T0137179

Lecture Notes in Computer Science

Lecture Notes in Artificial Intelligence **13916**

The series Lecture Notes in Artificial Intelligence (LNAI) was established in 1988 as a topical subseries of LNCS devoted to artificial intelligence.

The series publishes state-of-the-art research results at a high level. As with the LNCS mother series, the mission of the series is to serve the international R & D community by providing an invaluable service, mainly focused on the publication of conference and workshop proceedings and postproceedings.

Ning Wang · Genaro Rebolledo-Mendez · Noboru Matsuda · Olga C. Santos · Vania Dimitrova
Editors

Artificial Intelligence in Education

24th International Conference, AIED 2023
Tokyo, Japan, July 3–7, 2023
Proceedings

Editors
Ning Wang
University of Southern California
Los Angeles, CA, USA

Noboru Matsuda
North Carolina State University
Raleigh, NC, USA

Vania Dimitrova
University of Leeds
Leeds, UK

Genaro Rebolledo-Mendez
University of British Columbia
Vancouver, BC, Canada

Olga C. Santos
UNED
Madrid, Spain

ISSN 0302-9743 ISSN 1611-3349 (electronic)
Lecture Notes in Artificial Intelligence
ISBN 978-3-031-36271-2 ISBN 978-3-031-36272-9 (eBook)
https://doi.org/10.1007/978-3-031-36272-9

LNCS Sublibrary: SL7 – Artificial Intelligence

This Springer imprint is published by the registered company Springer Nature Switzerland AG
The registered company address is: Gewerbestrasse 11, 6330 Cham, Switzerland

Preface

The 24th International Conference on Artificial Intelligence in Education (AIED 2023) marks the 30th Anniversary of the International Artificial Intelligence in Education Society and the 24th edition of its International Conference. This year, the conference's theme was AI in Education for a Sustainable Society and it was held in Tokyo, Japan. It was conceived as a hybrid conference allowing face-to-face and online contributions. AIED 2023 was the next in a series of annual international conferences for presenting high-quality research on intelligent systems and the cognitive sciences for the improvement and advancement of education. It was hosted by the prestigious International Artificial Intelligence in Education Society, a global association of researchers and academics specialising in the many fields that comprise AIED, including, but not limited to, computer science, learning sciences, educational data mining, game design, psychology, sociology, and linguistics among others.

The conference hoped to stimulate discussion on how AI shapes and can shape education for all sectors, how to advance the science and engineering of intelligent interactive learning systems, and how to promote broad adoption. Engaging with the various stakeholders – researchers, educational practitioners, businesses, policy makers, as well as teachers and students – the conference set a wider agenda on how novel research ideas can meet practical needs to build effective intelligent human-technology ecosystems that support learning.

AIED 2023 attracted broad participation. We received 311 submissions for the main program, of which 251 were submitted as full-papers, and 60 were submitted as short papers. Of the full-paper submissions, 53 were accepted as full-papers (thus the full-paper acceptance rate is 21.11%) and another 14 were accepted as short-papers. Of the 60 short-paper submissions we received, 12 were accepted as short-papers. In addition to the 53 full-papers and 26 short-paper submissions, 65 were selected as posters. This set of posters was complemented with the other poster contributions submitted for the Poster and Late Breaking results track that is compiled in an adjunct volume.

The submissions underwent a rigorous double-masked peer-review process aimed to reduce evaluation bias as much as possible. The first step of the review process was done by the program chairs, who verified that all papers were appropriate for the conference and properly anonymized. Authors were also asked to declare conflicts of interest. After the initial revision, the program committee members were invited to bid on the anonymized papers that were not in conflict according to their declared conflicts of interest. With this information, the program chairs made the review assignment, which consisted of three regular members to review each paper plus a senior member to provide a meta-review. Some senior program committee members who are known as reliable from their previous involvement in the conference were not given any review load and were subsequently asked to make last-minute reviews when needed. The management of the review process (i.e., bidding, assignment, discussion, and meta-review) was done with the Easychair platform, which was configured so that reviewers of the same paper were also anonymous to each other. To avoid a situation where programme committee members would be involved in too many submissions, we originally requested committee members to withdraw from reviewing when they had more than two submissions. This led to having an insufficient number of reviewers so we invited back a select number of authors who could review papers, taking extra care that they would not review or meta-review their own papers.

As a result, each submission was reviewed anonymously by at least three members of the AIED community who are active researchers and then a discussion was led by a senior member of the AIED society. The role of the meta-reviewers was to assert and seek consensus to reach the final decision about acceptance and also to provide the corresponding meta-review. They were also asked to check and highlight any possible biases or inappropriate reviews. Decisions to accept/reject were taken by the two program chairs and confirmed with the general chairs. For borderline cases, the contents of the paper were read in detail before reaching the final decision. The decision of the papers submitted by one of the program chairs and one of the general chairs were looked at only by the other program and general chairs. In summary, we are confident that the review process assured a fair and equal evaluation for the submissions received without any bias as far as we are aware.

AIED 2023 offered other venues to present original contributions beyond the paper presentations and keynotes, including a Doctoral Consortium Track, an Industry and Innovation Track, Interactive Events, Posters/Late-Breaking Results, a Practitioner Track, and a track where published researchers from the International Journal of AIED could present their work. Since this year marks the 30th anniversary of the International AIED Society, the conference had a BlueSky Track that included papers reflecting upon the progress of AIED in the last 30 years and envisioning what's to come in the next 30 years. All of these additional tracks defined their own peer-review process. The conference also had a Wide AIED track where participants from areas of the World not typically present at the conference could present their work as oral presentations. All these contributions are compiled in an adjunct volume.

For making AIED 2023 possible, we thank the AIED 2023 Organizing Committee, the hundreds of Program Committee members, the Senior Program Committee members, and the AIED Proceedings Chairs, Irene-Angelica Chounta and Christothea Herodotou.

July 2023

Ning Wang
Genaro Rebolledo-Mendez
Vania Dimitrova
Noboru Matsuda
Olga C. Santos

Organization

Conference General Co-Chairs

Vania Dimitrova	University of Leeds, UK
Noboru Matsuda	North Carolina State University, USA
Olga C. Santos	UNED, Spain

Program Co-Chairs

Genaro Rebolledo-Mendez	University of British Columbia, Canada
Ning Wang	University of Southern California, USA

Doctoral Consortium Co-Chairs

Neil Heffernan	Worcester Polytechnic Institute, USA
Elaine Harada Teixeira de Oliveira	Federal University of Amazonas, Brazil
Kalina Yacef	University of Sydney, Australia

Workshop and Tutorials Co-Chairs

Martina Rau	University of Wisconsin - Madison, USA
Lei Shi	Newcastle University, UK
Sergey Sosnovsky	Utrecht University, The Netherlands

Interactive Events Co-Chairs

Ifeoma Adaji	University of British Columbia, Canada
Camila Morais Canellas	Sorbonne Université, CNRS, France
Manolis Mavrikis	University College London, UK
Yusuke Hayashi	Hiroshima University, Japan

Industry and Innovation Track Co-Chairs

Zitao Liu	TAL Education Group, China
Diego Zapata-Rivera	Educational Testing Service, USA

Posters and Late-Breaking Results Co-chairs

Marie-Luce Bourguet	Queen Mary University of London, UK
Carrie Demmans Epp	University of Alberta, Canada
Andrey Olney	University of Memphis, USA

Practitioner Track Co-Chairs

Susan Beudt	German Research Centre for Artificial Intelligence (DFKI), Germany
Berit Blanc	German Research Centre for Artificial Intelligence (DFKI), Germany
Diego Dermeval Medeiros da Cunha Matos	Federal University of the Alagoas, Brazil
Jeanine Antoinette DeFalco	Transfr, USA
Insa Reichow	German Research Centre for Artificial Intelligence (DFKI), Germany
Hajime Shirouzu	National Institution for Educational Policy Research, Japan

Local Organising Chair

Maomi Ueno	University of Electro-Communications, Japan

AIED Mentoring Fellowship Co-Chairs

Amruth Kumar	Ramapo College of New Jersey, USA
Maria Mercedes (Didith) T. Rodrigo	Ateneo de Manila University, Philippines

Diversity and Inclusion Co-Chairs

Seiji Isotani	Harvard University, USA
Rod Roscoe	Arizona State University, USA
Erin Walker	University of Pittsburgh, USA

Virtual Experiences Chair

Guanliang Chen	Monash University, Australia

Publicity Co-Chairs

Son T. H. Pham	Stephen F. Austin State University, USA
Miguel Portaz	UNED, Spain
Pham Duc Tho	Hung Vuong University, Vietnam

Volunteer Chair

Jingyun Wang	Durham University, UK

Proceedings Co-Chairs

Irene-Angelica Chounta	University of Duisburg-Essen, Germany
Christothea Herodotou	Open University, UK

Awards Chair

Tanja Mitrovic	University of Canterbury, New Zealand

Sponsorship Chair

Masaki Uto	University of Electro-Communications, Japan

Senior Program Committee Members

Giora Alexandron	Weizmann Institute of Science, Israel
Ivon Arroyo	University of Massachusetts Amherst, USA
Roger Azevedo	University of Central Florida, USA
Ryan Baker	University of Pennsylvania, USA
Stephen B. Blessing	University of Tampa, USA
Min Chi	North Carolina State University, USA
Mutlu Cukurova	University College London, UK
Carrie Demmans Epp	University of Alberta, Canada
Vania Dimitrova	University of Leeds, UK
Dragan Gasevic	Monash University, Australia
Sébastien George	Le Mans Université, France
Floriana Grasso	University of Liverpool, UK
Peter Hastings	DePaul University, USA
Neil Heffernan	Worcester Polytechnic Institute, USA
Seiji Isotani	Harvard University, USA
Irena Koprinska	University of Sydney, Australia
Vitomir Kovanovic	University of South Australia, Australia
Sébastien Lallé	Sorbonne University, France
H. Chad Lane	University of Illinois at Urbana-Champaign, USA
James Lester	North Carolina State University, USA
Shan Li	McGill University, Canada
Roberto Martinez-Maldonado	Monash University, Australia
Noboru Matsuda	North Carolina State University, USA
Manolis Mavrikis	University College London, UK
Gordon McCalla	University of Saskatchewan, Canada
Bruce McLaren	Carnegie Mellon University, USA
Eva Millan	University of Malaga, Spain
Tanja Mitrovic	University of Canterbury, New Zealand
Riichiro Mizoguchi	Japan Advanced Institute of Science and Technology, Japan
Kasia Muldner	CUNET, Canada
Roger Nkambou	Université du Québec à Montréal, Canada
Benjamin Nye	University of Southern California, USA
Andrew Olney	University of Memphis, USA
Jennifer Olsen	University of San Diego, USA
Luc Paquette	University of Illinois at Urbana-Champaign, USA
Kaska Porayska-Pomsta	University College London, UK
Thomas Price	North Carolina State University, USA
Kenneth R. Koedinger	Carnegie Mellon University, USA
Genaro Rebolledo-Mendez	University of British Columbia, Canada

Ido Roll	Technion - Israel Institute of Technology, Israel
Jonathan Rowe	North Carolina State University, USA
Nikol Rummel	Ruhr University Bochum, Germany
Vasile Rus	University of Memphis, USA
Olga C. Santos	UNED, Spain
Sergey Sosnovsky	Utrecht University, The Netherlands
Mercedes T. Rodrigo	Ateneo de Manila University, Philippines
Marco Temperini	Sapienza University of Rome, Italy
Vincent Wade	Trinity College Dublin, Ireland
Ning Wang	University of Southern California, USA
Diego Zapata-Rivera	Educational Testing Service, USA

Program Committee Members

Seth Adjei	Northern Kentucky University, USA
Bita Akram	North Carolina State University, USA
Burak Aksar	Boston University, USA
Laia Albó	Universitat Pompeu Fabra, Spain
Azza Abdullah Alghamdi	KAU, Saudi Arabia
Samah Alkhuzaey	University of Liverpool, UK
Laura Allen	University of Minnesota, USA
Antonio R. Anaya	Universidad Nacional de Educacion a Distancia, Spain
Tracy Arner	Arizona State University, USA
Ayan Banerjee	Arizona State University, USA
Michelle Barrett	Edmentum, USA
Abhinava Barthakur	University of South Australia, Australia
Sarah Bichler	Ludwig Maximilian University of Munich, Germany
Gautam Biswas	Vanderbilt University, USA
Emmanuel Blanchard	IDÛ Interactive Inc., Canada
Nathaniel Blanchard	Colorado State University, USA
Geoffray Bonnin	Université de Lorraine - LORIA, France
Nigel Bosch	University of Illinois at Urbana-Champaign, USA
Bert Bredeweg	University of Amsterdam, The Netherlands
Julien Broisin	Université Toulouse 3 Paul Sabatier - IRIT, France
Christopher Brooks	University of Michigan, USA
Armelle Brun	LORIA - Université de Lorraine, France
Jie Cao	University of Colorado Boulder, USA
Dan Carpenter	North Carolina State University, USA
Alberto Casas-Ortiz	UNED, Spain

Wania Cavalcanti	Universidade Federal do Rio de Janeiro COPPEAD, Brazil
Li-Hsin Chang	University of Turku, Finland
Penghe Chen	Beijing Normal University, China
Zixi Chen	University of Minnesota, USA
Ruth Cobos	Universidad Autónoma de Madrid, Spain
Cesar A. Collazos	Universidad del Cauca, Colombia
Maria de los Ángeles Constantino González	Tecnológico de Monterrey Campus Laguna, Mexico
Seth Corrigan	University of California, Irvine, USA
Maria Cutumisu	University of Alberta, Canada
Jeanine DeFalco	US Army Futures Command, USA
M. Ali Akber Dewan	Athabasca University, Canada
Konomu Dobashi	Aichi University, Japan
Tenzin Doleck	Simon Fraser University, Canada
Mohsen Dorodchi	University of North Carolina Charlotte, USA
Fabiano Dorça	Universidade Federal de Uberlandia, Brazil
Cristina Dumdumaya	University of Southeastern Philippines, Philippines
Yo Ehara	Tokyo Gakugei University, Japan
Ralph Ewerth	Leibniz Universität Hannover, Germany
Stephen Fancsali	Carnegie Learning, Inc., USA
Alexandra Farazouli	Stockholm University, Sweden
Effat Farhana	Vanderbilt University, USA
Mingyu Feng	WestEd, USA
Márcia Fernandes	Federal University of Uberlandia, Brazil
Carol Forsyth	Educational Testing Service, USA
Reva Freedman	Northern Illinois University, USA
Selen Galiç	Hacettepe University, Turkey
Wenbin Gan	National Institute of Information and Communications Technology, Japan
Michael Glass	Valparaiso University, USA
Benjamin Goldberg	United States Army DEVCOM Soldier Center, USA
Alex Sandro Gomes	Universidade Federal de Pernambuco, Brazil
Aldo Gordillo	Universidad Politécnica de Madrid (UPM), Spain
Monique Grandbastien	LORIA, Université de Lorraine, France
Beate Grawemeyer	University of Sussex, UK
Nathalie Guin	Université de Lyon, France
Jason Harley	McGill University, Canada
Bastiaan Heeren	Open University, The Netherlands
Laurent Heiser	Université Côte d'Azur, France

Wayne Holmes	University College London, UK
Anett Hoppe	TIB Leibniz Information Centre for Science and Technology; Leibniz Universität Hannover, Germany
Lingyun Huang	McGill University, Canada
Yun Huang	Carnegie Mellon University, USA
Ig Ibert-Bittencourt	Federal University of Alagoas, Brazil
Tomoo Inoue	University of Tsukuba, Japan
Paul Salvador Inventado	California State University Fullerton, USA
Mirjana Ivanovic	University of Novi Sad, Serbia
Stéphanie Jean-Daubias	Université de Lyon, France
Johan Jeuring	Utrecht University, The Netherlands
Yang Jiang	Columbia University, USA
Srecko Joksimovic	University of South Australia, Australia
David Joyner	Georgia Institute of Technology, USA
Akihiro Kashihara	University of Electro-Communications, Japan
Mizue Kayama	Shinshu University, Japan
Hieke Keuning	Utrecht University, The Netherlands
Yeojin Kim	North Carolina State University, USA
Kazuaki Kojima	Teikyo University, Japan
Amruth Kumar	Ramapo College of New Jersey, USA
Tanja Käser	EPFL, Switzerland
Andrew Lan	University of Massachusetts Amherst, USA
Mikel Larrañaga	University of the Basque Country, Spain
Hady Lauw	Singapore Management University, Singapore
Nguyen-Thinh Le	Humboldt Universität zu Berlin, Germany
Tai Le Quy	Leibniz University Hannover, Germany
Seiyon Lee	University of Pennsylvania, USA
Marie Lefevre	Université Lyon 1, France
Blair Lehman	Educational Testing Service, USA
Carla Limongelli	Università Roma Tre, Italy
Fuhua Lin	Athabasca University, Canada
Nikki Lobczowski	University of Pittsburgh, USA
Yu Lu	Beijing Normal University, China
Vanda Luengo	Sorbonne Université, France
Collin Lynch	North Carolina State University, USA
Sonsoles López-Pernas	Universidad Politécnica de Madrid, Spain
Aditi Mallavarapu	University of Illinois at Chicago, USA
Mirko Marras	University of Cagliari, Italy
Jeffrey Matayoshi	McGraw Hill ALEKS, USA
Kathryn McCarthy	Georgia State University, USA
Guilherme Medeiros-Machado	ECE Paris, France

Angel Melendez-Armenta	Tecnológico Nacional de México, Mexico
Wookhee Min	North Carolina State University, USA
Phaedra Mohammed	University of the West Indies, Trinidad and Tobago
Daniel Moritz-Marutschke	Ritsumeikan University, Japan
Fahmid Morshed-Fahid	North Carolina State University, USA
Ana Mouta	Universidad de Salamanca, Portugal
Tomohiro Nagashima	Saarland University, Germany
Huy Nguyen	Carnegie Mellon University, USA
Narges Norouzi	UC Berkeley, USA
Nasheen Nur	Florida Institute of Technology, USA
Marek Ogiela	AGH University of Science and Technology, Poland
Ranilson Paiva	Universidade Federal de Alagoas, Brazil
Radek Pelánek	Masaryk University Brno, Czechia
Eduard Pogorskiy	Open Files Ltd., UK
Elvira Popescu	University of Craiova, Romania
Miguel Portaz	UNED, Spain
Ethan Prihar	Worcester Polytechnic Institute, USA
Shi Pu	Educational Testing Service, USA
Mladen Rakovic	Monash University, Australia
Ilana Ram	Technion-Israel Institute of Technology, Israel
Sowmya Ramachandran	Stottler Henke Associates Inc, USA
Martina Rau	University of Wisconsin - Madison, USA
Traian Rebedea	University Politehnica of Bucharest, Romania
Carlos Felipe Rodriguez Hernandez	Tecnológico de Monterrey, Mexico
Rinat B. Rosenberg-Kima	Technion - Israel Institute of Technology, Israel
Demetrios Sampson	Curtin University, Australia
Sreecharan Sankaranarayanan	Carnegie Mellon University, USA
Mohammed Saqr	University of Eastern Finland, Finland
Petra Sauer	Beuth University of Applied Sciences, Germany
Robin Schmucker	Carnegie Mellon University, USA
Filippo Sciarrone	Universitas Mercatorum, Italy
Kazuhisa Seta	Osaka Prefecture University, Japan
Lele Sha	Monash University, Australia
Davood Shamsi	Ad.com, USA
Lei Shi	Newcastle University, UK
Aditi Singh	Kent State University, USA
Daevesh Singh	Indian Institute of Technology, Bombay, India
Sean Siqueira	Federal University of the State of Rio de Janeiro (UNIRIO), Brazil

Abhijit Suresh	University of Colorado Boulder, USA
Vinitra Swamy	EPFL, Switzerland
Michelle Taub	University of Central Florida, USA
Maomi Ueno	University of Electro-Communications, Japan
Josh Underwood	Independent, Spain
Maya Usher	Technion - Israel Institute of Technology, Israel
Masaki Uto	University of Electro-Communications, Japan
Rosa Vicari	Universidade Federal do Rio Grande do Sul, Brazil
Maureen Villamor	University of Southeastern Philippines, Philippines
Alessandro Vivas	UFVJM, Brazil
Alistair Willis	Open University, UK
Chris Wong	University of Technology Sydney, Australia
Peter Wulff	Heidelberg University of Education, Germany
Kalina Yacef	University of Sydney, Australia
Nilay Yalcin	University of British Columbia, Canada
Sho Yamamoto	Kindai University, Japan
Andrew Zamecnik	University of South Australia, Australia
Stefano Pio Zingaro	Universitá di Bologna, Italy
Gustavo Zurita	Universidad de Chile, Chile

International Artificial Intelligence in Education Society

Management Board

President

Vania Dimitrova	University of Leeds, UK

Secretary/Treasurer

Rose Luckin	University College London, UK
Bruce M. McLaren	Carnegie Mellon University, USA

Journal Editors

Vincent Aleven	Carnegie Mellon University, USA
Judy Kay	University of Sydney, Australia

Finance Chair

Ben du Boulay	University of Sussex, UK

Membership Chair

Benjamin D. Nye	University of Southern California, USA

Publicity Chair

Manolis Mavrikis	University College London, UK

Executive Committee

Diane Litman	University of Pittsburgh, USA
Kaska Porayska-Pomsta	University College London, UK
Manolis Mavrikis	University College London, UK
Min Chi	North Carolina State University, USA

Contents

Full Papers

Full Papers

Machine-Generated Questions Attract Instructors When Acquainted with Learning Objectives

Machi Shimmei[1(✉)], Norman Bier[2], and Noboru Matsuda[1]

[1] North Carolina State University, Raleigh, NC 27695, USA
{mshimme,Noboru.Matsuda}@ncsu.edu

[2] Carnegie Mellon University, Pittsburgh, PA 15213, USA
nbier@cmu.edu

Abstract. Answering questions is an essential learning activity on online courseware. It has been shown that merely answering questions facilitates learning. However, generating pedagogically effective questions is challenging. Although there have been studies on automated question generation, the primary research concern thus far is about if and how those question generation techniques can generate answerable questions and their anticipated effectiveness. We propose QUADL, a pragmatic method for generating questions that are aligned with specific learning objectives. We applied QUADL to an existing online course and conducted an evaluation study with in-service instructors. The results showed that questions generated by QUADL were evaluated as on-par with human-generated questions in terms of their relevance to the learning objectives. The instructors also expressed that they would be equally likely to adapt QUADL-generated questions to their course as they would human-generated questions. The results further showed that QUADL-generated questions were better than those generated by a state-of-the-art question generation model that generates questions without taking learning objectives into account.

Keywords: Question Generation · MOOCS · Learning Engineering

1 Introduction

Questions are essential components of online courseware. For students, answering questions is a necessary part of learning to attain knowledge effectively. The benefit of answering questions for learning (known as *test-enhanced learning*) has been shown in many studies [1, 2]. The literature suggests that answering retrieval questions significantly improves the acquisition of concepts when compared to just reading the didactic text [3]. Formative questions are also important for instructors. Students' answers for formative questions provide insight into their level of understanding, which, in turn, helps instructors enhance their teaching.

Despite the important role of questions on online courseware, creating large numbers of questions that effectively help students learn requires a significant amount of time and

N. Wang et al. (Eds.): AIED 2023, LNAI 13916, pp. 3–15, 2023.
https://doi.org/10.1007/978-3-031-36272-9_1

experience. To overcome this issue, researchers have been actively engaged in developing techniques for automated question generation [4]. However, in the current literature, most of the studies on question generation focus on linguistic qualities of generated questions like clarity and fluency. In other words, there has been a lack of research concern about the pedagogical value of the questions generated. It is therefore critical to develop a pragmatic technique for automatically generating questions that effectively help students learn.

The pedagogical value of questions can be discussed from multiple perspectives. In this study, we define pedagogical relevance as the degree to which a question helps students achieve learning objectives. Learning objectives specify goals that the students are expected to achieve, e.g., "*Explain the structure of the inner ear*." With this intention, the goal of the current study is to develop a machine-learning model that can generate questions that are aligned with given learning objectives. As far as we know, there has been no such question generation model reported in the current literature.

We hypothesize that if a key term or phrase related to a specific learning objective can be identified in a didactic text, then a verbatim question can be generated by converting the corresponding text into a question for which the key term or phrase becomes an answer. A verbatim question is a question for which an answer can be literally found in a related text. It is known that answering verbatim questions (even without feedback) effectively facilitates learning conceptual knowledge, arguably because doing so encourages students in the retrieval of relevant concepts [5].

Based on this hypothesis, we have developed a deep neural network model for question generation, called QUADL (**QU**iz generation with **A**pplication of **D**eep **L**earning). QUADL consists of two parts: the answer prediction model and the question conversion model. The answer prediction model predicts whether a given sentence is suitable to generate a verbatim question for a given learning objective. The output from the answer prediction model is a token index *<Is, Ie>* indicating a start and end of the target tokens, which are one or more consecutive words that represent key concepts in the given sentence. The question conversion model then converts the sentence into a question whose verbatim answer is the target tokens.

The primary research questions in this paper are as follows: Does QUADL generate questions that are pedagogically relevant to specific learning objectives? Is QUADL robust enough to apply to existing online courseware?

To answer those research questions, QUADL was applied to an existing online course (OLI[1] Anatomy & Physiology course). Then, in-service instructors were asked to evaluate the pedagogical relevance of generated questions through a survey (Sect. 4.2). In the survey, questions generated by QUADL were blindly compared to both those generated by Info-HCVAE [6] (a state-of-the-art question-generation system that does not take learning objective into account) and those generated by human experts.

The results show that QUADL questions are evaluated as on-par with human-generated questions, and remarkably better than the state-of-the-art question generation model in terms of the relevance to the learning objectives and the likelihood of adoption.

---

[1] Open learning Initiative (https://oli.cmu.edu).

From a broader perspective, QUADL has been developed as part of our effort to develop evidence-based learning engineering technologies that we call PASTEL (Pragmatic methods to develop Adaptive and Scalable Technologies for next-generation E-Learning) [7]. The primary goal for PASTEL is to assist instructional developers to build adaptive online courseware.

The major contributions of the current paper are as follows: (1) We developed and open sourced[2] a pragmatic question-generation model for online learning, QUADL, that can generate questions that are aligned with specific learning objectives, which is the first attempt in the current literature. (2) We demonstrated that instructors rated the pedagogical relevance of questions generated by QUADL as on-par with human-generated questions and higher than questions generated by a state-of-the-art model.

2 Related Work

There are two types of question generation models: answer-unaware models and answer-aware models. In *answer-unaware question generation models* (also known as answer-agnostic models), knowledge about the answer is not directly involved in the question generation pipeline either as an input or output (e.g., [8–10]). Given a source context (e.g., a sentence, a paragraph, or an article), the answer-unaware model generates a question(s) asking about a concept that appeared in the source context. However, the corresponding answer is not explicitly shown. Answer-unaware models are not suitable for QUADL because students' responses cannot be evaluated automatically without knowing correct answers.

In *answer-aware question generation models*, on the other hand, knowledge about the answer is explicitly involved in the question generation pipeline. In the current literature of question generation, the most actively studied models are answer-aware models for which the answer data are manually provided (e.g., [11–16]). There are also other answer-aware models that identify keywords (that become answers) by themselves and generate questions accordingly, called question-answer pair generation (QAG) models [6, 17–20]. There have been many QAG models specifically proposed for educational use [21–23]. QUADL is one such QAG model.

Some question generation models utilize extra knowledge in addition to the source context and/or an answer (though these models are not very common). For example, a model proposed by Pyatkin *et al.* [10] is given a predicate (e.g., "arrive") with a context, and produces a set of questions asking about all possible semantic roles of the given predicate (e.g., "start point", "entity in motion", etc.). Wang *et al.* [15] proposed a model whose input is a pair of a target and a paragraph. The target is a word or a phrase that specifies a topic of the question such as "size" or "sleep deprivation," but it is not an exact answer to the question. The model generates questions that ask about concepts relevant to the target in the paragraph.

QUADL also belongs to this type of question generation model. It takes a learning objective as the extra knowledge and generates questions that are suitable for the given

[2] The code and data used for the current study is available at https://github.com/IEClab-NCSU/QUADL.

learning objective. *As far as we are aware, no studies have been reported in the current literature that generate questions that align with specific learning objectives.*

3 Overview of *QUADL*

Figure 1 shows an overview of QUADL. Given a pair of a learning objective *LO* and a sentence *S*, <*LO, S*>, QUADL generates a question *Q* that is assumed to be suitable to achieve the learning objective *LO*. Examples of < *LO, S*> and *Q* are shown in Table 1 in Sect. 5.3. Notice that a target token is underlined in the sentence *S* and becomes the answer for the question.

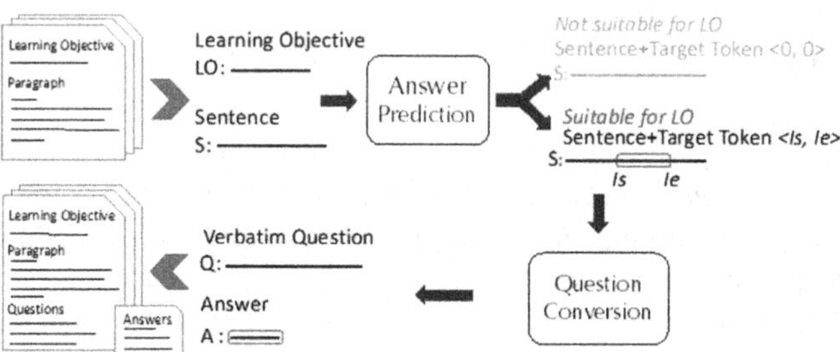

Fig. 1. An overview of QUADL. The answer prediction model identifies the start/end index <*Is, Ie*> of the target token (i.e., key term) in *S*. When *S* is not suitable for *LO*, it outputs <0,0>. The question conversion model converts *S* with target token to a verbatim question.

The input of the answer prediction model is a single sentence (or a *source sentence* for the sake of clarity) and a learning objective. In our application, each sentence *S* in a paragraph is paired with the learning objective *LO* as a single input <*LO, S*> to the model. The final output from the answer prediction model is a *target token index*, <*Is, Ie*>, where *Is* and *Ie* show the index of the start and end of the target token within the source sentence *S* relative to the learning objective *LO*. The models may output <$Is = 0, Ie = 0$>, indicating that the source sentence is not suitable to generate a question for the given learning objective. For the rest of the paper, we refer to source sentences that have non-zero indices (i.e., $Is \neq 0$ and $Ie \neq 0$) as *target sentences*, whereas the other source sentences that have the zero token index <0, 0> *non-target sentences*.

For the answer prediction model, we adopted Bidirectional Encoder Representation from Transformers (BERT) [24]. The final hidden state of the BERT model is fed to two single-layer classifiers. One of them outputs a vector of probabilities, $P_s(Is = i)$, indicating the probability of the *i-th* index in the sentence being beginning of the target token. Another model outputs a vector of probabilities for the end index, $P_e(Ie = j)$. To compute the final probability for being the target token, a normalized sum of $P_s(i)$ and $P_e(j)$ is first calculated as the joint probability $P(Is = i, Ie = j)$ for every possible span ($Is < Ie$) in the sentence. The probability $P(Is = 0, Ie = 0)$ is also computed, indicating a likelihood that the sentence is *not* suitable to generate a question for the learning objective. The pair <$Is = i, Ie = j$> with the largest joint probability becomes the final prediction.

As for the question conversion model, we hypothesize that if a target token is identified in a source sentence, a pedagogically valuable question can be generated by converting that source sentence into a verbatim question using a sequence-to-sequence model that can generate fluent and relevant questions. In the current implementation, we used the state-of-the-art technology, called ProphetNet [13], as a question conversion model. ProphetNet is an encoder-decoder pre-training model that is optimized by future n-gram prediction while predicting n-tokens simultaneously.

These two models in QUADL are domain and courseware independent. This section only describes their structures. The next section describes how those models were trained with the data from an existing online course for an evaluation study.

4 Evaluation Study

To evaluate the anticipated pedagogical value of the questions generated by QUADL, we conducted a human-evaluation study with in-service instructors who use an existing online course hosted by Open Learning Initiative (OLI) at Carnegie Mellon University.

4.1 Model Implementation

Answer Prediction Model. The answer prediction model was fine-tuned with courseware content data taken from the Anatomy and Physiology (A&P) OLI course. The A&P course consists of 490 pages and has 317 learning objectives. To create training data for the answer prediction model, in-service instructors who actively teach the A&P course manually tagged the didactic text as follows. The instructors were asked to tag each sentence *S* in the didactic text to indicate the target tokens relevant to specific learning objective *LO*. A total of 8 instructors generated 350 pairs of <*Lo, S*> for monetary compensation. Those 350 pairs of token index data were used to fine-tune the answer prediction model.

Since only a very small amount of data was available for fine-tuning, the resulting model was severely overfitted. We therefore developed a unique ensemble technique that we argue is an innovative solution for training deep neural models with extremely small data. A comprehensive description of the ensemble technique can be found in Shimmei *et al.* [25] Due to the space constraint, this paper briefly shows how the ensemble technique was applied to the answer prediction model. Note that the ensemble technique is not necessary if a sufficient amount of data are available.

To make an ensemble prediction, 400 answer prediction models were trained independently using the same training data, but each with a different parameter initialization. Using all 400 answer prediction models, an ensemble model prediction was made as shown below.

Recall that for each answer prediction model k ($k = 1, \ldots, 400$), two vectors of probabilities, the start index ${P_s}^k(i)$ and the end index ${P_e}^k(j)$, are output. Those probabilities were averaged across all models to obtain the ensemble predictions $P_s^*(i)$ and $P_e^*(j)$ for the start and end indices, respectively. The final target token prediction $P(Is = i, Ie = j)$ was then computed using P_s^* and P_e^* (see Sect. 3).

Subsequently, we applied rejection method. This method is based on the hypothesis that a reliable prediction has stable and high probabilities across models, while an unreliable prediction has diverse probabilities across models, which results in smaller values when averaged. Therefore, if either of $P_s^*(i)$ or $P_e^*(j)$ were below the pre-defined threshold $R \in (0,1)$, the model discarded the prediction $< Is = i, Ie = j >$. When all predictions were below the threshold, the model did not make any prediction for the sentence, i.e., the prediction is void.

Rejection increases the risk of missing target sentences but decreases the risk of creating questions from non-target sentences. For the sake of pedagogy, question quality is more important than the quantity. Rejecting target token prediction with a low certainty (i.e., below the threshold) ensures that the resulting questions are likely to be relevant to the learning objective. In the current study, we used a threshold of 0.4 for rejection. We determined the number of models and the threshold using the performance on the human-annotated test dataset.

Question Conversion Model. We used an existing instance of ProphetNet that was trained on SQuAD1.1 [26], one of the most commonly used datasets for question generation tasks. SQuAD1.1 consists of question-answer pairs retrieved from Wikipedia. We could train ProphetNet using the OLI course data. However, the courseware data we used for the current study do not contain a sufficient number of verbatim questions—many of the questions are fill-in-the-blank and therefore not suitable to generate a training dataset for ProphetNet.

4.2 Survey Study

Five instructors (the "participants" hereafter) were recruited for a survey study. The total of 100 questions used in the survey were generated by three origins: QUADL, Info-HCVAE, and human experts, as described below.

QUADL questions: 34 questions out of 2191 questions generated by QUADL using the method described above were randomly selected for the survey.
Info-HCVAE questions: 33 questions were generated by Info-HCVAE [6], a state-of-the-art question generation model that generates questions without taking the learning objective into account. Info-HCVAE extracts key concepts from a given paragraph and generates questions for them. We trained an Info-HCVAE model on SQuAD1.1. Info-HCVAE has a hyperparameter K that determines the number of questions to be generated from a paragraph. We chose K=5 because that was the average number of questions per paragraph in the A&P course. The Info-HCVAE model was applied to a total of 420 paragraphs taken from the courseware, and 2100 questions were generated. Questions with answers longer than 10 words or related to multiple sentences were excluded. Consequently, 1609 questions were left, among which 33 questions were randomly selected for the survey.
Human questions: 33 formative questions among 3578 currently used in the OLI A&P course were randomly collected. Most of the A&P questions are placed immediately after a didactic text paragraph. Only questions whose answers can be literally found in the didactic text on the same page where the question appeared were used because questions generated by QUADL are short-answer, verbatim questions. The course contains various types of formative questions such as fill-in-the-blank, multiple-choice, and short-answer

questions. Fill-in-the-blank questions were converted into interrogative sentences. For example, "The presence of surfactant at the gas-liquid interphase lowers the ____ of the water molecules." was changed to "The presence of surfactant at the gas-liquid interphase lowers what of the water molecules?". The multiple-choice questions were also converted into short-answer questions by hiding choices from participants.

Each survey item consists of a paragraph, a learning objective, a question, and an answer. Participants were asked to rate the prospective pedagogical value of proposed questions using four evaluation metrics that we adopted from the current literature on question generation [15, 21, 27]: answerability, correctness, appropriateness, and adoptability. Answerability refers to whether the question can be answered from the information shown in the proposed paragraph. Correctness is asking whether the proposed answer adequately addresses the question. Appropriateness is asking whether the question is appropriate for helping students achieve the corresponding learning objective. Adoptability is asking how likely the participants would adapt the proposed question to their class. Each metric was evaluated on a 5-point Likert scale.

Every individual participant rated all 100 questions mentioned above (i.e., 100 survey items). Five responses per question were collected, which is notably richer than any other human-rated study for question generation in the current literature that often involve only two coders.

5 Results

5.1 Instructor Survey

Inter-Rater Reliability: We first computed Krippendorff's alpha to estimate inter-rater reliability among study participants. There was moderate agreement for answerability, correctness, and appropriateness (0.56, 0.49, and 0.47 respectively). The agreement for adoptability was weak (0.34), indicating that there were diverse factors among participants that determine the adoptability of the proposed questions.

Overall Ratings: Figure 2 shows the mean ratings per origin (QUADL vs. Infor-HCVAE vs. Human) across participants for each metric. The plot shows that QUADL- and human-generated questions are indistinguishable on all four metrics, whereas questions generated by Info-HCVAE are clearly different.

Statistical tests confirmed the above observations. One-way ANOVA tests (when applied separately to each metric) revealed that origin (QUADL vs. Infor-HCVAE vs. Human) is a main effect for ratings on all four metrics; $F(2, 97) = 36.38, 24.15, 26.11$, and 25.03, for answerability, correctness, appropriateness, and adoptability, respectively, with $p < 0.05$ for each of them. A post hoc analysis using Tukey's test showed that there was a statistically significant difference between QUADL and Info-HCVAE at $p < .05$, but no significant difference between QUADL and human-generated questions for each of the four metrics.

Overall, the results suggest that in-service instructors acknowledged that questions generated by QUADL and humans had equal prospective pedagogical values when asked blindly. It is striking to see that *the in-service instructors suggested that they would be equally likely to adapt QUADL- and human-generated questions to their courses.*

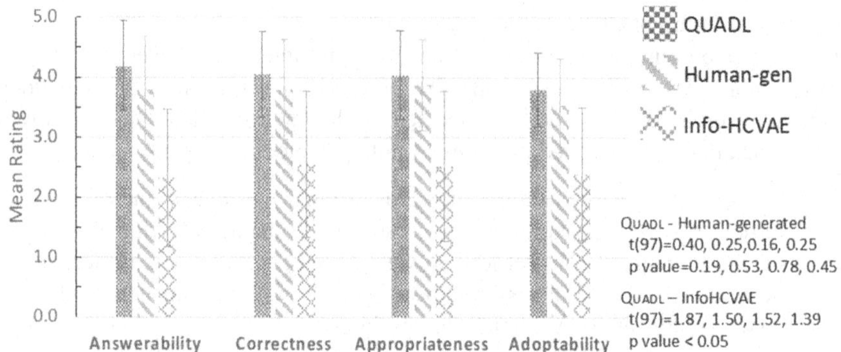

Fig. 2. Mean of the ratings for each metrics. The standard deviation is shown as an error bar. The value ranges from 1 as strongly disagree to 5 as strongly agree.

As for the comparison between question generation technologies with and without taking learning objectives into account, the results suggested that learning objective-specific questions (QUADL) were rated higher than learning objective-unaware questions (Info-HCVAE) on all four metrics. Since Info-HCVAE does not aim to generate questions for particular learning objectives, appropriateness might not be a fair metric. Yet, instructors showed clear hesitation to adapt questions generated by Info-HCVAE to their courses. The poor scores on answerability and correctness for Info-HCVAE questions might be mostly due to the poor performance of the question conversion model used (see Sect. 6 for the further discussion).

5.2 Accuracy of the Answer Prediction Model

To evaluate the performance of the answer prediction model, we operationalize the accuracy of the model in the target token identification task using two metrics: token precision and token recall. *Token precision* is the number of correctly predicted tokens divided by the number of tokens in the prediction. *Token recall* is the number of correctly predicted tokens divided by the number of ground truth tokens. For example, suppose a sentence "*The target tissues of the nervous system are muscles and glands*" has the ground truth tokens as "*muscles and glands*". When the predicted token is "*glands*", the token precision is 1.0 and token recall is 0.33.

We computed the average of token precision and token recall across the predictions for a test dataset that contains 25 target and 25 non-target sentences. The averages of token precision and token recall were 0.62 and 0.21, respectively. Even with the very limited amount of training data (350), the proposed ensemble-learning technique (Sect. 4.1) achieved 0.62 token precision with the cost of recall. The recall score is notably low because predictions with low certainty were rejected. For our primary purpose of making instructional questions, we value quality over quantity. Therefore, we consider the currently reported performance of the answer prediction model to be adequate in particular when considering the extremely low amount of data available for training. Of course, improving token recall is important future research.

5.3 Qualitative Analysis of Questions Generated by *QuADL*

There are questions generated by QuADL that were rated low by the instructors. How could QuADL be further improved to generate better questions? To answer this question, we sampled a few QuADL- and human-generated questions from the same paragraphs for the same learning objectives. Table 1 compares these questions, with their appropriateness scores as rated by participants.

We found that the QuADL questions had low appropriateness scores (<2) when they were not particularly specific and did not clearly require students to review the concepts mentioned in the corresponding learning objective. For example, Q2 in Table 1 is ambiguous because it does not mention that it is about the circulation of blood in the urinary tract. The human-generated question (Q1) has enough context and is clear enough to effectively encourage students to review important concepts in the learning objective. For Q2, the answer prediction model predicted a target sentence "Once filtered, the blood exits through the *renal vein.*" with "*renal vein*" as a target token. However, in the original paragraph just before the predicted target sentence, there were other relevant sentences: "The process begins with waste carrying blood entering each of the two kidneys through the renal artery. Urine is produced by the nephrons in the kidney." The question would be more appropriate if it included this context. Q4 better provides the information students need to answer the question and appropriately asks about the concept in the learning objective. It is rated as high as the human-generated question Q3.

Table 1. Example of generated questions. Human- and QuADL-generated questions from the same paragraph and learning objective are compared. ***S*** shows a source sentence of a question, and an answer is written in *italics*. For human-generated questions, the sentences that should be referred to in order to answer the question were retrieved from the paragraph.

Learning Objective: Identify gross and microscopic anatomy of the urinary tract.

Q1 (Human): When enough urine has been produced in the nephrons, it leaves through the ureters and urine is then transported to what? [Appropriateness: 4.4]

S: Each ureter transports the urine via peristalsis to the urinary *bladder*.

Q2 (QuADL): Through what part of the body does the blood exit? [Appropriateness: 2.2]

***S*:** Once filtered, the blood exits through the *renal vein.*

Learning Objective: Describe how the structure of these macromolecules allow the structures c the respiratory system to perform their functions.

Q3 (Human): The presence of surfactant at the gas-liquid interphase lowers what of the water molecules? [Appropriateness: 4.2]

S: At the gas-liquid interface of the alveoli cell membranes, surfactants found in the liquid surface layer lower ***surface tension*****.** Surface tension arises when water molecules hydrogen bond with each other.

Q4 (QuADL): What helps humidify and buffer the cells in direct contact with air? [Appropriateness: 4.2]

S: Throughout the entire respiratory system, *mucus* helps humidify and buffer the cells that are in direct contact with air.

To overcome this shortcoming, one of the solutions would be to change the unit-of-analysis for QUADL from sentence-level to paragraph-level. In the current study, the target sentence is always a single sentence. Capturing context more properly by taking an entire paragraph (or some relevant sentences) into account might reduce the ambiguity of the question and effectively encourage the retrieval of relevant concepts in the learning objective. Testing this hypothesis is an important future study.

6 Discussion

As far as we are aware, QUADL is the first model that generates pedagogical questions while taking learning objectives into account. The current study is also the first in the literature that evaluates the appropriateness of machine-generated questions relative to the learning objectives.

Through the current study, we encountered many obstacles related to education research. First, collecting datasets of adequate quality and size in order to train a deep neural network model is challenging. For the current study, even though some datasets for question generation are available [28–30], they did not necessarily meet our need—there are various domains, types of questions (e.g., multiple-choice vs. fill-in-the-blank), and difficulty levels. The current study demonstrated that the proposed ensemble method with rejection is a powerful solution for the issue of low data regime.

Second, with the lack of ground-truth data, comparing the machine- and human-generated questions is challenging. In the current study, to obtain human-generated questions, questions from an existing online course were retrieved using an automated process (as mentioned in Sect. 4). However, we noticed that some questions on the online course had extra text associated that provided information about the context. When those questions were converted into survey items, the information about the context was removed, which resulted in ambiguous questions that allowed different answers or were too general. We speculate that this technological complication might be a reason why some human-generated questions in the current survey study received low answerability and correctness scores by study participants.

Third, evaluating the validity and quality (or the "utility") of machine-generated questions is challenging. Strictly speaking, it requires a close-the-loop experiment where students are assigned to courseware with machine-generated questions, and their subsequent learning outcomes are measured. However, due to the cost of conducting a rigorously controlled study in an authentic learning environment, subjective evaluation by human experts is a common technique in the current literature [31]. Nonetheless, with the current promising results, we have been preparing a close-the-loop evaluation study with college students as an important next step towards the wide dissemination of the QUADL technology.

The results in Fig. 2 show that Info-HCVAE received notably low scores on answerability and correctness. This might be merely due to the inferior performance of the question conversion model that it uses. The current implementation of QUADL utilized an existing state-of-the-art question conversion module, ProphetNet. Due to this confounding factor, the current results do not allow us to draw a rigorous conclusion on how QUADL's feature of learning-objective awareness contributed to the better evaluation of

QuADL than Info-HCVAE. Technically speaking, the question of which of the models in QuADL—the answer prediction model or the question conversion model—lead to satisfactory performance has yet to be investigated. To achieve this goal, we plan to modify Info-HCVAE by replacing its question conversion model with ProphetNet. We acknowledge that a lack of conducting this rigorous comparison is a limitation of the current study and constitutes important future research.

The current study demonstrated the significant potential of QuADL as a pragmatic technology for creating next-generation evidence-based online courseware. As stated in the Introduction section, QuADL is part of PASTEL, a suite of evidence-based online learning engineering methods. The promising results reported in this paper encourage us to consider further extension of PASTEL. For example, it may be possible for a machine to learn concepts from texts on the internet and automatically generate didactic text for online courseware.

7 Conclusion

We found that a deep neural network model designed to generate questions that are aligned with the given learning objective, QuADL, performs on-par with human experts for the task of generating pedagogically valuable questions. QuADL is the first model in the literature that aims to generate questions that are suitable for attaining learning objectives. The results encourage us to further conduct an in-class evaluation study to measure students' learning with the machine-generated questions.

Acknowledgements. The research reported here was supported by National Science Foundation Grant No. 2016966 and No.1623702 to North Carolina State University.

References

1. Rivers, M.L.: Metacognition about practice testing: a review of learners' beliefs, monitoring, and control of test-enhanced learning. Educ. Psychol. Rev. **33**(3), 823–862 (2021)
2. Pan, S.C., Rickard, T.C.: Transfer of test-enhanced learning: meta-analytic review and synthesis. Psychol. Bull. **144**(7), 710 (2018)
3. Smith, M.A., Karpicke, J.D.: Retrieval practice with short-answer, multiple-choice, and hybrid tests. Memory **22**(7), 784–802 (2014)
4. Pan, L., et al.: Recent advances in neural question generation. arXiv preprint arXiv:1905. 08949 (2019)
5. Roediger, H.L., Karpicke, J.D.: The power of testing memory: basic research and implications for educational practice. Perspect. Psychol. Sci. **1**(3), 181–210 (2006)
6. Lee, D.B., et al.: Generating Diverse and Consistent QA pairs from Contexts with Information-Maximizing Hierarchical Conditional VAEs. Association for Computational Linguistics (2020)
7. Matsuda, N., et al.: PASTEL: Evidence-based learning engineering methods to facilitate creation of adaptive online courseware. In: Ouyang, F., et al. (eds.) Artificial Intelligence in STEM Education: The Paradigmatic Shifts in Research, Education, and Technology, pp. 1–16. CSC Press, New York, NY (in press)

8. Lewis, M., et al.: Bart: Denoising sequence-to-sequence pre-training for natural language generation, translation, and comprehension. arXiv preprint arXiv:1910.13461 (2019)
9. Du, X., et al.: Learning to ask: neural question generation for reading comprehension. In: Proceedings of the 55th Annual Meeting of the Association for Computational Linguistics. Association for Computational Linguistics (2017)
10. Pyatkin, V., et al.: Asking it all: generating contextualized questions for any semantic role. In: Proceedings of the 2021 Conference on Empirical Methods in Natural Language Processing. Association for Computational Linguistics (2021)
11. Bao, H., et al.: Unilmv2: pseudo-masked language models for unified language model pre-training. In: International Conference on Machine Learning. PMLR (2020)
12. Chan, Y.-H., Fan, Y.-C.: A recurrent BERT-based model for question generation. In: Proceedings of the 2nd Workshop on Machine Reading for Question Answering. Association for Computational Linguistics (2019)
13. Qi, W., et al.: ProphetNet: Predicting future N-gram for sequence-to-sequence pre-training. In: Findings of the Association for Computational Linguistics: EMNLP 2020. Association for Computational Linguistics (2020)
14. Wang, Z., et al.: QG-net: a data-driven question generation model for educational content. In: Proceedings of the Fifth Annual ACM Conference on Learning at Scale (2018)
15. Wang, Z., Valdez, J., Mallick, D.B., Baraniuk, R.G.: Towards human-like educational question generation with large language models. In: Rodrigo, M.M., Matsuda, N., Cristea, A.I., Dimitrova, V. (eds.) Artificial Intelligence in Education: 23rd International Conference, AIED 2022, Durham, UK, July 27–31, 2022, Proceedings, Part I, pp. 153–166. Springer International Publishing, Cham (2022). https://doi.org/10.1007/978-3-031-11644-5_13
16. Xiao, D., et al.: ERNIE-GEN: an enhanced multi-flow pre-training and fine-tuning framework for natural language generation. In: IJCAI (2020)
17. Du, X., Cardie, C.: Harvesting Paragraph-level Question-Answer Pairs from Wikipedia. Association for Computational Linguistics (2018)
18. Back, S., et al.: Learning to generate questions by learning to recover answer-containing sentences. In: Findings of the Association for Computational Linguistics (2021)
19. Subramanian, S., et al.: Neural Models for Key Phrase Extraction and Question Generation. Association for Computational Linguistics (2018)
20. Wang, B., et al.: Neural question generation with answer pivot. In: Proceedings of the AAAI Conference on Artificial Intelligence (2020)
21. Steuer, T., Filighera, A., Rensing, C.: Remember the facts? investigating answer-aware neural question generation for text comprehension. In: Bittencourt, I.I., Cukurova, M., Muldner, K., Luckin, R., Millán, E. (eds.) AIED 2020. LNCS (LNAI), vol. 12163, pp. 512–523. Springer, Cham (2020). https://doi.org/10.1007/978-3-030-52237-7_41
22. Willis, A., et al.: Key phrase extraction for generating educational question-answer pairs. In: Proceedings of the Sixth ACM Conference on Learning@ Scale (2019)
23. Qu, F., et al.: Asking questions like educational experts: Automatically generating question-answer pairs on real-world examination data. arXiv preprint arXiv:2109.05179 (2021)
24. Devlin, J., et al.: Bert: Pre-training of deep bidirectional transformers for language understanding. arXiv preprint arXiv:1810.04805 (2018)
25. Shimmei, M., Matsuda, N.: Can't inflate data? let the models unite and vote: data-agnostic method to avoid overfit with small data. In: 14th Inernational Conference on Educatinal Data Mining (to appear)
26. Rajpurkar, P., et al.: Squad: 100,000+ questions for machine comprehension of text. arXiv preprint arXiv:1606.05250 (2016)
27. Tamang, L.J., Banjade, R., Chapagain, J., Rus, V.: Automatic question generation for scaffolding self-explanations for code comprehension. In: Rodrigo, M.M., Matsuda, N., Cristea,

A.I., Dimitrova, V. (eds.) Artificial Intelligence in Education: 23rd International Conference, AIED 2022, Durham, UK, July 27–31, 2022, Proceedings, Part I, pp. 743–748. Springer International Publishing, Cham (2022). https://doi.org/10.1007/978-3-031-11644-5_77

28. Chen, G., et al.: LearningQ: a large-scale dataset for educational question generation. In: Twelfth International AAAI Conference on Web and Social Media (2018)
29. Lai, G., et al.: RACE: Large-scale ReAding Comprehension Dataset From Examinations. Association for Computational Linguistics (2017)
30. Welbl, J., et al.: Crowdsourcing multiple choice science questions. arXiv preprint arXiv:1707.06209 (2017)
31. Kurdi, G., et al.: A systematic review of automatic question generation for educational purposes. Int. J. Artif. Intell. Educ. **30**(1), 121–204 (2020)

SmartPhone: Exploring Keyword Mnemonic with Auto-generated Verbal and Visual Cues

Jaewook Lee(✉) and Andrew Lan

University of Massachusetts Amherst, Amherst, USA
jaewooklee@cs.umass.edu

Abstract. In second language vocabulary learning, existing works have primarily focused on either the learning interface or scheduling personalized retrieval practices to maximize memory retention. However, the learning content, i.e., the information presented on flashcards, has mostly remained constant. Keyword mnemonic is a notable learning strategy that relates new vocabulary to existing knowledge by building an acoustic and imagery link using a keyword that sounds alike. Beyond that, producing verbal and visual cues associated with the keyword to facilitate building these links requires a manual process and is not scalable. In this paper, we explore an opportunity to use large language models to automatically generate verbal and visual cues for keyword mnemonics. Our approach, an end-to-end pipeline for auto-generating verbal and visual cues, can automatically generate highly memorable cues. We investigate the effectiveness of our approach via a human participant experiment by comparing it with manually generated cues.

Keywords: Keyword Mnemonic · Vocabulary Learning · Large Language Models

1 Introduction

Learning vocabulary is key to learning second (mostly foreign) languages, but also a difficult task. One of the most well-known and effective methods is flashcards, i.e., writing the L2 (a second language word) word on the front and writing down the corresponding L1 word (a first or native language word) on the back, with content such as mnemonic or context. Moreover, one may manage flashcards by putting the cards in boxes to follow the Leitner system [13] to recall the word regularly following the forgetting curve [8]. However, both writing down every word and managing a bunch of cards require significant effort and can take a lot of effort from learners.

Technology advances have enabled vocabulary learning to shift from manually writing down the words to using software systems such as Anki [10] and Quizlet [21], which make language learning more efficient and engaging. Some systems use ideas behind intelligent tutoring systems to model the learner's knowledge state to intervene in the retrieval practice [18,22,23]. Many studies

N. Wang et al. (Eds.): AIED 2023, LNAI 13916, pp. 16–27, 2023.
https://doi.org/10.1007/978-3-031-36272-9_2

have shown that managing retrieval practice and designing personalized schedules using cognitive models can significantly improve learning efficiency [7,12]. Many systems also use gamified interfaces and enable learners to share decks with others, making the learning process more interactive and socially relevant [1,10,21]. However, despite these advances, the learning *content*, i.e., what is written on the flashcard, has mostly stayed the same throughout the years.

Regarding the content for second language learning, keyword mnemonic [3] is a notable memory encoding strategy that uses interactive visual imagery with a keyword that sounds like part of a foreign word. Forming the keyword-based interactive image takes a two-step approach: creating first an acoustic and then an imagery link. Imagine a native English speaker is learning the Spanish word *pato*, which means *duck*. The keyword that sounds like the word is *pot*. Using the keyword, the learner first creates an acoustic link between the keyword and the Spanish word. Then, the learner builds an imagery link that connects the sound and its meaning by using a verbal cue, such as "A duck wearing a pot on its head." By relating new information to existing knowledge, learners have an easier time memorizing the word and can retain it in memory for a longer time.

Previous studies on keyword mnemonics have shown their effectiveness compared with different learning strategies. Comparing keyword mnemonic with rote rehearsal and combining both strategies showed that the keyword group outperformed the other two groups [5]. Comparing the keyword mnemonic group with verbal and visual cues with mixed methods of contextual clues, word structure analysis, and opposite word pairs showed that the keyword group performed better in both short-term and long-term retention [20]. However, since the cues given in these studies are manually generated by experts, it is difficult to employ this approach at a large scale in the systems mentioned above.

In 2014, Savva et al. introduced an automatic keyword generation approach based on a cross-lingual system, TransPhoner [19]. It evaluates candidate keywords in the second language using the following measures for a given input word: imageability, phonetic similarity, orthographic similarity, and semantic similarity. The authors experimented on the effectiveness of TransPhoner using an evaluation set of 36 German words [9] with three other conditions: no keywords, randomly sampled keywords, and manually generated keywords. The result shows that the TransPhoner-generated condition achieved the highest score and the manually-generated keyword condition had no significant difference from randomly generated keywords.

Despite TransPhoner's success in automatically generating keywords as cues, other forms of richer verbal or visual cues that could further help learners build an imagery link cannot be automatically generated. The learner (or teacher) still needs to manually develop them to connect the keyword and the L1 word, which requires a lot of effort on their part. Moreover, it takes an expert to come up with an image as the visual cue that corresponds to the verbal cue. Using image APIs such as Google Image API, one can juxtapose images of a keyword and an L1 word, but doing is not as effective as showing both words together in a single image. To make keyword mnemonic scalable, we need an end-to-end solution that takes words as input and generates keyword, verbal and visual cues.

Contributions. In this paper, we detail a pipeline for automatically generating verbal and visual cues in one shot via text generator and text-to-image generator. Our contributions are as follows:

- We propose a large language model (LLM)-based pipeline that automatically generates highly memorable verbal and visual cues for an L1 word in language learning. We believe that our automated approach will significantly reduce content development costs by enhancing time efficiency and reducing manual generation effort. To the best of our knowledge, we are the first to apply LLMs in the context of keyword mnemonic.
- We implement a web application for human participant studies and use it to compare our approach with existing ones. We analyze the effectiveness of four approaches: automatically generated keyword only, automatically generated keyword with a verbal cue, automatically generated keyword with both verbal and visual cues, and manually generated keyword and verbal cues. We also outline avenues for future work that could stem from our approach.

2 Methodology

In this section, we detail our pipeline for automatically generating cues. Our work is driven by the following two research questions:

- Can we automatically generate human-level verbal cues for the keyword?
- Can we generate a visual cue that may facilitate building an imagery link that is described in a verbal cue?

We narrow the scope of automatically generating verbal and visual cues to the experiments conducted in previous studies [9,19] in this preliminary effort. We use the evaluation set of 36 German words and keywords from previous studies for both manually and automatically generated cues as baselines. Since verbal cues only exist for manually generated keywords, our task boils down to automatically generating verbal cues using TransPhoner-generated keywords and generating visual cues using verbal cues.

2.1 Pipeline for Auto-generating Verbal and Visual Cues

We propose a pipeline consisting of two LLMs that generate verbal and visual cues in two steps: First, we use a text generator to automatically generate a sentence containing the TransPhoner keyword as the verbal cue. Second, we use a text-to-image generator to generate an image as the visual cue. LLMs, pre-trained with massive datasets, have shown human-level performance on the tasks we described above through prompts. This is because LLMs are good for controllable text generation [17] and following instructions [16]. With proper prompts, models show their ability to solve the tasks with zero-shot or few-shot setups. We use zero-shot setup LLMs for both generating verbal and visual cues. We detail the pipeline through an example in Fig. 1 where we need to generate

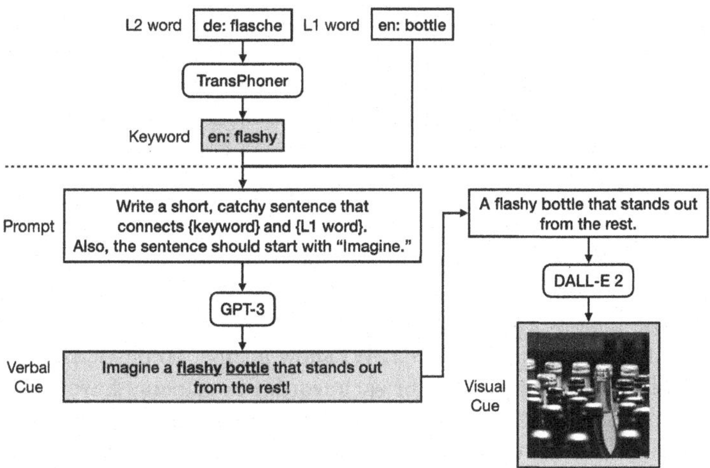

Fig. 1. Our end-to-end pipeline for automatically generating verbal and visual cues for an L2 word.

cues for the German word *flasche*, which means a *bottle*. The keyword generated by TransPhoner is *flashy*; Using the keyword and the meaning of the word, we create the prompt: "Write a short, catchy sentence that connects *flashy* and *bottle*." Additionally, we add a constraint on verbal cues to start with "Imagine" for two reasons. First, verbal cues in the previous study [9] are in that format. Since we are trying to answer whether we could achieve human-level verbal cues, we match the format. Second, we follow grammatical characteristics that come after the word imagine. After the word "Imagine", usually a noun or gerund comes out; we found that the generated verbal cue contains fewer ambiguous pronouns, which makes the cue more descriptive. This feature is key to linking the text generator and text-to-image generator within the same pipeline. Using the prompt, our text generator, GPT-3 [6] (text-davinci-003, temp = 0.5), generates the verbal cue. Then, we reuse the verbal cue as the prompt for our text-to-image generator, DALL-E 2 [15], by removing the word "Imagine". One can freely choose any LLMs to automatically generate these verbal and visual cues. We present the gray region in Fig. 1 to the participant as learning content.

3 Experimental Evaluation

In this section, we detail our experiments on presenting different content to different participants to explore whether automatically generated verbal and visual cues are effective in vocabulary learning.

3.1 Experimental Design

In the experiment, participants learn 36 German words and are tested on recalling both the German word (generation) and its English meaning (recognition).

The words are split into three sets, which means that each participant goes through the learning, recognition, and generation cycle three times. Words in each set are randomly shuffled for each participant. At the end of the experiment, we also ask participants to rate the helpfulness of the cues.

Learning and Testing. We provide each participant with both instructions on the study and the content that helps them learn the word; see Sect. 3.2 for details. Each word has a 30-s limit for the participant to memorize, and the participant can choose to move on to the next word after 15 s. After 30 s, we automatically move on to the next word. German words are pronounced twice, after 2 s and 7 s, respectively, after being displayed. We show a timer to participants to make them aware of the time remaining for each word. Participants have 15 s for both recognition and generation during testing. To avoid confusion between the two tests, we provide instructions such as "What is this in English?" and "What is this in German?". For generation, we also ask participants to use *a, o, u, s* instead of Umlaut *ä, ö, ü, ß*. We show a timer to participants as well. Words in both tasks are randomized in order.

Participants. We recruit participants from Amazon Mechanical Turk [2]. We require participants to be native English speakers with no German language experience. Considering the experiment takes about 40 min, we paid each participant $7.25 and added a bonus of $2.75 for those who got a score of over 70% on the final test. The bonus encourages participants to do their best. However, we acknowledge that some participants may cheat on tests to achieve a high score by using an external dictionary, which we have no control of.

Web Interface. We implement a React web application as our participant interface, which is designed based on the previous study [19]. We place an IRB-approved consent form on the front page and only participants who agree can participate in the experiment; the form explains in detail how the experiment is structured. We also show an example with a German word not in our evaluation set to clarify the procedure to participants. We collect metadata on time spent both during learning and testing, along with the responses to further investigate participant behavior.

3.2 Experimental Conditions

We first divide participants into two groups based on how the keyword was generated: automatically (auto-cue) and manually (manual-cue). Among many combinations of verbal and visual cues that can be presented to the participants, we choose conditions that enable both intra- and inter-group comparisons. We recruit a total of 80 participants for our study, with 20 in each condition.

As shown in Fig. 2, we show the example of our web interface on how the content is displayed in different conditions. For intra-group comparisons, we

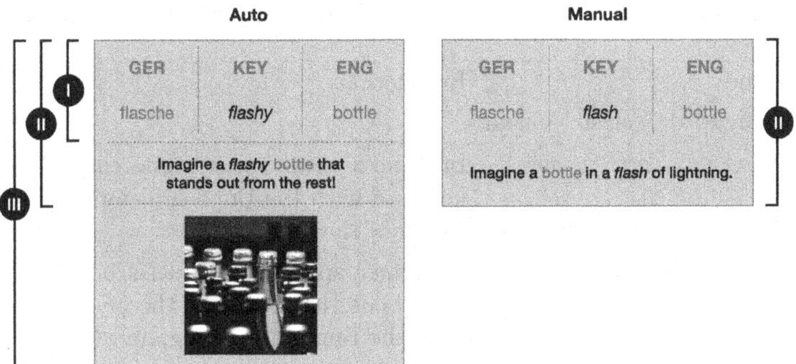

Fig. 2. A snapshot of our web interface shown to experiment participants.

further divide the auto-cue group into three conditions: Condition I is only provided with the TransPhoner-generated keyword, Condition II is provided with the keyword and the verbal cue generated by our pipeline, and Condition III is provided with the keyword and both the verbal and visual cues generated by our pipeline. For the inter-group comparisons, we provide both the auto-cue group and manual-cue group with information in Condition II. We note that the previous study [19] compared the groups with Condition I by not including verbal cues that were originally presented with the manually generated keywords [9]. The manually generated verbal cue and keyword should be considered as a whole since the keyword might have been chosen to provide a verbal cue with the best imageability among many keyword candidates.

We refer to these four conditions as Auto-I, Auto-II, Auto-III, and Manual-II. The instructions for each condition are shown in Table 1. We use the same instructions for Condition I from Savva et al. Our instructions for Condition II tell participants to create an imagery of a scene specified in a verbal cue. Our instructions for Condition III tell participants to remember the image, which is based on the verbal cue that describes a specific scene.

3.3 Evaluation Metrics

We use different metrics to score recognition and generation. For recognition, we use cosine similarity between the word embeddings [4] between the answer and the response. We also consider responses that miss "to" for "to"-infinitives to be correct. Unlike recognition, as a novice German learner, generation is bounded to the orthographic feature of vocabulary. Therefore, we use a standardized (subtracting 1 and normalizing to 1) Levenshtein distance to score generation, following previous studies [19]. We also ask participants to evaluate the helpfulness of the cues using a 5-point Likert scale, which is provided along with the entire 36 words and the cues.

Table 1. Cues and instructions we used for different experimental conditions.

Cond.	Cue			Instruction
	Keyword	Verbal	Visual	
I	yes	no	no	Imagine a visual scene connecting the given keyword with the English meaning, and the sound of the German word
II	yes	yes	no	Imagine a specific scene described in the verbal cue that connects the given keyword with the English meaning, and the sound of the German word
III	yes	yes	yes	Remember the image by following the verbal cue that connects the given keyword with the English meaning, and the sound of the German word

3.4 Results and Discussion

After we exclude participants who did not understand the experiment properly, such as those who wrote down the keyword when recalling the English meaning, we have a total of 72 participants: Auto-I (20) with an average age of 25.4 years (SD = 2.3), Auto-II (17) with an average age of 24.2 years (SD = 1.7), Auto-III (18) with an average age of 24.8 years (SD = 1.6), and Manual-II (17) with an average age of 25.3 years (SD = 1.1).

Figure 3 shows per-participant experimental data in box plots averaged among 36 German words. Learning time is time spent memorizing a word, while testing time is the average time on recognition and generation of the word. Similarly, the combined score is an average of recognition and generation scores. Learning time, testing time, and Likert scale are normalized by their maximum value.

The median of time spent on learning was 19.8, 18.9, 18.6, and 19.2 s, respectively, for the four conditions out of the 30 s time limit, which may suggest that cognitive load across different conditions is similar. The median of time spent on testing, i.e., the average time spent on recognition and generation, was 8.85, 9.75, 8.7, and 7.95 s out of the 15 s time limit. The median of the 5-point Likert scale was 4.2, 3.95, 4.25, and 4.4.

Now, we analyze the combined score based on the per-word combined score, as shown in Fig. 4. We perform a one-tailed Welch's t-test assuming unequal variances on the hypotheses of one condition being better than another. We set our level of significance to 5%. We detail each hypothesis below. Case A, B, and C in Fig. 4 are words we present with content generated through our pipeline for qualitative analysis.

Auto-I vs. Auto-II: Does a Verbal Cue Help Learning? We hypothesize that Auto-II, with additional verbal cues, will result in better recognition and generation scores than Auto-I, which uses only keywords. We define our null hypothesis (H_0) and alternate hypothesis (H_a) as follows:

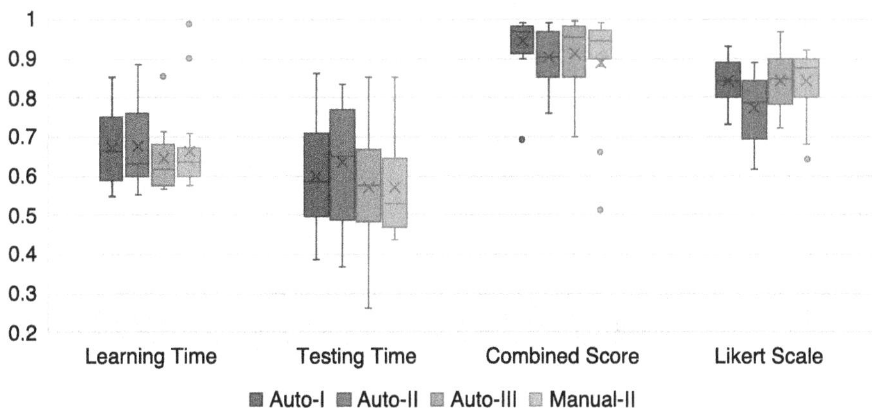

Fig. 3. Box plots of per-participant data for each experimental condition.

- H_0: $\mu_{Auto-II} \leq \mu_{Auto-I}$
- H_a: $\mu_{Auto-II} > \mu_{Auto-I}$

A right-tailed test shows there is no significant effect of verbal cues, $t(33) = -1.79, p = 0.96$; we cannot reject H_0. On the contrary, a left-tailed test shows statistical significance in favor of the keyword-only condition, $t(33) = 1.79, p = 0.04$. This result can be explained by several factors: The participants might have done rote rehearsals instead of building links as instructed in Table 1. Moreover, participants may come up with their own verbal cues that are more memorable than automatically generated ones. Personalized by default, participants' own verbal cues may be a better fit for each individual's own experience.

Auto-II vs. Manual-II: Are Automated Verbal Cues Effective? We hypothesize Manual-II to be an upper bound of Auto-II since the former cues are generated by experts in psycholinguistics. Therefore, we define our null hypothesis and alternate hypothesis as follows:

- H_0: $\mu_{Manual-II} \leq \mu_{Auto-II}$
- H_a: $\mu_{Manual-II} > \mu_{Auto-II}$

A right-tailed test shows that there is no significant difference between the two conditions, $t(24) = -0.32, p = 0.62$; we cannot reject H_0. In Fig. 4, we show three words where participants perform better in the Auto-II condition than Manual-II (case A) and otherwise (case B), respectively. Case A in Table 2 shows that auto-generated cues are more memorable than manual cues even with a grammatical error (risen should be raised) or not realistic (Reuben sandwich calling your name). Case B, on the other hand, contains keywords where auto-generated cues are not frequently used (Triton, frizzy) or making it hard to imagine (a wagon with stories). This result implies that although we can

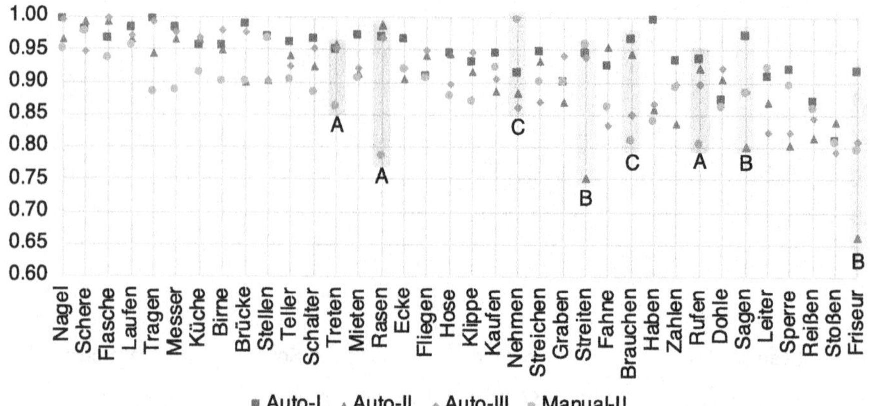

Fig. 4. Per-word combined score for all four experimental conditions, with three cases highlighting some words that work especially well with certain cues.

Table 2. Examples of automatically and manually generated verbal cues. A keyword is represented in *italic*, while a meaning is in **bold**.

Case	Word	Auto	Manual
A	Treten	Imagine **stepping** into *treason*, a treacherous path that can never be undone	Imagine you **step** on a stair *tread*
	Rasen	Imagine a *risen* **lawn** that is lush and green!	Imagine your **lawn** covered in *raisins*
	Rufen	Imagine *Reuben* **calling** out your name!	Imagine you **call** a friend to put a new *roof* on a cottage
B	Streiten	Imagine *Triton* and his trident **quarreling** with the waves	Imagine you **quarrel** about the Menai *straits*
	Sagen	Imagine a *wagon* full of stories just waiting to be **told**!	Imagine you **tell** someone *sago* is good for them
	Friseur	Imagine a **hairdresser** who can tame even the most *frizzy* hair!	Imagine your **hairdresser** inside a *freezer*
C	Nehmen	Imagine *Newman* taking the initiative to **take** action!	Imagine you **take** a *name* in your address book
	Brauchen	Imagine **needing** to fix a *broken* heart	Imagine *brokers* **need** much experience

automatically generate high-quality verbal cues, choosing appropriate keywords remains crucial. Therefore, we need to add keyword generation to the pipeline and evaluate the quality of both generated keywords and the verbal cue.

Auto-II vs. Auto-III: Does a Visual Cue Help Learning? We hypothesize better performance by Auto-III, which uses additional visual cues, than Auto-II. Therefore, we define our null hypothesis and alternate hypothesis as follows:

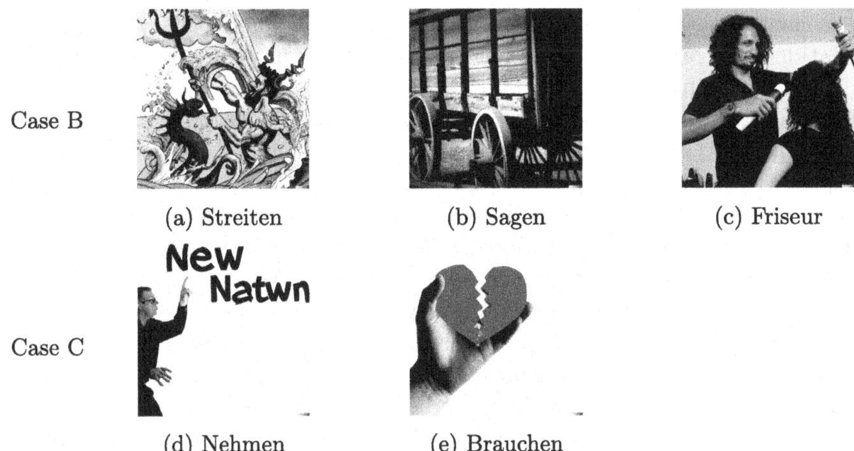

Fig. 5. Examples of visual cues generated by our pipeline in cases where they are helpful to participants and cases where they are not.

- H_0: $\mu_{Auto-III} \leq \mu_{Auto-II}$
- H_a: $\mu_{Auto-III} > \mu_{Auto-II}$

A right-tailed test shows that there is no significant difference between the two conditions, $t(32) = 0.39, p = 0.35$; we cannot reject H_0. In Fig. 4, we show three words for the cases where participants perform better in the Auto-III condition than in Auto-II (case B) and two for when it does not (case C), respectively. Case B shows that Auto-III, which has additional visual cues than Auto-II, performs similarly as Manual-II. Considering the previous comparison that Auto-II has a lower score than Manual-II, we see that Auto-III does somewhat outperform Auto-II. Therefore, we can conclude that visual cues help participant build the imagery link to some degree.

For a more qualitative analysis, Fig. 5 shows visual cues generated by our pipeline. Figure 5(a–c) shows that visual cues may be helpful in cases where keywords that lack imageability and are not frequently used (Triton, frizzy) or in cases where auto-generated verbal cues are hard to imagine (a wagon with stories). However, as shown in case C, visual cues for abstract words (to take, to need) do not help much. Figure 5(d–e) shows that in these cases the generated image is not descriptive enough to facilitate the imagery link. Interestingly, the Likert scale score was higher for Auto-III than Auto-II in every word except one. This result implies that participants think it is helpful to have additional visual cues. However, we cannot create effective visual cues for every word. Generating descriptive visual cues, especially for abstract words, remains a challenging task.

4 Conclusions and Future Work

In this paper, we explored the opportunity of using large language models to generate verbal and visual cues for keyword mnemonics. A preliminary human experiment suggested that despite showing some promise, this approach has limitations and cannot reach the performance of manually generated cues yet.

There are many avenues for future work. First, we need a larger-scale experiment in a real lab study, which provides us a controlled environment to test both short-term and long-term retention. Since we only tested short-term retention, it is possible that no approach can significantly outperform others. We also need more psycholinguistics perspectives on constraining time spent on learning and testing. By conducting the research in a more controlled environment, we can use additional information (e.g., demographics, language level) to help us conduct a deeper analysis of the results. We do clarify that using Amazon's Mechanical Turk to conduct experiments is standard in prior work, which is part of the reason why we chose this experimental setting. To track long-term retention, we likely have to resort to knowledge tracing models that handle either memory decay [11] or open-ended responses [14]. Second, we can extend our pipeline by generating the keyword automatically as well instead of using TransPhoner-generated keywords, which may make our approach even more scalable. One important aspect that must be studied is how to evaluate the imageability of the keywords and verbal cue that contains both keywords and vocabulary, which remains challenging. Third, we can generate personalized content for each participant. We may provide additional information in the text generator about the topic they are interested in that we could use to generate a verbal cue. Moreover, we can generate a story that takes all words into account. It is also possible to generate verbal cues in L2 as well, which may help learners by providing even more context. Fourth, instead of the pronunciation of the word, we can use other features in language to generate verbal cues. For example, when learning Mandarin, memorizing Chinese characters is as important as learning how to pronounce the word. The Chinese character 休 means *rest*, which is xiū in Mandarin. The character is called a compound ideograph, a combination of a person (人) and a tree (木), which represents a person resting against a tree. Combined with a keyword, *shoe*, for example, we could accomplish two goals with one verbal cue, "A person is *resting* by a tree, tying up their *shoe*." This way, we can make visual cues more descriptive for abstract words.

Acknowledgements. The authors thank the NSF (under grants 1917713, 2118706, 2202506, 2215193) for partially supporting this work.

References

1. von Ahn, L.: Duolingo. https://www.duolingo.com
2. Amazon: Amazon mechanical turk. https://www.mturk.com
3. Atkinson, R.C., Raugh, M.R.: An application of the mnemonic keyword method to the acquisition of a Russian vocabulary. J. Exp. Psychol. Hum. Learn. Mem. **1**(2), 126 (1975)
4. Bojanowski, P., Grave, E., Joulin, A., Mikolov, T.: Enriching word vectors with subword information. Trans. Assoc. Comput. Linguist. **5**, 135–146 (2017)

5. Brahler, C.J., Walker, D.: Learning scientific and medical terminology with a mnemonic strategy using an illogical association technique. Adv. Physiol. Educ. **32**(3), 219–224 (2008)
6. Brown, T., et al.: Language models are few-shot learners. In: Advances in Neural Information Processing Systems, vol. 33, pp. 1877–1901 (2020)
7. Carrier, M., Pashler, H.: The influence of retrieval on retention. Mem. Cogn. **20**, 633–642 (1992). https://doi.org/10.3758/BF03202713
8. Ebbinghaus, H.: Memory: a contribution to experimental psychology. Ann. Neurosci. **20**(4), 155 (2013)
9. Ellis, N.C., Beaton, A.: Psycholinguistic determinants of foreign language vocabulary learning. Lang. Learn. **43**(4), 559–617 (1993)
10. Elmes, D.: Anki. http://ankisrs.net
11. Ghosh, A., Heffernan, N., Lan, A.S.: Context-aware attentive knowledge tracing. In: Proceedings of the ACM SIGKDD, pp. 2330–2339 (2020)
12. Larsen, D.P., Butler, A.C., Roediger, H.L., III.: Repeated testing improves long-term retention relative to repeated study: a randomised controlled trial. Med. Educ. **43**(12), 1174–1181 (2009)
13. Leitner, S.: So lernt man lernen. Herder (1974). https://books.google.com/books?id=opWFRAAACAAJ
14. Liu, N., Wang, Z., Baraniuk, R., Lan, A.: Open-ended knowledge tracing for computer science education. In: Conference on Empirical Methods in Natural Language Processing, pp. 3849–3862 (2022)
15. OpenAI: Dall-e 2. https://openai.com/dall-e-2
16. Ouyang, L., et al.: Training language models to follow instructions with human feedback. In: Advances in Neural Information Processing Systems, vol. 35, pp. 27730–27744 (2022)
17. Prabhumoye, S., Black, A.W., Salakhutdinov, R.: Exploring controllable text generation techniques. arXiv preprint arXiv:2005.01822 (2020)
18. Reddy, S., Labutov, I., Banerjee, S., Joachims, T.: Unbounded human learning: optimal scheduling for spaced repetition. In: Proceedings of the 22nd ACM SIGKDD International Conference on Knowledge Discovery and Data Mining, pp. 1815–1824 (2016)
19. Savva, M., Chang, A.X., Manning, C.D., Hanrahan, P.: TransPhoner: automated mnemonic keyword generation. In: Proceedings of the SIGCHI Conference on Human Factors in Computing Systems, pp. 3725–3734 (2014)
20. Siriganjanavong, V.: The mnemonic keyword method: effects on the vocabulary acquisition and retention. Engl. Lang. Teach. **6**(10), 1–10 (2013)
21. Sutherland, A.: Quizlet. http://quizlet.com
22. Ye, J., Su, J., Cao, Y.: A stochastic shortest path algorithm for optimizing spaced repetition scheduling. In: Proceedings of the 28th ACM SIGKDD Conference on Knowledge Discovery and Data Mining, pp. 4381–4390 (2022)
23. Zylich, B., Lan, A.: Linguistic skill modeling for second language acquisition. In: LAK21: 11th International Learning Analytics and Knowledge Conference, pp. 141–150 (2021)

Implementing and Evaluating ASSISTments Online Math Homework Support At large Scale over Two Years: Findings and Lessons Learned

Mingyu Feng[1(✉)], Neil Heffernan[2], Kelly Collins[1], Cristina Heffernan[3], and Robert F. Murphy[4]

[1] WestEd, San Francisco, California 94107, USA
{mfeng,kelly.collins}@wested.org
[2] Worcester Polytechnic Institute, Worcester, Massachusetts 01609, USA
nth@wpi.edu
[3] The ASSISTments Foundation, Auburn, Massachusetts 01501, USA
cristina.heffernan@assistmetns.org
[4] LFC Research, Mountwin view, California 94043, USA
bfmurph@icloud.com

Abstract. Math performance continues to be an important focus for improvement. The most recent National Report Card in the U.S. suggested student math scores declined in the past two years possibly due to COVID-19 pandemic and related school closures. We report on the implementation of a math homework program that leverages AI-based one-to-one technology, in 32 schools for two years as a part of a randomized controlled trial in diverse settings of the state of North Carolina in the US. The program, called "ASSISTments," provides feedback to students as they solve homework problems and automatically prepares reports for teachers about student performance on daily assignments. The paper describes the sample, the study design, the implementation of the intervention, including the recruitment effort, the training and support provided to teachers, and the approaches taken to assess teacher's progress and improve implementation fidelity. Analysis of data collected during the study suggest that (a) treatment teachers changed their homework review practices as they used ASSISTments, and (b) the usage of ASSISTments was positively correlated with student learning outcome.

Keywords: ASSISTments · math learning · effective teaching · AI-based program · school implementation

1 Introduction

Math performance continues to be an important focus for improvement in the United States. Due to the promise of technology as a tool for improving mathematics education and closing the achievement gap, the use of educational technology in K-12 education has expanded dramatically in recent years, accelerated by the COVID-19 pandemic. The

N. Wang et al. (Eds.): AIED 2023, LNAI 13916, pp. 28–40, 2023.
https://doi.org/10.1007/978-3-031-36272-9_3

AIED and intelligent tutoring systems researchers and developers have built numerous technology-based learning platforms and programs, and many have been shown to be effective in a lab setting, or with a small number of closely monitored classrooms. Yet few of these products have been implemented at large scale in authentic school settings over an extended period. The challenges of wide adoption and effective implementation in schools come from several aspects, such as understanding school settings and meeting school priorities, availability of technology infrastructure to guarantee sufficient student/teacher access to equipment, integration of the program into established classroom practices, as well as training and continuous support for users to ensure sustained use with fidelity. In this paper we report on a large-scale efficacy randomized controlled trial (RCT) in diverse settings of the state of North Carolina in the U.S. In particular, we focus on the implementation of the ASSISTments platform and discuss how we have addressed each challenge to ensure faithful implementation.

The ASSISTments platform [1] is a technology-based, formative assessment platform for improving teacher practices and student math learning outcomes. As students work through problems and enter their answers into ASSISTments, the system provides immediate feedback on the correctness of answers and offers additional assistance in the form of hints or scaffolds. Students' performance on ASSISTments problems serves as an assessment of proficiency, enabling teachers to adjust classroom instruction and pacing to match the knowledge base of the class. ASSISTments was identified as effective at improving student learning and changing teacher's homework review practices during an efficacy study in Maine ([2, 3], effect size $g = .22, p < .01$). We recently conducted a large scale RCT to replicate the Maine study and see whether the found effects replicate in a heterogeneous population that more closely matches national demographics. The replication study to test the impact of the ASSISTments platform on students' math learning was guided by the following research questions:

- Student learning: *Do students in schools that use ASSISTments for homework learn more than students in schools that do homework without ASSISTments?*
- Impacts on classroom instruction: *Does using ASSISTments lead to adjustment in teachers' homework practices, and, if so, how they have done this?*
- Relationship between usage and student learning outcome: *What are the effects of implementation fidelity and dosage on learning?*

2 Background

2.1 The ASSISTments Program

Over the past two years, ASSISTments use in schools increased significantly, going from supporting 800 teachers to supporting 20,000 teachers and their 500,000 students. A rapid review that synthesizes existing evidence in online programs that promoted learning recommended ASSISTments as one of the few digital learning programs for use in response to the COVID pandemic [4]. A review of 29 studies that met rigorous standards of randomization [5] indicated that ASSISTments was one of only "Two interventions in the United States [that] stand out as being particularly promising."

The platform uses technology to give teachers new capabilities for assigning and reviewing homework and to give students additional support for learning as they do homework. Content in ASSISTments consists of mathematics problems with answers and hint messages. These mathematics problems are bundled into problem sets which teachers can use ASSISTments to assign to students in class or as homework. Students first do their assigned problems on paper and then enter their answers into ASSISTments to receive immediate feedback about the correctness of their answers, and/or hints on how to improve their answers or help separate multi-step problems into parts. One type of problem set is mastery-oriented "Skill Builders". Each skill builder provides opportunities for students to practice solving problems that focus on a targeted skill, until they reach a teacher-defined "mastery" threshold of proficiency (e.g., a streak of three correct answers on similar math problems). ASSISTments also implements the research-based instructional strategy of spaced practice [6] by automatically re-assessing students on skills and concepts that were "mastered" earlier at regular intervals, and providing further opportunities for students to hone those skills.

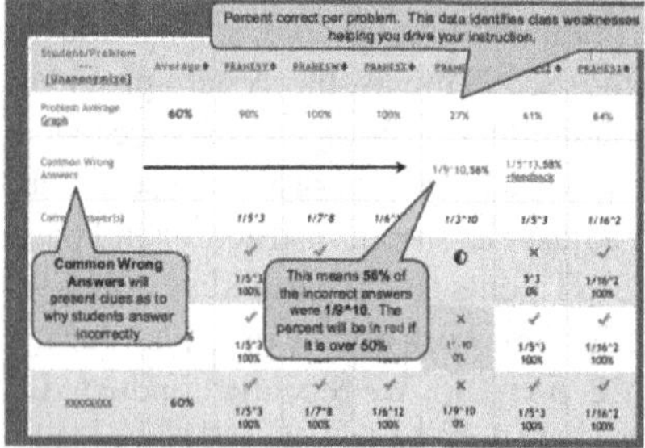

Fig. 1. An item report for teachers.

ASSISTments provides teachers with real-time, easily accessible reports that summarize student work for a particular assignment in a grid format (Fig. 1), which teachers can use to target their homework review in class and tailor instruction to their students' needs. These reports inform teachers about the average percent correct on each question, skills covered in the assignment, common wrong answers, and each student's answer to every question. The color-coded cells provide information on each student's performance on each problem, in a format that enables teachers to focus on the problems that were difficult for many students. Teachers use information about individual students to form instructional groups and address common difficulties.

2.2 Theoretical Framework

The design and development of ASSISTments aligns to theory- and empirically-based instructional practices of formative assessment [7] and skill development [8], and is built upon the following theoretical foundations and empirical research.

Immediate Feedback to Learners. Feedback has been identified as a powerful way to increase student learning [9]. An extensive set of studies (e.g., [10–12]) has found significant learning gains in response to feedback within computer-based instruction. A common form of feedback is correctness feedback, in which feedback informs the learner if their response is correct or incorrect, with no other information. Studies have shown positive results of correctness feedback [13, 14].

Formative Assessment. Formative assessment provides feedback to both teachers and students by informing teachers of student learning and progress and students of their own performance in relation to learning goals. Several decades of research have shown formative assessment, also known as assessment for learning, to be an effective way of improving both teaching and student learning [15–19].

Education technology can facilitate formative assessment and have a significant effect on improving math learning [20, 21]. Teachers benefit from the information and analytics on student learning and performance that these tools can provide, which help them modify instruction and better adapt to student learning needs [21, 22]).

Mastery Learning and Spaced Practice. In mastery learning, students only advance to a new learning objective when they demonstrate proficiency with the current one. Mastery learning programs lead to higher student achievement than more traditional forms of teaching [23, 24]. An Institute of Education Sciences (IES) Practice Guide [6] recommends spacing practice over time. Research has demonstrated that retention increases when learners are repeatedly exposed to content (see [25–27]).

Homework to Boost Learning. Homework is often required in middle schools and expected by students and parents, and a meaningful amount of instructional time is allocated to homework [28, 29]. Yet it is also somewhat controversial and perceived as needing improvement (e.g. [30, 31]). Homework provides a key opportunity for individual practice, which is undeniably important to mathematics learning. To "make homework work," teachers are recommended to streamline assignments and be intentional with each assignment, taking into account factors such as whether students understand the purpose of the assignment and the feedback the teacher will provide [32].

2.3 Research Design

Sample and Settings. The study[1] took place in North Carolina (NC), a state more demographically representative of the U.S. than Maine. Sixty-three schools from 41 different districts enrolled in the study. Demographic characteristics of the schools are shown in Fig. 2, comparing to the Maine study and the population in the U.S.. These schools served several different grade levels (6-8, 8-12, K-8) and were distributed across rural, town, suburban, and city communities (33 rural, 11 town, 8 suburban, and 11 city). Of the 63 schools, 18 were charter schools, 45 were public schools, and 48 received Title 1 funding. 102 7th grade math teachers and their classrooms enrolled in the study.

Baseline School Characteristics	Treatment Schools	Control Schools	NC Population	US Population	Sample in Original Study in Maine
Avg. % of White students	58	58	51	53	93
Avg. % Hispanic students	13	16	14	23	2
Avg. % Black students	20	21	27	16	2
Avg. % socioeconomically disadvantaged	53	48	59	54	39
% of Rural Schools	47	58	49	38	Over 50%

Fig. 2. Demographics of Participating Schools

[1] The study has been pre-registered on Registry of Efficacy and Effectiveness Studies (REES) https://sreereg.icpsr.umich.edu/framework/pdf/index.php?id=2064. The study has been approved by the Institutional Review Board at Worcester Polytechnic Institute and WestEd. Participating teachers all signed consent forms. Parents received a notification letter and opt-out form for their children.

Study Design. Schools within each district were paired based on their demographic characteristics and student prior performance on state math and English language arts (ELA) tests and then randomly assigned to a treatment or business-as-usual control condition. Overall, 32 schools and 60 teachers were assigned to the treatment condition and 31 schools and 52 teachers to the control condition. All seventh-grade math teachers in a school had the same assignment of condition.

The study was carried out during the 2018–19 and 2019–20 school years. During the study, ASSISTments was conceptualized as consisting of both teacher professional development (PD) and the use of the platform by teachers and students as illustrated by the theory of change in Fig 3.

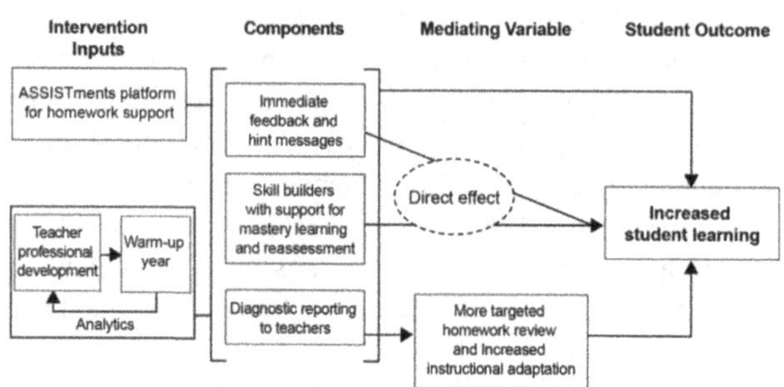

Fig. 3. ASSISTments Theory of Change

The program was implemented by all Grade 7 teachers in the treatment schools over two consecutive years—the first year was a ramp-up year, where teachers in the treatment condition learned and practiced using the tool with their students. The second year of implementation was the measurement year when treatment teachers continued to use ASSISTments with a new cohort of 7th grade students whose math learning outcome was measured in spring 2020. Treatment teachers received 2 days of PD during summer 2018 and 2019, and additional coaching and technical assistance distributed across the school year via webinars, video conferencing, and 2–3 times in-person visits each school year. Teachers had the full discretion to either select pre-built content from ASSISTments that aligns with the learning standards they teach in class, or request to have their own homework problems built in the system for them to assign.

Teachers in the control condition did not have access to ASSISTments or to the PD and coaching till summer 2020 ("delayed treatment"). They continued with their typical homework practices ("business-as-usual") including the possible use of other online homework platforms or supplemental programs in class. We used an end-of-year survey and daily instructional logs to collect information on their homework assignment and reviewing practices as well the online platforms control teachers have used to assign homework. Teachers reported using a variety of technology programs, including Kahoot,

Schoolnet, iXL, Khan Academy, iReady, Quizlet, Desmos, MobyMax, Prodigy, BrainPoP, SuccessMaker, Freckle, etc., as well as online resources provided by curriculum providers.

Influence of COVID. The onset of COVID-19 forced the closure of study schools March 2020 and the move to remote instruction until the end of the school year across all participating districts. This forced unplanned changes to the implementation of the study. It shortened the expected duration and lessened intensity of the use of ASSISTments by 2-3 months, which potentially could have affected the study's measured effect on student learning. ASSISTments use records indicated a significant drop in use coinciding with school closures. By the middle of April, 43% teachers in the treatment group resumed using ASSISTments to support their remote instruction, although at a reduced level compared to prior to school closures. Figure 4 summarizes usage before and after the outbreak of the pandemic during the 2019-20 school year. Additionally, the state End-Of-Grade test (EOG), which was planned to serve as the primary outcome measure of student learning, was canceled for spring 2020. We had to request teachers to administer a supplemental online assessment as a replacement.

Metrics	Pre.Covid/In Person				Post.Covid/Distance			
	mean	sd	min	max	mean	sd	min	max
% Problems Completed by Students out of the problems they started	0.94	0.12	0	1	0.91	0.23	0	1
% of Assignments Started by Students	0.63	0.27	0	1	0.45	0.38	0	1
Average % of Assignments Completed by Students out of the ones they were assigned to	0.49	0.29	0	1	0.38	0.38	0	1
Average # of days per week students worked in ASSISTments	1.54	0.60	0	3.7	0.80	0.88	0	5
Average minutes per week students worked in ASSISTments	17.92	12.18	0	92.6	11.53	17.79	0	208.4
% of assignment reports viewed (avg per teacher)	0.46	0.41	0	1	0.62	0.33	0	1

Fig. 4. Summary of ASSISTments Usage during 2019-20

3 Implementation of ASSISTments at Scale

The adoption and successful implementation of a technology-based program in authentic school settings for an extended period is influenced by numerous factors. Below we discuss in detail how we addressed some of them during multiple stages of the study.

3.1 Recruitment

Recruitment participants for large school-based studies, particularly RCTs, poses a significant challenge. Schools are often engaged in various initiatives at any given time. Given the dynamic nature of district and school-level initiatives, gaining buy-in for a research-based program like ASSISTments can be difficult, even though it is a free platform that has demonstrated rigorous empirical impact. Some schools may view ASSISTments as less sophisticated than similar commercial tools and may already be using paid tools for similar purposes. Moreover, schools may be reluctant to adopt a new free tool due to associated costs and the learning curve involved.

To address these challenges, researchers treated recruitment as an advertising campaign and identified factors that supported or inhibited participation, responding accordingly. Uncertainty associated with randomization hindered schools' willingness to participate, which was mitigated by offering control schools free access to ASSISTments in a non-study grade level plus delayed treatment for control schools. North Carolina schools varied in their technology adoption and teacher preferences for learning management systems (LMSs). To overcome these concerns, the recruitment was confined to districts that have implemented 1:1 access to computers. ASSISTments was developed as a versatile tool that can be used on any device without internet access and easily integrated with most common LMSs in North Carolina. The team examined district priorities based on public information and targeted recruitment efforts toward schools with a history of low math performance that prioritize the use of data to personalize student learning—a fit for the ASSISTments program.

3.2 Understanding School Context

Early during the study, we conducted interviews with school principals to identify factors that could affect the fidelity of implementation. Participants were asked about: their concerns regarding students' performance in math; the school's major priorities or areas of focus; the technology available to students at school, specifically for math, and at home; and their homework and data-use policy. Principals in the treatment group were also asked whether they foresaw of any obstacles to the use of ASSISTments or any additional supports that would help teachers use ASSISTments. The interviews helped us discover what could facilitate the use of ASSISTments (e.g., school initiative focusing on improvement and student identity; common planning time within grade level; established homework policy) and what could inhibit its use in the schools (e.g., limited access to technology). We then customized the communication and support for each school to ensure the program align with the school' context for easier implementation.

3.3 Training and Continuous Support

Research suggested when both PD and formative assessment technology are provided, teachers can learn more about their students and adapt instruction, resulting improvements in student outcomes. In this study, a former high school teacher and remediation program organizer coached treatment teachers on integrating ASSISTments into instruction and provided ongoing coaching and technical support throughout the school year via webinars, video conferencing, and in-person visits. During the PD, treatment teachers learned techniques such as encouraging students to rework problems they initially got wrong, focusing on incorrect homework problems, reviewing solution processes for difficult problems, and addressing common misconceptions. The PD training helped teachers to utilize the technology's reports to make informed decisions, such as identifying which problems or students to prioritize. As a result, teachers were able to personalize instruction for individual students, groups, or the entire class; for example, they could assign additional practice to individual students (e.g., by assigning skill builders) or adding other problems or hints to existing problem sets.

3.4 Specifying a Use Model and Expectation

To aid teachers in integrating ASSISTments into their existing instructional practices, the intervention was framed into a 4-step loop that aligns with common homework routines, as illustrated in Fig. 5. This approach helped teachers avoid the burden of figuring out how to utilize the program in their instruction and ensured consistency in program implementation. Per the use model, treatment teachers were expected to assign approximately 20–23 minutes of homework via the platform twice per week using a combination of textbook-based problems and pre-built content. Teachers were also expected to open reports for at least 50% of assignments. Opening the reports is a precursor to adapting instruction. Teachers were given autonomy in determining the amount and type of homework to assign. The clearly specified expectation served as an objective for sufficient dosage of implementation, which is crucial for the effectiveness of any intervention.

3.5 Monitoring Dosage and Evaluating Quality of Implementation

Corresponding to the four-step loop, the ASSISTments team created a rubric (Fig. 5) that was used to assess teacher's progress in implementing the four steps during the school year. During the in-person school visits, the local coach evaluated each teacher's status according to the rubric and recorded their evaluation in a coach log, which was used to track a teacher's progress over time and gauge whether additional support needed to be provided to the teacher. The color-coded log (one row per teacher, Fig. 6) made it easy for the coach to identify teachers who have improved from visit 1 to visit 3 (e.g., rows 4 or 5), or teachers who regressed over time (e.g., rows 1 or 6).

	2	1	0
Finding and Assigning Content	Teacher knows which content they want to use and where to find it.	Teacher is somewhat clear on what content to use and where to find it.	Teacher is unable to find content they want to use.
Students Doing Content	Students are completing at least 9 problems per week.	Students are doing content some of the time. (1-2 times per week)	Students are not doing content. (Less than 1 time per week)
Teachers Viewing Report Data	Teacher is checking reports for at least 75% of problem sets.	Teacher reviewed some data in the item report.	Teacher is not reviewing data in the report.
Teachers Sharing Report Data with Students	Teacher accomplishes the following during a report review: 3. Overviewing data on problems with which students struggled 4. Reteaching/direct instruction, student discussion or other method used to engage with the data.	Teacher is reviewing reports with students some of the time. (1-2 times per week)	Teacher is not reviewing the report with students. (Less than 1 time per week)

Fig. 5. 4-Step Rubric for Evaluating Progress

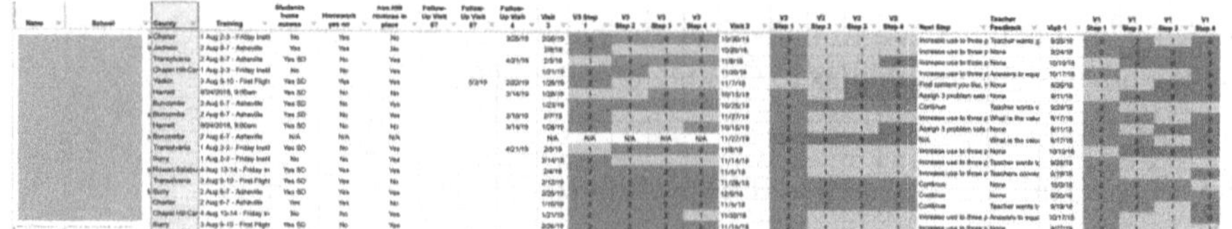

Fig. 6. A Part of Coach Logs of School Visits during the 2019-20 School Year

4 Data Collection

Student Learning. We obtained student demographic, enrollment and prior grade state test performance data from the state-wide database hosted by the North Carolina Education Research Data Center (NCERDC)[2]. Student demographic data accessed include gender, race/ethnicity, free and reduced-price lunch status, and individualized education program status. Students' 6th grade EOG math scores served as a measure of prior achievement. As a replacement for the state test, the study team requested teachers to administer the online Grade 8 Mathematics Readiness Test (MRT) early May during remote learning time. The test is aligned with the Common Core State Standards. An analysis found a strong alignment between MRT and the NC Course of Study Standard Topics. Analysis of the response data suggested a correlation of 0.68 between students' MRT scores and their 6th grade EOG test scores from spring 2019, and the reliability of the test (KR20 for dichotomous items) was 0.94.

Teacher Homework Practice Measure. An interview was conducted with a sample of treatment and control teachers to their homework practices and the kinds of instructional supports the school and district provide to teachers. A quantitative measure of teacher homework review practice-Targeted Homework Review-was created from teacher self-reports on weekly instructional logs. All teachers were asked to complete daily 10-minute online logs during 5 consecutive instructional days across 3 collection windows across the school year for a total 15 instructional days. On average, the log completion rate was 81 percent for control teachers and 96 percent for treatment teachers. Teachers were asked if they reviewed assignments with the whole class and the ways they select which problems to review with the class (e.g., *I reviewed all the problems, I reviewed a sample of problems that I think my students might have difficulty on, I asked my students what problems they wanted to go over*).

Student and Teacher Use of ASSISTments. The primary source of ASSISTments use data was the electronic use records collected by the ASSISTments system. All student and teacher actions on the platform are time-stamped. Student system use data includes their login time, duration of each session, the number of problems attempted, and answered correctly and incorrectly, whether an assignment was completed on time, and student-to-system interactions during problem-solving, such as requesting hint or commenting on a question. Teacher use data includes the dates each assignment was

[2] Our agreement with NCERDC doesn't permit sharing of obtained data with any third parties. Other data collected during the study has been deposited to the Open ICPSR data repository (https://www.openicpsr.org/openicpsr/project/183645/version/V1/view).

created, the kind of problems assigned, the type of assignment (e.g., whole class, or individualized), assignment due date, and whether the teacher opened a report. These data allowed us to assess the extent to which students used ASSISTments to complete their math homework and the extent to which teachers assigned problems and monitored students' homework performance according to the implementation model.

5 Analysis and Results

Student Learning Outcomes. For the impact analysis, we used a two-level hierarchical linear model (i.e., students nested within schools), controlling for school-level and student-level characteristics and students' prior achievement on their sixth-grade math state tests. The model didn't detect a significant difference between treatment and control students in their performance on the MRT test ($p > 0.05$).

Teacher Practice Outcomes. Analysis of teacher interview transcripts showed, all teachers in the treatment group noticed some changes in their instructional practices since they began using ASSISTments. Some of the changes identified by the teachers included: more thoughtful selection of problems; more one-on-one teaching; and being more efficient and more targeted in their instruction. Most of them have found the ASSISTments reports to be valuable and have shared them with students by projecting the anonymized reports on a screen in front of the class.

An HLM model with teacher and school levels was posited to look at the effect of ASSISTments on teacher practices, adjusting for the same school level covariates as in the student achievement analysis. The results indicated that after adjusting for school characteristics, treatment teachers reported a significantly percentage (17.4) of times when they applied targeted homework review practice ($p < 0.05$).

Relationship Between Usage and Learning Outcome. We conducted an exploratory analysis using the three-level hierarchical linear regression model using usage indicators to predict student achievement. Two usage indicators, *total number of problems completed by students*, and *total number of assignments at the classroom level*, are positively related to student performance on MRT. The results indicate that more homework assigned by the teacher through the ASSISTments platform is associated with higher math achievement ($p = .029$). More problems completed by a student is also associated with higher math achievement ($p < .001$).

6 Conclusion

Improving mathematics education is an important educational challenge. Many technology solutions have been developed and demonstrated evidence of promise during pilot studies, but many stayed in the research stage and very few have been effectively implemented in authentic school settings on a large scale for extended periods of time. In this paper, we reported on a large-scale efficacy study that explicitly adopted planned variation to test the impact of the ASSISTments program under a variety of setting and implementation characteristics. Particularly, we shared the procedures the team took to introduce the researcher-developed, free program to middle school teachers, and approaches to ensure consistency and faithful implementation of the program.

The results of our analyses suggested ASSISTments helped teachers practice the key aspects of rich in-depth formative assessment-driven instruction, and more use of the program was associated with higher student math achievement but didn't provide further evidence of the replicability of student learning impact on the selected measure.

The reported results are bound by several limitations related to the disruption of classroom instruction brought on by the onset of the COVID pandemic, including the unexpected alteration to the implementation model for ASSISTments and a shortening of the expected duration of the use of the intervention. The replacement outcome measure was not part of the state accountability assessment program and was administered at home during a time when teachers and schools were struggling to put together an instruction plan to engage students. Thus, the study experienced significant attrition in outcome data collection. Further analysis is ongoing to compare the sample, school context, implementation dosage, and counterfactuals between the original Maine study and the current study, and to better understand the nature of use of ASSISTments before and during the COVID pandemic.

Acknowledgement. This material is based on work supported by the Institute of Education Sciences of the U.S. Department of Education under Grant R305A170641. Any opinions, findings, and conclusions or recommendations expressed in this material are those of the authors and do not necessarily reflect the funders.

References

1. Heffernan, N., Heffernan, C.: The ASSISTments ecosystem: building a platform that brings scientists and teachers together for minimally invasive research on human learning and teaching. Int. J. Artif. Intell. Educ. **24**(4), 470–497 (2014). https://doi.org/10.1007/s40593-014-0024-x
2. Roschelle, J., Feng, M., Murphy, R.F., Mason, C.A.: Online mathematics homework increases student achievement. AERA Open **2**(4), 233285841667396 (2016). https://doi.org/10.1177/2332858416673968
3. Murphy, R., Roschelle, J., Feng, M., Mason, C.: Investigating efficacy, moderators and mediators for an online mathematics homework intervention. J. Res. Educ. Effectiveness **13**, 1–36 (2020). https://doi.org/10.1080/19345747.2019.1710885
4. Sahni, S.D., et al.: A What Works Clearinghouse Rapid Evidence Review of Distance Learning Programs. U.S. Department of Education (2021)
5. Escueta, M., Quan, V., Nickow, A.J., Oreopoulos, P.: Education Technology: An Evidence-Based Review. NBER Working Paper No. 23744. August 2017 JEL No. I20,I29,J24 (2017). http://tiny.cc/NBER
6. Pashler, H., Bain, P.M., Bottge, B.A., Graesser, A., Koedinger, K., McDaniel, M., et al.: Organizing instruction and study to improve student learning. IES practice guide (NCER 2007–2004). National Center for Education Research, Washington, D.C. (2007)
7. Heritage, M., Popham, W.J.: Formative Assessment in Practice: A Process of Inquiry and Action. Harvard Education Press, Cambridge, MA (2013)
8. Koedinger, K.R., Booth, J.L., Klahr, D.: Instructional complexity and the science to constrain it. Science **342**(6161), 935–937 (2013). https://doi.org/10.1126/science.1238056
9. Hattie, J., Timperley, H.: The power of feedback. Rev. Educ. Res. **77**(1), 81–112 (2007)

10. DiBattista, D., Gosse, L., Sinnige-Egger, J.A., Candale, B., Sargeson, K.: Grading scheme, test difficulty, and the immediate feedback assessment technique. J. Exp. Educ. **77**(4), 311–338 (2009)
11. Mendicino, M., Razzaq, L., Heffernan, N.T.: A comparison of traditional homework to computer supported homework. J. Res. Technol. Educ. **41**(3), 331–358 (2009)
12. Fyfe, E.R., Rittle-Johnson, B.: The benefits of computer-generated feedback for mathematics problem solving. J. Exp. Child Psychol. **147**, 140–151 (2016)
13. Kehrer, P., Kelly, K., Heffernan, N.: Does immediate feedback while doing homework improve learning? FLAIRS 2013. In: Proceedings of the 26th International Florida Artificial Intelligence Research Society Conference, pp. 542–545 (2013)
14. Kelly, K., Heffernan, N., Heffernan, C., Goldman, S., Pellegrino, J., Soffer-Goldstein, D.: Improving student learning in math through web-based homework review. In: Liljedahl, P., Nicol, C., Oesterle, S., Allan, D. (eds.) Proceedings of the Joint Meeting of PME 38 and PME-NA 36, vol. 3, pp. 417–424. PME, Vancouver, Canada (2014)
15. Black, P., Wiliam, D.: Classroom assessment and pedagogy. Assess. Educ.: Principles, Policy Pract. **25**(6), 551–575 (2018)
16. Bennett, R.E.: Formative assessment: a critical review. Assess. Educ.: Principles, Policy Pract. **18**(1), 5–25 (2011)
17. Dunn, K.E., Mulvenon, S.W.: A critical review of research on formative assessment: the limited scientific evidence of the impact of formative assessment in education. Pract. Assess. Res. Eval. **14**(7), 1–11 (2009)
18. OECD: Formative Assessment: Improving Learning in Secondary Classrooms. OECD Publishing, Paris (2005). https://doi.org/10.1787/9789264007413-en
19. Wiliam, D., Thompson, M.: Integrating assessment with learning: what will it take to make it work? In: Dwyer, C.A. (ed.) The Future of Assessment: Shaping Teaching and Learning, pp. 53–82. Lawrence Erlbaum Associates (2007)
20. Faber, J.M., Luyten, H., Visscher, A.J.: The effects of a digital formative assessment tool on mathematics achievement and student motivation: results of a randomized experiment. Comput. Educ. **106**, 83–96 (2017)
21. Koedinger, K.R., McLaughlin, E.A., Heffernan, N.T.: A quasi-experimental evaluation of an on-line formative assessment and tutoring system. J. Educ. Comput. Res. **43**(4), 489–510 (2010). https://doi.org/10.2190/EC.43.4.d
22. Sheard, M., Chambers, B., Elliott, L.: Effects of technology-enhanced formative assessment on achievement in primary grammar. Retrieved from the University of York website https://www.york.ac.uk/media/iee/documents/QfLGrammarReport_Sept2012.pdf (2012)
23. Anderson, J.R.: Learning and Memory: An Integrated Approach, 2nd edn. John Wiley and Sons Inc., New York (2000)
24. Koedinger, K., Aleven, V.: Exploring the assistance dilemma in experiments with Cognitive Tutors. Educ. Psychol. Rev. **19**, 239–264 (2007)
25. Rohrer, D.: The effects of spacing and mixing practice problems. J. Res. Math. Educ. **40**, 4–17 (2009)
26. Rohrer, D., Taylor, K.: The shuffling of mathematics practice problems improves learning. Instr. Sci. **35**, 481–498 (2007)
27. Rohrer, D., Pashler, H.: Increasing retention without increasing study time. Curr. Dir. Psychol. Sci. **16**, 183–186 (2007)
28. Fairman, J., Porter, M., Fisher, S.: Principals Discuss Early Implementation of the ASSISTments Online Homework Tutor for Mathematics: ASSISTments Efficacy Study Report 2. SRI International, Menlo Park, CA (2015)
29. Loveless, T.: The 2014 Brown Center Report on American Education: How well are American students learning? In: The Brown Center on Education Policy. The Brookings Institution, Washington, DC (2014)

30. Bennett, S., Kalish, N.: The Case Against Homework: How Homework is Hurting Our Children and What We Can do About it. Crown Publishers, New York, NY (2006)
31. Kohn, A.: The Homework Myth: Why our Kids Get too Much of a Bad Thing. Da Capo Press, Cambridge, MA (2006)
32. Pope, D.: Making Homework Work. ASCD (2020). Retrieved 20 Oct 2021

The Development of Multivariable Causality Strategy: Instruction or Simulation First?

Janan Saba[1] (✉), Manu Kapur[2], and Ido Roll[3]

[1] Hebrew University of Jerusalem, Jerusalem, Israel
janansaba3@gmail.com
[2] ETH-Zurich University, Zurich, Switzerland
manukapur@ethz.ch
[3] Technion, Haifa, Israel
roll@technion.ac.il

Abstract. Understanding phenomena by exploring complex interactions between variables is a challenging task for students of all ages. While the use of simulations to support exploratory learning of complex phenomena is common, students still struggle to make sense of interactive relationships between factors. Here we study the applicability of Problem Solving before Instruction (PS-I) approach to this context. In PS-I, learners are given complex tasks that help them make sense of the domain, prior to receiving instruction on the target concepts. While PS-I has been shown to be effective to teach complex topics, it is yet to show benefits for learning general inquiry skills. Thus, we tested the effect of exploring with simulations before instruction (as opposed to afterward) on the development of a multivariable causality strategy (MVC-strategy). Undergraduate students ($N = 71$) completed two exploration tasks using simulation about virus transmission. Students completed Task1 either before (*Exploration-first* condition) or after (*Instruction-first* condition) instruction related to multivariable causality and completed Task2 at the end of the intervention. Following, they completed transfer Task3 with a simulation on the topic of Predator-Prey relationships. Results showed that *Instruction-first* improved students' Efficiency of MVC-strategy on Task1. However, these gaps were gone by Task2. Interestingly, *Exploration-first* had higher efficiency of MVC-strategy on transfer Task3. These results show that while *Exploration-first* did not promote performance on the learning activity, it has in fact improved learning on the transfer task, consistent with the PS-I literature. This is the first time that PS-I is found effective in teaching students better exploration strategies.

Keywords: Multivariable causality strategy · exploratory learning · Interactive simulation

1 Introduction

In many authentic scientific phenomena, multiple factors act and interact with each other affecting outcomes. These variables often cannot be studied in isolation. For example, the impact of the rate of vaccination on pandemic spread depends on other factors related

N. Wang et al. (Eds.): AIED 2023, LNAI 13916, pp. 41–53, 2023.
https://doi.org/10.1007/978-3-031-36272-9_4

to the pandemic such as the transmission rate of the virus in the population, individuals' recovery rate, etc. This study focuses on evaluating and supporting learning of multivariable causality by studying the effects of multiple interacting relationships. The strategy of studying complex relationships is referred to as the multivariable causality strategy (MVC strategy) [20]. Using this strategy in educational settings is a challenging task for students of all ages [27, 30, 31]. The investigation of the multivariable causality of a phenomenon in real life can be very difficult, such as exploring how several factors simultaneously affect the transmission of a virus in the population. Therefore, interactive simulations are used by scientists and learners to investigate multiple relationships in complex phenomena. The use of simulation to support exploratory learning of complex phenomena is common [14, 28]. Students can actively explore these systems by systematically manipulating and changing multiple variables in the simulation and observing the outcomes using visual graphical representations [13, 22]. Studies have indicated the significant contribution of simulation-based learning in promoting students' conceptual understanding and learning transfer [4, 5]. However, an open question is the way in which simulations should be used in tandem with instruction. Thus, this study investigates applicability of Problem Solving before Instruction (PS-I) approach to this context. In PS-I, learners are given complex tasks that help them make sense of the domain, prior to receiving instruction on the target concepts [3]. While PS-I has been shown to be effective to teach complex topics, few studies have focused on the impact of exploration prior to instruction approach on the development of general inquiry skills [6, 10], but to the best of our knowledge, there is no study that tests the effectiveness of this learning approach on the development of MVC strategy. In the current work, we evaluate the effect of exploration with simulation prior to instruction on students' ability to apply and transfer the multivariable causality strategy.

2 Literature Review

2.1 Learning Multivariable Causality Strategy with Interactive Simulation

Much focus has been put on the acquisition of the Control Variables Strategy. Control Variables Strategy describes a sequence of trials in which all variables except the one being investigated are held constant (or "controlled") across experimental trials. Across multiple grades, mastering the Control Variables Strategy has become a standard requirement [20]. However, less attention is paid to the simultaneous effect of multiple factors on an outcome. Multivariable causality strategy (MVC strategy) is an important scientific inquiry skill. It focuses on investigating how multiple factors jointly affect an outcome. This is particularly true when exploring complex systems that involve interactive and nonlinear relationships [30].

The current study focuses on the development of MVC strategy, through learning with interactive simulations. Interactive simulations are visualized representations of complex phenomena that include events and processes. In educational settings, interactive simulations are designed to enable students to engage in scientific inquiry [22]. Using interactive simulations allows students to explore and interact with the system under investigation, by the means of manipulating elements, adjusting variables, and observing the outcome [15]. In STEM classrooms, interactive simulations have been increasingly accessible and used [8]. Several advantages are addressed for using interactive simulation in STEM classrooms. First, interactive simulations present simplified external visualizations of phenomena that cannot be easily observed otherwise. Second, situational feedback [1, 23] is provided by showing students the outcomes of their experiments [28] mainly graphical and numerical representations of the outcome. Third, they allow students to focus on key aspects of the phenomena and eliminate extraneous cognitive load [16]. Last, the use of these tools allows learners to engage in experiments that would otherwise be unsafe or impossible [4, 13]. For example, using interactive simulation to explore how several factors (e.g. transmission rate, recovery rate) can jointly affect the spread of a virus in a population could not be investigated in real life. However, interactive simulation can make it possible by allowing students to conduct several experiments, and test how different rates of these factors affect the spread of the virus. They enable students to provide an explanation and predict outcomes.

By using interactive simulation students engage in a meaningful learning environment, where they can explore key aspects of phenomena by setting their own exploration questions, conducting iterative experiments, and comparing and contrasting scenarios [4, 5, 16]. Several studies showed the benefits of using interactive simulations for advancing learning and knowledge transfer. For example, Finkelstein et al. found that in the same activity, students who learned through interactive simulation outperformed their peers who learned in a physical lab, even when the assessment was conducted in a physical lab [16]. Similarly, Saba et al. showed that learning chemistry by using interactive simulation better enhances students' conceptual learning and systems understanding compared to normative instruction using labs [5]. However, Roll and his colleagues found that, in the instruction context, learning with simulations assisted students to apply better inquiry strategies. But, these perquisites did not transfer to the following activity [2].

2.2 Problem Solving Prior to Instruction Approach to Learning

While simulations are effective, students often require additional support for their learning [4, 13]. This support is typically given in the form of direct instruction. However, the best timing of the instruction is unclear. Two approaches are addressed in the literature review. Several studies support the relative effectiveness of engaging in direct instruction [7, 18, 21]. They argue for providing students with high levels of support (in the form of instruction) prior to problem solving (referred to as Instruction followed by Problem-Solving, I-PS by [3]). For example, the Instruction phase may include a formal introduction to the target domain concepts and worked examples. Following, students move to a Problem-Solving phase where they engage in problem solving [26]. An alternative approach, which is tested in this study, is the approach of PS-I to learn of MVC Strategy [3]. PS-I approach to learning asks learners to develop solution attempts for problems related to new target concepts prior to the instruction phase that involves lectures and practice. The goal of the Problem-Solving phase is to prepare students for future learning from the Instruction phase [24, 25]. It was found in numerous studies that PS-I is effective mainly for teaching knowledge of various topics such as mathematics and physics [3, 6, 11], and prepares students better for "future learning"- the transfer of knowledge into a different domain [25]. Alas, this approach has rarely been investigated in the context of learning how to learn [10, 21]. Studies of Chase and Klahr and Matlen and Klahr [10, 21] used PS-I to teach the Control Variables Strategy. Results of these studies indicated no clear benefit to either approach. For example, Chase and Klahr worked on exploring the effect of PS-I on the development of Control Variables Strategy among elementary students using virtual labs. To assess the procedural and conceptual knowledge of Control Variables Strategy, two isomorphic variants of paper-and-pencil tests were used as pre- and posttests and a paper-and-pencil transfer test. The finding showed that there was no significant difference between the conditions (exploring prior to instruction versus instruction followed by exploration) related to learning the procedural and conceptual knowledge of Control Variables Strategy in both posttest and far transfer test. The lack of effect for PS-I in Chase and Klahr may be due to the student population being young, as PS-I is less effective for that age group. The exploration phase in Chase and Klahr study was based on conducting experiments in the virtual labs, but the assessments were conducted traditionally (paper-and-pencil). In the current study, we evaluate the PS-I approach for undergraduate students' learning MVC strategy using simulations [10].

The application of PS-I to learning with simulations requires some adaptation since scientific inquiry challenges are not traditional problem-solving tasks. In the context of learning MVC strategy with simulations, we adapt the Problem Solving to exploration, that is engaging in exploratory learning. It starts with the exploration phase where students participate in a simulation-based activity to answer a research question related to a certain phenomenon; they conduct several experiments in the simulation by manipulating the affecting variable and observing the outcome. Following is the Instruction phase, where students are taught the target concept. The I-PS approach starts with introducing and teaching the target concept, followed by an exploration phase where students use simulation to answer the research question based on what they were taught in the instruction phase. As an example, Fig. 1 illustrates the "Virus-Spread" simulation that was used in this study. Exploratory learning (based on PS-I) includes two phases, exploration phase followed by instruction phase. The exploration phase focuses on using the "Virus-Spread" simulation to explore the research question (for example: Use the "Virus-Spread" simulation to explore and describe how three factors of virus spread, transmission rate, recovery rate, and infected mortality rate, affect the number of dead people at the end of the pandemic?).

The alternative, the I-PS approach starts with introducing the conceptual meaning of MVC strategy (Instruction phase), followed by an exploration phase where students apply their understanding of MVC strategy to explore the same research question similar to the I-PS condition.

In the exploration phase, participating in simulation-based activities before instruction can be challenging, but it is engaging. Students can activate their prior knowledge to explore the relationship between variables [6]. Simulations provide feedback on how the interaction between variables has an impact on the observed outcome. In addition, simulations can assist students in identifying important characteristics of the phenomena and provide a wide range of problem spaces for students to explore [12]. A main challenge for exploratory learning is that students often fail to discover the target knowledge [19]. However, the provision of direct instruction following the exploration may mitigate this challenge and still benefit from the high agency offered by the exploration. Although

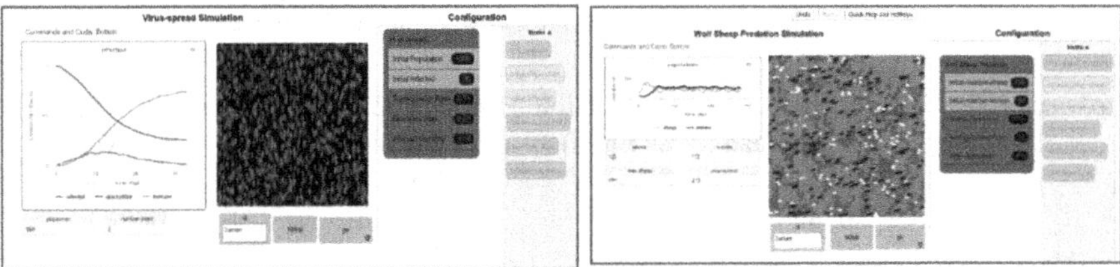

Fig. 1. Virus-spread simulation (left side); Wolf Sheep Predation simulation (right side) (Saba, 2022).

exploratory learning has been shown to improve conceptual understanding [12], few studies focused on the development of domain-general inquiry skills [6, 10].

This study seeks to examine the effectiveness of different learning approaches (*Exploration-first* versus *Instruction-first*) in promoting multivariable causality strategy (MVC strategy). To do that, the development of MVC strategy was assessed in the learning context, and in a transfer environment. The Transfer assessment evaluated future learning: How does instruction in one context affect students' ability to learn independently a new topic? Thus, this study has two research questions:

RQ1: what is the impact of engaging in exploration prior to instruction on the development of multivariable casualty strategy compared to engaging in instruction followed by exploration?
RQ2: to what extent does engaging in exploration before instruction contribute to transferring the multivariable casualty strategy and applying it in a new context compared to engaging in instruction before exploration?

3 Method

3.1 Participants

Seventy-one undergraduate students participated in the study. They were randomly assigned to either *Exploration-first* condition (n = 35) or *Instruction-first* condition (n = 36). The two conditions include the same materials but in a different order.

3.2 Design and Procedure

Students in both conditions engaged in two exploration tasks using the "Virus Spread" simulation (Task 1 and Task 2) and the instruction phase. Following the intervention, both groups engaged in a third exploration task (Task 3) using the "Wolf Sheep Predation" simulation. The difference between conditions was in the order of Task 1 and the Instruction. Table 1 illustrated the design of the study for both conditions.

3.3 Materials

Learning materials included three exploration tasks and instruction: Task 1 and Task 2 focused on exploring different problems related to virus transmission in a population and

Table 1. The sequence of learning design for each condition

Exploration-first	Task1- Virus Spread Exploration (20 min)	Instruction (15 min)	Task 2- Virus Spread Exploration (20 min)	Task 3- Transfer-Wolf Sheep Predation Exploration (20 min)
Instruction-first	Instruction (15 min)	Task 1- Virus Spread Exploration (20 min)	Task 2- Virus Spread Exploration (20 min)	Task 3- Transfer-Wolf Sheep Predation Exploration (20 min)

using "Virus Spread" simulation (https://technion.link/janan/Virus%20spread/). "Virus Spread" simulation is an online interactive simulation (Fig. 1) that is based on a combination between the "Virus-spread" NetLogo model [29] and the NetTango platform [17]. It enables students to explore how the interaction among several factors may affect the transmission of a virus in a certain population. Students can conduct iteratively several experiments by dragging and dropping factors' blocks and assigning them values; Pressing the "Setup" bottom to activate the determined value of each factor. Next, pressing the "Go" bottom to run the simulation. The simulation allows students to investigate the relationships between the three variables and how the interaction between them may affect the observed outcome. Task 3 targeted the exploration of a problem related to wolf and sheep predation within an ecosystem using the "Wolf Sheep Predation" simulation(https://technion.link/janan/Wolf%20sheep%20predation/). Similar to the "Virus-spread" simulation, it was developed by combining the "Wolf sheep predation" NetLogo model and the NetTango platform. Students were asked to explore the relationships and interactions among several factors inside an ecosystem that includes two populations, wolves and sheep (Fig. 1). The instruction phase illustrated phenomena consisting of multivariable interactions. It focused on exploring and reasoning about how nonlinear relationships between several variables give rise to the observed outcomes.

3.4 Data Sources and Analysis

Log files of students' interaction with simulations were collected for the three tasks. In this study, students explored how the interaction between three variables may affect the observed outcome. To view MVC strategy in students' log files, we looked for sets of experiments. A set of experiments contains at least three experiments which do not have to be sequential and characterized by controlling one variable (which has a fixed value in all the experiments) and exploring the effects of the other variables in tandem. Sets of experiments differed by the Efficiency of MVC strategy used in each set of experiments. Efficiency of a set of experiments has one out of four scores: Efficiency-score 0 refers to the lowest efficiency of MVC strategy of a set of experiments. It is an indication that the student randomly assigned values for the other two variables without relying on a specific strategy. Efficiency-score 1 refers to using the Control Variables Strategy, by controlling two variables (treating them as one combined variable) and testing only one variable at a time. Efficiency-score 2 refers to using MVC strategy in a set of experiments, where students control one variable and test interaction between the two other variables by applying at least twice the Control Variables Strategy, controlling the first variable and testing the second one, after then controlling the second variable and testing the first one. Efficiency-score 3 is an indication of using the highest efficiency of MVC strategy, but it differs from Efficiency-score 2 by the fact that a set of experiments of Efficiency-score 3 includes a shared baseline (using the same values of the variables when testing the interaction). Table 2 describes Efficiency scores and examples. Two components were created to evaluate the MVC strategy for each log file: (1) Quantity: the total number of sets of experiments students created in the task. (2) Efficiency: the sum of Efficiency scores of all the created sets of experiments divided by the number of experiments. As an example, one log file showed that a student conducted 17 experiments. Analyzing these experiments revealed 6 sets of experiments were created: 2 sets of experiments with an Efficiency-score 0, 2 sets of experiments with an Efficiency-score 1, and one set of experiments with an Efficiency-score 2. Thus, the Efficiency of MVC strategy was computed as the following: $(2 \times 0 + 2 \times 1 + 1 \times 2)/17 = 0.24$. Thus, for this student, the score for Quantity is 6, and the score for Efficiency is 0.24.

Table 2. Description of the four Efficiency -scores of MVC strategy for a set of experiments.

Efficiency-score	Description	Example[1]
0	Controlled one variable, without testing interactions between the other two variables	1 0.5 0.6 1 0.2 0.1 1 0.8 0.1
1	Using only Control Variables Strategy	1 0.8 0.6 1 0.8 0.1 1 0.8 0.5
2	Using MVC strategy without a shared baseline	1 0.5 0.6 1 0.5 0.1 1 0.8 0.1
3	Using MVC strategy with a shared baseline	1 0.5 0.6 1 0.5 0.5 1 0.6 0.5

[1]Example of Efficiency-score determined for a set of three experiments that has one controlled variable. Each line is one experiment that includes values of three variables

Repeated Measures ANOVA was conducted to analyze students' components of MVC strategy in Task 1 and Task 2 and to compare the two conditions on the task in which the instruction was given (RQ1). To evaluate transfer (RQ2), ANCOVAs for Task 3 were conducted, controlling for scores in Task 2 (Quantity and Efficiency). Task 2 shows students' ability to apply MVC Strategy at the end of the instructional sequence (Exploration followed by Instruction versus Instruction prior to Exploration). Thus, it serves as a measure of students' baseline knowledge following the intervention and can be used to evaluate Future Learning on Task 3, when no manipulation was given.

4 Results

Two repeated measures ANOVAs were conducted to measure the Quantity and Efficiency of the MVC strategy in Task 1 and Task 2 (see Table 3). Efficiency of MVC: The interaction between Task and Condition was significant with a medium effect size: $F(1.69) = 6.051$, $P = 0.016$, $\eta_p^2 = 0.081$. There was also a main effect for Task: $F(1,69) = 61.41$, $P = 0.000$, $\eta_p^2 = 0.47$. However, given the interaction, the main effects should not be interpreted. Quantity of MVC: There was no significant effect for Condition nor for its interaction with Task. There was a significant main effect for Task: $F(1,69) = 59.46$, $P = 0.000$, $\eta_p^2 = 0.46$.

Table 3. Statistical analysis of MVC strategy components, comparing between conditions in Task 1 and Task 2

Component					Statistical tests[5]					
	Task 1[1]		Task 2[2]		Task			(task × Condition)		
	Ex-I[3] *M(SD)*	I-Ex[4] *M(SD)*	Ex-I *M(SD)*	I-Ex *M(SD)*	F(1,69)	p	η_p^2	F(1,61)	p	η_p^2
Quantity	4.00 (2.65)	4.42 (2.51)	3.03 (2.18)	2.39 (1.85)	59.46	**0.000**[6]	0.46	1.801	0.184	0.03
Efficiency	0.53 (0.40)	0.61 (0.43)	0.50 (0.42)	0.49 (0.40)	61.41	**0.000**	0.47	6.051	**0.016**	.081

[1]Task 1 - Exploration phase with Virus Spread simulation; [2]Task 2 – After Intervention: Exploration with Virus Spread simulation; [3]Ex-I: Exploration- first condition (n = 35); [4]I-Ex: Instruction-first condition (n = 36); [5]Repeated Measure ANOVA; [6]Bolded item indicates significant *p*-value

Table 4 summarized the analysis of students' MVC strategy components (Quantity and Efficiency) in Task 3. Results showed significant differences between the conditions on students' Efficiency, with a large effect size: $F(1,69) = 4.096$, $P = 0.047$, $\eta_p^2 = 0.57$. Students from *Exploration-first* condition showed significantly higher Efficiency of MVC strategy compared to students from *Instruction-first* condition (Table 4). No significant differences were found between the two conditions in relation to the Quantity component: $F(1,69) = 1.208$, $P = 0.276$, $\eta_p^2 = 0.02$.

Table 4. Differences of Quantity and Efficiency components of MVC strategy between conditions in Task 3.

Component	Task 3[1]		Statistical tests		
	Ex-I[2] *M(SD)*	I-Ex[3] *M(SD)*	F(1,69)	p	$\eta_p{}^2$
Quantity	3.51 (2.69)	2.69 (2.17)	1.208	0.276	0.02
Efficiency	0.62 (0.40)	0.41 (0.31)	4.096	**0.047**[4]	0.57

[1]Task 3 – Transfer-Exploration with Predation simulation; [2]Ex-I: Exploration- first condition (n = 35); [3]I-Ex: Instruction-first condition (n = 36); [4]Bolded item indicates significant *p*-value; ANCOVA with covariates: Task 2: Quantity, and Efficiency; Task 2- Exploration phase with Virus Spread simulation

5 Discussion

This study focuses on learning MVC strategy using interactive simulation. We tested and compared the Quantity and Efficiency of MVC strategy of students in the two conditions (*Exploration-first* versus *Instruction-first*) at two different points of the learning sequence, Task 1 and Task 2. In addition, we evaluated Future Learning by looking at the transfer of MVC strategy across simulations and topics after invention. Significant differences were found between conditions related to the Efficiency of MVC strategy during the learning process. Thus, one main finding of this study is: Instruction helped in improving the Efficiency of MVC strategy when it comes prior to exploration only at the early stages of the learning. Right after instruction (Task 1), the condition who received it (Instruction prior to exploration) showed superior performance. However, these differences were all but eliminated by the time students in both conditions received the instruction (Task 2). These short-lived benefits for instruction were found previously when learning with simulations [2]. Also in that study, instruction helped students apply better inquiry strategies in the context of the instruction. Alas, these benefits did not transfer to the following activity. Few studies have focused on the development of general inquiry skills when engaging in exploration before instruction [6]. Chase & Klahr have found no significant effect of the sequence of learning on the development of the Control Variables Strategy in the posttest. Similar results were found in our study related to promoting MVC strategy. Despite this, our study differs from Chase & Klahr study in two aspects. First, we have compared the two conditions at two different points of the learning process, Chase & Klahr compared the two conditions based on students' posttests scores. Second, in our study data were collected and analyzed based on students' interaction with the simulation (log files), Chase & Klahr used a summative assessment in the form of paper-and-pencil [10]. Based on these two differences, the results of our study go one step further and provide an explanation for these results: Findings indicate that instruction prior to exploration can contribute to achieving immediate significant high-Efficiency MVC strategy only at the early stages of learning. But, the two different learning approaches achieve a very close Efficiency of MVC strategy at the end of the intervention. A second analysis of this study was conducted in relation to Future Learning. Particularly, a comparison was conducted between the two conditions related to the Quantity and Efficiency of MVC strategy when exploring a different simulation and different topic (Task 3). Results revealed a significant difference between the conditions in the Efficiency of MVC strategy. Students in *Exploration-first* condition significantly outperformed their counterparts in *Instruction-first* condition in the Efficiency of MVC strategy with a large effect size. Thus, a second main finding of this study is exploration prior to instruction approach contributes to the high Efficiency of MVC strategy when it is applied in a new context. Previous studies on exploration prior to instruction have demonstrated its significant contribution to the transfer of the conceptual understanding to different topics [12], and especially to learning of new material [25]. In this study, we found significant evidence of the contribution of the exploratory learning approach (PS-I paradigm, [3]) also on promoting general inquiry skills [6], namely the MVC strategy. This is particularly true when engaging in a new exploration task that contains a new simulation and a different topic. Our results are different from Chase & Klahr who found that students in both conditions (*Exploration-first* and *Instruction-first*) have similarly transferred their learning of Control Variables Strategy to another topic [10]. Our findings showed the significant impact of exploration prior to instruction on transferring the learning related to Efficiency of MVC strategy. This difference can be explained, as mentioned previously, in relation to the format of the transfer task, simulation-based task (in our study) versus summative assessment. When the transfer task includes a simulation, students can apply their new strategy in order to support future learning, rather than applying it without feedback [9]. Another explanation could be related to the age of the students who participate in the study. In Chase & Klahr study, middle-school students learned about Control Variables Strategy [10], in our study undergraduate students learned about MVC strategy. Similar to several works, such as Schwartz & Martin and Schwartz & Bransford [24, 25], we have found that exploration followed by instruction prepared students for future learning of MVC strategy as can be seen in the significant results of students' Efficiency of MVC strategy in the transfer Task 3 compared to students who learned about MVC before exploration. In relation to the Quantity of MVC strategy, no significant effects for Condition nor its interaction with Task were identified on any of the tasks: 1, 2, or 3. The lack of effect of instruction on the Quantity of experimental sets replicates earlier findings [4]. It seems that inquiry has a certain pace to it. While instruction can affect its content, it rarely affects its duration. The analysis further revealed that students in both conditions significantly decreased the Quantity of MVC strategy when exploring a second task with the same simulation at the end of the intervention. These results were not surprising. They can be explained by the fact that students of both conditions have experienced the Virus-Spread simulation and explored the interaction between the three factors in the previous exploration task (Task 1). Thus, they came to Task 2 knowing more about the domain, the simulation, and themselves as learners. Interestingly, students in the *instruction-first* condition decreased the Efficiency of their use of MVC strategy (from Task 1 to Task 2). Relatedly, students in the *explore-first* condition maintained their levels, but did not increase these. It may be that a second task on the same simulation invites a different pattern of exploration. Perhaps even the specific task was simpler and could be achieved with a lower-Efficiency MVC Strategy. Further studies should look at order effects of the tasks within the intervention.

6 Conclusions, Limitations, and Future Work

This study investigates the effectiveness of exploratory learning based on the PS-I approach to promoting the MVC strategy in the learning context, and in a transfer environment. Our findings show that in the learning context, providing learners with instruction first had immediate benefits on their learning process. However, by the end of the instructional sequence, no differences were found between the two learning approaches (PS-I versus I-PS) related to the MVC strategy. When looking at the big picture, a significant difference between conditions was found when engaging in a Future Learning task. This is the first study that found a significant contribution of the PS-I approach to prompting inquiry skills. Two factors contributed to this result: (1) The integration of interactive simulation in learning about MVC strategy as a vehicle to promoting the transfer of MVC strategy to another context; (2) Quantitively analyzing log files from students' interactions with the simulation to assess this general inquiry skill. One limitation of this work is the small sample size. This kind of analysis using students' log files usually requires a larger sample size to make our conclusions more robust. In a future study, a larger number of students would participate in the study. In addition, this study was conducted with undergraduate students. We believe that teaching about multivariable causality should be at earlier stages of high-school contexts. Finally, in this study, we used simulations from the domain of general biology. For future work, it is of interest to explore whether these results are also valid for using different simulations from different domains. Overall, this study makes several contributions. Theoretically, it helps to understand the process of learning, applying, and transferring MVC strategy in studying complex phenomena using simulations. Specifically, by applying the PS-I framework to this context, it provides the first evidence for the benefits of providing learners with opportunities for free exploration prior to instruction. The study also makes pedagogical contributions. The relevance of PS-I to the development of inquiry skills and their transfer has implications for task design and support. In particular, the study shows the effectiveness of PS-I in preparing students for future learning when engaging in new exploratory learning environments. Methodologically, the study identifies ways to quantify and evaluate the efficiency and quantity of students' exploration when studying complex phenomena using interactive simulations. Such a definition is important for a wide range of natural phenomena in which the traditional Control Variables Strategy does not suffice.

This work was supported by the Israeli Science Foundation award # 1573/21, and the Betty and Gordon Moore Foundation.

References

1. Roll, I., Yee, N., Briseno, A.: Students' adaptation and transfer of strategies across levels of scaffolding in an exploratory environment. In: Trausan-Matu, S., Boyer, K.E., Crosby, M., Panourgia, K. (eds.) ITS 2014. LNCS, vol. 8474, pp. 348–353. Springer, Cham (2014). https://doi.org/10.1007/978-3-319-07221-0_43
2. Roll, I., Baker, R.S.J.D., Aleven, V., Koedinger, K.R.: On the benefits of seeking (and avoiding) help in online problem solving environment. J. Learn. Sci. **23**(4), 537–560 (2014). https://doi.org/10.1080/10508406.2014.883977

3. Loibl, K., Roll, I., Rummel, N.: Towards a theory of when and how problem solving followed by instruction supports learning. Educ. Psychol. Rev. **29**(4), 693–715 (2017). https://doi.org/10.1007/s10648-016-9379-x
4. Roll, I., et al.: Understanding the impact of guiding inquiry: the relationship between directive support, student attributes, and transfer of knowledge, attitudes, and behaviors in inquiry learning. Instr. Sci. **46**(1), 77–104 (2018)
5. Saba, J., Hel-Or, H., Levy, S.T.:.Much.Matter.in.Motion: Learning by modeling systems in chemistry and physics with a universal programming platform. Interactive Learn. Env. **29**, 1–20 (2021)
6. Sinha, T., Kapur, M.: When problem solving followed by instruction works: evidence for productive failure. Rev. Educ. Res. **91**(5), 761–798 (2021)
7. Ashman, G., Kalyuga, S., Sweller, J.: Problem-solving or explicit instruction: Which should go first when element interactivity is high? Educ. Psychol. Rev. **32**, 229–247 (2020). https://doi.org/10.1007/s10648-019-09500-5
8. Blake, C., Scanlon, E.: Reconsidering simulations in science education at a distance: features of effective use. J. Comput. Assist. Learn. **23**(6), 491–502 (2007)
9. Bransford, J.D., Schwartz, D.L.: Rethinking Transfer: A Simple Proposal with Multiple Implications, vol. 24. American Educational Research Association, Washington DC (1999)
10. Chase, C.C., Klahr, D.: Invention versus direct instruction: for some content, it'sa tie. J. Sci. Educ. Technol. **26**(6), 582–596 (2017)
11. Darabi, A., Arrington, T.L., Sayilir, E.: Learning from failure: a meta-analysis of the empirical studies. Educ. Tech. Res. Dev. **66**(5), 1101–1118 (2018). https://doi.org/10.1007/s11423-018-9579-9
12. DeCaro, M.S., McClellan, D.K., Powe, A., Franco, D., Chastain, R.J., Hieb, J.L., Fuselier, L.: Exploring an online simulation before lecture improves undergraduate chemistry learning. International Society of the Learning Sciences (2022)
13. de Jong, T.: Learning and instruction with computer simulations. Educ. Comput. **6**, 217–229 (1991)
14. de Jong, T.: Technological advances in inquiry learning. Science **312**(5773), 532–533 (2006)
15. Esquembre, F.: Computers in physics education. Comput. Phys. Commun. **147**(1–2), 13–18 (2002)
16. Finkelstein, N.D., Adams, W.K., Keller, C.J., Kohl, P.B., Perkins, K.K., Podolefsky, N.S., et al.: When learning about the real world is better done virtually: a study of substituting computer simulations for laboratory equipment. Phys. Rev. Spec. Topics-Phys. Educ. Res. **1**(1), 10103 (2005)
17. Horn, M., Baker, J., Wilensky, U.: NetTango Web 1.0alpha. [Computer Software]. Evanston, IL. Center for Connected Learning and Computer Based Modeling, Northwestern University. http://ccl.northwestern.edu/nettangoweb/ (2020)
18. Hsu, C.-Y., Kalyuga, S., Sweller, J.: When should guidance be presented during physics instruction? Arch. Sci. Psychol. **3**(1), 37–53 (2015)
19. Kirschner, P., Sweller, J., Clark, R.E.: Why unguided learning does not work: an analysis of the failure of discovery learning, problem-based learning, experiential learning and inquiry-based learning. Educ. Psychol. **41**(2), 75–86 (2006)
20. Kuhn, D., Ramsey, S., Arvidsson, T.S.: Developing multivariable thinkers. Cogn. Dev. **35**, 92–110 (2015)
21. Matlen, B.J., Klahr, D.: Sequential effects of high and low instructional guidance on children's acquisition of experimentation skills: is it all in the timing? Instr. Sci. **41**(3), 621–634 (2013)
22. Moser, S., Zumbach, J., Deibl, I.: The effect of metacognitive training and prompting on learning success in simulation-based physics learning. Sci. Educ. **101**(6), 944–967 (2017)
23. Nathan, M.J.: Knowledge and situational feedback in a learning environment for algebra story problem solving. Interact. Learn. Environ. **5**(1), 135–159 (1998)

24. Schwartz, D.L., Bransford, J.D.: A time for telling. Cogn. Instr. **16**(4), 475–522 (1998)
25. Schwartz, D.L., Martin, T.: Inventing to prepare for future learning: The hidden efficacy of encouraging original student production in statistics instruction. Cogn. Instr. **22**, 129–184 (2004)
26. Stockard, J., Wood, T.W., Coughlin, C., Rasplica Khoury, C.: The effectiveness of direct instruction curricula: a meta-analysis of a half century of research. Rev. Educ. Res. **88**(4), 479–507 (2018). https://doi.org/10.3102/0034654317751919
27. Waldmann, M.R.: Combining versus analyzing multiple causes: how domain assumptions and task context affect integration rules. Cogn. Sci. **31**(2), 233–256 (2007)
28. Wieman, C.E., Adams, W.K., Perkins, K.K.: PhET: simulations that enhance learning. Science **322**(5902), 682–683 (2008)
29. Wilensky, U.: NetLogo models library [Computer software]. In Center for connected learning and computer-based modeling. Northwestern University (1999). http://cclnorthwestern.edu/netlogo/models/
30. Wu, H.K., Wu, P.H., Zhang, W.X., Hsu, Y.S.: Investigating college and graduate students' multivariable reasoning in computational modeling. Sci. Educ. **97**(3), 337–366 (2013)
31. Zohar, A.: Reasoning about interactions between variables. J. Res. Sci. Teach. **32**(10), 1039 (1995)

Content Matters: A Computational Investigation into the Effectiveness of Retrieval Practice and Worked Examples

Napol Rachatasumrit(✉), Paulo F. Carvalho, Sophie Li, and Kenneth R. Koedinger

Carnegie Mellon University, Pittsburgh, PA 15213, USA
{napol,koedinger}@cmu.edu, pcarvalh@cs.cmu.edu

Abstract. In this paper we argue that artificial intelligence models of learning can contribute precise theory to explain surprising student learning phenomena. In some past studies of student learning, practice produces better learning than studying examples, whereas other studies show the opposite result. We reconcile and explain this apparent contradiction by suggesting that retrieval practice and example study involve different learning cognitive processes, memorization and induction, respectively, and that each process is optimal for learning different types of knowledge. We implement and test this theoretical explanation by extending an AI model of human cognition — the Apprentice Learner Architecture (AL) — to include both memory and induction processes and comparing the behavior of the simulated learners with and without a forgetting mechanism to the behavior of human participants in a laboratory study. We show that, compared to simulated learners without forgetting, the behavior of simulated learners with forgetting matches that of human participants better. Simulated learners with forgetting learn best using retrieval practice in situations that emphasize memorization (such as learning facts or simple associations), whereas studying examples improves learning in situations where there are multiple pieces of information available and induction and generalization are necessary (such as when learning skills or procedures).

Keywords: Simulated Learners · Retrieval Practice · Worked Examples

1 Introduction

Retrieval practice - repeatedly trying to retrieve information by completing practice questions - has been shown to improve performance compared to re-studying [10,11]. Interestingly, re-study trials in the form of worked examples have also been shown to improve performance compared to answering more practice questions [9,14]. This apparent contradiction poses both theoretical and practical

N. Wang et al. (Eds.): AIED 2023, LNAI 13916, pp. 54–65, 2023.
https://doi.org/10.1007/978-3-031-36272-9_5

issues. Theoretically, to which degree do we have a complete understanding of the learning process if opposite approaches can yield similar results? Practically, when making suggestions for the application of cognitive science findings to educational contexts, practitioners are left wondering which approach to use and when.

Previous proposals to address this contradiction, have proposed that problem complexity was the critical dimension that defined whether retrieval practice or worked examples would improve learning [13,15]. The proposal is that worked examples improve learning of complex problems by reducing cognitive load, whereas practice improves learning of simpler materials that do not pose the same level of cognitive load. However, as Karpicke et al. pointed out, this explanation does not capture all the evidence [2]. For example, there is ample evidence that retrieval practice improves learning of complex texts [8]. Ultimately, problem complexity is hard to operationalize, and much of the discussion has centered around what constitutes complexity [2,7].

An alternative hypothesis follows directly from the Knowledge Learning Instruction framework (KLI) [4]. Based on analyses from over 360 *in vivo* studies using different knowledge content, KLI relates knowledge, learning, and instructional events and presents a framework to organize empirical results and make predictions for future research. A key premise of the KLI framework is that optimal Instructional design decisions depend on what Learning process the student must engage (e.g., is it verbatim memory, pattern induction, or sense making). In turn, the Learning process needed depends on the nature of the Knowledge that the student is to acquire (e.g., is it verbatim facts, problem-solving skills, or general principles). According to KLI, Instructional Events are activities designed to create learning. Textbooks, lectures and tests/quizzes are examples of commonly used Instructional Events. These Instructional Events in turn give rise to Learning Events - changes in cognitive and brain states associated with Instructional Events. KLI identifies three types of Learning Events: Memory processes, induction processes, and sense-making processes. These changes in cognitive and brain states influence and are influenced by the Knowledge Components (KCs) being learned. A KC is a stable unit of cognitive function that is acquired and modifiable. In short, KCs are the pieces of cognition or knowledge and are domain-agnostic. Although Learning Events and Knowledge Components cannot be directly observed, they can be inferred from Assessment Events, or outcome measures, such as exams and discussions.

KLI also offers a taxonomy for KCs based on how they function across Learning Events and relates differences in KCs with differences in Learning Events. In this way, KCs can be classified based on their application and response conditions. Facts such as "the capital of France is Paris" are constant application and constant response Knowledge Components (KCs) because there is only one single application of the KC and there is only one response. Conversely, skills such as equation solving are variable application and variable response KCs because there are multiple different problems that can elicit the same equation solving KC and there are multiple ways to apply this KC across different problems (e.g., solving an equation that one never saw, using a generalization extracted from studying many examples). Moreover, the KLI framework further suggests causal links between instructional principles (e.g., "retrieval practice", "worked-example study"), and changes in learner knowledge. For simple constant KCs such as facts, memory processes are more relevant. Conversely, for variable KCs such as skills, induction processes are more relevant. Thus, different types of KCs will interact with different types of Instructional Principles to create different learning.

In the context of facts ("What is the capital of France?"), learners need to successfully encode all of the information presented and be able to retrieve it later. Learning facts only requires learning the specific pieces of single practice items but does not require any synthesis across practice items. Conversely, in the context of skills ("Calculate the area of a rectangle with the following measurements"), learners need to generalize their knowledge across a series of studied instances. In this sense, learning skills requires identifying which pieces of the information are relevant for encoding and which are not. Put another way, when learning facts all presented information is critical and should be encoded, whereas when learning skills, only a subset of the presented information is relevant to forming an effective generalized skill. Finally, this theoretical proposal is also consistent with procedural differences between research on retrieval practice and example study. Research on retrieval practice generally tests learners' memory of the information presented in repeated trials, whereas research on worked examples generally uses different examples of the same concept in each trial.

Carvalho et al. [1] proposed that retrieval practice improves memory processes and strengthens associations, whereas studying worked examples improves inference processes and information selection for encoding. This proposal is consistent with previous work, showing that retrieval practice improves learning of associations, such as paired-associates or text that learners should try to retrieve either verbatim or by putting together several pieces of information [3]. Conversely, worked examples improve learning of knowledge requiring learners to infer or provide answers to multi-step problems or applying procedures (e.g., learning to calculate the area of a geometric solid [12].

To test this hypothesis, Carvalho et al. [1] conducted an experiment using a basic mathematical domain (calculating the area of geometrical shapes). Human participants (N = 95) were divided into 4 conditions: practice-only training of facts, study-practice training of facts, practice-only training of skills, and study-practice of skills. Carvalho et al. found a significant interaction between the type of concept studied and the type of training (see also Simulation Studies below).

In this study, we take a step further and provide a deeper understanding of the mechanisms behind the apparent contradiction between retrieval practice and worked examples in improving performance. To achieve this, we use an AI model of human learning to demonstrate the likelihood of these mechanisms generating such behavioral results similar to Carvalho et al.'s experiment and examine the extent to which the memory mechanism influences human learning. Specifically, we implemented two models, one without forgetting and another with forgetting.

Our results indicate that the simulated learners with forgetting match human results better and further analysis supports our proposed mechanism.

Furthermore, this study provides further evidence for the utility of computational models of human learning in the advancement of learning theory. As proposed by MacLellan et al. [5], the use of such models enables a bridge between learning theory and educational data, allowing for the testing and refinement of fundamental theories of human learning. This study extends this concept by demonstrating the ability of these models to contribute to evaluating theories that can explain even surprising student learning phenomena, for which existing learning theories may offer inconsistent explanations.

2 A Computational Model of Human Learning

Simulated learners are AI systems that learn to perform tasks through an interactive process, such as human demonstrations and feedback, usually with mechanisms that are intended to model how humans learn. In this work, we used the Apprentice Learner framework (AL), a framework for creating simulated learners based on different mechanistic theories of learning. Details on AL and its operation can be found elsewhere [5,16]; briefly, AL agents learn a set of production rules through an induction mechanism. The agents receive a set of states as input and search for the existing production rules that are applicable. If none of the existing production rules are applicable, AL agents will request a demonstration of a correct action and go through the induction process to construct a new rule for the current set of states. Later, when the agents encounter states that use the same production rule, the rule will get generalized or fine-tuned according to the examples they encounter. The learning process in AL is largely deterministic but some of the learning mechanisms have stochastic elements. For example, when multiple possible actions are possible, a stochastic probability matching process is used to select which one to execute.

The AL framework's production rules consist of two sets of conditions - the left-hand side (LHS) and right-hand side (RHS) - that include three essential components: where-part, when-part, and how-part. RHS is the action that AL thinks it should take in the form of a Selection-ActionType-Input (SAI) triple. The value of the SAI is calculated by a function composition called how-part, given the values extracted from the input states by the where-part. The where-part determines which state elements the production rule could be applied to. These sets of elements are called bindings and contain the elements to be used as arguments to the RHS of the skill and the selection of the state element to be acted upon. The when-part of a skill, a binary function (often in the form of a set of conditions), determines whether or not a particular skill should be activated given the current state of the system. Initially, the when-part and where-part may be either over-specific or over-general but will be refined to the appropriate level as the agent receives additional demonstrations or feedback on the skill. For instance, if the agent is given an example of calculating the area of a rectangle (e.g. given $l = 4$, $h = 3$, then the area is 12) and another for a triangle (e.g. given l = 5, h = 4, then the area is 10), it will learn two distinct skills (i.e. rectangle-area and triangle-area skills). However, since each skill is demonstrated only once, the conditions in the LHS remain unrefined. Consequently, when the agent is presented with another rectangle-area problem, it may mistakenly activate the triangle-area skill due to over-general conditions. Upon receiving negative feedback, AL will refine the when-part of the triangle-area skill to be more specific, such as only activating it when the shape is a triangle.

In previous work, AL agents have been shown to demonstrate human-like behaviors in learning academic tasks, such as fractions arithmetic, and multi-column addition [5]. Here, we used AL to test the mechanistic hypothesis that retrieval practice involves memory and retrieval processes, whereas studying examples involves induction processes. To do this, we developed a memory mechanism in AL and compared the performance of AL agents learning facts and skills in a setup similar to previous empirical results with humans (see also Simulation Studies below). We compare learning outcomes following training of facts and skills, using retrieval practice (practice-only) or worked examples (study-practice). In our study, we employed the same subject matter, but we altered the learning focus between fact acquisition (e.g., "What is the formula to calculate the area of a triangle?") and skill acquisition (e.g. "What is the area of the triangle below?").

Additionally, it is crucial to investigate the extent to which memory plays a role in this mechanistic hypothesis. From existing literature, it has been established that retrieval practice has a significant impact on memory. However, the implications of memory processes on the use of worked examples as a learning method are still unclear. For example, what is the potential impact of having perfect memory on these different modes of learning? Therefore, we also created a model without a forgetting mechanism and conducted the same experiment compared to the simulated learners with forgetting.

Our hypothesis predicts that the simulated learners with forgetting will perform similarly to human results, with better performance when learning facts and skills through retrieval practice and worked examples, respectively. The simulated learners without forgetting are not expected to match human results, with retrieval practice being less effective than worked examples in the acquisition of both facts and skills in the absence of a memory mechanism.

3 Simulation Studies

3.1 Data

The current work replicates the findings of Carvalho et al.'s [1] experiment on the effect of retrieval practice and worked examples on the different types of knowledge. In their studies, participants were divided into four groups: practice-only training of facts, study-practice training of facts, practice-only training of skills, and study-practice of skills. The participants learned how to calculate the area of four different geometrical shapes (rectangle, triangle, circle, and trapezoid) through a training phase that consisted of studying examples and practicing memorizing formulas or solving problems. Two multiple-choice tests were used as pre/posttests, with 16 questions in total, divided into two types of knowledge: fact-based ("What is the formula to calculate the area of the rectangle?") and skill-based ("What is the area of a rectangle that is 9 ft wide and 15 ft long?"). The training phase included study of examples and practice, with no feedback provided. A retention interval of trivia questions was used between the training and post-test. The order of the pre/posttest and the problems used were randomized for each participant.

To replicate the findings, our pre/post tests and study materials were adapted from the original study. Since the primary focus of our hypothesis is the interaction between types of training and types of knowledge, we decided to significantly simplify the encoding of the problems such that an agent is allowed to only focus on picking the correct production rule and selecting the relevant information from states presented. For example, instead of giving an agent a diagram, the input states include parsed information from the diagram, such as a shape type or lines' lengths, which allows the agent to focus only on picking the relevant information and the associated production rule instead of learning how to parse a diagram. This simplification is also plausible as parsing shapes is likely to be prior knowledge that human participants brought to the task. An example of the adaptation from human materials to agent input states is presented in Fig. 1.

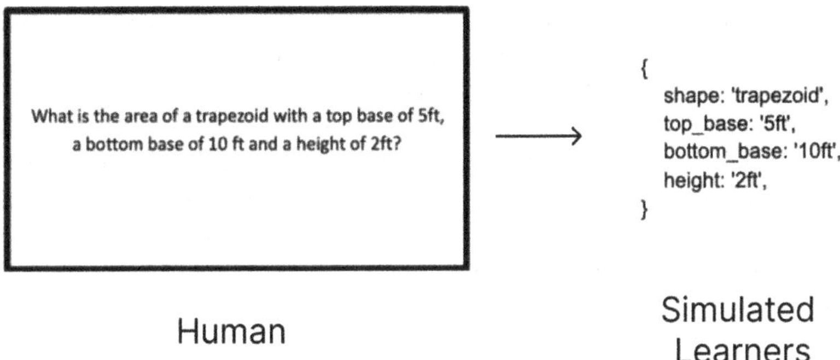

Fig. 1. Adaptation from human material to input states

The materials used during the study sessions were similar to those for the pre/posttests, but the agents were also given the solutions to the problems, in contrast to only questions without solutions or feedback in pre/posttests. For fact-based materials, the solutions were in the form of a corresponding string, while relevant operations, foci-of-attention, and numerical answers were provided for the skill-based materials. There were a total of 24 problems (six per geometrical shape) for each condition for the study session. The sample size and presentation order of the materials (i.e. pre/posttests and study materials) were kept consistent with the original study to ensure comparability.

3.2 Method

To evaluate our hypothesis that memory and forgetting processes are necessary for a learning benefit of retrieval practice, we leveraged the AL framework to create two models of human learning: a model with forgetting and a model without forgetting (i.e. having a perfect memory). Our memory mechanism implementation is based on Pavlik et al.'s memory model using ACT-R [6]:

$$m_n(t_{1...m}) = \beta + b_k + ln(\sum_{k=1}^{n} t_k^{-d_k})$$

In the simulated learners with forgetting, an activation strength (m_n) depends on the base activation (β), the strength of a practice type (b_k), ages of trials (τ_k), and decay rates (d_k). The decay rate for each trial depends on the decay scale parameter (c), the intercept of the decay function (α), and the activation strength of prior trials (m_{k-1}):

$$d_k(m_{k-1}) = ce^{m_{k-1}} + \alpha$$

The parameter values we used in the model in our study are shown in Table 1. These parameters were selected based on an initial parameter manual search. The activation strength of each production rule will be updated through the mathematical process described above, every time it is successfully retrieved both through demonstrations/examples or practice testing, but with different corresponding parameter values depending on the type of training. Then, the probability of a successful recall for a production rule will be calculated using the recall equation when simulated learners attempt to retrieve the rule.

In other words, the success of a production rule being activated for simulated learners depends on the model they are based on. In models without forgetting, the process is deterministic and the applicable rule will always be activated. However, in models with forgetting, the process is stochastic, with the probability of a successful being activated determined by the probability of a successful recall of the associated production rule.

Table 1. Memory model's parameters

α	τ_k	c	s	β	$b\_practice$	b_study
0.177	−0.7	0.277	1.4	5	1	0.01

There were a total of 95 AL agents, each agent matching a human participant in [1], assigned to one of four conditions: practice-only training of facts (N = 27), study-practice training of facts (N = 22), practice-only training of skills (N = 18), and study-practice training of skills (N = 28). Each agent went through the same procedure as human participants. It completed 16 pretest questions, 4 study sessions, and then completed 16 posttest questions after a waiting period.

In the study session, The agents were divided into two groups: the practice-only group, where they were trained with one demonstration (worked example) followed by three practice tests, and the study-practice condition, where they alternated between both types of training. During the practice tests, the agents were only provided with binary corrective feedback without the correct answer. The objective of the learning process for facts was for the agents to effectively link the appropriate constant (i.e. a formula string) to the specific state of the problem, as specified by the constant-constant condition (e.g. shape == trapezoid corresponds to the formula "$A = \frac{1}{2}(a + b) * h$"). The objective of learning skills, on the other hand, was for the agents to not only identify the appropriate formula to apply, but also to select the relevant variables from the given state of the problem, as outlined by the variable-variable condition (e.g. shape == square, base-length == 5, and diagonal-length == $\sqrt{50}$ corresponds to 5^2).

To account for human participants' prior knowledge, we pretrained each simulated learner to match pretest human performance [17]. In this experiment, we used average participants' pretest scores ($M = 0.59$ and 0.60, for facts and skills, respectively) as a target performance level for all conditions.

4 Results

4.1 Pretest

Overall, average agents' pretest performance was similar to human participants ($M = 0.56$ for facts and 0.58 for skills) ($t(958) = 0.310$, $p = 0.378$).

4.2 Learning Gain

Similar to the behavioral study in Carvalho et al. [1], we analyzed posttest performance controlling for pretest performance, for each type of trained concept (skills vs. facts) and training type (practice-only, vs. study-practice). A two-way ANOVA was performed to analyze the effects of type of training and type of concept studied on learning gains, and the results showed that there was a statistically significant interaction between the effects of type of training and type of concept in the simulated learners with forgetting ($F(1, 471) = 9.448$, $p = .002$), but none was found in the simulated learners without forgetting ($F(1, 471) = -3.843$, $p = 1$). Moreover, consistent with our prediction, simple main effects analysis showed that the type of training did have a statistically significant effect on learning gains in the simulated learners without memory ($F(1, 471) = 7.364$, $p = 0.007$), but not in the simulated learners with forgetting ($F(1, 471) = 0.845$, $p = 0.359$). On the other hand, the type of concept studied had a statistically significant effect on learning gains in both simulated learners with forgetting ($F(1, 471) = 13.052$, $p < 0.001$) and without forgetting ($F(1, 471) = 29.055$, $p < 0.001$). The similar pattern can also be seen in Fig. 2, comparing the learning gains for each condition between human participants (a), the simulated learners without forgetting (b), and the simulated learners with forgetting (c). The results

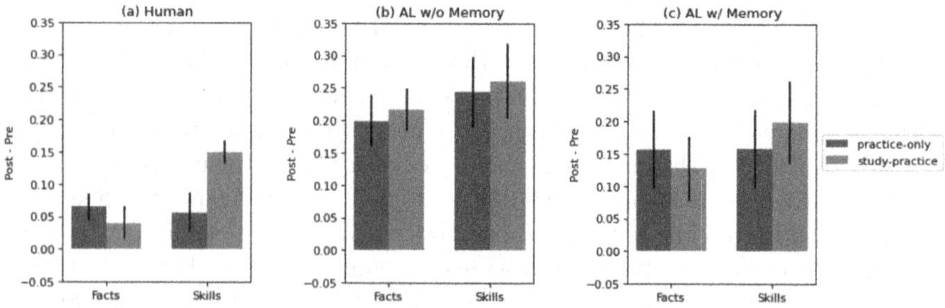

Fig. 2. Learning Gains Comparison between type of training and type of concept.

indicate that the simulated learners with forgetting in a study-practice condition led to higher learning gains for skills than a practice-only condition (19.9% vs 15.8%), $t(228) = -2.404$, $p = 0.009$, but the opposite was true for facts (12.7% vs 15.6%), $t(243) = 2.072$, $p = 0.020$. However, the simulated learners without forgetting led to higher learning gains for both skills (26.1% vs 24.4%), $t(228) = 1.106$, $p = 0.135$ and facts (21.6% vs 20.0%), $t(243) = -1.713$, $p = 0.044$, in the study-practice condition. These results suggest that the simulated learners with forgetting better align with human learning patterns.

4.3 Error Type

To further investigate the extent to which memory plays a role in this mechanistic hypothesis, we analyzed the types of errors made by simulated learners with forgetting at posttest (since the simulated learners without forgetting cannot commit a memory-based error, it would be unnecessary to conduct the analysis). We classified errors into two categories: memory-based and induction-based. Memory-based errors occurred when an applicable production rule was learned but not retrieved in the final test, whereas induction-based errors occurred when incorrect production rules were found or none were found. Table 2 displays the proportion of each error category, broken down by knowledge type, at the posttest stage. In general, the simulated learners committed more induction-based errors than memory-based errors (56.2% vs 43.8%). Additionally, the simulated learners in practice-only condition committed less memory-based errors compared to the ones from study-practice condition (41.8% vs 45.7%), but more induction-based errors (58.2% vs 54.2%), $t(466) = -1.467$, $p = 0.072$; even though, both groups exhibited similar proportions of both categories (82.7% vs 83.9% for induction-based errors and 17.3% vs 16.1% for memory-based errors) at pretest.

Table 2. Types of errors (memory-based and induction-based) in posttests for each training condition.

	Memory-based	Induction-based
practice-only	41.8% ± 27.9	58.2% ± 27.9
study-practice	45.7% ± 30.3	54.2% ± 30.3
overall	**43.8% ± 29.3**	**56.2% ± 29.3**

5 General Discussion

Our results indicate that the results of simulated learners with forgetting align well with human results, with retrieval practice being more effective for facts and worked examples being more effective for skills. In contrast, for simulated learners without forgetting, worked examples are more beneficial for both facts and skills, as the lack of a memory mechanism does not allow for the benefits of retrieval practice to be realized. These findings support our hypothesis that, according to the KLI framework, retrieval practice improves memory processes and strengthens associations, making it beneficial for learning facts where all presented information is important. Conversely, studying examples improves inference processes and information selection for encoding, making it beneficial for learning skills where only a subset of presented information is relevant.

Interestingly, the introduction of a memory mechanism in simulated learners with forgetting slightly decreases learning gains (22.2% for simulated learners without forgetting vs 13.7% for simulated learners with forgetting), $t(948) = 9.409$, $p < 0.0001$, but does not negate the benefits of worked examples over retrieval practice for skills. Furthermore, the breakdown of error categories revealed more induction-based errors than memory-based errors (59.1% vs 40.9%). This supports our hypothesis that skills learning involves more selectivity and inference, which are better aided by worked examples than by increased memory activation through retrieval practice.

In fact, the gap between retrieval practice and worked examples for skill learning increases even more in simulated learners with forgetting (1.7% vs 3.1%). A closer examination of these results suggested that this is because simulated learners with forgetting are better able to discard and "forget" incorrect production rules, which allows them to more effectively utilize the correct production rules. Simulated learners with forgetting were found to be more likely to select the correct production rules, due to stronger memory activation (because correct production rules are usually learned after incorrect ones, and not vice versa, allowing the correct production rules to have a stronger activation in memory). This offers a cursory insight into the importance of "forgetting" misconceptions for successful learning, but further research is required to fully understand the implications of its mechanism.

So far, we have presented evidence that computational models of human learning can be used to bridge the gap between learning theory and data. This approach allows for examination of learning theory in a variety of scenarios, including unexpected student learning phenomena. In particular, existing literature has shown inconsistencies in the effects of retrieval practices and worked examples on learning. To address the inconsistencies, Carvalho et al. [1] have proposed a plausible mechanism to explain the differences in these learning scenarios focusing on the selectivity of encoding of the tasks. Leveraging computational models of learning, we employed simulated learners as a means of validating the proposed theoretical framework. These simulated learners served as a valuable tool for investigating the mechanism of learning in greater depth, as we were able to analyze the types of errors made during the learning process. Additionally, by comparing the proposed theory to other existing theories, we were able to determine which theory best aligns with human data. Furthermore, an in-depth examination of the simulated learners revealed interesting insights, such as the potential benefits of forgetting in skill acquisition, which can serve as a guide for future research directions. Therefore, we have emphasized the possibilities that can be achieved through the use of computational models in education research.

6 Future Work

The current setup of our simulated learners has a limitation in that it is unable to transfer knowledge from skills to facts. This is a common phenomenon observed in human learning, where the learning gain on fact problems is increased when worked examples with skills are utilized (1.61% compared to 0.28% for retrieval practice). We hypothesize that when humans learn a skill through worked examples, they implicitly learn the associated fact, and are able to transfer that knowledge to other fact-based problems.

In future work, we plan to address the limitation in our simulated learners of transferring knowledge from skills to facts. We will investigate mechanisms for handling problem sets that combine both skills and facts, in order to improve the model's similarity to human learning. Additionally, we will explore increasing the complexity of problem encodings, such as incorporating the ability for the agent to parse diagrams, to better reflect real-world challenges in selectivity.

7 Conclusions

In summary, this study has highlighted the utility of computational models of human learning in bridging the gap between learning theory and data, as demonstrated through examination of unexpected learning phenomena. We started with an unexpected learning phenomena (inconsistencies in the effects of retrieval practices and worked examples on learning), and a proposed plausible mechanism (a mechanism focusing on the selectivity of encoding of the tasks). Then, through the use of computational models, we were able to not only confirm but also examine this proposed learning theory in more depth. Additionally, the ability of these models to examine different learning theories and identify which one best fits human learning, as well as provide valuable insights into the learning process, highlights the potential of computational models in the field of education research. Our findings demonstrate the potential for these models to inform the development of more effective teaching strategies and guide future research in this area.

Acknowledgements. This work was funded by NSF grant #1824257.

References

1. Carvalho, P.F., Rachatasumrit, N., Koedinger, K.R.: Learning depends on knowledge: the benefits of retrieval practice vary for facts and skills. In: Proceedings of the Annual Meeting of the Cognitive Science Society (2022)
2. Karpicke, J.D., Aue, W.R.: The testing effect is alive and well with complex materials. Educ. Psychol. Rev. **27**(2), 317–326 (2015)
3. Karpicke, J.D., Blunt, J.R.: Retrieval practice produces more learning than elaborative studying with concept mapping. Science **331**(6018), 772–775 (2011)
4. Koedinger, K.R., Corbett, A.T., Perfetti, C.: The knowledge-learning-instruction framework: bridging the science-practice chasm to enhance robust student learning. Cogn. Sci. **36**(5), 757–798 (2012)
5. Maclellan, C.J., Harpstead, E., Patel, R., Koedinger, K.R.: The apprentice learner architecture: closing the loop between learning theory and educational data. Int. Educ. Data Min. Soc. (2016)
6. Pavlik, P.I., Anderson, J.R.: Using a model to compute the optimal schedule of practice. J. Exp. Psychol. Appl. **14**(2), 101 (2008)
7. Rawson, K.A.: The status of the testing effect for complex materials: still a winner. Educ. Psychol. Rev. **27**(2), 327–331 (2015)
8. Rawson, K.A., Dunlosky, J.: Optimizing schedules of retrieval practice for durable and efficient learning: how much is enough? J. Exp. Psychol. Gen. **140**(3), 283 (2011)
9. Renkl, A.: The worked-out-example principle in multimedia learning. In: The Cambridge Handbook of Multimedia Learning, pp. 229–245 (2005)
10. Roediger, H.L., III., Agarwal, P.K., McDaniel, M.A., McDermott, K.B.: Test-enhanced learning in the classroom: long-term improvements from quizzing. J. Exp. Psychol. Appl. **17**(4), 382 (2011)
11. Roediger, H.L., III., Karpicke, J.D.: Test-enhanced learning: taking memory tests improves long-term retention. Psychol. Sci. **17**(3), 249–255 (2006)
12. Salden, R.J., Koedinger, K.R., Renkl, A., Aleven, V., McLaren, B.M.: Accounting for beneficial effects of worked examples in tutored problem solving. Educ. Psychol. Rev. **22**(4), 379–392 (2010)
13. Van Gog, T., Kester, L.: A test of the testing effect: acquiring problem-solving skills from worked examples. Cogn. Sci. **36**(8), 1532–1541 (2012)
14. Van Gog, T., Paas, F., Van Merriënboer, J.J.: Effects of process-oriented worked examples on troubleshooting transfer performance. Learn. Instr. **16**(2), 154–164 (2006)
15. Van Gog, T., Sweller, J.: Not new, but nearly forgotten: the testing effect decreases or even disappears as the complexity of learning materials increases. Educ. Psychol. Rev. **27**(2), 247–264 (2015)
16. Weitekamp, D., MacLellan, C., Harpstead, E., Koedinger, K.: Decomposed inductive procedure learning (2021)
17. Weitekamp, D., III., Harpstead, E., MacLellan, C.J., Rachatasumrit, N., Koedinger, K.R.: Toward near zero-parameter prediction using a computational model of student learning. Ann Arbor **1001**, 48105

Investigating the Utility of Self-explanation Through Translation Activities with a Code-Tracing Tutor

Maia Caughey and Kasia Muldner(✉)

Department of Cognitive Science, Carleton University, Ottawa, Canada
{maiacaughey,kasiamuldner}@cunet.carleton.ca

Abstract. Code tracing is a foundational programming skill that involves simulating a program's execution line by line, tracking how variables change at each step. To code trace, students need to understand what a given program line means, which can be accomplished by translating it into plain English. Translation can be characterized as a form of self-explanation, a general learning mechanism that involves making inferences beyond the instructional materials. Our work investigates if this form of self-explanation improves learning from a code-tracing tutor we created using the CTAT framework. We created two versions of the tutor. In the experimental version, students were asked to translate lines of code while solving code-tracing problems. In the control condition students were only asked to code trace without translating. The two tutor versions were compared using a between-subjects study ($N = 44$). The experimental group performed significantly better on translation and code-generation questions, but the control group performed significantly better on code-tracing questions. We discuss the implications of this finding for the design of tutors providing code-tracing support.

Keywords: Code Tracing Tutor · Self-Explanation · Translation from Python to English

1 Introduction

What skills do students need to be taught when learning to program? To date, the emphasis has been on program generation (i.e., teaching students how to write computer programs). To illustrate, the majority of work in the AIED community and/or computer science education focuses on supporting the process of program generation or aspects of it (e.g., [2,8,11]). Program generation is certainly a core skill, but it is only one of several competencies listed in theories of programming education [26]. A foundational skill proposed in Xie et al.'s framework [26] is code tracing. Code tracing involves simulating at a high level the steps a computer takes when it executes a computer program (including keeping track of variable values and flow of execution through the program). The high-level goal of our research is to design tutoring systems that help students

N. Wang et al. (Eds.): AIED 2023, LNAI 13916, pp. 66–77, 2023.
https://doi.org/10.1007/978-3-031-36272-9_6

learn to code trace. As a step in this direction, here we describe two alternative designs of a tutoring system supporting code tracing and the corresponding evaluation. We begin with the related work.

1.1 Code Tracing: Related Work

Code tracing helps students learn the mechanics of program execution, which promotes understanding of the constructs making up the program [19]. Tracing on paper is positively correlated with programming performance [16,17,24] and there is experimental evidence that teaching students to code trace first helps them to subsequently write programs [3]. However, many students find code tracing challenging. Some report not knowing where to start [27]. Others report avoiding code tracing altogether, resorting to suboptimal methods instead [7]. When students do code trace, their traces are often incomplete [6,9] and/or contain errors [9]. Thus, support for code tracing is needed.

One way to help students learn to code trace is with tutoring systems. In our prior work [13], we implemented CT-Tutor, which was designed to provide practice with code tracing through problems. To test the effect of scaffolding, we created two versions of the problem interface: (1) the high-scaffolding interface guided code tracing by requiring students to enter intermediate variable values and providing feedback on these entries; (2) the reduced-scaffolding interface only provided an open-ended scrap area. In both versions, for each problem CT-Tutor provided a corresponding example, shown either before the problem or after (based on condition). Students learned from the tutor but there was no significant effect of either scaffolding or example order. However, when the analysis included only students who learned from using the tutor, an interaction between scaffolding and example order emerged. When students were given the high-scaffolding interface, learning was highest when the example came *after* the problem, but the opposite was true for the reduced-scaffolding interface (more learning if the example came *before* the problem). CT-Tutor aimed to mirror the process of code tracing on paper. Another way to code trace is with a debugger (these are often included in program development environments). Nelson et al. [18] developed a tutoring system called PL-Tutor that like a traditional debugger showed the program state at each program-execution step, additionally prompting students to self-explain during various parts of the code-tracing process. PL-Tutor was evaluated by comparing learning from the tutor and Codecademy materials. There was no significant difference between the two conditions but students did learn from using PL-Tutor.

Other work focuses on evaluating debugging and/or program visualization tools. While access to debuggers can be beneficial [10,20], these are traditionally designed for students with some programming experience, since they use technical terms (e.g., stacks, frames). A potential limitation of these tools is that they do all the code-tracing work, and so students may fail to learn how to do it on their own without the help of the tool.

Other work focuses on the design of code-tracing examples (rather than problems as we do in the present work). Hosseini et al. [12] created a tutoring system that provided animations of code traces. The animations showed a visual trace of the program as well as the stack frame with values of variables. A second type of example included in the tutor was static, providing written explanations about code traces without animations. Students had access to both examples. Accessing animated examples was positively associated with higher course grades, while accessing static examples was associated with lower grades. Kumar [14] evaluated a tutoring system that presented a code-tracing example after an incorrect code-tracing solution was submitted. Students in the experimental group were prompted to self-explain the example; the control group did not receive prompts. There was no significant difference in gain scores between the prompted and unprompted groups, perhaps because tutor usage was not controlled (this study was done in a classroom context and students worked with the tutor on their own time, so some may have devoted little effort when answering the prompts).

As the summary above shows, obtaining significant effects related to code-tracing instructional manipulations is challenging, highlighting the need for more research. In general, existing support for code tracing focuses on showing the mechanics of program execution and state. However, in order to code trace a program, the programming language syntax and semantics need to be understood. This is not trivial for novices [21]. Accordingly, theories of programming instruction propose that translation is an essential component of learning to program [26]. Translation involves converting programming language syntax into an explanation of what the statement does in a human language. The translation grounds the code in more familiar language, which should free cognitive resources needed for performing the code trace. Translation can be characterized as the general mechanism of self-explanation [4], because it requires inferences over and beyond the instructional materials and because these inferences should be beneficial for performing the code trace.

Note that translation of individual programming statements or lines differs from describing what an entire program does. There is work on the latter aspect [17,25]. To illustrate, Whalley et al. [25] evaluated the reading comprehension skills of novice programmers. Participants were asked to explain what programs do at a high level. The prompts targeted entire programs or subsections that involved multiple program lines. In contrast, during code tracing, students read line-by-line to trace variable values and program behavior and so smaller program components are involved.

2 Current Study

To the best of our knowledge, work in AIED and beyond has not yet evaluated whether incorporating explicit support for line-by-line translations in a tutoring system promotes learning of code tracing. To fill this gap, we conducted a study to answer the following research question:

RQ: *In the context of a tutoring system supporting code tracing, does translation of programming language syntax into a human language (English in our work) prior to code tracing improve learning?*

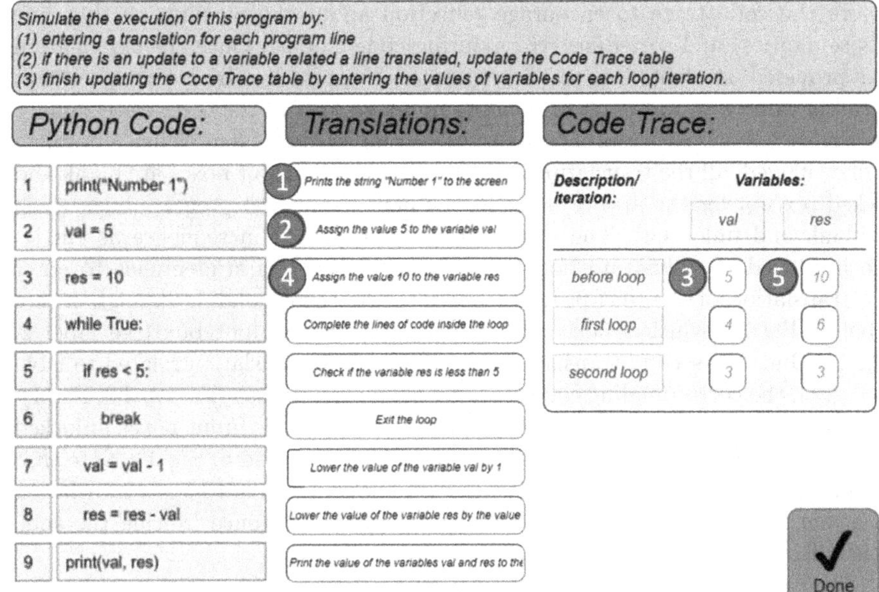

Fig. 1. The translation tutor interface with the full solution entered. The Python program to be code traced appears in the leftmost panel; the translations of the program from Python syntax to plain English are in the middle panel; the values of the program variables for a given loop iteration are in the right-most panel that contains the code-trace table. Notes: (1) the orange bubbles are for illustrative purposes and were not shown to students - see text for their description; (2) the solution to the translation, not shown here for space reasons, appears to the right of the code-trace table. (Color figure online)

To answer this question, we created two versions of an online tutoring system supporting code tracing of basic Python programs: (1) a translation tutor, and (2) a standard tutor. Both versions were created using CTAT [1]. CTAT is a tutor-building framework that facilitates tutor construction by providing tools to create the tutor interface and specify tutor behaviors. Both versions scaffolded the process of code tracing, but only the translation tutor required students to self-explain the meaning of the program being code traced.

2.1 Translation Tutor vs. Standard Tutor

The translation tutor presented one code-tracing problem per screen. Each screen included a brief Python program (Fig. 1, left), a translation panel used to enter plain English translations of the program (Fig. 1, middle), and a code-trace table used to input values of the program's variables during program execution (Fig. 1, right). To solve the problem, for each program line, students had to first self-explain the line by translating it into plain English. The translation was required before the code trace to encourage reflection about the meaning of that line (its semantics) and so reduce errors during code tracing. Once the explanation was provided, students had to code trace the program line by entering relevant variable value(s) into the corresponding input box(es) in the code-trace table.

Figure 1 shows the tutor interface with a completed problem. When a problem is first opened, all the translation and code-trace table input boxes are blank and locked (except for the first translation box next to the first program line, which is blank and unlocked). The input boxes are unlocked in sequence as entries are produced. To illustrate, for the problem in Fig. 1, a student has to enter the translation of the first program line into the corresponding input box (see label 1, Fig. 1), which unlocks the second translation input box (see label 2, Fig. 1). Once the second translation is produced (see translation related to label 2, Fig. 1), the corresponding code-trace table box is unlocked (see label 3, Fig. 1); this process continues, with the next translation and table input boxes unlocked after corresponding answers are generated (see labels 4 and 5, Fig. 1). Note that if a program includes a loop, as is the case for the program in Fig. 1, a translation of a line inside the loop body has to be produced only once (during the code trace of the first loop iteration).

The translation tutor provides immediate feedback for correctness on the entries in the code-trace table, by coloring them red or green for incorrect and correct entries, respectively. To guide solution generation, a correct table entry is required to unlock the next input box. For the translations, due to challenges with natural language parsing, immediate feedback is not provided, and any input submitted for a translation unlocks the next input box. Students can view a canonical translation solution after the entire code-trace problem is completed, by clicking on the purple box used to temporarily hide the translation solution (not shown in Fig. 1 for space reasons). At this point, students can revise their translations if they wish (all translation boxes are unlocked after the code trace-table is correctly completed).

In addition to feedback for correctness, the interface is designed to implicitly guide the code-tracing process in several ways, using tactics from our prior work [13]. First, as described above, the tutor requires students to enter the code trace step-by-step in the logical order dictated by the code-tracing process - cues are provided about this because the input boxes are locked (colored gray) until a step is correctly generated, at which point the color of the next input box changes to indicate it is unlocked and available for input. This design aims to encourage students to enter the entire solution step by step, motivated by the fact that when tracing on paper, students produce incomplete traces and/or skip steps when code tracing [6]. Second, because some students do not know which program elements to trace, the interface specifies the target variables to be traced (see code-trace table, Fig. 1).

The standard version of the tutor is identical to the translation version, except that its interface does not include the translation panel shown in Fig. 1 (middle). Thus, in the standard tutor, students only have to enter the code trace using the code-trace table. All other scaffolding included in the translation tutor is also provided by the standard tutor.

2.2 Participants

The study participants were 44 individuals (37 female, 6 male, 1 demi-femme) recruited using a range of methods: (1) class announcements in a first-year university programming class for cognitive science majors, (2) the SONA online recruitment system available to students in a first-year university class that provided a broad overview of cognitive science, (3) social media advertising via university Facebook groups, and (4) word of mouth. Participants either received a 2% bonus course credit for completing the study, or $25 compensation. To be eligible, participants either had to have no prior programming experience or at most one university-level programming course.

2.3 Materials

Both versions of the tutor were populated with four code-tracing problems. The same problems were used for the two versions; each problem showed a Python program with a while loop (e.g., see Fig. 1). A brief lesson was developed to provide an introduction and/or refresher to fundamental programming concepts. The lesson corresponded to a 20-min video showing a narrated slideshow. The lesson covered variable assignment, integer and string data types, basic conditional statements, and while loops.

A pretest and posttest were created to measure domain knowledge and learning. The tests were isomorphic, with the same number of questions and question content, but different variable names and values. There were three types of questions on the test, namely translation, code tracing, and code generation. Each test showed a series of six brief Python programs - for each program, students were asked to produce (1) a line-by-line translation of the program into plain English, and (2) a detailed code trace. The final question required the generation of a Python program. A grading rubric was developed, with the maximum number of points for each test equal to 44.5.

2.4 Experimental Design and Procedure

The study used a between-subjects design, with two conditions corresponding to the two tutor versions described in Sect. 2.1. Participants were assigned to the conditions in a round-robin fashion.

The study sessions were conducted individually through Zoom (one participant per session). The study took no more than two hours to complete. After informed consent was obtained, participants were given the link to the video lesson and were asked to watch it; they could take notes if they wished (20 min). After the lesson, participants completed a brief demographics questionnaire and the pretest (20 min. with a 5 min. grace period). Next, participants took a five-minute break. After the break, they were provided a link to the tutor (either the translation tutor or the standard tutor). After logging into the tutor website, they were given a five-minute introduction to the tutor by the researcher, including a demonstration on how to use the tutor (this was scripted for consistency between sessions). The demonstration used a problem from the video lesson. Next, participants used the tutor to solve four code-tracing problems. Participants were instructed to complete the problems at their own pace and told they would have 40 min to do so (with a 10-min grace period). After the tutor phase, participants were provided with a link to the posttest (20 min. with a five-minute grace period).

3 Results

The primary goal was to analyze if translation activities during code tracing with a tutoring system improved learning. Learning was operationalized by change from pretest to posttest (details below). The tests were graded out of 44.5 points, using a pre-defined grading scheme. Recall there were three types of questions on the test, namely translation, code tracing, and code generation (the latter was the transfer question, as the lesson and tutor did not cover program generation). We conducted separate analyses for each question type because each involved distinct concepts and lumping them together had potential to obscure findings. Initially, there were 22 participants in each condition. For a given question type, we removed from the analysis participants at ceiling at pretest for that question type, so the degrees of freedom will vary slightly.

The descriptive statistics are shown in Table 1. Collapsing across condition and question type, participants did learn from using the tutor, as indicated by the significant improvement from pretest to posttest (M_{gain} = 5.52%, SD = 10.24), $t(43) = 3.57$, $p < .001$, $d = 0.54$.

Prior to testing conditional effects, for each question type, we checked if there was a difference in the pretest scores between the two conditions using an independent samples t-test. Despite the assignment strategy, overall the translation-tutor group had slightly higher pretest scores, but this effect was not significant for any of the three question types (translation pretest scores: $t(42) = 1.21$, $p = .233$, $d = 0.37$; code-tracing pretest scores: $t(39) = 1.67$, $p = .103$, $d = 0.52$; code-generation pretest scores $t(42) = 1.49$, $p = .144$, $d = 0.50$).

To measure learning, we used normalized gain [5], calculated as follows:

$$\frac{posttest(\%) - pretest(\%)}{100\% - pretest(\%)}$$

Normalized gain characterizes how much a student learned (based on their raw gain from pretest to posttest), relative to how much they could have learned (based on their pretest score). This enables a more fair comparison between groups, particularly in situations where there are differences in pretest scores. This was the case with our data, i.e., the pretest scores for all three question categories were higher for the translation tutor group, albeit not significantly so.

Table 1. Mean and standard deviation for pretest, posttest, and raw gain (posttest – pretest) scores (by percentage) for each question type and condition.

Question Type		Standard Tutor $n = 22$	Translation Tutor $n = 22$
		M (SD)	M (SD)
Translation	Pretest (%)	66.99 (23.60)	74.88 (19.49)
	Posttest (%)	63.04 (28.71)	78.71 (22.58)
	Raw Gain (post-pre) (%)	−3.95 (15.70)	3.83 (8.11)
Code Trace	Pretest (%)	50.78 (23.43)	63.91 (26.96)
	Posttest (%)	66.75 (31.24)	64.81 (27.82)
	Raw Gain (post-pre)(%)	15.97 (20.93)	0.90 (8.35)
Code Generation	Pretest (%)	23.49 (28.48)	28.95 (31.34)
	Posttest (%)	31.44 (34.69)	51.32 (39.11)
	Raw Gain (post-pre)(%)	7.96 (21.59)	22.37 (28.06)

Normalized gain was calculated for each student separately. The average normalized gains for each question type and condition are shown in Fig. 2. A disadvantage for normalized gain is that it is arguably harder to interpret than raw gain (posttest - pretest). Thus, for the sake of completeness, we also report raw gain in Table 1. For the sake of parsimony we only report the inferential statistics for the normalized gain but the pattern of results for each question holds if raw gain is used as the dependent variable.

For the translation questions, the standard-tutor group performed slightly worse on the posttest than on the pretest as indicated by a negative normalized gain, while the translation-tutor group improved from pretest to posttest (see Fig. 2). This effect of tutor type on normalized gain was significant, $t(31.2) = 2.4$, $p = .025$ $d = 0.78$), with the translation-tutor group learning significantly more. The opposite pattern occurred for the code-trace questions, with only the standard-tutor group improving from pretest to posttest; the translation-tutor group had similar pretest and posttest scores (see Fig. 2). This effect of tutor type on normalized gain for the code-tracing scores was significant, $t(39) = 2.5$, $p = .016$, $d = 0.8$. For the code-generation transfer question, the translation-tutor group had higher normalized gain but not significantly so, $t(39) = 1.2$, $p = .251$, $d = 0.37$ (but see below after outliers were removed).

The results show that the effect of tutor type depended on question type (more translation learning with the translation tutor but more code-tracing learning with the standard tutor). This interaction between tutor type and question type was significant as indicated by a mixed ANOVA with question type as the within-subjects factor and tutor type as the between-subjects factor, $F(2, 74) = 6.57$, $p = .005$, ${\eta_p}^2 = .30$.

To ensure the validity of our results, we checked for the presence of outliers. There were no outliers due to data errors but there were several in each condi-

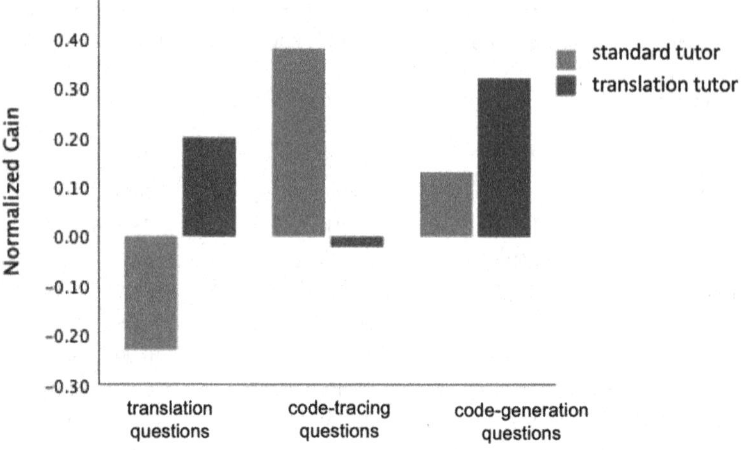

Fig. 2. Average normalized gain for each condition and question type (note that normalized gain is calculated differently from the raw gain shown in Table 1)

tion due to normal variation[1]. If outliers are a normal consequence of inherent individual differences, the advocated approach is to re-run the analysis without them and report if the original patterns change. This is advocated over removing outliers that are not due to errors as it portrays a more holistic view of the results. The removal of the outliers did not change the significant effect of tutor type for the translation and code-trace questions (translation: $t(41) = 2.09$, $p = .043$, $d = 0.64$; code-tracing $t(33.31) = 3.07$, $p = .04$, $d = 0.92$). For the code-generation transfer question, the outliers were influential. After they were removed, the translation-tutor group improved significantly more on the transfer question than the standard-tutor group, $t(27.43) = 2.72$, $p = .011$, $d = 0.88$.

4 Discussion and Future Work

The present paper describes the design and evaluation of a tutoring system supporting code-tracing activities. Overall students did learn from interacting with the tutoring system but the analysis of normalized gain scores broken down by question type revealed a nuanced pattern regarding the effect of tutor type.

We begin with the positive results. Students working with the translation tutor had to generate translations of programming language syntax into plain English. These students had significantly higher normalized gain on translation test questions, compared to the standard-tutor group not required to produce translations. Since translation is a form of self-explanation, our findings replicate prior work showing the benefits of self-explaining [4]. We acknowledge that the

[1] There was 1 outlier in the standard-tutor data for translation scores, 4 outliers in the translation-tutor data for code-tracing scores, and outliers for the code-generation scores (3 in the standard-tutor data and 1 in the translation-tutor data).

translation-tutor group had practice on translation, but explicit support does not always result in significant effects (e.g., [13,14,18]) and so this finding is encouraging. Notably, the standard-tutor group got slightly worse in their translation performance from pretest to posttest. This may have been due to the time gap between the lesson, which included translation information, and the posttest. Due to this gap, the standard-tutor group may have forgotten translation concepts from the lesson. In contrast, the translation-tutor group received support for translation with the tutor, which likely reinforced lesson concepts.

The translation-tutor group also had significantly higher normalized gain on the code-generation transfer question (after outlier removal). This group was scaffolded to generate explanations from Python to plain English. Because these explanations may have made the program more meaningful, this could have facilitated learning of code schemas (i.e., algorithms), which are useful for code generation [22].

We now turn to the unexpected result related to learning of code tracing, namely that the standard tutor group had significantly higher code-tracing normalized gains than the translation-tutor group. In fact, the translation-tutor group did not improve on code-tracing outcomes from pretest to posttest (normalized gain was close to zero). This surprising, as this group had opportunities for code-tracing practice, which based on prior work should increase learning [15,18]. One possible explanation for this finding is that the translation tutor required translation at each step and this may have increased cognitive load [23], which interfered with the code-tracing process. In particular, translations added an extra feature students had to pay attention to, and cognitive load theory predicts that splitting a student's attention between features can hinder learning. Other factors to explain why the translation-tutor group had low code-tracing performance include the timing of translations, effort, and feedback. The timing of the translations may have not been ideal - perhaps if the tutor prompted for translations at a different point in the instructional sequence, rather than concurrently during the code-trace activity, learning would have occurred. Moreover, writing translations requires effort. This increased effort may have led to fatigue, leaving less resources for the code trace. Finally, the feedback for the translations was delayed until after the code trace due to logistics reasons and came in the form of a canonical solution showing the "ideal" translations. This format requires students to invest effort into processing the feedback (e.g., comparing their translations to ones in the canonical solution), something that they may have been unmotivated to do.

In conclusion, the answer to our research question on whether translation activities improve learning depends on the outcome variable of interest (translation, generation, code tracing). Requiring translations to be completed concurrently with a code-trace activity produced a modest but significant boost for normalized gain related to translation questions; translation also improved normalized gain related to code generation, once outliers were removed. However, translation hindered learning of code tracing. To address this issue, future work needs to investigate alternative designs, including other instructional orderings.

For instance, perhaps students need to master the skill of translation from a given programming language to their native human language before *any* code tracing occurs. Another avenue for future work relates to providing guidance for translations. In the present study feedback in the form of a canonical solutions was provided after the code trace - perhaps novices require feedback earlier. Work is also needed to generalize our findings. In our study, the majority of participants were female. This was partly a function of the fact that about a third of the participants came from a first year programming class required for all cognitive science majors in our program (even ones not focused on computer science), and this class has a higher proportion of female students. In general, however, programming classes are increasingly required for non-computer science majors, and so research is needed on how to best support students from varied backgrounds.

Acknowledgements. This research was funded by an NSERC discovery grant.

References

1. Aleven, V., et al.: Example-tracing tutors: intelligent tutor development for non-programmers. Int. J. Artif. Intell. Educ. **26**(1), 224–269 (2016)
2. Anderson, J.R., Conrad, F.G., Corbett, A.T.: Skill acquisition and the Lisp tutor. Cogn. Sci. **13**(4), 467–505 (1989)
3. Bayman, P., Mayer, R.E.: Using conceptual models to teach basic computer programming. J. Educ. Psychol. **80**(3), 291 (1988)
4. Chi, M.T., Bassok, M., Lewis, M.W., Reimann, P., Glaser, R.: Self-explanations: how students study and use examples in learning to solve problems. Cogn. Sci. **13**(2), 145–182 (1989)
5. Coletta, V., Steinert, J.: Why normalized gain should continue to be used in analyzing preinstruction and postinstruction scores on concept inventories. Phys. Rev. Phys. Educ. Res. **16** (2020)
6. Cunningham, K., Blanchard, S., Ericson, B., Guzdial, M.: Using tracing and sketching to solve programming problems: replicating and extending an analysis of what students draw, pp. 164–172 (2017)
7. Cunningham, K., Ke, S., Guzdial, M., Ericson, B.: Novice rationales for sketching and tracing, and how they try to avoid it. In: Proceedings of the 2019 ACM Conference on Innovation and Technology in Computer Science Education, ITiCSE 2019, pp. 37–43. Association for Computing Machinery, New York (2019)
8. Fabic, G.V., Mitrovic, A., Neshatian, K.: Evaluation of parsons problems with menu-based self-explanation prompts in a mobile python tutor. Int. J. Artif. Intell. Educ. **29** (2019)
9. Fitzgerald, S., Simon, B., Thomas, L.: Strategies that students use to trace code: an analysis based in grounded theory. In: Proceedings of the First International Workshop on Computing Education Research, ICER 2005, pp. 69–80 (2005)
10. Hoffswell, J., Satyanarayan, A., Heer, J.: Augmenting code with in situ visualizations to aid program understanding. In: Proceedings of the 2018 CHI Conference on Human Factors in Computing Systems, CHI 2018, pp. 1–12 (2018)
11. Hosseini, R., et al.: Improving engagement in program construction examples for learning Python programming. Int. J. Artif. Intell. Educ. **30**, 299–336 (2020)

12. Hosseini, R., Sirkiä, T., Guerra, J., Brusilovsky, P., Malmi, L.: Animated examples as practice content in a Java programming course. In: Proceedings of the Technical Symposium on Computing Science Education, SIGCSE 2016, pp. 540–545 (2016)
13. Jennings, J., Muldner, K.: When does scaffolding provide too much assistance? A code-tracing tutor investigation. Int. J. Artif. Intell. Educ. **31**, 784–819 (2020)
14. Kumar, A.N.: An evaluation of self-explanation in a programming tutor. In: Trausan-Matu, S., Boyer, K.E., Crosby, M., Panourgia, K. (eds.) ITS 2014. LNCS, vol. 8474, pp. 248–253. Springer, Cham (2014). https://doi.org/10.1007/978-3-319-07221-0_30
15. Lee, B., Muldner, K.: Instructional video design: investigating the impact of monologue- and dialogue-style presentations, pp. 1–12 (2020)
16. Lister, R., Fidge, C., Teague, D.: Further evidence of a relationship between explaining, tracing and writing skills in introductory programming, pp. 161–165 (2009)
17. Lopez, M., Whalley, J., Robbins, P., Lister, R.: Relationships between reading, tracing and writing skills in introductory programming, pp. 101–112 (2008)
18. Nelson, G.L., Xie, B., Ko, A.J.: Comprehension first: evaluating a novel pedagogy and tutoring system for program tracing in CS1. In: Proceedings of the ACM Conference on International Computing Education Research, pp. 2–11 (2017)
19. Perkins, D.N., Hancock, C., Hobbs, R., Martin, F., Simmons, R.: Conditions of learning in novice programmers. J. Educ. Comput. Res. **2**(1), 37–55 (1986)
20. Pérez-Schofield, J.B.G., García-Rivera, M., Ortin, F., Lado, M.J.: Learning memory management with C-Sim: a C-based visual tool. Comput. Appl. Eng. Educ. **27**(5), 1217–1235 (2019)
21. Qian, Y., Lehman, J.: Students' misconceptions and other difficulties in introductory programming: a literature review. ACM Trans. Comput. Educ. **18**(1) (2017)
22. Soloway, E., Ehrlich, K.: Empirical studies of programming knowledge. IEEE Trans. Softw. Eng. **10**(5), 595–609 (1984)
23. Sweller, J.: Cognitive load theory. Psychol. Learn. Motiv. **55**, 37–76 (2011)
24. Venables, A., Tan, G., Lister, R.: A closer look at tracing, explaining and code writing skills in the novice programmer. In: Proceedings of the ACM Conference on International Computing Education Research, pp. 117–128 (2009)
25. Whalley, J.L., et al.: An Australasian study of reading and comprehension skills in novice programmers, using the Bloom and SOLO taxonomies. In: Proceedings of the 8th Australasian Conference on Computing Education, pp. 243–252 (2006)
26. Xie, B., et al.: A theory of instruction for introductory programming skills. Comput. Sci. Educ. **29**(2–3), 205–253 (2019)
27. Xie, B., Nelson, G.L., Ko, A.J.: An explicit strategy to scaffold novice program tracing. In: Proceedings of the ACM Technical Symposium on Computer Science Education, pp. 344–349 (2018)

Reducing the Cost: Cross-Prompt Pre-finetuning for Short Answer Scoring

Hiroaki Funayama[1,2](✉), Yuya Asazuma[1,2], Yuichiroh Matsubayashi[1,2], Tomoya Mizumoto[2], and Kentaro Inui[1,2]

[1] Tohoku University, Sendai, Japan
{h.funa,asazuma.yuya.r7}@dc.tohoku.ac.jp, {y.m,inui}@tohoku.ac.jp
[2] RIKEN, Tokyo, Japan
tomoya.mizumoto@a.riken.jp

Abstract. Automated Short Answer Scoring (SAS) is the task of automatically scoring a given input to a prompt based on rubrics and reference answers. Although SAS is useful in real-world applications, both rubrics and reference answers differ between prompts, thus requiring a need to acquire new data and train a model for each new prompt. Such requirements are costly, especially for schools and online courses where resources are limited and only a few prompts are used. In this work, we attempt to reduce this cost through a two-phase approach: train a model on existing rubrics and answers with gold score signals and finetune it on a new prompt. Specifically, given that scoring rubrics and reference answers differ for each prompt, we utilize key phrases, or representative expressions that the answer should contain to increase scores, and train a SAS model to learn the relationship between key phrases and answers using already annotated prompts (i.e., cross-prompts). Our experimental results show that finetuning on existing cross-prompt data with key phrases significantly improves scoring accuracy, especially when the training data is limited. Finally, our extensive analysis shows that it is crucial to design the model so that it can learn the task's general property. We publicly release our code and all of the experimental settings for reproducing our results (https://github.com/hiro819/Reducing-the-cost-cross-prompt-prefinetuning-for-SAS).

Keywords: Automated Short Answer Scoring · Natural Language Processing · BERT · domain adaptation · rubrics

1 Introduction

Automated Short Answer Scoring (SAS) is the task of automatically scoring a given student answer to a prompt based on existing rubrics and reference answers [9,14]. SAS has been extensively studied as a means to reduce the burden of manually scoring student answers in school education and large-scale examinations or as a technology for augmenting e-learning environments [7,10]. However, SAS in the practical application requires one critical issue to be addressed:

N. Wang et al. (Eds.): AIED 2023, LNAI 13916, pp. 78–89, 2023.
https://doi.org/10.1007/978-3-031-36272-9_7

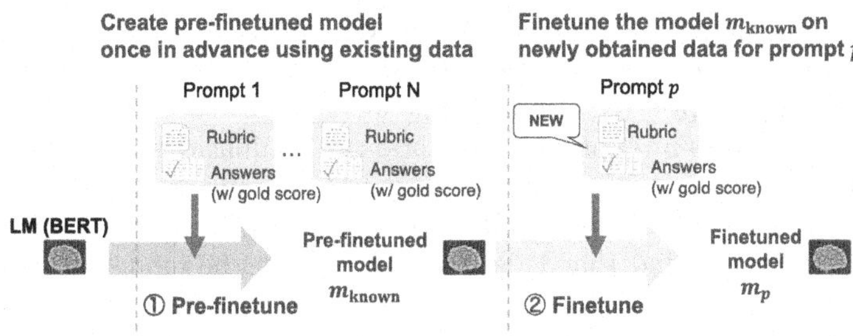

Fig. 1. Overview of our proposed method. We input key phrases, reference expressions, with an answer. We first pre-finetune the SAS model on already annotated prompts and then finetune the model on a prompt to be graded.

the cost of preparing training data. Data to train SAS models (i.e. students answers with human-annotated gold score signals) must be prepared for each prompt independently, as the rubrics and reference answers are different for each prompt [2].

In this paper, we address this issue by exploring the potential benefit of using *cross-prompt* training data, or training data consisting of different prompts, in model training. The cost of preparing training data will be alleviated if a SAS model can leverage cross-prompt data to boost the scoring performance with the same amount of in-prompt data. However, this approach imposes two challenges. First, it is not obvious whether a model can learn from cross-prompt data something useful for scoring answers to a new target prompt, as the new prompt must have different scoring criteria from cross-prompt data available a priori (*cross-prompt generalizability*). Second, in a real-world setting, cross-prompt data (possibly proprietary) may not be accessible when classrooms or e-learning courses train a new model for their new prompts (*data accessibility*). Therefore, we want an approach where one can train a model for a new prompt without accessing cross-prompt data while benefiting from cross-prompt training.

We address both challenges through a new two-phase approach: (i) train (pre-finetune) a model on existing rubrics and answers and (ii) finetune the model for a given new prompt (see Fig. 1). This approach resolves the data accessibility issue since the second phase (finetuning on a new prompt) does not require access to the cross-prompt data used in the first phase. Note that the second phase needs access only to the parameters of the pre-finetuned model. On the other hand, it is not obvious whether the approach exhibits cross-prompt generalizability. However, our experimental results indicate that a SAS model can leverage cross-prompt training data to boost the scoring performance if it is designed to learn the task's property shared across different prompts effectively.

Our contributions are as follows. (I) Through our two-phase approach to cross-prompt training, we conduct the first study in SAS literature to alleviate the need of expensive training data for training a model on every newly given prompt, while resolving the problem of limited accessibility to proprietary cross-prompt data. (II) We conduct experiments on a SAS dataset enriched with a large number of prompts (109), rubrics, and answers and show that a SAS model can benefit from cross-prompt training instances, exhibiting a considerable gain in score prediction accuracy on top of in-prompt training, especially in settings with less in-prompt training data. (III) We conduct an extensive analysis of the model's behavior and find that it is crucial to design the model so that it can learn the task's general property (i.e., a principle of scoring): an answer gets a high score if it contains the information specified by the rubric. (IV) In addition to our experiments, towards effective cross prompt SAS modeling which requires a large variety of prompts and answers, we added 1,0000 new data annotations (20 prompts with 500 answers each) to the RIKEN dataset [8], the only Japanese dataset available for ASAS.

We make our data annotations publicly available for academic purposes. We also publicly release our code and all of the experimental settings for reproducing our results[1].

2 Related Work

We position this study as a combination of the use of rubric and domain adoption using cross-prompt data.

To our knowledge, few researchers have focused on using rubrics. A study [16] used key phrases excerpted from rubrics to generate pseudo-justification cues, a span of the answer that indicates the reason for its score, and they showed the model performance is improved by training attention by pseudo-justification cues. [13] proposed a model that utilizes the similarity between the key concept described in rubrics and the answer. Following the utilization of rubrics in previous research, we also use the key phrases listed in the rubric, which are typical expressions used for achieving higher scores.

Domain adaptation in this field is also still unexplored. [15] further pre-trained BERT [4] on a textbook corpus to learn the knowledge of each subject (i.e. science) and report slight improvement. On the other hand, we pre-finetune the BERT on cross-prompt data to adopt the scoring task.

Since SAS has different rubrics and reference answers for each prompt, the use of cross-prompt data still remains an open challenge [6]. As far as we know, the only example is [12]. However, for each new prompt (i.e., target domain in their term), their model was required to be retrained with both in-prompt and cross-prompt data, which leaves the issue of limited accessibility to proprietary cross-prompt data. In contrast, our two-phase approach resolves the data accessibility issue since the second phase (i.e., finetuning on a new prompt) does not require access to the cross-prompt data used in the first phase. In our experiments, we modify the method used in [12] to adapt it to our task setting and compare the results of their method with our method.

[1] https://github.com/hiro819/Reducing-the-cost-cross-prompt-prefinetuning-for-SAS.

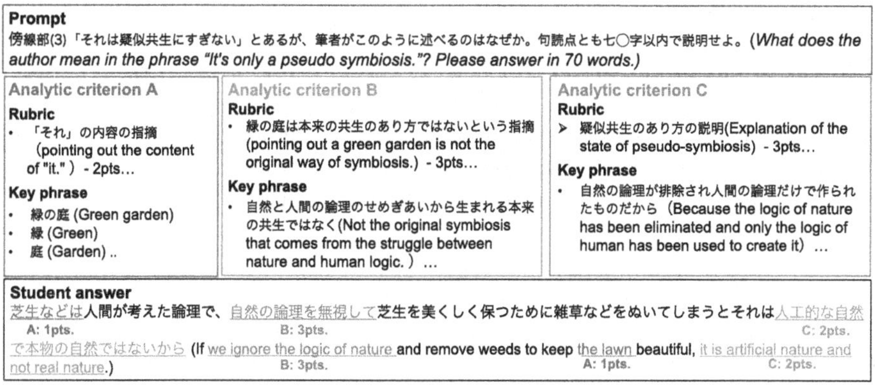

Fig. 2. Example of a prompt, scoring rubric, key phrase and student's answers excerpted from RIKEN dataset [8] and translated from Japanese to English. For space reasons, some parts of the rubrics and key phrase are omitted.

3 Preliminaries

3.1 Task Definition

Suppose X_p represents a set of all possible student answers for a prompt $p \in P$, and $\mathbf{x} \in X_p$ is an answer. Each prompt has a discrete integer score range $S = \{0, ..., N_p\}$, which is defined in the rubric. The score of each answer is chosen within the range S. Therefore, the SAS task is defined as assigning one of the scores $s \in S$ for each given input $\mathbf{x} \in X_p$.

In this study, we assume that every prompt is associated with a predefined rubric, which stipulates what information an answer must contain to get a score. An answer gets scored high if it contains the required information sufficiently and low if not. Figure 2 shows an example of a prompt with a rubric from the dataset we used in our experiments [8]. As in the figure, the required information stipulated by a rubric may also be presented by a set of key phrases to help human raters and students understand the evaluation criteria. Each key phrase gives an example of wording that gives an answer score high. In the dataset used in our experiments, every rubric provides a set of key phrases, and we utilize such key phrases in cross-prompt training.

In our cross-prompt training setting, we assume that we have some already graded prompts by human raters P_{known} and we then want to grade a new prompt p_{target} automatically. Within this cross-prompt setting, the model is required to score the answers having different score ranges. Therefore, we re-scale the score ranges of all P_{known} and p_{target} to $[0, 1]$, and as a result, the goal of the task is to construct a regression function $m : \bigcup_{p \in P_{\text{known}} \cup \{p_{\text{target}}\}} \{X_p\} \rightarrow [0, 1]$ that maps an student answer to a score $s \in [0, 1]$.

3.2 Scoring Model

A typical approach to construct a function m is to use deep neural networks. Suppose $\mathcal{D} = ((\mathbf{x}_i, s_i))_{i=1}^{I}$ is training data that consist of the pairs of an actually obtained student answer $\mathbf{x}_i$ and its corresponding human-annotated score s_i. I is the number of training instances. To train the model m, we attempt to minimize the Mean Squared Error loss on the training data $L_m(\mathcal{D})$:

$$m^* = \underset{m}{\operatorname{argmin}} \{L_m(\mathcal{D})\}, \quad L_m(\mathcal{D}) = \frac{1}{I} \sum_{(\mathbf{x},s)\in\mathcal{D}} (s - m(\mathbf{x}))^2, \tag{1}$$

where $m(\mathbf{x})$ is a score predicted by the model m for a given input $\mathbf{x}$. Once m^* is obtained, we can predict the score s of a new student answer as: $s = m^*(\mathbf{x})$.

We construct the model m as following. Let $\mathbf{enc}(\cdot)$ as the encoder, we first obtain a hidden vector $\mathbf{h}_\mathbf{x} \in \mathbb{R}^H$ from an input answer $\mathbf{x}$ as: $\mathbf{h}_\mathbf{x} = \mathtt{enc}(\mathbf{x})$.

Then, we feed the hidden vector $\mathbf{h}_\mathbf{x}$ to a linear layer with a sigmoid function to predict a score: $m(\mathbf{x}) = \mathtt{sigmoid}(\mathbf{w}^\top \mathbf{h}_\mathbf{x} + b)$, where $\mathbf{w} \in \mathbb{R}^H$ and $b \in \mathbb{R}$ are learnable parameters. In this paper, we used BERT [4], a widely used encoder in various NLP tasks, as the encoder.

4 Method

To leverage cross-prompt training data, we consider the following two-staged training process: (i) We first finetune the model with cross-prompt training instances so that it learns the task's general property (i.e., principles of scoring) shared across prompts, and (ii) we then further finetune the model with in-prompt training instances to obtain the model specific for the target prompt. We call the training in the first stage *pre-finetuning*, following [1]. The questions become what kind of general property can the model learn from cross-prompt training instances and how the model learns.

To address these questions, we first hypothesize that one essential property a SAS model can learn in pre-finetuning is the scoring principle: an answer generally gets a high score if it contains sufficient information specified by the rubric and gets a lower score if it contains less. The principle generally holds across prompts and is expected to be learned from cross-prompt training instances. To learn it, the model needs to have access to the information specified by the rubrics through pre-finetuning and finetuning. We elaborate on this below.

Key Phrases. As reference expressions for high score answers, we utilize key phrases described in the rubrics as shown in the middle part of Fig. 2. Key phrases are representative examples of the expressions that an answer must contain in order to gain scores.

Key phrases are clearly stated in each rubrics. We use those key phrases for each prompt p from the corresponding rubric, and generate a key phrase sequence k_p for p by enumerating multiple key phrases into a single sequence

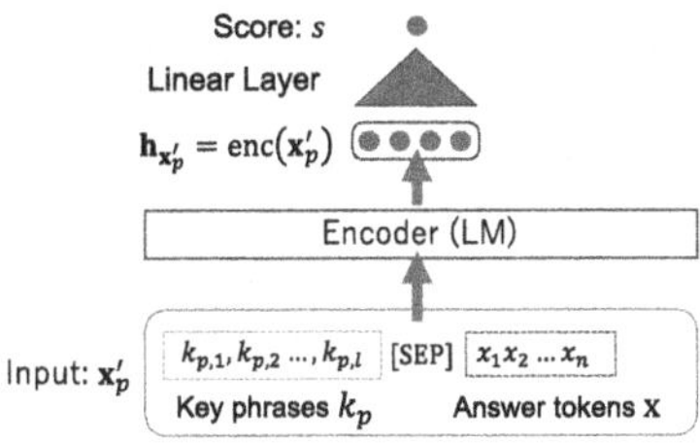

Fig. 3. Overall architecture of our model. We input key phrases and a student answer split by the [SEP] token.

with a comma delimiter. We then use the concatenated sequence $\mathbf{x}'_p$ of tokens k_p, [SEP], and $\mathbf{x}$ in this order, as our model input. For the model without using key phrases, we instead input a prompt ID to distinguish the prompt. We show the overall architecture of the model in Fig. 3.

Pre-finetuning. We utilize data from already annotated prompts P_{known} to train models for a new prompt p_{target}. For each prompt $p \in P$, there exists a key phrase sequence k_p. We create a concatenated input sequence $\mathbf{x}'_{p,i}$ for the i-th answer of the prompt p as: $\mathbf{x}'_{p,i} = \{k_p, [SEP], \mathbf{x}_{p,i}\}$. Then, we construct data for pre-finetuning as: $\mathcal{D}_{\text{known}} = \{(\mathbf{x}'_{p,i}, s_{p,i}) \mid p \in P_{\text{known}}\}_{i=1}^{I}$.

We pre-finetune the BERT-based regression model on this dataset $\mathcal{D}_{\text{known}}$ and obtain the model m_{known} : $m_{\text{known}} = \text{argmin}_m \{L_m(\mathcal{D}_{\text{known}})\}$. Next, we further finetune the pre-finetuned model m_{known} on $p \in P_{\text{target}}$ to obtain a model m_p for the prompt p as: $m_p = \text{argmin}_m \{L_m(\mathcal{D}_p)\}$.

5 Experiment

5.1 Dataset

RIKEN Dataset. We use the RIKEN dataset, a publicly available Japanese SAS dataset[2] provided in [8]. RIKEN dataset offers a large number of rubrics, prompts, and answers ideal for conducting our experiments. As mentioned in Sect. 1, we added 1,0000 new data annotations (20 prompts with 500 answers each) to the RIKEN dataset.

RIKEN dataset is a collection of annotated high school students' answers for Japanese Reading comprehension questions.[3] Each prompt in the RIKEN dataset has several scoring rubrics (i.e., analytic criterion [8]), and each answer is manually graded based on each analytic criterion independently (i.e., analytic score).

[2] https://aip-nlu.gitlab.io/resources/sas-japanese.

[3] Type of question in which the student reads a essay and answers prompts about its content.

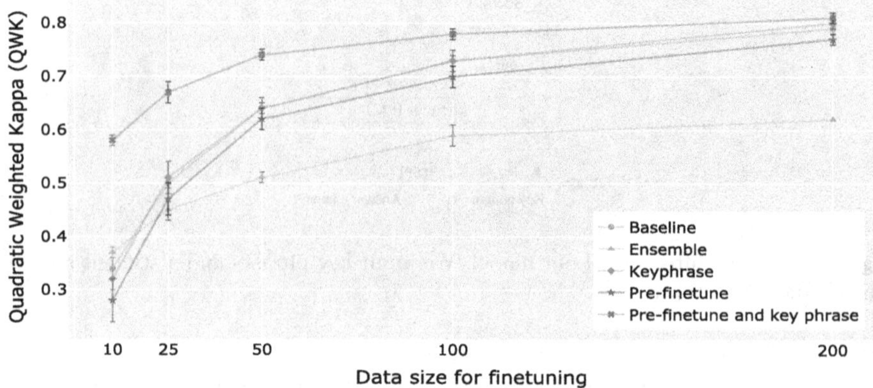

Fig. 4. QWK and standard deviation of four settings described in Sect. 5.2; `Baseline`, `Ensemble`, `Key phrase`, `Pre-finetune`, and `Pre-finetune & key phrase`. In the pre-finetuning phase, we use 88 prompts with 480 answers per prompt. We change the amount of data for finetuning as 10, 25, 50, 100, and 200.

In our experiment, we used 6 prompts (21 analytic criterion), same as [8], from RIKEN dataset as p_{target} to evaluate the effectiveness of pre-finetuning. We split answers for these 6 promts as 200 for train data, 50 for dev set and 250 for test set. For pre-finetuning, we used the remaining 28 prompts (88 analytic criterion), consisting of 480 answers per analytic criterion for training the model and 20 answers per analytic criterion as the dev set.

Following [5], we treat analytic criterion as an individual scoring task since each analytic score is graded based on each analytic criterion independently. For simplicity, we refer to each analytic criterion as a single, independent prompt in the experiments. Thus, we consider a total of 109 analytic criterion as 109 independent prompts in this dataset.

5.2 Setting

As described in Sect. 3.2, we used pretrained BERT [4] as the encoder for the automatic scoring model and use the vectors of CLS tokens as feature vectors for predicting answers.[4]

We train a model for 5 epochs in the pre-finetuning process. We then finetune the resulting model for 10 epochs. In the setting without pre-finetuning process, we finetune the model for 30 epochs. These epoch numbers were determined in preliminary experiments with dev set. During the finetuning process, we computed the QWK of the dev set at the end of each epoch and stored the best parameters with the maximum QWK.

Similar to previous studies [8,11], we use Quadratic Weighted Kappa (QWK) [3], the de facto standard evaluation metric in SAS, in the evaluation

[4] We used pretrained BERT models from https://github.com/cl-tohoku/bert-japanese for Japanese.

of our models. The scores were normalized to a range from 0 to 1 according to previous studies [8,11]. QWK was measured by re-scaling to the original range when evaluated on the test set.

To verify the effectiveness of cross-prompt pre-finetuning, we compare the following four settings in experiments; `Baseline`: Only finetune the BERT-based regression model for a target prompt without key phrases, the most straightforward way to construct a BERT-based SAS model. `Key phrase`: Only finetune BERT for a target prompt, we input an answer and key phrases to the model. `Pre-finetune`: Pre-finetune BERT on cross-prompt data, input only an answer. `Pre-finetune & key phrase`: Pre-finetune BERT on cross-prompt data, input an answer and key phrase pairs.

As mentioned in Sect. 2, [12] showed that ensembles of domain-specific and cross-domain models are effective. For comparison, we trained a cross-prompt model without a target prompt in advance since we cannot access the cross prompt data when creating a SAS model for a new prompt in our setting. We ensemble it with a prompt specific model trained only on the target prompt by averaging predicted scores of those models. We show results of the ensemble model as `Ensemble`.

5.3 Results

First, to validate the effectiveness of the pre-finetuning with key phrases, we examined the performance of the models for the five settings described in Sect. 5.2. Here, similar to [8], we experimented with 10, 25, 50, 100, and 200 training instances in the finetuning phase. The results are shown in Fig. 4.

We can see that pre-finetune without key phrases slightly lowers the model performance compared to Baseline. As expected, this result indicates that simply pre-finetuning on other prompts is not effective. Similarly, using only key phrases without pre-finetuning does not improve performance. QWK improves significantly only when key phrases are used and when pre-finetune is performed. Furthermore, we found that in our setting, where cross-prompt data is not accessible when adding a model for a new prompt, the method of ensembling cross-prompt and prompt specific models does not work effectively.

The gain was notably large when the training data was scarce, with a maximum improvement of about 0.25 in QWK when using 10 answers for finetuning compared to the Baseline. Furthermore, our results indicate that the pre-finetuning with key phrases can reduce the required training data by half while maintaining the same performance. On the other hand, the performance did not improve when we used 200 answers in training, which indicates that pre-finetuning does not benefit when sufficient training data is available. We note that the results of baseline models are comparable to the results of the baseline model shown in [8] (Fig. 4).

Impact of the Number of Prompts Used for Pre-finetuning. Next, we examined how changes in the number of prompts affect pre-finetuning: we fixed

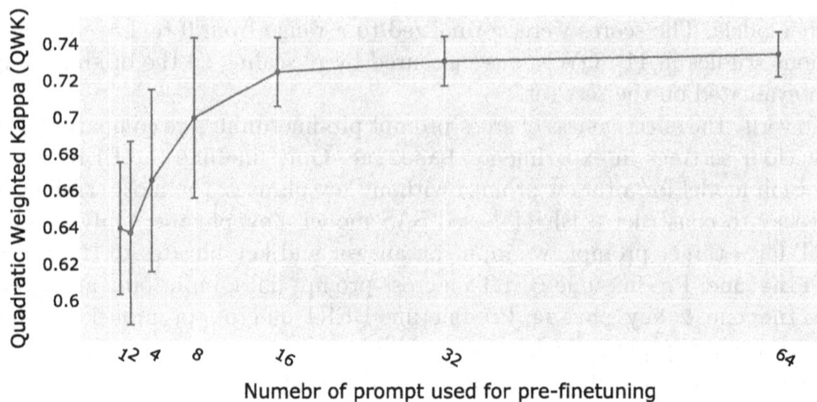

Fig. 5. QWK and standard deviation when the total number of answers used for pre-finetuning is fixed at 1,600 and the number of prompts used is varied from 1, 2, 4, 8, 16, 32, 64. For finetuning, 50 training instances were used.

the total number of answers used for the pre-finetuning at 1,600 and varied the number of prompts between 1, 2, 4, 8, 16, 32, 64. We performed finetuning using 50 answers for each prompt. The results are shown in Fig. 5. We see that the performance increases as the number of prompts used for pre-finetuning is increased. This result suggests that the more diverse the answer and key phrases pairs, the better the model understands their relationship. It also suggests that increasing the number of prompts is more effective for pre-finetuning than increasing the number of answers per prompts.

We can see the large standard deviation when the number of prompts used for pre-finetuning is small. We assume that the difference in the sampled prompts caused the large standard deviation, suggesting that some prompts might be effective for pre-finetuning while others are unsuitable for pre-finetuning. The result suggests that a certain number of prompts is needed for training in order to consistently obtain the benefits of cross-prompt learning for each new prompt.

5.4 Analysis: What Does the SAS Model Learn from Pre-finetuning on Cross Prompt Data?

We analyzed the behavior of the model in a zero-shot setting to verify what the model learned from pre-finetuning on cross-prompt data.

First, we examined the performance of the `Pre-finetune & key phrase` model in a zero-shot setting. As a result, we observed higher QWK scores for some prompts, as 0.81 and 079 points in the best-performing two prompts Y14_2-2_2_3-B and Y14_1-2_1_3-D, respectively. The results indicate that the model somewhat learns the scoring principle in our dataset through pre-finetune using the key phrases; i.e., an answer generally gets a high score if it contains sufficient information specified by the input key phrases.

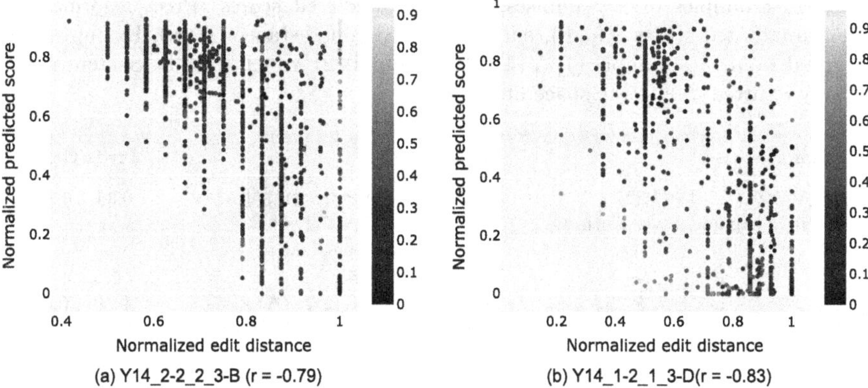

Fig. 6. Relationship between (x-axis) the normalized edit distance between the justification cue and key phrases in each answer to (a) Y14_2-2_2_3-B and (b) Y14_1-2_1_3-D and (y-axis) the predicted score in zero-shot settings. The color bars represent the absolute error between a predicted score and a gold score. r indicates the correlation coefficient.

Next, to examine how the key phrases contribute to the scoring, using the above two best-performing prompts, we examined the similarity between the key phrases and the manually annotated justification cues [8] (substrings of an answer which contributed to gain the score) in the student answer. For the similarity measure, we employ the normalized edit distance. Then we analyze the relationship between the edit distance and the predicted scores by the model.

The results for the two prompts with the highest QWK, Y14_2-2_2_3-B, Y14_1-2_1_3-D, are shown in Fig. 6. The color bars represent absolute error between a predicted score and a gold score. The correlation coefficients are -0.79 and -0.83, respectively, indicating a strong negative correlation between edit distance and predicted scores. This suggests that the more superficially distant the key phrases and answer, the lower the predicted model score will be for an answer. We also see that the model correctly predicts a variety of score points for the same edit distance values. We show some examples that have lower prediction error with high edit distance in Table 1. The examples indicate that the model predicts higher scores for answers that contain expressions that are semantically close to key phrases.

Those analysis indicate that the model partially grasp the property of the scoring task, in which an answer gains higher scores if the answer includes an expression semantically closer to the key phrases. Such a feature could contribute to the model's high performance, even when the model could not learn enough answer expression patterns from small training data.

Table 1. Examples of key phrases, answers, predicted scores (Pred.), normalized human annotated scores (Gold.), and normalized edit distance (Dist.). Examples are excerpted from the prompts (1) Y14_2-2_2_3-B and (2) Y14_1-2_1_3-D. Sentences are partially omitted due to the space limitation.

Key phrases	Answers	Pred.	Gold.
(1) 真実よりも幸福を優先する (Prioritize happiness over truth..)	幸福のためにはどうすれば良いかということについてばかり考える (.. only think about how to realize happiness.)	0.36	0.33
(2) 言葉を尽くして他人を説得する (Convince others with all my words)	説得に努まなければならない.. (..to try hard to convince others.)	0.50	0.50

6 Conclusion

In SAS, answers for each single prompt need to be annotated in order to construct a highly-effective SAS model specifically for that prompt. Such costly annotations are a major obstacle in deploying SAS systems into school education and e-learning courses, where resources are extremely limited. To alleviate this problem, we introduced a two-phase approach: train a model on cross-prompt data and finetune it on a new prompt. Given that scoring rubrics and reference answers are different in every single prompt, we cannot use them directly to train the model. Therefore, we utilized key phrases, or representative expression that answer should contain to gain scores, and pre-finetune the model to learn the relationship between key phrases and answers.

Our experimental results showed that pre-finetuning with key phrases greatly improves the performance of the model, especially when the training data is scarce (0.24 QWK improvement over the baseline for 10 training instances). Our results also showed that pre-finetuning can reduce the amount of required training data by half while maintaining similar performance.

Acknowledgments. We are grateful to Dr. Paul Reisert for their writing and editing assistance. This work was supported by JSPS KAKENHI Grant Number 22H00524, JP19K12112, JST SPRING, Grant Number JPMJSP2114. We also thank Takamiya Gakuen Yoyogi Seminar for providing invaluable data useful for our experiments. We would like to thank the anonymous reviewers for their insightful comments.

References

1. Aghajanyan, A., Gupta, A., Shrivastava, A., Chen, X., Zettlemoyer, L., Gupta, S.: Muppet: massive multi-task representations with pre-finetuning. In: EMNLP, pp. 5799–5811. Association for Computational Linguistics, Online and Punta Cana, Dominican Republic (2021). https://doi.org/10.18653/v1/2021.emnlp-main.468
2. Burrows, S., Gurevych, I., Stein, B.: The eras and trends of automatic short answer grading. Int. J. Artif. Intell. Educ. **25**(1), 60–117 (2015)

3. Cohen, J.: Weighted kappa: nominal scale agreement with provision for scaled disagreement or partial credit. Psychol. Bull. **70**(4), 213–220 (1968)
4. Devlin, J., Chang, M.W., Lee, K., Toutanova, K.: BERT: pre-training of deep bidirectional transformers for language understanding. In: NAACL-HLT, pp. 4171–4186 (2019). https://doi.org/10.18653/v1/N19-1423
5. Funayama, H., et al.: Balancing cost and quality: an exploration of human-in-the-loop frameworks for automated short answer scoring. In: Rodrigo, M.M., Matsuda, N., Cristea, A.I., Dimitrova, V. (eds.) AIED 2022. LNCS, vol. 13355, pp. 465–476. Springer, Cham (2022). https://doi.org/10.1007/978-3-031-11644-5_38
6. Haller, S., Aldea, A., Seifert, C., Strisciuglio, N.: Survey on automated short answer grading with deep learning: from word embeddings to transformers (2022)
7. Kumar, Y., et al.: Get it scored using autosas - an automated system for scoring short answers. In: AAAI/IAAI/EAAI. AAAI Press (2019). https://doi.org/10.1609/aaai.v33i01.33019662
8. Mizumoto, T., et al.: Analytic score prediction and justification identification in automated short answer scoring. In: BEA, pp. 316–325 (2019). https://doi.org/10.18653/v1/W19-4433
9. Mohler, M., Bunescu, R., Mihalcea, R.: Learning to grade short answer questions using semantic similarity measures and dependency graph alignments. In: ACL-HLT, pp. 752–762 (2011)
10. Oka, H., Nguyen, H.T., Nguyen, C.T., Nakagawa, M., Ishioka, T.: Fully automated short answer scoring of the trial tests for common entrance examinations for Japanese university. In: Rodrigo, M.M., Matsuda, N., Cristea, A.I., Dimitrova, V. (eds.) AIED 2022. LNCS, vol. 13355, pp. 180–192. Springer, Cham (2022). https://doi.org/10.1007/978-3-031-11644-5_15
11. Riordan, B., Horbach, A., Cahill, A., Zesch, T., Lee, C.M.: Investigating neural architectures for short answer scoring. In: BEA, pp. 159–168 (2017). https://doi.org/10.18653/v1/W17-5017
12. Saha, S., Dhamecha, T.I., Marvaniya, S., Foltz, P., Sindhgatta, R., Sengupta, B.: Joint multi-domain learning for automatic short answer grading. CoRR abs/1902.09183 (2019)
13. Sakaguchi, K., Heilman, M., Madnani, N.: Effective feature integration for automated short answer scoring. In: NAACL-HLT, Denver, Colorado, pp. 1049–1054. Association for Computational Linguistics (2015). https://doi.org/10.3115/v1/N15-1111
14. Sultan, M.A., Salazar, C., Sumner, T.: Fast and easy short answer grading with high accuracy. In: NAACL-HLT, San Diego, California, pp. 1070–1075. Association for Computational Linguistics (2016). https://doi.org/10.18653/v1/N16-1123
15. Sung, C., Dhamecha, T., Saha, S., Ma, T., Reddy, V., Arora, R.: Pre-training BERT on domain resources for short answer grading. In: EMNLP-IJCNLP, Hong Kong, China, pp. 6071–6075. Association for Computational Linguistics (2019). https://doi.org/10.18653/v1/D19-1628
16. Wang, T., Funayama, H., Ouchi, H., Inui, K.: Data augmentation by rubrics for short answer grading. J. Nat. Lang. Process. **28**(1), 183–205 (2021). https://doi.org/10.5715/jnlp.28.183

Go with the Flow: Personalized Task Sequencing Improves Online Language Learning

Nathalie Rzepka[1(✉)], Katharina Simbeck[1], Hans-Georg Müller[2], and Niels Pinkwart[3]

[1] University of Applied Sciences, Berlin, Germany
nathalie.rzepka@htw-berlin.de
[2] University of Potsdam, Potsdam, Germany
[3] Deutsches Forschungszentrum für Künstliche Intelligenz (DFKI), Berlin, Germany

Abstract. Machine learning (ML) based adaptive learning promises great improvements in personalized learning for various learning contexts. However, it is necessary to look into the effectiveness of different interventions in specific learning areas. We conducted an online-controlled experiment to compare an online learning environment for spelling to an ML based implementation of the same learning platform. The learning platform is used in schools from all types in Germany. Our study focuses on the role of different machine learning-based adaptive task sequencing interventions that are compared to the control group. We evaluated nearly 500,000 tasks using different metrics. In total almost 6,000 students from class levels 5 to 13 (ages from 11–19) participated in the experiment. Our results show that the relative number of incorrect answers significantly decreased in both intervention groups. Other factors such as dropouts or competencies reveal mixed results. Our experiment showed that personalized task sequencing can be implemented as ML based interventions and improves error rates and dropout rates in language learning for students. However, the impact depends on the specific type of task sequencing.

Keywords: Adaptive Learning · Task Sequencing · Learning Analytics

1 Introduction

In German schools, the use of online learning platforms has increased significantly, especially in the context of the COVID-19 pandemic. In particular for homework, these learning environments offer teachers the advantage that correction is automated, and students receive immediate feedback. The application of adaptive learning has already been shown to be useful in such online learning environments (Chen et al., 2005; Corbalan et al., 2008; Hooshyar et al., 2018; Mitrovic & Martin, 2004; van Oostendorp et al., 2014).

Adaptive learning is an overarching term, and by utilizing insights obtained from learning analytics and ML models, there are numerous ways to individually adjust a learning environment for a user. Which customization is most effective depends on the learning content, learning platform, and user group, among other factors. One out of a variety of adaptive learning interventions is adaptive task sequencing.

N. Wang et al. (Eds.): AIED 2023, LNAI 13916, pp. 90–101, 2023.
https://doi.org/10.1007/978-3-031-36272-9_8

In our work, we therefore investigate and compare different ML based adaptive task sequencing interventions in an online learning environment for the acquisition of German spelling. In order to do so, we transformed the existing platform into an adaptive environment based on machine learning.

Our article is structured as follows. First, we summarize related work on adaptive item sequencing. Afterwards, we briefly discuss research findings on adjusting the difficulty of tasks used in language platforms. In the following, we describe an online-controlled experiment design we conducted in a learning environment for German language learning. In addition, we explain different task sequencing processes that we compare. We then present our findings in chapter 4 and discuss them in chapter 5. In the end, we draw a conclusion and present limitations and future work.

2 Related Work

2.1 Adaptive Item Sequencing

Adaptive Item Sequencing, also known as Task Sequencing, is one method to implement curriculum learning. Curriculum learning as a form of transfer learning is based on the assumption that in order to perform a task, users have to train on source tasks and then transfer their learnings to the target task (Taylor & Stone, 2009). Task Sequencing is not solely an issue for digital learning, it also plays a role in conventional teaching and tutoring (Cock & Meier, 2012). Cock and Meier (2012) differentiate between four approaches to sequence different tasks. For all of them, the goal is to define a certain sequence of tasks (curriculum) that optimizes learning speed and performance on the target task (Narvekar et al., 2017). This follows the concept of flow theory. The flow theory assumes that a person experiences a satisfying state of focus and concentration when the challenges are neither too difficult, nor too easy (Csikszentmihalyi, 1990).

The Item Response Theory (IRT) provides explanations about the items' difficulty and learners' ability. IRT is a concept from psychometrics that is used in many contexts and includes educational measurements worldwide (Embretson & Reise, 2000, 4f). One model that is part of IRT is the Rasch Model. The Rasch measurement provides the opportunity to define item difficulty and user competency on the same scale. The Rasch Model has one ability parameter per learner and one parameter that defines the difficulty for each item (Wright, 1977). The model uses these parameters to estimate the probability of an individual being successful in answering an item. Equation 1 defines the Rasch model:

$$P_{vi} = \frac{e^{(\beta_v - \delta_i)}}{1 + e^{(\beta_v - \delta_i)}}, \tag{1}$$

where δ is the ability of a person i, β is the difficulty of an item v, and $P(vi)$ is the probability that person i answers item v correctly.

Adaptive Item Sequencing has been researched in various educational contexts in terms of learning outcomes and has yielded inconsistent results.

Hooshyar et al. compared an adaptive and a non-adaptive game for children and found that English reading skills improved more for the users of the adaptive game

(Hooshyar et al., 2018). More efficient learning through adaption is found by Corbalan et al., who conducted a 2 × 2 factorial design study (adaption or no adaption; program control or shared control) (Corbalan et al., 2008). In their study, Chen et al. proposed an e-learning system that implements individual learning paths based on the learners' ability and the item difficulty. They found that the satisfaction of learners using this system is very high. Evaluations showed that the system was able to provide learners with appropriate learning material, resulting in more efficient and effective learning (Chen et al., 2005). Increased efficiency was also found by van Oostendorp et al. when comparing an adaptive serious game to a non-adaptive serious game (van Oostendorp et al., 2014). Another positive example of adaptive task sequencing is the study of Mitrovic and Martin (2004). They compare two versions of a SQL-Tutor system. While in one system the problem complexity was defined by the teacher, in a second system it was set adaptively based on the student's knowledge. They found positive effects on the student's learning performance (Mitrovic & Martin, 2004).

In contrast, other studies did not find differences between groups in similar experiments. A learning system that assesses Algebra proficiency was evaluated with a focus on learning benefits through adaptive content (Shute et al., 2007). While the results showed enhanced student learning, it did not affect the accuracy of the assessment. Vanbecelaere et al. did not find any differences in spelling and reading improvement by comparing users from adaptive games and non-adaptive games (Vanbecelaere et al., 2020). They also found no differences in self-concept and motivation.

2.2 Individual Adjustment of Difficulty Levels in Language Learning

Several platforms that support first language learning already implement an individual adjustment of difficulty levels.

For example, Ghysels & Haelermans present an adaptive digital homework program that focuses on spelling skills (Ghysels & Haelermans, 2018). The program adapts the learning content to the user's previous performance and displays either repetitions or new modules. An experiment with Dutch 7th-grade students was conducted and found slightly better performance of the treatment group compared to the control group. The impact for low-performing students is higher than for high-performing students.

Further, Ronimus et al. evaluated an educational game designed specifically for dyslexic children and those with severe spelling and reading difficulties (Ronimus et al., 2020). They adjusted the difficulty level of the reading game according to the competency of the users. In their evaluation, they found no differences between the treatment and control groups.

Simpson and Park compared technology-assisted, individualized instruction to regular ones in school (Simpson & Park, 2013). In technology-assisted instruction, students received personalized instruction based on the Scantron achievement test scores. Results showed that the treatment group had significantly higher post-test scores compared to the control group.

As discussed, there is a lot of research already about difficulty adjustments in language learning as well as item sequencing in adaptive learning environments. However, the integration of both adaptive task sequencing and online language learning is not yet a research focus. Therefore, we formulate the following research questions:

RQ 1: What is the effect of ML based adaptive task sequencing on the number of mistakes made in an online platform for German spelling?
RQ 2: What is the effect of ML based adaptive task sequencing on the number of early dropouts on an online platform for German spelling?
RQ 3: What is the effect of ML based adaptive task sequencing on an online platform for German spelling on the users' competency in capitalization?
RQ 4: Are the effects dependent on the type of task sequencing?

3 Methodology

Orthografietrainer.net is an online learning platform to acquire German spelling skills for school children. The use is free of charge and is mainly used in blended classroom scenarios and for homework assignments. There are different areas of spelling to choose exercises from, for example capitalization, comma formation, or separated and combined spelling. The exercises are suitable for users from fifth grade upwards. While access numbers were already high in the past, the usage of the platform increased sharply since the COVID-19 pandemic. As of today, the platform has more than one million registered users and more than 11 million completed exercise sets.

The platform offers many advantages for both users and their teachers. First, all exercises are categorized by spelling area and difficulty. Each exercise set consists of ten spelling exercises the user must solve. After each exercise, the user receives immediate feedback as well as the possibility to display the appropriate spelling rule. Teachers benefit from the automated correction of the exercises and from a dashboard that summarizes the results of the class.

3.1 Online-Controlled Experiment

In our article, we present an online-controlled experiment that was carried out on the Orthografietrainer.net platform. To do so, the platform was transformed into an ML based adaptive learning platform. The experiment took place from 21.06.2022 to 31.10.2022 as was pre-registered at OSF[1] and the setup is described in detail in (Rzepka et al., 2022). Our focus was on the capitalization tasks and the user group students. During this time, each student that was starting a capitalization exercise was randomly assigned to one of three groups, one control group, and two task sequencing intervention groups. A user is always in the same group, which means that he or she experiences only one intervention or remains in the control group throughout the experiment. Users do not receive information on the group they are assigned to. Table 1 shows the number of users, sessions, and the number of sentences answered per intervention group. Users were evenly distributed between genders and were in grades 5 to 13 (ages 11–19). Most users were in grades 6–9 (ages 12–15), where a lot of spelling instruction takes place in school classrooms. Least users were in classes 10 and above.

[1] 10.17605/OSF.IO/3R5Y7.

Table 1. Experiment metrics per intervention group.

Intervention group	Number of users	Number of sessions	Number of answered sentences
Control	2,447	8,049	225,426
Intervention 1	1,818	5,733	148,565
Intervention 2	1,733	6,196	123,463

In both intervention groups, a prediction model was applied to determine whether the user is going to solve the task correctly or not and to calculate the class probabilities. For that purpose, we trained a decision tree classifier with a balanced data set of 200,000 records from the Orthografietrainer.net platform from the year 2020.

Task Sequencing

In our experiment we compared the control group with two machine learning-based task sequencing interventions. Figures 1, 2 and 3 describe the different task sequencing processes in detail.

Figure 1 shows the default task sequencing process of the Orthografietrainer.net platform that is experienced by the control group. Each session consists of 10 pre-defined sentences. If a user answers incorrectly for one of the sentences, the sentence is displayed again until the user answers it correctly. Following an incorrect answer, the user enters a loop to train the relevant orthographic rule on two additional sentences. These two sentences are pre-defined by the platform. With every mistake two sentences are added – that is also the case if the user gives an incorrect answer in the loop. After the two additional sentences are answered correctly, the previously incorrectly answered sentence is displayed again. If the user answers correctly, he or she continues with the remaining sentences from the exercise set. The session ends with a test phase in which all previously incorrectly answered sentences are repeated.

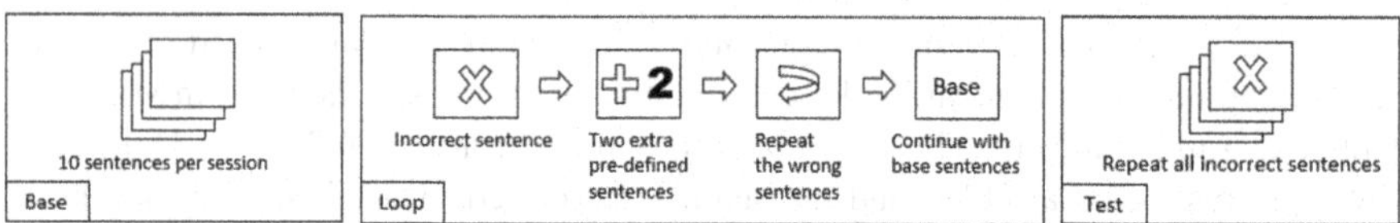

Fig. 1. Task sequence process in the control group.

Figure 2 shows the task sequencing process of the first intervention. As in the control group, the exercise set consists of 10 pre-defined sentences. If the user answers a sentence incorrectly, the loop is started. Sentences in the loop are, in contrast to the control group, not pre-defined. Instead, the machine learning-based prediction model calculates the probabilities of solving the sentences from a pool of sentences. The two sentences with the highest solving probability are displayed. After two additional sentences, the previously incorrectly answered sentence is repeated. If the user answers correctly, he or she continues with the remaining sentences from the exercise set. The session ends with a test phase in which all previously incorrectly answered sentences are repeated.

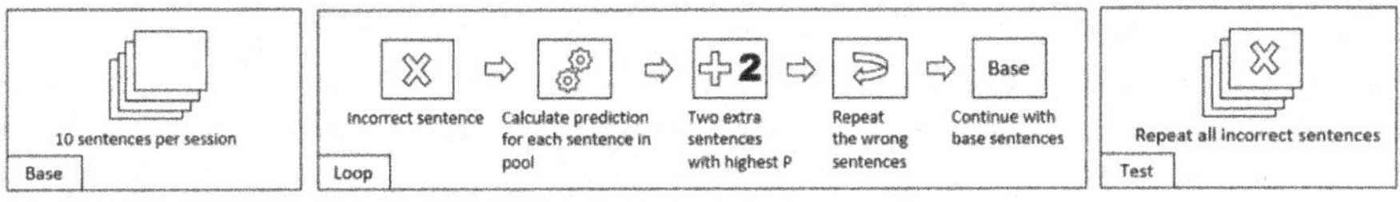

Fig. 2. Task sequence process in intervention 1.

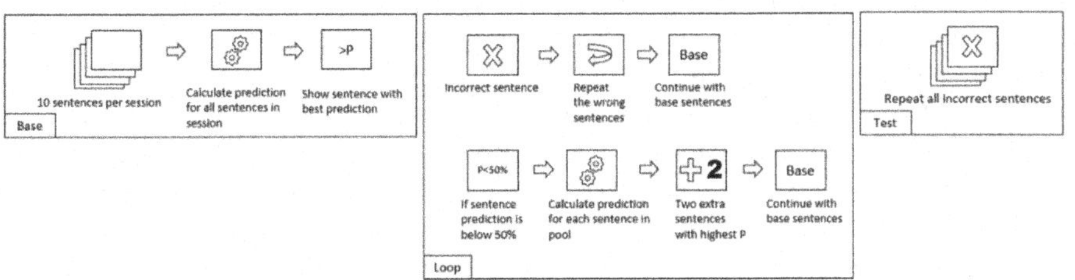

Fig. 3. Task sequence process in intervention 2.

The sentences selected from the pool to be displayed in the loop phase are deleted from it once they are shown. Thus, when the user enters the loop another time, two different sentences are added, again based on their prediction.

Figure 3 describes the second intervention's task sequencing process. The exercise set consists of 10 pre-defined sentences. The probability of the user answering correctly is calculated for all sentences in the exercise set. The sentence with the highest probability is displayed first. If the user answers the sentence correctly, the probability of solving each of the remaining sentences is calculated again and the sentence with the highest probability of solving is displayed next. If a user answers a sentence incorrectly, the user does not enter a loop phase, they only repeat the incorrect sentence and continue with a new sentence. After each submitted answer, the prediction is recalculated as it takes correct and incorrect answers into account. However, if the sentence with the highest probability of the user answering correctly has a likelihood of less than 50%, the user enters the loop phase. The loop phase is also entered if the user solved the previous sentence correctly. In the loop phase, the probability of solving all sentences from the pool is calculated by the prediction model. Then, the sentences with the two highest solving probabilities are added. After that, the user continues with the sentences from the exercise set. After each loop phase, a sentence from the exercise set is displayed, regardless of whether the probability of solving is less than 50%. That is implemented to hinder endless loop phases. The pedagogical concept of this task sequencing process is to present a preventive loop phase to avert the user from getting too difficult sentences and to allow the user to consolidate his knowledge before a more difficult sentence comes up.

Evaluation

To measure the effects and differences between the two intervention groups, we defined three overall evaluation criteria (OEC) which resulted in three hypotheses.

The first OEC is a relative number of incorrect answers. We calculate the number of incorrect answers in capitalization tasks for each user divided by the total number of capitalization tasks during the experimental phase. Here, all answers are included, also repetitions. We expect a decrease in the relative number of incorrect answers in the intervention groups compared to the control group.

The second OEC is the number of dropouts. If the set of ten sentences has not been completed in one session, it is classified as a dropout. After 45 min of inactivity, the user is automatically logged out of the platform. Therefore, we consider more than 45 min of inactivity as a dropout. The user has the possibility to continue with an exercise set at a later time, nevertheless, we label this as a dropout. We expect fewer dropouts in the intervention groups in comparison to the control group.

The third OEC is user competency. For each user, the competency in capitalization is calculated by the Rasch model by taking into account the users' correctly and incorrectly solved tasks (as described in Sect. 2.1). It calculates one user-sentence combination at a time, so if a sentence has been repeated, only the results of the last processed one is used. The Rasch model displays user competency and item difficulty on the same scale, making it easier to interpret. We expect an improved user competency in the intervention groups compared to the control group. The Rasch model is only used to determine the user competency; the adaptation is done with the prediction model described before.

The user groups are not perfectly equally distributed as they are randomly assigned to the groups. Before performing the statistical analysis, we therefore calculate the sample ration mismatch to determine if the groups are equal enough. For the statistical analysis, the data of each hypothesis is checked for homogeneity of variance and normal distribution. In all three cases, at least one assumption is not met, hence we proceed with non-parametric tests. We performed a Kruskal-Wallis-Test and, if the result is significant, we continue by performing a Wilcoxon-Mann-Whitney-Test. The level of significance is set at 0.05, but as we perform multiple tests, we carry out a Bonferroni correction. Therefore, the corrected level of significance is 0.017.

4 Results

4.1 H1 – Incorrect Answers

The first OEC describes the relative number of incorrect answers. As seen in Table 2, those numbers decreased significantly for both intervention groups in comparison to the control group. This indicates that both task sequencing interventions lead to a lower number of relative mistakes with an effect size of 3.00–4.05%. The task sequencing intervention 2 with preventive feedback instead of a traditional feedback loop improves slightly more. Both improvements are highly significant.

Furthermore, we found that effects are higher for girls than for boys in both interventions. Intervention 1 leads to a median of relative mistakes for girls of 8.65% while it is 10.00% for boys. The gap in intervention 2 is slightly smaller (7.68% for girls, 8.00% for boys). An analysis of effects concerning the grade level did not show great differences across grade levels. Additionally, we calculated the effects on low performers and high performers. We define low performers as the 10% of users with the highest number of relative mistakes. The high performers are defined as the 10% of users with the least number of relative mistakes. While there were no significant differences between

Table 2. Results of H1: Incorrect answers

Group	Median of relative mistakes	p-Value	Significance
Intervention 1	9.00%	2.8e−13	***
Intervention 2	7.95%	3.8e−42	***
Control	12.00%		

groups for high performers, we found a significant decrease in relative mistakes for the low performers for group 2 in comparison to the control group.

4.2 H2 – Dropout

The analysis of the second hypothesis led to inconsistent results (see Table 3). Intervention 1 had significantly higher dropouts than the control group with an effect size of 3.2%. On the other hand, intervention 2 led to significantly lower dropouts with an effect size of 1.5%. Concerning gender, we found a similar effect on girls and boys in intervention 1. However, intervention 2 was neither significant for boys nor girls. The analyses that compared grade levels did not reveal any differences.

Table 3. Results of H2: Dropouts

Group	Mean number of dropouts	p-Value	Significance
Intervention 1	0.1549	1.0e−07	***
Intervention 2	0.1088	0.0073	*
Control	0.1233		

4.3 H3 – User Competency

To evaluate H3 the overall competency in capitalization was calculated for each user by using the Rasch model. The last hypothesis expects a higher user competency from users in the intervention group. As seen in Table 4, intervention 1 resulted in significantly lower user competency. Intervention 2 is not significant.

Comparing the effect between gender, we found that competency decreases in the boy's cohort are larger than in the girl's one in intervention 1 (0.625 logits and 0.725 logits). Concerning grade levels, analyses showed that effects are only significant for lower grade levels up to grade level 9. No significant effects were found in grades 10 and above. We defined the top 10% and the bottom 10% of users in terms of competency, to compare high-performing and low-performing users. For high performers, differences between groups were not significant. However, intervention 1 showed a significant decrease in competency in comparison to the control group for low performers (p: 0.003, effect size: −0.305 logits).

Table 4. Results of H3: user competency

Group	Mean competency in logits	p-Value	Significance
Intervention 1	2.765	4.8e−14	***
Intervention 2	3.240	0.0443	n.s.
Control	3.535		

5 Discussion

We conducted an online-controlled experiment using an online platform for learning German spelling and grammar. We compared the traditional platform to two different machine learning based task sequencing processes. For that, we implemented a decision tree classifier to predict the users' success in processing a task. The first intervention sorts tasks according to their solving probability. Intervention 2 used the same approach as intervention 1, with the addition of preventive feedback loops. If the solving probability is less than 50%, similar tasks are added for practice before the user proceeds with more difficult tasks. In our analyses, we examined three hypotheses. First, we expect a lower relative number of mistakes. Secondly, we expect a lower dropout rate. Lastly, we expect an increase in user competency in capitalization. The user competency for the last hypothesis is calculated by the Rasch model.

Our results show that the relative number of mistakes decreases in both intervention groups, especially in intervention group 2. We interpret that task sequencing has a positive impact on users by showing the tasks in a sequence that is corresponds with the user's ability. The results were most evident for low performers. They benefit more from a personalized task sequence, while the benefit is not as substantial for high performers. Lower error rates can not only be explained by simpler sentences, as only the order of the sentences is adjusted; each user has to process the same 10 pre-defined sentences.

We found that intervention 1 led to more dropouts, while intervention 2 led to fewer dropouts. The increase in dropouts in intervention 1 could be explained by the simpler tasks in the beginning which might have bored the users. The decrease in dropouts in intervention 2 can easily be explained by the new task sequencing process outlined in chapter 3.2. Here, users only receive additional tasks if the prediction is below 50%. Therefore, users are no longer in long feedback loops and feedback loops are as needed. We interpret that a preventive feedback loop is beneficial for the users' endurance.

Interestingly, the competencies in intervention 1 decreased in comparison to the control group. While at first sight, this is contrary to the results of hypothesis 1, one must consider the relationship between the task sequencing process and the calculation of competence. Intervention 1 has a higher number of dropouts. Meaning, users from intervention 1 did not work on as many tasks as the users from the control group. That leads to fewer tasks and thus impacts the Rasch model. This suggests the relation between task sequencing processes and the Rasch model calculation must be considered in the interpretation. Competencies from intervention group 2 did not change significantly.

Concerning research question 4, one can say that the effects on the user's performance are dependent on the type of task sequencing. The decrease in errors is observed in both interventions, intervention 2 leads to fewer dropouts while the number of dropouts in intervention 1 is higher compared to the control group. In summary, intervention 2 seems to be more beneficial than intervention 1. Intervention 2 led to fewer errors, and fewer dropouts and does not impact the user's competency significantly.

Our study comes with several limitations that must be discussed thoroughly. First, it should be noted that the duration of the experiment includes summer vacations. Thus, it is conceivable that seasonal effects influence the results. Moreover, the OECs we used to assess the role of the ML based adaptive learning interventions are limited. With error rates, dropouts and competency, we only have a first insight in learning processes. More elaborate OECs to measure learning effectiveness (e.g., competency increase per time spent) should be used in further research. Further, the measurement of a dropout has its limitations. A dropout can of course happen when a user is frustrated or bored. However, there are many other reasons for a dropout over which the platform has no control. For example, the school lesson in which the platform is used may be over, or the student may end the exercise early for personal reasons. A dropout can therefore not be generalized as a negative event. Finally, the calculation and interpretation of the users' competency should be interpreted carefully. As described above, comparing different task sequencing processes means that we compare different numbers of tasks that influence the calculation. Furthermore, we calculated the competency for one point in time. However, it would be better to carry out a pre-and post-test to be able to interpret developments.

6 Conclusion

We present an online-controlled experiment that compares two adaptive machine learning based interventions in an online learning platform for spelling and grammar. Results of nearly 500.000 processed sentences were evaluated concerning the relative number of mistakes, the dropout rate, and the users' competency. In comparison to the control group, the intervention groups resulted in significantly lower error rates. Results concerning dropouts are inconsistent, which means that intervention 1 led to higher number of dropouts while intervention 2 led to fewer dropouts. We additionally found a decrease in user competencies in intervention 1 that we discussed thoroughly. In conclusion, adaptive and machine learning based task sequencing intervention can improve the student's success in online L1 learning platforms by decreasing the error rate of the users. We further conclude that interventions can also increase or decrease dropouts. The user competency we measured shows mixed results. Therefore, a trade-off must be made between competence measurement and task sequencing processes. Further research should include long running experiments, ideally a full school year. Furthermore, it might be possible to learn more about the reasons for dropouts, for example by conducting a survey when logging back in after the dropout. The calculation of the user competency might not be done within the actual exercise, but as separate pre- and posttests. Moreover, other measures could be used to evaluate the approaches, such as learning effectiveness per time spent.

References

Chen, C.-M., Lee, H.-M., Chen, Y.-H.: Personalized e-learning system using Item Response Theory. Comput. Educ. **44**(3), 237–255 (2005)

Cock, J., Meier, B.:. Task sequencing and learning. In: Seel, N.M. (ed.) Springer reference. Encyclopedia of the Sciences of Learning: With 68 Tables, pp. 3266–3269. Springer (2012). https://doi.org/10.1007/978-1-4419-1428-6_514

Corbalan, G., Kester, L., van Merriënboer, J.J.: Selecting learning tasks: effects of adaptation and shared control on learning efficiency and task involvement. Contemp. Educ. Psychol. **33**(4), 733–756 (2008). https://doi.org/10.1016/j.cedpsych.2008.02.003

Csikszentmihalyi, M.: Flow: The psychology of Optimal Experience (1990)

Embretson, S.E., Reise, S.P.: Item Response Theory for Psychologists. Multivariate Applications Book Series. Psychology Press (2000)

Ghysels, J., Haelermans, C.: New evidence on the effect of computerized individualized practice and instruction on language skills. J. Comput. Assist. Learn. **34**(4), 440–449 (2018). https://doi.org/10.1111/jcal.12248

Hooshyar, D., Yousefi, M., Lim, H.: A procedural content generation-based framework for educational games: toward a tailored data-driven game for developing early english reading skills. J. Educ. Comput. Res. **56**(2), 293–310 (2018). https://doi.org/10.1177/0735633117706909

Mitrovic, A., Martin, B.: Evaluating adaptive problem selection. In: De Bra, P.M.E., Nejdl, W. (eds.) AH 2004. LNCS, vol. 3137, pp. 185–194. Springer, Heidelberg (2004). https://doi.org/10.1007/978-3-540-27780-4_22

Narvekar, S., Sinapov, J., Stone, P.: Autonomous task sequencing for customized curriculum design in reinforcement learning. In: Proceedings of the Twenty-Sixth International Joint Conference on Artificial Intelligence. International Joint Conferences on Artificial Intelligence Organization (2017). https://doi.org/10.24963/ijcai.2017/353

Ronimus, M., Eklund, K., Westerholm, J., Ketonen, R., Lyytinen, H.: A mobile game as a support tool for children with severe difficulties in reading and spelling. J. Comput. Assist. Learn. **36**(6), 1011–1025 (2020). https://doi.org/10.1111/jcal.12456

Rzepka, N., Simbeck, K., Müller, H.-G., Pinkwart, N.: An Online controlled experiment design to support the transformation of digital learning towards adaptive learning platforms. In: Proceedings of the 14th International Conference on Computer Supported Education. SCITEPRESS – Science and Technology Publications (2022). https://doi.org/10.5220/0010984000003182

Shute, V., Hansen, E., Almond, R.: Evaluating ACED: The impact of feedback and adaptivity on learning. In: Luckin, R., Koedinger, K.R., Greer, J.E. (eds.) Frontiers in Artificial Intelligence and Applications, vol. 158. Artificial intelligence in Education: Building Technology Rich Learning Contexts that Work. IOS Press (2007)

Simpson, T., Park, S.: The effect of technology-supported, student-centered instruction on seventh grade students' learning in english/language arts. In: McBride, R., Searson, M. (eds.) SITE 2013, pp. 2431–2439. Association for the Advancement of Computing in Education (AACE) (2013). https://www.learntechlib.org/p/48467/

Taylor, M. E., Stone, P.: Transfer learning for reinforcement learning domains: a survey (2009). https://www.jmlr.org/papers/volume10/taylor09a/taylor09a.pdf?ref=https://codemonkey.link

van Oostendorp, H., van der Spek, E.D., Linssen, J.: Adapting the complexity level of a serious game to the proficiency of players. EAI Endorsed Trans. Game-Based Learn. **1**(2), e5 (2014). https://doi.org/10.4108/sg.1.2.e5

Vanbecelaere, S., van den Berghe, K., Cornillie, F., Sasanguie, D., Reynvoet, B., Depaepe, F.: The effectiveness of adaptive versus non-adaptive learning with digital educational games. J. Comput. Assist. Learn. **36**(4), 502–513 (2020)

Wright, B.D.: Solving measurement problems with the Rasch model. J. Educ. Meas. **14**(2), 97–116 (1977). http://www.jstor.org/stable/1434010

Automated Hand-Raising Detection in Classroom Videos: A View-Invariant and Occlusion-Robust Machine Learning Approach

Babette Bühler[1(✉)], Ruikun Hou[1,2(✉)], Efe Bozkir[2,4], Patricia Goldberg[1], Peter Gerjets[3], Ulrich Trautwein[1], and Enkelejda Kasneci[4]

[1] Hector Research Institute of Education Sciences and Psychology, University of Tübingen, Europastraße 6, 72072 Tübingen, Germany
{babette.buehler,ruikun.hou,patricia.goldberg, ulrich.trautwein}@uni-tuebingen.de

[2] Human-Computer Interaction, University of Tübingen, Sand 14, 72076 Tübingen, Germany
efe.bozkir@uni-tuebingen.de

[3] Leibniz-Institut für Wissensmedien, Schleichstraße 6, 72076 Tübingen, Germany
p.gerjets@iwm-tuebingen.de

[4] Human-Centered Technologies for Learning, Technical University of Munich, Arcisstraße 21, 80333 Munich, Germany
{efe.bozkir,enkelejda.kasneci}@tum.de

Abstract. Hand-raising signals students' willingness to participate actively in the classroom discourse. It has been linked to academic achievement and cognitive engagement of students and constitutes an observable indicator of behavioral engagement. However, due to the large amount of effort involved in manual hand-raising annotation by human observers, research on this phenomenon, enabling teachers to understand and foster active classroom participation, is still scarce. An automated detection approach of hand-raising events in classroom videos can offer a time- and cost-effective substitute for manual coding. From a technical perspective, the main challenges for automated detection in the classroom setting are diverse camera angles and student occlusions. In this work, we propose utilizing and further extending a novel view-invariant, occlusion-robust machine learning approach with long short-term memory networks for hand-raising detection in classroom videos based on body pose estimation. We employed a dataset stemming from 36 real-world classroom videos, capturing 127 students from grades 5 to 12 and 2442 manually annotated authentic hand-raising events. Our temporal model trained on body pose embeddings achieved an F_1 score of 0.76. When employing this approach for the automated annotation of hand-raising instances, a mean absolute error of 3.76 for the number of detected hand-raisings per student, per lesson was achieved. We demonstrate its application by investigating the relationship between hand-raising events

B. Bühler and R. Hou—Both authors contributed equally.

N. Wang et al. (Eds.): AIED 2023, LNAI 13916, pp. 102–113, 2023.
https://doi.org/10.1007/978-3-031-36272-9_9

and self-reported cognitive engagement, situational interest, and involvement using manually annotated and automatically detected hand-raising instances. Furthermore, we discuss the potential of our approach to enable future large-scale research on student participation, as well as privacy-preserving data collection in the classroom context.

Keywords: Hand-raising detection · Student Engagement · Educational Technologies · AI in Education

1 Introduction

Students' active participation in classroom discourse contributes to academic achievement in the school context [16]. To contribute verbally to the classroom discourse, students are required to raise their hands. Therefore, hand-raising in the classroom is an indicator of active participation and behavioral engagement, which is associated with achievement, cognitive engagement, perceived teacher emotional support [3], and motivation [2]. Further, a significant variation in the hand-raising frequency of eighth graders was due to situational interest in language art classes, and the self-concept in maths classes [2]. Results of such pioneering studies show the importance of hand-raising research to enable educators to understand and foster student engagement and active classroom participation, as well as the potential of employing hand-raising as an objective, low-inference behavioral engagement indicator. To study active participation, human observers often rate student behavior manually, which is time- and cost-intensive. Crowdsourcing strategies are often not applicable due to data protection regulations. This is part of the reason why studies of hand-raising have small sample sizes limited to certain grades, age groups, and school subjects [2,3], resulting in a lack of generalizability of results. Advances in computer vision and machine learning offer alternatives through automated recognition. This study aims to develop a robust approach to detect hand-raising events in classroom videos in an automated fashion and thus, develop a time- and cost-effective hand-raising assessment tool, replacing manual annotations to enable future large-scale research.

Previous research tackled automated hand-raising detection by either aiming to detect image patches of raised hands [17] or employing body pose estimation [19]. This research mostly focused on the real-time assessment of hand-raisings on the classroom level, i.e., as part of classroom monitoring systems [1]. However, research on hand-raising and its role in individual students' learning processes is still scarce. Further, previous approaches, mapping hand-raisings to individual students, are often not evaluated on real-world classroom videos containing authentic hand-raisings. Therefore, we propose a hand-raising detection approach, built and evaluated on diverse real-world classroom data, to identify individual students' hand-raisings for post hoc analysis in education research. To investigate the relation between hand-raising and learning activities, (1) we conduct a correlation analysis of manually annotated hand-raisings to cognitive engagement, involvement, and situational interest reported after each lesson. For enabling such research with automated action recognition in the classroom, one of the biggest challenges is that students are often filmed from diverse angles and might be partially occluded by classmates or learning materials. Therefore, (2) we propose a novel hand-raising gesture recognition approach based on view-invariant and occlusion-robust embeddings of body pose estimations and temporal classification. Since we are not directly working on the image stream, this approach allows student privacy to be protected by directly extracting poses in real-time and eliminating the need to store sensitive video data. Since, in addition to recognizing the gesture itself, identifying who and how often someone raised their hand is of particular interest for education research, (3) we apply and evaluate our classification approach for the automated annotation of hand-raising instances for individual students. We then conduct correlation analysis to learning-related activities comparing manually and automatically annotated hand-raisings.

2 Related Work

Initial work addressing the automatic recognition of hand-raising gestures formulated it as an object detection task, localizing raised hands and arms frame-by-frame. It investigated hand-raising in simple and staged settings with few people in the frame, focusing on techniques such as temporal and spatial segmentation, skin color identification, shape and edge feature analysis [21] or the geometric structure of edges [22]. However, such methods reach their limits when applied in a real classroom where a large number of students are visible, they occlude each other, and image resolution becomes lower.

Therefore, other works [14,17,18] aimed to overcome challenges of various gestures and low resolution, by introducing architectural adaptions of deep learning models. They achieved reliable results for detecting raised-hand image patches in realistic classroom scenes, with mean average precision (mAP) ranging between 85.2% [14] and 91.4% [18].

Object detection approaches are useful for measuring hand-raising rates at the class level, but it is challenging to analyze individual student behavior because raised hands cannot be easily attributed to specific students. To this end, a two-step approach combining object detection and pose estimation was performed by [8], heuristically matching the detected hand bounding box to a student based on body keypoints. On six real-world classroom videos, 83.1% detection accuracy was achieved. [11] chose the reverse approach, using pose estimation to obtain arm candidate areas and then classifying the corresponding image patch, achieving a F_1-score of 0.922 on a test video of college students.

Another strand of work directly employed pose estimation algorithms to detect hand-raising in the classroom. A classroom monitoring system by [1] used OpenPose [4] to extract eight upper body keypoints, for which direction unit vectors and distance between points were computed as geometric features. On scripted classroom scenes, a hand-raising prediction accuracy of 94.6% with a multi-layer perceptron was reported. The evaluation on real-world videos, only containing six hand-raising instances, yielded a recall of 50%. Furthermore, a classroom atmosphere management system for physical and online settings implemented rule-based hand-raising detection employing Kinect pose estimations [19]. The detection accuracy was however not evaluated on real-world classroom videos. Likewise, geometric features (i.e., normalized coordinates, joint distances, and bone angles) extracted from pose estimations were used by [12] for student behavior classification. Their approach to detecting four behaviors was evaluated on six staged classroom videos with a precision of 77.9% in crowded scenes.

As described above, utilizing pose estimation offers the advantage that hand-raisings can be attributed to a specific student, which is important when studying participation on the student level. One common limitation of previous research relying on pose estimation is that either the performance of models has only been extensively evaluated on scripted videos and small real-world datasets [1,12], or it implemented rule-based detection that was not evaluated on classroom videos at all [19]. How these approaches perform in a real-world classroom scenario, where hand-raisings are likely to be expressed with various, sometimes subtle gestures, is unknown. In conclusion, automated annotation of hand-raisings needs to be student specific and methodologically robust for real-world classroom scenarios. Therefore, we propose a recognition approach based on pose estimation trained and tested on a challenging dataset containing real-world school lessons with realistic hand-raising behaviors, representing a wide range of age groups (grades 5 to 12) and school subjects. Furthermore, our approach copes with the main technical challenges of such videos, such as camera angles and occlusions of students. To tackle those, we adapted and extended state-of-the-art view-invariant, occlusion-robust pose embeddings. Going beyond related works which detected hand-raising in a frame-by-frame fashion, we developed a classification model which integrates temporal information and is thus able to capture the large variety of hand-raisings expressed in their dynamic process.

3 Methodology

3.1 Data

Data Collection. The classroom videos utilized in the study were recorded in real-world lessons at a German secondary school, approved by the ethics committee of the University of Tübingen (Approval SPSVERBc1A2.5.4-097_aa). A total of 127 students, 56.3% male, from grades 5 to 12 were videotaped during 36 lessons across a wide variety of subjects (see Table 1). All recordings were captured by cameras (24 frames per second) mounted in the front left or right corner of the classroom. After each lesson, students completed a questionnaire on learning activities in the lesson, including self-reported involvement [6], cognitive engagement [15], and situational interest [10]. The employed scales have been shown to be related to engagement observer ratings and learning outcomes [7]. This resulted in video and questionnaire data for 323 student-lesson instances. Due to missing questionnaire information (i.e., item non-response), 18 instances had to be excluded for correlation analysis with learning-related activities.

Manual Annotation. Two human raters manually annotated hand-raisings in all 36 videos. The intra-class correlation coefficient of the two raters for the number of hand-raisings per student and lesson was 0.96, indicating very high inter-rater reliability. The number of hand-raisings averaged across the two observers was employed to analyze associations between student hand-raising and self-reported learning activities. A total of 2442 hand-raising events were annotated. Summary statistics on hand-raisings by students and lessons across grades are shown in Table 1. On average, one student raised their hand 7.4 times per lesson, while an average total amount of 64.6 hand-raisings occurred per lesson.

Half of the data was used to build the automated hand-raising detection model, requiring more fine-granular annotations. To this end, for 18 videos, hand-raising was annotated in a spatio-temporal manner using the VIA software [5], including the start and end time of hand-raisings as well as bounding boxes, with a joint agreement probability of 83.36%. To increase reliability, we combined the two observers' annotations by temporal intersection, resulting in 1584 hand-raising instances with an average duration of 15.6 s. The remaining 18 videos, only coded with regard to the number of hand-raisings of each student, in addition to four videos employed as a test set for the hand-raising classifier, then served to validate our developed model for automated annotation.

Table 1. Summary statistics of hand-raisings (N = 305).

Grade	Subjects	Lessons	Hand-raisings per student					Hand-raisings per lessons				
			M	SD	Md	Min	Max	M	SD	Md	Min	Max
5	B	2	15.093	6.091	15	5	26	203.750	6.010	204	199.5	208
6	B, L	3	8.625	7.934	6	0	33	126.500	89.705	85	66	230
7	E, I	2	9.420	12.991	5	0	60	117.750	64.700	118	72	164
8	E, F, G, H, IMP, L	8	5.079	4.860	4	0	23	44.438	20.711	40	8	74
9	E, G	2	6.053	5.550	5	0	24	57.500	23.335	58	41	74
10	G, P	3	8.912	7.997	5	0	26	50.500	14.309	58	34	60
11	A, ET, G, P, PS	6	7.813	7.203	6	0	24	52.083	17.019	53	33	83
12	C, H, M, P	10	5.169	5.061	4	0	31	36.700	22.671	40	7	89
Total		36	7.425	7.451	5	0	60	64.556	53.155	48	7	230

Subject abbreviations: B Biology, L Latin, E English, I Informatics, F French, G German, H History, IMP Informatics Math & Physics, P Physics, A Arts, ET Ethics, PS Psychology, C Chemistry, M Math

3.2 Skeleton-Based Hand-Raising Detection

This section presents our machine learning-based approach to automated hand-raising gesture detection. Figure 1 depicts the processing pipeline: generating sequential poses, extracting embeddings, and performing binary classification.

Preprocessing. To extract 2D human poses, we used the OpenPose library [4], estimating 25 body keypoints per student in each frame. Since students' lower body parts are often invisible due to occlusion, we focused on the 13 upper body keypoints representing the head, torso, and arms. To generate one skeleton

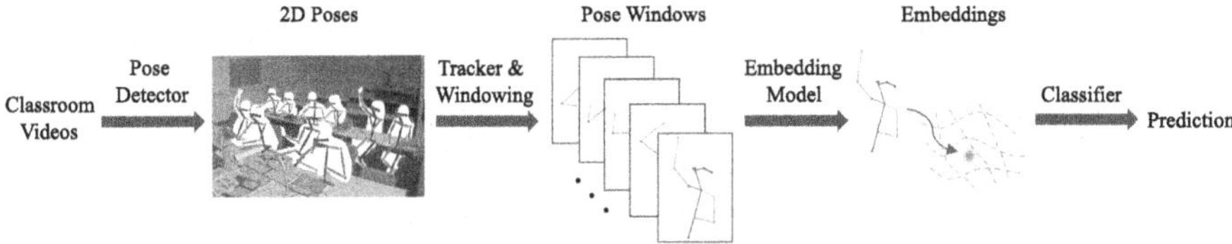

Fig. 1. Hand-raising detection pipeline.

tracklet for each student over time, we implemented an intersection-over-union tracker. Then, we used the 18 videos with spatio-temporal annotations (Sect. 3.1) to create a dataset for classifier development. Based on the annotations, tracklets were divided into subsequences labeled as "hand-raising" and "non-hand-raising" and then split into 2-second windows (48 frames) without overlap. This resulted in a highly imbalanced large-scale dataset of 243,069 instances, of which 12,839 (ca. 5%) represented hand-raisings.

Pose Embeddings. To handle issues of viewpoint change and partial visibility, we adopted a recent approach by Liu et al. [13], attempting to extract pose embeddings. First, the embedding space is view-invariant, where 2D poses projected from different views of the same 3D poses are embedded close and those from different 3D poses are pushed apart. Second, occlusion robustness is achieved by using synthetic keypoint occlusion patterns to generate partially visible poses for training. Third, the pose embeddings are probabilistic, composed of a mean and a variance vector, defining thus a Gaussian distribution that takes ambiguities of 2D poses during projection from 3D space into consideration. We tailored the training procedure to our classroom setup, training the embedding model from scratch on Human3.6M [9], a human pose dataset containing a large number of recordings from four camera views. We executed OpenPose on all images and neglected lower body keypoints to focus on learning similarity in the upper body poses. We employed the neck keypoint as the origin coordinate and normalized the skeletons by the neck-hip distance. To find a trade-off between performance and computational cost, the embedding dimension was selected as 32. This training process was independent of the classroom data, ensuring the generalization capability of our approach.

To leverage the probabilistic embeddings as input to a downstream classifier, we concatenated the two embedding outputs (mean and variance) into a 64D feature vector. Moreover, we extracted the following two types of geometric features as baselines to compare their recognition performance against pose embeddings: First, Lin et al. [12] distinguished between four student behaviors, by using eight keypoints and designing a 26D feature vector that concatenates normalized keypoint coordinates, keypoint distances, and bone angles. Second, Zhang et al. [20] proposed using joint-line distances for skeleton-based action recognition. We computed the corresponding distances on the basis of the 13 upper body keypoints, resulting in a 297D feature vector.

Classification. To achieve binary classification based on sequential inputs, we employed long short-term memory (LSTM) models. Similar to [20], we constructed three LSTM layers with 128 hidden units, followed by a fully connected layer applying a sigmoid activation function. Thus, the multi-layer model directly takes a 2-second sequence of frame-wise feature vectors as input, encodes temporal information, and estimates hand-raising probability. We trained the model using an empirically derived batch size of 512, binary cross-entropy loss, and Adam optimizer with a fixed learning rate of 0.001. To avoid overfitting, we held out 10% of training examples as validation data and set up an early stopping callback with the patience of 10 epochs according to validation loss. Additionally, we trained non-temporal random forest (RF) models for baseline comparisons. For model input, we generated an aggregated feature vector for each time window by calculating and concatenating the mean and standard deviation of all features. To handle imbalanced data, we compared the performance with and without class weighting while training both models. Class weights were set inversely proportional to class frequencies in the training set. We evaluated model performance in a video-independent manner, using 14 videos for training and 4 videos for testing. Our videos, stemming from un-staged classroom lessons, contain highly varying numbers of hand-raisings, making video-independent cross-validation infeasible due to unequal class distributions in each fold.

3.3 Automated Hand-Raising Annotation

Afterwards, we applied this hand-raising detection technique to estimate the number of hand-raisings of a student in class. To take full advantage of our data, we utilized 22 videos to evaluate the automated annotation performance, consisting of the second half of our videos and the four videos used to test classifiers. Following the pipeline in Fig. 1, we generated one tracklet for each student in each video. To achieve a more robust temporal detection, a tracklet was sliced into 48-frame sliding windows with a stride of 8 frames. Then, we extracted pose embeddings and applied the trained classifier to estimate the hand-raising probability for each window. When the average probability exceeded 0.5, a frame was assigned to hand-raising, and consecutive frames were combined into one hand-raising instance. We merged any two adjacent hand-raisings with an interval of fewer than 4 s and discarded those with a duration of less than 1 s, which only occur in less than 5% of the cases respectively according to the annotations of the training videos. This resulted in the number of estimated hand-raisings for each student in each video.

4 Results

4.1 Relation Between Hand-Raising and Self-reported Learning Activities

To demonstrate the importance of hand-raising analysis in classroom research, we examined the association between manually annotated hand-raisings and self-reported learning activities. The number of hand-raising instances of a student per lesson, annotated by human raters, is significantly positively correlated to self-reports of cognitive engagement ($r = 0.288$, $p < 0.001$), situational interest ($r = 0.379$, $p < 0.001$), and involvement ($r = 0.295$, $p < 0.001$) of students.

4.2 Hand-Raising Classification

To assess the performance of hand-raising classifiers, we utilized the F_1-score which is the harmonic mean of precision (i.e., the proportion of correct hand-raising predictions of all hand-raising predictions) and recall (i.e., the share of correctly predicted hand-raising instances of all hand-raising instances). We first tested different classification models with and without class weighting, using pose embeddings as input features. The results are shown in the upper part of Table 2. The greater penalty for misclassifying any hand-raising example when using weighting results in fewer false negatives and more false positives, i.e., a higher recall but lower precision. The best performance was obtained by the temporal LSTM model without class weighting, achieving a F_1-score of 0.76.

In the second step, we compared three pose representations, namely pose embeddings and two types of geometric features, employing LSTM classifiers. As shown in the lower part of Table 2, the pose embeddings generally yield better performance than the geometric features with respect to all three metrics, revealing the effectiveness of the data-driven approach. Notably, the pose embeddings achieve a noticeable increase in F_1-score over the features used in [12]. Moreover, despite the trivial improvement over the features used in [20], the pose embeddings benefit from less computational effort for inference, in comparison to the distance calculation over 297 joint-line combinations.

In order to gain a deeper understanding of our model, we investigated misclassified instances. Figure 2 depicts four misclassified examples, using the LSTM model on pose embeddings. They reveal poses are misclassified as hand-raising when they have a similar skeleton representation, e.g., when students scratch their heads (Fig. 2a) or rest their head on their hand (Fig. 2b). In turn, subtle hand-raisings that do not involve raising the hand or elbow above the shoulder are more difficult to recognize (Fig. 2c). Occasionally, students simply indicate hand-raisings by extending their index finger, which can not be represented in the skeleton we used. Furthermore, the classifier is prone to false predictions if keypoints of the hand-raising arm are not detected throughout the clip (Fig. 2d).

Fig. 2. Skeleton samples of misclassified windows using LSTM with pose embeddings. True label: (a, b) non-hand-raising; (c, d) hand-raising.

Table 2. Comparison of different classifiers and pose representations.

Classifier	Pose Representation	Precision	Recall	F_1-Score
RF	Pose embeddings	**0.950**	0.540	0.688
RF (w/ weighting)		0.763	0.695	0.727
LSTM		0.818	0.709	**0.760**
LSTM (w/ weighting)		0.570	**0.783**	0.660
LSTM	Geometric features [12]	0.816	0.576	0.676
	Geometric features [20]	0.812	0.694	0.748
	Pose embeddings	**0.818**	**0.709**	**0.760**

4.3 Automated Hand-Raising Annotation

After automatically annotating hand-raisings on a student level for the 22 validation videos (see Sect. 3.3), we compared the estimated hand-raising counts with the ground truth by calculating the mean absolute error (MAE). The automated annotation, employing the best-performing LSTM model achieved a MAE of 3.76. According to the ground truth, a student raised their hand 6.10 times in a lesson on average. Besides the moderate classification performance, this result can be attributed to the tracking capability. Some targets cannot be tracked continuously due to occlusion or failing pose estimation. Using a fragmentary tracklet can generate multiple sub-slots for one actual hand-raising instance, resulting in overestimation. By using the subset of tracklets that span at least 90% of the videos, the MAE decreased to 3.34. To demonstrate the application of such automated annotations, we did a re-analysis of the relation of manually and automatically annotated hand-raisings to self-reported learning activities on the validation videos. Table 3 shows that both annotations are significantly related to the three learning activities, showing comparable r values.

Table 3. Pearson correlations of different hand-raising annotations with self-reported learning activities in validation videos (N = 173).

Hand-raising annotation	Cognitive engagement		Situational interest		Involvement	
	r	p	r	p	r	p
Manually	0.222	0.003	0.326	0.000	0.213	0.005
Automated	0.222	0.003	0.288	0.000	0.201	0.008

5 Discussion

We found that the frequency of student hand-raisings is significantly related to self-reported learning activities in a diverse real-world classroom dataset covering grades five to twelve and a variety of subjects. These results are in line with previous research [2,3] and emphasize the important role of hand-raising as an observable cue for students' engagement in classroom discourse. To enable such analyses, we proposed a novel approach for the automated detection of hand-raising events in classroom videos. The employed view-invariant and occlusion-robust pose embeddings outperformed more simplistic geometric features used in previous research [12]. When applying the developed classification model for person-specific hand-raising instance annotation, the total amount of hand-raising instances per person was slightly overestimated, mainly stemming from discontinuous pose tracks. The comparison of correlation analysis between manually and automatically annotated hand-raising and learning-related activities showed comparable results for the two methods, despite the imperfect prediction accuracy, suggesting that automated annotation can be a useful proxy for research. It is important to stress that we employed a dataset containing un-staged real-world school lessons and authentic hand-raisings in naturalistic settings. This is why we employ and compare feature representations from previous research in our setting rather than directly comparing the performance results which were based on either scripted videos or extremely small sample sizes [1,12]. Furthermore, the data includes a wide span of age groups, ranging from 11- to 18-year-olds. As shown in Table 1, the frequency, as well as the manner of raising the hand and indicating the wish to speak in the classroom, differs substantially between those age groups.

These results strengthen the potential of automated hand-raising annotation for research investigating hand-raising as a form of participation in classroom discourse and an indicator of behavioral engagement. Replacing time-consuming manual coding with an automated process will allow cost-effective large-scale research on classroom interaction in the future. The approach is highly robust to camera viewpoints and therefore presumably generalizes well to new classroom setups. Further, the employment of pose estimation allows for privacy-preserving data collection, as poses can be extracted in real-time, thus eliminating the need to store video recordings. However, as students' body-pose information might be regarded as sensitive data, it is important to note that automated behavior detection should solely be used as a post hoc analytical tool for research purposes instead of real-time classroom monitoring.

One limitation of this work is that the absolute hand-raising frequency is still over- or underestimated by an average of almost four instances per person. Future research should continue to enhance the precision, by further improving the underlying pose estimation and tracking. This can either be achieved by optimizing camera angles to be most suitable for pose estimation as well as applying more advanced pose estimation techniques. Further room for improvement lies in the differentiation of difficult cases as shown in Fig. 2. Here hand-raisings that are more restrained, i.e., if students do not raise the hand above the head, are not recognized as such. A solution for this could be to include hand keypoint estimation, as the hand posture can possibly provide further information.

6 Conclusion

Our results indicate that hand-raising correlates to cognitive engagement, situational interest, and involvement and presents hence an important measure of active student participation in the classroom. We further show that the automated annotation is a viable substitute for manual coding of hand-raisings, opening up new possibilities for researching hand-raising in classroom discourse and student engagement on a large scale.

Acknowledgements. Babette Bühler is a doctoral candidate and supported by the LEAD Graduate School and Research Network, which is funded by the Ministry of Science, Research and the Arts of the state of Baden-Württemberg within the framework of the sustainability funding for the projects of the Excellence Initiative II. Efe Bozkir and Enkelejda Kasneci acknowledge the funding by the DFG with EXC number 2064/1 and project number 390727645. This work is also supported by Leibniz-WissenschaftsCampus Tübingen "Cognitive Interfaces" by a grant to Ulrich Trautwein, Peter Gerjets, and Enkelejda Kasneci. We thank Katrin Kunz and Jan Thiele for their excellent assistance.

References

1. Ahuja, K., et al.: Edusense: practical classroom sensing at scale. Proc. ACM Interact. Mob. Wearable Ubiquit. Technol. **3**(3), 1–26 (2019)
2. Böheim, R., Knogler, M., Kosel, C., Seidel, T.: Exploring student hand-raising across two school subjects using mixed methods: an investigation of an everyday classroom behavior from a motivational perspective. Learn. Instr. **65**, 101250 (2020)
3. Böheim, R., Urdan, T., Knogler, M., Seidel, T.: Student hand-raising as an indicator of behavioral engagement and its role in classroom learning. Contemp. Educ. Psychol. **62**, 101894 (2020)
4. Cao, Z., Hidalgo Martinez, G., Simon, T., Wei, S., Sheikh, Y.A.: OpenPose: realtime multi-person 2D pose estimation using part affinity fields. IEEE Trans. Pattern Anal. Mach. Intell. **43**, 172–186 (2019)
5. Dutta, A., Zisserman, A.: The via annotation software for images, audio and video. In: ACM International Conference on Multimedia, pp. 2276–2279 (2019)
6. Frank, B.: Presence messen in laborbasierter Forschung mit Mikrowelten: Entwicklung und erste Validierung eines Fragebogens zur Messung von Presence. Springer, Heidelberg (2014)
7. Goldberg, P., et al.: Attentive or not? Toward a machine learning approach to assessing students' visible engagement in classroom instruction. Educ. Psychol. Rev. **33**, 27–49 (2021)
8. Zhou, H., Jiang, F., Shen, R.: Who are raising their hands? Hand-raiser seeking based on object detection and pose estimation. In: Asian Conference on Machine Learning, pp. 470–485 (2018)
9. Ionescu, C., Papava, D., Olaru, V., Sminchisescu, C.: Human3.6M: large scale datasets and predictive methods for 3D human sensing in natural environments. IEEE Trans. Pattern Anal. Mach. Intell. **36**(7), 1325–1339 (2014)

10. Knogler, M., Harackiewicz, J.M., Gegenfurtner, A., Lewalter, D.: How situational is situational interest? Investigating the longitudinal structure of situational interest. Contemp. Educ. Psychol. **43**, 39–50 (2015)
11. Liao, W., Xu, W., Kong, S., Ahmad, F., Liu, W.: A two-stage method for hand-raising gesture recognition in classroom. In: International Conference on Educational and Information Technology. ACM (2019)
12. Lin, F.C., Ngo, H.H., Dow, C.R., Lam, K.H., Le, H.L.: Student behavior recognition system for the classroom environment based on skeleton pose estimation and person detection. Sensors **21**(16), 5314 (2021)
13. Liu, T., et al.: View-invariant, occlusion-robust probabilistic embedding for human pose. Int. J. Comput. Vis. **130**(1), 111–135 (2022)
14. Nguyen, P.D., et al.: A new dataset and systematic evaluation of deep learning models for student activity recognition from classroom videos. In: International Conference on Multimedia Analysis and Pattern Recognition. IEEE (2022)
15. Rimm-Kaufman, S.E., Baroody, A.E., Larsen, R.A., Curby, T.W., Abry, T.: To what extent do teacher-student interaction quality and student gender contribute to fifth graders' engagement in mathematics learning? J. Educ. Psychol. **107**(1), 170 (2015)
16. Sedova, K., et al.: Do those who talk more learn more? The relationship between student classroom talk and student achievement. Learn. Instr. **63**, 101217 (2019)
17. Si, J., Lin, J., Jiang, F., Shen, R.: Hand-raising gesture detection in real classrooms using improved R-FCN. Neurocomputing **359**, 69–76 (2019)
18. Liu, T., Jiang, F., Shen, R.: Fast and accurate hand-raising gesture detection in classroom. In: Yang, H., Pasupa, K., Leung, A.C.-S., Kwok, J.T., Chan, J.H., King, I. (eds.) ICONIP 2020. CCIS, vol. 1332, pp. 232–239. Springer, Cham (2020). https://doi.org/10.1007/978-3-030-63820-7_26
19. Yu-Te, K., Han-Yen, Y., Yi-Chi, C.: A classroom atmosphere management system for analyzing human behaviors in class activities. In: International Conference on Artificial Intelligence in Information and Communication. IEEE (2019)
20. Zhang, S., Liu, X., Xiao, J.: On geometric features for skeleton-based action recognition using multilayer LSTM networks. In: IEEE Winter Conference on Applications of Computer Vision, pp. 148–157 (2017)
21. Jie, Y., Cooperstock, J.R.: Arm gesture detection in a classroom environment. In: Sixth IEEE Workshop on Applications of Computer Vision (2002). ISBN 0769518583
22. Bo, N.B., van Hese, P., van Cauwelaert, D., Veelaert, P., Philips, W.: Detection of a hand-raising gesture by locating the arm. In: IEEE International Conference on Robotics and Biomimetics (2011). ISBN 9781457721380

Robust Educational Dialogue Act Classifiers with Low-Resource and Imbalanced Datasets

Jionghao Lin[1,2], Wei Tan[1(✉)], Ngoc Dang Nguyen[1(✉)], David Lang[3], Lan Du[1], Wray Buntine[1,4], Richard Beare[1,5], Guanliang Chen[1], and Dragan Gašević[1]

[1] Monash University, Clayton, Australia
{jionghao.lin1,wei.tan2,dan.nguyen2,lan.du,richard.beare, guanliang.chen,dragan.gasevic}@monash.edu
[2] Carnegie Mellon University, Pittsburgh, USA
[3] Stanford University, Stanford, USA
dnlang86@stanford.edu
[4] VinUniversity, Hanoi, Vietnam
wray.b@vinuni.edu.vn
[5] Murdoch Children's Research Institute, Melbourne, Australia
richard.beare@mcri.edu.au

Abstract. Dialogue acts (DAs) can represent conversational actions of tutors or students that take place during tutoring dialogues. Automating the identification of DAs in tutoring dialogues is significant to the design of dialogue-based intelligent tutoring systems. Many prior studies employ machine learning models to classify DAs in tutoring dialogues and invest much effort to optimize the classification accuracy by using limited amounts of training data (*i.e.,* low-resource data scenario). However, beyond the classification accuracy, the robustness of the classifier is also important, which can reflect the capability of the classifier on learning the patterns from different class distributions. We note that many prior studies on classifying educational DAs employ cross entropy (CE) loss to optimize DA classifiers on low-resource data with imbalanced DA distribution. The DA classifiers in these studies tend to prioritize accuracy on the majority class at the expense of the minority class which might not be robust to the data with imbalanced ratios of different DA classes. To optimize the robustness of classifiers on imbalanced class distributions, we propose to optimize the performance of the DA classifier by maximizing the area under the ROC curve (AUC) score (*i.e.,* AUC maximization). Through extensive experiments, our study provides evidence that (i) by maximizing AUC in the training process, the DA classifier achieves significant performance improvement compared to the CE approach under low-resource data, and (ii) AUC maximization approaches can improve the robustness of the DA classifier under different class imbalance ratios.

Keywords: Educational Dialogue Act Classification · Model Robustness · Low-Resource Data · Imbalanced Data · Large Language Models

N. Wang et al. (Eds.): AIED 2023, LNAI 13916, pp. 114–125, 2023.
https://doi.org/10.1007/978-3-031-36272-9_10

1 Introduction

One-on-one human tutoring has been widely acknowledged as an effective way to support student learning [16,25,26]. Discovering effective tutoring strategies in tutoring dialogues has been considered a significant research task in the design of dialogue-based intelligent tutoring systems [6] (*e.g.,* AutoTutor [13]) and the practice of effective tutoring [9,25]. A common approach of understanding the tutoring dialogue is to use dialogue acts (DAs), which can represent the intent behind utterances [9,17,25]. For example, a tutor's utterance (*e.g.,* "*Well done!*") can be characterized as the `Positive Feedback (FP)` DA [25]. To automate the identification of DAs in tutoring dialogues, prior research has often labeled a limited number of utterances and used these labeled utterances to train machine learning models to classify the DAs [9,14,16,17].

Prior research on educational DA classification has demonstrated promising classification accuracy in reliably identifying DAs [5,7,9,11,19]. However, the existing studies might overlook the impact of imbalanced DA class distribution in the classifier training process. Many of them (*e.g.,* [5,7,9,11]) have limited labeled DA datasets (*i.e.,* low-resource scenario [12]) and certain types of DA may be the minority DA class in the dataset (*i.e.,* imbalanced scenario [12]). For example, the DA about feedback is rarely seen in the labeled dataset in the work by Min *et al.* [11] but the provision of feedback is an important instructional strategy to support students. We argue that the issue of the imbalanced DA classes on the low-resource dataset might negatively impact the DA classification, especially for identifying crucial but underrepresented classes. To obtain more reliable classified DAs, it is necessary to investigate the approach to enhance the classifier robustness, which involves the capability of the classifier on learning the patterns from different class distributions under the low resources [12,23].

A robust classifier is able to maintain the performance when the distribution of classes in input data is varied [23]. The prior works in educational DA classification have optimized the DA classifiers by the *Cross Entropy* (CE) loss function which often tends to prioritize accuracy on the majority class at the expense of the minority class(es) and might not be robust enough to highly imbalanced class distribution [12]. Inspired by the recent advances in the model robustness literature, we propose to optimize the *Area Under ROC Curve* (AUC) of the DA classifier. AUC score is a metric that can measure the capability of the classifier in distinguishing different classes [12,23,27,30]. Maximizing the AUC score in the model training process has been shown to benefit the performance of classifiers with low-resource and highly-imbalanced scenarios in many domains (*e.g.,* medical and biological domains) [12,23]. However, the use of AUC maximization is still under-explored in the educational domain. With the intent to enhance the robustness of the DA classifier, we conducted a study to explore the potential value of the AUC maximization approaches on *(i) the classification performance of a DA classifier under the low-resource scenario* and *(ii) the robustness of the DA classifier on the highly-imbalanced data distribution.*

To demonstrate the effectiveness of AUC maximization, we adapted approaches of AUC maximization to replace the CE loss function for the DA classifier. We compared the classification performance of the DA classifier between the classifier optimized by CE and by AUC maximization on the low-resource and imbalanced DA dataset. Through extensive experiments, our studies demonstrated that (i) the adoption of the AUC Maximization approaches outperformed the CE on the low-resource educational DA classification, and (ii) the AUC maximization approaches were less sensitive to the impacts of imbalanced data distribution than the CE.

2 Background

2.1 Educational Dialogue Act Classification

Existing research in the educational domain has typically trained machine learning models on the labeled sentences from tutoring dialogues to automate the DA classification [2,5,7,9,11,18,19]. Boyer *et al.* [2] trained a Logistic Regression model on 4,806 labeled sentences from 48 tutoring sessions. Their work [2] achieved the accuracy of 63% in classifying 13 DAs. Samei *et al.* [18] trained Decision Trees and Naive Bayes models on 210 sentences randomly selected from their full tutoring dialogue corpus and they achieved the accuracy of 56% in classifying 7 DAs. In the later study, Samei *et al.* [19] trained a Logistic Regression model on labeled sentences from 1,438 tutorial dialogues and achieved an average accuracy of 65% on classifying 15 DAs. Though achieving satisfied performance on DA classification, we argue that these studies overlooked the imbalanced DA class distribution, which could negatively impact the DA classifier performance on minority but crucial classes. For example, in the dataset in [19], 23.9% sentences were labeled as `Expressive` (*e.g.*, "*Got it!*") while 0.3% labeled as `Hint` (*e.g.*, "*Use triangle rules*"). They obtained Cohen's κ score of 0.74 on identifying `Expressive` but 0.34 on `Hint`. It should be noted that the provision of a hint is an important instructional strategy in the tutoring process, but Cohen's κ score on identifying the hint DA was not sufficient. Therefore, it is necessary to enhance the robustness of the DA classifier on imbalanced data.

2.2 AUC Maximization on Imbalanced Data Distribution

The Area Under ROC Curve (AUC) is a widely used metric to evaluate the classification performance of machine learning classifiers on imbalanced datasets [12,23,27,30]. As discussed, many prior studies in educational DA classification have encountered the challenge of imbalanced DA class distribution [2,11,19]. To enhance the capability of the classifier to the imbalanced data distribution, machine learning researchers have begun employing the AUC maximization approaches to optimize classifiers towards the AUC score [12,23,27,30]. Specifically, the process of AUC maximization aims to optimize the AUC score via an AUC surrogate loss, instead of using the standard objective/loss function (*e.g.*, CE) which aims to optimize the classification accuracy of the classifier. The CE function can support the classifier in achieving sufficient classification performance when labeled training instances are sufficient. However, CE often tends to prioritize accuracy on the majority class at the expense of the minority class [12]. Thus, the CE loss function is vulnerable to highly imbalanced data distribution in the classifier training process [12]. It should be noted that most prior research in educational DA classification (*e.g.,* [2,9,19]) has used CE to optimize their classifiers on the imbalanced data distribution. As the classifier optimized by AUC maximization approaches might be less sensitive to imbalanced distribution than the CE, we propose that it is significant to explore the potential values of AUC maximization for educational DA classification.

3 Methods

3.1 Dataset

In this study, we obtained ethical approval from the Monash Human Research Ethics Committee under ethics application number 26156. The tutorial dialogue dataset used in the current study was provided by an educational technology company that offers online tutoring services. The dataset was collected with the consent of tutors and students for use in research. The dataset included records of tutoring sessions where tutors and students worked together to solve problems in various subjects (*e.g.,* math, physics, and chemistry) through chat-based communication. Our study adopted 50 tutorial dialogue sessions, which contained 3,626 utterances (2,156 tutor utterances and 1,470 student utterances. The average number of utterances per tutorial session was 72.52 (min = 11, max = 325) where tutors made an average of 43.12 utterances (min = 5, max = 183) per session and students made an average of 29.40 utterances (min = 4, max = 142) per session. We provided a sample dialogue in the digital appendix via https://github.com/jionghaolin/Robust.

3.2 Scheme for Educational Dialogue Act

To identify the dialogue acts (DAs), our study employed a pre-defined educational DA coding scheme introduced in [25]. This scheme has been shown to be effective for analyzing online one-on-one tutoring dialogues in many studies [8,9,24]. The DA scheme [25] was originally designed in a two-level structure. The second-level DA scheme, which included 31 DAs, could present more detailed information from the tutor-student dialogue. Thus, our study decided to use the second-level DA scheme to label the tutor-student utterances. Due to space reasons, we displayed the details of the full DA scheme in an electronic appendix at https://github.com/jionghaolin/Robust, and the scheme could also be found in [25]. Before the labeling, we first divided each utterance into multiple sentences as suggested by Vail and Boyer [25] and then we removed the sentences which only contained symbols or emojis. Two human coders were recruited to label the DA for each sentence, and a third educational expert was involved in resolving any disagreements. Two coders achieved Cohen's κ score of 0.77, which indicated a substantial level of agreement between the coders.

3.3 Approaches for Model Optimization

We aimed to examine the potential values of AUC maximization approaches to enhance the performance of educational DA classifiers. Inspired by the recent works in AUC Maximization [12,28–30], our study selected the approaches for training the DA classifier as follows:

- **Cross Entropy (CE):** CE is a commonly used loss function to optimize the classifier. The loss values of CE were calculated by comparing the predicted probability distribution to the true probability distribution. Our study used the DA classifier optimized by the CE approach as the baseline.
- **Deep AUC maximization (DAM):** Yuan *et al.* [29] proposed a robust AUC approach (*i.e.*, DAM) to optimize the performance of the DA classifier by optimizing it for the AUC surrogate loss. Optimizing the AUC score can significantly improve the performance of classifiers on imbalanced data.
- **Compositional AUC (COMAUC):** COMAUC proposed by Yuan *et al.* [30] involves minimizing a compositional objective function by alternating between maximizing the AUC score and minimizing the cross entropy loss. Yuan *et al.* [30] theoretically proved that COMAUC could substantially improve the performance of the classifier. Consequently, this is one of the methods we sought to implement to improve the performance of the DA classifier.

3.4 Model Architecture by AUC Maximization

We denote the training set as $D = \{(x_1, y_1), (x_2, y_2), \ldots, (x_n, y_n)\}$ where x_i represents the i-th input utterance with its corresponding label y_i, *i.e.*, $y_i \in \{\mathrm{NF}, \mathrm{FP}, \ldots, \mathrm{ACK}\}$ dialogue acts from [25]. n represents the size of our training set, which we assume to be small due to the challenging nature of obtaining data for the DA classification [14]. Additionally, we denote $\theta \in \mathbb{R}^d$ as the parameters of the utterance encoder f, and $\omega \in \mathbb{R}^{d\times 1}$ as the parameters of the class-dependent linear classifier g. In **Phase 1** (P_1 see Fig. 1), we input the i-th sentence, *i.e.*, x_i, along with its contextual sentences [10] x_{i-1} and x_{i-2}, *i.e.*, $\mathbf{x} = \{x_i, x_{i-1}, x_{i-2}\}$, to the utterance encoder. In **Phase 2** (P_2), we employed BERT [4] as the encoder due to its effectiveness in educational classification tasks [9,15,20]. The BERT model can learn the latent encoding representation of the input sentences and output these representations for the class-dependent linear classifier in **Phase 3** (P_3). In the following, we illustrate the process of generating the FP[1] prediction for the dialogue x_i:

$$z_{\mathrm{FP}} = g\left(\omega_{\mathrm{FP}}, f\left(\theta, \mathbf{x}\right)\right), \tag{1}$$

$$\Pr\left(y = \mathrm{FP}\right) = \sigma\left(z_{\mathrm{FP}}\right), \tag{2}$$

Since there are K classes[2], we need K linear classifiers, each to generate the prediction probability for its corresponding class via using the sigmoid function

[1] **FP**, `Positive Feedback` "*Well done!*", same abbreviation from [25].

[2] Our study has 31 dialogue acts as the classes to be classified.

$\sigma()$ in **Phase 4** (P_4). Lastly, to train the model with the surrogate AUC loss, we implement the deep AUC margin loss [29] for each of the sigmoid results from **Phase 4**, we demonstrate this surrogate loss for the **FP** class as follows:

$$\mathrm{AUC(FP)} = \mathbb{E}\left[\left(m - \sigma(z_{\mathrm{FP}}) + \sigma\left(z'_{\mathrm{FP}}\right)\right)^2\right], \tag{3}$$

where $\sigma(z_{\mathrm{FP}})$ represents the prediction probability for the sentence that is labeled as FP, and $\sigma(z'_{\mathrm{FP}})$ represents the prediction probability for the sentence that is not. The margin m, normally set as 1 [12,28–30], serves to separate the correct and incorrect prediction for the FP class, encouraging the correct and incorrect prediction to be distinguishable from each other. Then, the AUC losses of K classes are collected, summed up and back propagated in **Phase 5** (P_5) to tune the BERT encoder parameters θ and the class-dependent linear classifier parameters ω. We repeat the process until no further improvements can be achieved and obtain the optimal DA classifier for the DA classification.

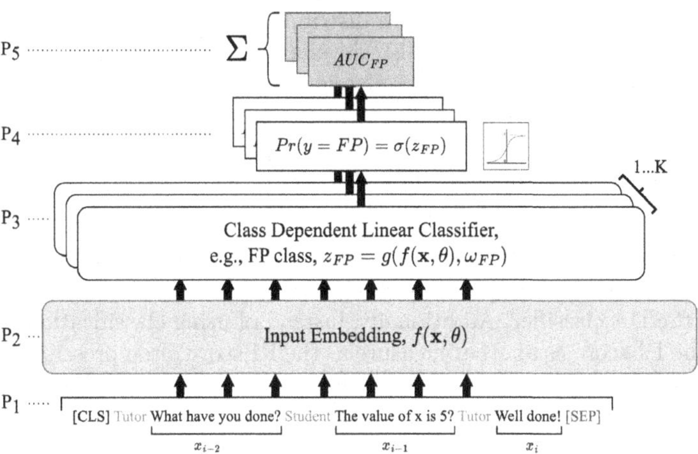

Fig. 1. Architecture of optimizing the classifier by AUC maximization approaches.

3.5 Study Setup

We aimed to evaluate the effectiveness of AUC maximization methods in two DA classification scenarios: (i) a low-resource scenario and (ii) an imbalanced distribution scenario. The dataset (50 dialogue sessions) was randomly split to *training* (40 sessions) and *testing set* (10 sessions) in the ratio of *80%:20%* where the *training set* contained 3,763 instances and *testing set* contained 476 instances.

Low-Resource Scenario. To examine the efficacy of AUC maximization approaches on the classification performance of the DA classifier, our study first investigated the impact of AUC approaches against the traditional CE for the multi-class DA classification to classify 31 DAs under the low-resource setting.

Inspired by [12], we simulated the experiments under the low-resource setting with a *training set* size of $\{25, 50, 100, 200, 400, 800\}$ randomly sampled from the full *training set*. Then, we evaluated the classification performance of the DA classifier optimized by the CE and AUC approaches on the full *testing set*. For each *training set* size, we trained the DA classifiers optimized by AUC maximization approaches and the CE baseline on 10 random training partitions and analyzed their average performance. This allowed us to investigate the performance of AUC maximization approaches under low-resource conditions.

Imbalanced Scenario. To study the robustness of the DA classifier to the imbalanced data distribution, we simulated two settings of imbalanced data distribution on the training and testing dataset. For each setting, we conducted binary classification on a specific type of DA and investigated the impact of imbalanced data distribution at different levels. First, we simulated the classifier performance for all approaches with imbalanced data distribution at different levels in the *training set*, which is also known as distribution shifting, *i.e.*, the data distribution on the training set does not match the distribution on testing [12,23]. Second, we simulated the data distribution shifting on both *training* and *testing sets*. For both setups, we also developed a random generator that creates imbalanced datasets for the DA classification tasks. This generator created *training* and *testing sets* by sampling sentence sets based on the percentage of specific feedback sentences. By evaluating the performance of AUC maximization methods on these different data distributions, we were able to assess its robustness under various imbalanced data conditions.

Evaluation Metrics. In line with the prior works in educational DA classification [9,14,17–19], our study also used Cohen's κ scores to evaluate the performance of the DA classifier. Additionally, instead of using classification accuracy, we used the F1 score as another measure as the F1 score often presents more reliable results on imbalanced distribution due to its nature of implicitly including both precision and recall performance [12].

4 Results

4.1 AUC Maximization Under Low-Resource Scenario

To evaluate the effectiveness of AUC approaches (*i.e.*, DAM and COMAUC), we investigated the proposed optimization approaches (*i.e.*, CE, DAM, and COMAUC) as described in Sect. 3.5. We run 10 different random seeds for each approach to minimize the impact of random variation and obtain reliable estimations of the model's performance. We plotted the averaged results with error bars for each approach in Fig. 2 where the green, blue and red lines represented the approaches CE, AUC, and COMAUC, respectively. Figure 2 shows that when the training set size was small (*e.g.*, 25, 50, 100, 200, 400 sentences), the averaged F1 score of the AUC approaches generally outperformed the CE approach; these differences were significant at the 95% level of confidence. It should be noted that the gap between AUC and CE approaches achieved the most significant difference at 100 sentences. Furthermore, when the training set size increased to 800 sentences, COMAUC outperformed both DAM and CE on average and demonstrated more stable performance as indicated by the error bars. These findings illustrated that COMAUC is an effective and reliable AUC maximization approach under low-resource scenarios.

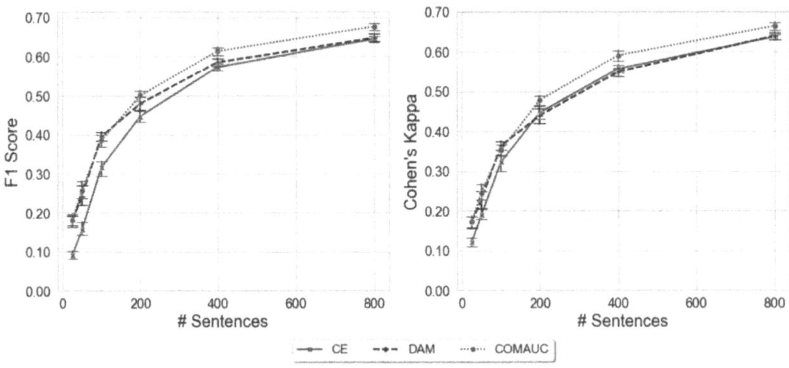

Fig. 2. Performance of DA classifiers with different optimization approaches.

4.2 AUC Maximization Under Imbalanced Scenario

To understand the extent to which AUC approaches can enhance the robustness of the DA classifier to the impact of imbalanced data, we conducted binary classification experiments on a DA from two perspectives as discussed in the **Imbalanced Scenario** in Sect. 3.5. We choose the dialogue act `Positive Feedback` (FP) as the candidate for analysis as it is widely used in tutoring dialogues.

Inspired by the result shown in Fig. 2 where the performance gap of F1 score between AUC and CE approaches demonstrated the most significant difference at 100 instances; thus, for the first setting in the imbalanced scenario, we decided to simulate the training set with 100 instances and explored the impact of the distribution shifting on FP class in the training set. As introduced in Sect. 3.5, we adjusted the ratio of FP in the training dataset from 1% to 80% and examined the classification performance of the DA classifier in different ratios. In Fig. 3, we found that the F1 score of the CE approach was lower than those of AUC approaches (*i.e.*, DAM and COMAUC) when the DA classifier was trained on the dataset with the ratio of 80% for the FP class in the training set. Though the F1 scores of the AUC approaches also decreased under these conditions, the decrease was less pronounced than that of CE. When scrutinizing Cohen's κ score, the DA classifier optimized by CE approach demonstrated vulnerable performance when the FP ratios were 60% and 80% in the training set whereas the Cohen's κ score of both AUC approaches maintained Cohen's κ above 0.60.

For the second setting in the imbalanced scenario, we also examined the classification performance of the DA classifier on the training set with 100 instances but a smaller testing set that contained 50 instances, which was designed for

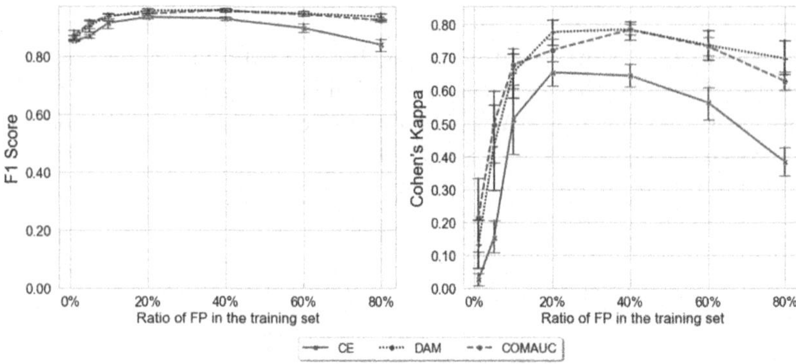

Fig. 3. Classification performance on FP class with different optimization approaches to the data shift in **training set**

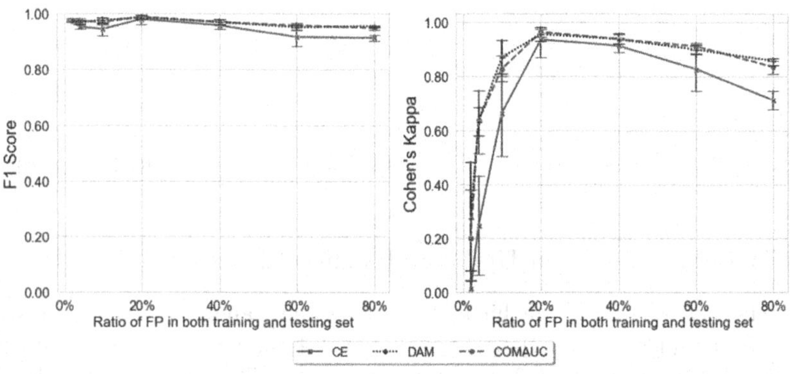

Fig. 4. Classification performance on FP class with different optimization approaches to the data shift in both **training** and **testing set**

facilitating the analysis. As introduced in Sect. 3.5, the same ratio of the specific DA class (*e.g.*, FP) was adjusted simultaneously in the training and testing sets. Figure 4 shows that the three selected approaches achieved promising F1 score in different ratios of FP but the AUC approaches were more stable than the CE approach when the ratios were 60% and 80%. When scrutinizing Cohen's κ scores, AUC approaches were less susceptible to changes in ratios compared to CE. Additionally, the CE approach demonstrated a higher variance of both the Cohen's κ scores and F1 score in Fig. 4 compared to the AUC approaches. These results indicate that the AUC approaches (*i.e.*, DAM and COMAUC) were more robust to the imbalanced data distribution when the ratio of FP was adjusted in the training and the testing set compared to the CE approach.

5 Discussion and Conclusion

Classification is a fundamental task in applying artificial intelligence and data mining techniques to the education domain. However, many educational classification tasks often encounter challenges of low-resource and imbalanced class distribution [1,3,9,31] where the low-resource dataset may not be representative of the population [21,22] and the classifier might overfit the majority class of the dataset [12]. A robust classifier can optimize the classification performance under low-resourced datasets and provide reliable results from various data distributions [12]. Thus, enhancing model robustness to low-resourced and imbalanced data is a crucial step to deploying machine learning algorithms in the educational domain. Our study investigated various approaches regarding the robustness of classifiers for the educational DA classification task. Through extensive experiments, our study provided evidence that AUC maximization approaches (*e.g.*, DAM and COMAUC) can enhance the classification performance of DA classifiers on the limited dataset (*i.e.*, low-resource scenario) and the robustness of the classifier on imbalanced data distribution (*i.e.*, imbalanced scenario) compared to the widely-used standard approach, *i.e.*, Cross Entropy.

Implications. *Firstly*, it is beneficial to adopt the AUC maximization approach for educational DA classification tasks where the training dataset is limited or low-resource. For example, the DA classification in the learning context of medical training [31] also encounters the low-resource issue. Additionally, many classification tasks in the educational domain also encounter the issue of low-resource annotation (*e.g.*, assessment feedback classification [3]). Driven by the findings in our study, we suggest adopting the AUC maximization approaches to the low-resource classification tasks in the educational domain. *Secondly*, in real-world tutoring dialogues, the data distribution about DAs from tutors and students is unavoidable to be imbalanced and changeable over time. To obtain a reliable DA classifier, researchers need to fine-tune the DA classifier when the new batch of the training instances is ready. Our results showed that the DA classifier optimized by the widely-used CE approach was brittle to the issues of imbalanced distribution. Additionally, the imbalanced data distribution widely exists in many educational classification tasks. For example, from a practical standpoint, one challenge in predicting student academic performance is the presence of students who are at high risk of failing was highly imbalanced compared to the students who have excellent or medium performance [1]. The results of our study call for future research on the potential of AUC approaches in identifying at-risk students from the highly imbalanced data distribution.

Limitations. We acknowledged that our study only simulated the imbalanced data distribution by pre-defined imbalanced ratios. It is necessary to investigate the imbalanced data in real-world tutoring dialogues.

References

1. Al-Luhaybi, M., Yousefi, L., Swift, S., Counsell, S., Tucker, A.: Predicting academic performance: a bootstrapping approach for learning dynamic Bayesian networks. In: Isotani, S., Millán, E., Ogan, A., Hastings, P., McLaren, B., Luckin, R. (eds.) AIED 2019. LNCS (LNAI), vol. 11625, pp. 26–36. Springer, Cham (2019). https://doi.org/10.1007/978-3-030-23204-7_3

2. Boyer, K., Ha, E.Y., Phillips, R., Wallis, M., Vouk, M., Lester, J.: Dialogue act modeling in a complex task-oriented domain. In: Proceedings of the SIGDIAL 2010 Conference, pp. 297–305 (2010)
3. Cavalcanti, A.P., et al.: How good is my feedback? A content analysis of written feedback. In: Proceedings of the LAK, LAK 2020, pp. 428–437. ACM, New York (2020)
4. Devlin, J., Chang, M.W., Lee, K., Toutanova, K.: BERT: pre-training of deep bidirectional transformers for language understanding. In: Proceedings of NAACL-HLT, Minneapolis, Minnesota, pp. 4171–4186. Association for Computational Linguistics (2019)
5. D'Mello, S., Olney, A., Person, N.: Mining collaborative patterns in tutorial dialogues. J. Educ. Data Min. **2**(1), 1–37 (2010)
6. Du Boulay, B., Luckin, R.: Modelling human teaching tactics and strategies for tutoring systems: 14 years on. Int. J. Artif. Intell. Educ. **26**(1), 393–404 (2016)
7. Ezen-Can, A., Boyer, K.E.: Understanding student language: an unsupervised dialogue act classification approach. J. Educ. Data Min. **7**(1), 51–78 (2015)
8. Ezen-Can, A., Grafsgaard, J.F., Lester, J.C., Boyer, K.E.: Classifying student dialogue acts with multimodal learning analytics. In: Proceedings of the Fifth LAK, pp. 280–289 (2015)
9. Lin, J., et al.: Is it a good move? Mining effective tutoring strategies from human–human tutorial dialogues. Futur. Gener. Comput. Syst. **127**, 194–207 (2022)
10. Lin, J., et al.: Enhancing educational dialogue act classification with discourse context and sample informativeness. IEEE TLT (under reviewing process)
11. Min, W., et al.: Predicting dialogue acts for intelligent virtual agents with multimodal student interaction data. Int. Educ. Data Min. Soc. (2016)
12. Nguyen, N.D., Tan, W., Buntine, W., Beare, R., Chen, C., Du, L.: AUC maximization for low-resource named entity recognition. In: Proceedings of the AAAI Conference on Artificial Intelligence (2023)
13. Nye, B.D., Graesser, A.C., Hu, X.: Autotutor and family: a review of 17 years of natural language tutoring. Int. J. Artif. Intell. Educ. **24**(4), 427–469 (2014)
14. Nye, B.D., Morrison, D.M., Samei, B.: Automated session-quality assessment for human tutoring based on expert ratings of tutoring success. Int. Educ. Data Min. Soc. (2015)
15. Raković, M., et al.: Towards the automated evaluation of legal casenote essays. In: Rodrigo, M.M., Matsuda, N., Cristea, A.I., Dimitrova, V. (eds.) AIED 2022. LNCS, vol. 13355, pp. 167–179. Springer, Cham (2022). https://doi.org/10.1007/978-3-031-11644-5_14
16. Rus, V., Maharjan, N., Banjade, R.: Dialogue act classification in human-to-human tutorial dialogues. In: Innovations in Smart Learning. LNET, pp. 183–186. Springer, Singapore (2017). https://doi.org/10.1007/978-981-10-2419-1_25
17. Rus, V., et al.: An analysis of human tutors' actions in tutorial dialogues. In: The Thirtieth International Flairs Conference (2017)
18. Samei, B., Li, H., Keshtkar, F., Rus, V., Graesser, A.C.: Context-based speech act classification in intelligent tutoring systems. In: Trausan-Matu, S., Boyer, K.E., Crosby, M., Panourgia, K. (eds.) ITS 2014. LNCS, vol. 8474, pp. 236–241. Springer, Cham (2014). https://doi.org/10.1007/978-3-319-07221-0_28
19. Samei, B., Rus, V., Nye, B., Morrison, D.M.: Hierarchical dialogue act classification in online tutoring sessions. In: EDM, pp. 600–601 (2015)
20. Sha, L., et al.: Is the latest the greatest? A comparative study of automatic approaches for classifying educational forum posts. IEEE Trans. Learn. Technol. 1–14 (2022)

21. Tan, W., Du, L., Buntine, W.: Diversity enhanced active learning with strictly proper scoring rules. In: Advances in Neural Information Processing Systems, vol. 34, pp. 10906–10918 (2021)
22. Tan, W., et al.: Does informativeness matter? Active learning for educational dialogue act classification. In: Wang, N., et al. (eds.) AIED 2023. LNAI, vol. 13916, pp. 176–188. Springer, Cham (2023)
23. Taori, R., Dave, A., Shankar, V., Carlini, N., Recht, B., Schmidt, L.: Measuring robustness to natural distribution shifts in image classification. In: Advances in Neural Information Processing Systems, vol. 33, pp. 18583–18599 (2020)
24. Vail, A.K., Grafsgaard, J.F., Boyer, K.E., Wiebe, E.N., Lester, J.C.: Predicting learning from student affective response to tutor questions. In: Micarelli, A., Stamper, J., Panourgia, K. (eds.) ITS 2016. LNCS, vol. 9684, pp. 154–164. Springer, Cham (2016). https://doi.org/10.1007/978-3-319-39583-8_15
25. Vail, A.K., Boyer, K.E.: Identifying effective moves in tutoring: on the refinement of dialogue act annotation schemes. In: Trausan-Matu, S., Boyer, K.E., Crosby, M., Panourgia, K. (eds.) ITS 2014. LNCS, vol. 8474, pp. 199–209. Springer, Cham (2014). https://doi.org/10.1007/978-3-319-07221-0_24
26. VanLehn, K., Graesser, A.C., Jackson, G.T., Jordan, P., Olney, A., Rosé, C.P.: When are tutorial dialogues more effective than reading? Cogn. Sci. **31**(1), 3–62 (2007)
27. Yang, T., Ying, Y.: AUC maximization in the era of big data and AI: a survey. ACM Comput. Surv. **55**(8) (2022). https://doi.org/10.1145/3554729
28. Ying, Y., Wen, L., Lyu, S.: Stochastic online AUC maximization. In: Advances in Neural Information Processing Systems, vol. 29 (2016)
29. Yuan, Z., Yan, Y., Sonka, M., Yang, T.: Large-scale robust deep AUC maximization: a new surrogate loss and empirical studies on medical image classification. In: 2021 IEEE/CVF ICCV, Los Alamitos, CA, USA, pp. 3020–3029. IEEE Computer Society (2021)
30. Yuan, Z., Guo, Z., Chawla, N., Yang, T.: Compositional training for end-to-end deep AUC maximization. In: International Conference on Learning Representations (2022). https://openreview.net/forum?id=gPvB4pdu_Z
31. Zhao, L., et al.: METS: multimodal learning analytics of embodied teamwork learning. In: LAK23: 13th International Learning Analytics and Knowledge Conference, pp. 186–196 (2023)

What and How You Explain Matters: Inquisitive Teachable Agent Scaffolds Knowledge-Building for Tutor Learning

Tasmia Shahriar(✉) and Noboru Matsuda

North Carolina State University, Raleigh, NC 27695, USA
{tshahri,noboru.matsuda}@ncsu.edu

Abstract. Students learn by teaching a teachable agent, a phenomenon called tutor learning. Literature suggests that tutor learning happens when students (who tutor the teachable agent) actively reflect on their knowledge when responding to the teachable agent's inquiries (aka knowledge-building). However, most students often lean towards delivering what they already know instead of reflecting on their knowledge (aka knowledge-telling). The knowledge-telling behavior weakens the effect of tutor learning. We hypothesize that the teachable agent can help students commit to knowledge-building by being inquisitive and asking follow-up inquiries when students engage in knowledge-telling. Despite the known benefits of knowledge-building, no prior work has operationalized the identification of knowledge-building and knowledge-telling features from students' responses to teachable agent's inquiries and governed them toward knowledge-building. We propose a Constructive Tutee Inquiry that aims to provide follow-up inquiries to guide students toward knowledge-building when they commit to a knowledge-telling response. Results from an evaluation study show that students who were treated by Constructive Tutee Inquiry not only outperformed those who were not treated but also learned to engage in knowledge-building without the aid of follow-up inquiries over time.

Keywords: Learning by teaching · teachable agents · tutor-tutee dialogue · knowledge-building · algebra equation solving

1 Introduction

A teachable agent (TA) is a computer agent that students can interactively teach. Literature affirms that students learn by teaching the TA, a phenomenon called the *tutor learning* [1–3]. In this paper, we address students who teach a TA as *tutors*. The current literature suggests that tutors learn by teaching when they reflect on their understanding [4], revisit the concepts, provide explanations to make sense of solution steps [5], and recover from misconceptions or knowledge gaps. These activities are defined as *knowledge-building* activities and are known to facilitate tutor learning better than *knowledge-telling* activities [6, 7] that merely rephrase known information and explanations with shallow reasoning.

N. Wang et al. (Eds.): AIED 2023, LNAI 13916, pp. 126–138, 2023.
https://doi.org/10.1007/978-3-031-36272-9_11

Despite the effectiveness of knowledge-building, tutors are often biased towards delivering what they know to their TA (i.e., knowledge-telling) [7–9]. Such a lack of cognitive effort weakens tutor learning. Roscoe *et al.* [1] further argued that tutors only engage in knowledge-building when they realize they do not know or understand something. Researchers reported that tutors are likely to realize their knowledge gaps and engage in knowledge building if the TA asks follow-up inquiries [6, 7, 10].

In this work, we implement a method for producing TA follow-up inquiries that is capable of engaging tutors in knowledge-building activities. We call it *Constructive Tutee Inquiry* (CTI). We investigate features of tutors' responses to identify if the tutor engaged in knowledge-building or knowledge-telling. This is the first work to operationalize the detection of knowledge-building and knowledge-telling activities from tutors' responses and generate follow-up inquiries to guide tutors toward knowledge-building when they fail to do so. To evaluate the effectiveness of the proposed CTI, we address the following research questions: **RQ1**: Can we identify knowledge-building and knowledge-telling from tutor responses to drive CTI? **RQ2:** If so, does CTI facilitate tutor learning? **RQ3:** Does CTI help tutors learn to engage in knowledge-building? An empirical evaluation study was conducted as a randomized controlled trial with two conditions to answer those research questions. In the treatment condition, the TA asked follow-up inquiries (i.e., CTI), whereas, in the control condition, the TA did not ask follow-up inquiries. The results show that treatment tutors outperformed control tutors and learned to engage in knowledge-building over time.

Our major contributions are: (1) we operationalized knowledge-building and knowledge-telling and developed a machine learning model to automatically identify knowledge-telling responses, (2) we developed CTI that guides tutors to commit knowledge-building responses, (3) we conducted an empirical study to validate CTI and demonstrated its effect on engaging tutors in knowledge-building, (4) we built and open-sourced the response classifier and the coded middle-grade tutors' response data available at: https://github.com/IEClab-NCSU/SimStudent.

2 SimStudent: The Teachable Agent

SimStudent [11] is our teachable agent (TA). It is shown at the bottom left corner of Fig. 1 which displays the entire user interface of the online learning environment called APLUS (Artificial Peer Learning Using SimStudent). In the current work, SimStudent is taught how to solve linear algebraic equations. A tutor may enter any equation in the tutoring interface (Fig. 1-a). Once the equation is entered, SimStudent consults its existing knowledge base and performs one step at a time. A step consists of either choosing one of four basic math transformations to apply (add, subtract, divide, and multiply) or executing a suggested transformation (e.g., computing a sum of two terms). The knowledge base consists of production rules automatically composed by SimStudent using inductive logic programming to generalize solutions demonstrated by the tutor [12]. When SimStudent suggests a step, the tutor must provide corrective (yes/no) feedback based on their opinion. When SimStudent gets stuck, the tutor needs to demonstrate the step. Over the course of a tutoring session, SimStudent continuously modifies the production rules in its knowledge base according to the tutor's feedback and demonstrations.

The tutor interacts with SimStudent using the chat panel (Fig. 1-b). Currently, SimStudent asks: (1) "Why did you perform *[transformation]* here?" when the tutor demonstrates a *transformation*, and (2) "I thought *[transformation]* applies here. Why do you think I am wrong?" when the tutor provides negative feedback to SimStudent's suggested *transformation*. The tutor provides a textual explanation in the response box (Fig. 1-c) to answer SimStudent's inquiries or may submit an empty response. In any case, SimStudent replies, "Okay!" and proceeds to the next solution step.

The tutor can give a new equation using the "New Problem" button, instruct the TA to solve a problem from the beginning using the "Restart" button, and undo using the "Undo" button (Fig. 1-f). Additionally, the tutor may quiz (Fig. 1-e) SimStudent anytime to check its knowledge status. Quiz topics include four levels (Fig. 1-g). The final challenge consists of equations with variables on both sides. The tutor can refresh their knowledge on solving equations using the resources like the Problem Bank, the Introduction Video, the Unit Overview, and worked-out Examples (Fig. 1-d). The teacher agent, Mr. Williams (Fig. 1-h), provides on-demand, voluntary hints on how to teach. For example, if the tutor repeatedly teaches one-step equations, Mr. Williams might provide the hint, "Your student failed on the two-step equation. Teaching similar equations will help him pass that quiz item".

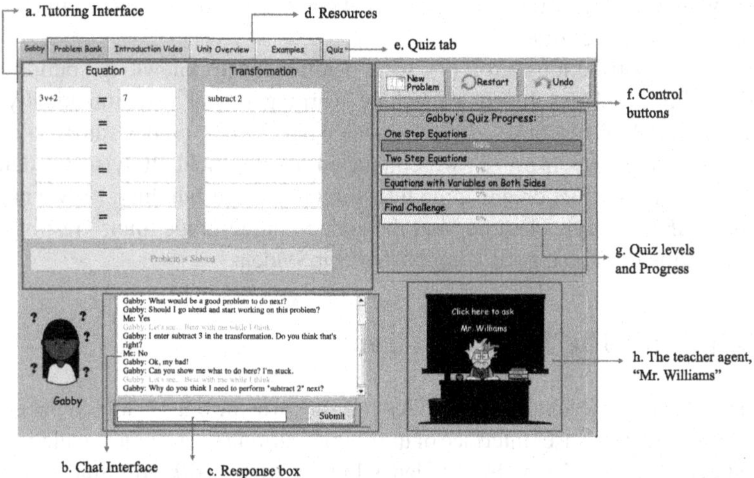

Fig. 1. APLUS interface with SimStudent named Gabby in the bottom left corner.

3 Constructive Tutee Inquiry

3.1 Motivation

The purpose of Constructive Tutee Inquiry is to help tutors generate knowledge-building responses to TA's inquiries. Existing literature argues that tutors switch from providing knowledge-telling to knowledge-building responses by hitting impasses or moments when they realize they do not know something or need to double-check their understanding [6, 7, 9]. Such moments are highly improbable to attain without proper scaffolding on the tutors' responses.

In the rest of the paper, we call the TA's first inquiry in an attempt to understand a topic the *initial inquiry*. For instance: "Why did you perform *[transformation]* here?" is an initial inquiry where SimStudent attempts to understand the reason behind applying the particular *transformation*. We call the TA's subsequent inquiries after the initial inquiry on the same topic the *follow-up inquiries.*

Constructive Tutee Inquiry (CTI) is a TA's follow-up inquiry to solicit knowledge-building responses. The conversational flow chart using the CTI engine is shown in Fig. 2. Two major technologies (together named as CTI Engine) that drive CTI are the *response classifier* and the *dialog manager*. The response classifier analyzes a tutor's response and classifies it as one of the pre-defined types that show what information the TA needs to seek in the next follow-up inquiry. The dialog manager selects the appropriate follow-up inquiry based on the tutor's response type with the intention to solicit a knowledge-building response. Figure 3 illustrates an example inquiry generated by the dialog manager based on the response classifier output class.

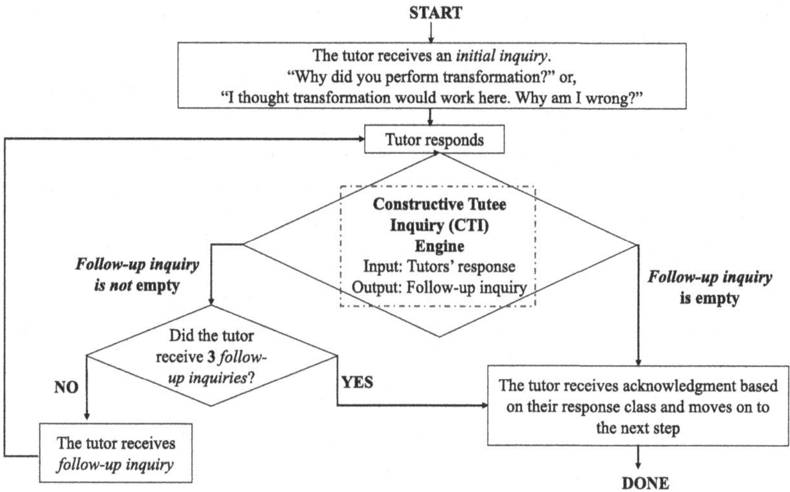

Fig. 2. A complete conversation flow chart using CTI Engine

3.2 Response Classifier

The goal of a response classifier is to identify knowledge-building and knowledge-telling responses and to determine what information to seek in the subsequent follow-up inquiry to solicit a knowledge-building response. The theoretical definition of knowledge-building and knowledge-telling responses (described earlier) inspired us to categorize

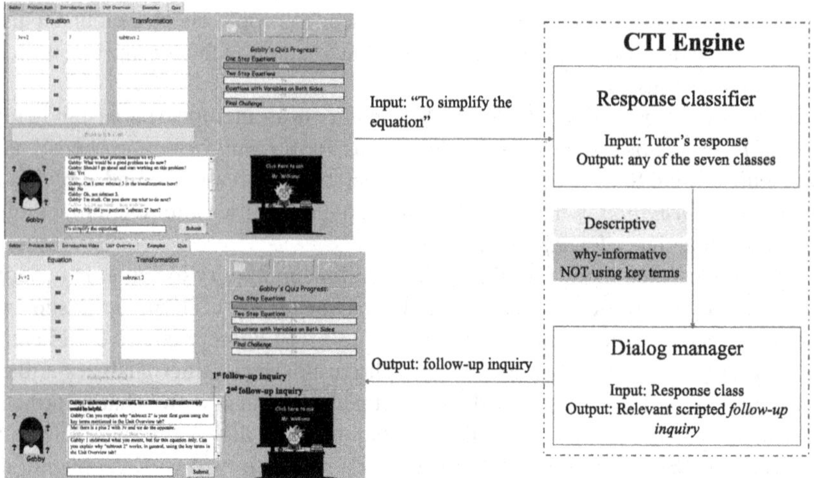

Fig. 3. How CTI Engine works with an example inquiry generated by Dialog Manager

responses in a top-down approach based on *sentence formation*, *relevancy*, *information content,* and *intonation.*

(i) *Sentence formation (ill-formed vs. well-formed)*: Responses that do not have syntactic or semantic meaning are *ill-formed* sentences. For example, responses with poor sentence structure, responses containing misspelled or unknown words, and responses consisting of incomplete sentences are ill-formed sentences. Responses that have syntactic or semantic meaning are *well-formed* sentences.

   TA inquiry**:** Why did you perform divide 2?
   Tutor response (ill-formed): jhjgjkgjk (unknown word)/coficient needs to go by dividing (poor sentence structure with a misspelled word)/the coefficient (incomplete sentence)
   Tutor response (well-formed): because it will cancel out 2.

(ii) *Relevancy (relevant vs irrelevant)*: Responses are *relevant* if they can independently reveal some information about the working domain and *irrelevant* if they could belong to any problem-solving domain.

   TA inquiry**:** Why did you perform divide 2?
   Tutor response (relevant): it will help you solve the equation.
   Tutor response (irrelevant): Do as I say/I kind of had a feeling that it may be right

(iii) *Information content* (*why* vs *what/how*): Responses are *why-informative* if it describes why a solution step or an alternative solution step is correct or incorrect. Responses are *what/how-informative* if it describes what solution step to perform or how a solution step is executed.

   TA inquiry**:** I thought "subtract 2" applies here. Why am I wrong?

Tutor response (why): It should not be subtract 2 because we have a negative constant with variable term and subtract 2 would make the equation worse./It should not be subtract 2 because it would make the equation worse.
Tutor response (what/how): you need to divide 2 instead. (what)/3v-2–2 = 3v-4 (how).

(iv) *Intonation (descriptive vs reparative)*: Responses are *descriptive* if someone explains their stance and *reparative* if someone acknowledges they made a mistake. Note that a response is not reparative if someone is repairing someone else's mistake.

TA inquiry: I thought "subtract 2" applies here. Why am I wrong?
Tutor response (descriptive): Because subtract 2 will not help you get rid of the constant.
Tutor response (reparative): Sorry, it should be subtract 2. I am wrong.

Our first hypothesis towards operationalization is that a knowledge-building or knowledge-telling response must be well-formed, relevant, and either descriptive or reparative. Therefore, any ill-formed, irrelevant tutor responses belong to the "other" class, which is neither knowledge-building nor knowledge-telling, for which the TA must seek a well-formed, relevant response from the tutor. Our second hypothesis is that the information content is what distinguishes knowledge-building responses from knowledge-telling responses. Since why-informative explanation promotes inferences [13, 14] and tutors are likely to realize their gaps by generating inferences [6], our third hypothesis is that "descriptive, why-informative responses" or "reparative, why-informative responses" are knowledge-building responses. Consequently, "descriptive, what/how-informative responses" or "reparative, what/how-informative responses" are considered to be knowledge-telling responses.

To empirically assess the quality of the why-informative response as an indication of knowledge-building, we analyzed 2676 responses from 165 tutors (79 seventh-grade, 51 eighth-grade, and 35 ninth-grade students) from 4 schools during our past studies. Our analysis revealed that high-gaining tutors with both high and low prior knowledge tend to include domain-dependent key terms (*constant*, *coefficient*, *like terms*, etc.) in their why-informative explanation, e.g. "I performed divide 2 because it would get rid of the coefficient and isolate the variable on its own". An ANCOVA analysis fitting the post-test score with the count of responses using key terms while controlling the pre-test score indicated that the count of responses using key terms is a reliable predictor of the post-test score; $F(1{,}162) = 8.2$, $p < 0.01$. Our analysis confirmed findings from another study that explanations using glossary terms facilitate learning [15]. Therefore, we define knowledge-building responses as "descriptive/reparative, why-informative responses using key terms. Consequently, knowledge-telling responses are defined as "descriptive/reparative, why-informative responses NOT using key terms" and "descriptive/reparative, what/how-informative responses." Figure 4 shows the seven classes of our response classifier.

Two human coders categorized 2676 responses into our defined seven classes. Based on Cohen's Kappa coefficient, the inter-coder reliability for this coding showed $\kappa = 0.81$. Disagreements were resolved through discussion. We used an open-source machine

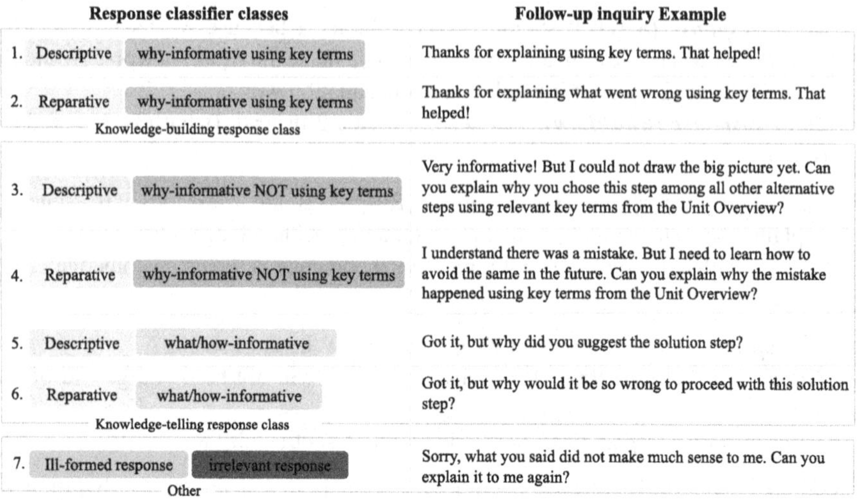

Response classifier classes		Follow-up inquiry Example
1. Descriptive	why-informative using key terms	Thanks for explaining using key terms. That helped!
2. Reparative	why-informative using key terms	Thanks for explaining what went wrong using key terms. That helped!
Knowledge-building response class		
3. Descriptive	why-informative NOT using key terms	Very informative! But I could not draw the big picture yet. Can you explain why you chose this step among all other alternative steps using relevant key terms from the Unit Overview?
4. Reparative	why-informative NOT using key terms	I understand there was a mistake. But I need to learn how to avoid the same in the future. Can you explain why the mistake happened using key terms from the Unit Overview?
5. Descriptive	what/how-informative	Got it, but why did you suggest the solution step?
6. Reparative	what/how-informative	Got it, but why would it be so wrong to proceed with this solution step?
Knowledge-telling response class		
7. Ill-formed response	irrelevant response	Sorry, what you said did not make much sense to me. Can you explain it to me again?
Other		

Fig. 4. Seven classes of response classifier and the corresponding follow-up inquiry generated by dialog manager. The features in a response based on which the classes are formed is shown using color coded box.

learning tool called LightSide [16] to train our learning model from text to identify our defined classes based on the *sentence formation*, *relevancy*, *information content, intonation*, and *presence of key terms.* Next, we describe the proposed scripted dialog manager that generates follow-up inquiries to nudge tutors to generate responses that are *descriptive* or *reparative, contain why-information and* use key terms.

3.3 Dialog Manager

We constructed a script to manage what follow-up inquiries to ask based on the tutors' response classes (shown in Fig. 4). We used an open-source XML script-based dialog manager called Tutalk [17], which uses the output from the response classifier for its decision-making to fashion dialog routes. The dialog manager asks at most three follow-up inquiries until the tutor provides a knowledge-building response, which is operationalized as *descriptive or reparative* responses that contain *why-information* and use domain-dependent key terms. If the knowledge-building response is not provided, the associated transformation (add, subtract, multiply, or divide) is cached. The TA temporarily moves on to the next solution step while informing the tutor that it will again ask the same initial inquiry (e.g., "*Hmm…I am still confused; I will get back to it later. Let's move on for now!*"). The TA again asks the same initial inquiry when the tutor demonstrates or provides negative feedback on the cached transformation.

4 Method

We have conducted an empirical evaluation study to validate the effectiveness of Constructive Tutee Inquiry (CTI). The experiment was conducted as a randomized controlled trial with two conditions: the treatment (CTI) condition, where SimStudent asks both initial inquiries and follow-up inquiries, and the control (NoCTI) condition, where SimStudent only asks initial inquiries.

33 middle school students (14 male and 19 female) of 6th-8th grade were recruited for the study and received monetary compensation. The study was conducted either in-person or online via Zoom, with 20 students attending in-person and 13 participating online. For the online sessions, APLUS was accessed through Zoom screen-sharing.

Students were randomly assigned between conditions ensuring an equal balance of their grades: 17 students (10 in-person and 7 online) for the CTI condition—7 sixth-graders, 7 seventh-graders, 3 eighth-graders—and 16 students (10 in-person and 6 online) for the NoCTI condition—5 sixth-graders, 8 seventh-graders, 3 eighth-graders.

Participants took a pre-test for 30 min on the first day of the study. The test consisted of 22 questions. Ten questions were solving-equation questions (with 2 one-step questions, 2 two-step questions, and 6 questions with variables on both sides), and 12 questions were multiple-choice questions used to measure the proficiency of algebra concepts (one question to formulate an equation from a word problem; seven questions to identify variable, constant, positive, negative, and like terms in an equation; and four questions to identify the correct state of an equation after performing a transformation on both sides). In the following analysis, we call the ten solving-equation questions the Procedural Skill Test (PST) and the 12 multiple-choice questions the Conceptual Knowledge Test (CKT).

The highest score any participant could achieve in the overall test, PST and CKT, are 22, 10, and 12, respectively. No partial marks were provided. In the analysis below, the test scores are normalized as the ratio of the score obtained by the participant to the maximum possible score.

A two-tailed unpaired t-test confirmed no condition difference on the pre-test. Overall test: $M_{\text{CTI}} = 0.60 \pm 0.18$ vs. $M_{\text{NoCTI}} = 0.63 \pm 0.24$; $t(30) = -0.35$, $p = 0.73$. PST: $M_{\text{CTI}} = 0.56 \pm 0.30$ vs. $M_{\text{NoCTI}} = 0.64 \pm 0.30$; $t(30) = -0.75$, $p = 0.46$. CKT: $M_{\text{CTI}} = 0.64 \pm 0.13$ vs. $M_{\text{NoCTI}} = 0.62 \pm 0.22$; $t(30) = 0.27$, $p = 0.79$.

Immediately after taking the pre-test, all participants watched a 10-min tutorial video on how to use APLUS. Participants were informed in the video that their goal was to help their synthetic peer (the teachable agent) pass the quiz. Participants were then free to use APLUS for three days for a total of 2 h or to complete their goal (i.e., passing the quiz), whichever came first. Upon completion, participants took a 30-min post-test that was isomorphic to the pre-test.

We controlled the time on task. A two-tailed unpaired t-test confirmed no condition difference on the minutes participants spent on our application $M_{\text{CTI}} = 292$ vs. $M_{\text{NoCTI}} = 243$; $t(30) = 1.23$, $p = 0.21$.

In the following analysis, we use the learning outcome data (the normalized pre-and post-test scores) along with process data (participants' activities while using APLUS) automatically collected by APLUS, including (but not limited to) interface actions taken by participants, the teachable agent inquiries, and participants' responses.

5 Results

5.1 RQ1: Can we Identify Knowledge-Building and Knowledge-Telling from Tutor Responses to Drive CTI?

If our operationalized features of knowledge-building response (KBR) truly indicates tutor's reflecting on their understanding, we must see a positive correlation between the frequency of KBR and learning gain. A regression analysis fitting normalized gain[1] and the frequency of KBR confirmed that KBR is a reliable predictor of learning gain; *normalized gain* = 0.17 + 0.01 * KBR; $F_{\mathrm{KBR}}(1, 31) = 5.08$, $p < .05$. The regression model implies that *providing one more descriptive or reparative responses containing why-information using key terms (i.e., KBR) results in a 1% increase of normalized gain.*

We also conducted the same analysis with knowledge-telling response (KTR), which, by definition, is *descriptive or reparative responses containing why-information NOT using key terms or* containing *what/how information* only. The result showed that frequency of KTR was not a predictor of learning gain; $F_{\mathrm{KTR}}(1, 31) = 2.66$, $p = 0.11$.

5.2 RQ2: Does CTI Facilitate Tutor Learning?

Figure 5 shows an interaction plot with normalized pre- and post-test score contrasting CTI and NoCTI conditions. An ANCOVA analysis with normalized post-test as a dependent variable and condition as an independent variable while controlling normalized pre-test revealed a main effect of condition, although the effect size was small; $M_{CTI} = 0.75 \pm 0.22$ vs. $M_{NoCTI} = 0.66 \pm 0.25$; $F_{\mathrm{Condition}}(1, 30) = 5.18$, $p < 0.05$, $d = 0.35$. We conducted the same analysis for PST ($M_{CTI} = 0.72 \pm 0.31$, $M_{NoCTI} = 0.65 \pm 0.33$; $F_{\mathrm{Condition}}(1, 30) = 3.54$, $p = 0.06.$, $d = 0.21$) and CKT ($M_{CTI} = 0.77 \pm 0.18$, $M_{NoCTI} = 0.68 \pm 0.23$; $F_{\mathrm{Condition}}(1, 30) = 2.94$, $p = 0.09$, $d = 0.48$). The current data suggests that *CTI tutors had a higher post-test score than the NoCTI tutors, and the discrepancy mostly came from the difference in their proficiency in solving equations.*

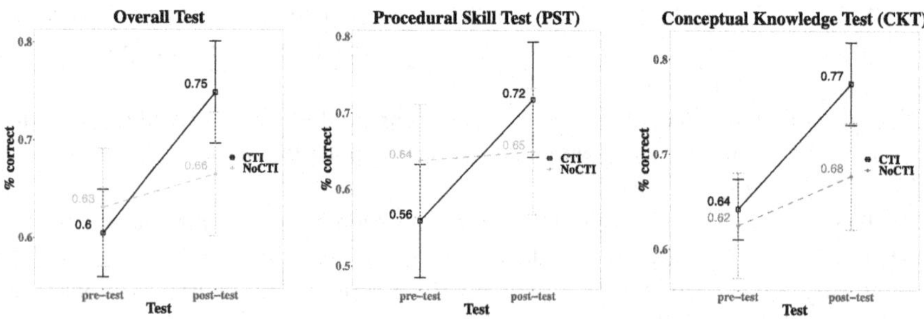

Fig. 5. Interaction plot of % correct from pre to post between conditions (CTI vs NoCTI)

[1] $\textit{Normalized gain} = \frac{\textit{Normalized Post score} - \textit{Normalized Prescore}}{1 - \textit{Normalized Prescore}}$

5.3 RQ3: Does CTI Help Tutors Learn to Engage in Knowledge-Building?

To understand if CTI tutors learned to engage in more knowledge-building responses, we first conducted a one-way ANOVA on the frequency of knowledge-building responses with condition as a between-subject variable. The result revealed a main effect for condition; $M_{CTI} = 19.06 \pm 10.21$, $M_{NoCTI} = 4.19 \pm 5.10$, $F_{Condition}(1, 30) = 27.04$, $p < 0.001$. The current data suggests that *CTI tutors generated significantly more knowledge-building responses than NoCTI tutors.*

We further hypothesized that receiving follow-up inquiries helped CTI tutors learn to provide knowledge-building responses (KBRs) to initial inquiries over time. To test this hypothesis, we calculated the ratio of KBR to the total responses generated by tutors on every initial-inquiry opportunity they receive. We only considered the first 30 initial inquiries because most of the tutors answered at least 30 initial inquiries before they could complete their goal, i.e., passing the quiz.

Figure 6 shows the chronological change of the ratio of KBR to the total responses provided by tutors on the y-axis and the chronological initial inquiry opportunity on the x-axis. For example, on the 5th initial inquiry, 35% of responses generated by CTI tutors were knowledge-building responses. A two-way ANOVA with ratio of KBR as the dependent variable and opportunity for the initial-inquiry and condition as the independent variable revealed a significant interaction between initial-inquiry opportunity (IIO) and condition: $F_{IIO:condition}(1, 52) = 10.52$, $p < 0.01$. *The current data suggests that CTI tutors shifted to provide knowledge-building responses at a higher frequency than NoCTI tutors over time.*

In the CTI condition, the teachable agent, SimStudent, was programmed to ask the same initial inquiry again (as described in Sect. 3.3) if the tutor failed to provide a KBR for a transformation in their previous try. Therefore, one possible reason for CTI tutors providing more KBR than NoCTI tutors might be that CTI tutors programmatically received more initial inquiries. To our surprise, there was no difference in the average number of initial inquiries tutors received in both conditions; $M_{CTI} = 26.5 \pm 15.15$ vs. $M_{NoCTI} = 30.5 \pm 17.46$, $t(110) = 1.30$, $p = 0.20$. *This finding implies that CTI tutors generated more knowledge-building responses than NoCTI tutors even when they received an equal number of initial inquiries.*

6 Discussion

We provided empirical support that an inquisitive teachable agent with constructive inquiries can be built even as a simple application of scripted dialogue, and its effect has been demonstrated. Our proposed Constructive Tutee Inquiry (CTI) helped tutors achieve a higher post-test score than tutors without CTI, arguably by helping them engage in knowledge-building, like reflecting on their own understanding [4] and recovering from their misconceptions or knowledge gaps [18] while tutoring. The gradual increase of the spontaneous knowledge-building responses on the initial inquiry also suggests that tutors who were incapable of engaging in knowledge-building at the beginning of the study eventually learned to do so over time.

Despite the cogent results on the effectiveness of CTI in the current study, two concerns remain that require scrutiny. First, the small effect size of the current study may

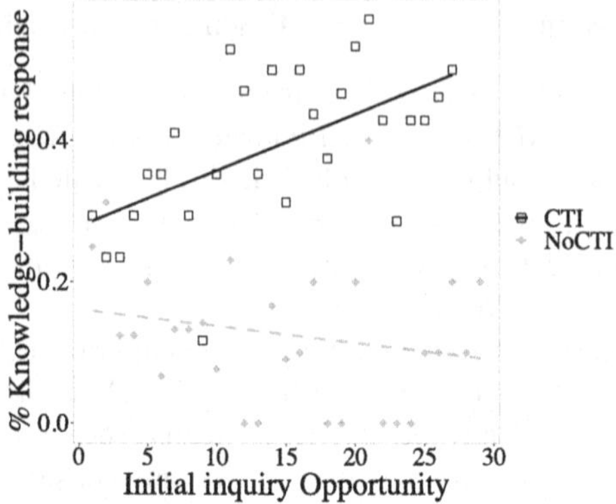

Fig. 6. Ratio of knowledge-building responses to the total responses provided by tutors on the subsequent initial inquiry opportunity

suggest room for system improvement. Second, the current data do not show evidence of students' learning gain on the conceptual knowledge test (CKT), even though tutors were prompted to explain solution steps using domain-dependent key terms. One probable reason could be a lack of internal validity for the CKT (i.e., not truly measuring the conceptual knowledge), which requires revision. Alternatively, it could be because CTI only helped tutors acquire shallow skills for solving equations using key terms without the need to deeply understand the meaning of those key terms or the interconnected relationship among them. Such formulation of shallow skills is argued to only facilitate procedural learning and not conceptual learning [19].

We investigated the nature of the knowledge-building responses (KBR) tutors generated by breaking them down into "descriptive vs. reparative" why-informative responses with key terms, as shown in Table 1. The majority of KBR tutors generated were descriptive responses; tutors rarely repaired their mistakes. One possible explanation for this could be that tutors realized their mistake only after exhaustively responding to all the follow-up inquiry turns, and no turns were left for prompting tutors to provide an explanation of why the mistake happened using key terms. Another reason could be tied to a lack of in-depth understanding of the key terms that might have affected their capability to realize mistakes. Then, even if they did realize their mistakes, it was harder to explain why the mistake happened using those terms.

Table 1. KBR broken down into descriptive vs reparative types

Condition	Avg # Knowledge-building responses (KBR)	
	Avg. # Descriptive	Avg. # Reparative
CTI	17.96	1.10
NoCTI	4.19	0

7 Conclusion

We found that guiding tutors towards knowledge-building using Constructive Tutee Inquiry (CTI) eventually helped tutors learn to perform knowledge-building over time without the aid of follow-up inquiries. It acted as implicit training for the tutor to learn to engage in knowledge-building. In the literature on learning by teaching, we are the first to develop a teachable agent that facilitates tutor learning with appropriate follow-up inquiries without consulting an expert domain knowledge.

Our analysis highlights response features that aid tutor learning. Furthermore, responding to CTI also helped the tutors enhance their equation-solving skills, which was reflected in their performance. However, it was not as effective in their conceptual learning. Our future interest is to understand the root cause behind this finding and to revise the follow-up inquiries to facilitate both procedural and conceptual learning.

Acknowledgment. This research was supported by the Institute of Education Sciences, U.S. Department of Education, through grant No. R305A180319 and National Science Foundation grants No. 2112635 (EngageAI Institute) and 1643185 to North Carolina State University. The opinions expressed are those of the authors and do not represent views of the Institute or the U.S. Department of Education and NSF.

References

1. Roscoe, R.D., Chi, M.T.H.: Understanding tutor learning: knowledge-building and knowledge-telling in peer tutors' explanations and questions. Rev. Educ. Res. **77**(4), 534–574 (2007)
2. Chi, M.T.H., et al.: Learning from human tutoring. Cogn. Sci. **25**, 471–533 (2001)
3. Graesser, A.C., Person, N.K., Magliano, J.P.: Collaborative dialogue patterns in naturalistic one-to-one tutoring. Appl. Cogn. Psychol. **9**(6), 495–522 (1995)
4. Butler, D.L.: Structuring instruction to promote self-regulated learning by adolescents and adults with learning disabilities. Exceptionality **11**(1), 39–60 (2003)
5. Hong, H.-Y., et al.: Advancing third graders' reading comprehension through collaborative Knowledge Building: a comparative study in Taiwan. Comput. Educ. **157**, 103962 (2020)
6. Roscoe, R.D., Chi, M.: Tutor learning: the role of explaining and responding to questions. Instr. Sci. **36**(4), 321–350 (2008)
7. Roscoe, R.D.: Self-monitoring and knowledge-building in learning by teaching. Instr. Sci. **42**(3), 327–351 (2013). https://doi.org/10.1007/s11251-013-9283-4
8. Roscoe, R.D., Chi, M.T.H.: The influence of the tutee in learning by peer tutoring. In: Forbus, K., Gentner, D., Regier, T. (eds.) Proceedings of the 26th Annual Meeting of the Cognitive Science Society, pp. 1179–1184. Erlbaum, Mahwah, NJ (2004)
9. Roscoe, R.D.: Opportunities and Barriers for Tutor Learning: Knowledge-Building, Metacognition, and Motivation. University of Pittsburgh (2008)
10. Shahriar, T., Matsuda, N.: "Can you clarify what you said?": Studying the impact of tutee agents' follow-up questions on tutors' learning. In: Roll, I., McNamara, D., Sosnovsky, S., Luckin, R., Dimitrova, V. (eds.) Artificial Intelligence in Education: 22nd International Conference, AIED 2021, Utrecht, The Netherlands, June 14–18, 2021, Proceedings, Part I, pp. 395–407. Springer International Publishing, Cham (2021). https://doi.org/10.1007/978-3-030-78292-4_32

11. Matsuda, N., et al.: Learning by teaching simstudent – an initial classroom baseline study comparing with cognitive tutor. In: Biswas, G., Bull, S., Kay, J., Mitrovic, A. (eds.) AIED 2011. LNCS (LNAI), vol. 6738, pp. 213–221. Springer, Heidelberg (2011). https://doi.org/10.1007/978-3-642-21869-9_29
12. Li, N., et al.: Integrating representation learning and skill learning in a human-like intelligent agent. Artif. Intell. **219**, 67–91 (2015)
13. Williams, J.J., Lombrozo, T., Rehder, B.: The hazards of explanation: overgeneralization in the face of exceptions. J. Exp. Psychol. Gen. **142**(4), 1006 (2013)
14. Rittle-Johnson, B., Loehr, A.M.: Eliciting explanations: constraints on when self-explanation aids learning. Psychon. Bull. Rev. **24**(5), 1501–1510 (2016). https://doi.org/10.3758/s13423-016-1079-5
15. Aleven, V.A., Koedinger, K.R.: An effective metacognitive strategy: learning by doing and explaining with a computer-based cognitive tutor. Cogn. Sci. **26**(2), 147–179 (2002)
16. Mayfield, E., Rosé, C.P.: LightSIDE: Open source machine learning for text. In: Handbook of Automated Essay Evaluation, pp. 146–157. Routledge (2013)
17. Carolyn, R.: Tools for authoring a dialogue agent that participates in learning studies. Artificial Intelligence. In: Education: Building Technology Rich Learning Contexts That Work, vol. 158, p. 43 (2007)
18. Cohen, J.: Theoretical considerations of peer tutoring. Psychol. Sch. **23**(2), 175–186 (1986)
19. Nilsson, P.: A framework for investigating qualities of procedural and conceptual knowledge in mathematics—An inferentialist perspective. J. Res. Math. Educ. **51**(5), 574–599 (2020)

Help Seekers vs. Help Accepters: Understanding Student Engagement with a Mentor Agent

Elena G. van Stee(✉), Taylor Heath, Ryan S. Baker, J. M. Alexandra L. Andres, and Jaclyn Ocumpaugh

University of Pennsylvania, Philadelphia, USA
evanstee@sas.upenn.edu

Abstract. Help from virtual pedagogical agents has the potential to improve student learning. Yet students often do not seek help when they need it, do not use help effectively, or ignore the agent's help altogether. This paper seeks to better understand students' patterns of accepting and seeking help in a computer-based science program called Betty's Brain. Focusing on student interactions with the mentor agent, Mr. Davis, we examine the factors associated with patterns of help acceptance and help seeking; the relationship between help acceptance and help seeking; and how each behavior is related to learning outcomes. First, we examine whether students accepted help from Mr. Davis, operationalized as whether they followed his suggestions to read specific textbook pages. We find a significant positive relationship between help acceptance and student post-test scores. Despite this, help accepters made fewer positive statements about Mr. Davis in the interviews. Second, we identify how many times students proactively sought help from Mr. Davis. Students who most frequently sought help demonstrated more confusion while learning (measured using an interaction-based ML-based detector); tended to have higher science anxiety; and made more negative statements about Mr. Davis, compared to those who made few or no requests. However, help seeking was not significantly related to post-test scores. Finally, we draw from the qualitative interviews to consider how students understand and articulate their experiences with help from Mr. Davis.

Keywords: Help Seeking · Help Acceptance · Pedagogical Agents

1 Introduction

Despite the growing body of research on the relationship of help seeking with various student characteristics (e.g., prior knowledge, self-regulatory skills, demographic characteristics) and learning outcomes in computer-based learning environments (CBLEs), we do not yet fully understand how all of these variables interact [1–3]. In particular, we still do not understand which factors lead students to accept help when it is initiated by the system (rather than by the student), and the conditions under which accepting such help contributes to better learning [4]. We also do not understand the degree to which help acceptance and help seeking are related behavioral patterns [1].

N. Wang et al. (Eds.): AIED 2023, LNAI 13916, pp. 139–150, 2023.
https://doi.org/10.1007/978-3-031-36272-9_12

The present study takes up these questions by investigating patterns of help acceptance and help seeking in a CBLE for science called Betty's Brain [5]. Many CBLEs, including Betty's Brain, offer help to learners. Such help can take the form of hints, reflection prompts, or directions to relevant information [1]. In some cases, CBLEs diagnose learners' needs and deliver unsolicited assistance, however, there are also situations where help must be proactively sought by the learner.

As in traditional classrooms, seeking help effectively in CBLEs requires cognitive and metacognitive skills [6]. Thus, there is a risk that students who stand to benefit most from the system's help may be least equipped to seek and apply it. Early studies found that students with less prior knowledge were not well-equipped to self-diagnose gaps in their knowledge [7]. Similarly, [8] found that low-performing students often ended their conversation with the mentor agent immediately after being asked what kind of help they needed, which may indicate that they were struggling to identify and articulate the gaps in their understanding [8].

Help seeking is not uniformly beneficial. For example, help seeking can be counterproductive to learning when students are overly reliant on high-level help [9], or when they game the system [10]. In contrast to the predictions of prior help-seeking theories and models, [11] found that avoiding help was associated with better learning than seeking help on steps for which students had low prior knowledge of the relevant skills. The relationship between help seeking and performance has also been found to vary across school demographic contexts [3]. Specifically, [3] found that higher hint usage was associated with higher math performance in urban schools but associated with lower math performance in suburban/rural schools. Other demographic categories (e.g., schools with high or low numbers of economically disadvantaged or limited English proficiency students) also showed differences.

Compared to proactively seeking help, accepting help prompted by the system requires less initiative from the student. Even so, struggling students may lack the knowledge or skills to respond productively after receiving system-initiated help. A recent study suggests that anxiety may inhibit frustrated students from accepting help delivered by a mentor agent [12]. This study used interviews to understand how trait-level science anxiety shaped students' behaviors after experiences of frustration; they reported that students with higher anxiety "seemed unable to process the help they were given" by the mentor agent, whereas less anxious students were more receptive.

Understanding the factors that lead students to accept system-initiated help is important because there is evidence that help acceptance is positively correlated with learning. [4] found that the degree to which students accepted a pedagogical agent's offer to provide help was a stronger predictor of learning gains than was their standardized test score. In this study, positive correlations to learning gains were seen both for students' willingness to receive advice from the pedagogical agent, and their actual compliance with the agent's advice.

This paper seeks to clarify the relationship between help acceptance, help seeking, and learning by identifying (a) student characteristics and experiences associated with help acceptance and help seeking, (b) whether there is an association between students' patterns of help acceptance and help seeking, and (c) how help acceptance and help seeking are related to learning outcomes. To do so, we combine interaction log and *in-situ* interview data collected from a sample of middle school students who used the Betty's Brain computer-based learning environment to learn about climate change. Betty's Brain offers an ideal setting to examine help acceptance and help seeking in tandem because it includes a mentor agent, Mr. Davis, who both initiates assistance and responds to student-initiated requests. Focusing on students' interactions with Mr. Davis, we examine how patterns of help seeking and help acceptance are related to one another; student characteristics and experiences while learning; and post-test scores. Finally, we complement these quantitative findings by using previously-collected interviews to consider how students themselves experience and describe their interactions with Mr. Davis.

2 Mr. Davis and Betty's Brain

In Betty's Brain [5], students construct a visual causal map that represents the relationships in complex scientific processes (i.e., climate change). Students use this map to teach a virtual agent named Betty. Students can gain information about the topic by reading the science resources book; learn strategies for teaching causal reasoning by reading the teacher's guide; check Betty's understanding by asking her questions; and evaluate their progress by having Betty take a quiz. Throughout these activities, the mentor agent Mr. Davis is available to provide on-demand help. Mr. Davis is presented to students as an experienced teacher whose role is to mentor them in the process of teaching Betty. Mr. Davis also grades Betty's quizzes and offers suggestions for ways to improve. His advice includes evaluations of the accuracy of the students' causal map links and suggestions to read specific pages in the science resources book or teacher's guide that contain the information necessary to fix incorrect links [13].

3 Methods

3.1 Participants

Data was obtained from a previously published study [14]. The data for this study were originally gathered from 92 sixth-grade students at an urban middle school in Tennessee, who spent four days (approx. 50 min/day) using Betty's Brain to learn about climate change in December 2018. The school has a diverse student population, with 60% White, 25% Black, 9% Asian, and 5% Hispanic students, and 8% of students enrolled in the free/reduced lunch program. The individual demographic information of the students was not collected. Throughout the unit, multiple forms of data were collected on the students' activity, experiences, and performance. After dropping 4 students who did not complete the anxiety survey from our analysis, our sample size was 88.

3.2 Interaction Log Data

Betty's Brain recorded the students' computer activity as interaction log data, which allows us to analyze various aspects of their virtual interactions, including the reading suggestions they received from Mr. Davis, whether they visited the pages he suggested, and the number of times each student initiated a conversation with Mr. Davis.

Help Acceptance. We operationalized help acceptance as a binary variable indicating whether a student ever followed Mr. Davis' reading suggestions. We consider the help to have been accepted if a student visited the page indicated by Mr. Davis in the period between the time that page suggestion was delivered and Mr. Davis' next page suggestion. We use a binary variable rather than a proportion because almost half (42%) of the students in the sample never followed a single page suggestion.

Help Seeking. For help seeking, we created an ordinal variable based on the total number of times each student began a conversation with Mr. Davis: low requesters (0–1 help requests), moderate requesters (2–4 requests), and high requesters (5–7 requests). The low and high request groups roughly correspond with the lowest and highest quintiles of requests, respectively.

Affective States. The interaction log data also enabled the tracking of students' affective states while using the program, using detection algorithms that had already been integrated into Betty's Brain [15]. The affect detectors for each of five epistemic emotions (boredom, confusion, engaged concentration, delight, and frustration) generated predictions (probabilities between 0 and 1) of the student's affective state every 20 s based on their activity in the program. In the following analyses, we average the affective probabilities at the student level across each student's entire history of interaction with the learning system.

3.3 *In-situ* Interviews

Additionally, members of the original research team conducted 358 short, *in-situ* interviews with students while they were participating in the program, gathering qualitative data about their experiences. Real-time monitoring of affective and behavioral sequences through an application called Quick Red Fox was used to prompt interviews, enabling the interviewers to delve into the cognition associated with crucial learning moments and changes in students' emotional states (see [16, 17]). The interviews were recorded, manually transcribed, and qualitatively coded (see [14] for details about the coding procedure). In the following analysis, we use variables indicating the proportion of interviews for each student that contained (a) positive and (b) negative statements about Mr. Davis [17].

3.4 Learning and Anxiety Measures

Student learning was assessed using pre- and post-test measures administered within the system, with a possible maximum of 18 points on each (see [18]). Additionally, trait-level science anxiety was measured using a revised version [12] of the Math Anxiety Scale [19]. The questions were modified to focus on science rather than math and were designed to elicit students' thoughts and feelings about science in general, not just their experiences in the current learning environment.

4 Results

In this paper we explore both the more passive construct of help acceptance and the more active construct of help seeking. For both help acceptance and help seeking, we discuss how the behavior measured is predicted by students' pre-test score, science anxiety, and affective states while learning, as well as how it is associated with students' perceptions of Mr. Davis. Finally, we analyze how both help acceptance and help seeking are related to learning outcomes. Using the help acceptance and help seeking measures previously constructed, we predict post-test scores among students.

4.1 Help Acceptance

To determine which students accepted help from Mr. Davis, we first identified the number of times each student followed Mr. Davis' suggestion to read a specific page in the science textbook or teacher's guide. Overall, we found that help acceptance was uncommon: on average, students followed help recommendations 6.33% of the time ($SD = 0.093$). Further, 42% of students never followed Mr. Davis' reading suggestions and even the most compliant student only followed 50% of the reading suggestions they received.

We then examine student-level factors—pre-test scores, trait-level science anxiety, and average incidence of affective state while learning—that might predict whether students followed Mr. Davis' reading suggestions. Using binary logistic regression, we estimate the log odds of students ever accepting help versus never accepting help (Table 1). None of the student characteristics or affective states we examined were significant predictors of help acceptance.

Table 1. Binary Logistic Regression Predicting Help Acceptance

	b (*SE*)	*p*-value
Pre-test Score	−0.003 (0.093)	0.976
Science Anxiety	0.022 (0.029)	0.455
Boredom	−2.282 (4.488)	0.611
Engaged Concentration	−1.545 (4.460)	0.729
Frustration	4.866 (3.025)	0.108
Confusion	−0.396 (3.183)	0.901
Delight	12.553 (13.948)	0.368
Constant	−10.121 (10.580)	0.339
N	88	
AIC	129.23	

Next, we examine how help acceptance was related to help seeking and students' perceptions of Mr. Davis using two-tailed T-tests (Table 2). Comparing the means for each group (help accepters vs. non-accepters), we find that help accepters made on average 4.3% fewer positive statements about Mr. Davis than non-accepters ($t(39.37) = 2.025$, $p = 0.050$). The difference in the mean number of conversation requests was not statistically significant between help accepters and non-accepters ($t(75.098) = -0.374$, $p = 0.710$). Finally, there was no statistically significant difference in the means for negative perceptions of Mr. Davis between help accepters and non-accepters ($t(83.071) = -0.132$, $p = 0.896$).

Table 2. T-Tests Comparing Means of Help-Seeking and Perceptions of Mr. Davis by Help Acceptance

	Never Accepted Help	Ever Accepted Help	Mean Difference	t	p-value
Conversation requests	2.892	3.039	−0.147	−0.374	0.710
Negative Davis comments	0.093	0.098	−0.005	−0.132	0.896
Positive Davis comments	0.050	0.007	0.043	2.025	0.050
N	88	88	88	88	88

4.2 Help Seeking

Next, we examine how many times each student initiated a conversation with Mr. Davis—a form of proactive help seeking behavior. We find that 93% of students used this function at least once, indicating that most of the class knew the feature was available. We use an ordinal variable to measure help seeking (low requesters, moderate requesters, and high requesters), as discussed above. On average, students made 2.977 conversation requests over the course of the unit ($SD = 1.800$).

Using ordinal logistic regression, we predict help seeking using the same student-level characteristics examined for help acceptance: pre-test score, science anxiety, and the five affective states in Table 3 [20]. In contrast to the findings for help acceptance, we found that help seeking was significantly associated with both science anxiety and confusion. A one point increase in a student's anxiety level is associated with an increase in the tendency towards higher help requests ($\beta = 0.071$ ($SE = 0.029$, $p = 0.014$)). A student being confused also corresponded to a $\beta = 6.400$ increase in the tendency towards higher help requests ($SE = 3.050$, $p = 0.036$). However, a student's level of help seeking was not statistically significantly associated with pre-test score or the other four affective states.

Finally, we perform a series of bivariate OLS regressions to examine how help seeking is related to help acceptance and students' perceptions of Mr. Davis (Table 4).

Regarding students' perceptions of Mr. Davis, we find that high requesters made 14.2% more negative statements about Mr. Davis in the interviews on average ($SE =$

Table 3. Ordinal Logistic Regression Predicting Help Seeking

	b (*SE*)	*p*-value
Pre-test Score	−0.008 (0.089)	0.933
Science Anxiety	0.071 (0.029)	0.014
Boredom	6.543 (4.481)	0.144
Engaged Concentration	−5.813 (4.213)	0.168
Frustration	−3.936 (2.890)	0.173
Confusion	6.400 (3.050)	0.036
Delight	−5.303 (11.408)	0.642
Low|Moderate	−2.625 (8.962)	
Moderate|High	0.544 (8.946)	
N	88	
Log-Likelihood	−76.42	
AIC	170.84	

Table 4. OLS Regression Coefficients Predicting Help Acceptance and Perceptions of Mr. Davis using Help Seeking Behavior

	Moderate Requesters		High Requesters		Constant	
	b (*SE*)	*p*-value	*b* (*SE*)	*p*-value	*b* (*SE*)	*p*-value
Reading compliance	−0.037 (0.025)	0.143	−0.029 (0.031)	0.363	0.091 (0.021)	< 0.001
Negative Davis comments	−0.018 (0.048)	0.716	0.142 (0.061)	0.023	0.081 (0.041)	0.055
Positive Davis comments	0.001 (0.024)	0.966	−0.011 (0.031)	0.728	0.026 (0.021)	0.207
N	88		88		88	

0.061), compared to low requesters (the reference group) ($p = 0.023$). The moderate request group made the fewest negative statements, but the difference between the low and moderate request groups was not statistically significant ($\beta = -0.018$, $SE = 0.048$, $p = 0.716$). As Table 4 demonstrates, there were no significant relationships between help seeking and positive perceptions of Mr. Davis, or between help seeking and help acceptance.

4.3 Learning Outcomes

Are patterns of help acceptance and help seeking related to learning outcomes in Betty's Brain? To examine this, we consider whether (a) the student ever accepting help from

Mr. Davis or (b) the frequency of seeking help from Mr. Davis is associated with post-test scores. We use OLS regression to predict post-test scores beginning with a simple model including only pre-test scores as a baseline, followed by models including our key independent variables (i.e., help acceptance and help seeking, measured categorically as stated above) and student-level characteristics: science anxiety, the five affective states, and positive and negative perceptions of Mr. Davis (Table 5).

Table 5. OLS Regression Coefficients Predicting Post-Test Scores

	b (*SE*)	*p*-value
Pre-test Score	0.733 (0.110)	< 0.001
Help Acceptance	1.146 (0.583)	0.053
Help Seeking		
Moderate	0.379 (0.716)	0.599
High	0.389 (0.949)	0.683
Science Anxiety	−0.052 (0.036)	0.151
Boredom	13.939 (5.454)	0.013
Engaged Concentration	−2.584 (5.239)	0.623
Frustration	−6.669 (3.653)	0.072
Confusion	1.109 (3.756)	0.769
Delight	−2.435 (15.623)	0.877
Negative Davis Comments	−2.635 (1.525)	0.088
Positive Davis Comments	1.145 (3.306)	0.730
Constant	10.467 (11.759)	0.376
N	88	
R2	0.4457	

Across models, pre-test score remains the most significant predictor of post-test score. Controlling for other factors, a one point increase in pre-test score corresponds to an increase in the post-test score by close to three-quarters of a point ($\beta = 0.733$, $SE = 0.110$, $p < 0.001$). Further, we find that help-acceptance is marginally related to post-test score after controlling for the other student-level measures. Students who accept help score $\beta = 1.15$ points higher on average than those who did not ($SE = 0.583$, $p = 0.053$). However, neither moderate ($\beta = 0.379$, $SE = 0.716$, $p = 0.599$) nor high ($\beta = 0.389$, $SE = 0.949$, $p = 0.683$) help seeking are significantly associated with post-test scores.

4.4 Insights from Qualitative Interviews

The results above suggest that students who use Mr. Davis the most appear to like him the least. Why is this the case? This is a particularly curious finding, given that students who accepted Mr. Davis's help also learned more (though we do not have evidence that this relationship was causal – it might also have been selection bias). To shed light on students' perceptions of Mr. Davis, we turned to the qualitative interviews.

One possible explanation is that students who use Mr. Davis the most have higher expectations for the help he should be providing and are thus dissatisfied with the level of support received. For example, one student who both accepted and sought help from Mr. Davis suggested that Betty's Brain could be improved if the developers "let Mr. Davis help a little more." This student expressed frustration that Mr. Davis did not provide more explicit guidance, explaining that "when I ask Mr. Davis [about a cause-and-effect relationship], he always says that I'll have to figure it out on my own."

Relatedly, if students seek help from Mr. Davis but are unable to understand or use his assistance effectively, they may direct their negative feelings towards him. For example, one high requester reported that he "started a conversation with Mr. Davis and [Mr. Davis] told me that I was wrong, and so I got confused." If help seeking produces further confusion rather than clarity, students may harbor negative feelings toward the person they asked for help.

5 Conclusions

This study aimed to advance understandings of the relationship between help seeking, help acceptance, and learning. Our results indicate that help acceptance and help seeking are distinct behavioral patterns within Betty's Brain: help acceptance was not significantly associated with help seeking in our sample. And whereas science anxiety and confusion were associated with an increased likelihood of being a high help requester, these measures did not predict help acceptance. Finally, while help acceptance was a marginally significant predictor of improved performance (observed as higher post-test scores), help seeking was not associated with learning outcomes.

Even so, we see a key point of similarity regarding help accepters' and help seekers' perceptions of Mr. Davis: we observed negativity toward Mr. Davis among the students who had the most interaction with him. In terms of help acceptance, students who followed Mr. Davis' reading suggestions made fewer positive statements about Mr. Davis than those who never followed his reading suggestions. In terms of help seeking, the highest requesters made more negative statements about Mr. Davis than low requesters. The interviews tentatively suggest that negativity toward Mr. Davis may stem from a desire for more guidance and/or the student's inability to respond productively to his assistance. We also found that negative perceptions of Mr. Davis were associated with lower post-test scores, net of help seeking, help acceptance, and all other variables.

This study has several limitations. First, the study's sample size was relatively small, and involved only a single school. Second, although previous research demonstrates that academic help-seeking patterns are associated with student demographic characteristics [1, 21, 22], we were unable to analyze variation by race, gender, or socioeconomic status because we did not receive individual-level data on these factors. We also did not have a large enough sample to use school-level demographic data as a proxy for individual-level demographic data [3]. Third, this study did not examine how students react to Mr. Davis' assessments of the accuracy of their causal maps. It is possible that the identification of a mistake may trigger different emotional responses and require different strategies to respond effectively, compared to receiving a suggestion to read a specific page.

Despite these limitations, our study takes a step towards understanding the relationship between help acceptance and help seeking within computer-based learning environments. Our results indicate that help acceptance and help seeking are distinct behaviors in Betty's Brain, and that only help acceptance is significantly related to learning. Consistent with prior research [8], we also found that help acceptance was relatively uncommon: 42% of our sample never followed Mr. Davis' reading suggestions, and even the most compliant student only followed half of the reading suggestions they received. Perhaps counterintuitively, we found that students who accepted help from Mr. Davis made fewer positive statements about him in the interviews. It is possible that this finding points to changes needed in his profile, some of which were addressed in a recent study [13]. Overall, we hope that the results presented here will spur further research to understand the factors associated with accepting and benefitting from system-initiated help.

Acknowledgments. This work was supported by NSF #DRL-1561567. Elena G. van Stee was supported by a fellowship from the Institute of Education Sciences under Award #3505B200035 to the University of Pennsylvania during her work on this project. Taylor Heath was supported by a NIH T32 Grant under award #5T32HD007242-40. The opinions expressed are those of the authors and do not represent views of the funding agencies.

References

1. Aleven, V., Stahl, E., Schworm, S., Fischer, F., Wallace, R.: Help seeking and help design in interactive learning environments. Rev. Ed. Res. **73**, 277–320 (2003). https://doi.org/10.3102/00346543073003277
2. Aleven, V., Roll, I., McLaren, B.M., Koedinger, K.R.: Help helps, but only so much: research on help seeking with intelligent tutoring systems. Int. J. Artif. Intell. Educ. **26**(1), 205–223 (2016). https://doi.org/10.1007/s40593-015-0089-1
3. Karumbaiah, S., Ocumpaugh, J., Baker, R.S.: Context matters: differing implications of motivation and help-seeking in educational technology. Int. J. Artif. Intell. Educ. **32**, 685–724 (2021). https://doi.org/10.1007/s40593-021-00272-0
4. Segedy, J.R., Kinnebrew, J.S., Biswas, G.: Investigating the relationship between dialogue responsiveness and learning in a teachable agent environment. In: Biswas, G., Bull, S., Kay, J., Mitrovic, A. (eds.) AIED 2011. LNCS (LNAI), vol. 6738, pp. 547–549. Springer, Heidelberg (2011). https://doi.org/10.1007/978-3-642-21869-9_97
5. Biswas, G., Segedy, J.R., Bunchongchit, K.: From design to implementation to practice a learning by teaching system: betty's brain. Int. J. Artif. Intell. Educ. **26**(1), 350–364 (2015). https://doi.org/10.1007/s40593-015-0057-9
6. Karabenick, S.A., Gonida, E.N.: Academic help seeking as a self-regulated learning strategy: current issues, future directions. In: Alexander, P.A., Schunk, D.H., Green, J.A. (eds.) Handbook of Self-Regulation of Learning and Performance, pp. 421–433. Routledge Handbooks Online (2017). https://doi.org/10.4324/9781315697048.ch27
7. Wood, H., Wood, D.: Help seeking, learning and contingent tutoring. Comput Educ. **33**, 153–169 (1999)

8. Segedy, J., Kinnebrew, J., Biswas, G.: Supporting student learning using conversational agents in a teachable agent environment. In: van Aalst, J., Thompson, K., Jacobson, M.J., Reimann, P. (eds.) The Future of Learning: Proceedings of the 10th International Conference of the Learning Sciences, pp. 251–255. International Society of the Learning Sciences, Sydney, Australia (2012)
9. Mathews, M., Mitrović, T., Thomson, D.: Analysing high-level help-seeking behaviour in ITSs. In: Nejdl, W., Kay, J., Pu, P., Herder, E. (eds.) AH 2008. LNCS, vol. 5149, pp. 312–315. Springer, Heidelberg (2008). https://doi.org/10.1007/978-3-540-70987-9_42
10. Baker, R.S., Corbett, A.T., Koedinger, K.R., Wagner, A.Z.: Off-task behavior in the cognitive tutor classroom: when students "game the system." In: Proceedings of ACM CHI 2004: Computer-Human Interaction, pp. 383–390 (2004)
11. Roll, I., Baker, R., Aleven, V., Koedinger, K.R.: On the benefits of seeking (and avoiding) help in online problem-solving environments. J. Learn. Sci. **23**(4), 537–560 (2014). https://doi.org/10.1080/10508406.2014.883977
12. Andres, J.M.A.L., Hutt, S., Ocumpaugh, J., Baker, R.S., Nasiar, N., Porter, C.: How anxiety affects affect: a quantitative ethnographic investigation using affect detectors and data-targeted interviews. In: Wasson, B., Zörgő, S. (eds.) Advances in Quantitative Ethnography: Third International Conference, ICQE 2021, Virtual Event, November 6–11, 2021, Proceedings, pp. 268–283. Springer International Publishing, Cham (2022). https://doi.org/10.1007/978-3-030-93859-8_18
13. Munshi, A., Biswas, G., Baker, R., Ocumpaugh, J., Hutt, S., Paquette, L.: Analysing adaptive scaffolds that help students develop self-regulated learning behaviours. J. Comput. Assist. Learn. **39**, 351–368 (2022). https://doi.org/10.1111/jcal.12761
14. Hutt, S., Ocumpaugh, J., Andres, J.M.A.L., Munshi, A., Bosch, N., Paquette, L., Biswas, G., Baker, R.: Sharpest tool in the shed: Investigating SMART models of self-regulation and their impact on learning. In: Hsiao, I.-H., Sahebi, S., Bouchet, F., Vie, J.-J. (eds.) Proceedings of the 14th International Conference on Educational Data Mining (2021)
15. Jiang, Y., et al.: Expert feature-engineering vs. deep neural networks: which is better for sensor-free affect detection? In: Penstein Rosé, C., et al. (eds.) AIED 2018. LNCS (LNAI), vol. 10947, pp. 198–211. Springer, Cham (2018). https://doi.org/10.1007/978-3-319-93843-1_15
16. Hutt, S., et al.: Quick red fox: an app supporting a new paradigm in qualitative research on AIED for STEM. In: Ouyang, F., Jiao, P., McLaren, B.M., Alavi, A.H. (eds.) Artificial Intelligence in STEM Education: The Paradigmatic Shifts in Research, Education, and Technology (In press)
17. Ocumpaugh, J., et al.: Using qualitative data from targeted interviews to inform rapid AIED development. In: Rodrigo, M.M.T., Iyer, S., and Mitrovic, A. (eds.) 29th International Conference on Computers in Education Conference, ICCE 2021 – Proceedings. pp. 69–74 (2021)
18. Hutt, S., et al.: Who's stopping you? – using microanalysis to explore the impact of science anxiety on self-regulated learning operations. In: Proceedings of the Annual Meeting of the Cognitive Science Society, vol. 43 (2021)
19. Mahmood, S., Khatoon, T.: Development and validation of the mathematics anxiety scale for secondary and senior secondary school students. Br. J. Sociol. **2**, 170–179 (2011)
20. McCullagh, P.: Regression models for ordinal data. J. R. Stat. Soc. Ser. B Stat. Methodol. **42**, 109–142 (1980)

21. Ryan, A.M., Shim, S.S., Lampkins-uThando, S.A., Kiefer, S.M., Thompson, G.N.: Do gender differences in help avoidance vary by ethnicity? An examination of African American and European American students during early adolescence. Dev. Psychol. **45**, 1152 (2009). https://doi.org/10.1037/a0013916
22. Calarco, J.M.: "I need help!" social class and children's help-seeking in elementary school. Am. Soc. Rev. **76**, 862–882 (2011). https://doi.org/10.1177/0003122411427177

Adoption of Artificial Intelligence in Schools: Unveiling Factors Influencing Teachers' Engagement

Mutlu Cukurova[1(✉)], Xin Miao[2], and Richard Brooker[2]

[1] University College London, London, UK
m.cukurova@ucl.ac.uk
[2] ALEF Education, Abu Dhabi, UAE

Abstract. Albeit existing evidence about the impact of AI-based adaptive learning platforms, their scaled adoption in schools is slow at best. In addition, AI tools adopted in schools may not always be the considered and studied products of the research community. Therefore, there have been increasing concerns about identifying factors influencing adoption, and studying the extent to which these factors can be used to predict teachers' engagement with adaptive learning platforms. To address this, we developed a reliable instrument to measure more holistic factors influencing teachers' adoption of adaptive learning platforms in schools. In addition, we present the results of its implementation with school teachers (n = 792) sampled from a large country-level population and use this data to predict teachers' real-world engagement with the adaptive learning platform in schools. Our results show that although teachers' knowledge, confidence and product quality are all important factors, they are not necessarily the only, may not even be the most important factors influencing the teachers' engagement with AI platforms in schools. Not generating any additional workload, increasing teacher ownership and trust, generating support mechanisms for help, and assuring that ethical issues are minimised, are also essential for the adoption of AI in schools and may predict teachers' engagement with the platform better. We conclude the paper with a discussion on the value of factors identified to increase the real-world adoption and effectiveness of adaptive learning platforms by increasing the dimensions of variability in prediction models and decreasing the implementation variability in practice.

Keywords: Artificial Intelligence · Adoption · Adaptive Learning Platforms · Factor Analysis · Predictive Models · Human Factors

1 Introduction

AI in Education literature has an increasing amount of research that shows the positive impact of using AI in adaptive learning platforms to support students' academic performance (VanLehn, Banerjee, Milner, & Wetzel, 2020), their affective engagement (D'Mello et al., 2007), and metacognitive development (Azevedo, Cromley, & Seibert,

N. Wang et al. (Eds.): AIED 2023, LNAI 13916, pp. 151–163, 2023.
https://doi.org/10.1007/978-3-031-36272-9_13

2004) in controlled studies. Several learning platforms including OLI learning course (Lovett, Meyer, & Thille, 2008), SQL-Tutor (Mitrovic & Ohlsson, 1999), ALEKS (Craig et al., 2013), Cognitive Tutor (Pane, Griffin, McCaffrey, & Karam, 2014) and ASSISTments (Koedinger, McLaughlin, & Heffernan, 2010), have also been shown to have statistically significant positive impacts on student learning in real-world settings. It is important to note that some of these are conducted as multi-state studies using matched pairs of schools across the USA and still showed positive impact (i.e. Pane et al., 2014). These results are particularly significant, as more general studies examining the positive impact of educational interventions are notoriously hard to reach statistical significance (see also, Du Boulay, 2016). Despite such strong impact results, the adoption of adaptive AIED platforms is slow in many schools at best, or simply doesn't exist in many others across the globe. There is an increasing concern about how to promote the adoption of AI-based adaptive learning platforms in schools to enhance students' learning performance.

The slow adoption of AIED systems in real-world settings might, in part, be attributable to the frequent neglect of a range of other factors associated with complex educational systems. These include but are not limited to understanding and deliberately considering the learners' and the teachers' preferences (e.g. Zhou et al., 2021), why and how exactly in the system the tool will be used by the teachers (e.g. Buckingham Shum et al., 2019), the social contexts in which the tools will be used and the perceived/actual support offered to teachers from such social contexts in their practice, physical infrastructure, governance, and leadership of the schools (Benavides et al., 2020), as well as the ethical (e.g. Holmes et al., 2021) and societal implications related to fairness, accountability and transparency of the system (e.g. Sjoden, 2020). Within this complex ecosystem of factors, existing research tends to ignore the relevance of such holistic factors or focus mainly on investigating individual teacher-level factors with generic technology adoption considerations (e.g. perceived usefulness and ease of use (Davis, 1989)). This teacher-centric way of looking at the educational ecosystems is useful but partial in that it ignores where the power lies in introducing such tools into educational institutions in the first place, be it government or local authority policy, the leaders of the school, senior teaching staff in the institution, school infrastructure and culture, community and collaboration opportunities among individual teachers or even the learners themselves. Besides, generic technology acceptance evaluations like Technology Acceptance Model tend to overlook the peculiarities of a specific technology like AI which has different implications for users' agency, accountability and transparency of actions. Likewise, they miss the issue of who builds the tools and thus the role of the technical, social and market forces within which educational technology developers must operate. Perhaps, this is, at least to a certain extent, due to the limited scope of AIED solutions focusing mainly on the technical and pedagogical aspects of delivery in a closed system, rather than taking a "mixed-initiative" approach (Horvitz, 1999) aiming to combine human and machine factors in complex educational systems with considerations of more holistic factors (Cuban, Kirkpatrick, & Peck, 2001). There is an urgent need for understanding the factors influencing the adoption of artificial intelligence solutions in real-world practice. In order to fill in this gap, in this research paper we present,

1) The development of a reliable survey instrument to measure more holistic factors influencing the adoption of adaptive learning platforms in schools.
2) The results of a country-level implementation of the survey and compute the extent to which they predict teacher engagement with adaptive learning platforms.

Based on the results observed, we provide suggestions for scaled adoptions of AI-based adaptive learning platforms in schools more widely.

2 Context and the Adaptive Learning Platform Studied

Adoption of AI-based adaptive learning platforms, or any particular artefact as a matter of fact, is first and foremost likely to be influenced by the particular features of the artefact itself. Hence, building upon Vandewaetere and Clarebout's (2014) four-dimensional framework to structure the diversity of adaptive learning platforms in schools, we first present the details of the particular adaptive learning platform we investigated. The specific adaptive platform we studied is called the Alef Platform, which is a student-centred adaptive learning system developed by Alef Education. The platform has been implemented and used as the primary learning resource across six core curricula from Grade 5 to Grade 12 in all K-12 public schools in the United Arab Emirates. By design, the platform is a student-centred adaptive learning system which allows learners to self-regulate learning through adaptive tests, bite-sized multimedia content and analytics that provides feedback on cognitive and behavioural performance. Adaptation is dynamic and covers both the content and the feedback. The platform allows shared control from both students and teachers. So, teachers can control lessons assigned to students for classroom management, curriculum pacing and behavioural management purposes. In addition, students, who lack training in self-regulated learning and struggle to engage with adaptive content, can override the platform's suggestions. By providing control and continuous analytics feedback about what students are working on and how they are performing, teachers can support, and intervene at the right point in a student's learning process in classrooms regardless of system suggestions. Additionally, school leaders are also provided with learning analytics dashboards to monitor school-level performance, identify gaps for intervention and support teachers' weekly professional development. Currently, the platform only adapts to cognitive and behavioural learner parameters based on student answers to items and engagement behaviours but does not consider any affective or meta-cognitive dimensions in adaptation.

3 Methodology

Based on previously established methodologies published (e.g. Nazaretsky et al., 2022a), the development process included the following four steps: 1) Top-down design of the categories and items based on the literature review of factors influencing technology adoption in schools and bottom-up design of additional items based on discussions with teachers and experts, (2) Pilot with a small group of teachers for polishing the initially designed items, (3) Exploratory Factor Analysis for bottom-up analysis of emerged factors, (4) Reliability analysis of the emerged factors to verify their internal validity.

In the first step, we have undertaken a semi-systematic literature review on factors that influence the adoption of AI in schools. Similar to Van Schoors et al., (2021), we run the following search string on the web of science database (TS = ((adaptiv* OR adjust* OR personal* OR individual* OR tailor* OR custom* OR intelligent* OR tutor*) AND (learn* OR education* OR class* OR school OR elementary OR primary OR secondary) AND (digital* OR online OR computer*) AND (adopt* OR use* OR interve* OR experiment*))) and we filtered papers from the last 25 years. Based on the review of the identified papers, it became clear that the change connected to the adoption of AI in schools does not merely happen by means of introducing a new digital tool for learning at the classroom level, rather, it is profoundly connected to existing teaching practices, school leadership, teachers' vision and perception of the platform, as well as the infrastructure and technical pedagogical support. In this respect, in order to understand and measure factors influencing AI adoption in schools, the need for an instrument that counts for the different dimensions of schools that are engaged when innovative transformations take place became clear. Based on the summary of the identified papers we categorised indicators of adoption at the school level into four: a) digital maturity and infrastructure, b) pedagogical, c) governance and d) teacher empowerment and interaction. These dimensions are selected because they broadly cover the essential elements relevant to the adoption of AI in schools, as discussed in research studies of the relevant literature. Secondly, unlike available technology adoption instruments in the literature, these aspects do not *only focus* on the technical aspects of the platform features, neither they are only at an individual teacher level. Rather they consider multidimensional factors at the whole-school level which are crucial to the successful adoption of AI-based Adaptive Platforms in Schools. These aspects involve conceptual constructs of the framework from which the initial instrument items were built. It is important to note that these four aspects should not be perceived as isolated entities that do not transverse and correlate. Rather, these aspects are interrelated and influence one another.

To summarise the identified dimensions briefly,

a) Digital maturity is understood as an aspect that sheds light on the organizational aspect of digital adoption, technical infrastructure and teachers' knowledge and confidence in the use of adaptive learning platforms. Adopting a whole-school approach towards technology implementation necessitates examining the level at which the school as an organizational entity is already integrated with digital technologies (e.g., the infrastructure, processes and interactions). Moreover, it is crucial to evaluate the level at which the staff, including teachers and leaders trust, feel at ease and confident with using and deploying AI tools in their daily practices.
b) The pedagogical aspect mainly aims to get a sense of to what extent adaptive learning platforms are actively embedded within the pedagogical practices of the school and require changes from the traditional practices of teacher's practice. This dimension is also associated with the school's curriculum alignment, teachers' workload, and assessment cultures of the schools.

c) The governance dimension requires an account of the school as a structural organization in which a shared vision about the role of AI tools is put into practice. The governance is particularly significant, as with any organizational change needing to be stabilized and supported at the management level. In this respect, the governance aspect sheds light on the vision of different school actors on the use of adaptive learning platforms and more specifically focuses on the vision of school leaders and principals about the direction they envisage for their school with respect to the adoption of AI tools.
d) At last, the teacher empowerment and interaction dimension aims to generate insights into the extent to which individual teachers help each other out and create learning communities for sharing and interacting with their experiences of a particular AI tool. Moreover, this aspect revisits the role of students as active contributors to adoption through shared school-level practices.

After the literature review was completed, we ran in-depth discussions and interviews with the creators of the platform to cross-check survey items to be aligned with the key design principles for the platform as well as their professional experience in implementing adaptive learning platforms in schools. Then, the initial set of items was piloted with a group of 10 experienced teachers. The participants were informed about the purpose of the survey and its development methodology, and then asked to fill in the survey individually and discuss any feedback they considered relevant. As a result, the survey items were finalized and translated as required. Relevant ethical and practical approvals were obtained from school principals and local education governing agencies before the large-scale implementation. The resulting instrument consisted of 30 five-level Likert items to measure potential factors influencing the adoption of adaptive learning platforms in K12 public schools and 8 items on socio-demographics (The instrument can be seen in the Appendix link below after the references).

To explore the factorial structure of the collected data all items were subjected to exploratory factor analysis (EFA) using the principle factor extraction (PFE) approach. Since our theoretical and experimental results indicated correlations between potential factors we chose a dimensionality reduction method that does not assume that the factors are necessarily orthogonal (e.g. principal axis factoring). The analysis was completed using relevant packages in *Python*. Initially, data was examined for outliers and erroneous responses were removed, such as respondents with incorrect email addresses or invalid ages. Then, for the remaining 529 responses, the Kaiser-Meyer-Olkin (KMO) Measure of Sampling Adequacy (MSA) was used to verify the sampling adequacy for the analysis ($KMO = 0.913 > 0.9$). Bartlett's test of sphericity ($\chi 2 = 9819.2$, $p < 0.001$) indicated that the correlation structure is adequate for meaningful factor analysis. Using the *FactorAnalyzer* package we ran the EFA with the PFE approach and the results showed seven factors with eigenvalues over 1. The seven-factor model also met all the quality criteria suggested by Bentler and Bonett (1980) and was chosen as the best model for explaining the response data. For deciding on the mapping of items to factors, we used a threshold value of .75. We didn't have any items with double loading thanks to the high threshold value set and items with no loading (not explained by any of the factors) were qualitatively judged to decide whether they should be retained or dropped. Then, we examined the internal consistency of the assignment of items to a factor and labelled each factor accordingly. The factors identified and labelled are presented below (Table 1).

Factor 1. Presented strong positive correlations with 'No Additional Analytics', 'No Additional Switching Tools', 'No Additional Classroom Behavioural Management', 'No Additional Balancing of Learning' items which are all related to not adding any additional workload to teachers. So, we labelled this factor as the workload factor.

Factor 2. Showed strong positive correlations with 'Believe in Success', 'Trust in the Platform', 'Defending the Platform' items which are all related to measuring the extent to which teachers have ownership of the platform and trust it. We labelled this factor as the teachers' ownership and trust factor.

Factor 3. Had strong positive correlations with 'Part of Community', 'Helpful Professional Development', 'Access to Help', 'Sufficient Guidance' items which all relate to the extent to which teachers feel that they are supported and get help from a community. Therefore, we labelled this factor as teachers' perceived support factor.

Factor 4. Had strong positive correlations with 'Knowledge and Skills to Use' and 'Self-Efficacy' items so we labelled this factor as teachers' perceived knowledge and confidence factor.

Factor 5. Showed strong positive correlations with 'Maintaining Engagement', and 'Finding Balance' items which are related to teachers' sense of control in the use of the platform in their classroom practice. So, we labelled this factor as the classroom orchestration factor.

Factor 6. Presented strong positive correlations with 'Great Content', 'Efficiency', and 'Satisfaction with the Platform' items which all relate to teachers' perceived quality of digital learning content and platform features. Therefore, we labelled this product quality factor.

Factor 7. Had a strong negative correlation with the 'No Privacy Concerns' item so we labelled it as the ethical factor.

Table 1. Seven Factors and Variance Explained

Factors	Sum of Squared Loadings	Average Var Extracted	Cumulative Variance
Factor 1	3.270	0.093	0.093
Factor 2	3.119	0.089	0.183
Factor 3	2.915	0.083	0.266
Factor 4	1.997	0.057	0.323
Factor 5	1.894	0.054	0.377
Factor 6	1.519	0.043	0.420
Factor 7	1.172	0.033	0.454

4 Results

4.1 Teachers' Responses to the Items

The instrument was double-translated and delivered bilingually through Survey Monkey. It was administered to 792 UAE K12 public school teachers from Grade 5 to Grade 12, teaching 6 subjects including English, Math, and Science. Teachers' age ranged from 22 to 55, with the average age of 44. Among the teachers, only 2% were novice teachers with 1–2 years of teaching experience while above 80% of teachers had at least 6 years of teaching experience. With regards to their experience using adaptive learning platforms, 6.4% had less than a year's experience, 13% had 1–2 years of experience, 30% had 2–3 years of experience and about 50% had 3–5 years of experience. The responded teachers consisted of a balanced gender distribution. In terms of results, we first used the instrument to measure the distribution of teachers' responses to each of the factors that resulted from the analysis. Per factor, the score of the teacher on that factor was computed by averaging the scores of the items belonging to that factor. The score of the individual item regarding the ethical concern was taken as it is. The resulting distributions of the scores by the chosen subcategories are presented in Fig. 1.

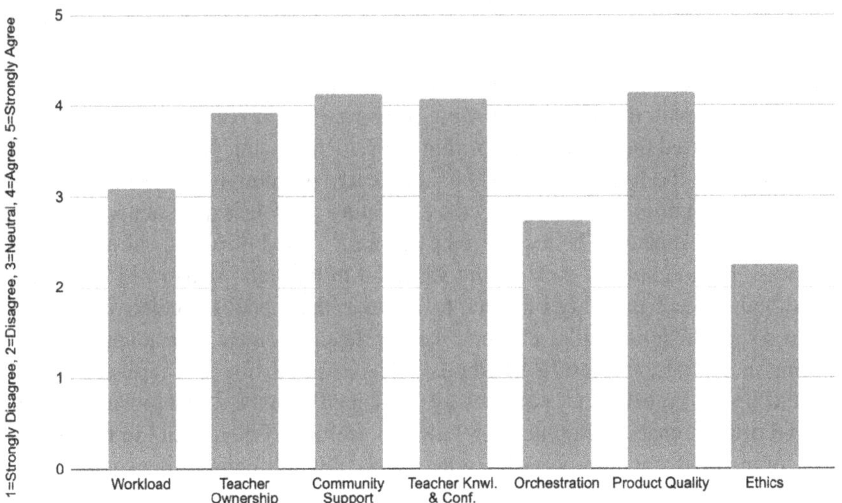

Fig. 1. Teachers' mean response values to seven factors of the survey.

Based on the mean values provided, the mean value for the workload factor was 3.09, indicating that the teachers didn't think that the adaptive learning platform added to or decreased their workload. The mean value for the teacher ownership factor was 3.92, 4.13 for the community support factor, and 4.07 for the teacher knowledge and confidence factor. These indicate that teachers overall agreed that they had enough knowledge and confidence to use the adaptive learning platform, they received enough support for its effective use and they have a high amount of ownership of the platform. The mean value for the product quality factor was 4.14, indicating that teachers on average agreed that the platform was of good quality and were satisfied with it. On the other hand, the mean value for the orchestration factor was 2.74, indicating a somewhat disagreement that teachers can find a balance between the other activities of teaching and the use of the platform as well as struggling to keep students engaged with the platform during classroom activities. Finally, the mean value for the ethics factor was 2.25, indicating that teachers might have some concerns about privacy, safety, and data security while using the platform.

4.2 Predicting Teachers' Engagement with the Adaptive Learning Platform

Furthermore, we studied the correlations between the items included in the seven factors identified above and teachers' active days on the adaptive learning platform. Adaptive learning platform engagement days were calculated using the platform's log data for the first half of the 2022–2023 academic year. As summarized in Table 2, most items that correlated statistically significantly with teachers' engagement with the adaptive learning platform were related to the seven factors discussed above. Those that are different are the items that had double loadings on multiple factors. For instance, the extent to which teachers think that the platform provides 'Helpful Feedback' is related to both ownership and community support factors, or 'content alignment' with national curricula item is related to ownership and product quality factors, 'Easy to Use' item is loaded on ownership, community and quality factors.

At last, we used the data from the survey to build a predictive model of teachers' engagement with the adaptive learning platform in their real-world practice. More specifically, we used the *catboost* algorithm with the following parameters: 1000 iterations, depth of 2, the learning rate of 0.01, and early stopping rounds set to 30. These parameters were chosen based on previous literature and our experiments to optimize the model's performance. The results indicated an R^2 value of 0.244, showing that the model was able to explain ~%24 of the variation in teachers' real-world engagement data. Although there is plenty of space to improve the model's predictive power, for complex social phenomena like teachers' real-world engagement with adaptive learning platforms in schools, the results can be considered promising. Perhaps more importantly, the highest co-efficiency values were associated with the extent to which teachers perceived professional development provided as useful and their belief in the potential success of the platform to improve learning which further highlights the significance of these items for predicting teachers' real-world engagement with the adaptive learning platform in schools.

5 Discussion

The use of artificial intelligence in the field of education has greatly contributed to the design and development of effective adaptive learning platforms. However, as these platforms move out of our labs to schools, most researchers have come to realize that many important and complex problems regarding their adoption exceed the scope of the system features. Change connected to any innovation adoption does not merely happen by means of introducing a new digital tool for learning at the classroom level, rather, it is profoundly

Table 2. Significantly correlating items with teachers' real-world engagement with the adaptive learning platform

	Pearson Correlation Coefficient	P-Value	Adjusted-P-Value
Helpful CPD	0.246061	p = 0.00 < .001	p = 0.00 < .001
Believe in Success	0.225836	p = 0.00 < .001	p = 0.00 < .001
Satisfied w/ Platform	0.200222	p = 0.00 < .001	p = 0.00 < .001
Helpful Feedback	0.190905	p = 0.00 < .001	p = 0.00 < .001
Defend the Platform	0.190418	p = 0.00 < .001	p = 0.00 < .001
Great Content	0.182100	p = 0.00 < .001	p = 0.00 < .001
Platform is Efficient	0.180516	p = 0.00 < .001	p = 0.00 < .001
Trust in the Platf	0.178809	p = 0.00 < .001	p = 0.00 < .001
Part of Community	0.154447	p = 0.00 < .001	p = 0.00 < .01
Content Alignment	0.153034	p = 0.00 < .001	p = 0.00 < .01
Easy to Use	0.144989	p = 0.00 < .001	p = 0.00 < .01
Sufficient Guidance	0.142695	p = 0.00 < .001	p = 0.00 < .01
Leadership's Believe	0.139671	p = 0.00 < .01	p = 0.00 < .01
Access to Help	0.134303	p = 0.00 < .01	p = 0.00 < .01
Can Find Balance	0.131564	p = 0.00 < .01	p = 0.01 < .01
No Ext. Lesson Planning	0.112758	p = 0.01 < .01	p = 0.02 < .05
Self-Efficacy	0.108566	p = 0.01 < .05	p = 0.03 < .05
Opportunity to Share	0.106320	p = 0.01 < .05	p = 0.03 < .05

connected to teaching practices, school leadership, teachers' vision and perception as well as the infrastructure and technical pedagogical support (Ilomäki & Lakkala, 2018). Recent research in AIED literature highlights the importance of such factors and shows that the combination of human mentoring and AI-driven computer-based tutoring can have a positive impact on student performance, with several studies demonstrating the promise of this approach (Chine et al., 2022). However, teachers' acceptance level of AI technologies influences the integration of such human-AI teaming approaches (Ifinedo et al., 2020). In this respect, for human-AI personalisation approaches to work, there is a need for an approach towards innovation adoption that accounts for the different dimensions of schools that are relevant when innovative transformations take place. To scale the adoption of artificial intelligence platforms in schools, a multitude of other relevant factors should work together to encourage successful implementations. In this study, we developed a reliable instrument to measure some of these important factors in the adoption of adaptive learning platforms and built a predictive model of teachers' real-world engagement using the survey data as input.

Our results highlighted a set of influential factors that play an important role in the adoption of adaptive learning solutions by teachers. First of all, it is important that the adaptive platform does not lead to an increase in teachers' existing workload (even if it does not lead to any decrease in it). Teachers' workload is already a significant concern in many education systems and any adaptive platform system implementation should aim to protect the *status quo* or decrease the workload rather than adding to it by any means. Our results indicate that no additional workload to use platform analytics, no additional workload of switching between different tools, no additional workload of classroom management, and no additional workload of switching between different pedagogical practices during the implementation of the adaptive platforms are desired. Therefore, understanding teachers' existing practices in schools and thinking about when and how the use of adaptive platforms should be implemented requires careful thinking. Second, a significant amount of effort should be put into increasing teachers' trust and ownership of the adaptive platform. The more teachers believe in the potential of the platform and trust in its success, the more likely they are to adopt it in their practice. Involving teachers in the iterative research and design process of the adaptive learning platform, and making sure it is genuinely addressing teacher needs might help with teacher trust (Nazaretski *et al.*, 2022b) and ownership (Holstein et al., 2019). In addition, the extent to which individual teachers receive guidance, professional development, and support regarding the use of the adaptive platform, as well as the increased opportunities for teachers to help each other and create learning communities for sharing and interacting with their experiences of the platform, appear to play a significant role in their likelihood of engaging with the adaptive learning platforms. As well-evidenced in previous work regarding the adoption of technology in education in general, ease of use, the perceived quality of the platform as well as teachers' skills and confidence in using them were also identified as key elements in gaining and maintaining teachers' engagement with adaptive learning platforms. In addition, the dimensions of the orchestration of the platform in classroom settings and privacy concerns were highlighted in our results. Therefore, designing lesson plans that help teachers to more effectively switch between the use of the adaptive platform and other pedagogical activities, providing teachers guidance on how to increase their students' engagement with the platform during class time as well as ensuring that the platform does not cause any ethical and privacy-related concerns can increase the teachers' adoption of adaptive learning platforms.

It was interesting to observe that the importance and need for strong investment in the technical infrastructure to ensure that teachers have access to reliable hardware, software, and support when they are implementing the adaptive learning platforms were not specifically highlighted in our results. The significance of such factors associated with the infrastructure is well-reported in the literature (Ifinedo et al., 2020). However, this surprising result is likely due to the fact that in the context of our study, for schools we studied there is a system-level top-down approach to digitising teaching and learning, providing schools with all the material aspects of technology needed (e.g. laptops, charging, storage, internet, etc.) as well as the capacity within schools to communicate potential technical challenges are part of the product and services offered.

At last, our prediction model of teachers' real-world engagement with the adaptive learning platform achieved an R^2 value of 0.244. This is a humble but promising result to predict a complex and dynamic phenomenon like teachers' real-world engagement. AI has proven its potential to provide applications for adaptive learning to support students' academic performance (VanLehn, et al., 2020), yet another valuable potential of AI models is to help us increase the dimensions of variability in our system-level models of complex constructs like the adoption of adaptive platforms. The results of this study not only can contribute to practices of wider adoption of adaptive learning platforms in schools through the suggestions provided above, but also can help us build highly dimensional models for better predicting teachers' real-world engagement behaviours. How to promote the broad adoption of AI in education is one of the main goals of our community and an important step we need to take to progress in our mission of using AI to contribute to a world with equitable and universal access to quality education at all levels (UN's SDG4). If we are to scale the use and adoption of artificial intelligence in schools, we need to better understand, measure and model system-level variables identified in this study.

6 Conclusion

The scaled adoption of artificial intelligence platforms has the potential to revolutionize the way students learn and teachers teach. While previous research has predominantly focused on developing technologies that are both pedagogically and technologically sound, the factors influencing real-world adoption have received limited attention. To the best of our knowledge, there is no previously established instrument for measuring school-level factors influencing the adoption of adaptive platforms. Hence, the contribution of this research paper is twofold. First, it introduces a new instrument to measure school-level factors influencing teachers' adoption of adaptive platforms in schools and presents seven factors with their internal reliability and validity. Second, it presents a large-scale implementation of the survey and develops a predictive model of teachers' real-world engagement using this survey data. These improve our understanding of the factors influencing the adoption of artificial intelligence in schools and can be used to increase further adoption of AI in real-world school settings.

Acknowledgement. This research was partially funded by ALEF Education and the European Union's Horizon 2020 research and innovation programme under grant agreement No 101004676.

Appendix

The full survey used can be accessed here https://docs.google.com/document/d/1qTsbbjN7_v5w34UyNp3FMKcCLwUhf2w2tYKUq-udrxo/edit?usp=sharing.

References

Azevedo, R., Cromley, J.G., Seibert, D.: Does adaptive scaffolding facilitate students' ability to regulate their learning with hypermedia? Contemp. Educ. Psychol. **29**(3), 344–370 (2004)

Benavides, L.M.C., Tamayo Arias, J.A., Arango Serna, M.D., Branch Bedoya, J.W., Burgos, D.: Digital transformation in higher education institutions: a systematic literature review. Sensors **20**(11), 3291 (2020)

Bentler, P.M., Bonett, D.G.: Significance tests and goodness of fit in the analysis of covariance structures. Psychol. Bull. **88**(3), 588 (1980)

Buckingham Shum, S., Ferguson, R., Martinez-Maldonado, R.: Human-centred learning analytics. J. Learn. Anal. **6**(2), 1–9 (2019)

Chine, D.R., et al.: Educational equity through combined human-ai personalization: a propensity matching evaluation. In: Rodrigo, M.M., Matsuda, N., Cristea, A.I., Dimitrova, V. (eds.) Artificial Intelligence in Education: 23rd International Conference, AIED 2022, Durham, UK, July 27–31, 2022, Proceedings, Part I, pp. 366–377. Springer International Publishing, Cham (2022). https://doi.org/10.1007/978-3-031-11644-5_30

Craig, S.D., Hu, X., Graesser, A.C., Bargagliotti, A.E., Sterbinsky, A., Cheney, K.R., et al.: The impact of a technology-based mathematics after-school program using ALEKS on student's knowledge and behaviors. Comput. Educ. **68**, 495–504 (2013)

Cuban, L., Kirkpatrick, H., Peck, C.: High access and low use of technologies in high school classrooms: Explaining an apparent paradox. Am. Educ. Res. J. **38**(4), 813–834 (2001)

Davis, F.D.: Perceived usefulness, perceived ease of use, and user acceptance of information technology. MIS Q. **13**(3), 319 (1989). https://doi.org/10.2307/249008

D'Mello, S., Picard, R.W., Graesser, A.: Toward an affect-sensitive AutoTutor. IEEE Intell. Syst. **22**(4), 53–61 (2007)

Du Boulay, B.: Recent meta-reviews and meta-analyses of AIED systems. Int. J. Artif. Intell. Educ. **26**(1), 536–537 (2016)

Holmes, W., et al.: Ethics of AI in education: towards a community-wide framework. Int. J. Artif. Intell. Educ. **32**, 504–526 (2021)

Holstein, K., McLaren, B.M., Aleven, V.: Co-designing a real-time classroom orchestration tool to support teacher–AI complementarity. J. Learn. Anal. **6**(2), 27–52 (2019)

Horvitz, E.: Principles of mixed-initiative user interfaces. In: Proceedings of the SIGCHI Conference on Human Factors in Computing Systems, pp. 159–166 (1999)

Ilomäki, L., Lakkala, M.: Digital technology and practices for school improvement: innovative digital school model. Res. Pract. Technol. Enhanc. Learn. **13**(1), 1–32 (2018). https://doi.org/10.1186/s41039-018-0094-8

Ifinedo, E., Rikala, J., Hämäläinen, T.: Factors affecting Nigerian teacher educators' technology integration: Considering characteristics, knowledge constructs, ICT practices and beliefs. Comput. Educ. **146**, 103760 (2020)

Koedinger, K.R., McLaughlin, E.A., Heffernan, N.T.: A quasi-experimental evaluation of an online formative assessment and tutoring system. J. Educ. Comput. Res. **43**(4), 489–510 (2010)

Lovett, M., Meyer, O., Thille, C.: JIME - The open learning initiative: measuring the effectiveness of the OLI statistics course in accelerating student learning. J. Interact. Media Educ. **2008**(1), 13 (2008). https://doi.org/10.5334/2008-14

Mitrovic, A., Ohlsson, S.: Evaluation of a constraint-based tutor for a database language. Int. J. Artif. Intell. Educ. **10**, 238–256 (1999)

Nazaretsky, T., Cukurova, M., Alexandron, G.: An instrument for measuring teachers' trust in AI-based educational technology. In: LAK22: 12th International Learning Analytics and Knowledge Conference, pp. 56–66 (2022a)

Nazaretsky, T., Ariely, M., Cukurova, M., Alexandron, G.: Teachers' trust in AI-powered educational technology and a professional development program to improve it. Br. J. Edu. Technol. **53**(4), 914–931 (2022)

Pane, J.F., Griffin, B.A., McCaffrey, D.F., Karam, R.: Effectiveness of cognitive tutor algebra I at scale. Educ. Eval. Policy Anal. **36**(2), 127–144 (2014)

Sjödén, B.: When lying, hiding and deceiving promotes learning - a case for augmented intelligence with augmented ethics. In: Bittencourt, I.I., Cukurova, M., Muldner, K., Luckin, R., Millán, E. (eds.) AIED 2020. LNCS (LNAI), vol. 12164, pp. 291–295. Springer, Cham (2020). https://doi.org/10.1007/978-3-030-52240-7_53

Vandewaetere, M., Clarebout, G.: Advanced technologies for personalized learning, instruction, and performance. In: Spector, J.M., Merrill, M.D., Elen, J., Bishop, M.J. (eds.) Handbook of research on educational communications and technology, pp. 425–437. Springer, New York (2014). https://doi.org/10.1007/978-1-4614-3185-5_34

VanLehn, K., Banerjee, C., Milner, F., Wetzel, J.: Teaching Algebraic model construction: a tutoring system, lessons learned and an evaluation. Int. J. Artif. Intell. Educ. **30**(3), 459–480 (2020)

Van Schoors, R., Elen, J., Raes, A., Depaepe, F.: An overview of 25 years of research on digital personalised learning in primary and secondary education: A systematic review of conceptual and methodological trends. Br. J. Edu. Technol. **52**(5), 1798 (2021)

Zhou, Q., et al.: Investigating students' experiences with collaboration analytics for remote group meetings. In: Roll, I., McNamara, D., Sosnovsky, S., Luckin, R., Dimitrova, V. (eds.) AIED 2021. LNCS (LNAI), vol. 12748, pp. 472–485. Springer, Cham (2021). https://doi.org/10.1007/978-3-030-78292-4_38

The Road Not Taken: Preempting Dropout in MOOCs

Lele Sha[1(✉)], Ed Fincham[1], Lixiang Yan[1], Tongguang Li[1], Dragan Gašević[1,2], Kobi Gal[2,3], and Guanliang Chen[1(✉)]

[1] Centre for Learning Analytics, Faculty of Information Technology, Monash University, Clayton, Australia
{lele.sha,lixiang.Yan,tongguang.Li,dragan.gasevic, guanliang.chen}@monash.edu, ed.fincham@proton.me
[2] School of Informatics, University of Edinburgh, Edinburgh, UK
kgal@exseed.ed.ac.uk
[3] Department of Software and Information Systems, Ben-Gurion University, Be'er Sheva, Israel

Abstract. Massive Open Online Courses (MOOCs) are often plagued by a low level of student engagement and retention, with many students dropping out before completing the course. In an effort to improve student retention, educational researchers are increasingly turning to the latest Machine Learning (ML) models to predict student learning outcomes, based on which instructors can provide timely support to at-risk students as the progression of a course. Though achieving a high prediction accuracy, these models are often "black-box" models, making it difficult to gain instructional insights from their results, and accordingly, designing meaningful and actionable interventions remains to be challenging in the context of MOOCs. To tackle this problem, we present an innovative approach based on Hidden Markov Model (HMM). We devoted our efforts to model students' temporal interaction patterns in MOOCs in a transparent and interpretable manner, with the aim of empowering instructors to gain insights about actionable interventions in students' next-step learning activities. Through extensive evaluation on two large-scale MOOC datasets, we demonstrated that, by gaining a temporally grounded understanding of students' learning processes using HMM, both the students' current engagement state and potential future state transitions could be learned, and based on which, an actionable next-step intervention tailored to the student current engagement state could be formulated to recommend to students. These findings have strong implications for real-world adoption of HMM for promoting student engagement and preempting dropouts.

Keywords: Hidden Markov Models · MOOCs Dropout · Student Engagement

1 Introduction

Over the last decade, the proliferation of MOOCs has provided millions of students with unprecedented access to open educational resources. However, from their conception, MOOCs have attracted widespread criticism due to students' limited interactions with the course materials and high dropout rates [17]. Driven by this, an extensive body

N. Wang et al. (Eds.): AIED 2023, LNAI 13916, pp. 164–175, 2023.
https://doi.org/10.1007/978-3-031-36272-9_14

of research has developed accurate predictive models of student dropout [3,9]. Despite achieving a high accuracy (around 90% AUC [17]) with increasingly more advanced Machine Learning (ML) models, these models provide a limited understanding of the underlying phenomenon, i.e., how a student disengages from the course which ultimately leads to a dropout. In providing course instructors with relevant insights, on the one hand, educational researchers have attempted to represent students' interactions as a set of "prototypical", richly descriptive patterns of student activity [12,15], from which a high-level description of students' engagement with the course material may emerge (e.g., "on-track", "behind" or "disengaged" [12]). However, studies of this strand of research often failed to provide actionable insights to instructors/students, i.e., how to persist student in an on-track state, or how to prevent students from being disengaged. On the other hand, researchers have foregone the understanding of student engagement and have instead aimed to directly predict a future learning path towards successful completion of a course [16,25]. Though achieving a high prediction accuracy (up-to 80%), these approaches relied on "black-box" models, making it difficult for instructors/students to embrace such suggestions as actionable. In an effort to bridge these two disparate fields (i.e., understanding student engagement and providing actionable learning suggestions), and develop models that offer both interpretable insights as well as actionable suggestions about improving student retention in the future, we posit that it is essential to consider students' engagement patterns as a *non-static* phenomenon which may change with time, e.g., an "on-track" student may be demotivated and become less engaged with the course materials [2]. By taking such *temporal dynamics* of engagement (i.e., how students' engagement evolves as the course progresses) into account, meaningful future actions could be inferred directly based on the current engagement state for improving student retention.

As such, we investigated the following **Research Questions** (RQs): **RQ 1** – To what extent can *temporal engagement patterns* be identified that are reflective of student retention in MOOCs? **RQ 2** – How can these patterns be used for *promoting engagement* and *preempting dropout*? To address RQ 1, we took temporal dynamics into account by utilising a fully transparent and interpretable model for analyzing sequential data (e.g., students' interaction with different course materials over time), namely, a Hidden Markov Model (HMM). The HMM consists of a set of latent states, which represent discrete points in time, along with a transition matrix that governs the switching dynamics between them. So we may capture the underlying latent states behind students' interactions, and the potential future transitions between different states. To address RQ 2, we further explored the use of HMM-identified patterns and transitions as a basis to suggest next-step actionable learning activities to students (i.e., the learning material to be accessed at the next step) based on their current engagement states. Through extensive evaluation on the two MOOCs datasets, we demonstrated that HMM may capture not only richly descriptive patterns about student interactions (termed as engagement patterns) in the hidden states, but also salient information regarding the future engagement trend (e.g., whether a student may continue to engage with the course). This indicates that HMM may be a valuable tool for researchers and practitioners to understand the temporal dynamics of student engagement. Besides, in a simulated dropout study, the HMM-suggested next-step learning activity is the only one that can achieve simultaneously a high contribution to student retention and a high probability of being performed by a student given their current engagement state, compared to the two other baselines, thereby providing strong motivation for future real-world evaluation and adoption.

2 Related Work

Interpreting Interactions; Illuminating Engagement. While students' engagement may be opaque, the engagement state (i.e., the underlying state which drives how a student interacts with course material) generally manifests itself as a mediating role in students' subsequent behaviour [3,15]. This behaviour is observable and may be directly modeled [3,12,18,21]. For instance, Herskovits et al. [8] proposed that diverse interaction patterns reflect high levels of engagement. To quantify this diversity, the authors used principal component analysis at regular time intervals to project student's interactions along the top three principal components. These interactions were then clustered into common trajectories. In a similar vein, Kizilcec et al. [12] adopted a simple but effective k-means clustering approach and showed that high-level prototypical engagement trajectories consistently emerged from three different computer science MOOCs. Then, the later work of Coleman et al. [1] rejected the curation of rigidly defined feature sets and instead used Latent Dirichlet Allocation (LDA) to discover behavioural patterns, or "topics", directly from students' interactions (in an unsupervised manner). By representing students as a mixture of these latent types, students may be characterised by a distribution over the set of course resources (termed as "use-cases"), e.g., a shopping use case (where students only access the free components of the course) vs. a disengaging use case (where students interaction gradually attenuated as the course progresses). The shortcoming of these prior approaches is that they treat student engagement patterns mostly as a static phenomenon, where user activity is represented as a "bag-of-interactions", which ignores the temporal dependency that is inherent to the data – a student may be, for example, actively engaged for the first week of a course, before transitioning into a state of disengagement in the following weeks. We argue that such a temporal factor could bring valuable insights about how and when a student first started to disengage, which subsequently may inform relevant engagement strategies to preempt their dropout.

Promoting Engagement; Preempting Dropout. While the combination of richly descriptive models of students' interactions with the improvement of learning outcomes (e.g., dropout) has been largely overlooked by the literature, a number of fields come close. For instance, the literature surrounding learning strategies (defined as "thoughts, behaviours, beliefs, or emotions that facilitate the acquisition, understanding, or later transfer of new knowledge and skills" [8]), identify common interaction patterns which are associated with learning outcomes [4,11,13,14,20]. While these studies generally found that students' learning strategies/trajectories were associated with their learning outcome, they did not explicitly model this relationship, so were unable to evaluate the extent to which different learning strategies are contributing to a dropout/non-dropout of a student. In a different vein, researchers have forgone the notion of student engagement and instead, attempted to generate a learning path (i.e., the learning materials/resources a student accesses at the next step in order to achieve a certain learning outcome) directly from the latest black-box ML models. For instance, Mu at al. [16] attempted to suggest learning resources based on a student's past learning experience using a XGBoost coupled with an explainer model. So the past activity with the highest feature importance (based on model explanation) and an undesirable learning outcome were diagnosed as important actions to be performed by students. Similarly, Zhou et al. [25] recommended learning resources based on students' expected future performance, which are predicted using a deep neural network-based model. Although achieving a high accuracy (up-to 80%) at recommending the correct resources, these approaches were limited from the perspective of a learner's dynamic engagement state, as the learning paths are separately modelled from how a student may actually interact with the course material given their current engagement state (e.g., recommending highly effective strategies to a "shopping" student who merely intended to browse the free component of the course is considered pointless). Besides, black-box models lack transparency and interpretability, which can make it challenging to gain educational insights from the results of these models and subsequently limit actual adoption in real-world educational scenarios. Driven by this, we adopted a fully transparent HMM model to analyse student engagement dynamics, which we then use for informing an engaging next-step learning activities based on a student's current engagement state. To our knowledge, our approach is the first to suggest actionable activities based on HMM.

3 Method

3.1 Dataset

To ensure a robust analysis on student engagement in MOOCs, we adopted two MOOCs datasets in the evaluation. The first dataset **Big data in Education** (denote as **Big data**) was a course offered by Columbia University in 2013 and delivered on the Coursera platform. In this course, students learned a variety of educational data-mining methods and applied them to research questions related to the design and improvement of interventions within educational software and systems. The course material consisted of lecture videos, formative in-video quizzes, and 8 weekly assignments. The second dataset **Code Yourself** was offered by the University of Edinburgh in 2015 on the Coursera platform. The course was designed to introduce teenagers to computer programming, while also covering basic topics in software engineering and computational thinking. The course materials consisted of lectures, videos, formative in-video quizzes, peer-reviewed programming projects, and 5 weekly assignments. Although the two courses had initial enrolments of over 45,000 and 59,900, respectively (as shown in the statistics Table 1), a large proportion of these students did not actively participate, and only 18,222 and 26,514 accessed at least a single item in the course; we restrict our study to these students. Of these active students, only 3,600 and 10,183 was active the entire course, and only about 2% of all students have successfully completed the course. This low completion rate can be attributed to a high rate of student attrition – common across MOOC environments – which, although significantly attenuated after the first weekly assignment, remained substantial throughout the course.

Table 1. The descriptive statistics of the two MOOCs datasets used in this study. The fractions within brackets indicate the percentage of No. enrolled.

Dataset	Duration	No. enrolled	No. dropout	No. active learners		
				At least once	Entire course	Completed
Big data	8 weeks	45,000	41,400 (92.00%)	18,222 (40.49%)	3,600 (8.00%)	638 (1.42%)
Code yourself	5 weeks	59,900	49,717 (83.00%)	26,514 (44.26%)	10,183 (17.00%)	1,591 (2.66%)

3.2 Modeling Student Engagement by HMM

Hidden Markov Model (HMM) is a well-established unsupervised machine learning approach that helps to make inferences about the latent variables (i.e., hidden states) through analyzing the observable actions [6,15]. Importantly, we opted for a variant known as sticky HMM [6]. The "sticky" assumption in the Markov model is that once a student has adopted a particular state, they persist in that state for as long as possible until a new state is required to describe their actions. Not only does this assumption represent many scenarios in real-world data, where states persist through time [6], but it also helps combat the unrealistically rapid switching dynamics that are present in models without this state-persistence bias [5]. The sticky HMM requires us to specify the number of states. However, to mitigate the impact this has on our model, we place a non-parametric prior over the state space [6]. The implementation uses a weak-limit approximation to the hierarchical Dirichlet process [24], which approximates the unbounded state space by a truncated representation with τ states, where we specify $\tau = 10$ [15]. The prior places a diminishing probability mass on infrequent states and thus focuses the majority of the probability mass on a relatively small number of major states in the model. For brevity, we refer to the sticky HMM as an HMM in this paper.

RQ 1: Understanding student engagement. In our engagement analysis, we used student interaction in MOOC as observable actions to model the student engagement state as hidden states. As we were interested in the temporal relation of student interaction, we represented students' interactions on a *weekly* basis. Specifically, we created a 10 element vector. The first three elements of the vector represented whether the student took actions for each type (i.e., assignment a, lecture l, quiz q) from the past week (denote as a_p, l_p and q_p respectively); the next three elements represented actions from the current week; the next three elements represented actions from the future week; and finally, the last element indicated whether the student interacted with any resources at all during the current week (denote as o). Therefore, during each week, a student G_i is represented by the 10 binary-element vector: $G_i = [a_p, l_p, q_p, a_c, l_c, q_c, a_f, l_f, q_f, o]$. By doing so, we can model temporal engagement and understand how a student engagement from the perspective of their interaction with the past, current, and future weeks.

RQ 2: Recommending learning path. Based on the HMM-identified student engagement states and their transitions, we further explored the use of these states and transitions to suggest important actions to be performed in future weeks in order for students to transition/persist in the engaging state. Specifically, at a given week w, based on a student's performed actions μ_w at w (e.g., accessing lecture slides from $w - 1$), we identified a student's current engagement state S_w by finding the state with the highest probability of emitting μ. Next, we utilised the state transition probability matrix T to calculate a favorable state at next week S_{w+1}, where S_{w+1} was the top-3 most likely state to be transitioned from S_w and had the highest probability to persist in engaging states (determined via a descriptive analysis shown in Sect. 4) in the following weeks. To illustrate this, we present an example in Fig. 1: suppose **Learner** started a 5-week course. After finishing week 1, based on their interactions with course material, we identify their current engagement state ① (i.e., the state with the highest probability of emitting μ). The aim is to decide which state should **Learner** proceed to next in order *to maximize their chance of staying engaged* with the course. Therefore, we identify the top-3 most probable transitions from ① based on T: ⑤ (which is on Path a), ③ (Path b) and ② (Path c) respectively. Since ⑤ is not an engaging state, we move on to Path b (on the left side of Fig. 1). To calculate the total probability of staying engaged on Path b (denote as E_b), we multiply all transition probabilities between engaging state pairs step-wise on the path, e.g., the first step being ③ $\rightarrow$ ① is determined by T(③,①). Then $E_b = T$(③,①) $\times T$(①,②) $\times T$(②,③), similarly, the engaging probability of $E_c = T$(②,③) $\times T$(③,①) $\times T$(①,②) (Path c shown on the right side of Fig. 1). Lastly, the higher probability between the E_b and E_c is selected as the engaging path, and learning activities are recommended based on activities with the highest emission probability from the state of first step e.g., if E_b has a higher engaging probability, then activities with the highest emission probability of state ③ are selected from the activity vector as instruction to students. Since the student may still not behave as instructed, we recalculate the path at each week to ensure the engaging path is up-to-date with the student current engagement state.

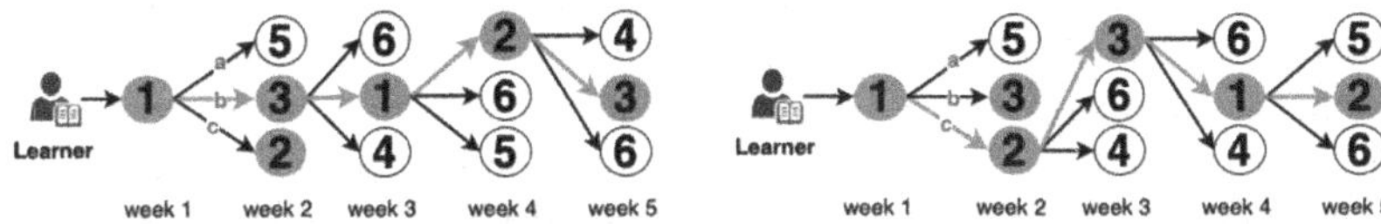

Fig. 1. An example illustration of a recommended learning path.

Evaluation Approach. We assessed the effectiveness of HMM-recommended next-step learning actions from the perspective of student dropout [3,9], i.e., by taking the learning activity in the recommended path, how likely would a student dropout in the following week. In line with previous studies [3,7], the *dropout* was a binary label and is set to 1 if a student takes no further actions in the following weeks, and set to 0 otherwise [3,17]. Given that the latest Deep Learning (DL) dropout predictor can achieve a high prediction accuracy (around 90% AUC [17]), we utilise the predictive power by building a LSTM model – one of the representative DL approaches widely adopted to predict dropout in MOOCs [16,17,25]. We input students' temporal activity representation (as detailed in Sect. 3.2) and their dropout labels to the LSTM model, which output whether a student would dropout on a weekly basis. Importantly, for each week's prediction, we changed the last week's activity to the HMM-recommended activity (i.e., assuming the student followed the recommendation) to assess the impact on a dropout prediction. For instance, suppose we were assessing the impact of HMM-recommended activity in week 3, the input to the LSTM model consisted of $[\mu_1, \mu_2, \mu_3]$, we replaced

μ_3 with HMM-recommended activity resulting in $[\mu_1, \mu_2, \mu_3']$, and performed subsequent dropout prediction. Although we could simply compare the predicted dropout label between the original $[\mu_1, \mu_2, \mu_3]$ and HMM-suggested $[\mu_1, \mu_2, \mu_3']$, to understand specific impact of μ_3', we adopted a widely-used black-box explainer model LIME [19] to examine the importance of μ_3' towards a dropout/non-dropout prediction. Given that LIME is more efficient than other black-box explainers (e.g., SHAP [23]), LIME can be used to explain a large number of student instances in MOOCs. LIME produce a feature importance score in the interval $[-100, 100]$, where 100 indicates a high importance to a dropout prediction, and -100 indicates high importance to a non-dropout prediction. To demonstrate the effectiveness of our approach, we included in total three baseline approaches, inspired from previous literature:

- `Random` samples from the learning activities of those who successfully completed the course randomly as the recommended learning action to a learner.
- `Diagnose Past Performance (DPP)` [16] utilised a XGBoost model to recommend missing learning activities from previous weeks that has the highest contribution (measured in SHAPLEY value) to a dropout label.
- `Future Performance Prediction (FPP)` [25] utilised a LSTM model to predict the optimal learning activities to be performed, which would result in the lowest dropout probability based on a learner's past activity.

3.3 Study Setup

Pre-processing. Given the goal of our study was to analyse student engagement with the MOOCs, we restricted our study to students who accessed the course material at least once, i.e., the *At least once* column in Table 1. In line with [15], the training and testing results of HMM and LSTM models are reported based on the 5-fold cross-validation. For all experiments, HMM and LSTM were trained using the same training set, and we ensured that the testing set was not used in the training phase. Given the high class imbalance ratio of dropout vs. non-dropout labels inherently in a MOOC course, which may cause the LSTM to over-classify the majority class (i.e., dropouts), we applied random under-sampling to ensure the training set of the LSTM model has an equal distribution of dropout and non-dropout samples.

Model Implementation. We implement HMM by Python package `bayesian-hmm`[1]. To train HMM, we used the Markov Chain Monte Carlo (MCMC) sampling estimation with an iteration of 2000 steps. The LSTM model was implemented using the Python package `keras`. In line with the prior dropout prediction literature [17], the model was composed by two LSTM layers of size 32 and 64 and a Dense layer (with sigmoid activation) with a hidden size of 1. During training, we used a batch size of 32 with an Adam optimiser and set the learning rate and training epoch parameters to 0.001 and 20, respectively. Lastly, LIME was implemented using the Python `LIME` package[2]. In response to the call for increased reproducibility in educational research [7], which benefit practitioners, we open-source code used in this study[3]

[1] https://pypi.org/project/bayesian-hmm/.
[2] https://github.com/marcotcr/lime.
[3] https://github.com/lsha49/hmm-engagement.

Evaluation Metrics. Given that the feature importance of HMM-suggested activity (to a non-dropout label) depends on the accuracy performance of the LSTM dropout prediction, we adopted two commonly adopted accuracy metrics that are robust against class imbalance: AUC and F1 scores.

4 Results

To investigate student engagement, we generated plots illustrating their parameters: state and state transition. For the sake of brevity, these plots are only displayed for the larger course under study (Big Data in Education). Although we observed similar results across both MOOCs datasets.

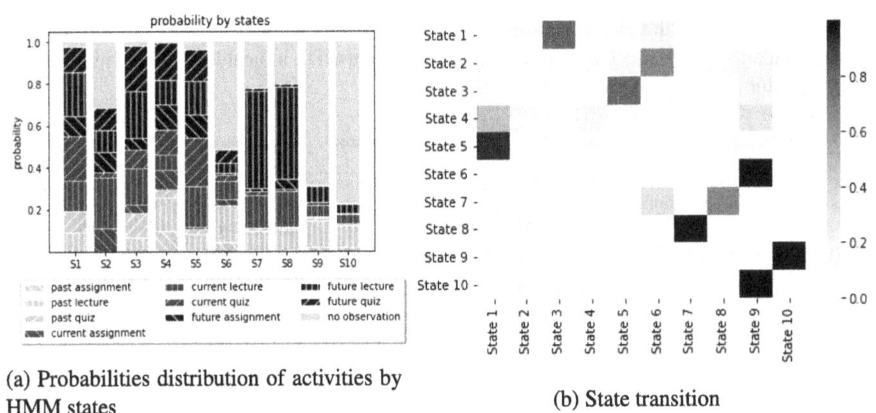

(a) Probabilities distribution of activities by HMM states

(b) State transition

Fig. 2. The HMM states and transition illustration of Big data in education.

RQ1: Understanding student engagement. To answer our first research question, we generated plots illustrating the hidden states (Fig. 2a) and transitions between different states (Fig. 2b) in Fig. 2. Among all states, students belonged to *state 1–5* generally tended to interact with the current week (denoted in blue) and the next week's (in dark blue) learning materials. As such, these states could also be given semantic labels in keeping with those found in the literature – "on-track" [12]. By contrast, *state 6–10* were characterised mostly by inactivity. For instance, *state 9–10* were mostly non-active (i.e., no observation as denoted in grey), while *state 7–8* were mostly only viewing lectures, thus may be categorised as "disengaged". Lastly, we noted that *state 6* was more active than *state 9–10*, and performed all the action types (as opposed to just viewing lectures). However, students in *state 6* tended to focus on past activities (denoted in light blue), compared to activities in the current or future weeks, so we assign it a "behind" label. It is important to note that, while some states appeared to have similar observable patterns, e.g., between the on-track *state 4* and *state 5* or between the disengaged *state 7* and *state 8*, their transitions may have been different, as detailed in Fig. 2b. For instance, while *state 5* almost exclusively transitioned to another on-track *state 1*, *state 4* may have transitioned to a quite different on-track *state 2* and a disengaged *state 9*, which highlighted the importance of closely monitoring the state transitions beyond the static patterns. Interestingly, students who were in a disengaged or behind state almost exclusively kept transitioning to disengaged states. This indicated that, once students started to disengage from the course material, they were unlikely to revert back to an engaging state and complete the course, which corroborated with previous findings [2,22]. In comparison, though the "on-track" *state 1–5* in general, tended to transition to another "on-track" state, they may have also transitioned to a disengaged state – especially *state 2* and *state 4* (except for *state 5*), indicating the importance of persisting students in an engaging state despite that they are already engaging.

Table 2. The LIME feature importance score (denoted as L_{score}) and transition probability (denoted as P_{path}) of the HMM-suggested path. The results in row `Actual` were calculated by using students' original activity. The signs ↑ and ↓ are used to indicate whether a higher (↑) or lower (↓) value was more preferred in a metric.

Dataset	Path	week 2		week 3		week 4		week 5		week 6		week 7	
		↓ L_{score}	↑ P_{path}	↓ L_{score}	↑ P_{path}	↓ L_{score}	↑ P_{path}	↓ L_{score}	↑ P_{path}	↓ L_{score}	↑ P_{path}	↓ L_{score}	↑ P_{path}
Big data	Actual	0.04	0.72%	0.11	**0.82%**	0.12	1.24%	−0.03	1.33%	−0.01	0.61%	0.01	**1.18%**
	Random	−0.93	0.01%	−0.54	0.75%	−2.76	1.69%	−5.76	1.56%	−0.03	1.23%	0.02	0.65%
	DPP	1.02	0.02%	−1.31	0.55%	−2.24	0.46%	1.35	0.28%	0.01	0.42%	**−0.05**	0.39%
	FPP	5.89	0.06%	**−6.51**	0.57%	0.74	0.47%	−0.12	0.28%	**−0.31**	0.45%	0.14	0.39%
	HMM	**−1.25**	**1.18%**	−1.6	0.76%	**−4.59**	**2.22%**	**−6.40**	**2.40%**	−0.07	**1.88%**	0.02	0.65%
Code yourself	Actual	−1.61	9.79%	−2.30	2.29%	6.03	**3.82%**	–		–		–	
	Random	4.66	4.51%	−4.43	0.91%	−2.36	0.81%	–		–		–	
	DPP	−5.43	**10.17%**	3.64	2.29%	3.65	3.82%	–		–		–	
	FPP	**−10.28**	2.68%	**−15.50**	2.30%	−2.22	3.82%	–		–		–	
	HMM	2.85	6.25%	−3.89	**2.71%**	**−2.64**	2.29%	–		–		–	

RQ2: Promoting students onto engaging paths. Given the above findings, we further explored whether the proposed HMM-based approach could be used to promote student engagement and preempt dropout. To this end, we utilised HMM to generate engaging paths and test the feature importance score of the path via a dropout predictor (as detailed in Sect. 3.2). Given that the validity of this approach relied on using an accurate dropout predictor, we first evaluated the accuracy of the LSTM model (measured in F1 score and AUC). Overall, we observed that the model achieved a high F1 (0.85–0.91) and AUC (0.82–0.92) in weeks 3–7 for Big data and a high F1 and AUC (0.80) in week 3 for Code yourself, and thus could serve as a strong basis to evaluate the generated path. (We have included a complete report of F1 and AUC results in digital appendix[4]) Given that the rest of the weeks achieved lower F1 and AUC (below 0.80), we restricted our analysis of the feature importance score in weeks 3–7 for Big data and week 3 for Code yourself. We summarise the LIME importance score towards a dropout/non-dropout prediction in Table 2. To take into account the scenario where the model recommended a path consisting of highly engaging learning activities, but were difficult to follow (e.g., recommending students to perform all activities), we included a

[4] https://docs.google.com/document/d/1hntPaZ1NfSJF4t2I9x83CKtAt4C0jRB9Ma8OfdYj2_Q/edit?usp=sharing.

measure of the probability of students performing the recommended activity on a path (i.e., P_{path}), based on the HMM transition matrix (detailed in Sect. 3.2). Overall, we observed that, for both datasets, FPP achieved the best performance in terms of LIME score (L_{score}) that contributed to a non-dropout (3 out of 6 in weeks 3–7 for Big data and week 3 for Code yourself), closely followed by the proposed HMM approach (2 out of 6). Though, HMM achieved the highest path transition probability P_{path} (4 out of 6), even higher than the Actual activity (i.e., student's original action), while none of the other baseline approaches managed to surpass Actual. In particular, while FPP generated activity performed the best in terms of LIME score, it had the lowest or the second lowest path probability in 5 out of 6 instances (i.e., week 3–7 in Big data). This indicated that, FPP tended to recommend highly engaging learning activities that were unlikely to be performed by the learners. In comparison, HMM-recommended activities had a more balanced performance between the LIME score (top-2) and transition probability (top-1), indicating that the recommended engaging paths were more likely to be adopted by learners (compared to other path-generating strategies) and as such are more likely to preempt students' dropout.

5 Discussion and Conclusion

This paper investigated the effectiveness of a HMM-based approach to analyse student temporal engagement patterns. Compared to previous approaches, HMM was shown to be able to not only capture students' static engagement patterns (represented as "state"), but also the transitions between different states. So, we were able to uncover insights about the temporal dynamics about student engagement. In particular, we found that while disengaged students were unlikely to transition to an engaging state, the engaging students may either stayed engaged or became disengaged in future weeks. This highlighted the importance of persisting students' engagement state in order to preempt their dropout. Driven by this, we further explored the use of the student engagement state and their transitions to inform a learning path that can promote students' engagement. Through extensive experimentation, we demonstrated that, not only our HMM-suggested learning activities were effective at preempting student dropout, but these activities were also more likely to be performed by students given their current engagement state (compared to other baseline approaches). These results provide a strong motivation for adoption of the proposed HMM approach in a real-world MOOC setting.

Implications. Firstly, our evaluations demonstrated the effectiveness of HMM as a fully transparent and interpretable approach for modeling student-temporal interactions and identifying hidden patterns and their transitions. All this provides useful insights about students' current engagement states (e.g., whether they are on-track for completion) and future engagement trends (e.g., whether they are likely to stay engaged in the future with the course material), making it a valuable tool for educators and researchers potentially beyond a MOOC setting (e.g., to model student interactions in a hybrid learning environment). Secondly, given that students who have already been disengaged are almost never going to transition back to an engaging state in a MOOC setting, we posit that providing guidance/intervention at the time of dropout or immediately before dropout (by utilising predictive models) may be futile and laborious. Instead, educators and practitioners may repurpose such efforts onto persisting the engagement of those "on-track" students and preempt their dropout. Thirdly, we showed that HMM-generated learning path is the only approach that can achieve simultaneously a high contribution to non-dropout prediction and a high probability of being performed by a student given their current engagement state. This implies that students may be inclined to follow the suggested activity and stay engaged with minimal human intervention. Given the low teacher-student ratio in MOOCs settings [10], the HMM-based approach may be implemented to persist student engagement and support teaching at scale.

Limitations. We acknowledge the following limitations of our study. First, the analysis involved only two MOOCs datasets, which enabled us to model student engagement using three different student interaction types (i.e., assignment, lecture, and quiz). To further increase the generalizability of our findings, we plan on adopting additional datasets, potentially beyond MOOC settings, and with more interactions (e.g., forum activity) so as to further explore HMM's capability in modeling student engagement. Second, we have not conducted a human evaluation on the real-world impact of HMM-suggested engagement paths. Given the promising findings, for future work, we plan on implementing the proposed approach in a real-world MOOC setting and evaluate the effectiveness of preempting student dropouts.

References

1. Coleman, C.A., Seaton, D.T., Chuang, I.: Probabilistic use cases: discovering behavioral patterns for predicting certification. In: Proceedings of the Second ACM Conference on Learning@ Scale, pp. 141–148 (2015)
2. Deng, R., Benckendorff, P., Gannaway, D.: Learner engagement in MOOCs: scale development and validation. Br. J. Edu. Technol. **51**(1), 245–262 (2020)
3. Fei, M., Yeung, D.Y.: Temporal models for predicting student dropout in massive open online courses. In: 2015 IEEE International Conference on Data Mining Workshop (ICDMW), pp. 256–263. IEEE (2015)
4. Fincham, E., Gašević, D., Jovanović, J., Pardo, A.: From study tactics to learning strategies: an analytical method for extracting interpretable representations. IEEE Trans. Learn. Technol. **12**(1), 59–72 (2018)
5. Fox, E.B., Sudderth, E.B., Jordan, M.I., Willsky, A.S.: An HDP-HMM for systems with state persistence. In: Proceedings of the 25th International Conference on Machine Learning, pp. 312–319 (2008)
6. Fox, E.B., Sudderth, E.B., Jordan, M.I., Willsky, A.S.: A sticky HDP-HMM with application to speaker diarization. Ann. Appl. Stat. 1020–1056 (2011)
7. Gardner, J., Yang, Y., Baker, R.S., Brooks, C.: Modeling and experimental design for MOOC dropout prediction: a replication perspective. Int. Educ. Data Min. Soc. (2019)
8. Hershcovits, H., Vilenchik, D., Gal, K.: Modeling engagement in self-directed learning systems using principal component analysis. IEEE Trans. Learn. Technol. **13**(1), 164–171 (2019)
9. Jiang, S., Williams, A., Schenke, K., Warschauer, M., O'dowd, D.: Predicting MOOC performance with week 1 behavior. In: Educational Data Mining 2014 (2014)
10. Joksimović, S., et al.: How do we model learning at scale? A systematic review of research on MOOCs. Rev. Educ. Res. **88**(1), 43–86 (2018)

11. Jovanović, J., Gašević, D., Dawson, S., Pardo, A., Mirriahi, N., et al.: Learning analytics to unveil learning strategies in a flipped classroom. Internet High. Educ. **33**(4), 74–85 (2017)
12. Kizilcec, R.F., Piech, C., Schneider, E.: Deconstructing disengagement: analyzing learner subpopulations in massive open online courses. In: Proceedings of the Third International Conference on Learning Analytics and Knowledge, pp. 170–179 (2013)
13. Lin, J., Lang, D., Xie, H., Gašević, D., Chen, G.: Investigating the role of politeness in human-human online tutoring. In: Bittencourt, I.I., Cukurova, M., Muldner, K., Luckin, R., Millán, E. (eds.) AIED 2020. LNCS (LNAI), vol. 12164, pp. 174–179. Springer, Cham (2020). https://doi.org/10.1007/978-3-030-52240-7_32
14. Lin, J., et al.: Is it a good move? Mining effective tutoring strategies from human-human tutorial dialogues. Futur. Gener. Comput. Syst. **127**, 194–207 (2022)
15. Mogavi, R.H., Ma, X., Hui, P.: Characterizing student engagement moods for dropout prediction in question pool websites. arXiv preprint: arXiv:2102.00423 (2021)
16. Mu, T., Jetten, A., Brunskill, E.: Towards suggesting actionable interventions for wheel-spinning students. Int. Educ. Data Min. Soc. (2020)
17. Prenkaj, B., Velardi, P., Stilo, G., Distante, D., Faralli, S.: A survey of machine learning approaches for student dropout prediction in online courses. ACM Comput. Surv. (CSUR) **53**(3), 1–34 (2020)
18. Raković, M., et al.: Towards the automated evaluation of legal casenote essays. In: Rodrigo, M.M., Matsuda, N., Cristea, A.I., Dimitrova, V. (eds.) AIED 2022. LNCS, vol. 13355, pp. 167–179. Springer, Cham (2022). https://doi.org/10.1007/978-3-031-11644-5_14
19. Ribeiro, M.T., Singh, S., Guestrin, C.: "Why should I trust you?" explaining the predictions of any classifier. In: Proceedings of the 22nd ACM SIGKDD International Conference on Knowledge Discovery and Data Mining, pp. 1135–1144 (2016)
20. Saqr, M., López-Pernas, S., Jovanović, J., Gašević, D.: Intense, turbulent, or wallowing in the mire: a longitudinal study of cross-course online tactics, strategies, and trajectories. Internet High. Educ. 100902 (2022)
21. Sha, L., et al.: Assessing algorithmic fairness in automatic classifiers of educational forum posts. In: Roll, I., McNamara, D., Sosnovsky, S., Luckin, R., Dimitrova, V. (eds.) AIED 2021. LNCS (LNAI), vol. 12748, pp. 381–394. Springer, Cham (2021). https://doi.org/10.1007/978-3-030-78292-4_31
22. Sinclair, J., Kalvala, S.: Student engagement in massive open online courses. Int. J. Learn. Technol. **11**(3), 218–237 (2016)
23. Štrumbelj, E., Kononenko, I.: Explaining prediction models and individual predictions with feature contributions. Knowl. Inf. Syst. **41**(3), 647–665 (2014)
24. Teh, Y., Jordan, M., Beal, M., Blei, D.: Sharing clusters among related groups: hierarchical Dirichlet processes. In: NeurIPS, vol. 17 (2004)
25. Zhou, Y., Huang, C., Hu, Q., Zhu, J., Tang, Y.: Personalized learning full-path recommendation model based on LSTM neural networks. Inf. Sci. **444**, 135–152 (2018)

Does Informativeness Matter? Active Learning for Educational Dialogue Act Classification

Wei Tan[1], Jionghao Lin[1,2(✉)], David Lang[3], Guanliang Chen[1], Dragan Gašević[1], Lan Du[1], and Wray Buntine[1,4]

[1] Monash University, Clayton, Australia
{wei.tan2,jionghao.lin1,guanliang.chen,dragan.gasevic,lan.du}@monash.edu
[2] Carnegie Mellon University, Pittsburgh, USA
[3] Stanford University, Stanford, USA
dnlang86@stanford.edu
[4] VinUniversity, Hanoi, Vietnam
wray.b@vinuni.edu.vn

Abstract. Dialogue Acts (DAs) can be used to explain what expert tutors do and what students know during the tutoring process. Most empirical studies adopt the random sampling method to obtain sentence samples for manual annotation of DAs, which are then used to train DA classifiers. However, these studies have paid little attention to sample informativeness, which can reflect the information quantity of the selected samples and inform the extent to which a classifier can learn patterns. Notably, the informativeness level may vary among the samples and the classifier might only need a small amount of low informative samples to learn the patterns. Random sampling may overlook sample informativeness, which consumes human labelling costs and contributes less to training the classifiers. As an alternative, researchers suggest employing statistical sampling methods of Active Learning (AL) to identify the informative samples for training the classifiers. However, the use of AL methods in educational DA classification tasks is under-explored. In this paper, we examine the informativeness of annotated sentence samples. Then, the study investigates how the AL methods can select informative samples to support DA classifiers in the AL sampling process. The results reveal that most annotated sentences present low informativeness in the training dataset and the patterns of these sentences can be easily captured by the DA classifier. We also demonstrate how AL methods can reduce the cost of manual annotation in the AL sampling process.

Keywords: Informativeness · Active Learning · Dialogue Act Classification · Large Language Models · Intelligent Tutoring Systems

1 Introduction

Traditional one-on-one tutoring involves human participants (*e.g.*, a course tutor and a student), which has been widely acknowledged as an effective form of

N. Wang et al. (Eds.): AIED 2023, LNAI 13916, pp. 176–188, 2023.
https://doi.org/10.1007/978-3-031-36272-9_15

instruction [16,24]. Understanding what expert tutors do and students know during the tutoring process is a significant research topic in the field of artificial intelligence in education [5], which can contribute to the design of dialogue-based intelligent tutoring systems (*e.g.,* AutoTutor [14]) and the practice of human tutoring [12,24]. A typical method used to understand the tutoring process is to map the sentences from tutors and students onto dialogue acts (DAs) which can manifest the intention behind sentences [16,24]. For example, a tutor's sentence (*e.g., "No, it is incorrect!"*) can be coded as the `Negative Feedback` DA. The mapping process often needs a pre-defined DA scheme developed by educational experts, and the DA scheme is used to annotate the sentences with the DAs [16,17]. However, manually annotating the DAs for millions of tutoring dialogues is impractical for conducting research related to educational dialogue since the annotating process is often time-consuming and costly [15]. Therefore, researchers typically annotate a small amount of sentences from tutoring dialogues and further use the annotated sentences to train machine learning classifiers to automate the annotation process for DAs [12,16].

Though previous works demonstrated the potential of automating the DA classification [3,6,12,15,17], little attention has been given to investigate the extent to which the DA classifiers can learn the patterns from the annotated sentences, which is important to build a robust classifier to facilitate tutoring dialogue analysis. The ability of the DA classifiers that learn the relation between the model inputs (*e.g.,* sentences) and outputs (*e.g.,* DAs) is defined as learnability [10,21]. To improve the learnability of a DA classifier, it is necessary to train the DA classifier on the instances that can help the classifier capture the underlying patterns and further improve the generalization of the DA classifier on the full dataset [10]. These instances are considered highly informative samples in the dataset [10]. It should be noted that the informativeness level might be different among the samples, and the classifier can learn the patterns of the low informative samples with a small amount of training data [10,21]. Therefore, annotating more high informative samples and less low informative samples could further save human annotation costs. However, few works investigated sample informativeness on the task of educational DA classification, which motivated us to focus on the **first research goal**, *i.e., investigate the sample informativeness levels of annotated DAs.* Additionally, the previous works annotated the sentences by randomly sampling from the archived tutoring dialogue corpora [12,15–17], which might not provide sufficient high informative samples for training the DA classifier. We argue that blindly annotating the dialogue sentences to train the DA classifier might consume excessive human annotation costs and not enhance the classifier's learnability on the full dataset [19]. Instead, a promising solution is to use the statistical active learning (AL) methods, which aim to select highly informative samples to train the classifier [19]. To provide details about the efficacy of AL methods, the **second research goal** of the current study was to *investigate the extent to which statistical AL methods support the DA classifier training on the DAs with different informativeness levels.*

Our results showed that most annotated instances from the training samples presented low informativeness, which indicated that the patterns of these samples could be sufficiently captured by our DA classifier. Additionally, the DA classifier needs to train on more high informativeness samples to improve the classification performance on the unseen dataset, which further requires support by statistical AL methods. We compared the state-of-the-art AL method (*i.e.*, CoreMSE [22]) with other commonly-used AL methods (*i.e.*, Least Confidence and Maximum Entropy) and random baseline. We found that the CoreMSE AL method could select more high informative instances for the DA classifier at the early stage of the classifier training process and gradually reduced the selection of the low informative instances, which could alleviate the cost of manual annotation.

2 Related Work

2.1 Educational Dialogue Act Classification

To automate classifying educational DAs, previous studies in the educational domain have employed machine learning models to train on annotated sentences [1,12,15,17,18]. For example, Boyer *et al.* [1] employed a Logistic Regression model to train on linguistic features (*e.g.*, N-grams) from the annotated sentences in 48 programming tutoring sessions. They [1] achieved 63% accuracy on classifying thirteen DAs. Later on, Samei *et al.* [18] used Decision Trees and Naive Bayes models to train on linguistic features and contextual features (*e.g.*, the speaker's role of previous sentences) from 210 annotated tutoring sentences on the science-related topic. Their work [18] achieved the accuracy of 56% in classifying seven educational DAs. These previous works trained DA classifiers by using the sentences which were randomly sampled from the larger tutoring dialogue corpora. Though these works demonstrated the feasibility of automating DA classification [1,12,15,17,18], it still remains largely unknown whether the DA classifier learned the patterns sufficiently from the randomly selected sentences, which is important for building robust and reliable DA classifiers.

2.2 Sample Informativeness

Training a DA classifier on a sufficient amount of annotated sentences can help the classifier achieve satisfactory classification performance. However, manual annotation is time-consuming and expensive [12,15,17]. To mitigate the human annotation cost and also help the classifier achieve satisfied classification performance, Du *et al.* [4] suggested annotating the most informative samples, which can reduce the generalization error and uncertainty of the DA classifier on the unseen data. A recent work [21] proposed the *Data Maps* framework which used **Confidence**, **Variability**, and **Correctness** to measure the informativeness. **Confidence** denotes the averaged probability of the classifier's prediction on the correct labels across all training epochs; **Variability** denotes the spread of probability across the training epochs, which capture the uncertainty of the classifier; **Correctness** denotes the fraction of the classifier correctly predicted labels over all training epochs [21]. Building upon the work introduced in [21],

Karamcheti *et al.* [10] further used the **Correctness** to categorize the samples into four groups: *Easy*, *Medium*, *Hard*, and *Impossible* [21] and these groups indicated the extent to which the classifier can learn the patterns from the annotated instances. Inspired by [4,10,21], we argue that when scrutinizing the sampling process for training the DA classifier, if samples are randomly selected, there will be redundant for overly sampling the *Easy* samples and insufficient for sampling high informative samples, which might consume the human annotation budget and lead to poor generalizability of the DA classifier. As a remedy, it is important to select the most suitable samples for training the classifier. AL can offer promising methods to select the most suitable samples.

2.3 Statistical Active Learning

Recent studies on educational DA classification [12,15] have agreed that the high demand for the annotated dataset was still an issue for the DA classification task. To alleviate this issue, a promising solution is to use AL methods which can select the high informative samples from the unlabeled pool and send them to Oracle (*e.g.*, human annotator) for annotation [19]. Traditionally, there are three typical scenarios of AL methods: 1) *membership query synthesis*, which focuses on generating artificial data-point for annotation rather than sampling the data-point from the real-world data distribution, 2) *stream-based sampling*, which focuses on scanning through a sequential stream of non-annotated data-points and make sampling query decision individually, and 3) *pool-based sampling* which focuses on selecting the most informative samples from the non-annotated data pool and send them to the oracle for annotation [19].

As the annotated DAs were available in our study and the dataset was not collected in a sequential stream, we consider our study fits well with the *pool-based sampling* scenario. The pool-based AL methods can both reduce the computational cost of model training and maintain the performance of the model trained on the annotated dataset [19]. Many studies employed the pool-based AL methods (*e.g.*, Least Confidence [8,20] and Maximum Entropy [11]) on the educational tasks (*e.g.*, student essay classification [8], and educational forum post classification [20]) and their results have demonstrated the promise of AL methods on alleviating the demand for annotated datasets. However, it still remains largely unknown about the extent to which AL methods can select the informative samples to support the automatic classification of educational DA.

3 Methods

3.1 Dataset

The current study obtained ethics approval from the Monash University Human Research Ethics Committee under application number 26156. The dataset used in our study was provided by an educational technology company that operated online tutoring services and collected the data from tutors and students along with informed consent allowing the use of the de-identified dataset for research. The dataset contained detailed records of the tutoring process where tutors and students collaboratively solve various problems (the subjects including mathematics, physics, and chemistry) via textual message. In total, our dataset contained 3,626 utterances (2,156 tutor utterances and 1,470 student utterances) from 50 tutorial dialogue sessions. The average number of utterances per tutorial dialogue session was 72.52 (min = 11, max = 325); tutors averaged 43.12 utterances per session (min = 5, max = 183), and students 29.40 utterances per session (min = 4, max = 142). We provided a sample dialogue in the digital appendix via https://github.com/jionghaolin/INFO.

3.2 Educational Dialogue Act Scheme and Annotation

Identifying the DAs in tutorial dialogues often relies on a pre-defined educational DA coding scheme [9]. By examining the existing literature, we employed the DA scheme introduced in [24] whose effectiveness in analyzing online one-on-one tutoring has been documented in many previous studies (*e.g.*, [7,12,23]). The DA scheme developed in [24] characterizes the DAs into a two-level structure. To discover more fine-grained information from tutor-student dialogue, in our study, we decided to annotate the tutoring dialogues by using the second-level DA scheme. Notably, some utterances in dialogues contained multiple sentences, and different sentences can indicate different DAs. To address this concern, Vail and Boyer [24] suggested partitioning the utterances into multiple sentences and annotating each sentence with a DA. After the utterance partition, we then removed the sentences which only presented meaningless symbols or emoji. Lastly, we recruited two human coders to annotate DAs, and we obtained Cohen's κ score of 0.77 for the annotation. The annotations achieved a substantial agreement between the two coders, and we recruited a third educational expert to resolve the inconsistent cases. The full DA scheme can be found in a digital appendix via https://github.com/jionghaolin/INFO, which contains 31 DAs. Due to the space limit, we only presented students' DAs in Table 1.

Table 1. The DA scheme for annotating student DAs. The DAs were sorted based on their frequency (*i.e.*, the column of **Freq.**) in the annotated dataset.

Dialogue Acts (DAs)	Sample Sentences	Freq.
Confirmation Question	*"So that'd be 5?"*	4.93%
Request Feedback by Image	*[Image]*	4.34%
Understanding	*"Oh, I get it"*	1.46%
Direction Question	*"Okay what do we do next?"*	1.20%
Information Question	*"Isn't there a formula to find the nth term?"*	1.06%
Not Understanding	*"I don't know."*	0.24%
Ready Answer	*"Yep, ready to go."*	0.07%

3.3 Identifying Sample Informativeness via Data Maps

To identify the informativeness of annotated instances, we need to train the annotated dataset on a classifier. Building upon our previous work [13], we used ELECTRA [2] as the backbone model for classifying 31 DAs, which is effective in capturing nuanced relationships between sentences and the DAs. The dataset (50 dialogue sessions) was randomly split to *training* (40 sessions) and *testing set* (10 sessions) in the ratio of *80%:20%* for training the classifier. The classifier achieved accuracy of 0.77 and F1 score of 0.76 on the testing set. Then, we applied the *Data Maps*[1] to the DA classifier to analyze the behaviour of the classifier on learning individual instance during the training process. Following the notation of *Data Maps* in [21], the training dataset denotes $Dataset = \{(x, y^*)_i\}_{i=1}^{N}$ across E epochs where the N denotes the size of the dataset, ith instance is composed of the pair of the observation of x_i and true label y_i^* in the dataset. **Confidence** ($\hat{\mu_i}$) was calculated by $\hat{\mu_i} = \frac{1}{E}\sum_{e=1}^{E} p_{\theta^{(e)}}(y_i^*|x_i)$ where $p_{\theta^{(e)}}$ is the model's probability with the classifier's parameters $\theta^{(e)}$ at the end of the eth epoch. **Variability** ($\hat{\sigma_i}$) was calculated by the variance of $p_{\theta^{(e)}}(y_i^*|x_i)$ across epochs: $\hat{\sigma_i} = \sqrt{\frac{\sum_{e=1}^{E}(p_{\theta^{(e)}}(y_i^*|x_i) - \hat{\mu_i})^2}{E}}$. **Correctness** ($Cor$) was the fraction of the classifier correctly annotated instance x_i over all epochs. A recent study [10] further categorized **Correctness** into *Easy* ($Cor \geq 0.75$), *Medium* ($0.75 > Cor \geq 0.5$), *Hard* ($0.5 > Cor \geq 0.25$), and *Impossible* ($0.25 > Cor \geq 0$). In line with the studies [10,21], we mapped the training instances along two axes: the $y-axis$ indicates the confidence and the $x-axis$ indicates the variability of the samples. The colour of instances indicates **Correctness**.

3.4 Active Learning Selection Strategies

We aimed to adopt a set of AL methods to examine the extent to which AL methods support the DA classifier training on the DAs with different informativeness levels. We employed the state-of-the-art AL method (CoreMSE [22]) to compare with two commonly used AL methods (*i.e.*, Least Confidence [19] and Maximum Entropy [25]) and the random baseline. Following the algorithms in the papers, we re-implemented them in our study. **Random Baseline** is a sampling strategy that samples the number of instances uniformly at random from the unlabeled pool [19]. **Maximum Entropy** is an uncertainty-based method that chooses the instances with the highest entropy scores of the predictive distribution from the unlabeled pool [25]. **Least Confidence** is another uncertainty-based method that chooses the instances with the least confidence scores of the predictive distribution from the unlabeled pool [19]. **CoreMSE** is a pool-based AL method proposed by [22]. The method involves both diversity and uncertainty measures via the sampling strategy. It selects the diverse samples with the highest uncertainty scores from the unlabeled pool, and the uncertainty scores are estimated by the reduction in classification error. These diverse samples can provide more information to train the model to achieve high performance.

[1] https://github.com/allenai/cartography.

3.5 Study Setup

In the AL training process, the DA classifier was initially fed with 50 training samples randomly selected from the annotated dataset. For each AL method, the training process was repeated six times using different random seeds. A batch size of 50 was specified. We set the maximum sequence length to 128 and fine-tuned it for 30 epochs. The AdamW optimizer was used with the learning rate of 2e-5 to optimize the training of the classifier. All experiments were implemented on RTX 3090 and Intel Core i9 CPU processors.

4 Results

4.1 Estimation of Sample Informativeness

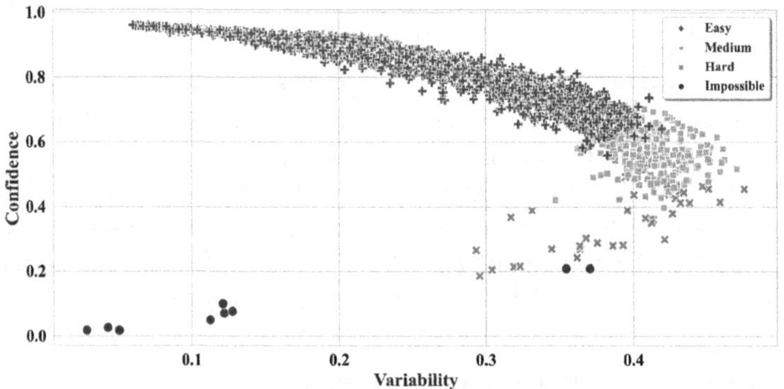

Fig. 1. The sample informativeness on the annotated sentences.

To estimate the sample informativeness level of each training instance, we employed the *Data Maps*. In Fig. 1, the y-axis represents the **Confidence** measure, the x-axis represents **Variability**, and the colours of instances represent **Correctness**, which are detailed in Sect. 3.3. The samples in the *Easy* group (*i.e.*, blue scatters) always presented a high confidence level, which indicates that most sample points could be easily classified by our DA classier. Then, the samples in the *Medium* and *Hard* groups presented generally lower confidence and higher variability compared with the *Easy* group, which indicates that the samples in *Medium* and *Hard* were more informative than those in *Easy*. Lastly, the samples in the *Impossible* group always presented low confidence and variability. According to [21], the samples in *Impossible* could be the reasons for mis-annotation or insufficient training samples (*e.g.*, insufficient training samples for a DA can impede the classifier's ability to capture its classification pattern) for the DA classifier. Most training samples were considered *Easy* to be classified, and only a small number of samples are *Impossible*. The distribution of *Easy* and *Impossible* points in Fig. 1 indicates that our dataset presented high-quality annotation, which is generally considered learnable for the DA classifier.

We further investigated the distribution of informativeness level for the annotated DAs[2]. As described in Sect. 3.2, we mainly present the distribution of the informativeness level for each DA in Fig. 2. We sorted the DAs based on their frequency in Table 1. We observed that most samples in the `Confirmation Question` and `Request Feedback by Image` DAs were in the *Easy* group, which indicates that both DAs are easy for the DA classifier to learn. Then, in the middle of Fig. 2, the `Understanding`, `Direction Question`, and `Information Question` DAs had roughly 1% frequency in Table 1. Compared with the first two DAs in Fig. 2, the middle three DAs had a higher percentage of *Medium*, which indicates that the DA classifier might be less confident in classifying these DAs. Lastly, the DAs `Not Understanding` and `Ready Answer` had the frequency lowered than 1% in Table 1. The samples in `Not Understanding` and `Ready Answer` were considered *Hard* and *Impossible* to be classified, respectively. The reasons might be that the annotated samples of `Not Understanding` and `Ready Answer` in our dataset were insufficient for the DA classifier to learn the patterns.

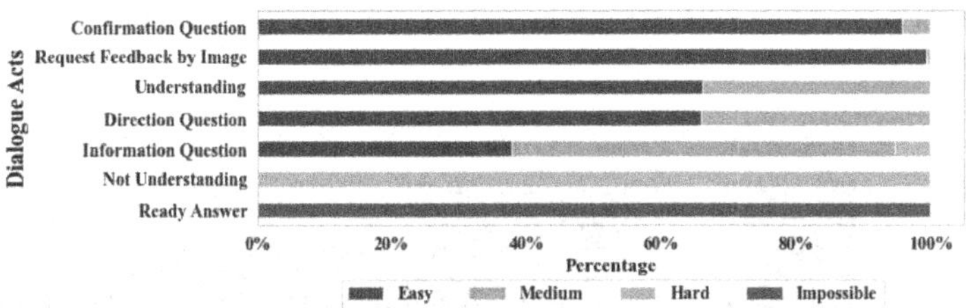

Fig. 2. Distribution of informativeness level for each DA (Student only)

4.2 Efficacy of Statistical Active Learning Methods

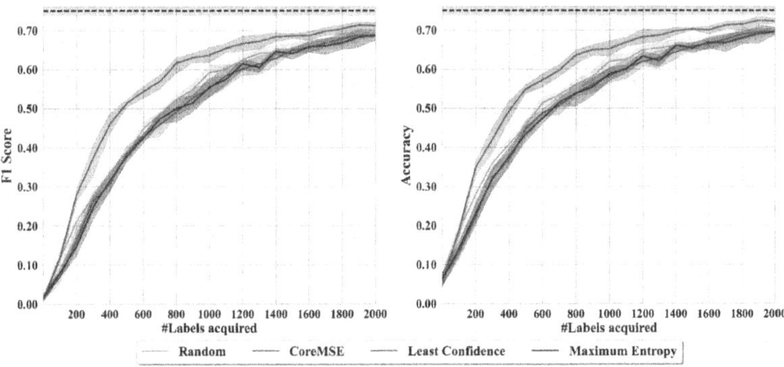

Fig. 3. Learning Curve of AL methods. The sampling batch for each AL method is 50. The classification performance was measured by F1 score and accuracy.

[2] Due to the space limit, we only present a part of the analysis results, and the full results can be accessed at https://github.com/jionghaolin/INFO.

To investigate how AL methods select different informative samples for training the DA classifier, we first evaluated the overall performance of the DA classifier with the support of AL methods. In Fig. 3, the x-axis represents the training sample size after selecting samples and the y-axis represents the classification performance measured by F1 score and classification accuracy. We compared the CoreMSE method to the baseline methods, *i.e.*, Maximum Entropy (ME), Least Confidence (LC), and Random. The results in Fig. 3 demonstrate the learning curves of the models where the CoreMSE method could help the DA classifier achieve better performance with fewer training samples compared with the baseline methods. For example, when acquiring 600 samples from the annotated data pool, the CoreMSE method could achieve roughly the F1 score of 0.55, which was equivalent to the classifier performance training on 900 samples with the use of baseline methods. It indicates that the CoreMSE method could save 30% human annotation costs compared to the baseline methods. Whereas the efficacy of LC and ME methods was similar to that of the random baseline in both F1 score and accuracy value; this indicates that the traditional uncertainty-based AL methods (*i.e.*, LC and ME) might not be effective on our classification task.

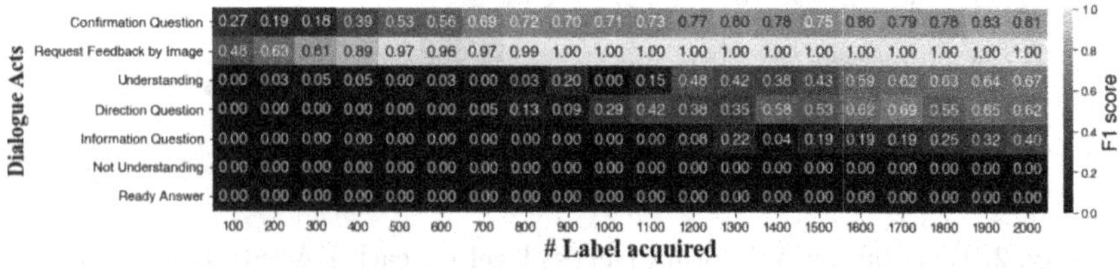

Fig. 4. F1 score for the dialogue acts only specific to students (**Random**)

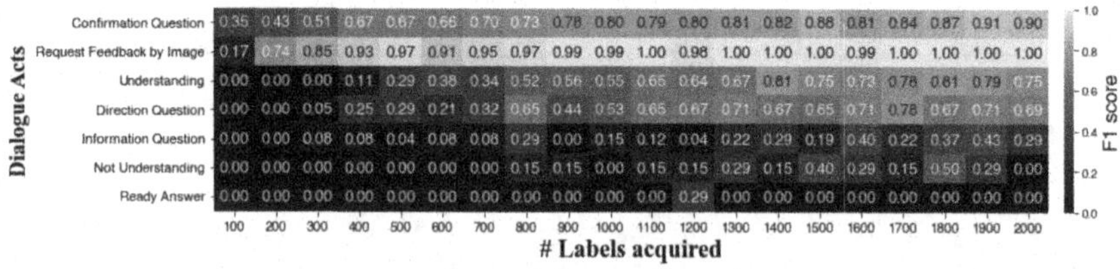

Fig. 5. F1 score for the dialogue acts only specific to students (**CoreMSE**)

Based on the results in Fig. 3, we observed that the random baseline performed slightly better than the ME and LC methods. In the further analysis of each DA, we decided to compare the efficacy between CoreMSE and the random baseline[3]. As shown in Fig. 4 and Fig. 5, we sorted DAs based on their frequency in the dataset. Figure 4 shows the changes of F1 scores for each DA with the use

[3] Due to the space limit, we documented our full results in a digital appendix, which is accessible via https://github.com/jionghaolin/INFO.

of the random baseline; this indicates that the DA classifier performed quite well for classifying the `Confirmation Question` and `Request Feedback by Image` DAs as the training sample size increased. Regarding the `Understanding` and `Direction Question`, the DA classifier needed more than 1,600 annotated samples to achieve a decent performance. For the DAs `Information Question`, `Not Understanding`, and `Ready Answer`, the DA classifier almost failed to learn the patterns from the samples selected by the Random method. In comparison, Fig. 5 shows the classification performance for each DA with the support of the CoreMSE method. The results indicate that CoreMSE could support the DA classifier to make accurate predictions for the DAs (e.g., `Understanding`) when acquiring fewer annotation samples than the random baseline. Then, though the classification performance for the DAs `Information Question` and `Not Understanding` was not sufficient, CoreMSE demonstrated the potential to improve the accuracy for both DAs as more labels were acquired. Lastly, both Random and CoreMSE methods failed to support the DA classifier to make an accurate prediction on the `Ready Answer` DA. The reason might be the fact that the `Ready Answer` DA was rare in our annotated dataset.

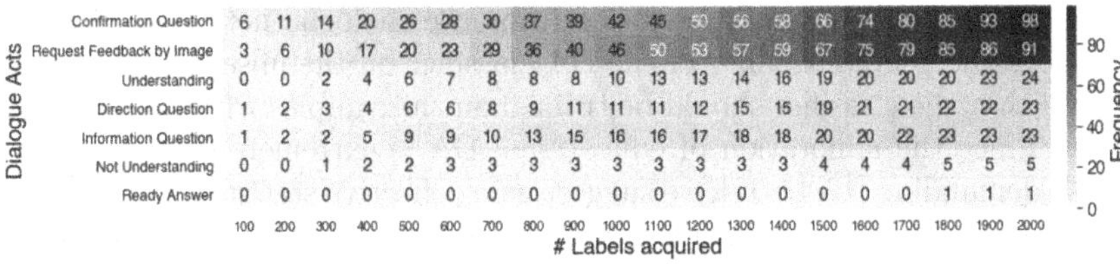

Fig. 6. The distribution of sampling frequency for each dialogue acts (**Random**)

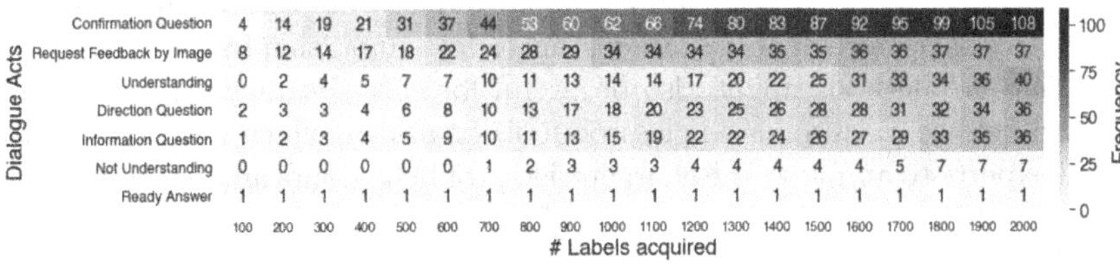

Fig. 7. The distribution of sampling frequency for each dialogue acts (**CoreMSE**)

Next, we investigated the sampling process between the CoreMSE and random methods to learn more details about how the CoreMSE method saved the annotation budget. For each sampling batch, we counted the cumulative frequency for each DA in Figs. 6 and 7. For example, in the first 200 acquired annotations (Fig. 6), the random method acquired 11 `Confirmation Question`, among which 6 of them were from the first 100 samples and 5 from the subsequent 100 acquired labels. We observed that compared with the random baseline (Fig. 6), the CoreMSE method (Fig. 7) gradually reduced sampling `Request`

`Feedback by Image` instances from 700 labels acquired. This result indicates that the random baseline retained sampling the DA `Request Feedback by Image` instances even the F1 score achieved satisfactory performance, which might consume the budget for the human manual annotation, whereas, the CoreMSE method could alleviate the manual annotation cost when the DA classifier sufficiently learned the patterns. Then, compared with the random baseline (Fig. 6), the CoreMSE method (Fig. 7) generally selected more instances of `Understanding`, `Direction Question`, and `Information Question` DAs for training the DA classifier across the sampling process, which could explain the reason why the CoreMSE method supported the classifier achieving better performance than the random baseline.

5 Conclusion

Our study demonstrated the potential value of using a well-establish framework *Data Maps* [21] to evaluate the informativeness of instances (*i.e.,* the tutorial sentences annotated with dialogue acts) in automatic classification of educational DAs. We found that most instances presented low informativeness in the training dataset, which was easy-to-learn for the dialogue act (DA) classifier. To improve the generalizability of the DA classifier on the unseen instances, we proposed that the classifier should be trained on the samples with high informativeness. Since the annotation of educational DA is extremely time-consuming and cost-demanding [12,15,17], we suggest using effective statistical AL methods for annotation. Our study provided evidence of how the state-of-the-art statistical AL methods (*e.g.,* CoreMSE) could select informative instances for training the DA classifier and gradually reduce selecting the easy-to-learn instances, which can alleviate the cost of manual annotation. We acknowledged that some instances were quite rare in our original annotation, so we plan to employ the AL methods to select more of these rare instances for annotation in future research. Lastly, the AL methods might also be useful for costly evaluation tasks in the education field (*e.g.,* automated classroom observation), which requires the educational experts to annotate the characteristics of behaviours among the students and teachers. Thus, a possible extension of the current work would be to develop an annotation dashboard for human experts to be used in broader educational research.

References

1. Boyer, K., Ha, E.Y., Phillips, R., Wallis, M., Vouk, M., Lester, J.: Dialogue act modeling in a complex task-oriented domain. In: Proceedings of the SIGDIAL 2010 Conference, pp. 297–305 (2010)
2. Clark, K., Luong, M.T., Le, Q.V., Manning, C.D.: Electra: pre-training text encoders as discriminators rather than generators. In: ICLR (2019)
3. D'Mello, S., Olney, A., Person, N.: Mining collaborative patterns in tutorial dialogues. J. Educ. Data Min. **2**(1), 1–37 (2010)

4. Du, B., et al.: Exploring representativeness and informativeness for active learning. IEEE Trans. Cybern. **47**(1), 14–26 (2015)
5. Du Boulay, B., Luckin, R.: Modelling human teaching tactics and strategies for tutoring systems: 14 years on. IJAIED **26**(1), 393–404 (2016)
6. Ezen-Can, A., Boyer, K.E.: Understanding student language: an unsupervised dialogue act classification approach. JEDM **7**(1), 51–78 (2015)
7. Ezen-Can, A., Grafsgaard, J.F., Lester, J.C., Boyer, K.E.: Classifying student dialogue acts with multimodal learning analytics. In: Proceedings of the Fifth LAK, pp. 280–289 (2015)
8. Hastings, P., Hughes, S., Britt, M.A.: Active learning for improving machine learning of student explanatory essays. In: Penstein Rosé, C., et al. (eds.) AIED 2018. LNCS (LNAI), vol. 10947, pp. 140–153. Springer, Cham (2018). https://doi.org/10.1007/978-3-319-93843-1_11
9. Hennessy, S., et al.: Developing a coding scheme for analysing classroom dialogue across educational contexts. Learn. Cult. Soc. Interact. **9**, 16–44 (2016)
10. Karamcheti, S., Krishna, R., Fei-Fei, L., Manning, C.D.: Mind your outliers! Investigating the negative impact of outliers on active learning for visual question answering. In: Proceedings of the 59th ACL, pp. 7265–7281 (2021)
11. Karumbaiah, S., Lan, A., Nagpal, S., Baker, R.S., Botelho, A., Heffernan, N.: Using past data to warm start active machine learning: does context matter? In: LAK21: 11th LAK, pp. 151–160 (2021)
12. Lin, J., et al.: Is it a good move? Mining effective tutoring strategies from human-human tutorial dialogues. Futur. Gener. Comput. Syst. **127**, 194–207 (2022)
13. Lin, J., et al.: Enhancing educational dialogue act classification with discourse context and sample informativeness. IEEE TLT (in press)
14. Nye, B.D., Graesser, A.C., Hu, X.: Autotutor and family: a review of 17 years of natural language tutoring. IJAIED **24**(4), 427–469 (2014)
15. Nye, B.D., Morrison, D.M., Samei, B.: Automated session-quality assessment for human tutoring based on expert ratings of tutoring success. Int. Educ. Data Min. Soc. (2015)
16. Rus, V., Maharjan, N., Banjade, R.: Dialogue act classification in human-to-human tutorial dialogues. In: Innovations in Smart Learning. LNET, pp. 183–186. Springer, Singapore (2017). https://doi.org/10.1007/978-981-10-2419-1_25
17. Rus, V., et al.: An analysis of human tutors' actions in tutorial dialogues. In: The Thirtieth International Flairs Conference (2017)
18. Samei, B., Li, H., Keshtkar, F., Rus, V., Graesser, A.C.: Context-based speech act classification in intelligent tutoring systems. In: Trausan-Matu, S., Boyer, K.E., Crosby, M., Panourgia, K. (eds.) ITS 2014. LNCS, vol. 8474, pp. 236–241. Springer, Cham (2014). https://doi.org/10.1007/978-3-319-07221-0_28
19. Settles, B.: Active Learning. Synthesis Digital Library of Engineering and Computer Science. Morgan & Claypool, San Rafael (2012)
20. Sha, L., Li, Y., Gasevic, D., Chen, G.: Bigger data or fairer data? Augmenting BERT via active sampling for educational text classification. In: Proceedings of the 29th COLING, pp. 1275–1285 (2022)
21. Swayamdipta, S., et al.: Dataset cartography: mapping and diagnosing datasets with training dynamics. In: Proceedings of EMNLP. ACL, Online (2020)
22. Tan, W., Du, L., Buntine, W.: Diversity enhanced active learning with strictly proper scoring rules. In: NeurIPS, vol. 34 (2021)
23. Vail, A.K., Grafsgaard, J.F., Boyer, K.E., Wiebe, E.N., Lester, J.C.: Predicting learning from student affective response to tutor questions. In: Micarelli, A., Stam-

per, J., Panourgia, K. (eds.) ITS 2016. LNCS, vol. 9684, pp. 154–164. Springer, Cham (2016). https://doi.org/10.1007/978-3-319-39583-8_15
24. Vail, A.K., Boyer, K.E.: Identifying effective moves in tutoring: on the refinement of dialogue act annotation schemes. In: Trausan-Matu, S., Boyer, K.E., Crosby, M., Panourgia, K. (eds.) ITS 2014. LNCS, vol. 8474, pp. 199–209. Springer, Cham (2014). https://doi.org/10.1007/978-3-319-07221-0_24
25. Yang, Y., Loog, M.: Active learning using uncertainty information. In: Proceedings of 23rd International Conference on Pattern Recognition, pp. 2646–2651 (2016)

Can Virtual Agents Scale Up Mentoring?: Insights from College Students' Experiences Using the CareerFair.ai Platform at an American Hispanic-Serving Institution

Yuko Okado[1(✉)], Benjamin D. Nye[2], Angelica Aguirre[1], and William Swartout[2]

[1] California State University, Fullerton, Fullerton, CA 92831, USA
{yokado,anaguirre}@fullerton.edu

[2] University of Southern California Institute for Creative Technologies, Playa Vista, CA 90094, USA
{nye,swartout}@ict.usc.edu

Abstract. Mentoring promotes underserved students' persistence in STEM but is difficult to scale up. Conversational virtual agents can help address this problem by conveying a mentor's experiences to larger audiences. The present study examined college students' ($N = 138$) utilization of CareerFair.ai, an online platform featuring virtual agent-mentors that were self-recorded by sixteen real-life mentors and built using principles from the earlier MentorPal framework. Participants completed a single-session study which included 30 min of active interaction with CareerFair.ai, sandwiched between pre-test and post-test surveys. Students' user experience and learning gains were examined, both for the overall sample and with a lens of diversity and equity across different, potentially underserved demographic groups. Findings included positive pre/post changes in intent to pursue STEM coursework and high user acceptance ratings (e.g., expected benefit, ease of use), with under-represented minority (URM) students giving significantly higher ratings on average than non-URM students. Self-reported learning gains of interest, actual content viewed on the CareerFair.ai platform, and actual learning gains were associated with one another, suggesting that the platform may be a useful resource in meeting a wide range of career exploration needs. Overall, the CareerFair.ai platform shows promise in scaling up aspects of mentoring to serve the needs of diverse groups of college students.

Keywords: Virtual Agents · Mentoring · Dialog Systems · STEM Outreach · Hispanic-Serving

1 Introduction

Nearly every successful science, technology, education, and mathematics (STEM) professional was inspired and guided by different mentors across their career.

N. Wang et al. (Eds.): AIED 2023, LNAI 13916, pp. 189–201, 2023.
https://doi.org/10.1007/978-3-031-36272-9_16

Mentoring is particularly important for under-represented minority (URM) students' engagement and persistence in STEM fields [3]. As URM students belong to racial/ethnic groups that are not proportionally represented in STEM fields, they often have less exposure to STEM careers and have a greater need for mentoring. Unfortunately, mentors from backgrounds similar to URM students' will also be under-represented and often over-burdened [5]. Moreover, despite its effectiveness, traditional person-to-person mentoring is difficult to scale up, as mentors are limited by their schedules, and they hold only partial information about careers based on their own experiences. Even with the advent of and growth in online virtual mentoring programs [10], they typically differ only in modality (e.g., teleconference) and face similar problems with scaling.

To address this challenge, AI could be leveraged to help share STEM mentors' experiences on a wider scale. An earlier project for high school STEM outreach (MentorPal [11,12]) demonstrated that it was feasible to video-record real-life mentors answering questions often asked by mentees and use these to generate conversational virtual agent-mentors. Students could pick from suggested questions or type/speak their own questions, where natural language understanding would respond with the mentor's best-match answer. Building on this approach, the CareerFair.ai project is studying how a wide array of virtual mentors in a "virtual career fair" may increase interest and persistence in STEM-based career pathways, especially among students at Minority-Serving Institutions (MSIs). Unlike in prior research, real mentors could self-record and generate their virtual mentors in an online platform, enabling a wider set of mentors to be created. A mixed-methods study was conducted with students at a large U.S. Hispanic-Serving Institution (HSI), to understand their user experience and learning gains with CareerFair.ai, both overall and with a lens of diversity and equity.

2 CareerFair.ai Design

The CareerFair.ai platform provides two distinct capabilities: 1) Mentor Publishing: a self-recording web-based platform was developed so mentors could record, edit, and publish their own virtual mentors and 2) Virtual Mentoring: a portal where students can find and chat with virtual mentors and mentor panels.

Background. This research builds on findings with virtual agents showing that conversational agents using recorded human videos can compellingly convey personal experiences [16]. These and other types of interactive digital narratives better increase learning and engagement compared to traditional learning formats, such as readings or didactic presentations [9]. As a milestone in this area, the New Dimensions in Testimony project enabled museum visitors to converse with hologram recordings of Holocaust survivors to learn about their personal experiences, producing strong engagement in survivors' stories and lives [17]. As noted earlier, the MentorPal project developed and tested a small cohort of virtual STEM mentors with high school students, which showed immediate (pre/post-test) gains in students' knowledge about and interest in STEM careers

[12]. However, this earlier work indicated the need for expanding mentors, as students requested more occupations and more diversity to be represented.

Mentor Publishing. Considering the large number of career fields and their intersection with the diverse identities of URM learners, it was recognized that scaling up virtual mentoring must not be constrained by specific recording equipment or pre-scheduled recording sessions with research staff. To address this, a mentor publishing portal was built, to allow an unlimited number of mentors to record, edit, and preview their mentors flexibly. While it is not possible to describe this process in detail due to space limitations, this allows new mentors to opt-in to the CareerFair.ai platform on an ongoing basis and also to return to improve their virtual mentor at any time. While mentors can record any question, they are recommended to first answer questions from a carefully curated "STEM Careers" question set with 256 questions, organized by topic. The set combines questions that students *should* ask, based on existing research and professional insights, and questions that students said they *wanted* to ask, based on needs assessments conducted at the HSI where the current study was conducted ($N =$ 1197). Mentors were invited to the project to allow students to explore different STEM careers, including those identified via needs assessments, and to increase exposure to mentors from under-represented or underserved backgrounds.

Virtual Mentoring for Students. On the student-facing CareerFair.ai platform, students are presented with an interface where they can view profiles of individual mentors as well as mentor panels, or a roundtable-style panel of mentors (Fig. 1). The header and footer show logos with links to the students' home institution and collaborating outreach organizations. When students click on a desired mentor or mentor panel, the "virtual mentor" interface is shown (Fig. 2), where students can pose questions free-form in a text box or choose questions

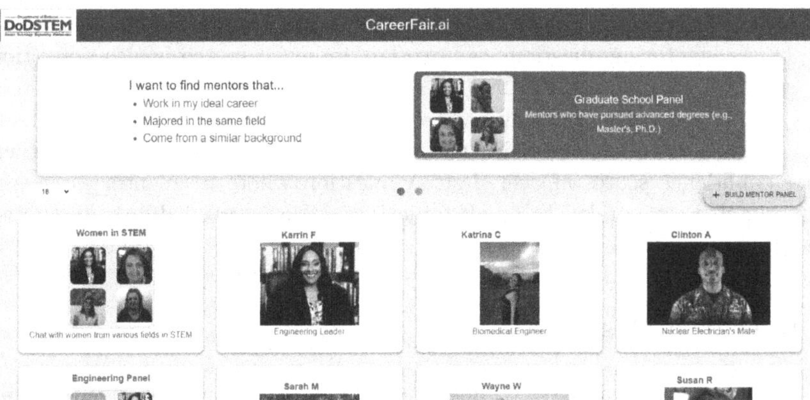

Fig. 1. The CareerFair.ai Home Page.

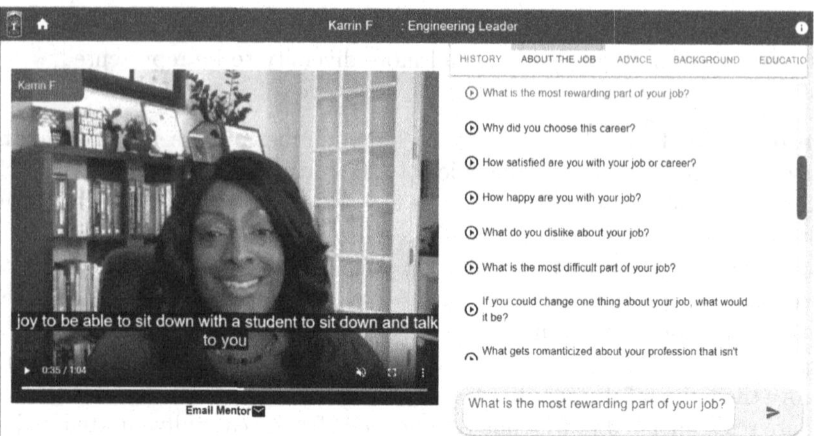

Fig. 2. The Virtual Mentor Chat Interface.

from sets of "suggested questions" grouped by topic on the right side of the screen. In response to the question, a videotaped response classified as the most relevant answer based on the natural language question answering model would play back on the left [11,12]. In the mentor panel format, the most relevant answer for each mentor on the panel would play back, one at a time, similar to a roundtable format where mentors take turns answering the same question.

Question Answering. User-entered questions are classified using a logistic regression classifier. Similar to MentorPal [12], answer classification is based on a feature vector of sentence embeddings for the input question. However, while MentorPal used an average of individual Word2Vec embeddings, the current system instead uses Sentence-BERT [14]. As S-BERT captures relationships between words, this increased accuracy on our benchmark mentor test set from 73% ideal answers [12] to 82% ideal answers (i.e., exact match to a human expert). As mentoring questions can overlap, an exact match is also not strictly necessary, and in practice over 90% of inputs receive "reasonable answers" (i.e., an expert would rate the answer as responsive to the question). The mentor panel dialog controller also helps to improve answer quality, as mentors with higher confidence scores answer first. A question where no mentor has a confident match is responded to as off-topic (e.g., "I'm sorry, I didn't understand your question...").

3 Research Design

The present study examined students' user experience and perceived learning gains as a result of using the CareerFair.ai platform, both for the overall sample and with a lens of diversity and equity across different demographic groups that may be underserved. Analyses included self-reported quantitative ratings and write-in descriptions, as well as from logged activity on the CareerFair.ai platform. Exploratory in nature, the study examined:

1. User experience and acceptance of the platform
2. Perceived impact as a result of using the platform
3. Anticipated and actual information learned and explored by users
4. Whether any of the above differed for underserved groups of users.

Participants. A sample of 138 students (35.8% STEM majors; 96.3% undergraduate, 36.5% first-generation college student, 45.7% Hispanic, 54.3% underrepresented minority per the National Science Foundation definition, 65.2% non-male) was recruited from California State University, Fullerton (CSUF), a large public HSI in the United States. The demographic characteristics of the sample were consistent with the student population at the institution.

Procedures. Participants in the study had access to three mentor panels (CSUF Alumni; Women/Womxn in STEM; Engineering) and up to 16 virtual mentors (range = 10–16; 44% of color and 44% non-male). Participants were recruited for an online Qualtrics protocol using word-of-mouth, referrals from collaborating campus organizations, and the Psychology research pool. After providing informed consent, they completed a pre-test survey, after which they were redirected to the CareerFair.ai platform. Participants were asked to actively interact with the platform for 30 min, after which a pop-up emerged to redirect them to the post-test survey. Data were checked for effort and valid responding by trained research assistants. All procedures were approved by the Institutional Review Board at CSUF.

Measures. Usability and acceptance of the platform were measured using a version of a Unified Theory of Acceptance and Use of Technology (UTAUT) measure [18], with nine items on ease of use, acceptance, and intent to use the platform (e.g., "I found CareerFair.ai easy to use", "Using CareerFair.ai is a good idea", "I would recommend CareerFair.ai to other students") rated on a 6-point scale ranging from 1 (*completely disagree*) to 6 (*completely agree*). Participants also rated: "I learned more about a career I am interested in" and "I learned more about new career opportunities that I would be interested in," using a 5-point Likert scale ranging from 1 (*strongly disagree*) to 5 (*strongly agree*).

Two measures were administered at both pre-test and post-test. Participants' perceived value of and expectations for success in STEM fields were assessed using the Value-Expectancy STEM Assessment Scale (VESAS [1]), which included 15 items (e.g., "STEM is an important field for me", "I feel I have what it takes to succeed in a STEM-related job") also rated on a 5-point scale for agreement. Scores for two subscales, perceived value of STEM fields and expectations for success in STEM careers, were obtained. Participants also rated three statements on how likely they were to major, minor, or take additional classes in STEM on a 5-point scale ranging from 1 (*very unlikely*) to 5 (*very likely*).

Regarding mentor selection, participants indicated which mentor they interacted with the most, which was coded for mentor being of color or non-male. Participants also indicated the mentor panel(s) used via a multiple-answer item.

Four write-in items were also administered and analyzed. At pre-test, participants described their expected learning gains by answering: "What do you most hope to learn or gain from using the CareerFair.ai website?" and "What kinds of questions do you hope to get answered by the (virtual) mentors?". These were coded for correspondence with available content or features on the CareerFair.ai platform (0 = *Not addressed*, 1 = *Addressed*). At post-test, participants described what they had learned and explored, in response to: "What were some things you learned from using the CareerFair.ai website?" and "Please describe main things that you explored on the CareerFair.ai site." Responses were coded for recurring themes, using inductive coding procedures with the phenomenological approach [15], and also deductively for correspondence with desired learning gains at pre-test (0 = *Does not correspond*, 1 = *Corresponds at least partially*).

The content accessed by participants on the platform was also logged. On the CareerFair.ai system, the mentor's response to each question is tagged with a Topic (e.g., "What is a typical day like on the job?" is associated with the Topic, "About the Job"). Topics for all the responses that were played back or viewed by the participant for at least 3 s were tallied for each user's session.

Analyses. Descriptive statistics were obtained, and changes in scores between pre-test and post-test were tested using repeated-measures analysis of variance (ANOVA). Point-biserial correlations were used to examine associations between themes (codes) found for the post-test write-in items and the frequency with which responses in each Topic were viewed. Potential differences by demographic groups were explored using chi-square tests of independence (for categorical variables) and ANOVA (for quantitative variables).

4 Results

User Acceptance. Participants rated the platform highly on the UTAUT measure, with overall mean across the nine items in the "Agree" range ($M = 5.15$, $SD = 0.58$; $Median = 5.11$, $Mode = 5$). URM students gave significantly higher UTAUT ratings ($M = 5.26$, $SD = 0.52$) compared to non-URM students ($M = 5.01$, $SD = 0.62$), $F(1, 136) = 6.90, p = .01$. For all but one item, average ratings fell in the "Agree" to "Completely Agree" range (means, medians, and modes between 5 to 6). The two highest rated statements were, "I found CareerFair.ai easy to use" and "Using CareerFair.ai is a good idea."

A few global items asked the participants about their overall impression of the platform. Consistent with the UTAUT ratings, a vast majority (96.4%) of the participants indicated that they would recommend the platform to others.

This did not differ by most demographic groups, though non-male students were significantly more likely to recommend the platform (98.9%) than male students (91.7%), $\chi^2(1, N = 137) = 4.42, p = .04$.

Mean ratings for learning about a career of interest (M = 3.80, SD = 1.15) and new career opportunities of interest (M = 3.98, SD = 0.88) fell in the Neutral to Agree range. The ratings generally did not differ by demographic groups, though STEM majors expressed higher agreement that they learned about a career they were interested in (M = 4.12, SD = 0.95), compared to non-STEM majors (M = 3.63, SD = 1.22), $F(1, 137) = 6.12, p = .02$.

Impact. Small but positive increases were observed between pre-test and post-test in participants' intent to major (M = 3.21, SD = 1.39 pre-test and M = 3.41, SD = 1.40 post-test), $F(1, 137) = 8.86, p = .003$, partial $\eta^2 = .06$, minor (M = 3.07, SD = 1.19 pre-test and M = 3.32, SD = 1.16 post-test), $F(1, 135) = 12.02, p < .001$, partial $\eta^2 = .08$, and take additional courses in STEM (M = 3.25, SD = 1.16 pre-test and M = 3.53, SD = 1.12 post-test), $F(1, 136) = 13.34, p < .001$, partial $\eta^2 = .09$. No differences were found for these effects across different demographic groups.

Similar to these findings, perceived value of STEM fields as assessed by VESAS evidenced a modest change from pre-test (M = 25.66, SD = 2.71) to post-test (M = 26.35, SD = 2.89), $F(1, 135) = 5.74, p = .018$, and a significant difference for this effect was found between first-generation and non-first generation students, $F(1, 133) = 6.47, p = .01$, partial $\eta^2 = .05$. Post-hoc analyses indicated that only first-generation college students, not non-first-generation students, evidenced a gain ($\Delta = 1.35$ points on average) over time. By contrast, there was no change in the subscale score for expectations for success in STEM careers, $F(1, 135) = 0.01, p = .91$.

Mentor Selection. From the CareerFair.ai home page, participants had an option to interact with individual mentors or with "mentor panels." Approximately two-thirds (n = 96; 69.1%) of the sample elected to use at least one mentor panel. In terms of selection and behavior, STEM majors (n = 49) were more likely than non-STEM majors (n = 88) to skip mentor panels (36.7%) or use Engineering mentor panels (20.4%), whereas non-STEM majors were more likely than STEM majors to use Alumni (20.5%) or Women/Womxn (27.3%) mentor panels, $\chi^2(4, N = 137) = 12.48, p = .01$. Non-male students ($n$ = 90) were more likely than male students (n = 48) to use Women/Womxn (27.8%, as opposed to 12.5% among male students) or multiple mentor panels (27.8%, as opposed to 4.2%), whereas male students were more likely than non-male students to use the Engineering (29.2% vs. 2.2%) mentor panel, $\chi^2(4, N = 138) = 33.01, p < .001$.

In terms of the mentor most utilized, 49.3% of participants reported interacting the most with a mentor of color, whereas 76.1% reported doing so with a non-male mentor. No demographic differences emerged in focusing on a mentor of color, thus students from different groups were equally likely to focus on a mentor of color. Non-male participants selected a non-male mentor at a significantly higher rate (87.4%) than male participants (55.3%), $\chi^2(1, N = 134) = 17.23, p < .001$, otherwise focusing on a non-male mentor was equally likely across different groups.

Of note, mentor selection did not generally influence self-reported user experience of using the platform, but those that interacted with a non-male mentor gave a higher overall UTAUT rating ($M = 5.21$, $SD = 0.57$), $F(1, 132) = 4.62, p = .03$, and rating that mentor panel better facilitated learning about careers compared to one-on-one interactions with mentors, rated on a 1 (strongly disagree) to 5 (strongly agree) scale ($M = 4.14$, $SD = 0.73$), $F(1, 87) = 7.24, p = .01$, on average compared to those that interacted with a male mentor ($M = 4.96$, $SD = 0.60$ and $M = 3.60$, $SD = 1.00$, respectively).

Self-reported Learning. Table 1 summarizes the recurrent themes found in participants' descriptions of what they learned from and explored on the CareerFair.ai platform. The most frequently mentioned themes were related to the mentor's career path, general advice or approach to careers, and specifics about the mentor's job. URM students and non-male students both reported learning about mentors' personal career paths (49.3% and 48.9%, respectively) more than their counterparts (32.2% and 27.7%, respectively), $\chi^2(1, N = 137) = 4.07, p = .04$ for URM comparisons, $\chi^2(1, N = 137) = 5.73, p = .02$ for gender comparisons. Additionally, first-generation college students more frequently reported learning about educational requirements for jobs (24%) compared to their counterparts (10.5%), $\chi^2(1, N = 136) = 4.44, p = .04$. Upperclassmen (21.3%) were more likely than others (0.0–7.3%) to explore different jobs on the platform, $\chi^2(2, N = 134) = 6.24, p = .04$. STEM majors were more likely to explore pragmatic strategies in career building (12.5%), $\chi^2(1, N = 135) = 5.77, p = .02$, and platform features themselves (20.8%), $\chi^2(1, N = 135) = 5.77, p = .02$ than non-STEM majors (2.3% and 6.9%, respectively), $\chi^2(1, N = 135) = 5.75, p = .02$. URM students explored personal challenges experienced by mentors more often (6.8%) than non-URM students (0%), $\chi^2(1, N = 136) = 4.35, p = .04$. Surprisingly, male students reported exploring work-life balance at a greater rate (12.8%) than non-male students (2.2%). $\chi^2(1, N = 136) = 6.15, p = .01$.

Associations with Content Accessed on the Platform. Participants asked an average of 20.71 questions ($SD = 12.38$) and a total of 2917 responses were viewed. More advanced undergraduate and graduate students asked more questions than lower-level undergraduates, $Welch'sF(2, 10.45) = 4.37, p = .04$. The most commonly queried Topics were About the Job (35%), Background (21%), Advice (14%), and Education (12%). The codes for self-reported learning gains and explorations were compared against questions logged by CareerFair.ai.

Questions were classified by Topic under which it belongs and each student's questions were counted by Topic. Point-biserial correlations between the coded qualitative data and frequency of question-topics were examined and are presented in Table 1. Of the 18 codes, a majority (11 codes; 61%) showed significant links to actual questions asked, with most of those associations being logically linked (e.g., self-reported learning and explorations related to career advice were both linked to asking more questions within the Advice Topic). The following codes were not significantly associated with specific Topics: In learning, General Career Approach (reported by 38%), Job-Specific Details (29%), Possible Jobs in STEM (25%); in exploring, Demographic-Specific Information (15%), Platform Features (12%), Possible Jobs (12%), and Obstacles Overcome (4%).

In examining participants' desired and actual learning gains, 77.6% of the participants ($n = 104$) had at least partial correspondence between desired (pre-test) and reported (post-test) gains. Of note, 91.9% of the participants at pre-test mentioned desired learning gains that could be addressed by the platform.

Table 1. Themes coded for self-reports and their correlation with question Topics

Theme/Code (*Description*)	Occurrence	Topics for Questions Asked
"What were some things you learned?"		
Mentor's Path (*Career and Personal History*)	42%	Development ($r = .17$), Education ($r = .19$)
Pragmatic/Strategic Advice	23%	Advice ($r = .36$)
Education	15%	Education ($r = .20$), Graduate School ($r = .41$), Lifestyle ($r = .18$)
Work-Life Balance	4%	For Fun ($r = .26$), Lifestyle ($r = .38$)
Platform (*Features or Content*)	4%	Motivation and Vision ($r = .21$)
"Main things that you explored"		
Job Details	41%	About the Job ($r = .26$)
Mentor's Path (*Career and Personal History*)	36%	Background ($r = .33$)
Advice	18%	Advice ($r = .42$)
Education-Related Information	15%	Education ($r = .20$), Graduate School ($r = .28$)
Work-Life Balance	6%	Computer Science ($r = .23$), For Fun ($r = .18$), Lifestyle ($r = .19$)
Pragmatics (*How-To's*)	6%	Advice ($r = .31$)

5 Discussion

The CareerFair.ai platform shows promise in scaling up some of the benefits of mentoring and providing diverse sets of students with an interactive and personalized way to learn about different career paths in STEM. The results suggest high levels of user acceptance and positive impact on such outcomes as interest in pursuing STEM coursework and perceived value of STEM fields. Of note, URM students gave higher user acceptance ratings than non-URM students, which suggests that the platform could be a welcome resource for diverse sets of students, possibly by addressing unmet needs that occur more frequently among

URM students [2] and critical needs related to diversity, equity, and inclusion, as discussed below. Outcomes related to intent to pursue STEM coursework and perceived value of STEM fields showed positive increases between pre-test and post-test assessments, with outcomes generally not being moderated by demographic factors; thus, the CareerFair.ai platform has the potential to positively influence interest and persistence in STEM for students from all backgrounds. Moreover, the only way in which demographic factors moderated these key outcomes was that first-generation college students showed a greater increase in perceived value of STEM fields after using the platform.

Based on participants' selection of mentors, we found that gender influenced the way participants selected mentors. Non-male students were more likely to interact with the "Women/Womxn in STEM" mentor panel and/or a non-male mentor one-on-one. Moreover, the UTAUT ratings for the platform and for mentor panels were also higher among those who focused on interacting with a non-male mentor. A greater proportion of non-male participants expressed the willingness to recommend platform to others. These findings indicate the importance of increasing students with access to gender-minority mentors, a known need [4], as their insights are in demand especially among non-male students. Otherwise, minimal differences emerged across subgroups in choosing mentor panels or focusing on mentors of color, though STEM and non-STEM majors did show some differences in their utilization of available mentor panels.

Notably, participants' self-reported descriptions of the content they learned from and explored on the platform corresponded with both their desired learning gains at pre-test and with the actual content they had viewed on the CareerFair.ai platform. Consistent with earlier research on MentorPal [11], students engaged with specifics about the mentor's job, sought out advice, or asked questions related to education. The Topics associated with the actual questions asked correlated, by-and-large, in a very logical way. Thus, participants appeared to retain content from the platform, at least in some topic areas. These findings show promise for using an AI-based virtual agent to provide users with content that is more personally relevant and efficient to access than searching uncurated content online, scheduling appointments with career counselors or other parties that may only hold partial knowledge, or attending roundtables or other events where little of the content is tailored to the individual student. Thus, this tool may complement in-person mentoring by providing detailed follow-up, preparatory exploration, or interview practice.

In addition to the findings, the platform has further implications related to diversity, equity, and inclusion. One, because representation matters in providing students with role models that share similar backgrounds or values [6], a web platform like CareerFair.ai can support not just career mentoring but also a sense of belongingness by incrementally growing to host a broader and more diverse set of mentors than students can easily access in any one location. Second, students from underserved backgrounds may face additional barriers to accessing mentors, including stereotype threat [8] and limitations in resources (e.g., time, networks, travel) to find and schedule meetings with mentors. The CareerFair.ai platform provides students with a free-of-charge resource where they can efficiently sample different mentors' perspectives and guidance without facing those types of barriers inherent in finding individual mentors. A recent needs analysis of URM students conducted by another research group [7] also noted the importance of rapport-building and suggesting resources for students of color; as the CareerFair.ai platform includes questions intended to build rapport (e.g., conversational and personal mentor recordings) and hyperlinks to external resources, these suggest convergence toward suggested features for effective "virtual mentoring."

The study had several limitations. While the number of mentors is growing, an earlier usability study indicated demand for additional careers [13]. Further research with larger sample sizes and at different minority-serving institutions is also recommended, to corroborate the findings. To best examine the naturalistic utilization of the CareerFair.ai platform, it may be useful to allow users to exit the platform when they wish, though the current usage time limit of 30 min was supported by prior work involving the MentorPal technology [12]. Moreover, longitudinal studies examining further use of the platform, retention of any information learned, and effects of the virtual mentors on participants' actual career exploration and planning behaviors are needed. Finally, because the virtual mentors play back pre-recorded responses, they are limited in their ability to address user-specific situations (e.g., a participant asks for advice specific to their own circumstance), and the platform is more likely to focus on informational aspects of mentoring and representation than social-emotional aspects of mentoring.

6 Conclusions and Future Directions

The CareerFair.ai platform – which features virtual agent-mentors that were self-recorded and published by real-life mentors – may help improve access to career information, guidance, and mentoring for a wide range of students. Students reported high levels of acceptance and evidenced good correspondence across what they wanted to learn from the platform, content that they accessed on the platform, and their self-reported takeaways. The platform provides a promising approach to scale up mentoring, particularly for first-generation and URM students. As noted, future research should examine the virtual mentoring under more naturalistic conditions and in coordination with live mentors; it may be appropriate to conceptualize the CareerFair.ai platform as a stepping stone towards direct interaction with a mentor and expand this role beyond its current capabilities (e.g., the "Email Mentor" button). Future research is needed to clarify to what extent the virtual mentors can disseminate different facets of mentoring activities and could be integrated with additional, face-to-face mentoring.

Acknowledgments. This material is based upon work supported by the National Defense Education Program (NDEP) for Science, Technology, Engineering, and Mathematics (STEM) Education, Outreach, and Workforce Initiative Programs under Grant

No. HQ0034-20-S-FO01. The views expressed in this publication do not necessarily reflect the official policies of the Department of Defense nor does mention of trade names, commercial practices, or organizations imply endorsement by the U.S. Government. We thank the CareerFair.ai project mentors, research assistants, and study participants.

References

1. Appianing, J., Van Eck, R.N.: Development and validation of the value-expectancy stem assessment scale for students in higher education. Int. J. STEM Educ. **5**(24) (2018)
2. Chelberg, K.L., Bosman, L.B.: The role of faculty mentoring in improving retention and completion rates for historically underrepresented STEM students. Int. J. High. Educ. **8**(2), 39–48 (2019)
3. Chemers, M.M., Zurbriggen, E.L., Syed, M., Goza, B.K., Bearman, S.: The role of efficacy and identity in science career commitment among underrepresented minority students. J. Soc. Issues **67**(3), 469–491 (2011)
4. Dawson, A.E., Bernstein, B.L., Bekki, J.M.: Providing the psychosocial benefits of mentoring to women in STEM: careerwise as an online solution. N. Dir. High. Educ. **2015**(171), 53–62 (2015)
5. Domingo, C.R., et al.: More service or more advancement: institutional barriers to academic success for women and women of color faculty at a large public comprehensive minority-serving state university. J. Divers. High. Educ. **15**(3), 365–379 (2022)
6. Fealing, K.H., Lai, Y., Myers, S.L., Jr.: Pathways vs. pipelines to broadening participation in the STEM workforce. J. Women Minorities Sci. Eng. **21**(4), 271–293 (2015)
7. Mack, N.A., Cummings, R., Huff, E.W., Gosha, K., Gilbert, J.E.: Exploring the needs and preferences of underrepresented minority students for an intelligent virtual mentoring system. In: Stephanidis, C., Antona, M. (eds.) HCII 2019. CCIS, vol. 1088, pp. 213–221. Springer, Cham (2019). https://doi.org/10.1007/978-3-030-30712-7_28
8. Martin-Hansen, L.: Examining ways to meaningfully support students in STEM. Int. J. STEM Educ. **5**(1), 1–6 (2018)
9. McQuiggan, S.W., Rowe, J.P., Lee, S., Lester, J.C.: Story-based learning: the impact of narrative on learning experiences and outcomes. In: Woolf, B.P., Aïmeur, E., Nkambou, R., Lajoie, S. (eds.) ITS 2008. LNCS, vol. 5091, pp. 530–539. Springer, Heidelberg (2008). https://doi.org/10.1007/978-3-540-69132-7_56
10. Neely, A.R., Cotton, J., Neely, A.D.: E-mentoring: a model and review of the literature. AIS Trans. Hum.-Comput. Interact. **9**(3), 220–242 (2017)
11. Nye, B., Swartout, W., Campbell, J., Krishnamachari, M., Kaimakis, N., Davis, D.: Mentorpal: Interactive virtual mentors based on real-life STEM professionals. In: Interservice/Industry Simulation, Training and Education Conference (2017)
12. Nye, B.D., et al.: Feasibility and usability of mentorpal, a framework for rapid development of virtual mentors. J. Res. Technol. Educ. **53**(1), 1–23 (2020)
13. Okado, Y., Nye, B.D., Swartout, W.: Student acceptance of virtual agents for career-oriented mentoring: a pilot study for the careerfair.ai project. In: American Educational Research Association Annual Meeting (2023)
14. Reimers, N., Gurevych, I.: Sentence-BERT: sentence embeddings using Siamese BERT-networks. In: 9th EMNLP-IJCNLP Conference, pp. 3982–3992 (2019)

15. Saldana, J.: Fundamentals of Qualitative Research. Oxford University Press, Oxford (2011)
16. Swartout, W., et al.: Virtual humans for learning. AI Mag. **34**(4), 13–30 (2013)
17. Traum, D., et al.: New dimensions in testimony: digitally preserving a holocaust survivor's interactive storytelling. In: Schoenau-Fog, H., Bruni, L.E., Louchart, S., Baceviciute, S. (eds.) ICIDS 2015. LNCS, vol. 9445, pp. 269–281. Springer, Cham (2015). https://doi.org/10.1007/978-3-319-27036-4_26
18. Venkatesh, V., Morris, M.G., Davis, G.B., Davis, F.D.: User acceptance of information technology: toward a unified view. MIS Q. 425–478 (2003)

Real-Time AI-Driven Assessment and Scaffolding that Improves Students' Mathematical Modeling during Science Investigations

Amy Adair[1(✉)], Michael Sao Pedro[2], Janice Gobert[1,2], and Ellie Segan[1]

[1] Rutgers University, New Brunswick, NJ 08901, USA
amy.adair@gse.rutgers.edu
[2] Apprendis, Berlin, MA 01503, USA

Abstract. Developing models and using mathematics are two key practices in internationally recognized science education standards, such as the Next Generation Science Standards (NGSS) [1]. However, students often struggle at the intersection of these practices, i.e., developing mathematical models about scientific phenomena. In this paper, we present the design and initial classroom test of AI-scaffolded virtual labs that help students practice these competencies. The labs automatically assess fine-grained sub-components of students' mathematical modeling competencies based on the actions they take to build their mathematical models within the labs. We describe how we leveraged underlying machine-learned and knowledge-engineered algorithms to trigger scaffolds, delivered proactively by a pedagogical agent, that address students' individual difficulties as they work. Results show that students who received automated scaffolds for a given practice on their first virtual lab *improved* on that practice for the next virtual lab on the same science topic in a different scenario (a near-transfer task). These findings suggest that real-time automated scaffolds based on fine-grained assessment data can help students improve on mathematical modeling.

Keywords: Scaffolding · Intelligent Tutoring System · Science Practices · Performance Assessment · Formative Assessment · Science Inquiry · Mathematical Modeling · Developing and Using Models · Virtual Lab · Online Lab · Pedagogical Agent · Next Generation Science Standards Assessment

1 Introduction

To deepen students' understanding of scientific phenomena and ensure that students are fully prepared for future careers related to science and mathematics [2], students must become proficient at key science inquiry practices, i.e., the ways in which scientists study phenomena. Standards, such as the Next Generation Science Standards (NGSS) [1], define several such practices, including NGSS Practice 2 (Developing and Using Models) and Practice 5 (Using Mathematics and Computational Thinking). However,

N. Wang et al. (Eds.): AIED 2023, LNAI 13916, pp. 202–216, 2023.
https://doi.org/10.1007/978-3-031-36272-9_17

the difficulties that students experience with the scientific practices related to using mathematics and developing models can be barriers for students' access to and success in high school science coursework and future STEM careers [3–5]. Specifically, students often have difficulties developing mathematical models (i.e., graphs) with quantitative data in science inquiry contexts [6] because they struggle to properly label the axes of their graphs [7], interpret variables on a graph [8], make connections between equations and graphs [4], or choose the functional relationship to create a best-fit line or curve [9, 10]. Thus, students need resources capable of formatively assessing and scaffolding their competencies in a rigorous, fine-grained way as they work [12, 13] so that they can develop these critical competencies and, in turn, transfer them across science contexts [11].

In this paper, we evaluate the design of virtual labs in the Inquiry Intelligent Tutoring System (Inq-ITS), which are instrumented to automatically assess and scaffold students' competencies as they conduct investigations and develop mathematical models to represent and describe science phenomena [14–16]. To do so, we address the following research question: Did individualized scaffolding, triggered by automated assessment algorithms, help students improve on their mathematical modeling competencies from the first virtual lab activity to a second virtual lab activity on the same topic in a different scenario (i.e., a near-transfer task)?

1.1 Related Work

Some online environments seek to assess and support students' competencies related to mathematical modeling for science, such as constructing and exploring computational models (e.g., Dragoon) [17], drawing qualitative graphs of science phenomenon (e.g., WISE) [18], and physics problem solving (e.g., Andes) [19]. However, these environments do not assess students' mathematical modeling competencies within the context of a full science inquiry investigation. Further, they do not provide AI-driven real-time scaffolding on the full suite of other NGSS practices (e.g., Planning and Conducting Investigations), all of which are needed for conducting an authentic investigation that uses mathematical models to make inferences about science phenomena.

Scaffolding in online learning environments for both math and science has yielded student improvement on competencies by breaking down challenging tasks into smaller ones [20], providing hints on what to do next for students who are stuck on a task [21], and reminding students about the progress and steps taken thus far [20]. While scaffolding strategies have been applied to the online learning environments for modeling in science [19, 22], there are no studies, to our knowledge, that investigate the efficacy of AI-driven scaffolds for *mathematical modeling associated with science inquiry*, as envisioned by the practices outlined in the NGSS. Thus, the goal of the current study is to evaluate the use of real-time automated scaffolding in the Inq-ITS labs to improve students' competencies on science inquiry and mathematical modeling practices.

2 Methods

2.1 Participants and Materials

Participants included 70 students across four eighth grade science classes taught by the same teacher from the same school in the northeastern region of the United States during Fall 2022. Thirty-one percent of students qualify for free or reduced-price lunch; 71% identify as White, 16% as Hispanic, and 6% as two or more races.

All students completed two Inq-ITS mathematical modeling virtual labs on the disciplinary core idea of Forces and Motion (NGSS DCI PS2.A). Both labs were augmented with automated scaffolding. Students completed the two labs during their regularly scheduled classes. In these labs, students used simulations to collect data and develop mathematical models to demonstrate the relationship between the roughness/friction of a surface and the acceleration of a moving object on that surface (see Sect. 2.2 for more details). Inq-ITS automatically assessed students' competencies using previously validated educational data-mined and knowledge-engineered algorithms [14, 16, 23], which triggered scaffolds to students as they worked.

2.2 Inq-ITS Virtual Lab Activities with Mathematical Modeling

The virtual labs consisted of six stages that structured the investigation and captured different aspects of students' competencies at several NGSS practices (Table 1, Fig. 1). The goal of each activity was to develop a mathematical model (i.e., a best-fit curve represented by a graph and corresponding equation) that can explain how changing one factor (e.g., roughness of a ramp/road) impacted an outcome (e.g., acceleration of a sled sliding down that ramp, or acceleration of the truck on the road). Descriptions of each stage and how each stage aligned to NGSS practices are shown in Table 1.

We consider the tasks presented in both labs as isomorphic, near-transfer tasks [24, 25], since they consisted of the same stages and focused on the same physical science concept (i.e., the relationship between friction/roughness of a surface and acceleration of a moving object on that surface). However, the scenarios depicted in the simulations differed. In the first lab (Truck), students investigated the mathematical relationship between the roughness/friction of a *flat road* and the acceleration of the *truck* on that road (Fig. 2, left). In the second lab (Ramp), students investigated the mathematical relationship between the roughness/friction of a *ramp* and the ending acceleration of a *sled* sliding down the ramp (Fig. 2, right). In both cases, students learn that, when they only change the roughness/friction of the surface (i.e., road/ramp) and keep all other variables constant, there is a negative linear relationship between the friction of the surface and the acceleration of the object moving along that surface.

The design of the lab focuses on students' competencies with inter-related practices including collecting controlled data [26], plotting/graphing the data [27], and determining the informal line/curve of best fit [9] *without* deriving the algebraic equations, which shifts the focus to modeling the phenomenon rather than completing rote "plug-and-chug" methods often taught in physics problem-solving contexts [28]. This design not only helps students more readily identify the similarities in the mathematical and scientific relationship between the variables in the two scenarios (i.e., the friction of the road/ramp vs. the acceleration of the truck/sled), but also helps students develop more sophisticated understandings of the scientific meaning in the graphs, a task with which students often struggle [29].

Table 1. Stages of the Inq-ITS Mathematical Modeling Virtual Lab Activity

Stage	Primary Related NGSS Practice(s)	Description of Stage
Stage 1: Hypothesizing/Question Formation	Practice 1: Asking Questions & Defining Problems	Students form a question about the mathematical relationship between an independent and dependent variable based on a given goal (e.g., If I change the roughness of the ramp, then I will be able to observe that the roughness of the ramp and the acceleration of the sled at the end of the ramp have a linear relationship).
Stage 2: Collecting Data	Practice 3: Planning & Carrying Out Investigations	Students collect data using a simulation that can be used to investigate the relationship between the variables outlined in their hypothesis (e.g., roughness of the ramp and acceleration of the sled at the end of the ramp). The data that they collect are automatically stored in a data table.
Stage 3: Plotting Data	Practice 2: Developing & Using Models Practice 5: Using Mathematics & Computational Thinking	Students select trials from their data table to plot on a graph and select the variables to place on the x-axis and y-axis of their graph. Ideally, students should place their independent variable (e.g., roughness of the ramp) on the x-axis and their dependent variable (e.g., acceleration of the sled at the end of the ramp) on the y-axis, and students should only plot *controlled* data.

(*continued*)

Table 1. (*continued*)

Stage	Primary Related NGSS Practice(s)	Description of Stage
Stage 4: Building Models	Practice 2: Developing & Using Models Practice 5: Using Mathematics & Computational Thinking	Students select the type of mathematical relationship that best fits the shape of the plotted data (linear, inverse, square, inverse square, horizontal). Students also determine the coefficient and constant for the equation of the best-fit curve/line as well as check the fit (i.e., coefficient of determination, R^2), which is automatically calculated and stored in their table along with a snapshot of their graph and equation. Ideally, students should create a model that fits the data points and demonstrates the mathematical relationship between the two variables. Students are *not* expected to calculate the coefficient and constants for the equation of their model, but rather they are expected to use the slider to create a best-fit curve/line.
Stage 5: Analyzing Data	Practice 4: Analyzing & Interpreting Data	Students interpret the results of their graphs by making a claim about the relationship between the variables, identifying if it was the relationship that they had initially hypothesized, and selecting the graphs and corresponding equations that best demonstrated this relationship.
Stage 6: Communicating Findings	Practice 6: Constructing Explanations	Students write an explanation of their findings in the claim, evidence, and reasoning (CER) format.

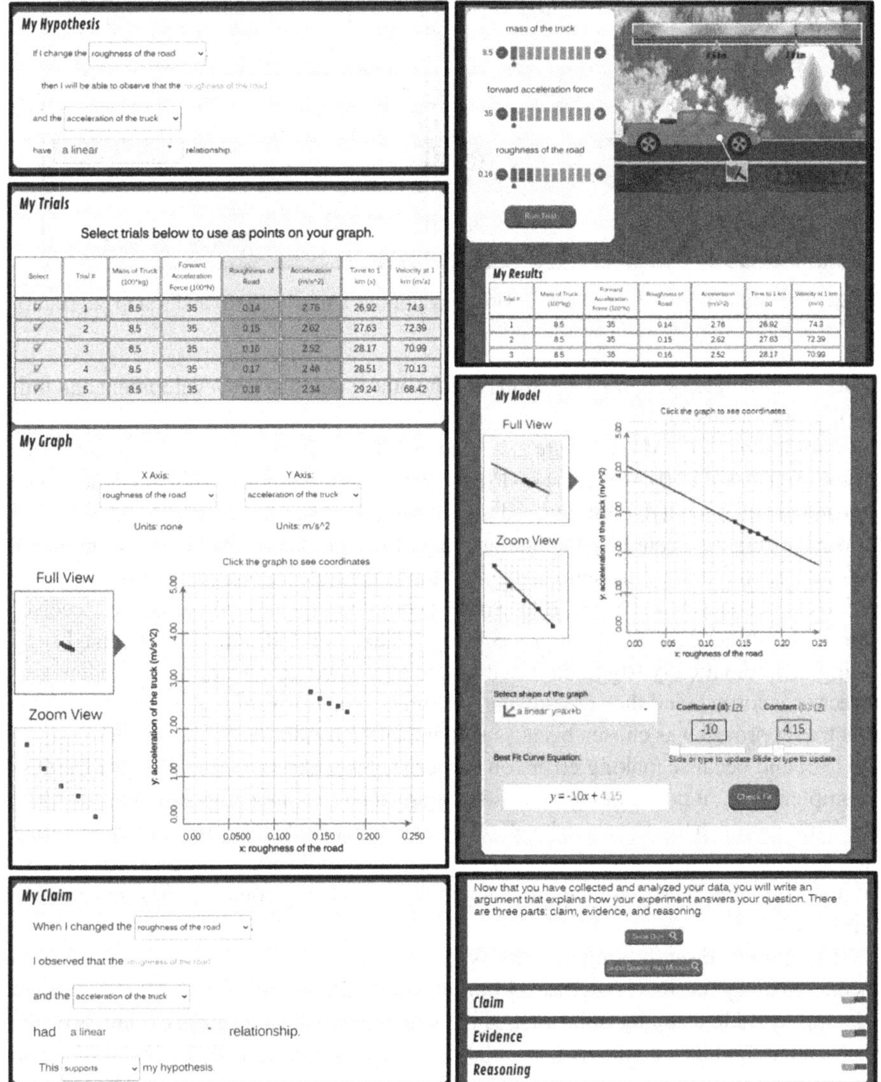

Fig. 1. Screenshots of Inq-ITS mathematical modeling virtual lab; stages include (1) Hypothesizing (top left), (2) Collecting Data (top right), (3) Plotting Data (middle left), (4) Building Models (middle right), (5) Analyzing Data (bottom left), (6) Communicating Findings (bottom right).

2.3 Approach to Automated Assessment and Scaffolding of Science Practices

Inq-ITS automatically assesses and scaffolds their competencies on fine-grained components, or "sub-practices," of the related NGSS practices elicited in each stage of the lab activity (Table 2). For this study, the automated scoring algorithms were active for the first four stages of the lab (Hypothesizing, Collecting Data, Plotting Data, and Building

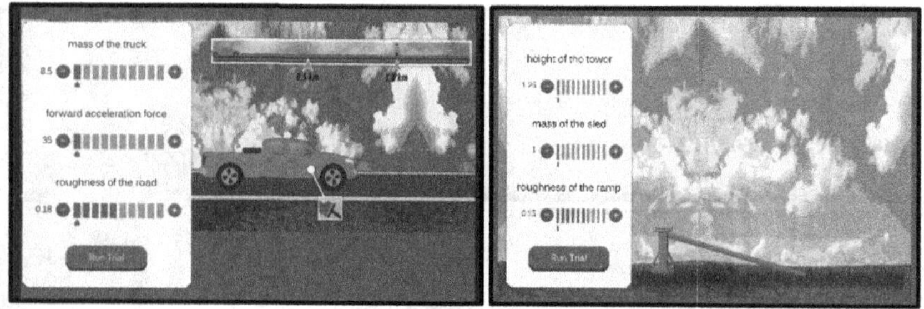

Fig. 2. The simulation in the Collecting Data stage of the Truck lab (left) and Ramp lab (right).

Models). Automated scoring algorithms for the other stages are in development and thus out of scope of this study.

Assessment and scaffolding are executed as follows. Each sub-practice is automatically scored as either 0 (incorrect) or 1 (correct) using previously validated educational data-mined and knowledge-engineered algorithms [14, 23]. The algorithms take as input the work products created by the student (e.g., their graphs or mathematical models), and/or distilled features that summarize the steps they followed (e.g., the processes they used to collect data) [14–16, 23]. If the student completes the task correctly (i.e., receives 1 for all sub-practices), they can proceed to the next stage. If not, individualized scaffolding is automatically triggered based on the sub-practices on which the student was correct or incorrect, and they are prevented from moving forward to the next stage. This proactive approach was chosen because students often cannot recognize when to ask for help [30] and because making errors on earlier stages make subsequent stages fruitless to complete (e.g., it does not make sense to graph data that are completely confounded) [16]. This approach has shown to be effective in helping students learn and transfer other science inquiry competencies [31] even after many months [32]; however, to date, we had not tested this approach with the mathematical modeling competencies described in this study.

The automated scaffolding appears as an on-screen pop-up message delivered from a pedagogical agent, Rex. Rex scaffolding messages are specifically designed to orient and support students on the sub-practice for which they are struggling, explain how the sub-practice should be completed, and elaborate on why the sub-practice is completed in that way [30–33]. Students also have the option to ask further predefined questions to the agent to receive definitions for key terms and further elaborations on how to complete the sub-practice. If students continue to struggle, the student will eventually receive a bottom-out hint [30, 33] stating the actions they should take within the system to move forward in the activity. If the student needs support on multiple sub-practices, the scaffolds are provided in the priority order that was determined through discussions with domain experts and teachers familiar with the task and the Inq-ITS system. For example, if a student is struggling with both the "Good Form" and "Good Fit" sub-practices for the "Building Models" stage (Table 2), the student will receive scaffolding on the "Good Form" sub-practice *first* since the student must be able to identify the shape of the data before fitting the model to the data.

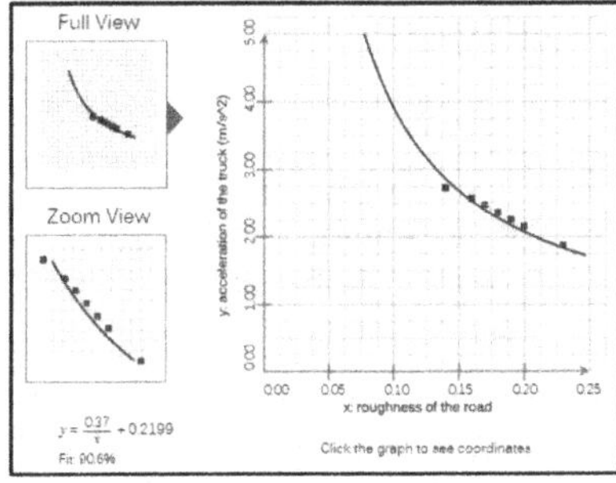

Fig. 3. Example screenshot of a student struggling with "Math Model has Good Form" sub-practice, but not with "Math Model has Good Fit" sub-practice (left; note: the student selected an *inverse* relationship when they should have chosen a *linear* relationship, given the variables on their graph); the first scaffold the student would receive to remediate this difficulty (right).

To illustrate, consider a student who is struggling with choosing the mathematical functional form that best demonstrates the relationship between variables, a common difficulty for students [9, 10, 15]. In this case, the student creates a mathematical model that appears to fit the data points plotted on the graph, but the function chosen for the model does *not* best represent the shape of the data in the graph (Fig. 3, left). When the student chooses to move on to next stage, the Inq-ITS assessment algorithms use features of the student's mathematical model, including the shape of the mathematical model (e.g. linear, square), the numerical values chosen for their coefficients and constants, and their fit scores, to determine that the student built a mathematical model with a "good fit" but *not* a "good form" (see Table 2 for sub-practice criteria). Rex then provides feedback to help the student ensure their model has the correct functional form expected between the variables selected for their graph. In this example, the first scaffold the student receives from Rex states, "Your mathematical model won't help you determine if your hypothesis is supported or not. Even though it fits the data points closely, its shape does not represent the trend in your data points." (Fig. 3, right). If the student continues to struggle on this sub-practice, the student will receive the next level of scaffold (i.e., a procedural hint), stating "Let me help you some more. Look at what kind of shape your data points make. Then, when you select the shape of the graph, choose the option that looks most like the shape your data points make." If the student continues to struggle after receiving the first two scaffolds, the student will receive a bottom-out hint stating, "Let me help you some more. The shape of your data looks most like linear." As illustrated, scaffolds are designed to support students in building their mathematical modeling competencies by focusing on the *fine-grained* sub-practice (e.g., choosing the correct functional form) with which the student is struggling in that moment.

2.4 Measures and Analyses

To measure students' competencies, students' stage scores are calculated as the average of the sub-practice scores for that stage (Table 2) before scaffolding was received (if any), as has been done in previous studies [32]. Because students may receive multiple scaffolds addressing different sub-practices on a single stage and the effect of those

Table 2. Operationalization of Automatically Scored Sub-Practices in the Inq-ITS Virtual Lab

Stage	Sub-Practice	Criteria
Stage 1: Hypothesizing	Hypothesis IV	A variable that can be *changed* by the experimenter was chosen as the IV in the hypothesis drop-down menu
	Hypothesis IV Goal-Aligned	The *goal-aligned* IV (the IV from the investigation goal) was chosen as the IV in the hypothesis drop-down menu
	Hypothesis DV	A dependent variable that will be *measured* was chosen as the DV in their hypothesis drop-down menu
	Hypothesis DV Goal-Aligned	The *goal-aligned* DV (the DV from the investigation goal) was chosen as the DV in the hypothesis drop-down menu
Stage 2: Collecting Data	Data Collection Tests Hypothesis	The student collected controlled data that can be used to develop a mathematical model demonstrating the relationship between the IVs and DVs stated in the investigation goal. Assessed by EDM algorithm [23]
	Data Collection is Controlled Experiment	The student collected controlled data that can be used to develop a mathematical model demonstrating the relationship between *any* of the changeable variables and the DV stated in the investigation goal. Assessed by EDM algorithm [23]
	Data Collection has Pairwise-IV CVS	The student collected at least two trials, where only the goal-aligned IV changes and all other variables are held constant (i.e., controlled variable strategy; CVS)
Stage 3: Plotting Data	Graph's X-Axis is an IV & Y-Axis is a DV	Using the drop-down menus, the student selected one of the potential IVs for the x-axis of their graph and one of the potential DVs for the y-axis of their graph

(*continued*)

Table 2. (*continued*)

Stage	Sub-Practice	Criteria
	Axes of Graph are Goal Aligned	Using the drop-down menus, the student selected the *goal-aligned* IV for x-axis and the *goal-aligned* DV for y-axis
	Axes of Graph are Hypothesis Aligned	The student selected the *hypothesis-aligned* IV (i.e., the IV that the student chose in hypothesis) as the x-axis of their graph and the *hypothesis-aligned* DV (i.e., the DV that the student chose in hypothesis) as the y-axis of their graph
	Graph Plotted Controlled Data	The student only plotted controlled data with respect to the variable chosen for the x-axis
	Graph Plotted Minimum for Trend	The student plotted controlled data with 5 unique values for the variable chosen for the x-axis. This number is sufficient to see mathematical trends for Inq-ITS' simulation designs
Stage 4: Building Models	Math Model has Good Form	The student built a model with the correct mathematical relationship, based on the variables selected for the graph's axes
	Math Model has Good Fit	The student built a model that fits the plotted data with at least 70% fit. This minimum score balances between students spending too much effort maximizing fit, and not having a useful model. It represents a reasonably strong fit to the data

(*continued*)

Table 2. (*continued*)

Stage	Sub-Practice	Criteria
	Math Model has Good Fit *and* Form	The student built a model that *both* has the correct mathematical relationship based on the variables selected for the axes of the graph *and* fits the plotted data with at least 70% fit. If the student has one model with good fit but not good form and another model with good form but not good fit, the student does not get credit

scaffolds may be entangled, we use the measures of students' overall competencies at the stage level for this study's analyses.

To determine the impact of the real-time AI-driven scaffolding, we analyzed how the scaffolded students' competencies from the first virtual lab activity (Truck) to the second virtual lab activity (Ramp). We note that students who received scaffolding on one stage (e.g., Collecting Data) did not necessarily receive scaffolding on another stage (e.g., Plotting Data). As such, we examined students' competencies on each stage separately to determine students' improvement on the competency for which they were helped. Furthermore, to isolate whether each type of scaffolding improved students' performance on the respective competency, we ran four two-tailed, paired samples *t*-tests with a Bonferroni correction (i.e., one for each competency to account for the chance of false-positive results when running the multiple *t*-tests).

We recognize that our analytical approach does not account for the effects of scaffolding on one competency possibly leading to improvements on other competencies (despite the student only having received scaffolding on one of the competencies). For example, a student may receive scaffolding on the Plotting Data stage, which in turn potentially impacts their performance with fitting the mathematical model to the plotted data on the subsequent Building Models stage [16]. However, unpacking the correlation between competencies and how the scaffolding may affect performance on multiple competencies was outside the scope of this study.

3 Results

We found that, for all stages, the scaffolded students' competencies *increased* from the first lab (Truck) to the second (Ramp; Fig. 2). With a Bonferroni corrected alpha (0.05/4 =.0125), the differences were significant for all four stages (i.e., Hypothesizing, Collecting Data, Plotting Data, Building Models; Table 3). Further, the effect sizes (Cohen's d) were large, suggesting that the automated scaffold was effective at helping students to improve at those competencies within the Inq-ITS labs.

Table 3. Average inquiry practice scores across activities and results of paired samples t-tests

Stage	N	Lab 1: M (SD)	Lab 2: M (SD)	Within-Subjects Effects
Hypothesizing	27	.41 (.30)	.74 (.27)	$t(26) = -5.45, p < .001, d = 1.05$
Collecting Data	37	.58 (.22)	.84 (.24)	$t(36) = -6.05, p < .001, d = 1.00$
Plotting Data	24	.63 (.21)	.86 (.21)	$t(23) = -4.12, p < .001, d = .84$
Building Models	31	.24 (.20)	.59 (.43)	$t(30) = -4.40, p < .001, d = .79$

4 Discussion

Students' competencies with mathematical modeling practices during science inquiry are critical for deep science learning [1, 2] and for future STEM courses and careers [4, 5]. However, students have difficulties with many aspects of mathematical modeling crucial to analyzing scientific phenomena [6] including those of focus in this study (e.g., identifying the functional form in plotted data [9, 10]). When students struggle with constructing and interpreting graphs in mathematics, it hampers their ability to transfer those competencies to science contexts [4, 34]. Further, even though these mathematical modeling competencies are necessary for developing deep understanding of science phenomena [6, 22, 29], they are not often addressed in science classrooms [7]. Thus, there is a need for resources that provide immediate, targeted support on the specific components for which students struggle, when it is optimal for learning [30, 31].

In this study, we found that students who received AI-driven real-time scaffolds during a virtual lab improved their mathematical modeling competencies when completing a near-transfer (i.e., isomorphic; [24]) task on the same physical science topic in a different scenario. These results suggest that scaffolds that address the sub-practices associated with each of the four stages in the virtual lab (i.e., Hypothesizing, Collecting Data, Plotting Data, and Building Models) are beneficial for students' learning and transfer of their mathematical modeling competencies. We speculate that students improve because the lab design operationalized the mathematical modeling practices (e.g., NGSS Practices 2 & 5) into *fine-grained* sub-practices. More specifically, students were given support based on their *specific* difficulties with these fine-grained sub-practices (e.g., labeling the axes of the graph, identifying the functional form in a graph, etc.). Furthermore, by addressing concerns raised by others who have articulated the lack of specificity in the NGSS for assessment purposes [11], we have evidence that our approach toward operationalizing, assessing, and scaffolding the sub-practices associated with the NGSS practices can positively impact students' learning.

Though promising, to better understand the generalizability of students' improvement as well as whether the improvement occurred because of the scaffolding or because of the practice opportunities, a randomized controlled experiment with a larger sample size comparing students' improvement with scaffolding versus without scaffolding in the virtual labs will be conducted. Future work will also disentangle how the scaffolding on one practice can impact students' competencies on other practices and examine students' ability to transfer their mathematical modeling competencies across physical science topics and assessment contexts outside of Inq-ITS, all of which are critical to achieve the vision of NGSS [1].

Acknowledgements. This material is based upon work supported by an NSF Graduate Research Fellowship (DGE-1842213; Amy Adair) and the U.S. Department of Education Institute of Education Sciences (R305A210432; Janice Gobert & Michael Sao Pedro). Any opinions, findings, and conclusions or recommendations expressed are those of the author(s) and do not necessarily reflect the views of either organization.

References

1. Next Generation Science Standards Lead States: Next Generation Science Standards: For States, By States. National Academies Press, Washington (2013)
2. National Science Board: Science and engineering indicators digest 2016 (NSB-2016-2). National Science Foundation, Arlington, VA (2016)
3. Gottfried, M.A., Bozick, R.: Supporting the STEM pipeline: linking applied STEM course-taking in high school to declaring a STEM major in college. Educ. Fin. Pol. **11**, 177–202 (2016)
4. Potgieter, M., Harding, A., Engelbrecht, J.: Transfer of algebraic and graphical thinking between mathematics and chemistry. J. Res. Sci. Teach. **45**(2), 197–218 (2008)
5. Sadler, P.M., Tai, R.H.: The two high-school pillars supporting college science. Sci. Educ. **85**(2), 111–136 (2007)
6. Glazer, N.: Challenges with graph interpretation: a review of the literature. Stud. Sci. Educ. **47**, 183–210 (2011)
7. Lai, K., Cabrera, J., Vitale, J.M., Madhok, J., Tinker, R., Linn, M.C.: Measuring graph comprehension, critique, and construction in science. J. Sci. Educ. Technol. **25**(4), 665–681 (2016)
8. Nixon, R. S., Godfrey, T. J., Mayhew, N. T., Wiegert, C. C.: Undergraduate student construction and interpretation of graphs in physics lab activities. Physical Review Physics Education Research 12(1), (2016).
9. Casey, S.A.: Examining student conceptions of covariation: a focus on the line of best fit. J. Stat. Educ. **23**(1), 1–33 (2015)
10. De Bock, D., Neyens, D., Van Dooren, W.: Students' ability to connect function properties to different types of elementary functions: an empirical study on the role of external representations. Int. J. Sci. Math. Educ. **15**(5), 939–955 (2017)
11. Penuel, W.R., Turner, M.L., Jacobs, J.K., Van Horne, K., Sumner, T.: Developing tasks to assess phenomenon-based science learning: challenges and lessons learned from building proximal transfer tasks. Sci. Educ. **103**(6), 1367–1395 (2019)
12. Furtak, E.M.: Confronting dilemmas posed by three-dimensional classroom assessment. Sci. Educ. **101**(5), 854–867 (2017)
13. Harris, C.J., Krajcik, J.S., Pellegrino, J.W., McElhaney, K.W.: Constructing Assessment Tasks that Blend Disciplinary Core Ideas, Crosscutting Concepts, and Science Practices for Classroom Formative Applications. SRI International, Menlo Park, CA (2016)
14. Gobert, J.D., Sao Pedro, M., Raziuddin, J., Baker, R.S.: From log files to assessment metrics: measuring students' science inquiry skills using educational data mining. J. Learn. Sci. **22**(4), 521–563 (2013)
15. Dickler, R., et al.: Supporting students remotely: Integrating mathematics and sciences in virtual labs. In: International Conference of Learning Sciences, pp. 1013–1014. ISLS (2021)

16. Olsen, J., Adair, A., Gobert, J., Sao Pedro, M., O'Brien, M.: Using log data to validate performance assessments of mathematical modeling practices. In: Artificial Intelligence in Education. Posters and Late Breaking Results, Workshops and Tutorials, Industry and Innovation Tracks, Practitioners' and Doctoral Consortium: 23rd International Conference, AIED 2022, Durham, UK, July 27–31, 2022, Proceedings, Part II, pp. 488–491. Springer International Publishing, Cham (2022). https://doi.org/10.1007/978-3-031-11647-6_99
17. Vanlehn, K., Wetzel, J., Grover, S., Sande, B.: Learning how to construct models of dynamic systems: an initial evaluation of the Dragoon intelligent tutoring system. IEEE Trans. Learn. Technol. **10**(2), 154–167 (2016)
18. Matuk, C., Zhang, J., Uk, I., Linn, M.C.: Qualitative graphing in an authentic inquiry context: how construction and critique help middle school students to reason about cancer. J. Res. Sci. Teach. **56**(7), 905–936 (2019)
19. VanLehn, K., et al.: The Andes physics tutoring system: lessons learned. Int. J. Artif. Intell. Educ. **15**(3), 147–204 (2005)
20. Koedinger, K.R., Anderson, J.R.: The early evolution of a Cognitive Tutor for algebra symbolization. Interact. Learn. Environ. **5**(1), 161–179 (1998)
21. Aleven, V., McLaren, B.M., Roll, I., Koedinger, K.R.: Help helps, but only so much: research on help seeking with intelligent tutoring systems. Int. J. Artif. Intell. Educ. **26**(1), 205–223 (2016)
22. Fretz, E.B., Wu, H.K., Zhang, B., Davis, E.A., Krajcik, J.S., Soloway, E.: An investigation of software scaffolds supporting modeling practices. Res. Sci. Educ. **32**(4), 567–589 (2002)
23. Sao Pedro, M., Baker, R., Gobert, J., Montalvo, O., Nakama, A.: Leveraging machine-learned detectors of systematic inquiry behavior to estimate and predict transfer of inquiry skill. User Model. User-Adap. Inter. **23**, 1–39 (2013)
24. Bassok, M., Holyoak, K.J.: Interdomain transfer between isomorphic topics in algebra and physics. J. Exp. Psychol. **15**(1), 153–166 (1989)
25. Bransford, J.D., Schwartz, D.L.: Rethinking transfer: a simple proposal with multiple implications. Rev. Res. Educ. **24**(1), 61–100 (1999)
26. Siler, S., Klahr, D., Matlen, B.: Conceptual change when learning experimental design. In: International Handbook of Research on Conceptual Change, pp.138–158. Routledge (2013)
27. Koedinger, K.R., Baker, R.S., Corbett, A.T.: Toward a model of learning data representations. In: Proceedings of the Twenty-Third Annual Conference of the Cognitive Science Society, pp. 45–50. Erlbaum, Mahwah, NJ (2001)
28. Uhden, O., Karam, R., Pietrocola, M., Pospiech, G.: Modelling mathematical reasoning in physics education. Sci. Educ. **21**(4), 485–506 (2012)
29. Jin, H., Delgado, C., Bauer, M., Wylie, E., Cisterna, D., Llort, K.: A hypothetical learning progression for quantifying phenomena in science. Sci. Educ. **28**(9), 1181–1208 (2019)
30. Aleven, V., Koedinger, K.R.: Limitations of student control: do students know when they need help? In: Gauthier, G., Frasson, C., VanLehn, K. (eds.) Intelligent Tutoring Systems, pp. 292–303. Springer Berlin Heidelberg, Berlin, Heidelberg (2000). https://doi.org/10.1007/3-540-45108-0_33
31. Sao Pedro, M., Baker, R., Gobert, J.: Incorporating scaffolding and tutor context into Bayesian knowledge tracing to predict inquiry skill acquisition. In: Educational Data Mining, pp. 185–192 (2013)
32. Li, H., Gobert, J., Dickler, R.: Evaluating the transfer of scaffolded inquiry: what sticks and does it last? In: Isotani, S., Millán, E., Ogan, A., Hastings, P., McLaren, B., Luckin, R. (eds.) AIED 2019. LNCS (LNAI), vol. 11626, pp. 163–168. Springer, Cham (2019). https://doi.org/10.1007/978-3-030-23207-8_31

33. Wood, H., Wood, D.: Help seeking, learning and contingent tutoring. Comput. Educ. **33**, 153–169 (1999)
34. Rebello, N.S., Cui, L., Bennett, A.G., Zollman, D.A., Ozimek, D.J.: Transfer of learning in problem solving in the context of mathematics and physics. In: Learning to Solve Complex Scientific Problems, pp. 223–246. Routledge, New York (2017)

Improving Automated Evaluation of Student Text Responses Using GPT-3.5 for Text Data Augmentation

Keith Cochran[1(✉)], Clayton Cohn[2], Jean Francois Rouet[3], and Peter Hastings[1]

[1] DePaul University, Chicago, IL 60604, USA
kcochr11@depaul.edu
[2] Vanderbilt University, Nashville, TN 37240, USA
[3] Université de Poitiers, 86073 Poitiers Cedex 9, France

Abstract. In education, intelligent learning environments allow students to choose how to tackle open-ended tasks while monitoring performance and behavior, allowing for the creation of adaptive support to help students overcome challenges. Timely feedback is critical to aid students' progression toward learning and improved problem-solving. Feedback on text-based student responses can be delayed when teachers are overloaded with work. Automated evaluation can provide quick student feedback while easing the manual evaluation burden for teachers in areas with a high teacher-to-student ratio. Current methods of evaluating student essay responses to questions have included transformer-based natural language processing models with varying degrees of success. One main challenge in training these models is the scarcity of data for student-generated data. Larger volumes of training data are needed to create models that perform at a sufficient level of accuracy. Some studies have vast data, but large quantities are difficult to obtain when educational studies involve student-generated text. To overcome this data scarcity issue, text augmentation techniques have been employed to balance and expand the data set so that models can be trained with higher accuracy, leading to more reliable evaluation and categorization of student answers to aid teachers in the student's learning progression. This paper examines the text-generating AI model, GPT-3.5, to determine if prompt-based text-generation methods are viable for generating additional text to supplement small sets of student responses for machine learning model training. We augmented student responses across two domains using GPT-3.5 completions and used that data to train a multilingual BERT model. Our results show that text generation can improve model performance on small data sets over simple self-augmentation.

Keywords: data augmentation · text generation · BERT · GPT-3.5 · educational texts · natural language processing

N. Wang et al. (Eds.): AIED 2023, LNAI 13916, pp. 217–228, 2023.
https://doi.org/10.1007/978-3-031-36272-9_18

1 Introduction

Researchers in educational contexts investigate how students reason and learn to discover new ways to evaluate their performance and provide feedback that promotes growth. Intelligent learning environments (ILEs) for K-12 students are designed to incorporate inquiry-based, problem-solving, game-based, and open-ended learning approaches [17,21,23]. By allowing students to choose how they approach and tackle open-ended tasks [37], they can utilize the resources available in the environment to gather information, understand the problem, and apply their knowledge to solve problems and achieve their learning objectives. At the same time, ILEs monitor students' performance and behavior, allowing for the creation of adaptive support to help students overcome challenges and become more effective learners [2,7,33].

Some research in this field aims to understand the factors that impact learning in various contexts. One area of study is centered on national and international literacy standards [1], which mandate that students should be able to think critically about science-related texts, understand scientific arguments, evaluate them, and produce well-written summaries. This is crucial for addressing societal issues such as bias, "fake news," and civic responsibility. However, achieving deep comprehension of explanations and arguments can be difficult for teenage students [25]. Additionally, research in discourse psychology suggests that students' reading strategies are shaped by their assigned reading task and other contextual dimensions [8]. For example, prior research has shown that students generate different types of inferences when reading as if to prepare for an exam compared to reading for leisure [9]. Similarly, students' writing is influenced by their perception of the audience [12].

Student responses in educational settings usually have a specific structure or purpose, which aligns with the grading criteria and demonstrates the student's level of understanding of the material. Natural Language Processing (NLP) techniques like sentence classification can be used to analyze student performance and provide feedback quickly [19]. *BERT-based* models have revolutionized the NLP field by being pre-trained on large datasets such as Wikipedia and BooksCorpus [15], giving them a deep understanding of language and how words are used *in context*. These models can then be fine-tuned for specific tasks by adding an output layer and training it with a smaller labeled dataset.

One common approach to improve models' performance with limited data is data augmentation [30]. This technique is commonly used in other fields of AI, such as computer vision. Attempts have been made to apply data augmentation techniques to text data [11], but it is more challenging because small changes in the text can produce bigger changes in the meaning, leading to errors in model training. Some current data augmentation techniques for text data involve modifying original responses, such as misspelling words or replacing them with similar words [34].

In this paper, we investigate text generation using three different "temperatures" and compare the results to a baseline measurement and a self-augmentation method, where the original data set is replicated to increase the training data. This technique of self-augmentation has been successful in previous research [13], and similar methods have been applied to computer vision with improved model performance [29]. We aim to determine the appropriate level of augmentation and establish a baseline measurement for comparison when additional augmentation techniques are applied.

2 Background and Research Questions

Data sets in educational contexts can sometimes be large, but when they are comprised of students' hand-generated responses, they tend to be on the order of at most few hundred responses. The amount of data obtained is sometimes a function of the nature of the tests. Modern machine learning models come pre-trained on various data sets. However, in order to improve performance on a given downstream task, these models need to be fine-tuned using labeled data [36]. Although some of these models can be good at zero-shot or few-shot learning [35], they are designed to allow further fine-tuning to improve performance for specific tasks when sufficient training data in both quantity and quality is available [18].

These educational data sets are also often imbalanced, meaning each label does not have equal representation. Machine learning models perform better when the data is close to being balanced across labels [28]. Data augmentation has improved model performance in image processing [30]. However, that process does not translate directly to text-based models. Simple replication of the data can be used and is referred to as self-augmentation. Looking at techniques beyond self-augmentation, [6] describes a taxonomy and grouping for data augmentation types. Cochran et al. showed that augmentation using masking, noise, and synonyms can improve classification performance [14]. This study continues that research by exploring augmentation using a generative AI method.

Recent studies have used text generation to improve classifier performance by augmenting data to create additional training data artificially [27,31]. The intent is to address the imbalance in data sets and allow smaller data sets to acquire larger data volumes to aid model training. Several survey papers on text augmentation break down the various types of data augmentation currently being researched [5,16,22]. In the generative method of text augmentation, artificial student responses are generated using a predictive model that predicts the response given a text prompt as input.

The OpenAI API performs NLP tasks such as classification or natural language generation given a prompt. OpenAI provides an interface for the Generative Pretrained Transformer 3.5 (GPT-3.5) [10], one of the most powerful language models available today [26]. A recent study has shown improvement for short text classification with augmented data from GPT-3.5, stating that it can be used without additional fine-tuning to improve classification performance [3]. Additionally, [6] note that GPT-3.5 is the leading augmentation method among recent papers and may even be able to replicate some instances whose labels are left out of the data set (zero-shot learning).

The student response data sets contain labels for each response corresponding to a hand-graded value on a grading rubric. Transformer-based NLP models, such as BERT [15] and GPT-3.5 [10], are now the industry standard for modeling many NLP tasks. Previous research by [14] shows that BERT-based transformers work well for text classification of student responses to STEM questions. Therefore, we are using GPT-3.5 for augmentation and continue to use BERT-based models for classification. Since we have two data sets in two languages, English and French, we use a BERT-based multilingual model as the classifier of choice.

RQ 1: Can artificially-generated responses improve base model classification performance? Our hypothesis **H1** is that additional augmented data will improve model performance for smaller data sets. Determining how large a data set needs to be before it would not require data augmentation is out of the scope of this study. Here, we are determining if augmentation will work for these data sets at all (i.e. can we reject the null hypothesis that the models perform the same with and without augmented data).

RQ 2: Can artificially-generated responses outperform self-augmentation when used for training models for sentence classification? Our hypothesis **H2** is that artificially-generated responses will outperform the self-augmentation method because they are not simple copies of the data, so more of the domain is likely to be filled with unique examples when creating the augmented data space.

RQ 3: Does temperature sampling of the artificially-generated student responses affect model performance? Recall the temperature variable for the OpenAI API allows for altering the probability distribution for a given pool of most likely completions. A lower value creates responses almost identical to the prompt text. A higher value (up to a maximum of 1) allows the model to choose more "risky" choices from a wider statistical field. **H3** proposes that augmenting the data with slightly more risky answers, equating to a temperature of 0.5, will provide the best performance in general. Until we test, we do not want to speculate if a number closer to 1.0 would improve performance or not. We hypothesize the temperature setting of 0.5 would be on the low side.

RQ 4: Does performance ultimately degrade when the model reaches a sufficient level of augmentation? It can be assumed that any augmentation would encounter overfitting, where model performance begins to degrade at some point [13]. **H4** is that the performance will degrade with additional augmentation after a peak is reached. **H5** is that the performance will degrade more slowly with higher temperature augmented data sets and thus support the idea that more risk involved in generated responses is better for larger amounts of augmentation.

3 Methods

3.1 Data Sets

Two data sets were obtained for this study. The first data set is from a discourse psychology experiment at a French university where 163 students were given an article describing links between personal aggression and playing violent video games. The participants were asked to read the article and write a passage to either a friend in a "personal" condition or a colleague in an "academic" condition. Our evaluation was around whether or not they asserted an opinion on the link between violent video games and personal aggression. The label quantities from the data set are shown in Table 1. The majority label quantity, "No Opinion", is shown in bold. The rightmost column gives the Entropy measure for the data, normalized for the four possible outcomes. A dataset which is balanced across labels would have an entropy value near 1.

Table 1. French Student Response Data Split for the Opinion Concept

Label	No Opinion	Link Exists	No Link	Partial Link	Normalized Entropy
Count	**118**	7	13	25	0.619

The second data set was obtained from a study [4,20,24,37] on students learning about rainwater runoff with responses from 95 6th-grade students in the southeastern United States. Responses were given in the English language.

Each of the six concepts was modeled individually as a binary classification task. Student responses that included the corresponding concept were coded as **Present**. Responses were otherwise coded as **Absent**. As previously mentioned, many small educational data sets are imbalanced. Table 2 shows the label quantities indicating the scarcity of data and the degree of imbalance in the dataset.

Table 2. Rainwater Runoff Student Response Data Split per Question

Concept	Absent	Present	Entropy
1	10	**85**	0.485
2a	25	**70**	0.831
2b	**64**	31	0.911
3a	44	**51**	0.996
3b	**73**	22	0.895
3c	**57**	38	0.971

3.2 Augmentation Approach

The label quantities shown in Tables 1 and 2, along with the normalized entropy, have the **majority quantity of reference** for that particular data set shown in bold. A balanced data set would have equal quantities across all labels and normalized entropy values at or near 1.0. We define an augmentation level of 1x when all labels have the same quantity as the majority quantity of reference for that data set. All data sets after that was augmented in multiples of the majority quantity up to 100x, or 100 times the majority quantity of reference for that data set. We generated data using GPT-3.5 (model "text-curie-001") with the prompt "paraphrase this sentence" and inserted an actual student response to fill in the rest of the language prompt. The data was generated, stored, and used directly in fine-tuning the BERT-based language models. The only modification was to add BERT's "special" [CLS] and [SEP] tokens so the model could process the text.

The OpenAI API provides a method for varying the degree of "aggressiveness" in generating text by adjusting temperature sampling. In this study, we performed tests at temperature values of 0.1, 0.5, and 0.9 to determine if temperature is an important factor in text generation such that it affects model performance.

After GPT-3.5 was used to create artificial student responses to augment small data sets, those augmented data sets were then used to determine if model performance using sentence classification improves or degrades.

3.3 Model Classification

Since we had data sets in two different languages, we chose a multilingual model to compare the use of language when performing fine-tuning. We chose the Microsoft *Multilingual L12 H384* model as a basis for all testing due to its performance gains over the base BERT model and its improved ability for fine-tuning [32]. We fine-tuned it using a combination of original data and augmented data for training. Data was held out from the original data set for testing purposes. A separate BERT model was fine-tuned for each concept and augmentation type to classify the data by adding a single feed-forward layer. This resulted in 28 separate *BERT-based* models that were fine-tuned and evaluated for this study. We used the micro-F_1 metric as the performance measurement. The models were trained and evaluated ten times, with each training iteration using a different seed for the random number generator, which partitions the training and testing instances. The train/test split was 80/20.

3.4 Baseline Evaluation

We evaluated two different baseline models for each concept. The *a priori* model chose the majority classification for each concept. For our *unaugmented* baseline, we applied BERT prototypically without data augmentation or balancing. The baseline performance results are shown in Table 3.

4 Results

Table 3 presents a summary of the results. Each row corresponds to a concept for the English data set, with one row for the French Data set. The leftmost data column shows the percentage of the answers for each concept marked with the majority label. The following two columns present the baseline results. On the right are the maximum F_1 scores for each concept using either self-augmented or generated data and indicating the augmentation level used to achieve that maximum performance. The highest performance for each data set is shown in bold, indicating which method or data set was used to achieve that score.

Table 3. Performance (micro-F_1) of baseline vs all augmented models

Concept	% Maj. Label	Baseline		Max Performance		
		a priori	Unaug.	Self	GPT-3.5	Aug.
French	73	**0.720**	0.575	0.636	0.612	21x
C1	89	**0.940**	0.735	0.789	0.815	0.6x
C2a	73	0.850	0.757	**0.931**	0.921	8x
C2b	67	0.670	0.547	0.852	**0.874**	55x
C3a	54	0.700	0.532	0.726	**0.815**	55x
C3b	77	0.770	0.684	0.926	**0.952**	55x
C3c	60	0.600	0.568	0.747	**0.832**	89x

Figure 1 illustrates how each of the four augmentation methods affected model performance as more augmentation was used to train the model. The "self" label on the chart indicates the self-augmentation method of creating multiple copies of the original data. The numbers 0.1, 0.5, and 0.9 indicate the temperature setting used on the GPT-3.5 API to provide varied responses, as previously discussed. Note that as augmentation increases, the "self" method peaks and begins to decline in performance with additional training data added, where the augmented models using GPT-3.5 do not drop off as much. This indicates the model is more tolerant of this generated data than continuing to fine-tune on the same small data set, copied multiple times.

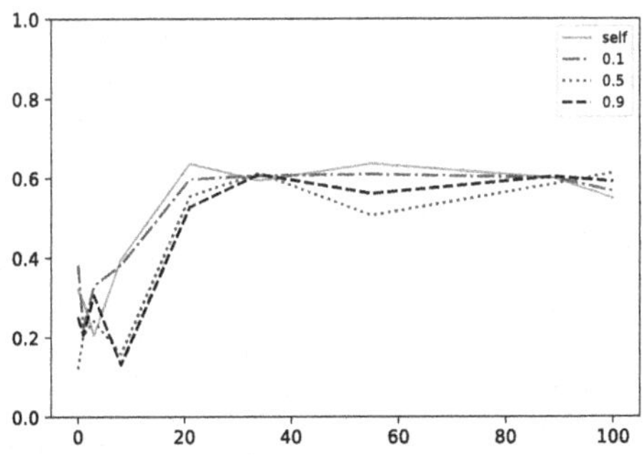

Fig. 1. French Model Performance (micro-F_1) per Augmentation Type. (Note: The x-axis shows the level of augmentation applied from 0x to 100x.)

Figure 2 shows how each of the model's (one for each concept) performances varied with training data using different augmentation types of self, and the other lines indicate each of the three temperatures used to generate text. Note that self-augmentation peaks early while the other types continue improving performance.

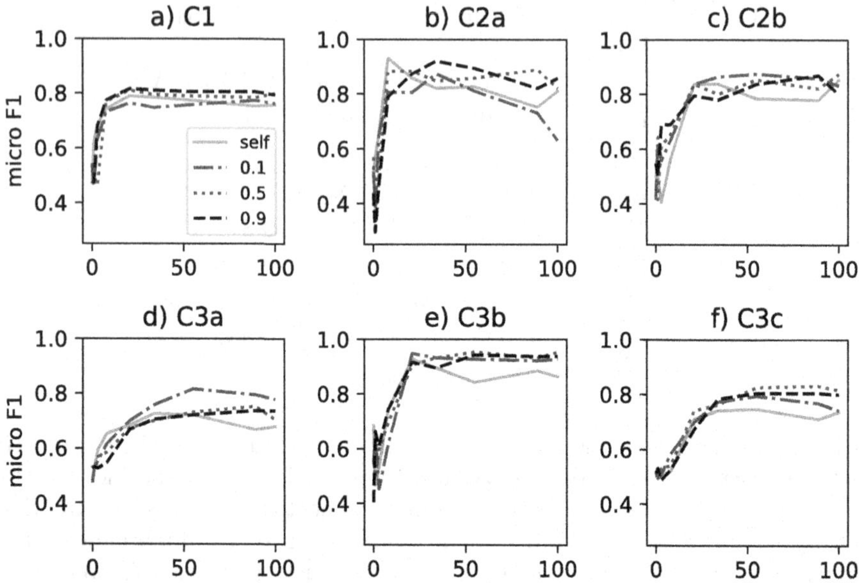

Fig. 2. Rainwater Runoff Model Performance per Augmentation Type. (Note: The x-axis shows the level of augmentation applied from 0x to 100x.)

5 Discussion

Recall **RQ 1**, which asks if artificially-generated responses improve base model performance. This research shows that the augmented model outperformed the unaugmented model in all seven concepts. However, the *a priori* computation which always selects the majority label won on two of the data sets. Our hypothesis **H1** stated that additional augmented data would improve model performance for smaller data sets, and that was shown to be supported by this data. The base model testing without augmentation was always improved upon with augmented data. However, two data sets, C1 and the French Data, were heavily imbalanced. Their entropy values were far from ideal, as shown in Tables 1 and 2. In these cases where entropy is low, guessing the majority label performed better than machine learning models to predict the label for the student response. Determining augmentation methods that improve performance on low-entropy data sets is an area of further research.

RQ 2 asks if artificially-generated responses outperform self-augmentation when training models for classification. This study shows that the maximum performance was achieved using artificially-generated responses in four out of seven concepts. Our hypothesis **H2** stating that artificially-generated responses will outperform simple replication of existing data as in the self-augmentation method was partially supported. This experiment shows that although simply replicating given data might produce good performance, generating new examples usually produced the best performance. However, this research also shows that the performance degrades faster when the only data augmentation is from self-augmentation. This needs further research to determine if this is a consistent way to get the model training jump-started before adding other types of augmentation into the mix.

Next, **RQ 3** asks if temperature sampling of the artificially-generated student responses affects model performance. Examining the maximum performance at each augmentation level did not reveal a single winner among the three temperatures used for student response generation. **H3** proposed that augmenting the data with slightly more risky answers, equating to a temperature of 0.5, will provide the best performance in general. In the charts presented in Figs. 1 and 2, each data set augmented by the three temperatures varied in performance but were similar to each other. Toward higher values of augmentation, the more risky generation using a temperature of 0.9 continued to increase in performance, indicating reduced overfitting during training. Temperature variation did not significantly alter performance but should be investigated further to see if that is true in general, or only in these specific data sets.

Finally, **RQ 4** ponders if performance ultimately degrades when the model reaches a sufficient level of augmentation. In all models tested, performance peaks and degrades after 55x to 89x augmentation. Table 3 shows the augmentation level at which performance peaked and began to fade. **H4** states that the performance will degrade with additional augmentation, which was supported by all the models tested. **H5** further states that the performance will degrade more slowly with higher temperature and thus more risky generated responses. When examining the performance changes in Figs. 1 and 2, the highest temperature, 0.9, rose in performance similar to other temperatures but decreased at a slower pace than the other temperatures, especially at higher augmentation levels. Due to this observation, this hypothesis is supported by the data.

6 Conclusion

This study intended to determine if GPT-3.5 was a viable solution to generate additional data to augment a small data set. We used one multilingual *BERT-based* model, trained it using two different data sets in two languages augmented by two different methods, and compared that result to baseline models against one using self-augmentation and three with GPT-3.5 augmentations. In four out of seven cases, a model augmented with GPT-3.5 generated responses pushed the performance beyond what could be achieved by other means.

Another objective of this study was to determine if setting the temperature or riskiness in GPT-3.5 response generation would affect performance. Our data shows that while it may not achieve peak results, the higher temperature generated text has more longevity because the models could take on more augmented data and maintain stability than other temperatures or self-augmentation.

These empirical tests show that augmentation methods such as self-augmentation and text generation with GPT-3.5 drastically improve performance over unaugmented models. However, the performance achieved by these models leveled off quickly after augmentation amounts of around fifty times the amount of original data. In addition, two data sets with severely imbalanced data did not improve performance enough to overcome their *a priori* computed values.

Using a higher temperature value when generating data from the GPT-3.5 model did not yield the highest-performing results but came very close. The added benefit of using the higher temperature is that the generated student responses seemed more diverse, allowing the model to prevent overfitting, even at higher augmentation levels.

7 Future Work

In this study, we showed how self-augmentation rises quickly but then levels off and degrades performance. Further research must be done to increase the diversity in generated student responses and research combinations of different augmentation techniques, including GPT-3.5 temperature that might be introduced at different augmentation levels.

References

1. Achieve Inc.: Next Generation Science Standards (2013)
2. Azevedo, R., Johnson, A., Chauncey, A., Burkett, C.: Self-regulated learning with MetaTutor: advancing the science of learning with metacognitive tools. In: Khine, M., Saleh, I. (eds.) New Science of Learning, pp. 225–247. Springer, New York (2010). https://doi.org/10.1007/978-1-4419-5716-0_11
3. Balkus, S., Yan, D.: Improving short text classification with augmented data using GPT-3. arXiv preprint arXiv:2205.10981 (2022)
4. Basu, S., McElhaney, K.W., Rachmatullah, A., Hutchins, N., Biswas, G., Chiu, J.: Promoting computational thinking through science-engineering integration using computational modeling. In: Proceedings of the 16th International Conference of the Learning Sciences (ICLS) (2022)
5. Bayer, M., Kaufhold, M.-A., Buchhold, B., Keller, M., Dallmeyer, J., Reuter, C.: Data augmentation in natural language processing: a novel text generation approach for long and short text classifiers. Int. J. Mach. Learn. Cybern. **14**, 135–150 (2022). https://doi.org/10.1007/s13042-022-01553-3
6. Bayer, M., Kaufhold, M.A., Reuter, C.: A survey on data augmentation for text classification. arXiv preprint arXiv:2107.03158 (2021)

7. Biswas, G., Segedy, J.R., Bunchongchit, K.: From design to implementation to practice a learning by teaching system: Betty's brain. Int. J. Artif. Intell. Educ. **26**(1), 350–364 (2016)
8. Britt, M.A., Rouet, J.F., Durik, A.M.: Literacy Beyond Text Comprehension: A Theory of Purposeful Reading. Routledge (2017)
9. van den Broek, P., Tzeng, Y., Risden, K., Trabasso, T., Basche, P.: Inferential questioning: effects on comprehension of narrative texts as a function of grade and timing. J. Educ. Psychol. **93**(3), 521 (2001)
10. Brown, T.B., et al.: Language models are few-shot learners. arXiv preprint arXiv:2005.14165 (2020)
11. Chen, J., Tam, D., Raffel, C., Bansal, M., Yang, D.: An empirical survey of data augmentation for limited data learning in NLP. arXiv preprint arXiv:2106.07499 (2021)
12. Cho, Y., Choi, I.: Writing from sources: does audience matter? Assess. Writ. **37**, 25–38 (2018)
13. Cochran, K., Cohn, C., Hastings, P.: Improving NLP model performance on small educational data sets using self-augmentation. In: Proceedings of the 15th International Conference on Computer Supported Education (2023, to appear)
14. Cochran, K., Cohn, C., Hutchins, N., Biswas, G., Hastings, P.: Improving automated evaluation of formative assessments with text data augmentation. In: Rodrigo, M.M., Matsuda, N., Cristea, A.I., Dimitrova, V. (eds.) AIED 2022. LNCS, vol. 13355, pp. 390–401. Springer, Cham (2022). https://doi.org/10.1007/978-3-031-11644-5_32
15. Devlin, J., Chang, M.W., Lee, K., Toutanova, K.: BERT: pre-training of deep bidirectional transformers for language understanding. arXiv preprint arXiv:1810.04805 (2018)
16. Feng, S.Y., et al.: A survey of data augmentation approaches for NLP. arXiv preprint arXiv:2105.03075 (2021)
17. Geden, M., Emerson, A., Carpenter, D., Rowe, J., Azevedo, R., Lester, J.: Predictive student modeling in game-based learning environments with word embedding representations of reflection. Int. J. Artif. Intell. Educ. **31**(1), 1–23 (2020). https://doi.org/10.1007/s40593-020-00220-4
18. Gururangan, S., et al.: Don't stop pretraining: adapt language models to domains and tasks. arXiv preprint arXiv:2004.10964 (2020)
19. Hastings, P., Hughes, S., Britt, A., Blaum, D., Wallace, P.: Toward automatic inference of causal structure in student essays. In: Trausan-Matu, S., Boyer, K.E., Crosby, M., Panourgia, K. (eds.) ITS 2014. LNCS, vol. 8474, pp. 266–271. Springer, Cham (2014). https://doi.org/10.1007/978-3-319-07221-0_33
20. Hutchins, N.M., et al.: Coherence across conceptual and computational representations of students' scientific models. In: Proceedings of the 15th International Conference of the Learning Sciences, ICLS 2021. International Society of the Learning Sciences (2021)
21. Käser, T., Schwartz, D.L.: Modeling and analyzing inquiry strategies in open-ended learning environments. Int. J. Artif. Intell. Educ. **30**(3), 504–535 (2020)
22. Liu, P., Wang, X., Xiang, C., Meng, W.: A survey of text data augmentation. In: 2020 International Conference on Computer Communication and Network Security (CCNS), pp. 191–195. IEEE (2020)
23. Luckin, R., du Boulay, B.: Reflections on the Ecolab and the zone of proximal development. Int. J. Artif. Intell. Educ. **26**(1), 416–430 (2016)

24. McElhaney, K.W., Zhang, N., Basu, S., McBride, E., Biswas, G., Chiu, J.: Using computational modeling to integrate science and engineering curricular activities. In: Gresalfi, M., Horn, I.S. (eds.) The Interdisciplinarity of the Learning Sciences, 14th International Conference of the Learning Sciences (ICLS) 2020, vol. 3 (2020)
25. OECD: 21st-Century Readers. PISA, OECD Publishing (2021). https://doi.org/10.1787/a83d84cb-en. https://www.oecd-ilibrary.org/content/publication/a83d84cb-en
26. Pilipiszyn, A.: GPT-3 powers the next generation of apps (2021)
27. Quteineh, H., Samothrakis, S., Sutcliffe, R.: Textual data augmentation for efficient active learning on tiny datasets. In: Proceedings of the 2020 Conference on Empirical Methods in Natural Language Processing (EMNLP), pp. 7400–7410. Association for Computational Linguistics (2020)
28. Schwartz, R., Stanovsky, G.: On the limitations of dataset balancing: the lost battle against spurious correlations. arXiv preprint arXiv:2204.12708 (2022)
29. Seo, J.W., Jung, H.G., Lee, S.W.: Self-augmentation: generalizing deep networks to unseen classes for few-shot learning. Neural Netw. **138**, 140–149 (2021). https://doi.org/10.1016/j.neunet.2021.02.007. https://www.sciencedirect.com/science/article/pii/S0893608021000496
30. Shorten, C., Khoshgoftaar, T.M.: A survey on image data augmentation for deep learning. J. Big Data **6**(1), 1–48 (2019)
31. Shorten, C., Khoshgoftaar, T.M., Furht, B.: Text data augmentation for deep learning. J. Big Data **8**(1), 1–34 (2021)
32. Wang, W., Wei, F., Dong, L., Bao, H., Yang, N., Zhou, M.: MiniLM: deep self-attention distillation for task-agnostic compression of pre-trained transformers. In: Advances in Neural Information Processing Systems, vol. 33, pp. 5776–5788 (2020)
33. Winne, P.H., Hadwin, A.F.: nStudy: tracing and supporting self-regulated learning in the Internet. In: Azevedo, R., Aleven, V. (eds.) International Handbook of Metacognition and Learning Technologies. SIHE, vol. 28, pp. 293–308. Springer, New York (2013). https://doi.org/10.1007/978-1-4419-5546-3_20
34. Wu, L., et al.: Self-augmentation for named entity recognition with meta reweighting. arXiv preprint arXiv:2204.11406 (2022)
35. Xia, C., Zhang, C., Zhang, J., Liang, T., Peng, H., Philip, S.Y.: Low-shot learning in natural language processing. In: 2020 IEEE Second International Conference on Cognitive Machine Intelligence (CogMI), pp. 185–189. IEEE (2020)
36. Yogatama, D., et al.: Learning and evaluating general linguistic intelligence. arXiv preprint arXiv:1901.11373 (2019)
37. Zhang, N., Biswas, G., McElhaney, K.W., Basu, S., McBride, E., Chiu, J.L.: Studying the interactions between science, engineering, and computational thinking in a learning-by-modeling environment. In: Bittencourt, I.I., Cukurova, M., Muldner, K., Luckin, R., Millán, E. (eds.) AIED 2020. LNCS (LNAI), vol. 12163, pp. 598–609. Springer, Cham (2020). https://doi.org/10.1007/978-3-030-52237-7_48

The Automated Model of Comprehension Version 3.0: Paying Attention to Context

Dragos Corlatescu[1,2], Micah Watanabe[3], Stefan Ruseti[1], Mihai Dascalu[1,4(✉)], and Danielle S. McNamara[3]

[1] University Politehnica of Bucharest, 313 Splaiul Independentei, 060042 Bucharest, Romania
{dragos.corlatescu,stefan.ruseti,mihai.dascalu}@upb.ro
[2] CrowdStrike, 201 Str. Barbu Văcărescu, 020285 Bucharest, Romania
dragos.corlatescu@crowdstrike.com
[3] Department of Psychology, Arizona State University, PO Box 871104, Tempe, AZ 85287, USA
{micah.watanabe,dsmcnama}@asu.edu
[4] Academy of Romanian Scientists, Str. Ilfov, Nr. 3, 050044 Bucharest, Romania

Abstract. Reading comprehension is essential for both knowledge acquisition and memory reinforcement. Automated modeling of the comprehension process provides insights into the efficacy of specific texts as learning tools. This paper introduces an improved version of the Automated Model of Comprehension, version 3.0 (AMoC v3.0). AMoC v3.0 is based on two theoretical models of the comprehension process, namely the Construction-Integration and the Landscape models. In addition to the lessons learned from the previous versions, AMoC v3.0 uses Transformer-based contextualized embeddings to build and update the concept graph as a simulation of reading. Besides taking into account generative language models and presenting a visual walkthrough of how the model works, AMoC v3.0 surpasses the previous version in terms of the Spearman correlations between our activation scores and the values reported in the original Landscape Model for the presented use case. Moreover, features derived from AMoC significantly differentiate between high-low cohesion texts, thus arguing for the model's capabilities to simulate different reading conditions.

Keywords: Reading Comprehension · Automated Model of Comprehension · Language Model

1 Introduction

Reading comprehension is a foundational skill for both knowledge acquisition [9] and memory reinforcement [1]. Successful reading comprehension depends on both individual factors (e.g., decoding ability, goals, motivation, and prior knowledge) and features of the text (e.g., syntactic complexity and cohesion).

N. Wang et al. (Eds.): AIED 2023, LNAI 13916, pp. 229–241, 2023.
https://doi.org/10.1007/978-3-031-36272-9_19

Importantly, these two factors (individual features and text features) have interactive effects. For example, less skilled readers benefit from highly cohesive texts, whereas skilled readers benefit from less cohesive texts [11]. While the theoretical effects of individual and text features on comprehension have been documented, predicting the effects across different texts depends on effective modeling of the comprehension process. Therefore, modeling the comprehension process across numerous texts can provide information on the efficacy of specific texts as learning tools prior to their use in research and classrooms.

The Automated Model of Comprehension (AMoC) provides teachers and researchers the ability to model the reading comprehension process of different texts. AMoC leverages the general frameworks provided by the Construction-Integration (CI) Model and Landscape Model and integrates modern language models and natural language processing tools in order to simulate the readers' mental representation during reading. When a text is analyzed with AMoC, a concept graph is produced, which represents the generation of concepts (nodes) and inferences (edges). While previous iterations of AMoC have been tested and published, the current version uses state-of-the-art language models that are superior to those used in previous versions.

The Automated Model of Comprehension has its roots in the Construction-Integration model [10] and the Landscape Model [1]. The CI model describes comprehension as a two-step process. Readers first *construct* a mental representation of a text's meaning by generating inferences, and then *integrate* the new, text-based information with their own existing knowledge. The Landscape Model simulates the activation of concepts in a similar way to the CI Model; however, it is further used to simulate the fluctuation of activated concepts across time. Within the Landscape Model, prior knowledge can be activated through two different mechanisms: cohort activation, which is the passive linking of concepts related to the reader's mental representation through the formation of associative memory traces, and coherence-based retrieval, which represents the importance of a word in relation to the text (such as causal, temporal, or spatial connections). The coherence parameter is smaller when the reading process is more superficial. Importantly, both the CI model and the Landscape model were automated, but the parameters and connections were entirely hand-sewn - meaning the experimenter listed the nodes and connections, as well as the assumed knowledge activated by the reader, and the models simply'put' the pieces together. Our objective is to automate the entire reading process, including which nodes are activated, their connections, and whatever outside knowledge is activated during the reading process.

Two versions of the Automated Model of Comprehension have been published (i.e., v1.0 [5] and v2.0 [3]), but recent developments in text analysis have enabled improvement in AMoC's internal modeling and output. The third version of AMoC uses the Transformer [16] architecture for taking into account context and having a better representation of words. We publicly released our model and all corresponding code on GitHub[1]. In contrast, AMoC v2.0 [3] employed a

[1] https://github.com/readerbench/AMoC.

combination of word2vec [13] and Wordnet [14] to generate nodes and edges. The main feature of the Transformer is the self-attention mechanism which enables a better understanding of the relations between words in a sentence across a wider span. Figure 1 introduces attention edges generated using BertViz [17] from the "knight" token in the second sentence to its previous occurrence, as well as lexical associations to other words. Additionally, Transformers are faster in both the training phase and inference phase in comparison to Recurrent Neural Networks due to their parallel nature. This architecture has been successfully employed on a variety of NLP tasks for which it holds state-of-the-art performance.

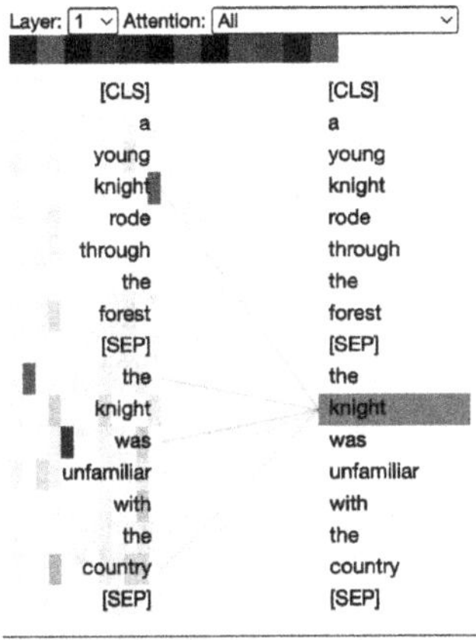

Fig. 1. BertViz self-attentions.

Generative Pre-training Transformer (GPT) is a large Transformer-based generative language model [15]. Instead of focusing on representing word embeddings such as the BERT [6] encoder, GPT's goal is to generate text with cursive ideas that map well with the context and are syntactically correct. As such, the end goal can be considered the generation of text that is indistinguishable from human-generated writing. The training process involved predicting the next word in a sequence given the previous words. Additionally, GPT can be customized to slightly guide its prediction - for example, the length in tokens can be set by the user. N-grams of different lengths can be set not to repeat in order to have a more diverse generation. Sampling can also be used when choosing the next word in a generation to ensure a more diverse output. Multiple GPT models have been released at the time of this writing, namely versions 1–4. GPT-3 and 4 [2] require extensive resources and are available only via API. Since we want to make AMoC available to everyone and deployable on most computers with average specifications, we decided to train and test AMoC v3.0 using GPT-2.

2 Method

AMoC mimics human comprehension by using the generative capabilities of state-of-the-art language models. When analyzing a text, AMoC generates a concept graph. The concept map contains nodes and edges. Each node represents a concept (i.e., a noun, verb, adjective, or adverb) and includes information about the word lemma and whether the concept is text-based or inferred. The edges are weighted links, the weight being derived from the attention scores of the GPT-2 model in different contexts. Here, we consider the mean attention score between 2 words across all layers from GPT-2. The AMOC v3.0 workflow is depicted in Fig. 2 as follows.

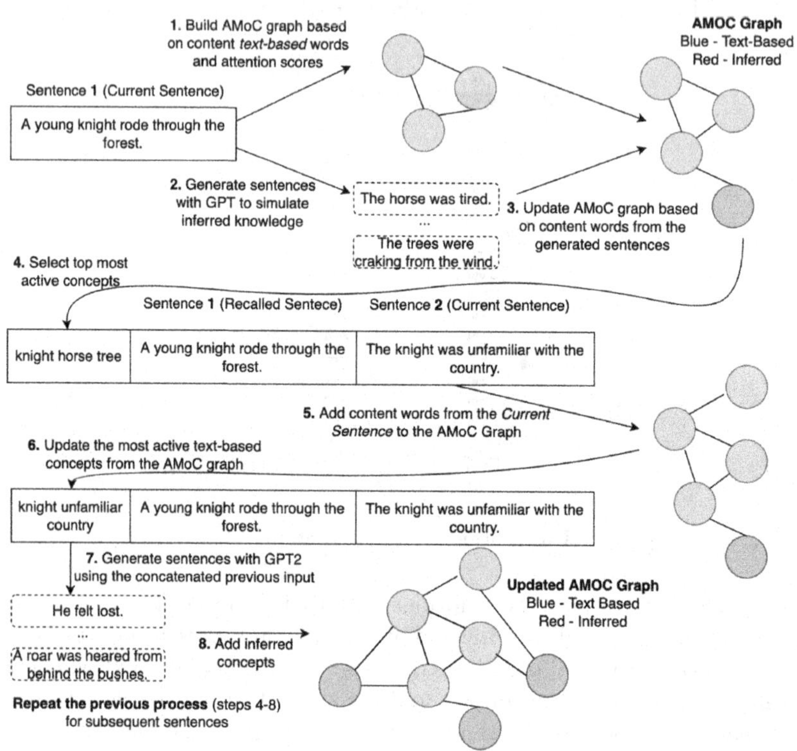

Fig. 2. AMoC v3 processing flow.

2.1 Processing Flow

The text analysis begins with an empty concept graph. There are no previous sentences in the working memory of the model nor any concepts in the AMoC graph. In the first step, the content words from the sentence (i.e., nouns, verbs, adjectives, and adverbs) are added to the graph as nodes with edges connecting them. The weights of the edges are given by the attention scores of the *Model*. In the second step, sentences are generated using the *Model*, while the content words from the generated sentences are added to the graph, along with the edges connecting the generated content words to the text-based content words.

Note that this first sentence scenario is just a sub-scenario of a general analysis presented below.

For each sentence after the first sentence, the model retains active concepts from the *PrevSent* sentences as nodes in the concept graph. In addition, the model retains edges and utilizes a decay factor to mimic reading over time (see below). The *PrevSent* sentences, top *Active* concepts from the concept graph, and the current sentence are given to the model. New nodes are generated, and edges are added based on the attention scores of the content words from the current sentence and the other content words.

New sentences are generated using the *Model* that receives as input the current sentence, past *PrevSent* sentences, and top *Active* nodes from the AMoC graph that are text-based. Note that in the generation process, we do not consider only the concepts that are text-based (i.e. not the inferred concepts). The generation can take multiple forms using the *Imagin* (parameter) to make them more diverse or more clustered on the same idea. We can also vary the number of sentences generated as well as the length of tokens from the generation sequence.

The top *DictExp* number of words that have a maximum attention score greater than a *AttThresh* threshold are added to the graph. This process limits the number of inferred concepts from the generation process as well as imposes a condition that the words should have a strong connection (attention score) with at least one of the words from the "source" text.

When analyzing the next sentence in the text (i.e., simulating the reading of another sentence), the model supports a decay factor (i.e., forgetting the previous sentences) (parameter *WD*) such that all the edges in the graph are decayed with a specified percentage. In addition, the importance of the nodes can be scaled with their Age of Exposure [4] scores (parameter *UseAOE*).

The process repeats with subsequent sentences, past *PrevSent* sentences as a sliding window, top *Active* concepts active from the graph, and corresponding method to update the AMoC graph, generate follow-up sentences and add new inferred concepts to the AMoC graph.

2.2 Features Derived from AMoC

The following list introduces the graph metrics that consider either the entire AMoC graph or only the active nodes from the AMoC graph. In order to obtain the final values, the graph after each sentence was saved, and the metrics were computed on each of the intermediate graphs as the mean value of the targeted nodes. Then, the mean of all of these mean values per sentence was computed for the entire AMoC graph.

- Closeness centrality - a local measure of centrality per node that takes into account the minimum distance from a node to all other nodes;
- Betweenness centrality - a local measure reflecting the number of shortest paths that go through a node;
- Degree centrality - a local measure computed as the sum of the weights to adjacent nodes;

- Density - a global measure of the AMoC graph, which is the ratio between the number of edges and the maximum possible number of edges in the AMoC graph;
- Modularity - a global measure of how difficult it is to split the AMoC graph into multiple subgraphs.

2.3 Experimental Setup

The model considers the following Parameters presented in Table 1. Each of the parameters aligns with assumptions regarding the reader's skill in parsing, understanding, and remembering text, their knowledge of words, and their tendency to go beyond the text using elaboration and imagery. Note that the default values were chosen based on an expert judgment that follows an analogy to the manner in which other comprehension models were evaluated (i.e., the CI and Landscape models).

Table 1. AMoC v3.0 parameters.

Parameter	Description	Default value
Model	Normal (more complex) versus Distilled (less complex, faster) GPT	gpt2
PrevSent	Number of sentences that are remembered in their raw form (max 3)	1
Active	Number of active concepts in the reader's mind	5
DictExp	Maximum dictionary expansion	7
AttThresh	Attention score threshold for new concepts	0.3
Imagin	Values from 1 to 4 denoting potential elaboration and imagery by the reader (1 less imaginative, 4 more imaginative)	1
WD	Percentage of decay of each edge from sentence to sentence	0.1
UseAOE	Binary value indicative of whether to scale by AOE scores when computing the importance of the nodes	1 (True)

2.4 Comparison Between AMoC Versions

The main differences between the three AMoC versions are displayed in Table 2. AMoC v3.0 is implemented in Python like AMoC v2.0 [3], while AMoC v1.0 [5] was in Java. spaCy [8] is also used in both the 2nd and 3rd versions to apply basic processing on the text (i.e., lemmatization, POS tagging for the extraction of content words). The 1st and the 2nd versions of AMoC use word2vec to provide weights for the edges in the concept graph, while in comparison, the 3rd version uses GPT2 attention scores. As mentioned in the introduction section, Transformer models (including GPT2) obtained state-of-the-art results in multiple fields due to the self-attention mechanism, along with other aspects. In addition, the similarity scores obtained by Transformer models are context-related. Therefore, these scores should be of higher quality than other similarity scores that are general and do not take context into account (e.g., scores provided by word2vec). The inference process also differs in AMoC v3.0 compared to the previous versions. AMoC v3.0 uses GPT2 text generation to extract new concepts that are more related to the context of the text, rather than just picking similar words from a language model (word2vec) or WordNet. Lastly, AMoC v3.0 is highly customizable with 8 parameters that can be manually configured to serve the user's needs, surpassing by far the 3 customizable parameters from AMoC v2.0.

Table 2. Comparison between AMoC versions.

Feature	AMoC v1.0	AMoC v2.0	AMoC v3.0
Programming Language	Java	Python	Python
Sentence Segmentation	Stanford Core NLP v3.8	SpaCy v2.3.5	SpaCy v3.4
Tokenization	Stanford Core NLP v3.8	SpaCy v2.3.5	SpaCy v3.4
Language model used for weighting text-based concepts	Word2Vec	Word2Vec	GPT2 attention scores
Inferring new concepts	WordNet + Word2Vec	WordNet + Word2Vec	GPT2 text generation
Adding newly inferred concepts	PageRank	PageRank	GPT2 attention scores
Number of parameters	3	3	8

3 Results

We begin this section by showcasing the stages of the AMoC v3.0 processing flow using real text. Next, we present two experiments conducted on both AMoC v2.0 and AMoC v3.0 to validate the models and also provide a performance comparison between them. The first experiment involves computing correlations between the Landscape Model word scores and the activation scores from AMoC models, while the second experiment analyzes how related are the AMoC graph centrality metrics to the cohesion of the text.

3.1 Use Case

We used the text from the original Landscape Model paper to mimic how AMoC v3.0 works. We showcase the first and the last sentences with the corresponding outputs of our model (see Fig. 3). The nodes in the graph have the following colors: dark blue - text-based and active; light blue - text-based and not active; dark red - inferred and active; light red - inferred and not active. Also, note that there are maximum 10 nodes displayed, even though there might be more. This decision was made so that the concepts could be easily followed. The parameters used were the default ones, as presented in Table 1.

- First text sentence: A young knight rode through the forest
  • Top Graph Concepts: None
  • Modcl Generated sentences (selection of 2):
    * The girl who rode it was an apprentice knight, and she wore a light pink dress. "This one is my daughter's second bride."
    * At this moment he stopped. He looked down at the black sword which he held in his hand.
- Last text sentence: She married the knight
  • Top Graph Concepts: dragon red knight princess king
  • Model Generated sentences (selection of 2):
    * He married her by making him the best knight he could be. Because of that, his knights would be able to make the top of this league, even without the royal crown.
    * "The man in charge of the palace."

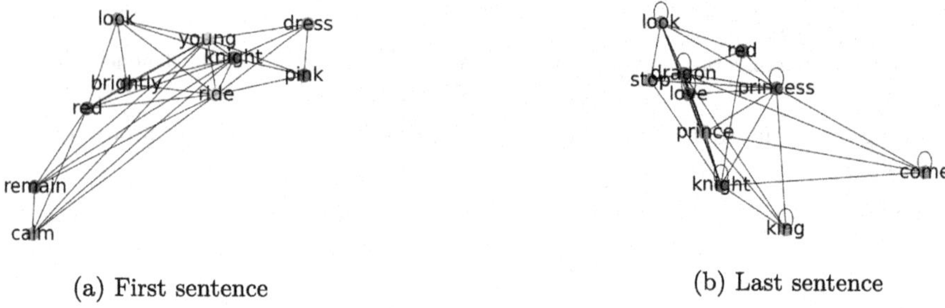

(a) First sentence (b) Last sentence

Fig. 3. AMoC graphs after each selected sentence.

3.2 Correlations to the Landscape Model

The first experiment involves analyzing the correlation between the activation scores presented in the Landscape Model paper and the activation scores obtained by AMoC v2.0 and AMoC v3.0. The scores from the Landscape Model are from 0 to 5 for each sentence, where 0 denotes that the word is not present at all in the memory, while 5 is the highest level of activation. The scores are for both text-based and inferred concepts. However, since inferring new concepts differs from method to method, we decided to consider only text-based words in this analysis.

The setup of the experiment was the following: we performed a grid-search for two parameters *maximum active concepts* and *maximum dictionary expansion*, while keeping the other parameters fixed for both models, namely: for AMoC v2.0, the *activation score* was set to 0.3; for AMoC v3.0, the *Model* was GPT-2, the *PrevSent* was 1, the *AttThresh* was 0.3, the *Imagin* was 1, the *WD* was 0.1, and the *UseAOE* was True. The *maximum active concepts* was varied between 2 and 5, and the *maximum dictionary expansion* was varied between 3 and 7. Then, for each case, the Spearman correlation between AMoC models was computed with the scores from the Landscape model. Two types of Spearman correlations were considered: correlation per word with respect to the sentences and correlation per sentence with respect to the words.

Table 3 shows the Spearman correlation scores. The scores argue that, in general, there is a correlation between the two approaches, even though they are approaching the problem from different starting points. While inspecting the word correlation, AMoC v3.0 is performing better than its previous version. AMoC v2.0 is performing slightly better in the sentence correlation category, but only with a marginal difference of .03. Figure 4 depicts surface plots with plots scores obtained from the grid search as a surface plot; the views are consistent with the mean scores.

Table 3. Mean Correlation Scores.

Spearman Correlation	AMoC v2.0	AMoC v3.0
Per Word	.566	.656
Per Sentence	.791	.764

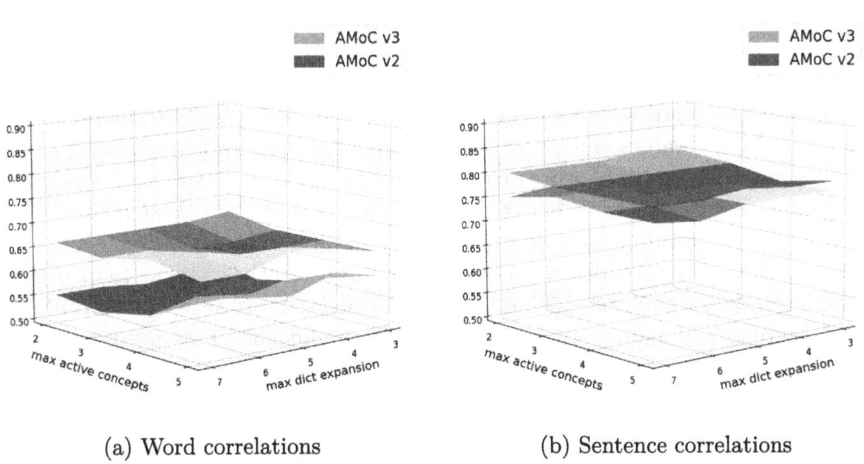

(a) Word correlations (b) Sentence correlations

Fig. 4. Surface plots denoting correlations between both AMoC models and the Landscape Model.

An important observation is to be made about AMoC v3.0 compared to AMoC v2.0 and the Landscape model, as the last two give a high weight to the text-based words and also to their absence. For example, in the last two sentences from the text (*"The princess was very thankful to the knight. She married the knight."*), the activation score in the Landscape Model for the word "dragon" is 0, which means that the reader forgot everything about the dragon. In a similar manner, when the knight and the dragon fought, the princess also had 0 as activation. We believe that those scores should be lower, but not 0, since we are referring to the central characters of the story. As such, AMoC v3.0 offers activation of approximately 0.8 (out of 1) for "dragon" in the last sentences; nevertheless, this also has a negative impact on the overall correlations with the Landscape model results.

3.3 Diffentiating Between High-Low Cohesion Texts

This experiment was conducted to assess the extent to which the features derived from AMoC differentiate between high-low cohesion texts. Moreover, we evaluate whether cohesion differences have a lower impact on an individual with higher knowledge by changing the simulation parameters for AMoC.

The considered dataset consists of 19 texts in two forms [12], one having high cohesion and one having low cohesion. The (initial) low cohesion texts were modified by expert linguists to have a higher cohesion, but the same ideas were retained in both versions. In general, the texts are of medium length, with a mean length of approximately forty sentences. The high cohesion texts tend to be a little longer both in the number of words and of sentences, but this property did not affect the experiment.

The Linear Mixed Effects [7] statistical test was employed to analyze how well AMoC models differentiate between the same texts, one with a high and the other with low cohesion. The parameters were similar for the two models, namely: the maximum dictionary expansion was set to 9, while the maximum active concepts were set to 7. For AMoC v3.0, all the other parameters were given the default values since it supports multiple configurable parameters.

The majority of the features computed with AMoC v3.0 were statistically significant in differentiating between high-cohesion and low-cohesion texts (see Table 4). The centrality measurements are relevant to this situation because a high-cohesion text should have its concepts more tightly coupled than a low-cohesion text. In contrast, AMoC v2.0 did not perform as well - except for one feature, the others were not statistically significant. This argues for the superiority of the current version of AMoC. Also, AMoC v2.0 does not compute metrics for the active concepts, but nonetheless, the all-nodes statistics are a good indicator that the newer version is superior.

Table 4. Differences between low and high cohesion text in AMoC v2.0 and v3.0 configurations.

Feature	All/Active	AMoC v2.0		AMoC v3.0	
		F	p	F	p
Closeness centrality	active	–	–	18.37	<.001
	all	1.79	.197	6.05	.024
Betweenness centrality	active	–	–	13.06	<.001
	all	0.11	.741	0.90	.353
Degree centrality	active	–	–	10.61	<.001
	all	1.49	.237	1.95	.179
Density	active	–	–	10.61	<.001
	all	1.49	.237	1.95	.179
Modularity	active	–	–	3.71	.070
	all	4.39	.050	5.55	.029

4 Conclusions and Future Work

In this paper, we introduced the third version of the Automated Model of Comprehension. AMoC v3.0 improves on past versions by using modern Natural Language Techniques (i.e., large language models) that produce better language representation and more relevant relations between words. A detailed comparison was presented between the current version and the previous two versions, where we showed that AMoC v3.0 has improved performance and customizability compared to previous versions. These improvements enable users of AMoC to better model the reading comprehension process across texts.

Two experiments were conducted in order to evaluate AMoC and compare v2.0 to v3.0. The first experiment showed that AMoC v3.0 has higher correlations with the text-based concepts from the Landscape Model than AMoC v2.0 at the word level while maintaining a similar correlation at the sentence level. The second experiment was an analysis of the centrality graph metrics extracted from the concept graphs generated by AMoC v2.0 and AMoC v3.0 and demonstrated that AMoC v3.0 is superior. In addition, AMoC is the only existing model that simulates potential inferences drawn by the reader and provides automated predictions of readers' comprehension and text quality with no experimenter-driven decisions. The better modeling outcomes of AMoC v3.0, combined with the simulation of inferences and predicted comprehension, have implications for the design of curricula. For example, a teacher can use the generated inferences to design more personalized assessments of reading comprehension.

Future work on AMoC v3.0 will explore its capabilities by testing more complex reading outcomes, such as reading times. Additionally, we anticipate using AMoC to develop texts and reading assessments. Applying the text analysis provided by AMoC to educational settings affords the opportunity to improve the specifications of the model and assess its utility as a tool for curriculum design.

Acknowledgements. This work was supported by the Ministry of Research, Innovation, and Digitalization, project CloudPrecis, Contract Number 344/390020/06.09.2021, MySMIS code: 124812, within POC, the Institute of Education Sciences (NSF R305A130124, R305A190063), the U.S. Department of Education, and the National Science Foundation (NSF REC0241144; IIS-0735682).

References

1. Van den Broek, P., Young, M., Tzeng, Y., Linderholm, T.: The landscape model of reading: inferences and the online construction of a memory representation. The Construction of Mental Representations During Reading, pp. 71–98 (1999)
2. Brown, T., et al.: Language models are few-shot learners. In: Advances in Neural Information Processing Systems, vol. 33, pp. 1877–1901 (2020)
3. Corlatescu, D.-G., Dascalu, M., McNamara, D.S.: Automated model of comprehension V2.0. In: Roll, I., McNamara, D., Sosnovsky, S., Luckin, R., Dimitrova, V. (eds.) AIED 2021. LNCS (LNAI), vol. 12749, pp. 119–123. Springer, Cham (2021). https://doi.org/10.1007/978-3-030-78270-2_21
4. Dascalu, M., McNamara, D.S., Crossley, S.A., Trausan-Matu, S.: Age of exposure: a model of word learning. In: 30th AAAI Conference on Artificial Intelligence, pp. 2928–2934. AAAI Press (2016)
5. Dascalu, M., Paraschiv, I.C., McNamara, D.S., Trausan-Matu, S.: Towards an automated model of comprehension (AMoC). In: Pammer-Schindler, V., Pérez-Sanagustín, M., Drachsler, H., Elferink, R., Scheffel, M. (eds.) EC-TEL 2018. LNCS, vol. 11082, pp. 427–436. Springer, Cham (2018). https://doi.org/10.1007/978-3-319-98572-5_33
6. Devlin, J., Chang, M.W., Lee, K., Toutanova, K.: BERT: pre-training of deep bidirectional transformers for language understanding. In: Conference of the North American Chapter of the Association for Computational Linguistics: Human Language Technologies, pp. 4171–4186. Association for Computational Linguistics (2019). https://doi.org/10.18653/v1/n19-1423
7. Gałecki, A., Burzykowski, T.: Linear mixed-effects model. In: Gałecki, A., Burzykowski, T. (eds.) Linear Mixed-Effects Models Using R, pp. 245–273. Springer, New York (2013). https://doi.org/10.1007/978-1-4614-3900-4_13
8. Honnibal, M., Montani, I.: spaCy 2: natural language understanding with bloom embeddings. Convolutional Neural Netw. Incremental Parsing **7**(1) (2017)
9. Kintsch, W.: The role of knowledge in discourse comprehension: a construction-integration model. Psychol. Rev. **95**(2), 163 (1988)
10. Kintsch, W., Welsch, D.M.: The construction-integration model: a framework for studying memory for text, pp. 367–385. Lawrence Erlbaum Associates Inc., Hillsdale (1991)
11. McNamara, D.S., Kintsch, W.: Learning from texts: effects of prior knowledge and text coherence. Discourse Process. **22**(3), 247–288 (1996)
12. McNamara, D.S., Louwerse, M.M., McCarthy, P.M., Graesser, A.C.: Coh-metrix: capturing linguistic features of cohesion. Discourse Process. **47**(4), 292–330 (2010)
13. Mikolov, T., Chen, K., Corrado, G., Dean, J.: Efficient estimation of word representation in vector space. In: Workshop at ICLR (2013)
14. Miller, G.A.: WordNet: a lexical database for English. Commun. ACM **38**(11), 39–41 (1995)

15. Radford, A., et al.: Language models are unsupervised multitask learners. OpenAI Blog **1**(8), 9 (2019)
16. Vaswani, A., et al.: Attention is all you need. In: Advances in Neural Information Processing Systems, vol. 30, pp. 5998–6008. Curran Associates Inc. (2017)
17. Vig, J.: BertViz: a tool for visualizing multihead self-attention in the BERT model. In: ICLR Workshop: Debugging Machine Learning Models (2019)

Analysing Verbal Communication in Embodied Team Learning Using Multimodal Data and Ordered Network Analysis

Linxuan Zhao[1(✉)], Yuanru Tan[2], Dragan Gašević[1], David Williamson Shaffer[2], Lixiang Yan[1], Riordan Alfredo[1], Xinyu Li[1], and Roberto Martinez-Maldonado[1]

[1] Monash University, Clayton, VIC 3108, Australia
linxuan.zhao@monash.edu
[2] University of Wisconsin-Madison, Madison, WI, USA

Abstract. In embodied team learning activities, students are expected to learn to collaborate with others while freely moving in a physical learning space to complete a shared goal. Students can thus interact in various team configurations, resulting in increased complexity in their communication dynamics since unrelated dialogue segments can concurrently happen at different locations of the learning space. This can make it difficult to analyse students' team dialogue solely using audio data. To address this problem, we present a study in a highly dynamic healthcare simulation setting to illustrate how spatial data can be combined with audio data to model embodied team communication. We used ordered network analysis (ONA) to model the co-occurrence and the order of coded co-located dialogue instances and identify key differences in the communication dynamics of high and low performing teams.

Keywords: Collaborative Learning · Multimodality · Communication

1 Introduction

Learning to effectively work in teams in co-located settings remains irreplaceable in many professional sectors, even though online teamwork practices have become more common [29]. Embodied team learning is one such setting where students can freely move around a physical learning space to learn how to interact effectively with resources and other students [5]. This is common practice in high-risk sectors where inadequate teamwork skills have been linked to failures and safety issues [15] such as in firefighting [2] and healthcare [9]. The conditions in such settings can lead to high complexity in communication dynamics, as dialogue segments of team members can happen in parallel at different physical locations with varied team member configurations. Figure 1 illustrates this situation in the context of nursing simulation, where a team of students may need to temporarily split themselves into two sub-teams to complete various tasks.

N. Wang et al. (Eds.): AIED 2023, LNAI 13916, pp. 242–254, 2023.
https://doi.org/10.1007/978-3-031-36272-9_20

The dialogue within those two sub-teams can be unrelated, resulting in dialogue segments being spatially distributed. In such a situation, teachers can find it difficult to assess the team's performance as critical events can happen simultaneously. Students can also struggle to reflect on their team dynamics since they cannot easily have a comprehensive view of their activity as a whole [21].

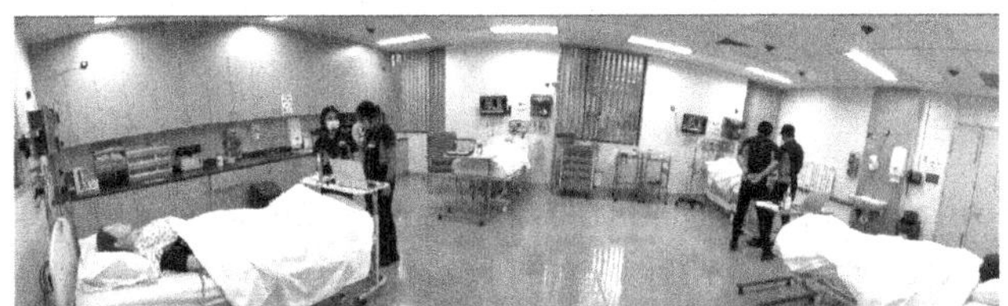

Fig. 1. Embodied teamwork in immersive healthcare simulation where students can temporarily split into sub-teams to complete tasks in parallel towards a shared goal

Rapid advancements in multimodal sensing [14] and artificial intelligence (AI) innovations are enabling new opportunities to automatically model dialogue in education. For instance, previous studies have demonstrated that teacher-student classroom communication can be modelled to provide feedback to teachers on the authority level of their questions [6] and their classroom discourse skills [22]. It has also been proposed that modelling student-student communication can enable the analysis of teamwork [34] and collaborative problem-solving skills [1]. Recent works have demonstrated the value of rendering student verbal communication data visible using word-based interfaces to support the assessment of group collaboration [16] and interfaces showing key communication events to help teachers identify groups of students that may require more help [28].

These previous works suggest that the automated analysis of the content of group communication is becoming feasible and can be highly valuable to study and support learning. However, in these studies, students have been expected to work physically together, therefore generating one dialogue segment throughout a learning session. None of these works targeted highly dynamic embodied team learning situations where students can freely create sub-teams, which can lead to distributed dialogue segments, making it hard to extract meaning from students' logged dialogue. The only previous work studying such a highly dynamic learning situation was presented by Zhao et al. [34] in which authors combined spatial data with audio to enable extracting meaning from distributed dialogue segments. However, the authors did not consider the order of occurrence of key constructs that may augment the meaningfulness of the dialogue analysis. As suggested by the most of other previous works, it has been critical to consider such order for analysing verbal communication [6,16,17,22]. Against this gap in the literature, we formulate the following research question (RQ): *To what extent can the order of occurrence of key high-order team constructs be modelled from students' distributed coded dialogue to identify effective team learning practices?*

Our work addresses this question, and goes beyond previous work, by presenting a study that illustrates how audio and spatial data of students can be used to model critical high-order constructs of embodied team communication. The study was conducted in the context of a highly dynamic healthcare simulation setting. We used ordered network analysis (ONA) to model the order of occurrence of coded co-located dialogue data and identify effective team learning practices by analysing key differences between high and low performing teams.

2 Methods

Learning Context. The study involved a series of immersive team simulations, held over four weeks in 2021, which were part of the regular activities of an undergraduate course of the Bachelor of Nursing at *Undisclosed University*. These were conducted in a specialised classroom space equipped with four patient manikins and medical equipment simulating a hospital ward (see Fig. 1). Multimodal data (details provided below) from 228 consenting students (aged 20 to 23), grouped in 57 teams of four students, were collected. Due to limitations of the microphone hardware (e.g., signal interference) and other practical challenges (e.g., students accidentally turning microphones off), only the high-quality data from 60 students grouped in 15 teams were used in this study. Three nursing teachers, who designed the simulations, monitored the simulation from a control room and assessed the students' performance based on their observations.

The learning design included the following four phases. (1) An **initial handover**, in which the first two students would enter the room and listen to an introduction conveyed by the doctor (played by a teacher). (2) The **initial assessment**, in which the same students would make a plan and would need to react to an unexpected event pre-programmed by the teachers, involving the identification of a serious problem suffered by the patient (in bed 4) and escalating the situation by calling for help. (3) **Resolving emergency**, in which the two other students in the team would enter the room, receive handover information and collaborate to help the patient at risk (defined by teachers as the **primary task**) while completing tasks for the three other patients (the **secondary tasks**). (4) **Emergent diagnosis**, in which an emergency doctor would enter the room and students would provide an update of the situation.

Each simulation was between 15 and 30 min long (avg = 20.25 min.; st. dev. = 8.13 min). Teachers assessed students' team performance in phases 2 and 3, since students' team dynamics mostly happened in these two phases.

Apparatus. Portable wireless (Xiaokoa) headset microphones were provided to consenting students to capture their voices. A multi-channel (TASCAM US-16×08) audio interface was used to synchronise the audio streams and store them into individual files. For spatial data, waist bags each containing a positioning sensor (Pozyx) were provided to each student. These data included each student's body orientation and their x-y spatial coordinates. The team assessment results were collected using a questionnaire filled by the teachers who evaluated the simulation. The questionnaire assessed their teamwork effectiveness based on a

7-point Likert scale. All data collection devices were synchronised automatically. Ethical approval was obtained from the *Undisclosed University*.

Modelling Multimodal Sensor Data. Utterance intervals (when individual speech started and ended) were automatically extracted from audio signals captured by each microphone using a voice activity detection script created via the Python library *py-webrtcvad*. The utterance intervals were used to organise students' utterances in the sequence of their turns of talking, and utterance content would be coded using the coding scheme described below. The utterance content was transcribed using a third-party transcription service.

Since students could be chatting at completely different locations and have conversations in parallel, spatial data was needed to organise students' dialogue into corresponding dialogue segments. We used the body orientation and spatial coordinates to detect dialogue segments by adopting the f-formations theory [23]. An f-formation appears whenever multiple people sustain a spatial and orientational relationship for collaboration in close proximity [23]. As a previous study in healthcare [27] suggested, communication between healthcare professionals commonly happens in close proximity (less than 1.5 m), we applied an f-formation detection algorithm [35] to differentiate dialogue segments. Furthermore, the spatial data were also used to detect Spaces of Interests (SoIs), namely the spaces of *primary tasks* (bed 4) and *secondary tasks* (beds 1–3).

Coding Scheme and Procedure. To analyse the content of dialogue segments, a coding scheme was designed for embodied teamwork communication in this context by adapting previous coding schemes [13,19] and a team theory framework [20]. The coding scheme includes four higher-order teamwork constructs and nine communication behaviours for coding the dialogue. The first construct, *shared leadership*, captures instances when students assigned others or themselves to specific tasks or provided handover information to bring others on board as no student was formally appointed as a leader. The second construct, *situation awareness*, captures the communication related to identification and reaction to patients' emergency state [19]. The third construct, *shared mental model*, captures the communication for establishing a shared understanding of the current situation and a potential plan to tackle this [32]. The last construct, *closed-loop communication*, refers to the double-checking of information or acknowledging the receipt of information [13]. More details about the definitions and examples for each communication behaviour are provided in Table 1.

The coding was done at an utterance level and each utterance could have multiple codes. Two researchers coded 20% of the dialogue data. Cohen's kappa was used to measure inter-rater reliability (0.6 was the threshold for acceptance) [12]. The kappa for each code was greater than 0.7 (see column 5 in Table 1). The coding of the remaining 80% of the data was completed by one researcher.

Ordered Network Analysis. As suggested by theories of communication in healthcare teams [15], the sequential order of communication is essential to demonstrate effective healthcare teamwork. For example, the order of communication from *information sharing* to *task allocation* can illustrate assigning tasks

Table 1. Teamwork communication coding scheme and corresponding definition, example, and inter-rater reliability.

Teamwork Constructs	Communication behaviours (codes)	Definition	Example	Kappa
Shared leadership	Task allocation [19]	a student explicitly assigns a task to others or proactively self-allocates a task	"You do the medical observation, and I will do the discharge for the bed three patient."	0.744
	Provision of handover information [15]	a student updates to others regarding a task to which they have not been exposed	"She is day-one post total hysterectomy. She has got a history of heart disease..."	0.853
Situation Awareness	Escalation [7]	a student informs others that the situation goes beyond their capabilities and they need extra help	"I think we need to call the emergency team for help."	0.747
Shared mental model	Planning [32]	a student lists several tasks remaining to be done for provoking subsequent task allocation	"She is due for antibiotics and pain meds, and we also need to call her family."	0.781
	Information sharing [15]	a student proactively shares information that was not asked by others	"Her wound is dry and intact. There is no concern now."	0.744
	Information request [15]	a student asks someone else a question to get information	A: "Is the IV necessary for this patient?" (Information request) B: "She does not need it" (Responding to request)	0.794
	Responding to request [15]	a student provides information responding to a previously asked question		0.804
Closed-loop communication	Acknowledgement [4]	a student acknowledges receipt of information or instructions from others	"Yes", "I agree", "Okay"	0.858
	Checking-back [4]	a student double-checks the information or instructions from others	A: "Can you give her 1ml IV fluid?" B: "1ml IV fluid" (Check-back)	0.922

based on clinical evidence, which can be indicative of an effective team coordination strategy [15]. Thus, we used ordered network analysis (ONA) [30].

ONA is a technique to quantify and visualise the directed connection within coded data. The ONA algorithm employs similar functions and procedures as Epistemic Network Analysis (ENA), a widely used network analysis technique for the modeling and comparison of learning phenomena [3,24,25]. The key difference is that ONA accounts for the order of connections during the modeling and visualising processes. ONA starts processing the data by accumulating directed connections within *units of analysis* (i.e., subjects of research interest, such as a team) as high-dimensional vectors. These accumulations operate on the *line* level (i.e., the fundamental unit of meaning in the data, such as an utterance) by counting the order of codes' co-occurrences within *stanza windows* (i.e., the temporal context formed by span of *lines*). Then, the ordered co-occurrences are aggregated across each *conversation* (i.e., collections of meaningfully related *stanza windows*, such as all *stanza windows* of a sub-team) for each *unit of analysis* to obtain cumulative connection vectors. Next, a dimensional reduction via Singular Value Decomposition (SVD) or Means Rotation (MR) is applied to the collection of those cumulative vectors to project them as points in a two dimensional space.

In our study, we applied MR to maximise the variances between two groups of units on the x-axis of the space. Then, ONA networks are visualised in this two-dimensional space using two coordinated representations for each unit of analysis: (1) a projected point, which represents the location of its network in the low-dimensional projected space (shown as red or blue points in our study), and (2) a directed weighted network where nodes correspond to the codes, and edges reflect the relative frequency of connection between two codes. Specifically, the node size is proportional to the frequency of its represented code being connected with other codes; and the size of the coloured inner circle of a node is proportional to self-connections, i.e., the frequency of the code making a connection with itself. Between each pair of nodes, the edge consists of a pair of triangles with varied sizes to illustrate the frequency of directed connections. The bigger and darker a triangle is, the higher frequency of connections is. A black chevron is placed on the more frequent side of an edge to support recognizing the direction of connections. For example, a chevron pointing from node A to node B represents that the directed connection from A to B is more frequent.

To conduct ONA, several parameters need to be specified, namely: *lines*, *conversations*, *stanza windows*, *units of analysis*, and *codes*. In our study, students' utterances were used as *lines* and co-located dialogue segments as *conversations*. We used a *stanza window* size of three lines to accumulate connections, since we tested multiple configurations of *stanza window* and found this setting provided the highest variance in the ONA model [26]. The *units of analysis* in this study were the unique combinations of phase (i.e., phases 2 and 3) and SoIs (i.e., primary tasks and secondary tasks). Regarding *codes*, we included all codes in Table 1 except for *checking-back*, since it had extremely low frequency (n = 105) compared to other codes (mean = 791.7, st. dev. = 644.3). We omitted this code to maximise the clarity of analysis as suggested in [31]. We also excluded *provision of handover information* code in the secondary task phase 3 model for a similar reason. We built ONA models using the R implementation of ONA [10].

To address our RQ, we divided the 15 teams into seven low-performing and eight high-performing teams based on the median score of their team performance. Using these two groups, we can identify prominent differences of team communication in the mean ONA networks of high and low performing teams. To demonstrate the prominent differences, we created ONA mean network subtractions by subtracting the two groups' mean networks. We also conducted Mann-Whitney U tests on the distribution of projected points to compare if the differences between the groups were statistically significant.

3 Results

3.1 Primary Tasks

Primary Task: Phase 2. As shown in Fig. 2 (left), for the case of dialogue data at bed 4 (the primary task), several directed connections appeared in high-

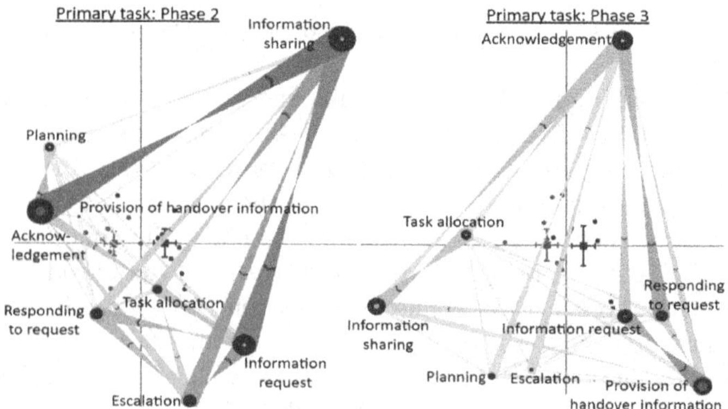

Fig. 2. Mean network subtractions for low (red edges) and high (blue edges) performing teams in spaces for primary tasks during the initial assessment, phase 2 (left); and resolving emergency, phase 3 (right) (Color figure online)

performing teams, while the connection from *planning* to *acknowledgement* was the only major directed connection in low-performing teams. This is not a surprising result since the primary task in phase 2 was to identify the patient at risk and announce escalation, so team behaviours related to information exchange and escalation were expected to be more frequent for high-performing teams. The Mann-Whitney U tests also showed the differences between the high-performing teams (N = 8, Mdn = 0.40, Q1 = 0.19, Q3 = 0.37) and the low-performing teams (N = 7, Mdn = −0.301, Q1 = −0.34, Q3 = −0.20) on the first dimension (U = 54, $p = 0.001$, $r = 0.58$) were significant.

Specifically, although both groups of teams showed connections to *information request*, a key difference is that high-performing teams requested information more frequently after contributing with new information (*sharing information*) or after announcing an emergency (*escalation*). In contrast, *information requests* were limited in low-performing teams and happened only after *planning*, Considering that the directed connection from *planning* to *acknowledgement* is stronger than from *planning* to *information request*, we can conclude that low-performing teams demonstrated inefficient communication while planning. Moreover, evidence of the effective dynamics of high-performing teams is also illustrated by the directed connections to *responding to request*. For example, *information request*, and *escalation* are pointing towards *responding to request*. This suggests that students in high-performing teams exchanged information related to the critical task (announcing escalation) more frequently which was critical in this phase. Although similar connections occurred for low-performing teams, the strength of those connections was relatively weak, as indicated by the faded red edges.

Primary Task: Phase 3. The primary task in phase 3 mainly consisted of providing handover information to new team members and then offering emergent medical support to the patient at risk (in bed 4) while taking care of stable patients in the other beds. As shown in Fig. 2 (right), dialogue from the high-performing teams showed stronger connections among a small set of nodes on the right side of the x-axis; while low-performing teams had stronger connections on the left side. The Mann-Whitney U tests showed the difference between the high-performing teams (N = 8, Mdn = 0.152, Q1 = 0.02, Q3 = 0.31) and the low-performing teams (N = 7, Mdn = −0.13, Q1 = −0.19, Q3 = −0.10) on the first dimension (U = 52, $p = 0.004$, $r = 0.52$) was significant.

Although both groups of teams showed various connections to the node *provision of handover information*, the connections to other nodes were different. The blue circle in the node *provision of handover information* shows that the high-performing teams repeatedly provided handover information, and also when team members explicitly requested such information (see directed connection from *information request* to *provision of handover information*). This shows that the high-performing teams could provide handover information fluently and respond to questions from others. In contrast, the low-performing teams communicated about handover information infrequently, as indicated by the faded red edges connected to *provision of handover information*. This indicates that some team members in the low-performing teams may have not had complete information relevant for the primary task. Additionally, the red edges show that the low-performing teams communication focused on *task allocation*, *information sharing*, *planning*, and *escalation* announcement to gather information, which are constructs that were expected to occur in the previous phase, rather than focusing on *provision of handover information* which was critical in this phase.

3.2 Secondary Tasks

The secondary tasks were related to the three patients with stable physical conditions. Students were expected to put less effort into the secondary tasks but still perform them as they would prioritise aiding the patient at risk.

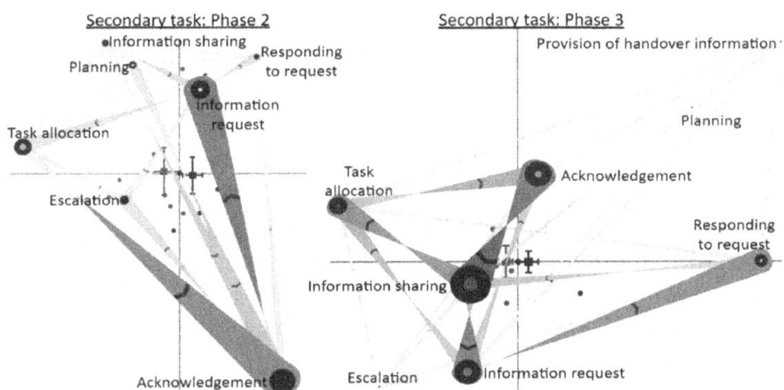

Fig. 3. Mean network subtractions for low (red edges) and high (blue edges) performing teams in spaces for secondary tasks during the initial assessment – phase 2 (left) and resolving the emergency – phase 3 (right) (Color figure online)

Secondary Tasks: Phase 2 and Phase 3. As shown in Fig. 3 (left), the majority of edges are in red. This is expected since the teams were expected to put less effort into secondary tasks, and teams rated as low-performing were not aligned with this expectation. The Mann-Whitney U tests showed a significant difference between the high-performing teams (N = 8, Mdn = 0.17, Q1 = 0.01, Q3 = 0.29) and the low-performing teams (N = 7, Mdn = −0.12, Q1 = −0.32, Q3 = −0.02) on the first dimension (U = 50, $p = 0.012$, $r = 0.53$).

The high-performing teams only made one strong connection from *acknowledgement* to *information request*. This suggests that the high-performing teams frequently asked questions in the middle of a dialogue segment, which may demonstrate their engagement in team communication. Yet, the larger number of red edges suggests that low-performing teams may have over-emphasised the secondary tasks. The node *escalation* also reveals important differences between the two groups. Specifically, regarding all the edges connected to *escalation*, the ones pointed towards *escalation*, namely from *task allocation*, *information request*, and *acknowledgement*, are all from the low-performing teams. This suggests that the low-performing teams sometimes allocated tasks or exchanged information and then announced escalation. Since this happened in the space of the secondary tasks, the low-performing teams may have incorrectly announced escalation on stable patients. In contrast, the edges pointing away from *escalation* are all from the high-performing teams. This shows that the high-performing teams sometimes announced escalation and then arranged tasks or exchanged information. This may happen in a situation that the student working on primary tasks went to the space of secondary tasks to inform the student working in this space about escalation of the patient at risk and asked a student for help.

Regarding the communication in the secondary tasks' space in phase 3, the Mann-Whitney U tests also showed a significant difference between the high-performing teams (N = 8, Mdn = 0.013, Q1 = 0.01, Q3 = 0.16) and the low-performing teams (N = 7, Mdn = −0.13, Q1 = −0.26, Q3 = −0.08) on the first dimension (U = 56, $p = 0.001$, $r = 0.50$). The main finding is that low-performing teams still overemphasised the secondary tasks in phase 3, as the majority of edges in Fig. 3 (right) appeared in low-performing teams.

4 Discussion

To address our RQ, we used multimodal data and ONA to model the ordered communication behaviours in embodied team learning where dialogue segments were spatially distributed. From the analysis of mean network subtractions, the effective team learning practice in this setting involves the following behaviours. (1) Prioritising primary tasks. Through the network subtractions for primary tasks and secondary tasks, we found that high-performing teams prioritised the primary tasks while low-performing teams prioritised secondary tasks. This is aligned with the healthcare literature that suggests that developing effective patient prioritisation skills is critical in this kind of learning settings [7]. (2) Timely and correctly performing critical tasks (i.e., escalation and providing handover information). The results illustrate that high-performing teams announced escalation on the correct patient timely (in phase 2) and provided abundant handover information. In contrast, low-performing teams more frequently announced escalation on the wrong (stable) patients or late in phase 3, putting less effort into providing handover information. This suggests that timely and correctly performing critical tasks is an effective team practice, as announcing escalation correctly and timely is related to the effective application of medical resources to maximise patient safety [7], and providing handover information can contribute to effective team coordination [15]. (3) Coordinating the team efficiently. All four network subtractions demonstrated that low-performing teams put more effort into communicating about team coordination (stronger connections to *planning* and *task allocation*). This suggests that low-performing teams were less efficient in coordinating their teams, so they had to invest more effort on coordination, resulting in failure to demonstrate other key effective team behaviours.

This study has several **implications** for research in embodied teamwork education and co-located collaborative learning. We demonstrated how ONA graphs can be useful to analyse communication in an embodied team learning setting by modelling multimodal data. Other researchers can adapt this method to model and analyse verbal communication in similar learning settings, such as collaborative problem-solving [17] and laboratory classroom teaching [11]. We also considered the implication for students and teachers. However, current ONA visualisations can be hard for them to understand due to their complexity. Yet, future work can explore ways to effectively communicate the insights from ONA in ways that students or teachers can understand.

Regarding **limitations and future work**, the first limitation is the generalisability of the findings. The sample size in this study is limited (60 students in 15 teams) and the interpretations of the results are based on a specific learning design, so the findings are not meant to be generalisable. Another limitation results from the manual transcription and coding of the communication contents. This currently limits the portability and scalability [33] of this method. Overcoming this limitation is our future work. We plan to fully automate the transcription and coding procedures using speech-to-text (e.g., the recently released OpenAI whisper [18]) and natural language processing techniques [8]. Another future work is to design a method to convey the findings in ONA to students and teachers in an intuitive way to enable practical application.

Regarding **ethical considerations**, although spatial data in this study was anonymous, the audio recordings can lead to privacy concerns. However, it is a common practice to record audio data in similar studies [16,17]. To minimise privacy concerns, we used colours to represent the students and never collected any identity information. Furthermore, we controlled the access to our dataset to prevent any unintended use [33].

In **conclusion**, we presented a method to extract communication behaviours in embodied learning settings from multimodal data and analyse them using ONA. We illustrated the capability of this method to identify the key factors for differentiating high and low-performing teams in a team-based embodied healthcare simulation. The method in this paper can benefit practitioners to support their teaching and researchers studying embodied teamwork.

References

1. Abitino, A., Pugh, S.L., Peacock, C.E., D'Mello, S.K.: Eye to eye: gaze patterns predict remote collaborative problem solving behaviors in triads. In: Rodrigo, M.M., Matsuda, N., Cristea, A.I., Dimitrova, V. (eds.) AIED 2022. LNCS, vol. 13355, pp. 378–389. Springer, Cham (2022). https://doi.org/10.1007/978-3-031-11644-5_31
2. Adrot, A., Bia Figueiredo, M.: "Lost in digitization": a spatial journey in emergency response and pragmatic legitimacy. In: de Vaujany, F.-X., Adrot, A., Boxenbaum, E., Leca, B. (eds.) Materiality in Institutions. TWG, pp. 151–181. Springer, Cham (2019). https://doi.org/10.1007/978-3-319-97472-9_6
3. Bowman, D., et al.: The mathematical foundations of epistemic network analysis. In: Ruis, A.R., Lee, S.B. (eds.) ICQE 2021. CCIS, vol. 1312, pp. 91–105. Springer, Cham (2021). https://doi.org/10.1007/978-3-030-67788-6_7
4. Härgestam, M., Lindkvist, M., Brulin, C., Jacobsson, M., Hultin, M.: Communication in interdisciplinary teams: exploring closed-loop communication during in situ trauma team training. BMJ Open **3**(10) (2013)
5. Ioannou, M., Georgiou, Y., Ioannou, A., Johnson, M.: On the understanding of students' learning and perceptions of technology integration in low-and high-embodied group learning. **1**, 304–311 (2019)
6. Kelly, S., Olney, A.M., Donnelly, P., Nystrand, M., D'Mello, S.K.: Automatically measuring question authenticity in real-world classrooms. Educ. Res. **47**(7), 451–464 (2018)
7. Liaw, S.Y., Rethans, J.J., Scherpbier, A., Piyanee, K.Y.: Rescuing a patient in deteriorating situations (rapids): a simulation-based educational program on recognizing, responding and reporting of physiological signs of deterioration. Resuscitation **82**(9), 1224–1230 (2011)
8. Lin, J., et al.: Is it a good move? Mining effective tutoring strategies from human-human tutorial dialogues. Futur. Gener. Comput. Syst. **127**, 194–207 (2022)
9. Ludlow, K., Churruca, K., Ellis, L.A., Mumford, V., Braithwaite, J.: Decisions and dilemmas: the context of prioritization dilemmas and influences on staff members' prioritization decisions in residential aged care. Qual. Health Res. **31**(7), 1306–1318 (2021)
10. Marquart, C., Tan, Y., Cai, Z., Shaffer, D.W.: Ordered network analysis (2022). https://epistemic-analytics.gitlab.io/qe-packages/ona/cran/
11. Martinez-Maldonado, R., Echeverria, V., Schulte, J., Shibani, A., Mangaroska, K., Buckingham Shum, S.: Moodoo: indoor positioning analytics for characterising classroom teaching. In: Bittencourt, I.I., Cukurova, M., Muldner, K., Luckin, R., Millán, E. (eds.) AIED 2020. LNCS (LNAI), vol. 12163, pp. 360–373. Springer, Cham (2020). https://doi.org/10.1007/978-3-030-52237-7_29
12. McHugh, M.L.: Interrater reliability: the kappa statistic. Biochemia medica **22**(3), 276–282 (2012)
13. Miller, K., Riley, W., Davis, S.: Identifying key nursing and team behaviours to achieve high reliability. J. Nurs. Manag. **17**(2), 247–255 (2009)
14. Mitri, D.D., et al. (eds.): Proceedings of the First International Workshop on Multimodal Artificial Intelligence in Education. CEUR Workshop Proceedings (2021)
15. O'Daniel, M., Rosenstein, A.H.: Professional communication and team collaboration. In: Patient Safety and Quality: An Evidence-Based Handbook for Nurses (2008)
16. Praharaj, S., Scheffel, M., Schmitz, M., Specht, M., Drachsler, H.: Towards collaborative convergence: quantifying collaboration quality with automated co-located

collaboration analytics. In: 12th International Learning Analytics and Knowledge Conference, pp. 358–369 (2022)
17. Pugh, S.L., Subburaj, S.K., Rao, A.R., Stewart, A.E., Andrews-Todd, J., D'Mello, S.K.: Say what? Automatic modeling of collaborative problem solving skills from student speech in the wild. In: 14th International Conference on Educational Data Mining, pp. 55–67 (2021)
18. Radford, A., Kim, J.W., Xu, T., Brockman, G., McLeavey, C., Sutskever, I.: Robust speech recognition via large-scale weak supervision. OpenAI Blog (2022)
19. Riley, W., Hansen, H., Gürses, A.P., Davis, S., Miller, K., Priester, R.: The nature, characteristics and patterns of perinatal critical events teams. In: Advances in Patient Safety: New Directions and Alternative Approaches (2008)
20. Salas, E., Sims, D.E., Burke, C.S.: Is there a "big five" in teamwork? Small Group Res. **36**(5), 555–599 (2005)
21. Salas, E., Stevens, R., Gorman, J., Cooke, N.J., Guastello, S., von Davier, A.A.: What will quantitative measures of teamwork look like in 10 years? In: Proceedings of the Human Factors and Ergonomics Society Annual Meeting, vol. 59, no. 1, pp. 235–239 (2015)
22. Schlotterbeck, D., Uribe, P., Jiménez, A., Araya, R., van der Molen Moris, J., Caballero, D.: TARTA: teacher activity recognizer from transcriptions and audio. In: Roll, I., McNamara, D., Sosnovsky, S., Luckin, R., Dimitrova, V. (eds.) AIED 2021. LNCS (LNAI), vol. 12748, pp. 369–380. Springer, Cham (2021). https://doi.org/10.1007/978-3-030-78292-4_30
23. Setti, F., Lanz, O., Ferrario, R., Murino, V., Cristani, M.: Multi-scale F-formation discovery for group detection. In: 2013 IEEE International Conference on Image Processing, pp. 3547–3551. IEEE (2013)
24. Shaffer, D.W., Collier, W., Ruis, A.R.: A tutorial on epistemic network analysis: analyzing the structure of connections in cognitive, social, and interaction data. J. Learn. Anal. **3**(3), 9–45 (2016). https://doi.org/10.18608/jla.2016.33.3
25. Shaffer, D.W., et al.: Epistemic network analysis: a prototype for 21st-century assessment of learning. Int. J. Learn. Media **1**(2) (2009)
26. Siebert-Evenstone, A.L., Irgens, G.A., Collier, W., Swiecki, Z., Ruis, A.R., Shaffer, D.W.: In search of conversational grain size: modeling semantic structure using moving stanza windows. J. Learn. Anal. **4**(3), 123–139 (2017)
27. Sorokowska, A., et al.: Preferred interpersonal distances: a global comparison. J. Cross Cult. Psychol. **48**(4), 577–592 (2017)
28. Southwell, R., et al.: Challenges and feasibility of automatic speech recognition for modeling student collaborative discourse in classrooms. In: 15th International Conference on Educational Data Mining, pp. 302–315 (2022)
29. Takizawa, P.A., Honan, L., Brissette, D., Wu, B.J., Wilkins, K.M.: Teamwork in the time of Covid-19. FASEB BioAdvances **3**(3), 175–181 (2021)
30. Tan, Y., Ruis, A.R., Marquart, C., Cai, Z., Knowles, M.A., Shaffer, D.W.: Ordered network analysis. In: Damşa, C., Barany, A. (eds.) ICQE 2022. CCIS, vol. 1785, pp. 101–116. Springer, Cham (2023). https://doi.org/10.1007/978-3-031-31726-2_8
31. Wang, Y., Swiecki, Z., Ruis, A.R., Shaffer, D.W.: Simplification of epistemic networks using parsimonious removal with interpretive alignment. In: Ruis, A.R., Lee, S.B. (eds.) ICQE 2021. CCIS, vol. 1312, pp. 137–151. Springer, Cham (2021). https://doi.org/10.1007/978-3-030-67788-6_10
32. Westli, H.K., Johnsen, B.H., Eid, J., Rasten, I., Brattebø, G.: Teamwork skills, shared mental models, and performance in simulated trauma teams: an independent group design. Scand. J. Trauma Resuscitation Emerg. Med. **18**(1), 1–8 (2010)

33. Yan, L., Zhao, L., Gasevic, D., Martinez-Maldonado, R.: Scalability, sustainability, and ethicality of multimodal learning analytics. In: 12th International Learning Analytics and Knowledge Conference, New York, NY, USA, pp. 13–23 (2022)
34. Zhao, L., et al.: METS: multimodal learning analytics of embodied teamwork learning. In: 13th International Learning Analytics and Knowledge Conference (2023, in press)
35. Zhao, L., et al.: Modelling co-located team communication from voice detection and positioning data in healthcare simulation. In: 12th International Learning Analytics and Knowledge Conference, New York, NY, USA, pp. 370–380 (2022)

Improving Adaptive Learning Models Using Prosodic Speech Features

Thomas Wilschut[1(✉)], Florian Sense[2], Odette Scharenborg[3], and Hedderik van Rijn[1]

[1] Department of Experimental Psychology, University of Groningen, Groningen, The Netherlands
{t.j.wilschut,d.h.van.rijn}@rug.nl
[2] InfiniteTactics, LLC, Beavercreek, USA
florian.sense@infinitetactics.com
[3] Department of Multimedia and Computing, Delft University of Technology, Delft, The Netherlands
o.e.scharenborg@tudelft.nl

Abstract. Cognitive models of memory retrieval aim to describe human learning and forgetting over time. Such models have been successfully applied in digital systems that aid in memorizing information by adapting to the needs of individual learners. The memory models used in these systems typically measure the accuracy and latency of *typed* retrieval attempts. However, recent advances in speech technology have led to the development of learning systems that allow for spoken inputs. Here, we explore the possibility of improving a cognitive model of memory retrieval by using information present in speech signals during spoken retrieval attempts. We asked 44 participants to study vocabulary items by spoken rehearsal, and automatically extracted high-level prosodic speech features—patterns of stress and intonation—such as pitch dynamics, speaking speed and intensity from over 7,000 utterances. We demonstrate that some prosodic speech features are associated with accuracy and response latency for retrieval attempts, and that speech feature informed memory models make better predictions of future performance relative to models that only use accuracy and response latency. Our results have theoretical relevance, as they show how memory strength is reflected in a specific speech signature. They also have important practical implications as they contribute to the development of memory models for spoken retrieval that have numerous real-world applications.

Keywords: Adaptive Learning · Cognitive Modeling · Automatic Speech Recognition · Machine learning · Speech prosody · Pitch · Speaking Speed · Intensity

1 Introduction

Model-based adaptive learning systems optimize learning by tailoring learning procedures to the needs of the individual learner [11, 14, 22]. To this end, such systems aim to estimate and predict the extent to which a learner has successfully memorized information. As memory strength cannot be observed directly, models of memory retrieval use

N. Wang et al. (Eds.): AIED 2023, LNAI 13916, pp. 255–266, 2023.
https://doi.org/10.1007/978-3-031-36272-9_21

behavioral proxies, such as accuracy scores and response latencies to make informed predictions [1,15,26]. Here, we will build upon a number of recent studies showing that human speech contains information on a speaker's emotional state, confidence, and the accuracy of the spoken response [7,10]. The aim of this study is to explore the theoretical and practical feasibility of using the information present in spoken retrieval attempts to improve cognitive models of memory retrieval that can be applied in adaptive learning systems. To that end, we will here examine (a) whether the results of earlier studies, which identified a specific prosodic speech signature associated with accuracy and speaker confidence, generalize to a learning paradigm specifically, and (b) whether the information present in spoken retrieval attempts can be used to improve predictions of memory retrieval success in an applied learning setting.

As the extent to which a learner has successfully memorized an item is a latent state that cannot be measured directly, cognitive models of learning and forgetting use behavioral proxies. Response accuracy is a logical candidate, as it indicates whether or not the learner could successfully retrieve the memorandum. Correspondingly, accuracy is used in many models that predict performance [15,23]. However, using (only) accuracy as a behavioral index of memory strength results in a number of issues. First, accuracy-based models of forgetting have difficulties accounting for the passage of time between events (for example, in early Bayesian knowledge tracing models, information was never forgotten once an item flipped to the "known" state after an accurate response [13]). Second, using accuracy as a proxy of memory strength does not allow for meaningful discrimination within correct responses, and as a consequence, accurate performance predictions require many incorrect responses [19]. Because of these limitations, some models use the latency of a response in addition to its accuracy to predict performance. A core assumption these models rely on is the link between response latency and memory strength [4]. An abundance of experimental data supports this link: Faster responses are generally associated with more accurate responses and a stronger association between cue and response compared to slower responses[21]. Using response latency in addition to accuracy to predict learner performance has been proven to be successful in a range of adaptive learning applications [11,15,19,21].

An example of a successful approach to modeling forgetting using accuracy and response latency is the MemoryLab adaptive scheduling system www.memorylab.nl/en. The system is based on the ACT-R theory's declarative memory framework [1]. In this cognitive model, individual facts are represented as a memory chunk that has a certain activation, which corresponds to the fact's strength in memory. It gets boosted every time a fact is rehearsed and gradually decays over time if an item is not repeated. As some facts are more difficult to learn than other facts, and as some learners forget facts at a different rate than other learners, all facts are given a unique decay rate parameter for all learners, which allows the model to vary the rate at which items are forgotten between learners and between facts. The model relies on the above-mentioned link between retrieval speed and the memory strength—or activation—of that fact. Combining the link between response latency and memory activation with its activation decay functions, the model can calculate expected response times for retrieval attempts at any future point in time. Observed discrepancies between expected and observed response times and accuracy scores can consequently be used to update decay parameters, resulting in a system that can accurately capture and predict forgetting over time [20,24].

Accuracy and latency comprise only a part of all potential sources of information that may be used to inform models that aim to predict forgetting. In this study, we will focus on spoken language, which contains prosodic speech features (PSFs). PSFs are high-level properties of units of speech longer than individual phonemic segments, such as syllables, words or sentences [17]. PSFs can be roughly divided into three categories. First, **intonation** is the melodic pattern of an utterance and refers to the dynamics in pitch over the duration of a speech segment. Second, **rhythm** is defined by dynamics in timing, or speaking speed, over the duration of the speech segment. Third, **stress** refers to intensity (loudness) given to a syllable of speech, resulting in changes in relative intensity. Prosodic information usually reflects information that is not necessarily present in grammar or choice of vocabulary, such as the emotional state of the speaker, emphasis, or the form of utterance (e.g., question versus statement versus command) [27,29].

Of particular importance for the current study is the idea that speakers implicitly convey their levels of certainty about a response through speech prosody. For example, in many languages, speakers end a sentence or word with raising pitch and lower intensity when asking a question or to express uncertainty about a response [10]. Listeners are able to decode these prosodic speech signatures across languages and cultures [7]. Next to the idea that speech prosody conveys subjective speaker confidence, a recent study by Goupil and Aucouturier demonstrated that objective accuracy is distinctly reflected in the speech signal [6]. In their study, participants were instructed to complete a visual detection task: A word was briefly presented on the screen, followed by a visual mask. After a short interval, participants needed to verbally choose which word they saw from a number of alternatives and rate their confidence in the response. The results showed that the participants' speaking speed and intonation were associated with the objective accuracy of the response, and the intensity of the response was associated with subjective confidence. To date, it is unclear if the above mentioned effects generalize to the context of a learning paradigm.

Overall, earlier findings suggest that a speech signal for spoken retrieval attempts may contain information that can be used to improve models of memory retrieval in two ways. On one hand, as they are associated to objective accuracy, they may directly carry information about the latent memory strength of a response. At the same time, they may carry information about memory strength indirectly, as they are associated to subjective confidence in the response. In the current study, we aim to explore the hypotheses that these PSFs are indeed informative in a learning context and that they can be used to improve models of memory retrieval.

To pursue these research questions, we will rely on recent advancements in speech technology which have led to the development of adaptive learning systems that allow for spoken input. Such systems allow users to efficiently practice pronunciation, which is an important part of language acquisition that is largely omitted in traditional approaches [5]. Furthermore, they can be applied in situations where users do not have the ability (for example, young children) or opportunity (for example, while driving a car) to type. Finally, spoken learning allows for the extraction of PSFs in the spoken utterances. Compared to more traditional approaches to automatic speech processing, that often rely on deep learning–based classification of spectral components of the speech signal [12], extracting PSFs from the speech signal is computationally relatively

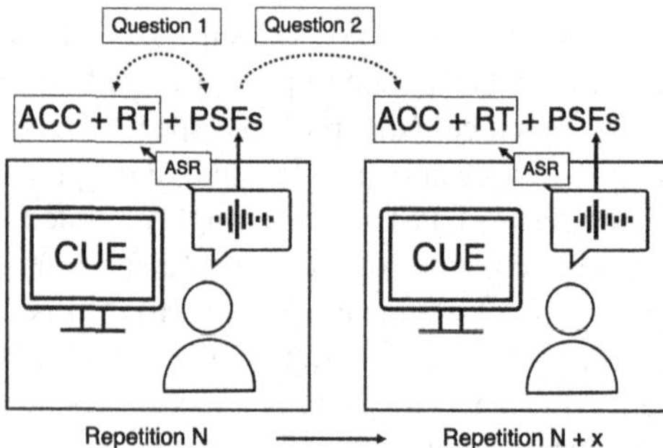

Fig. 1. Design and research questions. Participants saw a cue and responded using speech. Using automatic speech recognition (ASR), the accuracy (ACC) and response time (RT) of the response is determined. The first research question examines if PSFs derived from the speech signal are associated to accuracy and response time on the same trial. The second research question considers if current repetition (N) PSFs can be used to improve predictions for future repetition (N + X) accuracy and response time for the same item.

inexpensive [25]. In short, advances in speech technology have made the implementation of speech-based learning systems, in which PSFs are automatically extracted, practically feasible.

Given the growing popularity and practical feasibility of speech-controlled learning applications, we here explore the possibility of exploiting information present in the speech signal to improve cognitive models of memory retrieval used in personalized learning applications. To that end, we first aim to extend earlier studies that find a specific PSF signature associated with accuracy and with subjective confidence to the context of a memory retrieval paradigm specifically (see Fig. 1, Question 1). To foreshadow the results, we indeed found an association between PSFs and accuracy and response latency for spoken retrieval attempts. Second, in this study, we are the first to explore the possibility of improving predictions of future retrieval performance using a model that incorporates PSFs for previous attempts compared to a model uses accuracy and response latency only (see Fig. 1, Question 2). To foreshadow the results, we found that using previous repetition PSFs *in addition to accuracy and response latency* substantially improves predictions of future retrieval performance.

2 Methods

The analyses reported in this study are based on data from the experiment reported in [28], which demonstrated that latency-based adaptive learning algorithms can improve learning efficiency in speech-based learning systems. That study used four within-subjects conditions: two learning modality conditions (typing-based and speaking-based learning) and two item scheduling conditions (fully adaptive MemoryLab learning, based on the learners' response times and accuracy scores, and less adaptive flashcard-inspired learning, using the learners' accuracy only). All analyses reported here are based on the speech conditions of the experiment only. For the first research question (exploring the extent to which different PSFs can be used to explain performance on the same trial) data from both scheduling conditions was included. The second research question explored whether using PSFs can result in increased performance predictions compared to adaptive scheduling systems that used response times and accuracy scores only. Therefore, only data from the MemoryLab adaptive scheduling condition was included. We will briefly reiterate the relevant details of the study here, but for more details, see [28]. Materials, analysis scripts, data and code to recreate the current experiment and analyses are available on https://osf.io/dfexp/.

2.1 Participants

In total, data from 44 participants was available for analysis. This sample size was chosen based on previous studies [28]. Participants were first-year Psychology students that received course credit for participants. Participants were 17–29 years of age, and 73% female. Participants were native Dutch speakers and indicated that they were fluent in English. Participants gave informed consent and the study was approved by the ethical committee of the department of Psychology at the University of Groningen (study approval code: PSY-2021-S-0025).

2.2 Design and Procedure

Participants were asked to study the English translation of Swahili vocabulary items. For the first presentation of an item, participants saw a Swahili word on the computer screen in text, together with the written English translation of this word. Additionally, the spoken English translation of the word was presented through headphones. In all subsequent presentations, only the Swahili word was shown, and participants were instructed to speak the correct English translation, after which they received corrective written and auditory feedback. Voice utterances were transcribed to text automatically and in real time using the Google Web Speech API to provide corrective feedback. To prevent that minor transcription, tense or number errors would result in scoring the response as incorrect, responses were considered correct if Levenshtein's edit distance from response to answer [30] was equal to or less than 2. Response times were defined as the time elapsed between the start of the presentation of the item and the time at which the participant started speaking (voice onset time). In the MemoryLab adaptive scheduling conditions, items were scheduled based on the accuracy and latency of previous responses (see Introduction). For a more detailed description of the model, see [18,23].

2.3 Materials

For details on the materials used (word lists, exemplar pronunciations) as well as the setup (software and hardware) used in the experiment, see [28].

2.4 Speech Feature Extraction

PSFs were extracted after data collection using Praat 6.2.07 [3]. Before extracting PSFs, all silences were trimmed based on zero-crossings and short term intensity. We selected PSFs to include in the analyses based on previous literature (pitch slope, intensity, speaking speed) and on an exploratory basis (average pitch, jitter, shimmer). Subsequently, we extracted the following PSFs for each utterance: (1) the fundamental frequency for each 0.01 s window in the speech signal; (2) the average frequency over the full duration of the speech signal; (3) the average intensity over the duration over the speech signal; (4) the speaking speed defined as the average number of syllables uttered per second; (5) mean local jitter and (6) mean local shimmer. Finally, (7) we used least squares linear regression on all pitch observations in an utterance to compute the pitch slope. All PSFs were standardized within participants.

2.5 Data and Statistical Analyses

Statistical analyses were conducted in R 3.4.1 [16]. Trials containing PSFs that were more than 5 standard deviations above or below the mean value, as well as trials with response latencies below 200ms, were considered outliers and were removed from the data set before analyses. In total, 1.02% of all trials were excluded based on these criteria. After preprocessing, the data set contained a total of 7,334 learning trials from 44 participants.

To examine the association between PSFs and response accuracy and latency on the same trial, we computed Pearson's correlations. In addition, we fitted four mixed effects regression models to explain accuracy (logistic mixed effects regression models) and response latency (linear mixed effects regression models) using PSFs. In these models, standardized PSFs were added as fixed effects, and we controlled for by-item and by-participant variation by adding these factors as random effects to the models [2].

To explore if PSFs can be used to improve the adaptive learning model predictions, we trained and tested two variations of two regression models to predict (a) current trial response latency and (b) accuracy. Because regression analyses with a large number of predictor variables have a tendency to over-fit the data (reducing out-of-sample predictive properties) we used elastic-net penalized regression [31]. We used 10-fold cross validation to choose the optimal elastic net hyperparameters for mixture (alpha) and penalty (lambda). Both sets of models contained predictions based on the memory activation estimated during learning by the MemoryLab model as predictor. The models differed by their inclusion of the PSFs: only one set of models contained PSFs on previous trials. We included PSFs for up to five repetitions back for the same item. All models were trained and tested using a second 10-fold cross validation procedure, in which the models were trained on 90% of the data and tested on the remaining 10% of data 10 times. Because of the unbalanced nature of the sample (more correct than incorrect responses), we report average test classification precision, recall and F1-score metrics for the models that predict accuracy [8]. We report average test R-squared values, as well as root mean square error (RMSE) values, for the models predicting response latencies.

3 Results

3.1 The Association Between Speech Prosody and Memory Retrieval Performance

The first main goal of this study is to test the reliability of earlier studies that find an association between some PSFs and speaker confidence and accuracy [6,7,9], and to examine the exact way in which these effects can be found in a learning paradigm. Figure 2A shows Pearson's correlations between accuracy scores, response times, and PSFs for spoken retrieval attempts during learning. The top two rows of Fig. 2A show that there are several PSFs that correlate significantly with accuracy and response latencies. First, in line with earlier literature [6], we find that standardized pitch slope positively correlates with response times, and negatively correlates with correctness, indicating that longer response times and lower accuracy are associated with rising pitch. Average standardized intensity is negatively associated with response times and positively associated with correctness, indicating more accurate and faster responses are, on average, louder. Finally, speaking speed is negatively associated to response times and positively associated to correctness indicating that faster and more accurate responses are associated with higher speaking speed.

To further corroborate the correlational analyses reported above, we fitted four mixed effects regression models. Table 1.1 shows the results of the model predicting current trial accuracy from standardized pitch slope, speaking speed, intensity, jitter, shimmer, and average pitch[1]. The results show that pitch slope, speaking speed, and average intensity significantly explain accuracy: The lower the pitch slope, the higher the speaking speed and the higher the intensity, the higher the accuracy of the same response. Shimmer, jitter, and average pitch do not explain accuracy. When response times are added to the model (Table 1.2), pitch slope and intensity still significantly explain accuracy.

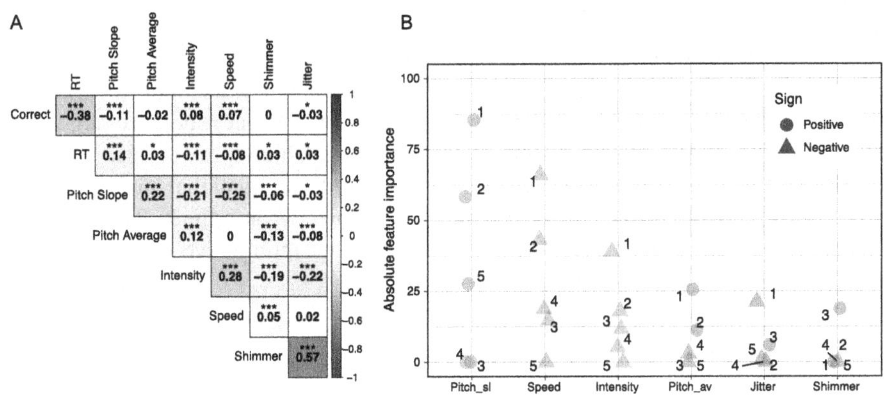

Fig. 2. A. Pearson's correlations between accuracy, response latencies, and PSFs for the same spoken retrieval attempt. Note: *p<.05; **p<.01; ***p<.001. **B**. Absolute PSF importance (penalized regression coefficient (ms)) in predicting current repetition recall performance (response latency) from previous repetition PSFs. Black numbers indicate how many trials back the PSF was recorded. Circles show positive regression coefficients, triangles show negative coefficients.

[1] The logistic regression coefficients in Table 1.1 and 1.2 can be converted to probabilities using an inverse logit transform. For example, in Table 1.1, a one standard deviation increase in pitch slope was associated with a decrease in accuracy from $e^{(1.885)}/(1 + e^{(1.885)}) = 0.868$ to $e^{(1.885-0.260)}/(1 + e^{(1.885-0.260)}) = 0.835$.

Table 1. The association between PSFs and retrieval performance.

	Accuracy		Response latency (ms)	
	(1. without RT)	(2. with RT)	(3. without accuracy)	(4. with accuracy)
Response latency (ms)		−0.001***		
Accuracy (cor./incor.)				−1541.371***
Pitch slope (z)	−0.260***	−0.196***	148.827***	99.541***
Speaking speed (z)	0.099***	0.058	−90.516***	−70.819***
Average intensity (z)	0.120**	0.084*	−62.560**	−39.173*
Jitter (z)	−0.049	−0.065	16.411	4.902
Shimmer (z)	0.050	0.087	38.286	46.826*
Average pitch (z)	−0.030	−0.028	20.423	13.315
Constant	1.885***	3.226***	2264.926***	3569.319***
Observations	7,334	7,334	7,334	7,334

*p<.05; **p<.01; ***p<.001

Table 1.3 shows that pitch slope, speaking speed and intensity explain variation in response times (also see Fig. 2B): The higher the pitch slope, the higher the response times. The higher the speaking speed, the lower the response time. The louder the learner gave the answer, the higher correctness. Finally, the higher the mean local shimmer, the longer the response latency. Table 1.4 shows the model in which accuracy is also included. In this case, pitch slope, speaking speed and intensity remain significantly associated with response latency.

Overall, the mixed effects models corroborate what is apparent in the correlational analyses reported above: Three features show the strongest association to memory strength or ease-of-retrieval. First, rising pitch (positive pitch slope) utterances are associated to low accuracy and long response latencies. Second, high speaking speed is associated with high accuracy and low response times. Third, we found that loud responses (high intensity) are associated to high accuracy and low response times. These results indicate that in the context of a learning task, PSFs are informative of the extent to which a learner has successfully memorized an item.

3.2 Improving Predictions of Future Performance Using Speech Prosody

The results discussed in the previous section demonstrate that a number of PSFs recorded during spoken retrieval attempts are associated with retrieval accuracy and response times, and therefore carry information about the extent to which the learner has successfully memorized an item. Given the reliability of these effects in the context of a learning paradigm, we will now test whether they can be used to improve out-of-sample predictions of future retrieval success in applied learning settings. We used penalized regression analyses to predict retrieval performance (accuracy and response times) from speech prosody during earlier retrieval attempts for the same item.

Table 2 shows the average results of four penalized regression models, two predicting current trial accuracy and two predicting current trial response times. For each outcome variable, we fitted one model with the original MemoryLab algorithm's estimations as predictor (right columns, 'without PSFs') and one model that contained both the original MemoryLab model estimations, as well as the six PSF recordings for the five preceding repetitions of the same item (left columns, 'with PSFs'). The results show that using previous repetition PSFs in addition to the original model resulted in increased test classification accuracy, precision, recall, and F1-score values. The F1-score increased by 13.5%, indicating substantial improvements in the precision and robustness of the model by adding PSFs. Using previous repetition PSFs also resulted in an increase in explained variance in response times (6.9%), and a reduction in overall Root Mean Square Error (RMSE).

Table 2. Improving model predictions using prosodic speech features.

Accuracy	With PSFs	Without PSFs	RT	With PSFs	Without PSFs
Accuracy	0.882	0.864	Test R^2	72.689	67.953
Precision	0.684	0.657	Test RMSE	700.484	749.035
Recall	0.339	0.286			
F1-score	0.453	0.399			

Note: values represent averages for 10-fold cross validated test predictions.

Figure 2B shows the importance of each of the PSFs included the in elastic net regression analyses. The average absolute feature importance (defined as the penalized regression coefficient) is shown on the x-axis. The black numbers indicate how many trials back the PSF was recorded. The best predictor of response times is pitch slope, followed by speaking speed and intensity. For most PSFs, the previous two repetitions were the most important predictors of current trial response time. PSFs recorded more than two repetitions in the past are generally less informative for current trial performance (more than two repetitions back, most speech features have very low coefficients). For average pitch, jitter and shimmer, we found low coefficients and inconsistent signs over preceding repetitions, underlining the lack of evidence for the informative value of these features. Overall, these results show that using pitch slope, speaking speed and intensity for the two repetitions preceding the current repetition most substantially improve predictions of future recall performance relative to accuracy and response latency.

4 Discussion

In this study, we (a) explored the association between high-level prosodic features in speech and recall performance in a learning paradigm, and (b) examined the possibility of using prosodic speech information to improve a cognitive model of memory retrieval that can be used to for item scheduling in an applied learning system. To this end, we analysed speech recordings from over 7,000 retrieval attempts for 44 participants and automatically extracted six high-level PSFs. We will reiterate and interpret the results for (a) and (b) in turn.

Correlational and mixed effects generalized linear regression analyses revealed that accuracy during spoken retrieval attempts is negatively associated to pitch slope, and positively associated to speaking speed and intensity. In other words, for incorrect retrieval attempts, participants were more likely to speak with rising pitch, lower speaking speeds, and lower intensity than for correct retrieval attempts. In addition, we found that response latency is positively associated to pitch slope, and negatively associated to speaking speed and intensity. These results replicate and extend the findings of Goupil and Acouturier [6], who recently reported that higher intensity was associated with higher retrieval accuracy. Intuitively, our results also align with earlier research on the association between *confidence* and prosody, in which rising pitch and lower speaking speeds were tied to uncertainty about a response [6, 10]. While subjective confidence in the response was not directly measured in this study, it is reasonable to assume that overall confidence in accurate and fast retrieval attempts was higher than confidence in inaccurate or slower responses. Finally, we found no clear effects of average pitch, average jitter and average shimmer in an utterance. These features were included in our analyses on an exploratory basis (we are not aware of any relevant studies that report an association between accuracy or subjective confidence in a response). We conclude that the latter features are not useful indices of memory strength in a learning context.

We found that using previous repetition PSFs, in addition to previous repetition response latencies and accuracy scores, to predict current repetition accuracy resulted in increased classification performance. The F1-score of the model increased with 13.5%, indicating that the balanced predictive performance of the model improved substantially. Similarly, we show that utilizing previous repetition PSFs increases the explained variance of the model predicting current repetition response latencies by 6.9%. Overall, these results show that using previous repetition PSFs in addition to response times and accuracy can result in a substantial improvement of overall adaptive learning model performance. These results demonstrate that PSFs provide information in addition to accuracy and response times that can be used as a behavioral proxy of latent memory strength in adaptive learning models.

Our results lead to various suggestions for future work. First, although we find that information gathered through prosodic speech analyses can improve model predictions, our results do not show *why* a specific speech signature is associated with better or worse recall performance. More specifically, pitch slope, speaking speed, and intensity could reflect the objective memory strength for a vocabulary item, much like the assumed relationship between memory strength and response times [4]. Alternatively, they could be a reflection of subjective confidence in the response. Future studies should shed light on this issue by directly assessing the speakers' subjective confidence in each response. Second, this study used single words as cues and response options. Arguably, more or stronger prosodic information could be extracted if combinations of multiple words, or short sentences are used, because longer utterances give the speaking more opportunity to vary intonation, rhythm and stress [17]. Finally, in the current study, PSFs were extracted offline after completion of the experiment. In future projects, PSFs should be extracted to update predictions of future performance in real time to optimize scheduling or feedback presentation in an adaptive learning session. A technical challenge associated with this approach is that for the current study, PSFs were standardized by each learner using the complete learning history for that learner, which would not be possible in a real-time setting. Future studies should consider the practical feasibility of standardising PSFs based on the first couple of responses only.

In addition, it is important to consider issues related to the specificity of participant sample used in the current study. In an additional set of analyses, we found no significant effects of age or gender on the effects reported in this study. In addition, earlier research suggests that universal PSFs are found in speech throughout languages and cultures [7, 17, 29]. Nevertheless, we acknowledge that it is of key importance to further examine if the results found in this study, which can be construed as a proof-of-concept, generalize to other groups of learners throughout the world.

Despite these open questions, our results have important implications. First, they have fundamental relevance as they are the first to couple a specific prosodic speech signature (falling pitch, high speaking speed, high vocal intensity) to high accuracy and fast responses in a learning paradigm, suggesting that PSFs may be used as a measure of speaker confidence during memory recall. More generally, although more research into the exact nature of the relationship between PSFs and memory performance is necessary, PSFs may prove to be a valuable new tool in the further exploration of important open research questions (e.g., about speaker certainty/confidence or feeling-of-knowing and a range of other meta-memory judgements). Second, our results are educationally relevant because they can contribute to improving cognitive models of memory retrieval that are used in real world learning settings: The results of this study lead to specific recommendations on *how* to use PSFs in adaptive learning models.

In conclusion, we show that spoken retrieval attempts contain information about the extent to which a learner has memorized an item, and that PSFs can be used to improve model predictions for learner performance on future trials. As such, they are a promising candidate to be used in learning research and in educationally relevant speech-based learning applications.

References

1. Anderson, J.R., Bothell, D., Lebiere, C., Matessa, M.: An integrated theory of list memory. J. Mem. Lang. **38**(4), 341–380 (1998)
2. Baayen, R.H., Davidson, D.J., Bates, D.M.: Mixed-effects modeling with crossed random effects for subjects and items. J. Mem. Lang. **59**(4), 390–412 (2008)
3. Boersma, P.: Praat: doing phonetics by computer (2006). http://www.praat.org/
4. Byrne, M.D., Anderson, J.R.: Perception and action. Atomic Comp. Thought **16**, 23–28 (1998)
5. Golonka, E.M., Bowles, A.R., Frank, V.M., Richardson, D.L., Freynik, S.: Technologies for foreign language learning: a review of technology types and their effectiveness. Comput. Assist. Lang. Learn. **27**(1), 70–105 (2014)
6. Goupil, L., Aucouturier, J.J.: Distinct signatures of subjective confidence and objective accuracy in speech prosody. Cognition **212**, 104661 (2021)
7. Goupil, L., Ponsot, E., Richardson, D., Reyes, G., Aucouturier, J.J.: Listeners' perceptions of the certainty and honesty of a speaker are associated with a common prosodic signature. Nat. Commun. **12**(1), 1–17 (2021)
8. Goutte, C., Gaussier, E.: A probabilistic interpretation of precision, recall and *F*-score, with implication for evaluation. In: Losada, D.E., Fernández-Luna, J.M. (eds.) ECIR 2005. LNCS, vol. 3408, pp. 345–359. Springer, Heidelberg (2005). https://doi.org/10.1007/978-3-540-31865-1_25

9. Jiang, N.: Lexical development and representation in a second language. Appl. Linguis. **21**(1), 47–77 (2000)
10. Jiang, X., Pell, M.D.: The sound of confidence and doubt. Speech Commun. **88**, 106–126 (2017)
11. Lindsey, R.V., Shroyer, J.D., Pashler, H., Mozer, M.C.: Improving students' long-term knowledge retention through personalized review. Psychol. Sci. **25**(3), 639–647 (2014)
12. Liu, Z.T., Rehman, A., Wu, M., Cao, W.H., Hao, M.: Speech emotion recognition based on formant characteristics feature extraction and phoneme type convergence. Inf. Sci. **563**, 309–325 (2021)
13. Nedungadi, P., Remya, M.: Incorporating forgetting in the personalized, clustered, Bayesian knowledge tracing (pc-BKT) model. In: 2015 International Conference on Cognitive Computing and Information Processing (CCIP), pp. 1–5. IEEE (2015)
14. Papousek, J., Pelánek, R., Stanislav, V.: Adaptive practice of facts in domains with varied prior knowledge. In: Educational Data Mining 2014, pp. 6–13 (2014)
15. Pavlik, P.I., Anderson, J.R.: Using a model to compute the optimal schedule of practice. J. Exp. Psychol. Appl. **14**(2), 101 (2008)
16. R Core Team: R: A Language and Environment for Statistical Computing. R Foundation for Statistical Computing, Vienna, Austria (2020)
17. Reed, B.S.: Analysing Conversation: An Introduction to Prosody. Macmillan International Higher Education (2010)
18. Sense, F., Behrens, F., Meijer, R.R., Van Rijn, H.: An individual's rate of forgetting is stable over time but differs across materials. Top. Cogn. Sci. **8**(1), 305–321 (2016)
19. Sense, F., Meijer, R.R., Van Rijn, H.: Exploration of the rate of forgetting as a domain-specific individual differences measure. Front. Educ. **3**(112) (2018)
20. Sense, F., van der Velde, M., Van Rijn, H.: Predicting university students' exam performance using a model-based adaptive fact-learning system. J. Learn. Anal. **8**, 1–15 (2021)
21. Settles, B., Brust, C., Gustafson, E., Hagiwara, M., Madnani, N.: Second language acquisition modeling. In: Proceedings of the Thirteenth Workshop on Innovative Use of NLP for Building Educational Applications, pp. 56–65 (2018)
22. Settles, B., Meeder, B.: A trainable spaced repetition model for language learning. In: Proceedings of the 54th Annual Meeting of the Association for Computational Linguistics (Volume 1: Long Papers), pp. 1848–1858 (2016)
23. Van Rijn, H., van Maanen, L., van Woudenberg, M.: Passing the test: improving learning gains by balancing spacing and testing effects. In: Proceedings of the 9th International Conference of Cognitive Modeling, vol. 2, pp. 7–6 (2009)
24. Van Rossum, G., Drake, F.L.: Introduction To Python 3: Python Documentation Manual Part 1. CreateSpace (2009)
25. Ververidis, D., Kotropoulos, C.: Sequential forward feature selection with low computational cost. In: 2005 13th European Signal Processing Conference, pp. 1–4. IEEE (2005)
26. Walsh, M.M., et al.: Mechanisms underlying the spacing effect in learning: a comparison of three computational models. J. Exp. Psychol. Gen. **147**(9), 1325 (2018)
27. Wennerstrom, A.: The Music of Everyday Speech: Prosody and Discourse Analysis. Oxford University Press, Oxford (2001)
28. Wilschut, T., Sense, F., van Rijn, H.: Speaking to remember: model-based adaptive vocabulary learning using automatic speech recognition. Available at SSRN 4227060 (2022)
29. Xu, Y.: Speech prosody: a methodological review. J. Speech Sci. **1**(1), 85–115 (2011)
30. Yujian, L., Bo, L.: A normalized Levenshtein distance metric. IEEE Trans. Pattern Anal. Mach. Intell. **29**(6), 1091–1095 (2007)
31. Zou, H., Hastie, T.: Regularization and variable selection via the elastic net. J. R. Stat. Soc.: Ser. B (Stat. Methodol.) **67**(2), 301–320 (2005)

Neural Automated Essay Scoring Considering Logical Structure

Misato Yamaura(✉), Itsuki Fukuda, and Masaki Uto(✉)

The University of Electro-Communications, Chofu, Tokyo 182-8585, Japan
{yamaura,fukuda,uto}@ai.lab.uec.ac.jp

Abstract. Automated essay scoring (AES) models based on deep neural networks (DNN) have recently achieved high accuracy. However, conventional neural AES models cannot explicitly consider the logical structure of each essay. Explicitly considering the logical structure in neural AES models is expected to improve scoring accuracy because logical structure is an important factor affecting essay quality. Accordingly, this study proposes a neural AES method that incorporates information about logical structure. First, the proposed method estimates the logical structure of each essay using the argument mining method, which is a machine learning method for extracting the logical structure from texts. Then, the logical structure is processed using a newly developed neural architecture, which we formulate as a transformer-based DNN model with modified self-attention, and a distributed representation of the logical structure is output. Finally, the proposed method integrates that distributed representation into conventional neural AES models to predict the essay score. We demonstrate the effectiveness of the proposed method through experiments using benchmark data for AES.

Keywords: automated essay scoring · deep neural networks · argument mining · natural language processing · educational measurement

1 Introduction

Essay writing tests have been widely used in various educational contexts to measure higher-order abilities, such as logical thinking and expression skills [1,10]. However, there are several issues associated with manual essay scoring, including the difficulty of ensuring fairness and its increased costs [26]. Automated essay scoring (AES), which utilizes natural language processing (NLP) and machine learning techniques to grade essays automatically, has been attracting attention in recent years as a means of resolving these problems [6,25]. AES is also important for writing education because its ability to provide immediate and accurate feedback is effective for developing students' writing skills [12,13,27]. Thus, AES

This work was supported by JSPS KAKENHI Grant Numbers 19H05663, 20K20817, and 21H00898.

N. Wang et al. (Eds.): AIED 2023, LNAI 13916, pp. 267–278, 2023.
https://doi.org/10.1007/978-3-031-36272-9_22

has been used with various kinds of writing instruction and support tools, including formative assessment systems and analytical scoring systems [9,20].

There are two major approaches for AES. One is the *feature-engineering* approach, which predicts essay scores using machine learning models, such as linear regression models and decision trees, from manually designed features, such as the number of words and grammatical errors [6,17]. The other is the *automatic feature extraction* approach [14,25] based on deep neural networks (DNN), where the input is a word sequence of a target essay and the output is a predicted score. In recent years, AES models using pre-trained transformer-based DNN models, including BERT (Bidirectional Encoder Representations from Transformers) [2], have achieved high accuracy [23,26].

Conventional neural AES models can consider the meaning of words and the relationships among them, but they cannot directly consider the logical structure of each essay [4,5]. Explicitly considering the logical structure in neural AES models is expected to improve scoring accuracy because the logical structure is an important factor that affects essay quality. To do so, an AES method was proposed that took logical structure into account [15]. Specifically, this method estimates the logical structure of each essay using the argument mining method, which is a machine learning method for extracting the logical structure from texts [16,18,22], and predicts scores using an AES model with a feature-engineering approach that uses manually designed features extracted from the logical-structure, such as the number of components in the logical structure and the number of edges between those components. However, that study reported that adding the logical-structure features did not improve scoring accuracy much. One reason for this conclusion might be that the previous method only used relatively superficial, manually designed features corresponding to the logical structure, and thus failed to capture deep and meaningful features.

To overcome this limitation, we propose a neural AES method that takes the logical structure into account without converting it into manually designed features. We begin by proposing a new transformer-based DNN model with modified self-attention to process the logical structure of each essay, which is estimated by an argument mining method. This neural architecture outputs a distributed representation of the logical structure, a fixed-dimensional real-valued vector aggregating the information about a given logical structure. The proposed AES method integrates the distributed representation into conventional neural AES models to predict an essay score. We demonstrate its effectiveness through experiments using benchmark data for AES.

2 Conventional Neural AES Model Using BERT

In this section, we introduce a conventional neural AES model using BERT [2], which has often been used as a baseline model in recent years.

BERT is a pre-trained transformer-based DNN model released by the Google AI Language Team in 2018. BERT has achieved state-of-the-art results in a variety of NLP tasks. In recent years, many variants of BERT, such as RoBERTa [11], ALBERT [7], and DistilBERT [19], have been proposed.

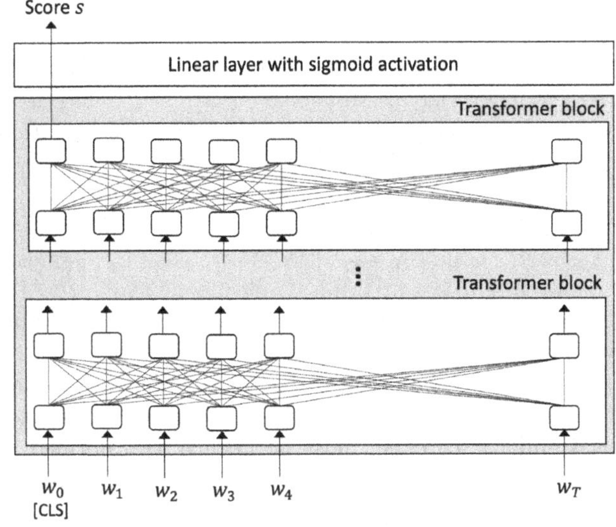

Fig. 1. BERT-based AES model.

BERT consists of 12 stacked transformer architectures, each containing a self-attention mechanism. The self-attention mechanism is the key component of BERT that captures the semantic relationships among words and outputs a contextualized word embedding vector, a fixed-dimensional real-valued vector, for each word. BERT can understand the context of a given text by capturing the meaning of each word and the relationships among words in the text through multi-layered self-attention.

BERT is trained in *pre-training* and *fine-tuning* steps. Pre-training is performed on large amounts of text data over two unsupervised learning tasks: *masked language model*, a task for predicting randomly masked words, and *next sentence prediction*, a task for determining whether two given sentences are adjacent. The pre-trained BERT can be applied to various downstream tasks by fine-tuning it, including AES. Fine-tuning is conducted with a task-specific supervised dataset after adding task-specific output layers and initializing BERT parameters to pretrained values.

The model structure for applying BERT to an AES task is shown in Fig. 1. When using BERT for AES, input essays require preprocessing, namely, adding a special token [CLS] to the beginning of each input. Specifically, if w_t represents the t-th word in a given essay, then the input is defined as the word sequence $\{w_0, w_1, \ldots, w_T\}$, where w_0 indicates the [CLS] token and T is the essay length. The BERT output vector corresponding to the [CLS] token can be used as a distributed representation for a given essay. Thus, letting $\boldsymbol{x}_0$ be the BERT output vector for the [CLS] token, we can predict an essay score s by inputting $\boldsymbol{x}_0$ to a linear layer with sigmoid activation, which is given as

$$s = \sigma(\boldsymbol{W}\boldsymbol{x}_0 + b), \tag{1}$$

where σ is the sigmoid function, and $\boldsymbol{W}$ and b are the weight vector and bias parameters.

Fine-tuning of the model is performed by the back–propagation algorithm using the following mean squared error (MSE) loss function.

$$\frac{1}{N}\sum_{n=1}^{N}(s_n - s_n^*)^2, \tag{2}$$

where s_n and s_n^* represent the predicted score and the gold-standard score for the n-th essay, and N is the number of essays in the training data.

Note that the predicted scores from the AES model take values between 0 and 1 because the sigmoid function is applied. Therefore, in the model training phase, the gold-standard scores s_n^* must be linearly transformed to the range $[0, 1]$. Furthermore, in the prediction phase, the predicted scores s_n must be linearly transformed to the gold-standard score scale.

Conventional neural AES methods, including the BERT-based AES model, can take into account the meaning of words and the relationships between them but cannot directly do so for the logical structure of each essay. In this study, we aim to overcome this limitation by integrating the information about the logical structure into the AES models. We use the argument mining method to estimate the logical structure.

3 Argument Mining

Argument mining consists of the following four subtasks, shown in Fig. 2.

1. **Component identification** Sentences or phrases corresponding to the components of logical structure (hereafter referred to as *logical components*) are extracted from a text.
2. **Component classification** Each logical component is classified into *Premise*, *Claim*, and *Major Claim*.

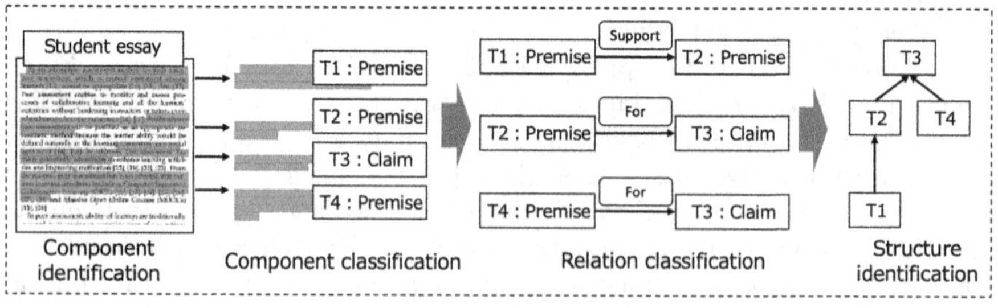

Fig. 2. Outline of argument mining.

3. **Relation classification** The logical relationships between logical components are classified into *Support*, *Attack*, *For*, and *Against* while considering the results of the component classification task.
4. **Structure identification** The optimal logical structure satisfying the tree structure constraints is estimated, taking into account the results of the relation classification task.

The most traditional argument mining method solves these four subtasks separately one by one [22], while recent DNN-based methods solve them in an end-to-end manner, allowing them to achieve higher performance [3,8,21,28]. In this study, we use the DNN-based argument mining model proposed by Eger et al. [3], which was trained using an existing dataset for argument mining [22], to estimate the logical structure of essays.

4 Proposed Method

The DNN model we developed processes the logical structure estimated by the argument mining method. We present this model and propose a new neural AES method that incorporates the processed logical structure information.

4.1 DNN Model for Processing Logical Structure

Our DNN model processes logical structures by extending BERT. Specifically, we extend the self-attention in BERT by incorporating a *visible matrix*, which controls visibility among arbitrary words. Given a logical structure estimated by the argument mining method, we construct a visible matrix where only words within logical components with a logical relationship are allowed to be referred to each other. The visible matrix is defined as follows.

Visible Matrix. Suppose that a target essay consists of P logical components $\{\mathbf{C}_p \mid p \in \{1, \ldots, P\}\}$. Each logical component $\mathbf{C}_p$ is a word sequence $\{w_i, w_{i+1}, \ldots, w_{I_p}\}$, where i is the starting position of the logical component in the essay and I_p is the number of words in the logical component. Furthermore, we define the logical relationship between logical components by a $P \times P$ matrix $\mathbf{R} = \{r_{uv} \mid u, v \in \{1, \ldots, P\}\}$, where $r_{uv} = r_{vu} = 1$ if the u-th logical component $\mathbf{C}_u$ and the v-th logical component $\mathbf{C}_v$ are logically related on the estimated logical structure, and $r_{uv} = r_{vu} = 0$ otherwise. Using these definitions, we define the visible matrix as a $(T+1) \times (T+1)$ matrix $\mathbf{M} = \{m_{tt'} \mid t, t' \in \{0, \ldots, T\}\}$, where $m_{tt'} = 0$ if words w_t and $w_{t'}$ have a logical relationship, and $m_{tt'} = -\infty$ otherwise. Specifically, satisfying the following two conditions makes $m_{tt'} = 0$.

1. The estimated logical structure has a logical component $\mathbf{C}_p$ consisting of the t-th word w_t and a logical component $\mathbf{C}_{p'}$ consisting of the t'-th word $w_{t'}$.
2. Two components $\mathbf{C}_p$ and $\mathbf{C}_{p'}$ have a logical relationship in the estimated logical structure, namely, $r_{pp'} = 1$.

Note that we defined the visible matrix as a $(T+1) \times (T+1)$ matrix because we assume that the input for our BERT extension model is the word sequence of a targeted essay with the [CLS] token, as in the conventional BERT-based AES model. Furthermore, we aim to aggregate all information into the output vector for the [CLS] token. To do so, we set all elements in the 0-th row of the visible matrix $\mathbf{M}$ to be 0, namely, $m_{0t} = 0; \forall t$.

BERT Using Visible Matrix. Our proposed method extends self-attention in BERT by using the visible matrix, which is created based on an estimated logical structure of a targeted essay. Hereafter, we refer to this model as *BERT-LS*.

BERT-LS calculates the l-th self-attention layer in BERT as

$$\mathbf{H}_l = \mathbf{A}_l \cdot \mathbf{V}_l, \tag{3}$$

where $\mathbf{H}_l$ is the $(T+1) \times D$ matrix whose t-th row corresponds to the distributed representation of the t-th word in the given essay, and D represents the dimension of the distributed representation. $\mathbf{V}_l$ is the $(T+1) \times D$ matrix calculated based on the previous self-attention layer output $\mathbf{H}_{l-1}$. $\mathbf{A}_l$ is the $(T+1) \times (T+1)$ matrix corresponding to the attention weights, where the component in the t-th row and t'-th column, $a_{ltt'}$, represents the degree of relevance between words w_t and $w_{t'}$. Given the visible matrix $\mathbf{M}$, BERT-LS calculates the attention weight matrix as

$$\mathbf{A}_l = softmax(\frac{\mathbf{Q}_l \cdot \mathbf{K}_l^{\top}}{\sqrt{d}} + \mathbf{M}), \tag{4}$$

where $\mathbf{Q}_l$ and $\mathbf{K}_l$ are the $(T+1) \times D$ matrices calculated based on the previous self-attention layer output $\mathbf{H}_{l-1}$, and d is a scaling factor.

Through the softmax operation, the attention weight $a_{ltt'}$ becomes 0 when the corresponding element in the visible matrix $m_{tt'}$ is $-\infty$, while it becomes a positive value when $m_{tt'} = 0$. When the attention weight $a_{ltt'} = 0$, the self-attention operation defined in Eq. (3) ignores the relation between words w_t and $w_{t'}$. Therefore, BERT-LS can process the input essay while considering only the relations among words within logical components with a logical relationship, which enables text processing while emphasizing the logical structure. In BERT-LS, we can regard the output vector corresponding to the [CLS] token in the last layer as a distributed representation focusing on the logical structure because we designed the visible matrix to aggregate all information into the output vector for the [CLS] token, as explained above. Note that when the visible matrix is the zero matrix, BERT-LS is consistent with the original BERT, meaning that BERT-LS is an extension of BERT.

Figure 3 shows an example of the process for creating a visible matrix and a conceptual diagram of the self-attention where the visible matrix is applied. This example is for an input text that has a logical structure consisting of two logically related components.

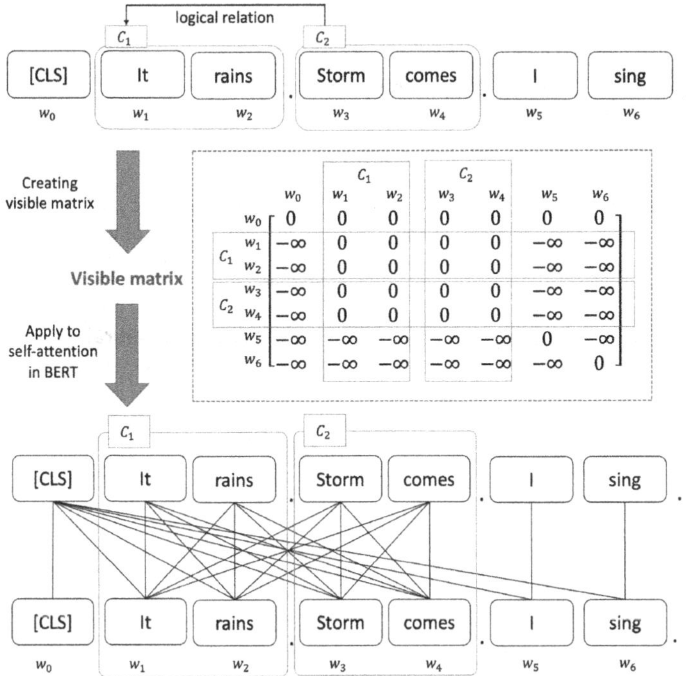

Fig. 3. Outline of visible matrix creation and self-attention with the visible matrix.

4.2 Neural AES Model Considering Logical Structure

In this section, we present the neural AES model incorporating BERT-LS. The architecture of the AES model is illustrated in Fig. 4. It consists of two DNN models, BERT-LS and a conventional neural AES model, including the BERT-based AES model introduced in Sect. 2.

Both models receive the word sequence together with the [CLS] token and output the distributed representation corresponding to the [CLS] token. The two distributed representation vectors are then concatenated, and the essay score is predicted using the concatenated vector. Specifically, letting $\boldsymbol{x}_0'$ be the output distributed representation from BERT-LS and $\boldsymbol{x}_0$ be that from the conventional neural AES model, we map the concatenated vector $[\boldsymbol{x}_0; \boldsymbol{x}_0']$ to a scalar score value s through the linear layer with sigmoid activation defined in Eq. (1). The training of the model is performed by a back–propagation algorithm using the MSE loss function defined in Eq. (2).

Note that BERT-LS ignores the relations among words that have no logical relation. Accordingly, the method combines BERT-LS and a conventional neural AES model, allowing it to use both the information focusing on the logical structure and the information aggregating the whole text.

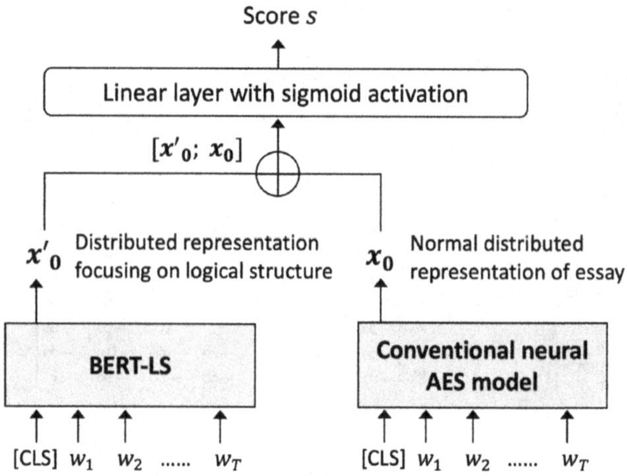

Fig. 4. Architecture of the proposed method.

5 Experiment

In this section, we demonstrate the effectiveness of our method through experiments using a benchmark dataset for AES.

5.1 Setup

We used Automated Student Assessment Prize (ASAP) for our dataset, the most widely used benchmark dataset for AES studies. ASAP consists of essays with gold-standard scores that were written by native English-speaking U.S. students for eight different prompts (essay tasks). Scores are provided based on ordered categories with different value ranges. Each prompt corresponds to one of three essay types: argumentative, source-dependent, and narrative. See Table 1 for detailed statistics and types for each prompt.

Table 1. Statistics of the ASAP Dataset.

Prompt	# of essays	Avg. length	Score range	Essay type
1	1,783	350	2–12	Argumentative
2	1,800	350	1–6	Argumentative
3	1,726	150	0–3	Source-Dependent
4	1,772	150	0–3	Source-Dependent
5	1,805	150	0–4	Source-Dependent
6	1,800	150	0–4	Source-Dependent
7	1,569	250	0–30	Narrative
8	723	650	0–60	Narrative

Using the ASAP dataset, we evaluated scoring accuracy in each prompt based on five-fold cross-validation with the quadratic-weighted Kappa (QWK) and the correlation coefficient as the evaluation metrics. The QWK, a metric showing agreement between predicted scores and ground truth, was the common evaluation metric used in the ASAP competition.

In this experiment, we examined several variants of the proposed method that use different neural AES models as the *conventional neural AES model* shown in Fig. 4. Specifically, we examined four pre-trained transformer-based neural AES models, BERT, RoBERTa [11], ALBERT [7], and DistilBERT [19], and one traditional neural AES model [24] based on long short-term memory (LSTM). We compared the scoring accuracy of these models to those without the BERT-LS architecture (corresponding to conventional AES models) to evaluate the effectiveness of incorporating the logical structure information extracted by the BERT-LS architecture.

The models used in this experiment were implemented with Pytorch and the *Transformers* library in *Hugging Face.* For BERT, RoBERTa, ALBERT, and DistilBERT, we used the base-size models. For the LSTM-based model, the hyperparameters were the same as those in the original paper [24]. We used the AdamW optimizer with mini-batch sizes of 16 and 30 as the maximum epochs.

5.2 Experimental Results

Tables 2 and 3 show the results, the model with the higher performance being given in bold. The p-value columns in the tables show the results of the paired samples t-tests for the average accuracy of the proposed and the conventional models using the same base AES model.

The tables indicate that in all cases our proposed method achieves higher averaged performances in terms of both QWK and correlation than the conventional method. Furthermore, the improvements were significant ($p < .05$) in many cases. These results show that using the information in the logical structure processed by BERT-LS is effective at improving AES accuracy.

Table 2. QWK values of the proposed and conventional methods.

		Prompt								Avg.	p-value
		1	2	3	4	5	6	7	8		
BERT	Proposed	**.815**	**.672**	**.693**	**.816**	**.809**	**.814**	.829	**.717**	**.771**	<.01
	Conventional	.795	.669	.671	.794	.803	.806	**.832**	.713	.760	
RoBERTa	Proposed	**.823**	**.682**	**.679**	**.823**	**.824**	.820	**.837**	**.736**	**.778**	<.01
	Conventional	.812	.673	.674	.814	.812	**.821**	.825	.717	.769	
ALBERT	Proposed	**.817**	**.679**	.676	**.812**	**.804**	.807	**.835**	**.723**	**.769**	.07
	Conventional	.792	.665	**.676**	.808	.800	**.811**	.834	.722	.763	
DistilBERT	Proposed	**.817**	**.679**	**.691**	**.808**	**.802**	.807	.827	**.735**	**.771**	<.05
	Conventional	.798	.674	.653	.803	.801	**.810**	**.829**	.726	.762	
LSTM	Proposed	**.817**	**.687**	**.686**	**.804**	**.806**	**.801**	**.827**	**.740**	**.771**	<.01
	Conventional	.804	.637	.656	.772	.796	.800	.739	.654	.732	

Table 3. Correlation values of the proposed and conventional methods.

		Prompt								Avg.	*p*-value
		1	2	3	4	5	6	7	8		
BERT	Proposed	**.843**	**.744**	**.723**	**.846**	**.847**	**.854**	.839	**.766**	**.808**	<.01
	Conventional	.829	.728	.710	.842	.842	.850	**.841**	.747	.799	
RoBERTa	Proposed	**.846**	**.754**	**.720**	**.854**	**.851**	.860	**.850**	.772	**.813**	.06
	Conventional	.837	.751	.717	.850	.847	**.862**	.844	**.776**	.811	
ALBERT	Proposed	**.847**	**.743**	**.724**	**.847**	.844	**.852**	.844	**.762**	**.808**	<.05
	Conventional	.829	.739	.713	.842	**.846**	.850	**.845**	.749	.802	
DistilBERT	Proposed	**.838**	**.744**	**.723**	**.843**	**.843**	**.850**	.836	**.765**	**.805**	<.05
	Conventional	.828	.739	.704	.842	.839	.847	**.838**	.750	.798	
LSTM	Proposed	**.838**	**.747**	**.721**	**.841**	**.848**	**.846**	**.833**	**.757**	**.804**	<.05
	Conventional	.826	.702	.715	.814	.844	.840	.749	.680	.771	

5.3 Analysis

As described in Table 1, the ASAP dataset consists of three essay types: argumentative, source-dependent, and narrative. The logical structure is expected to be important in the argumentative type, which asks students to discuss and justify their opinions on a specific topic, and the narrative type, which asks students to narrate a story about a specific topic. This is because written essays for these types are expected to be longer and more complex. However, in the source-dependent type, which asks students to respond to a question about a given text, the importance of the logical structure should be relatively low because the essay texts are relatively simple and short, as shown in Table 1.

To test this, we looked at the number of logical relationships in the logical structure of each essay and calculated the average number of them for each prompt. Table 4 shows the results, and we can see that the number of logical relationships tends to be larger for argumentative-type and narrative-type prompts (prompts 1, 2, 7, and 8) than for source-dependent-type prompts (prompts 3, 4, 5, and 6). In Table 4, we show the average improvements in QWK and correlation coefficient by the proposed method over all of the base AES models. The results show that the improvement in scoring accuracy with the proposed method tends to be large for prompts with a large number of logical relationships. These results support our expectations and suggest that our method succeeded in capturing the information contained in the logical structure.

Table 4. Average number of logical relations and the average improvements in QWK and correlations by the proposed method.

	Prompt							
	1	2	3	4	5	6	7	8
Avg. Num. of Logical Relations	9.63	6.39	0.31	0.35	0.41	0.57	0.72	4.66
Avg. Improvement in QWK	.018	.016	.019	.014	.007	.000	.019	.024
Avg. Improvement in Correlation	.012	.015	.010	.008	.003	.003	.017	.024

6 Conclusion

In this study, we proposed a new neural AES method that focuses on the logical structure of essays. We first developed a new DNN method that processes the logical structure estimated by an argument mining technique without converting it into manually designed features. Our AES method functions by integrating the distributed representation of the logical structure obtained from the proposed neural model into the conventional neural AES models. Experiments with benchmark data showed that incorporating the logical structure information with our method is effective at improving the scoring accuracy of AES.

For future work, we would like to analyze in greater detail the cases for which our method is especially effective. Additionally, we will investigate different ways of incorporating logical structures to determine how further improvements can be made.

References

1. Abosalem, Y.: Assessment techniques and students' higher-order thinking skills. Int. J. Second. Educ. **4**, 1–11 (2016)
2. Devlin, J., Chang, M.W., Lee, K., Toutanova, K.: BERT: pre-training of deep bidirectional transformers for language understanding. In: Proceedings of Conference of the North American Chapter of the Association for Computational Linguistics: Human Language Technologies, pp. 4171–4186 (2019)
3. Eger, S., Daxenberger, J., Gurevych, I.: Neural end-to-end learning for computational argumentation mining. In: Proceedings of the Annual Meeting of the Association for Computational Linguistics, pp. 11–22 (2017)
4. Hohenecker, P., Lukasiewicz, T.: Ontology reasoning with deep neural networks. J. Artif. Intell. Res. **68**, 503–540 (2020)
5. Janda, H.K., Pawar, A., Du, S., Mago, V.: Syntactic, semantic and sentiment analysis: the joint effect on automated essay evaluation. IEEE Access **7**, 108486–108503 (2019)
6. Ke, Z., Ng, V.: Automated essay scoring: A survey of the state of the art. In: Proceedings of the International Joint Conferences on Artificial Intelligence Organization, vol. 19, pp. 6300–6308 (2019)
7. Lan, Z., Chen, M., Goodman, S., Gimpel, K., Sharma, P., Soricut, R.: ALBERT: a lite BERT for self-supervised learning of language representations. In: International Conference on Learning Representations (2020)
8. Lawrence, J., Reed, C.: Argument mining: a survey. Comput. Linguist. **45**(4), 765–818 (2020)
9. Liu, M., Li, Y., Xu, W., Liu, L.: Automated essay feedback generation and its impact on revision. IEEE Trans. Learn. Technol. **10**(4), 502–513 (2017)
10. Liu, O.L., Frankel, L., Roohr, K.C.: Assessing critical thinking in higher education: current state and directions for next-generation assessment. ETS Res. Rep. Ser. **2014**(1), 1–23 (2014)
11. Liu, Y., et al.: RoBERTa: a robustly optimized BERT pretraining approach. arXiv (2019)
12. Ljungman, J., et al.: Automated grading of exam responses: an extensive classification benchmark. In: Soares, C., Torgo, L. (eds.) DS 2021. LNCS (LNAI), vol. 12986, pp. 3–18. Springer, Cham (2021). https://doi.org/10.1007/978-3-030-88942-5_1

13. Lu, C., Cutumisu, M.: Integrating deep learning into an automated feedback generation system for automated essay scoring. In: Proceedings of the International Conference on Educational Data Mining, pp. 573–579 (2021)
14. Nadeem, F., Nguyen, H., Liu, Y., Ostendorf, M.: Automated essay scoring with discourse-aware neural models. In: Proceedings of the Workshop on Innovative Use of NLP for Building Educational Applications, pp. 484–493 (2019)
15. Nguyen, H., Litman, D.: Argument mining for improving the automated scoring of persuasive essays. In: Proceedings of the Association for the Advancement of Artificial Intelligence Conference on Artificial Intelligence, vol. 32, pp. 5892–5899 (2018)
16. Palau, R.M., Moens, M.F.: Argumentation mining: the detection, classification and structure of arguments in text. In: Proceedings of the International Conference on Artificial Intelligence and Law, pp. 98–107 (2009)
17. Phandi, P., Chai, K.M.A., Ng, H.T.: Flexible domain adaptation for automated essay scoring using correlated linear regression. In: Proceedings of the Conference on Empirical Methods in Natural Language Processing, pp. 431–439 (2015)
18. Rinott, R., Dankin, L., Alzate Perez, C., Khapra, M.M., Aharoni, E., Slonim, N.: Show me your evidence - an automatic method for context dependent evidence detection. In: Proceedings of the Conference on Empirical Methods in Natural Language Processing, pp. 440–450 (2015)
19. Sanh, V., Debut, L., Chaumond, J., Wolf, T.: DistilBERT, a distilled version of BERT: smaller, faster, cheaper and lighter. arXiv (2019)
20. Shibata, T., Uto, M.: Analytic automated essay scoring based on deep neural networks integrating multidimensional item response theory. In: Proceedings of the International Conference on Computational Linguistics, pp. 2917–2926 (2022)
21. Shnarch, E., et al.: Will it blend? blending weak and strong labeled data in a neural network for argumentation mining. In: Proceedings of the Annual Meeting of the Association for Computational Linguistics, pp. 599–605 (2018)
22. Stab, C., Gurevych, I.: Parsing argumentation structures in persuasive essays. Comput. Linguist. **43**(3), 619–659 (2017)
23. Stab, C., Miller, T., Schiller, B., Rai, P., Gurevych, I.: Cross-topic argument mining from heterogeneous sources. In: Proceedings of the Conference on Empirical Methods in Natural Language Processing, pp. 3664–3674 (2018)
24. Taghipour, K., Ng, H.T.: A neural approach to automated essay scoring. In: Proceedings of the Empirical Methods in Natural Language Processing, pp. 1882–1891 (2016)
25. Uto, M.: A review of deep-neural automated essay scoring models. Behaviormetrika **48**(2), 459–484 (2021). https://doi.org/10.1007/s41237-021-00142-y
26. Uto, M., Okano, M.: Learning automated essay scoring models using item-response-theory-based scores to decrease effects of rater biases. IEEE Trans. Learn. Technol. **14**(6), 763–776 (2021)
27. Wang, J., Brown, M.: Automated essay scoring versus human scoring: a comparative study. J. Technol. Learn. Assess. **6**, 1–22 (2007)
28. Ye, Y., Teufel, S.: End-to-end argument mining as biaffine dependency parsing. In: Proceedings of the Conference of the European Chapter of the Association for Computational Linguistics, pp. 669–678 (2021)

"Why My Essay Received a 4?": A Natural Language Processing Based Argumentative Essay Structure Analysis

Bokai Yang[1(✉)], Sungjin Nam[2], and Yuchi Huang[2]

[1] University of Wisconsin-Eau Claire, Eau Claire, WI 54701, USA
bokai@email.sc.edu
[2] ACT, Iowa City, IA 52243, USA
{sungjin.nam,yuchi.huang}@act.org

Abstract. Writing argumentative essays is a critical component of students' learning. Previous works on automatic assessments on essay writing often focused on providing a holistic score for the input essay, which only summarized the essay's overall quality. However, to provide more pedagogical value and equitable educational opportunities for all students, an automatized system needs to provide detailed feedback on students' essays. To address this issue, we developed an essay argumentative structure feedback system to support educators and students. We employed natural language processing (NLP) and data mining techniques to explore the association between argumentative structure and essay scores. First, we proposed a cross-prompt, sentence-level ensemble model to classify the argumentative elements and extract the argumentative structures from the essay. The model worked across multiple datasets and achieved high performance. Second, after applying the classification model on the ACT writing tests, we performed a sequential mining process to extract representative argumentative structures. Our findings highlight the role of organizational argumentative structure in essay scoring. Furthermore, we found a common argumentative structure used by the high-scored essays. Finally, with the knowledge of argumentative elements and structures used in the previous essays, we proposed a feedback tool design to complement the current AES systems and help students improve their argument writing skill.

Keywords: Natural language processing · Educational data mining · Argument mining · Teaching analytics · Automated feedback

1 Introduction

Argumentative writing is an essential part of students' learning [18]. Several educational organizations evaluate students' writing skills in their examinations. However, manually grading each essay can be a non-trivial task. For example, given the current student-teacher ratio and the teacher shortage [2], the manual evaluation process can consume a significant amount of time from individual educators [3]. In the purpose of saving

N. Wang et al. (Eds.): AIED 2023, LNAI 13916, pp. 279–290, 2023.
https://doi.org/10.1007/978-3-031-36272-9_23

manual grading effort, automatic essay scoring (AES) aims to grade students' essay without human interference. Many recent AES research focused on tools that provide a holistic score [1]. Usually, an AES system takes a student's essay text as an input and assigns a score for the essay [5–7]. As a result, these AES systems can only provide a wholistic score of the essay.

Although the AES tools can provide a holistic score for students essays and save essay grading efforts, they are not enough to be applied in classroom settings [1]. Both teachers and students expect feedback from the AES tools. Teachers demand details of the graded essay and students need feedback to improve their essay.

Some early research in AES applied feature engineering and included features that might influence the essay score in their development of the AES systems [4, 8, 9]. Several features have been proved to affect the score, for example, the coherence and relevance of the essays [4], the effective use of evidence [9], the strength of the arguments [8]. However, there are still some possible influential features that remain unexplored. Evaluating an essay by including all the features that might affect its score remains a challenge. In addition, most of these studies have been limited to applying feature engineering to predict a holistic score for the essay. The assigned score of the essay still cannot be fully explained.

Argumentative structure is one of the dominant features that could influence the score of an argumentative essay [8]. Argument mining has been applied to identify the claims and premises in argumentative essays [25, 26]. With the ability to automatically identify and extract the argumentative structure of an essay, argument mining has been applied into education research [17]. Recent evidence suggests that argument mining could be used to support the AES tools [13]. However, there has been little discussion about what kind of argumentative structure contributes to the score.

This paper aims to capture the argumentative structures of essays and identify the patterns in low/high scored essays. To extract the argumentative elements of an essay, we applied an argument mining approach and proposed a cross-prompt, sentence-level ensemble model to classify the argumentative elements. In addition, a sequential pattern mining process was applied to extract the argumentative structures of the previous essays. In order to provide solid feedback, we seek to understand what kinds of structure contribute to the score. Thus, in this paper, we will answer the following research questions in the later sections:

RQ1: How reliable are the argument elements classification models when applied to datasets with different prompts?
RQ2: What is the most frequently used argumentative structure from the essays?
RQ3: Do essays with the same score share similar lengths and structures?

2 Literature Review

2.1 Automatic Essay Scoring

The early AES studies applied feature engineering and traditional machine learning algorithms to provide a score for the input essay [8, 9, 16]. For example, Rahimi and colleagues [9] proposed a score prediction method based on students' effective use of evidence and their organization of ideas and evidence to support their claims. They provided several features with rubric to predict the score of the evidence and organization dimensions. In their research, a random forest model was used as the classifier for scores. However, they identified evidence through a simple word-matching algorithm with a manually provided list of keywords for each topic. In a following study, Zhang and colleagues [16] further applied this evidence rubric method and provided feedback on how students used evidence in their writing. Persing and Ng [8] attempted a sentence-level, support vector machine (SVM) based AES model to score the essays through an argument strength dimension. The model included various predictive features, including the number of claims and supports, transitional phrases, coreference and so on. However, their argumentative elements identification process was still based on a string-matching approach with a sentence labeling rule. Recently, researchers have shown an increased interest in applying deep learning models to predict essay scores [5, 7]. Dong and colleagues [5] applied a recurrent convolutional neural network to learn text representation and provided a holistic score for the input essay. Although the deep learning approaches can learn deeper semantic relationship between essay texts and scores without feature engineering, these AES tools still only provide limited information on submitted essay responses to educators and students.

The recent advancement in deep-learning-based language models like BERT [19] and GPT-3 [24] have achieved state-of-the-art results in many NLP tasks. The focus of the AES research community also moves from traditional machine learning and early deep learning approaches to transformer-based models like BERT [19]. For example, Yang and colleagues [14] showed that their BERT-based model outperformed the previous state-of-the-art neural models. With the help of pre-trained BERT, Yang and Zhong [4] proposed a hierarchical structured model to extract semantic features at both sentence-level and document-level. The extracted semantic features were then used to evaluate coherence and relevance in the essays and compute the final score. In a recent study, a hierarchical BERT-based transfer learning approach was carried out by Xue, Tang and Zheng [15]. Instead of providing a single holistic score for the essay, their model provided scores for different features of the essay including grammar, lexicon, and supporting ideas. Although these studies showed interesting and breakthrough findings, with the long-term goal of producing feedback for students and educators, the model cannot provide details on why the essay received a certain score for each category. Regardless of state-of-the-art score prediction performance, it is still necessary for the novel AES systems to describe what is influencing the final score of an essay.

2.2 Argument Mining

Argument mining is a research area within NLP, aiming at extracting and identifying argumentative elements and structures from text [17]. Stab and Gurevych [10] proposed an SVM-based approach and classified sentences into four categories: major claim, claim, premise and none. In a following study, Nguyen and Litman [13] improved the argument component identification model developed by Stab and Gurevych [10]. They used argumentative features like number and fraction of argument components over total number of sentences, number of claims, number of premises and so on to improve the AES systems. Their results showed that the argument mining could improve the performance of the holistic-score-based AES systems. Deep learning massive language models have also been applied into argument mining [11, 12]. Wang and his colleagues [11] used pre-trained BERT model to mine three argumentation components: major claim, claim, and premise. Their approach considered both essay-level, paragraph-level, and word-level classification. In another study, Tariq and Debanjan [12] applied a token-level classification to identify claim and premise token of an essay. During the experiment, they noticed that a multitask model can identify additional cases that were missed by the single task model. In addition, they suggested that some sentences may contain multiple claims which caused misclassification. It has been proved that the performance of the AES model can be improved by applying the argument mining [13]. However, previous research still focused on predicting a final score for the input essay instead of providing feedback. As a result, the implementations of previous AES tools cannot provide detailed feedback.

3 Data

3.1 Feedback Prize Dataset

To train the argumentative elements classification model, we used the dataset of the Kaggle competition "Feedback Prize – Evaluating Student Writing" provided by [18]. The dataset consisted of 15,594 argumentative essays from about 15 prompts, written by United States students in grades 6–12. The essays were annotated for seven commonly used elements, including Lead, Position, Claim, Counterclaim, Rebuttal, Evidence and Concluding Statements. The explanations of each label are shown in Appendix Table 1. The overall inter-rater reliability of the dataset was 0.73.

Table 1. Labels and examples for argument elements

Labels	Examples
Position	I think that kids should have a grade B average
Claim	The cost of a life shouldn't be worth a text message, Snapchat or Instagram DM (Direct Message)
Evidence	BMW has a car that they call "Traffic Jam Assistant", this car can handle driving speeds up to 25 mph, but drivers still need to keep their hands on the wheel and when the car does not know what to do then vibrations go through the seat to notify the driver

According to the explanations on the competition webpage, counterclaim and rebuttal are also claims. In addition, as mentioned by the competition host, counterclaims and rebuttals had the lowest reliability. These two elements were often labeled as claims. Thus, in our experiments, we merged Claim, Counterclaim and Rebuttal to a single label Claim. We also found that there were overlaps between the labels. For example, the data labeled as lead and concluding statements consisted of positions, claims, and evidence. Since we applied a sentence-level prediction and the prediction process did not consider the location of the sentence, we excluded the sentences labeled as lead and concluding statements from the analysis. In conclusion, for this study, we used three labels: Position, Claim, Evidence. Some examples of the data are shown in Table 1.

3.2 ACT Writing Test Dataset

ACT offers optional essay exams used by colleges and universities for admissions decisions. To further analyze the association between the argumentative structure of essay and its score, we also obtained a large number of essays from the previous ACT tests. The ACT tests essays contained 13,990 essays from 27 prompts, collected from the previous ACT writing tests with dates ranging from September 2020 to March 2021. Students who took the tests come from more than 50 countries and over 500 regions. To test the generalizability of the model, we also applied a testing set which contained 30 essays (723 sentences) with scores ranging from 2 to 12.

4 System Design

In this section, we describe the proposed argumentative essay structure feedback providing tool and the datasets used to train and test the model. The system contained two parts: the ensemble model block and the essay analysis block. The system overview is shown in Fig. 1. First, we developed an ensemble model to classify the argumentative elements based on several pre-trained deep learning massive language models. To achieve the final goal of providing feedback for future students' essays, we need to understand the associations among the use of argumentative elements, the use of argumentative structure and essay scores in the essays. Thus, we further applied the ensemble model to classify the argumentative elements and extract argumentative structures from the ACT writing essays. Finally, the feedback could be provided based on the length and essay structure difference between the input essay and the high scored essays in previous tests.

4.1 Datasets

After preprocessing the Feedback Dataset, we obtained 120,630 data with three labels: Position, Claim and Evidence. We randomly split more than 110 K data into the training set and the validation set to train the models. The training set for the multiclass classification model contains more than 100K data and the validation set contains 10K data. To include more positions and evidence into the training set, we further increased the size of the training set and decreased the size of the validation set for the binary classification models used to predict position and evidence. The testing set from Feedback Prize dataset contained 10K randomly selected data. To test the generality of the model, 723 sentences (30 essays) from the ACT writing dataset were annotated by two experts on argumentative essay grading. The inter-rater reliability of the dataset was 0.615 (Weighted Kappa). In total, we built two training sets, two validation sets and two testing sets to train and test the ensemble model.

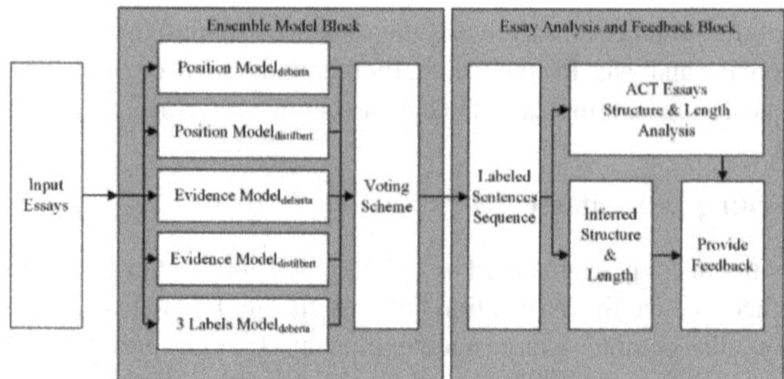

Fig. 1. System overview

4.2 Ensemble Model Block

We first built a multiclass classification model to classify all three kinds of argumentative elements. The model was built based on the pretrained model: DeBERTa [20]. We also built four binary classification models to classify positions and evidence using DeBERTa and DistilBERT [21]. We used these two pre-trained models for the binary classifiers since these models were trained on different corpus and provided the best results in our preliminary analysis.

The Voting Scheme. To synthesize the results of the five different models, a voting scheme algorithm was applied to decide the final label of a sentence. We want to emphasize that our goal is to find as many positions and evidence as we can. The design of the algorithm of the voting scheme was shown in Appendix Algorithm 1.

Evaluation Metrics. To evaluate the models, we applied the following evaluation metrics: accuracy, Macro-F1 and recall. The recall value showed our model's sensitivity to the positions and evidence sentences. Since the goal of this paper is to extract the argumentative structure of an essay, we tried to find as many positions and evidence as we could. Thus, the recall value was also included as an evaluation metric.

4.3 Essay Analysis Block

After applying the ensemble classification model on the ACT previous test essays, we could further analyze the elements and structures used in the previous essays. We first analyzed the association between essay length and score. Then we assigned a label of each sentence from the previous essays and calculated the average number of positions, evidence used and their proportions. Finally, to extract the argumentative structures of the previous essays, we applied a sequential pattern mining process on the label sequences of the previous essays.

Sequential Pattern Mining Process. The sequential pattern mining process was done by the TKS (top-k sequential pattern mining) [23] algorithm in SPMF package [22]. The TKS algorithm was designed to find the frequent subsequences. To better extract the argumentative structure, the label position and evidence were chosen to be the required items for each subsequence.

5 Results

5.1 Ensemble Model Results

To answer RQ1, we present our experiment results of the ensemble model in this section. The results of the models are shown in Table 2. As we can see from the results, the triple classification model achieved a better overall accuracy and F1 but lower recalls on the Feedback Prize testing set. The ensemble model received better results on almost all the evaluation metrics on the ACT testing set. Further explanations and discussions of the model results are shown in Sect. 6.

Table 2. Model evaluation results

Models	Feedback Prize								*ACT* Writing							
	Accuracy/F1		Recall/F1 (Position)		Recall/F1 (Evidence)		Recall/F1 (Claim)		Accuracy/F1		Recall/F1 (Position)		Recall/F1 (Evidence)		Recall/F1 (Claim)	
Position*	**0.943**	**0.877**	**0.811**	**0.786**	N/A		N/A		**0.936**	**0.736**	**0.351**	**0.465**	N/A		N/A	
Position^	0.94	0.864	0.747	0.762	N/A		N/A		0.935	0.728	0.333	0.447	N/A		N/A	
Evidence*	**0.913**	**0.906**	N/A		**0.862**	**0.881**	N/A		**0.881**	**0.791**	N/A		**0.521**	**0.629**	N/A	
Evidence^	0.909	0.901	N/A		0.85	0.874	N/A		0.871	0.773	N/A		0.5	0.601	N/A	
3 labels*	**0.865**	**0.848**	0.778	**0.792**	0.87	**0.879**	**0.884**	**0.874**	0.819	0.679	0.351	0.476	0.493	0.603	**0.956**	0.888
Ensemble	0.854	0.838	**0.836**	0.773	**0.895**	0.878	0.827	0.858	**0.835**	**0.73**	**0.544**	**0.59**	**0.679**	**0.7**	0.909	**0.893**

(The highest accuracy/recall/F1 are highlighted in bold. *: DeBERTa-base, ^: DistilBERT-base models).

5.2 ACT Tests Essays Analysis Results

To answer RQ2 and RQ3, we first analyzed the previous essays through their length. Most of the previous essays were between 8 and 35 in length (Appendix Fig. 1). After that, we analyzed the length, number of positions used, number of claims used, and number of evidence used for essays with different scores. The results can be found in Appendix Table 2 and Appendix Fig. 1. From our results, longer essays tended to receive higher scores. In addition, higher scored essays always contained more positions and evidence.

As for RQ2, the sequential pattern mining results showed that the most frequently used structure in the previous essays was position-claims-evidence-claims (PCEC). Around 40% of the students had this structure in their essay. This result also indicated that a huge number of the students did not include a position or evidence in their essays and their essays did not have an organizational structure.

Finally, we compared the top three frequent sequential patterns from essays with different scores. The results of the sequential pattern mining (top three frequent sequences) are shown in Table 3. We note that there are sequence overlaps between different mining results. As we can see from the sequential pattern mining results, essays with score 10 to 12 used a different structure which was *evidence-claims-position* (ECP). All the essays with lower scores tended to be the *position-claims-evidence-claims* (PCEC) structure.

Moreover, the lower scored essays tended to have no extracted structures in their writings. For example, for essays with a score less than 4, only 3% of them had an organizational structure. On the contrary, approximately 70% of the essays scored 12 used *evidence-claims-position* structures. These results showed that there were distinctive patterns in argumentative structures among essays with different scores.

Table 3. Sequential pattern mining results

Scores	Top Three Frequent Sequences	Counts and Proportions
12	evidence | claim | claim | claim | claim | claim | claim | position	59 (70.24%)
	claim | evidence | claim | claim | claim | claim | claim | position	58 (69.05%)
	evidence | claim | claim | claim | claim | claim | position | claim	58 (69.05%)
10–12	claim | evidence | claim | claim | claim | claim | claim | position	807 (56.91%)
	claim | claim | evidence | claim | claim | claim | claim | position	796 (56.14%)
	evidence | claim | claim | claim | claim | claim | claim | position	786 (55.43%)
7–9	position | claim | claim | claim | claim | evidence | claim | claim	2982 (40.38%)
	position | claim | claim | claim | evidence | claim | claim | claim	2978 (40.33%)
	position | claim | claim | claim | claim | claim | evidence | claim	2938 (39.79%)
4–6	position | claim | claim | claim | evidence	1348 (28.37%)
	position | claim | claim | evidence | claim	1344 (28.28%)
	position | claim | evidence | claim | claim	1318 (27.74%)
2–3	position | claim | claim | claim | evidence	14 (3.21%)
	position | claim | claim | evidence | claim	14 (3.21%)
	position | claim | evidence | claim | claim	13 (2.98%)

5.3 The Feedback Proving Process

Based on the knowledge of the ACT tests essays, we introduce a design mockup for a feedback tool (Fig. 2). We expect the tool can provide suggestions for students to improve their essay writing skills in three dimensions: length, argumentative elements, and essay structures.

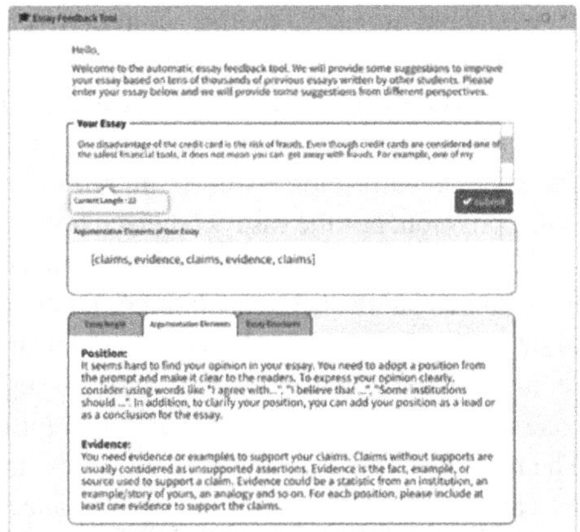

Fig. 2. An example screenshot for the feedback tool

6 Discussion

The purpose of this study is to capture and analyze argumentative structures from students' essays. This paper applied an NLP-based ensemble model to classify argumentative elements and extract argumentative structures. The ensemble model achieved high overall accuracy on both datasets with different prompts. The significance of this paper is that it highlighted the associations among argumentative elements, structures, and essay scores. Our study demonstrated how specific argumentative structures can contribute to higher or lower essay scores.

One intuitive finding from the analysis was that longer essays tended to receive higher scores, which was consistent with the finding of [8] that a short essay had less potential to make a strong argument. In addition, our analysis confirmed the findings from the previous studies [8, 9] that higher scored essays tend to use more positions and evidence argumentative components in their writings. These findings suggest that including more positions and evidence to support the claims can facilitate a stronger argumentative essay.

In previous studies, several factors were reported to affect essay score, including coherence, relevance, and the strength of the arguments [4, 8]. Our study further supports that certain argumentative structures can have positive impacts on essay scores. For example, Sect. 5.2 showed that the most frequently used structure in the ACT essay tests responses was position-claims-evidence-claims (PCEC). This is a basic structure that starts with a position, followed by claims and evidence to support this position. However, only around 40% of essays applied this structure. The sequential pattern mining results of essays with different scores further indicated that the PCEC structure was frequently used in essays with scores between 7 and 9. According to the ACT essay scoring rubric, the essays within this score range are usually considered as exhibiting a basic organizational structure.

From the analysis, we could identify another common organizational strategy, evidence-claims-positions (ECP) structure, from the high scored essays (10 to 12). Surprisingly, about 70% of the essays with a score of 12 used the ECP structure. After manually inspecting some examples, we found that these high-scored essays usually started with stories or facts. For example, a story of the author was used as a lead and the essay was concluded with a position. For the essays received lower scores (i.e., score 2 to 3), as illustrated in Table 3, only around 3% essays applied the position-claims-evidence structure. For the essays received scores between 4 and 6, less than 30% of them show common organizational structures. In other words, lack of organizational structures could indicate lower scores in argumentative essays. These findings suggested that the organizational structure is one of the dominant features that could influence the score of an argumentative essay. The results were consistent with other research suggested that the argumentation structure can be used to improve the performance of AES [13]. Based on these results, we can conclude that lower scored essays tended to lack positional and evidential components, do not apply an organizational argumentative structure, and are short in overall length.

To identify the argumentative elements and extract the argumentative structure of the essays, this study proposed an ensemble learning model. There were some differences in prediction performance of the ensemble models' components between Feedback Prize data and ACT writing test data. With the Feedback Prize data, all model variations were able to learn the tasks and perform accurate classification results (Table 2). However, with our ACT testing set data, the ensemble model outperformed the base models by a large margin. The ensemble model was able to find nearly 20% more positions and evidence compared to the base models. Overall, the result further confirmed the ability of the ensemble model to work across multiple datasets and achieve sufficient accuracy. Although the ensemble model achieved an overall accuracy of .835 on the ACT testing set, the recall and F1 scores for positions and evidence were not as good as the results on the Feedback Prize dataset. These inconsistencies might be caused by the cross-prompt implementation or different training corpus of the pre-trained language models. We believe the performance of the ensemble can be improved by adding more training sentences from the targeted prompts. In addition, the declined model performance for positions and evidence may also be caused by the sentence-level approach. After inspecting some misclassified cases, we found that our experts also have disagreement on the labels of these cases. Sometimes, an extremely long sentence may contain multiple argumentative components. Although our ensemble learning approach overcame this weakness to a certain degree, a border detection approach could be considered for follow-up studies.

While this study has limitations, we offer suggestions for future research implications. First, the classification models for argumentative elements can employ prompt-specific training processes to further improve accuracy. Second, for the analysis, we simplified seven argument elements into three. However, good arguments usually explore multiple perspectives and critically reflect the complexity of the given topic. Future work on classifying more detailed categories of argument elements, including lead, counterclaim, rebuttal, and evidence, would provide a deeper understanding of writing more effective argumentative essays. Lastly, automatic grading tools might have grading bias for minority student groups [27], such as students who use English as a second language. Although our ACT dataset contains essays from diverse populations, we did not analyze the structural differences between essays from culturally and linguistically diverse students. Further studies could analyze the structural difference among diverse student essays, address the bias, and provide more pedagogical value and equitable educational opportunities for all students.

7 Conclusion

In this study, we present a deep-learning-based ensemble model to extract the argumentative structures in students' essays and analyze distinctive argumentative structures for high-scored and low-scored essays. The findings of the study highlight that different argumentative structure usage is a significant factor in predicting essay scores. In future, we believe that our study can be further expanded to include other important factors in essay writing to provide richer feedback to students and improve automatic essay scoring engine's performance.

Appendix

Can be found at https://tinyurl.com/AIED2023.

References

1. Ke, Z., Ng, V.: Automated essay scoring: a survey of the state of the art. In: IJCAI, vol. 19, pp. 6300–6308 (2019)
2. Sutcher, L., Darling-Hammond, L., Carver-Thomas, D.: Understanding teacher shortages: an analysis of teacher supply and demand in the United States. Educ. Policy Anal. Arch. **27**, 35 (2019). https://doi.org/10.14507/epaa.27.3696
3. Uto, M.: A review of deep-neural automated essay scoring models. Behaviormetrika **48**(2), 459–484 (2021). https://doi.org/10.1007/s41237-021-00142-y
4. Yang, Y., Zhong, J.: Automated essay scoring via example-based learning. In: Brambilla, M., Chbeir, R., Frasincar, F., Manolescu, I. (eds.) ICWE 2021. LNCS, vol. 12706, pp. 201–208. Springer, Cham (2021). https://doi.org/10.1007/978-3-030-74296-6_16
5. Dong, F., Zhang, Y., Yang, J.: Attention-based recurrent convolutional neural network for automatic essay scoring. In: Proceedings of the 21st Conference on Computational Natural Language Learning (CoNLL 2017), pp. 153–162 (2017)
6. Süzen, N., Gorban, A.N., Levesley, J., Mirkes, E.M.: Automatic short answer grading and feedback using text mining methods. Procedia Comput. Sci. **169**, 726–743 (2020)
7. Taghipour, K., Ng, H.T.: A neural approach to automated essay scoring. In: Proceedings of the 2016 conference on empirical methods in natural language processing, pp. 1882–1891 (2016)
8. Persing, I., Ng, V.: Modeling argument strength in student essays. In: Proceedings of the 53rd Annual Meeting of the Association for Computational Linguistics and the 7th International Joint Conference on Natural Language Processing (vol. 1: Long Papers), pp. 543–552 (2015)
9. Rahimi, Z., Litman, D., Correnti, R., Wang, E., Matsumura, L.C.: Assessing students' use of evidence and organization in response-to-text writing: using natural language processing for rubric-based automated scoring. Int. J. Artif. Intell. Educ. **27**(4), 694–728 (2017)

10. Stab, C., Gurevych, I.: Identifying argumentative discourse structures in persuasive essays. In: Proceedings of the 2014 conference on empirical methods in natural language processing (EMNLP), pp. 46–56 (2014)
11. Wang, H., Huang, Z., Dou, Y., Hong, Y.: Argumentation mining on essays at multi scales. In: Proceedings of the 28th International Conference on Computational Linguistics, pp. 5480–549 (2020)
12. Alhindi, T., Ghosh, D.: "Sharks are not the threat humans are": Argument Component Segmentation in School Student Essays. arXiv preprint arXiv:2103.04518 (2021)
13. Nguyen, H., Litman, D.: Argument mining for improving the automated scoring of persuasive essays. In: Proceedings of the AAAI Conference on Artificial Intelligence, vol. 32, no. 1 (2018)
14. Yang, R., Cao, J., Wen, Z., Wu, Y., He, X.: Enhancing automated essay scoring performance via fine-tuning pre-trained language models with combination of regression and ranking. In: Findings of the Association for Computational Linguistics: EMNLP 2020, pp. 1560–1569 (2020)
15. Xue, J., Tang, X., Zheng, L.: A hierarchical BERT-based transfer learning approach for multi-dimensional essay scoring. IEEE Access **9**, 125403–125415 (2021)
16. Zhang, H., et al.: eRevise: Using natural language processing to provide formative feedback on text evidence usage in student writing. Proc. AAAI Conf. Artif. Intell. **33**(01), 9619–9625 (2019). https://doi.org/10.1609/aaai.v33i01.33019619
17. Cabrio, E., Villata, S.: Five years of argument mining: a data-driven analysis. In: IJCAI, vol. 18, pp. 5427–5433 (2018)
18. Crossley, S., Tian, Y.: Argumentation features and essay quality: exploring relationships and incidence counts. J. Writ. Res. **14**(1), 1–34 (2022). https://doi.org/10.17239/jowr-2022.14.01.01
19. Devlin, J., Chang, M. W., Lee, K., Toutanova, K.: Bert: pre-training of deep bidirectional transformers for language understanding. arXiv preprint arXiv:1810.04805 (2018)
20. He, P., Liu, X., Gao, J., Chen, W.: Deberta: Decoding-enhanced bert with disentangled attention. arXiv preprint arXiv:2006.03654 (2020)
21. Sanh, V., Debut, L., Chaumond, J., Wolf, T.: DistilBERT, a distilled version of BERT: smaller, faster, cheaper and lighter. arXiv preprint arXiv:1910.01108 (2019)
22. Fournier-Viger, P., et al.: The SPMF open-source data mining library version 2. In: Berendt, B., Bringmann, B., Fromont, É., Garriga, G., Miettinen, P., Tatti, N., Tresp, V. (eds.) ECML PKDD 2016. LNCS (LNAI), vol. 9853, pp. 36–40. Springer, Cham (2016). https://doi.org/10.1007/978-3-319-46131-1_8
23. Fournier-Viger, P., Gomariz, A., Gueniche, T., Mwamikazi, E., Thomas, R.: TKS: efficient mining of top-k sequential patterns. In: Motoda, H., Wu, Z., Cao, L., Zaiane, O., Yao, M., Wang, W. (eds.) Advanced Data Mining and Applications, pp. 109–120. Springer Berlin Heidelberg, Berlin, Heidelberg (2013). https://doi.org/10.1007/978-3-642-53914-5_10
24. Brown, T., et al.: Language models are few-shot learners. Adv. Neural. Inf. Process. Syst. **33**, 1877–1901 (2020)
25. Wan, Q., et al.: Automated Claim Identification Using NLP Features in Student Argumentative Essays. International Educational Data Mining Society (2021)
26. Persing, I., Ng, V.: End-to-end argumentation mining in student essays. In: Proceedings of the 2016 Conference of the North American Chapter of the Association for Computational Linguistics: Human Language Technologies, pp. 1384–1394 (2016)
27. Litman, D., Zhang, H., Correnti, R., Clare Matsumura, L., Wang, E.: A Fairness Evaluation of Automated Methods for Scoring Text Evidence Usage in Writing. In: Roll, I., McNamara, D., Sosnovsky, S., Luckin, R., Dimitrova, V. (eds.) Artificial Intelligence in Education: 22nd International Conference, AIED 2021, Utrecht, The Netherlands, June 14–18, 2021, Proceedings, Part I, pp. 255–267. Springer International Publishing, Cham (2021). https://doi.org/10.1007/978-3-030-78292-4_21

Leveraging Deep Reinforcement Learning for Metacognitive Interventions Across Intelligent Tutoring Systems

Mark Abdelshiheed(✉), John Wesley Hostetter, Tiffany Barnes, and Min Chi

North Carolina State University, Raleigh, NC 27695, USA
{mnabdels,jwhostet,tmbarnes,mchi}@ncsu.edu

Abstract. This work compares two approaches to provide metacognitive interventions and their impact on preparing students for future learning across Intelligent Tutoring Systems (ITSs). In two consecutive semesters, we conducted two classroom experiments: Exp. 1 used a classic artificial intelligence approach to *classify* students into different metacognitive groups and provide *static* interventions based on their classified groups. In Exp. 2, we leveraged Deep Reinforcement Learning (*DRL*) to provide *adaptive* interventions that consider the dynamic changes in the student's metacognitive levels. In both experiments, students received these interventions that taught *how* and *when* to use a backward-chaining (BC) strategy on a logic tutor that supports a default forward-chaining strategy. Six weeks later, we trained students on a probability tutor that only supports BC without interventions. Our results show that adaptive DRL-based interventions closed the metacognitive skills gap between students. In contrast, static classifier-based interventions only benefited a subset of students who knew *how* to use BC in advance. Additionally, our DRL agent prepared the experimental students for future learning by significantly surpassing their control peers on both ITSs.

Keywords: Reinforcement Learning · Artificial Intelligence · Intelligent Tutoring Systems · Metacognitive Interventions · Metacognitive Skills

1 Introduction

A challenging desired aspect of learning is being continuously prepared for future learning [9]. Our incremental knowledge is the evidence that preparation for future learning exists yet is hard to predict and measure [10]. Considerable research has found that one factor that facilitates preparing students for future learning is their metacognitive skills [7,26]. We focus on two types of metacognitive skills related to problem-solving strategies: *strategy-awareness* [5,20,22] and *time-awareness* [6,8], that are respectively, *how* and *when* to use each strategy.

Substantial work has demonstrated that metacognitive interventions of domain knowledge or problem-solving strategies accelerate preparation for future

N. Wang et al. (Eds.): AIED 2023, LNAI 13916, pp. 291–303, 2023.
https://doi.org/10.1007/978-3-031-36272-9_24

learning [7,19,22] and promote strategy- and time-awareness [12,23,26]. Such interventions included hints, feedback, prompted nudges, worked examples, and direct strategy presentation. However, these interventions were *static or hard-coded* into the learning environment despite the fact that students often acquire and master metacognitive skills as they learn [17]. Reinforcement Learning (RL) [24] is one of the most effective approaches for providing adaptive support and scaffolding across Intelligent Tutoring Systems (ITSs) [14,16,27]. The deep learning extension of RL, known as Deep RL (DRL), has been commonly utilized in pedagogical policy induction across ITSs [15,21] due to its higher support of model sophistication. As far as we know, no prior work has leveraged DRL to provide *adaptive* metacognitive interventions and investigated their impact on preparation for future learning across ITSs.

In this work, we conducted two consecutive experiments to compare two approaches for providing metacognitive interventions and their impact on preparing students for future learning on ITSs. In Exp. 1, we utilized a Random Forest Classifier (RFC) to classify students into different metacognitive groups and then provide static interventions based on their classification, while in Exp. 2, we leveraged a DRL-based approach for adaptive interventions that consider the dynamic changes in the student's metacognitive levels. Based on strategy- and time-awareness, our prior work classified students into those who are *both* strategy- and time-aware (*StrTime*), those who are *only* strategy-aware (*StrOnly*), and the rest who follow the default strategy (*Default*) [5,6]. We found that only *StrTime* students were prepared for future learning, as they learned significantly better than their peers across different deductive domains. Motivated by such findings, we designed metacognitive interventions to teach students *how* and *when* to use a backward-chaining (BC) strategy on a logic tutor that supports a default forward-chaining strategy. After six weeks, students were trained on a probability tutor that only supports BC without receiving interventions. Our results showed that *Default* and *StrOnly* students benefited equally from our DRL policy and surprisingly outperformed *StrTime* students. However, the RFC-based approach only helped *StrOnly* students to catch up with *StrTime*.

2 Background and Related Work

2.1 Metacognitive Interventions for Strategy Instruction

Metacognition indicates one's awareness of their cognition and the ability to control and regulate it [13]. Strategy- and time-awareness are two metacognitive skills that address *how* and *when* to use a problem-solving strategy, respectively [6,8]. Much prior work has emphasized the role of strategy awareness in preparation for future learning [5,22] and the impact of time awareness on academic performance and planning skills [8,11].

Considerable research has shown that metacognitive interventions promote strategy- and time-awareness [12,23,26]. We focus on two metacognitive interventions: directly presenting the strategy [12,23] and prompting nudges to use it [7,19,26]. Spörer et al. [23] found that students who were explicitly instructed on comprehensive reading strategies surpassed their peers, who were taught by the instructors' text interactions, on a transfer task and follow-up test. They also understood how, when, and why to use each reading strategy.

Zepeda et al. [26] demonstrated that metacognitive interventions impact learning outcomes, strategy mastery, and preparation for future learning. The experimental condition who received tutoring nudges and worked examples performed significantly better on a physics test than their control peers. They also made better metacognitive judgments and demonstrated mastery of knowing how and when to use physics strategies. As an example of preparation for future learning, the experimental students performed better on a novel self-guided 'control of variables' learning task than their control peers.

Despite much prior work on metacognitive interventions, the interventions were either not adaptive, not applied to ITSs, or had no preparation for future learning assessment. In our work, we conducted two experiments to provide metacognitive interventions and investigated their impact on preparing students for future learning across ITSs. Specifically, we first attempted static metacognitive interventions using a classifier-based approach, then compared it against a DRL-based approach for adaptive metacognitive interventions.

2.2 Reinforcement Learning in Intelligent Tutoring Systems

Reinforcement Learning (RL) is a popular machine learning branch ideal in environments where actions result in numeric rewards without knowing a ground truth [24]. Due to its aim of maximizing the cumulative reward, RL has been widely used in educational domains due to the flexible implementation of reward functions [14,15,21]. Deep RL (DRL) is a field that combines RL algorithms with neural networks; for instance, Deep Q-Learning is the neural network extension of Q-Learning, where a neural network is used to approximate the Q-function [18]. Substantial work has used RL and DRL in inducing pedagogical policies across ITSs [15,21,27]. Zhou et al. [27] utilized hierarchical RL to improve the learning gain on an ITS. They showed that their policy significantly outperformed an expert and a random condition.

Ju et al. [15] presented a DRL framework that identifies the critical decisions to induce a critical policy on an ITS. They evaluated their critical-DRL framework based on two success criteria: *necessity* and *sufficiency*. The former required offering help in **all** critical states, and the latter required offering help **only** in critical states. Their results showed that the framework fulfilled both criteria. Sanz-Ausin et al. [21] conducted two consecutive classroom studies where DRL was applied to decide whether the student or tutor should solve the following problem. They found that the DRL policy with simple explanations significantly improved students' learning performance more than an expert policy.

Despite the wide use of RL and DRL on ITSs, the attempts to combine either with metacognitive learning have been minimal [16]. Krueger et al. [16] used RL to teach the metacognitive skill of knowing how much to plan ahead (Deciding How to Decide). Their metacognitive reinforcement learning framework builds on the semi-gradient SARSA algorithm [24] that was developed to approximate

Markov decision processes. They defined a meta Q-function, known as Q_{meta}, that takes the meta state of the environment and the planning horizon action. They evaluated their framework on two planning tasks, where the authors defined constrained reward functions, and the rewards could be predicted many steps ahead to facilitate forming a plan.

In our work, we induced and deployed a DRL policy of metacognitive interventions on a logic tutor and investigated its impact on preparing students for future learning on a subsequent probability tutor. Additionally, we did not override any DRL mathematical definition, such as the Q-function. Instead, our DRL algorithm's metacognitive aspect resides in our interventions' nature.

3 Logic and Probability Tutors

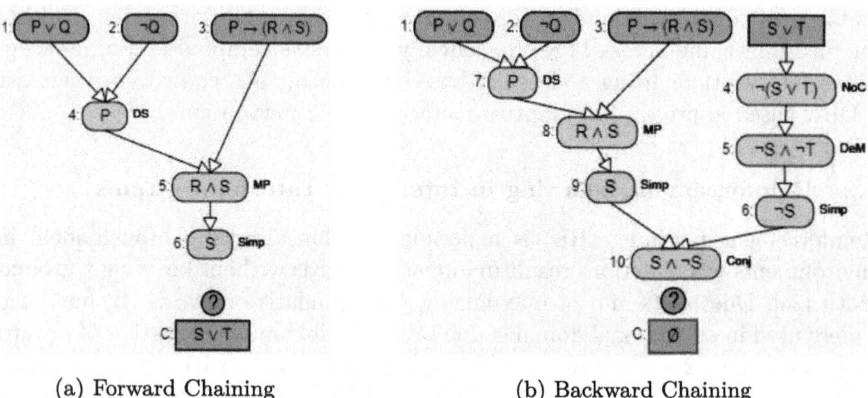

(a) Forward Chaining (b) Backward Chaining

Fig. 1. Logic Tutor Problem-Solving Strategies

Logic Tutor: It teaches propositional logic proofs by applying valid inference rules such as Modus Ponens through the standard sequence of pre-test, training and post-test. The three phases share the same interface, but training is the *only* one where students can seek and get help. The pre-test has two problems, while the post-test is harder and has six problems; the first two are isomorphic to the pre-test problems. Training consists of five ordered levels with an *incremental degree of difficulty*, and each level consists of four problems. Every problem has a score in the $[0, 100]$ range based on the accuracy, time and solution length.

The *pre-* and *post-test* scores are calculated by averaging their pre- and post-test problem scores. A student can solve any problem throughout the tutor by either a *forward-chaining* or a *backward-chaining (BC)* strategy. Figure 1a shows that for *forward chaining*, one must derive the conclusion at the bottom from givens at the top, while Fig. 1b shows that for *BC*, students need to derive a contradiction from givens and the *negation* of the conclusion. Problems are presented by *default* in forward chaining, but students can switch to BC by clicking a button in the tutor interface.

Probability Tutor: It teaches how to solve probability problems using ten principles, such as the Complement Theorem. The tutor consists of a textbook, pre-test, training, and post-test. Like the logic tutor, training is the only section for students to receive and ask for hints, and the post-test is harder than the pre-test. The textbook introduces the domain principles, while training consists of 12 problems, each of which can *only* be solved by *BC* as it requires deriving an answer by *writing and solving equations* until the target is ultimately reduced to the givens.

In pre- and post-test, students solve 14 and 20 open-ended problems, where each pre-test problem has an isomorphic post-test problem. The answers are graded in a double-blind manner by experienced graders using a partial-credit rubric, where grades are based *only* on accuracy in the $[0, 100]$ range. The *pre-* and *post-test* scores are the average grades in their respective sections.

4 Methods

As students can choose to switch problem-solving strategies *only* on the logic tutor, our interventions are provided in the logic training section [1–3,5]. It was shown that *StrTime* students frequently follow the desired behavior of switching ***early*** to *BC* on the logic tutor, their *StrOnly* peers switch ***late***, and the *Default* students make ***no*** switches and stick to the default strategy [4,6]. Additionally, we found that providing metacognitive interventions that recommend switching to BC —referred to as Nudges— or present problems directly in BC —known as Direct Presentation— for *Default* and *StrOnly* students cause them to catch up with their *StrTime* peers [2,3,5]. Therefore, we conducted two experiments to investigate different ways to present such metacognitive interventions for *Default* and *StrOnly* students.

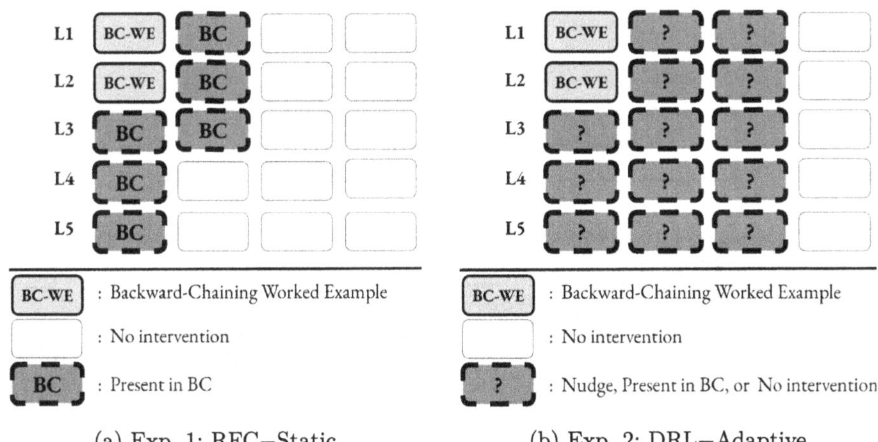

(a) Exp. 1: RFC−Static

(b) Exp. 2: DRL−Adaptive

Fig. 2. Training on the Modified Logic Tutor

4.1 Experiment 1: RFC−Static

We utilized a RFC that could early predict a student's metacognitive group —*Default*, *StrOnly* or *StrTime*— based on the incoming competence of the logic tutor. The RFC was previously shown to be 96% accurate [5].

After early prediction, we trained *StrTime* students on the original logic tutor with all problems presented by default in forward chaining, while *Default* and *StrOnly* students were assigned the modified tutor, as shown in Fig. 2a. Specifically, two worked examples (WE) on BC were provided, where the tutor showed a step-by-step solution, and six problems were presented in BC by default (Direct Presentation). We expected the WEs and six problems to teach students *how* and *when* to use BC. Note that we selected the colored problems in Fig. 2a based on our data's historical switches to BC [6]. This experiment provided a static metacognitive intervention —Direct Presentation— which was preferred to Nudges due to its prior success with *Default* and *StrOnly* students [3].

4.2 Experiment 2: DRL−Adaptive

We leveraged DRL to provide adaptive metacognitive interventions —*Nudge*, *Direct Presentation*, or *No Intervention*— regardless of the RFC's metacognitive group prediction. We trained *Experimental* students on the modified tutor shown in Fig. 2b. Figure 3 shows an example of a nudge, which is prompted after a number of seconds sampled from a probability distribution of prior students' switch behavior [6]. Since our interventions included the no-intervention option, we intervened in as many problems as possible. The WEs from Experiment 1 were kept, as they are vital for teaching students ***how*** to use BC. We did not intervene in the last training problem at each level, as it is used to evaluate the student's improvement on that level.

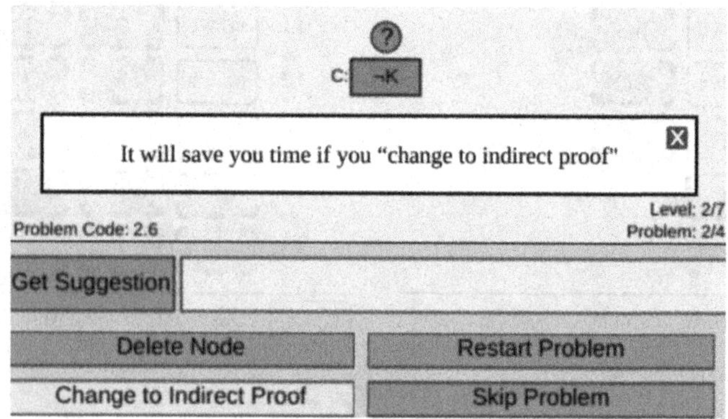

Fig. 3. Strategy Switch Nudge

Training Corpus and Policy Induction: To train our DRL agent, we utilized data collected from four previous studies consisting of 867 students [2,3,5,6] and performed a $80-20$ train-test split. The dataset consisted of a record per each student on a logic training problem represented as (**state**, **action**, **reward**). The *state* is the feature vector comprising 152 features that capture temporal, accuracy-based and hint-based behaviors. The *action* is either Nudge, Direct Presentation, or No Intervention. The *reward* is the immediate problem score on the logic tutor, as stated earlier in Sect. 3.

Our goal is to show that DRL works with our metacognitive interventions **rather** than *which* DRL algorithm is better with our interventions. We preferred DRL to RL due to its prevailing success in educational domains [15,21]. To select our DRL algorithm, we had to avoid a relatively simple algorithm such as Deep Q-Network (DQN), which overestimates action values [18] and may result in underfitting. Furthermore, we needed to avoid sophisticated DRL algorithms, such as autoencoders and actor-critic approaches, so that DRL does not overshadow the impact of our metacognitive interventions. In other words, a sophisticated DRL algorithm yielding an optimal policy would be acknowledged likely for its sophistication rather than for the metacognitive interventions it provided. Thus, we leveraged Double-DQN (DDQN), which solves the overestimation issue in DQN by **decoupling** the action *selection* from the action *evaluation* in two different neural networks [25]. The resulting modified Bellman equation becomes:

$$Q(s,a;\boldsymbol{\theta}) = r + \gamma\, Q(s', argmax_{a'} Q(s', a', \boldsymbol{\theta}); \boldsymbol{\theta}^{-}) \tag{1}$$

where r is the reward; γ is the discount factor; s and s' refer to the current and next states; a and a' denote the current and next actions. Specifically, DDQN uses the **main** ($\boldsymbol{\theta}$) neural network to *select* the action with the highest Q-value for the next state and then *evaluates* the Q-value of that action using the **target** ($\boldsymbol{\theta}^{-}$) neural network. After hyperparameter tuning, we picked the model with the lowest mean squared error loss. The deployed policy had two hidden layers with 16 neurons each, 1e-3 learning late, 9e-1 discount factor, 32 batch size, a synchronization frequency of 4 steps between main and target neural networks, and was trained until convergence ($\approx$2000 epochs).

5 Experiments Setup

The two experiments took place in an undergraduate Computer Science class at North Carolina State University in the Spring and Fall of 2022, respectively. The participants were assigned each tutor as a class assignment and told that completion is required for full credit. In both experiments, students were assigned to the logic tutor following the standard procedure of pre-test, training (Fig. 2a for Exp. 1 and Fig. 2b for Exp. 2), and post-test. We trained students on the probability tutor six weeks later, where no interventions were provided. Each probability training problem was randomly assigned for the student to solve on their own, for the tutor to present it as a worked example (WE), or both to collaborate in the form of collaborative problem-solving. Note that on both tutors, the problem order is the same for all students.

Exp. 1 (RFC) Participants: A total of 121 students finished both tutors and were classified by the RFC into 47 *Default*, 48 *StrOnly* and 26 *StrTime*. *Default* and *StrOnly* were randomly assigned to *Experimental (RFC)* and *Control (Ctrl)* conditions. *Experimental* received interventions on logic training (Fig. 2a), while *Control* and *StrTime* received no interventions. We had 24 $Default_{RFC}$, 25 $StrOnly_{RFC}$, 23 $Default_{Ctrl}$, 23 $StrOnly_{Ctrl}$ and 26 *StrTime* students.

Exp. 2 (DRL) Participants: A total of 112 students finished both tutors and were randomly assigned to *Experimental (DRL)* and *Control (Ctrl)* conditions. *Experimental* received interventions on logic training (Fig. 2b), while *Control* received no interventions. To investigate whether DRL would help students with different incoming metacognitive skills, we used the *RFC to ensure even distribution across conditions and metacognitive groups for comparison purposes.* We found that our DRL policy provided no interventions for $StrTime_{DRL}$ students 94% of the time. As a result, we combined $StrTime_{DRL}$ and $StrTime_{Ctrl}$ into *StrTime*. We had 22 $Default_{DRL}$, 24 $StrOnly_{DRL}$, 22 $Default_{Ctrl}$, 22 $StrOnly_{Ctrl}$ and 22 *StrTime* students.

6 Results

6.1 Experiment 1: RFC–Static

Table 1. Experiment 1 (RFC–Static) Results

	Experimental (RFC)		*Control*		
	$Default_{RFC}$ ($N = 24$)	$StrOnly_{RFC}$ ($N = 25$)	$Default_{Ctrl}$ ($N = 23$)	$StrOnly_{Ctrl}$ ($N = 23$)	*StrTime* ($N = 26$)
Logic Tutor					
Pre	57.8 (20)	56.5 (17)	55.9 (20)	55.7 (21)	57.4 (20)
Iso. Post	76.5 (14)	**83.9** (**9**)*	72.3 (16)	73 (15)	81.8 (11)*
Iso. NLG	0.18 (.15)	**0.32** (**.11**)*	0.14 (.29)	0.16 (.32)	0.31 (.17)*
Post	73.6 (13)	**80.5** (**9**)*	69.1 (13)	71.6 (11)	79.8 (10)*
NLG	0.16 (.12)	**0.31** (**.12**)*	0.12 (.31)	0.13 (.3)	0.28 (.16)*
Probability Tutor					
Pre	75.5 (16)	74.2 (15)	73.8 (14)	74.8 (16)	76.1 (16)
Iso. Post	75.2 (17)	**93.6** (**5**)*	70.9 (14)	72.6 (15)	91.2 (7)*
Iso. NLG	0.04 (.34)	**0.35** (**.16**)*	-0.03 (.29)	-0.02 (.31)	0.28 (.18)*
Post	74.6 (19)	**92.7** (**7**)*	69.5 (16)	70.9 (17)	90.3 (8)*
NLG	0.02 (.37)	**0.32** (**.18**)*	-0.07 (.32)	-0.04 (.36)	0.25 (.17)*

In a row, bold is for the highest value, and asterisk means significance over no asterisks.

Table 1 compares the groups' performance in Experiment 1. We show the mean and standard deviation of pre- and post-test scores, isomorphic scores, and the learning outcome in terms of the normalized learning gain (NLG) [6,14] defined as ($NLG = \frac{Post-Pre}{\sqrt{100-Pre}}$), where 100 is the maximum test score. We refer to pre-test, post-test and NLG scores as *Pre*, *Post* and *NLG*, respectively. The RFC was 97% accurate in classifying students who received no interventions—$Default_{Ctrl}$, $StrOnly_{Ctrl}$ and $StrTime$. On each tutor, a one-way ANOVA using group as factor found no significant difference on *Pre*: $F(4, 116) = 0.21$, $p = .93$ for logic and $F(4, 116) = 0.38$, $p = .82$ for probability.

To measure the students' improvement on isomorphic problems, repeated measures ANOVA tests were conducted (one for each group on each tutor) using {*Pre*, *Iso. Post*} as factor. We found that $StrOnly_{RFC}$ and *StrTime* learned significantly with $p < .0001$ on both tutors, while $Default_{RFC}$ and the control groups did not perform significantly higher on *Iso. Post* than *Pre* on either tutor. These findings verify the RFC's accuracy, as *StrTime* learned significantly on both tutors, while *Control* did not, despite each receiving no interventions.

On both tutors, a one-way ANCOVA using *Pre* as covariate and group as factor found a significant effect on *Post*: $F(4, 115) = 7.4$, $p < .0001$, $\eta^2 = 0.56$. for logic and $F(4, 115) = 8.2$, $p < .0001$, $\eta^2 = 0.63$. for probability. Follow-up pairwise comparisons with Bonferroni adjustment ($\alpha = .05/10$) showed that $StrOnly_{RFC}$, $StrTime > Default_{RFC}$, $Default_{Ctrl}$, $StrOnly_{Ctrl}$ on both tutors. Similar patterns were found on *NLG* using ANOVA. In essence, while no significant difference was found between $StrOnly_{RFC}$ and *StrTime*, each significantly outperformed the remaining three groups.

6.2 Experiment 2: DRL−Adaptive

Table 2. Experiment 2 (DRL−Adaptive) Results

	Experimental (DRL)		*Control*		
	$Default_{DRL}$ ($N = 22$)	$StrOnly_{DRL}$ ($N = 24$)	$Default_{Ctrl}$ ($N = 22$)	$StrOnly_{Ctrl}$ ($N = 22$)	*StrTime* ($N = 22$)
			Logic Tutor		
Pre	55.6 (21)	56.1 (21)	55.2 (19)	56.4 (23)	58.2 (19)
Iso. Post	91.9 (5)*	**92.1** (**4**)*	72.7 (18)	74.1 (17)	83.4 (12)*
Iso. NLG	0.46 (.12)*	**0.48** (**.09**)*	0.18 (.3)	0.14 (.27)	0.35 (.11)*
Post	**87.7** (**5**)*	87.6 (5)*	70 (15)	69.7 (16)	80.2 (11)*
NLG	**0.45** (**.12**)*	0.44 (.14)*	0.16 (.33)	0.1 (.31)	0.31 (.15)*
			Probability Tutor		
Pre	76.9 (15)	74.6 (16)	75.2 (15)	76.7 (13)	78.6 (14)
Iso. Post	94.5 (5)*	**96.1** (**3**)*	73.9 (10)	71.4 (14)	89.1 (7)*
Iso. NLG	0.36 (.11)*	**0.43** (**.13**)*	-0.02 (.18)	-0.06 (.21)	0.24 (.15)*
Post	94.1 (6)*	**95.6** (**4**)*	71.8 (11)	68.6 (17)	87.7 (8)*
NLG	0.34 (.13)*	**0.39** (**.16**)*	-0.07 (.24)	-0.1 (.25)	0.22 (.19)*

In a row, bold is for the highest value, and asterisk means significance over no asterisks.

Table 2 compares the groups' performance in Experiment 2. The RFC was 98% accurate in classifying students who received no interventions —$Default_{Ctrl}$, $StrOnly_{Ctrl}$ and *StrTime*. A one-way ANOVA using group as factor found no significant difference on *Pre*: $F(4, 107) = 0.18$, $p = .95$ for logic and $F(4, 107) = 0.45$, $p = .77$ for probability. We conducted repeated measures ANOVA for each group on each tutor using {*Pre*, *Iso. Post*} as factor. We found that $Default_{DRL}$, $StrOnly_{DRL}$ and *StrTime* learned significantly with $p < .0001$ on both tutors, unlike $Default_{Ctrl}$ and $StrOnly_{Ctrl}$.

A one-way ANCOVA using *Pre* as covariate and group as factor found a significant effect on *Post* on both tutors: $F(4, 106) = 9.3$, $p < .0001$, $\eta^2 = 0.61$ for logic and $F(4, 106) = 10.6$, $p < .0001$, $\eta^2 = 0.74$ for probability. Subsequent Bonferroni-corrected analyses ($\alpha = .05/10$) revealed that $Default_{DRL}$, $StrOnly_{DRL}$ > *StrTime* > $Default_{Ctrl}$, $StrOnly_{Ctrl}$ on both tutors. For instance, $Default_{DRL}$ had significantly higher *Post* than *StrTime* ($t(42) = 3.9$, $p < .001$, $d = 0.88$ for logic and $t(42) = 3.7$, $p < .001$, $d = 0.91$ for probability) and $Default_{Ctrl}$ ($t(42) = 6.6$, $p < .0001$, $d = 1.6$ for logic and $t(42) = 7.1$, $p < .0001$, $d = 2.5$ for probability). Similar patterns were observed using ANOVA on *NLG*. In brief, the two DRL groups benefited equally from our policy, significantly outperformed their control peers, and surprisingly surpassed *StrTime* students.

6.3 Post-hoc Analysis

To compare the results between the two experiments, we performed a Shapiro-Wilk normality test for each metric for each group in Tables 1 and 2. The results showed no evidence of non-normality ($p > .05$). Therefore, we conducted independent samples t-test for every two identical groups between Tables 1 and 2. Specifically, we found no significant difference between the *StrTime* groups across the two tables[1]. Similarly, no such difference was observed between the $Default_{Ctrl}$ groups or between their $StrOnly_{Ctrl}$ peers.

The main objective was to compare our *static RFC-based* and *adaptive DRL-based* interventions. First, we compared the interventions' distribution within each experiment. The RFC experimental students received static interventions; hence, the distribution is identical between $Default_{RFC}$ and $StrOnly_{RFC}$. For DRL students, $Default_{DRL}$ received 94(33%) Nudges, 65(23%) Direct Presentation and 127(44%) No Intervention, while $StrOnly_{DRL}$ received 82(26%) Nudges, 74(24%) Direct Presentation and 156(50%) No Intervention. A chi-square test showed no significant difference in the distribution of interventions between the experimental DRL groups: $\chi^2(2, N = 598) = 3.2$, $p = .2$.

Second, we compared the learning performance on both tutors. On the logic tutor, a two-way ANCOVA using *Pre* as covariate, and condition {*RFC*, *DRL*} and metacognitive group {*Default*, *StrOnly*} as factors, found a significant interaction effect on *Post*: $F(1, 90) = 37.9$, $p < .0001$, $\eta^2 = 0.78$. There was also a main effect of condition: $F(1, 90) = 28.4$, $p < .0001$, $\eta^2 = 0.59$

[1] This holds for all metrics, such as *Pre*, *Post* and *NLG*.

in that the *DRL* groups significantly outperformed their *RFC* peers. Follow-up Bonferroni-corrected analyses ($\alpha = .05/6$) confirmed that $Default_{DRL}$, $StrOnly_{DRL} > Default_{RFC}$, $StrOnly_{RFC}$. For example, $StrOnly_{DRL}$ had significantly higher *Post* than $StrOnly_{RFC}$: $t(47) = 3.6$, $p < .001$, $d = 0.98$. Similar results were found using ANOVA on *NLG*.

On probability, a two-way ANCOVA using *Pre* as covariate and the same two factors found a significant interaction effect on *Post*: $F(1, 90) = 29.1$, $p < .0001$, $\eta^2 = 0.46$. Subsequent pairwise analyses with Bonferroni adjustment ($\alpha = .05/6$) revealed that $Default_{DRL}$, $StrOnly_{DRL}$, $StrOnly_{RFC} > Default_{RFC}$. For instance, $StrOnly_{RFC}$ had significantly higher *Post* than $Default_{RFC}$: $t(47) = 6.4$, $p < .0001$, $d = 1.3$. Similar results were observed on *NLG* using ANOVA.

7 Discussions and Conclusions

We showed that leveraging DRL to provide adaptive metacognitive interventions closed the gap between metacognitive groups and caused them to surpass their control peers. Surprisingly, our DRL policy allowed the experimental students to outperform *StrTime* students significantly. On the other hand, using a RFC-based approach to provide static interventions only benefited a subset of students —*StrOnly*— who caught up with their *StrTime* peers.

It is also evident that DRL prepared the experimental students for future learning [9] by outperforming their control peers on both tutors. In other words, the experimental students outperformed their peers on probability based on interventions they received on logic. This finding suggests that they acquired backward-chaining skills in logic and transferred them to probability, where they received no interventions.

Limitations and Future Work. There are at least two caveats in our work. First, splitting students into experimental and control conditions resulted in relatively small sample sizes. Second, the probability tutor supported only one strategy, which restricted our intervention ability to the logic tutor. The future work involves comparing the RFC-based and DRL-based approaches within one study. Additionally, we aim to make the probability tutor support forward chaining, like the logic tutor.

Acknowledgments. This research was supported by the NSF Grants: 1651909, 1660878, 1726550 and 2013502.

References

1. Abdelshiheed, M., Hostetter, J.W., Barnes, T., Chi, M.: Bridging declarative, procedural, and conditional metacognitive knowledge gap using deep reinforcement learning. In: CogSci (2023)
2. Abdelshiheed, M., Hostetter, J.W., Shabrina, P., Barnes, T., Chi, M.: The power of nudging: exploring three interventions for metacognitive skills instruction across intelligent tutoring systems. In: CogSci, pp. 541–548 (2022)

3. Abdelshiheed, M., Hostetter, J.W., Yang, X., Barnes, T., Chi, M.: Mixing backward- with forward-chaining for metacognitive skill acquisition and transfer. In: Rodrigo, M.M., Matsuda, N., Cristea, A.I., Dimitrova, V. (eds.) AIED 2022. LNCS, vol. 13355, pp. 546–552. Springer, Cham (2022). https://doi.org/10.1007/978-3-031-11644-5_47
4. Abdelshiheed, M., Maniktala, M., Barnes, T., Chi, M.: Assessing competency using metacognition and motivation: the role of time-awareness in preparation for future learning. In: Design Recommendations for Intelligent Tutoring Systems, vol. 9, pp. 121–131. US Army CCDC Soldier Center (2022)
5. Abdelshiheed, M., Maniktala, M., Ju, S., Jain, A., Barnes, T., Chi, M.: Preparing unprepared students for future learning. In: CogSci, pp. 2547–2553 (2021)
6. Abdelshiheed, M., Zhou, G., Maniktala, M., Barnes, T., Chi, M.: Metacognition and motivation: the role of time-awareness in preparation for future learning. In: CogSci, pp. 945–951 (2020)
7. Belenky, D.M., et al.: Examining the role of manipulatives and metacognition on engagement, learning, and transfer. J. Probl. Solving **2**(2), 6 (2009)
8. de Boer, H., et al.: Long-term effects of metacognitive strategy instruction on student academic performance: a meta-analysis. Educ. Psychol. Rev. **24**, 98–115 (2018)
9. Bransford, J.D., Schwartz, D.L.: Rethinking transfer: a simple proposal with multiple implications. Rev. Res. Educ. **24**(1), 61–100 (1999)
10. Detterman, D.K., Sternberg, R.J.: Transfer on Trial: Intelligence, Cognition, and Instruction. Ablex Publishing (1993)
11. Fazio, L.K., et al.: Strategy use and strategy choice in fraction magnitude comparison. J. Exp. Psychol. Learn. Mem. Cogn. **42**(1), 1 (2016)
12. Fellman, D., et al.: The role of strategy use in working memory training outcomes. J. Mem. Lang. **110**, 104064 (2020)
13. Flavell, J.H.: Metacognition and cognitive monitoring: a new area of cognitive-developmental inquiry. Am. Psychol. **34**(10), 906 (1979)
14. Hostetter, J.W., Abdelshiheed, M., Barnes, T., Chi, M.: A self-organizing neuro-fuzzy q-network: Systematic design with offline hybrid learning. In: AAMAS (2023)
15. Ju, S., Zhou, G., Abdelshiheed, M., Barnes, T., Chi, M.: Evaluating critical reinforcement learning framework in the field. In: Roll, I., McNamara, D., Sosnovsky, S., Luckin, R., Dimitrova, V. (eds.) AIED 2021. LNCS (LNAI), vol. 12748, pp. 215–227. Springer, Cham (2021). https://doi.org/10.1007/978-3-030-78292-4_18
16. Krueger, P.M., Lieder, F., Griffiths, T.: Enhancing metacognitive reinforcement learning using reward structures and feedback. In: CogSci (2017)
17. Kuhn, D.: Metacognitive development. Curr. Dir. Psychol. Sci. **9**(5), 178–181 (2000)
18. Mnih, V., et al.: Human-level control through deep reinforcement learning. Nature **518**(7540), 529–533 (2015)
19. Richey, J.E., Zepeda, C.D., Nokes-Malach, T.J.: Transfer effects of prompted and self-reported analogical comparison and self-explanation. In: CogSci, vol. 37 (2015)
20. Roberts, M.J., Erdos, G.: Strategy selection and metacognition. Educ. Psychol. **13**, 259–266 (1993)
21. Sanz Ausin, M., Maniktala, M., Barnes, T., Chi, M.: Exploring the impact of simple explanations and agency on batch deep reinforcement learning induced pedagogical policies. In: Bittencourt, I.I., Cukurova, M., Muldner, K., Luckin, R., Millán, E. (eds.) AIED 2020. LNCS (LNAI), vol. 12163, pp. 472–485. Springer, Cham (2020). https://doi.org/10.1007/978-3-030-52237-7_38

22. Shabrina, P., Mostafavi, B., Abdelshiheed, M., Chi, M., Barnes, T.: Investigating the impact of backward strategy learning in a logic tutor: Aiding subgoal learning towards improved problem solving. IJAIED (2023)
23. Spörer, N., et al.: Improving students' reading comprehension skills: effects of strategy instruction and reciprocal teaching. Learn. Instr. **19**(3), 272–286 (2009)
24. Sutton, R.S., et al.: Reinforcement Learning: An Introduction. MIT Press, Cambridge (2018)
25. Van Hasselt, H., Guez, A., Silver, D.: Deep reinforcement learning with double Q-learning. In: AAAI, vol. 30 (2016)
26. Zepeda, C.D., et al.: Direct instruction of metacognition benefits adolescent science learning, transfer, and motivation. J. Educ. Psychol. **107**(4), 954 (2015)
27. Zhou, G., Azizsoltani, H., Ausin, M.S., Barnes, T., Chi, M.: Hierarchical reinforcement learning for pedagogical policy induction. In: Isotani, S., Millán, E., Ogan, A., Hastings, P., McLaren, B., Luckin, R. (eds.) AIED 2019. LNCS (LNAI), vol. 11625, pp. 544–556. Springer, Cham (2019). https://doi.org/10.1007/978-3-030-23204-7_45

Enhancing Stealth Assessment in Collaborative Game-Based Learning with Multi-task Learning

Anisha Gupta[1(✉)], Dan Carpenter[1], Wookhee Min[1], Bradford Mott[1], Krista Glazewski[2], Cindy E. Hmelo-Silver[2], and James Lester[1]

[1] North Carolina State University, Raleigh, NC, USA
{agupta44,dcarpen2,wmin,bwmott,lester}@ncsu.edu
[2] Indiana University, Bloomington, IN, USA
{glaze,chmelosi}@indiana.edu

Abstract. Collaborative game-based learning environments offer the promise of combining the strengths of computer-supported collaborative learning and game-based learning to enable students to work collectively towards achieving problem-solving goals in engaging storyworlds. Group chat plays an important role in such environments, enabling students to communicate with team members while exploring the learning environment and collaborating on problem solving. However, students may engage in chat behavior that negatively affects learning. To help address this problem, we introduce a multidimensional stealth assessment model for jointly predicting students' out-of-domain contributions to group chat as well as their learning outcomes with multi-task learning. Results from evaluating the model indicate that multi-task learning, which simultaneously performs the multidimensional stealth assessment, utilizing predictive features extracted from in-game actions and group chat data outperforms single-task variants and suggest that multi-task learning can effectively support stealth assessment in collaborative game-based learning environments.

Keywords: Stealth Assessment · Multi-Task Learning · Computer-Supported Collaborative Learning · Collaborative Game-Based Learning · Game-Based Learning Environments

1 Introduction

Computer-supported collaborative learning has been shown to effectively foster students' collaborative problem-solving skills with pedagogical strategies including inquiry learning and problem-based learning [1–3]. Collaborative game-based learning combines the benefits of both computer-supported collaborative learning, which focuses on social aspects of learning, and game-based learning environments that create engaging learning experiences. Collaborative game-based learning enables students to discuss topics, ask questions, and collectively brainstorm towards achieving shared goals with their team in a game environment. Unlike single-player game-based learning environments, students can play an active role in guiding their peers while also receiving help from

N. Wang et al. (Eds.): AIED 2023, LNAI 13916, pp. 304–315, 2023.
https://doi.org/10.1007/978-3-031-36272-9_25

teammates towards mastering in-game concepts together [4]. In particular, collaborating in small groups helps students learn better in problem-based collaborative game-based learning, where students actively participate in engaging problem solving and collaboratively work towards learning and completing the game, while a teacher plays the role of a facilitator who provides guidance for each team [2, 5].

Collaborative game-based learning environments are often equipped with online chat interfaces for students to communicate with each other [5–7]. Although this is a useful tool for students to exchange ideas while solving a shared problem, some students may engage in out-of-domain chat [6, 8]. Out-of-domain messages sent in the chat could be a source of distraction to students and could negatively impact their shared learning experience. While some out-of-domain chat behavior may be constructive, it could also transition students to negative affective states such as frustration or perhaps lead to unproductive learning outcomes when it is associated with certain affective states such as confusion [9]. It would thus be desirable to detect participation in out-of-domain chat early in collaborative game-based learning environments and allocate the necessary pedagogical support to students with the aim of helping them regulate their behavior. To address this challenge, we introduce a stealth assessment model that dynamically predicts students' out-of-domain chat contributions as well as their learning outcomes. Stealth assessment, which is grounded in evidence-centered design [10], uses a stream of students' interaction data within a game-based learning environment to make inferences about their competencies [11]. While assessing individual students' knowledge has often been a focus of stealth assessment, it can be extended to support assessments in collaborative game-based learning by inferring conclusions regarding both learning and collaboration. In collaborative game-based learning environments, it can be beneficial to identify which students are engaging in out-of-domain chat behaviors during the collaborative learning experience. While some level of rapport building is expected in collaborative game-based learning environments, it is important to ensure that such interactions do not distract students from the primary learning objective. Our goal is not to eliminate all out-of-domain chat, but rather to provide a tool that can help teachers identify and address out-of-domain messages that are not conducive to learning. Accurate and early prediction of learning outcomes and contribution to out-of-domain messages can be especially useful for providing targeted pedagogical assistance to students at the early phase of students' interactions with the learning environment. For example, if certain members of a group are predicted to engage in out-of-domain behavior, then they may be impeding the learning process for the entire group, at which point the stealth assessment model can inform a facilitator or an automated scaffolding system to intervene in the online chat and redirect the focus of the conversation in the early phase of collaborative learning.

This work presents a multidimensional stealth assessment model that explores multi-task learning for simultaneously predicting post-test performance and out-of-domain contribution of students by dynamically analyzing their interactions with a collaborative game-based learning environment for teaching ecosystem science to middle school students. We explore the benefits of leveraging information from messages exchanged in the group chat, the representations of which are driven by a transformer-based language model, in addition to game trace logs, as input features to our multi-task learning stealth assessment model. To effectively model these multimodal features, we adopt a late fusion approach that concatenates chat data representations and in-game action representations generated using two separate variants of recurrent neural networks in our multidimensional stealth assessment model. A competitive baseline using random forest models was created to evaluate our deep learning-based stealth assessment models with respect to predictive accuracy of both post-test performance and out-of-domain contributions as well as the models' early prediction capacity for both predictive tasks.

2 Related Work

Stealth assessment models have been extensively explored in the context of game-based learning environments [11]. Stealth assessment is an application of evidence-centered design (ECD) [10], which performs inference of higher-level student competencies based on task-level evidence. ECD consists of three core components: a competency model that models students' knowledge and skills, an evidence model that relates observations of student behavior to the competency variables, and a task model that provides the students with problems suited to demonstrate their competency and gather evidence of their knowledge. Stealth assessment implements ECD in the context of game-based learning by using fine-grained action-level information to model student competencies without interrupting gameplay or disrupting engagement [12, 13]. Stealth assessment has been extensively explored for unobtrusively measuring student learning outcomes such as problem solving [14], creativity [15], and computer science skills [16] in game-based learning environments. Multi-task learning [17], a technique for inducing shared representations by modeling multiple related machine learning tasks and leveraging them to improve predictive performance, has been investigated for stealth assessment with a goal to improve predictive performance of stealth assessment models. Gupta et al. used multi-task learning for predicting post-test scores and quality of written reflections in single-player game-based learning environments [18].

Collaboration has been shown to be highly associated with students' learning outcomes in collaborative learning environments [19]. For example, Sung et al. found a significant difference in learning outcomes between groups of students who collaborated in a game-based learning environment and students who played the game individually [20]. Given this relationship between learning outcomes and constructive collaboration, we present a stealth assessment model that simultaneously predicts post-test performance and out-of-domain chat contribution of students in collaborative game-based learning environments with multi-task learning, utilizing action logs and distributed representations of group chat messages.

Prior work has also explored natural language processing techniques to obtain embeddings of student-generated text for improving the predictive performance of stealth assessment models [18]. Pre-trained embeddings, which are vector representations of text such as words and phrases that are learned from large amounts of text data, have also been examined to represent chat messages exchanged by students in small group-based collaborative game-based learning environments. Carpenter et al. investigated Word2Vec, ELMo, and BERT embeddings for identifying out-of-domain messages in collaborative chats [8], reporting best predictive performance using pre-trained BERT embeddings.

Park et al. showed that an LSTM-based model trained on pre-trained BERT embeddings of collaborative group chats outperformed other baseline models for disruptive talk detection [6]. In our current study, we present a model that predicts out-of-domain contribution over the entire course of a students' gameplay, based on evidence of previous actions and chat records, in addition to predicting their post-test performance after gameplay. Our work is first to investigate a stealth assessment model to evaluate students' collaboration and learning using multi-task learning for collaborative game-based learning environments.

3 Dataset

Our experiments were conducted on data collected from two studies using CRYSTAL ISLAND: ECOJOURNEYS (Fig. 1), a collaborative game-based learning environment for middle school ecosystem science. For both the studies, students completed a pre-test measuring ecosystems content knowledge, and then they played the game for 6 classroom periods, followed by attempting a post-test assessment with the same questions as the pre-test. In the game students are tasked with identifying an illness that is affecting the fish population on a remote island. Students in the class are divided into teams that work together towards solving the problem in the game. A team typically consisted of four students, with a minimum of three students. Each member of the team explores the game environment to gather information scattered across the map. Students gather at a virtual whiteboard integrated in the game to discuss and organize their findings to collaboratively figure out what might be plaguing the fish on the island. Each team is also equipped with an in-game group chat interface, where students communicate with their teammates to discuss their findings and collaborate on ideas. A facilitator, who is a domain-expert research team member, regulated the chat for each group, prompting for group discussion when needed, taking a poll on ideas, and occasionally nudging the focus of the team back to the in-game problem solving if students were distracted.

In our current work, we used a dataset consisting of a total of 8,350 chat messages, of which 2,000 messages were sent by the facilitator and the remaining messages were sent by 72 consented and assented middle school students (11–12 years old, 31 females and 41 males) divided into 18 groups. Three students who did not complete the post-test assessment were excluded from the dataset, but their group chat messages were included as context for predicting out-of-domain contribution and learning outcomes of other students in the dataset. On average, each group chat consisted of 463.89 messages (SD = 285.26), and an average of 92.78 messages were sent by each student across all the groups (SD = 78.45).

3.1 Out-of-Domain Labeling

Our chat data comprises messages collected from both studies using the collaborative game-based learning environment. Chat messages from the first study were previously labeled by two researchers [21], where the inter-rater reliability measured with Cohen's kappa is 0.751, indicating substantial agreement. The chat messages from the second study were labeled by two researchers, including one rater who labeled the dataset from

Fig. 1. CRYSTAL ISLAND: ECOJOURNEYS collaborative game-based learning environment

the first study. Cohen's kappa of 0.878 was achieved between the two raters on the chat messages from the second study, indicating almost perfect agreement. All three researchers involved in labeling the chat messages for both the studies labeled 20% of the first dataset in common and achieved Fleiss' kappa of 0.90, indicating very good agreement. A total of 2,259 out of 8,350 chat messages were labeled as out-of-domain in the combined dataset ($M = 132.88$, $SD = 107.75$ per group; $M = 28.96$, $SD = 42.21$ per student). The rubric used for labeling the chat dataset for both the studies is shown in Table 1 [21].

Table 1. Rubric for labeling chat dataset as on-task or out-of-domain.

Label	Rubric	Example
On-task	Text that contributes to the science discussion in the group, demonstrates or addresses a relevant affective state, fosters collaboration, or asks a relevant question	*"Im at the whiteboard now, what do I do?"* *"Tilapia need warm temperatures in their water, dissolved oxygen, and food (Cant remember what food)"*
Out-of-domain	Text that is meaningless or unintelligible, or off-topic conversation that is unrelated to learning outcomes and fails to address affective states of students in the group	*"REEEEEEEEE"* *"are u boy or girl"* *"WHATS A TURTLELE"* *"gonna be as mean as possible"*

3.2 Post-test Assessment

As noted above, we utilize data collected from two separate studies that used CRYSTAL ISLAND: ECOJOURNEYS. These studies had slightly different pre- and post-test assessment questionnaires. The first study had a total of 42 questions. Based on the outcome of the first study, item analysis was performed on the test content to shorten its length and the second study was conducted with 32 assessment questions that were directed towards the same learning goals. For the post-test performance prediction task, we predict performance labels driven by the normalized score for each student. The distribution of normalized post-test scores across data collected from both studies is shown in Fig. 2.

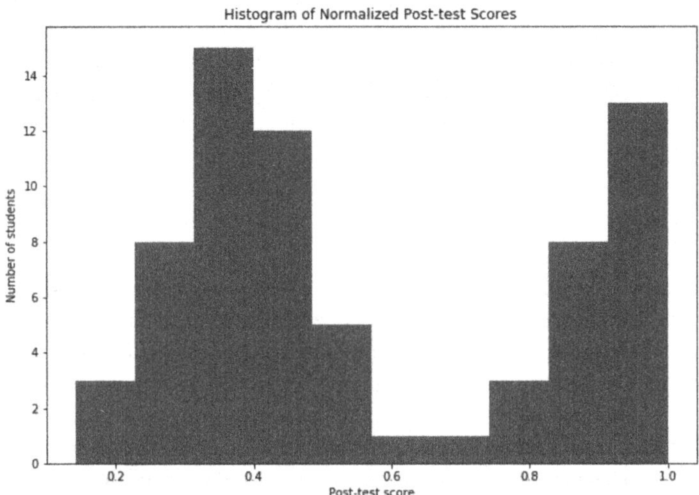

Fig. 2. Histogram of normalized post-test score distribution across both studies

The post-test questions were presented to the students in various formats such as multiple choice, interpreting charts and tables, and classification problems. Questions included problems directly based on in-game text, such as classifying components of an ecosystem as biotic or abiotic, and completing incomplete definitions of technical terms including respiration and decomposition. The post-test assessments also included questions that tested understanding of in-game concepts, presenting students with case studies that reported findings, requiring them to interpret the reported values from charts and tables and suggest the most likely explanation for the observations.

3.3 Feature Extraction

The game trace logs of the students comprise of 21 distinct action types, including *SpokeTo*, *Activated*, *MovedTo*, *ReceivedChatMessage* and *SentChatMessage.* At any given timestamp, we construct a 21-dimensional, count-based representation (i.e., the number of times a student performed each action type). These representations are designed to support sequential predictions from the beginning of gameplay to the end with a goal to enable our stealth assessment models to make early predictions and inform adaptive interventions as early as possible. The counts for each action type were normalized using z-score normalization for effective modeling. In addition, pre-trained

DistilBERT embeddings [22] were extracted for each message in the chat. DistilBERT, which was trained on Toronto Book Corpus and English Wikipedia (same as BERT), is a lighter version of BERT that maintains 95% of BERT's predictive performance while being 60% faster than BERT and 120% faster than ELMo and BiLSTM, lending itself to be a suitable technique for processing potential high volumes of chat messages exchanged by students in cases where our stealth assessment model is deployed to perform real-time assessments. The rubric for labeling the out-of-domain dataset is case-agnostic, so we use the uncased version of the DistilBERT embeddings to represent each chat message. Given the limited number of students in the dataset compared to the high dimensionality (768 dimensions) of the pre-trained DistilBERT embeddings, we perform principal component analysis (PCA) to reduce the dimensionality of the DistilBERT embeddings to 32 dimensions to be appropriate for training our stealth assessment models while preserving variance in the dataset.

3.4 Class Labeling

In our work, we cast each predictive task as a binary classification task so that our stealth assessment model can provide actionable feedback in the early phase of students' collaborative learning by achieving high predictive capacity. We define out-of-domain contribution of a student as the ratio of the number of out-of-domain messages sent by a student to the total number of out-of-domain messages sent in the group. If no out-of-domain messages were sent in the group, the out-of-domain contribution of all students in the group is considered to be zero. A higher score for the out-of-domain contribution metric indicates that the student negatively contributed to the group discussion by sending more out-of-domain chats. We performed median splits to determine high and low classification labels for each of the prediction tasks. The threshold for out-of-domain chat contribution was determined to be 20.3% (34 students in low category, 35 students in high category). Similarly, we performed a median split on the normalized post-test scores to determine binary class labels. The threshold used for labeling to post-test performance was 47.62% (33 students in low category, 36 students in high category).

4 Model Architecture

In this section, we present our multi-task learning stealth assessment model for predicting out-of-domain contribution to chat and post-test performance for students in a collaborative game-based learning environment. Based on a preliminary analysis of the data, our model is designed to make a prediction utilizing the past 20 actions performed by the student in the game, and 25 recent messages exchanged in the group chat for every new action that the student takes. The preliminary analysis suggested that using 20 actions and 25 recent messages as input to the stealth assessment models not only captures useful game and dialogue context but also addresses data sparsity issues compared to utilizing the entire history of the actions and chats as input.

The model architecture embeds two recurrent neural networks as subnetworks to separately model game-trace logs and group chats, respectively. The subnetwork that processes game action features comprises a long short-term memory (LSTM) network

(8 hidden units, 0.1 dropout, 0.01 L2 kernel, recurrent, and bias regularization factors), followed by a dropout layer with a dropout rate of 0.2. The subnetwork that processes recent group chats comprises a gated recurrent unit (GRU) network (8 hidden units, 0.1 dropout, 0.01 L2 kernel, recurrent, and bias regularization factors), followed by a dropout layer with a dropout rate of 0.2. The outputs from each of these subnetworks are concatenated to construct a combined representation (i.e., late fusion), which is fed into two separate dense layers (each with 1 output unit and sigmoid activation function) for predicting post-test performance and out-of-domain contribution, respectively. Our stealth assessment model was trained using the Adam optimizer [23] (learning rate = 0.01), and binary cross-entropy was used as the loss function. This model architecture is shown in Fig. 3. For single-task models, we only preserved layers specific to the prediction task, while removing all other layers added to the model architecture for the other prediction task. We maintain a non-neural random forest baseline that utilizes all available features to evaluate with our deep, multi-task learning-based stealth assessment model.

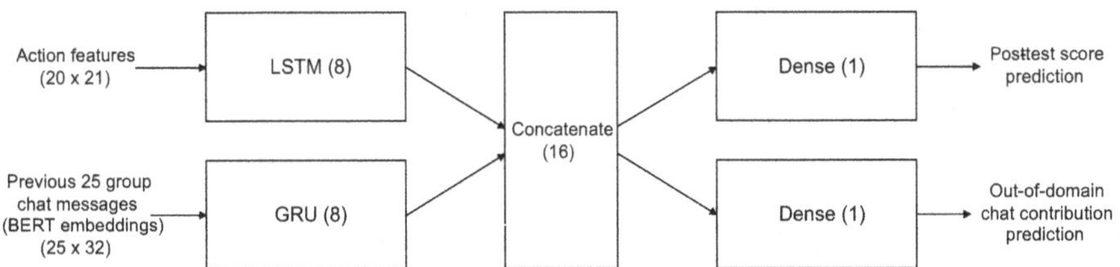

Fig. 3. Multi-task deep learning-based stealth assessment model architecture

5 Results

The model was evaluated using group-level 5-fold nested cross-validation. For each fold of cross-validation, we performed an 80–20 split on the dataset to create a training set and a separate validation set that is used to optimize the models. Accuracy and two early prediction metrics—standardized convergence point (SCP) [24] and convergence rate (CR) [25]—were used as evaluation metrics for our model. SCP measures how much in advance of the end of students' gameplay the model can consistently make accurate predictions about the outcome. The metric considers the average of the percentage of gameplay that elapses before a model can consistently and accurately predict the correct outcome, while penalizing non-converged sequences (i.e., student interaction sequences that did not yield a final correct prediction). CR indicates the percentage of converged student interaction sequences, which supplements SCP to demonstrate early prediction robustness of predictive models. In short, a higher CR and a lower SCP indicate a better early prediction performance.

The average of the results from each cross-validation fold is reported in this paper. We perform ablation studies on our stealth assessment model to determine the importance of game trace and group chat features for each of the prediction tasks and also determine if the prediction tasks benefit from multi-task learning of these two modalities. The results of our evaluation are presented in Tables 2 and 3 for post-test performance and out-of-domain contribution predictions, respectively.

Table 2. Single-task and multi-task learning results for predicting post-test performance. RF is the random forest baseline, DL is our deep learning-based stealth assessment model, A represents action features and G represents group chat features. (The best score is marked in bold for each metric.)

ML (Features)	Single-Task Learning				Multi-Task Learning			
	Macro Accuracy (%)	Micro Accuracy (%)	CR (%)	SCP (%)	Macro Accuracy (%)	Micro Accuracy (%)	CR (%)	SCP (%)
RF (A,G)	**67.5**	**66.5**	68.19	**52.26**	N/A	N/A	N/A	N/A
DL (A)	63.84	65.47	**70.84**	54.34	68.3	68.46	**70.68**	**49.53**
DL (G)	55.47	54.79	52.08	91.87	56.14	56.33	60.2	90.27
DL (A,G)	66.1	66.35	65	55.41	**70.37**	**71.01**	70.5	54.8

Table 3. Single-task and multi-task learning results for predicting out-of-domain performance. (The best score is marked in bold for each metric.)

ML (Features)	Single-Task Learning				Multi-Task Learning			
	Macro Accuracy (%)	Micro Accuracy (%)	CR (%)	SCP (%)	Macro Accuracy (%)	Micro Accuracy (%)	CR (%)	SCP (%)
RF (A,G)	63.34	62.78	69.86	59.08	N/A	N/A	N/A	N/A
DL (A)	**70.64**	**68.77**	71.53	**53.64**	67.67	68.51	74.21	49.44
DL (G)	46.39	46.38	41.54	96.13	47.34	47.42	41.34	88.14
DL (A,G)	68.03	67.36	**75.97**	54.22	**75.27**	**74.84**	**75.97**	**47.19**

6 Discussion

From Tables 2 and 3, we observe that our multi-task, deep learning-based stealth assessment model outperforms the random forest baseline for both prediction tasks, while the majority class baselines are 57.93% and 58.17% (macro accuracy) for post-test performance and out-of-domain contribution predictions, respectively. The best predictive accuracy is obtained using multi-task learning for predicting post-test performance

(70.37% macro accuracy, 71.01% micro accuracy) and out-of-domain chat contribution (75.27% macro accuracy, 74.84% micro accuracy) simultaneously, outperforming stealth assessment models based on single-task learning as well as random forest. The best early prediction results for both the post-test performance prediction task (49.53% SCP) and out-of-domain contribution prediction task (47.19% SCP) is also obtained in a multi-task learning setting. It should be noted that multi-task learning of game trace logs is especially effective for improving predictive performance of post-test performance, both in terms of accuracy and early prediction. This suggests that the out-of-domain contribution prediction task helped induce effective shared representations from the game trace logs subnetwork that is also beneficial for the post-test performance prediction task. From the early prediction results, we observe that the SCP for the out-of-domain contribution prediction task improves when there is multi-task learning with the multimodal data (SCP for the out-of-domain contribution prediction task in a single-task learning setting is 54.22%, compared to 47.19% when multi-task learning is applied). This could indicate that multi-task learning helps the model achieve stability in its predictions for out-of-domain contribution early, resulting in more consistent predictions with better convergence scores.

For single-task learning models for predicting post-test performance, we obtain the best result (66.1% macro accuracy, 66.35% micro accuracy, 65% convergence rate and 55.41% standardized convergence point) using a combination of game trace logs and group chat features, indicating that including group chat evidence is beneficial for the task of predicting post-test performance. Predictive accuracy of the single-task learning model for out-of-domain contribution is reduced when group chat features are used in addition to game trace log features, which suggests that the model is unable to effectively model recent group chat features when they are combined with game interaction features, whereas our multi-task stealth assessment models achieve the highest predictive performance as well as the best early prediction performance for out-of-domain contribution prediction, when the multimodal features are used.

7 Conclusion

Collaborative game-based learning environments are promising platforms for students to participate in team learning and collaborate on problem solving. Chat systems integrated in these environments serve as discussion forums and enable communication between groups of students. However, students may engage in out-of-domain discussions in the forum, which may be disruptive to the overall learning in the group. In this work, we present a deep learning-based stealth assessment model that supports early prediction of students' post-test performance and out-of-domain contribution in the group chat using multi-task learning. The multi-task learning-based stealth assessment models outperform the non-neural random forest baseline for both prediction tasks. We obtained the best prediction and early prediction results using both game trace logs and recent group chats as input features in a multi-task learning setting, suggesting that predicting post-test performance and out-of-domain contribution of students in collaborative game-based learning environments are related stealth assessment tasks that can benefit from multi-task learning.

In future work, it will be interesting to enrich chat message representations with sentiment scores, number of unique participants in the group chat context window, similarity of chat messages to in-game text, and the frequency of a student's contribution in recent chat history. Misspelled words in the chat can be corrected using a text normalization system, which could further improve robustness of stealth assessment models. Other pre-trained embeddings obtained from large language models can also be explored. Furthermore, it will also be important to investigate other modalities, such as posture and eye gaze data, for predicting learning outcomes and out-of-domain contribution of students in collaborative game-based learning environments. Finally, it will be important to incorporate the stealth assessment model alongside student-adaptive scaffolding to evaluate the impact on learning outcomes and processes within collaborative game-based learning environments.

Acknowledgements. This research was supported by the National Science Foundation under Grants DRL-1561655, DRL-1561486, IIS-1839966, and SES- 1840120. Any opinions, findings, and conclusions expressed in this material are those of the authors and do not necessarily reflect the views of the National Science Foundation.

References

1. Dillenbourg, P., Järvelä, S., Fischer, F.: The evolution of research on computer-supported collaborative learning. In: Technology-enhanced learning, pp. 3–19. Springer, Dordrecht (2009)
2. Hmelo-Silver, C.E., Chernobilsky, E.: Understanding collaborative activity systems: the relation of tools and discourse in mediating learning. In: Embracing Diversity in the Learning Sciences: Proceedings of the Sixth International Conference of the Learning Sciences, p. 254. Psychology Press (2004) October)
3. Jeong, H., Hmelo-Silver, C.E., Jo, K.T.: Ten years of computer-supported collaborative learning: a meta-analysis of CSCL in STEM education during 2005–2014. Educ. Res. Rev. **28**, 100284 (2019)
4. Engle, R.A., Conant, F.R.: Guiding principles for fostering productive disciplinary engagement: explaining an emergent argument in a community of learners classroom. Cogn. Instr. **20**(4), 399–483 (2002)
5. Saleh, A., Chen, Y., Hmelo-Silver, C.E., Glazewski, K.D., Mott, B.W., Lester, J.C.: Coordinating scaffolds for collaborative inquiry in a game-based learning environment. J. Res. Sci. Teach. **57**(9), 1490–1518 (2020)
6. Park, K., et al.: Detecting disruptive talk in student chat-based discussion within collaborative game-based learning environments. In: LAK21: 11th International Learning Analytics and Knowledge Conference, pp. 405–415. (2021) April
7. Jeong, H., Hmelo-Silver, C.: Technology supports in CSCL (2012)
8. Carpenter, D., et al.: Detecting off-task behavior from student dialogue in game-based collaborative learning. In: International Conference on Artificial Intelligence in Education, pp. 55–66. Springer, Cham (2020) July
9. Sabourin, J.L., Rowe, J.P., Mott, B.W., Lester, J.C.: Considering alternate futures to classify off-task behavior as emotion self-regulation: a supervised learning approach. J. Educ. Data Min. **5**(1), 9–38 (2013)
10. Mislevy, R.J., Steinberg, L.S., Almond, R.G.: Focus article: on the structure of educational assessments. Meas. Interdisc. Res. Perspect. **1**(1), 3–62 (2003)

11. Shute, V.J.: Stealth assessment in computer-based games to support learning. Comput. Games Instruction **55**(2), 503–524 (2011)
12. Henderson, N., et al.: Enhancing stealth assessment in game-based learning environments with generative zero-shot learning. International Educational Data Mining Society (2022)
13. Kim, Y.J., Almond, R.G., Shute, V.J.: Applying evidence-centered design for the development of game-based assessments in physics playground. Int. J. Test. **16**(2), 142–163 (2016)
14. Zhao, W., Shute, V., Wang, L.: Stealth assessment of problem-solving skills from gameplay. In: Interservice/Industry Training, Simulation, and Education Conference (I/ITSEC), (15212) (2015)
15. Shute, V.J., Rahimi, S.: Stealth assessment of creativity in a physics video game. Comput. Hum. Behav. **116**, 106647 (2021)
16. Min, W., et al.: DeepStealth: game-based learning stealth assessment with deep neural networks. IEEE Trans. Learn. Technol. **13**(2), 312–325 (2019)
17. Zhang, Y., Yang, Q.: A survey on multi-task learning. IEEE Transactions on Knowledge and Data Engineering (2021)
18. Gupta, A., Carpenter, D., Min, W., Rowe, J.P., Azevedo, R., Lester, J.C.: Multimodal multi-task stealth assessment for reflection-enriched game-based learning. In MAIED@ AIED, pp. 93–102 (2021)
19. Dillenbourg, P., Fischer, F.: Computer-supported collaborative learning: the basics. Zeitschrift für Berufs-und Wirtschaftspädagogik **21**, 111–130 (2007)
20. Sung, H.Y., Hwang, G.J.: A collaborative game-based learning approach to improving students' learning performance in science courses. Comput. Educ. **63**, 43–51 (2013)
21. Carpenter, D., et al.: Detecting off-task behavior from student dialogue in game-based collaborative learning. In: Artificial Intelligence in Education: 21st International Conference, AIED 2020, Ifrane, Morocco, July 6–10, 2020, Proceedings, Part I 21 pp. 55-66. Springer International Publishing (2020)
22. Sanh, V., Debut, L., Chaumond, J., Wolf, T.: DistilBERT, a distilled version of BERT: smaller, faster, cheaper and lighter. arXiv preprint arXiv:1910.01108 (2019)
23. Kingma, D.P., Ba, J.: Adam: a method for stochastic optimization. arXiv preprint arXiv:1412.6980 (2014)
24. Min, W., et al.: A generalized multidimensional evaluation framework for player goal recognition. In: Twelfth Artificial Intelligence and Interactive Digital Entertainment Conference (2016) September)
25. Blaylock, N., Allen, J.: Corpus-based, statistical goal recognition. In: IJCAI, vol. 3. pp. 1303–1308 (2003) August

How Peers Communicate Without Words-An Exploratory Study of Hand Movements in Collaborative Learning Using Computer-Vision-Based Body Recognition Techniques

Qianru Lyu[1](✉), Wenli Chen[1], Junzhu Su[1], Kok Hui John Gerard Heng[1], and Shuai Liu[2]

[1] National Institute of Education, Nanyang Technological University, Singapore 637616, Singapore
Nie20.lq@e.ntu.edu.sg

[2] Tianjin University, Beiyang Avenue 300072, Tianjin, China

Abstract. Accumulating research in embodied cognition highlights the essential role of human bodies in knowledge learning and development. Hand movement is one of the most applied body motions in the collaborative ideation task when students co-construct knowledge with and without words. However, there is a limited understanding of how students in a group use their hand movements to coordinate understandings and reach a consensus. This study explored students' hand movement patterns during the different types of knowledge co-construction discourses: quick consensus-building, integration-oriented consensus building, and conflict-oriented consensus building. Students' verbal discussion transcripts were qualitatively analyzed to identify the type of knowledge co-construction discourses. Students' hand motion was video-recorded, and their hand landmarks were detected using the machine learning tool MediaPipe. One-way ANOVA was conducted to compare students hand motions in different types of discourses. The results found there were different hand motion patterns in different types of collaboration discourses. Students tended to employ more hand motion during conflict-oriented consensus building discourses than during quick consensus building and integration-oriented consensus building discourses. At the group level, the collaborating students were found to present less equal hand movement during quick consensus-building than integration-oriented consensus building and conflict-oriented consensus building. The findings expand the existing understanding of embodied collaborative learning, providing insights for optimizing collaborative learning activities incorporating both verbal and non-verbal language.

Keywords: Hand Movement · Collaborative Learning · Gesture · Body Recognition Techniques

N. Wang et al. (Eds.): AIED 2023, LNAI 13916, pp. 316–326, 2023.
https://doi.org/10.1007/978-3-031-36272-9_26

1 Introduction

Recognized as a twenty-first-century educational trend, collaborative learning becomes an important instructional practice to support higher-level learning [1]. From the socio-cultural learning perspective, collaborative learning is a "coordinated and synchronous activity" [2] when students work as a group to solve an ill-structured complex problem through knowledge exchange and negotiation [3, 4]. Students are expected to co-construct knowledge via externalizing and negotiating ideas, building upon each other's ideas, and reaching a shared understanding [5]. In the context of university-level engineering classrooms, efficient collaborative learning is particularly required when student engineers analyze design problems and ideate alternative solutions to solve real-world problems [6]. However, student engineers were found to have challenges engaging in successful collaborative learning and deserve more research effort to explore how to support collaborative learning in engineering classrooms [7, 8].

To achieve a nuanced understanding of students' collaborative learning, researchers demonstrated the multimodal nature of learning interactions in classrooms [9], when students communicate with both verbal and non-verbal cues such as gestures, and facial expressions [10]. More recently, embodied cognition theory was introduced to collaborative learning studies to make sense of the multimodal knowledge co-construction process [11]. Among the various bodily engagements, hand movement was found to support peer interaction in a notable way, especially in engineering classrooms when students need to externalize and comprehend their abstract ideas [12]. More recently, the advancement of computer vision opened a new window for moment-by-moment detection and tracking of body movement [13, 14]. These trained deep neural networks provide valuable opportunities for investigating the role of hand movement in the collaborative learning context [15]. Supported by the computer vision-based body recognition tools, a comprehensive analysis of both verbal and non-verbal cues can be implemented, helping with finetuning the design and instructions for collaborative learning practices in the engineering classroom.

To realize this goal, we will introduce the theoretical underpinnings, review the relative empirical findings, and identify the research questions to be addressed. The method and results sessions illustrate how we addressed the research questions and the relative findings. Theoretical contributions and implications will be included in the discussion session.

2 Literature Review

Collaborative learning studies in recent years stress a close examination of students' interaction process as important learning evidence. Taking a process-oriented perspective, researchers highlighted three important consensus-building discourses that represent the different knowledge co-construction quality: quick consensus-building, conflict-oriented consensus-building, and integration-oriented consensus-building [16]. There have been accumulating studies employing this framework to conduct utterance analysis during collaborative learning [17, 18]. Though utterance analysis-based research adds value to the interpretation of the collaborative learning outcome quality, it has some limitations in view that the nature of human learning and interactions is multimodal [19]. Multimodality in learning stresses the multiplicity of modes of communication including not only verbal cues but also gesture, action, and visual cues [20]. In the face-to-face collaborative learning context, gestures and actions naturally emerge in the meaning-making process as students employ their bodies to externalize ideas and understand each other [15]. In recent years, this link between cognition and body has been discussed in embodied cognition research.

Embodied cognition theory emphasizes the role of the body in forming cognitive representations and understanding human cognition with a meaningful connection with the concrete world [21, 22]. Embodied cognition theory also states that bodily behavior influences the way we think [11]. From this perspective, cognition constitutes body-based sensorimotor information that takes place both inside and outside the human brain such as gesture, action, and analogical mapping [23]. The embodied cognition theories, therefore, provide insights for a more comprehensive understanding of students' communication and interactions during collaborative learning incorporating bodily engagements and nonverbal behaviors [11].

Among the different embodiments, hand movement is one of the most useful nonverbal behaviors in the classroom [24] and can convey important information [25, 26] as co-thought and/or co-speech gestures. Students were found to use their hands to express their understanding of a new concept even before they were able to express it verbally [27]. Though hand movement was demonstrated useful in multiple domains such as mathematics, science, and foreign language vocabulary [28], few empirical studies looked into its role in design activities, a key collaborative learning practice in engineering classrooms. Researchers argued that hands contribute to the description, mutual understanding, and negotiation of design concepts for student engineers during collaborative learning [29]. Hand movement can provide a connection between representations and design ideas, make abstract concepts more concrete, and direct learner attention [12]. Therefore, a nuanced understanding of student engineers' hand movement patterns during collaborative learning could provide insights for optimizing their design process with both cognitive and bodily engagement.

Existing studies of hand movement in collaborative learning contexts applied manual coding of students' hand movements such as pointing, beating, mirroring, and echoing [30–32]. This kind of qualitative analysis takes significant manpower and makes it difficult to quantify students' hand movement for further statistical analysis, limiting the implications for real-world collaborative learning practices. With the advancement of computer vision techniques in recent years, some trained deep neural networks can be deployed to track human body gesture data based on standard web images or videos, such as OpenPose [13], DensePose [14], and MediaPipe [33]. Education researchers have applied these tools to detect bodily engagement and highlighted its feasibility and usability in collaborative learning contexts without the need for expensive wearables [15]. Building on the latest research, this study adopted the computer vision-based model to quantify students' hand movements in collaborative learning activities.

This study aims to fine-tune the existing collaborative learning practices in the engineering design context incorporating embodied cognition theories. A review of the existing literature showed that though hand movement was found to play a key role in collaborative learning tasks in engineering classrooms, how students employed hands to co-construct knowledge remains under-studied. In particular, it is unclear whether and how students engaged in the different knowledge co-construction discourses with different hand movement patterns. This question is significant for realizing a holistic understanding of how students collaborate with both verbal and body cues and expanding the application of embodied cognition theories to support real-world collaborative learning practices. To bridge the gap, this study conducted a moment-by-moment analysis of students' hand movements during different types of face-to-face knowledge co-construction discourse. To understand how students applied hand movement during different discourses, this study quantified individual students' hand movements throughout the different discourses (RQ1). To explore how the students within the same group employed hand motions synchronously, this study compared the collaborating students' hand movements within each discourse (RQ2). Two research questions are proposed in this study:

1. Were there differences in how students employ hand movement between different collaborative learning discourses?
2. What were the hand movement patterns of student groups during the different collaborative learning discourses?

3 Method

This exploratory study applied a mixed-method design. Both quantitative and qualitative research methods were employed to address the research questions. This study is approved by the university ethical committee of Nanyang Technological University. Informed consent forms to participate in the study were collected from all the participants.

3.1 Participants and Learning Context

A total of 5 pairs of 4th-year engineering students (average age = 25, female = 1) from a Singapore university participated in this study. Participants were asked to finish a design problem in dyads within one hour. The task required them to design an autonomous robot that is able to perform certain functions in the given arena including navigating the arena, collecting the ball, and delivering the ball to a specific place. Each pair of student engineers were expected to collaboratively analyze the design problem, identify the required functions, and generate alternative robot designs. The learning environment is face-to-face collaborative learning when students can see each other's upper body during discussion.

3.2 Data Collection

The data collected include an audio recording of students' verbal discussion and a video recording of students' collaborative learning. Each audio recording file was 1 h long from the beginning of the task till the end. Students' discussion was transcribed to text for further analysis of knowledge co-construction discourse. Each video file was 1 h in length, capturing two collaborating students' upper bodies throughout the task. The recording angle remained the same from the task beginning till the end.

3.3 Data Analysis and Instruments

To identify students' knowledge co-construction discourses, the authors conducted a content analysis of students' discussion transcripts based on the coding scheme (see Table 1). The unit of analysis for coding was one utterance of a distinctive idea conveyed by students in the same group. One utterance usually contains several sentences together that conveyed a complete idea. Each knowledge co-construction discourse was coded with its category and timestamps (beginning time and ending time). The three knowledge co-construction discourses are defined by the argumentative knowledge co-construction framework [16], a powerful analytical lens to unpack the collaborative learning process [34]. The three consensus-building approaches are widely adopted to reach a nuanced understanding of students' reasoning about, elaborating on, modifying conceptions, and building common grounds [35]. In the end, a total of 48 quick consensus-building, 30 integration-oriented consensus building, and 14 conflict-oriented consensus building were identified. The inter-rater reliability was considered satisfactory (Cohen's kappa = 0.872).

To investigate students' hand movement during their knowledge co-construction discourses, MediaPipe, an open deep learning-based library was applied to detect and extract the human body poses data at every single frame (see Figs. 1 and 2). For each frame, the MediaPipe outputs the location of the 75 landmarks for each human body in the scene, including 33 pose landmarks (Pose Landmark Model) and 21 hand landmarks (Hand Landmark Model) for each hand. Each landmark will be reported as a variable with columns for x, y, and z coordinates in the CSV file. With an interest in hand movement, the authors focused on 7 landmarks representing both hands of each person (Landmark ID #16, #17, #18, #19, #20, #21 in Fig. 1). These landmarks were chosen mainly because of they were relatively more stably captured compared to the hand landmark model. Considering the collaborative learning context, we set the framerate as 10 to process in the MediaPipe, meaning 10 observations for each second for each individual.

The movement quantity of each hand landmark was calculated by adding up the difference between its coordinates at one observation and the coordinates at the next observation within one knowledge co-construction discourse. To answer the first research question, each individual student's hand motion value within each knowledge co-construction discourse was calculated by adding up all 7 landmarks' movement quantity. To answer the second research question, we calculated the ratio of the two collaborating students' hand movement quantity at the group level (the higher student hand movement quantity/the lower student hand movement quantity) within each same knowledge co-construction discourse. The closer the ratio value was to 1, the more equally the collaborating students applied their hand motions during the knowledge co-construction discourse. The knowledge co-construction discourses containing missing values were excluded for further analysis. To compare students' hand movement quantity in different knowledge co-construction discourses, one-way ANOVA was conducted for both the individual level and group level.

Table 1. Coding scheme of knowledge co-construction discourses.

Knowledge co-construction discourses	Explanation	Example
Quick Consensus Building	Accept a peer contribution without any modification	"-What else? -Good cable management maybe? -Ah okay."
Integration-oriented Consensus Building	Take over the perspective of their learning peers and/or integrate different perspectives	"-I think we can use the long-range sensor to like localize where it is…Wait we only have 1 sensor? -Yeah we only got 1 short range sensor.. The mid-range sensor -Then what about 1 on top for the height? -Okay then we have like 1 sensor for each corner
Conflict-oriented Consensus Building	Reject and/or repair contributions of their learning peers with further replacement, modification, and/or supplementation	"-I want the robot to be seen -Err… I don't think it matters at this moment -Well…I mean, it's not to be optimized for, it's more like just adding some colors to the styrofoam ball.. Shape like the tennis and then we put at our strong point -Then maybe we put it with an explanation -Okay."

4 Results

To understand how much individual students applied hand movement during different types of knowledge co-construction discourses (RQ1), the one-way ANOVA was conducted. The distribution of the hand movement quantities of different knowledge co-construction discourses is shown in Fig. 3. The width of the column represents the frequency of a certain movement quantity. The one-way ANOVA results indicated that students applied significantly different hand movement quantities in different types of knowledge co-construction discourses, $F(3, 181) = 7.779$, $p < 0.01$. The Tukey's HSD Test reported that students' hand movement quantity in conflict-oriented consensus building ($M = 52.74$, $SD = 43.39$) was significantly higher than that in quick consensus building ($M = 28.09$, $SD = 26.62$, $p < 0.001$) and integration-oriented consensus building ($M = 31.53$, $SD = 29.24$, $p = 0.005$). There was no statistically significant difference in mean hand movement quantity between quick consensus-building and integration-oriented consensus building ($p = 0.773$). This result indicated that students

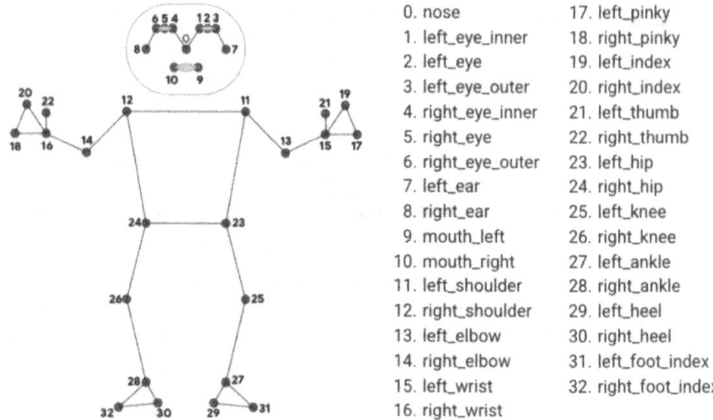

Fig. 1. Pose landmark model (https://mediapipe.dev/).

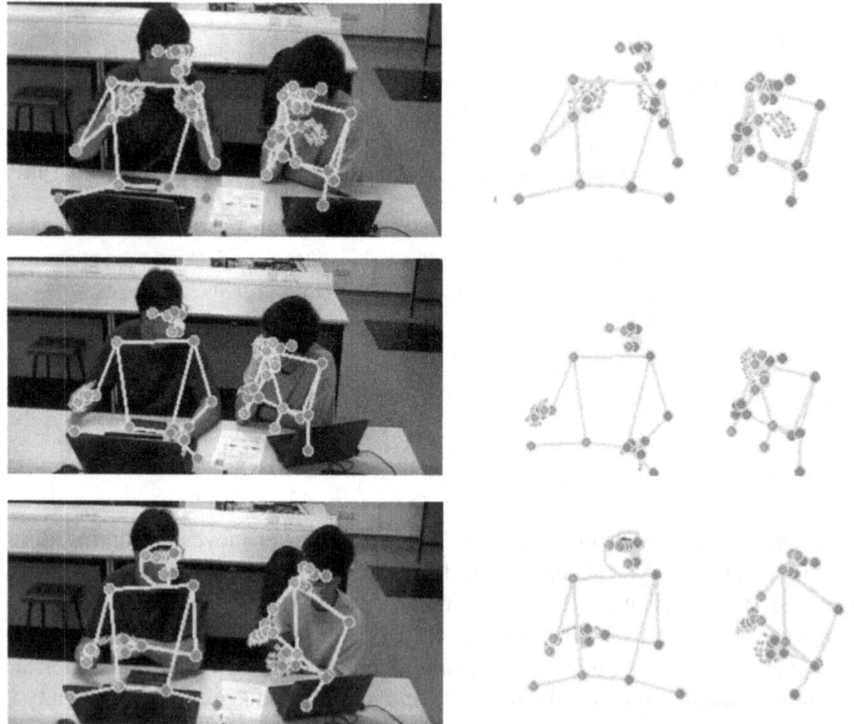

Fig. 2. Moment-by-moment extracted landmarks from video recording.

in the conflict-oriented consensus building discourses applied significantly more hand motions than in the other two knowledge co-construction discourses.

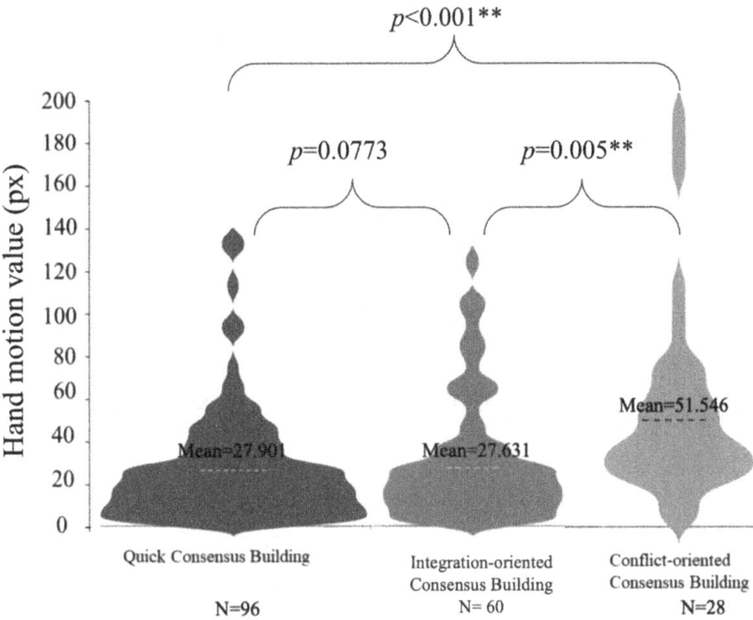

Fig. 3. Hand motion value distribution in different types of knowledge co-construction discourses.

To understand how students at the group level applied hand movement during the different knowledge co-construction discourses (RQ2), the one-way ANOVA was conducted for the ratio value of each knowledge co-construction discourse. The results are visualized in Fig. 4. The closer the ratio value was to 1, the more equally the collaborating students applied their hand movement during the knowledge co-construction discourse. A higher ratio value indicated that the collaborating students applied more different quantities of hand movement during the knowledge co-construction discourse. The one-way ANOVA results showed there was a significantly different ratio value among the three knowledge co-construction discourses, $F(2, 89) = 7.233$, $p < 0.01$. Tukey's HSD Test reported that the ratio value of conflict-oriented consensus building ($M = 1.26$, $SD = 0.43$) was significantly lower than quick consensus building ($M = 2.22$, $SD = 0.99$, $p = 0.001$). There was no significant difference in the ratio value between quick consensus building and integration-oriented consensus or between integration-oriented consensus and conflict-oriented consensus building. The results indicated that the collaborating students tended to employ different quantities of hand movement within the same group. During quick consensus building discourses, it was likely that one student employed much more hand movement than his/her peer. In comparison, students tended to employ similar quantities of hand motions during conflict-oriented consensus building discourses.

In summary, the results indicate different hand movement patterns during different types of knowledge co-construction discourses. Students engaging in conflict-oriented consensus building discourses tended to employ significantly more hand movement than the other two discourses. Students from the same group applied a similar quantity of

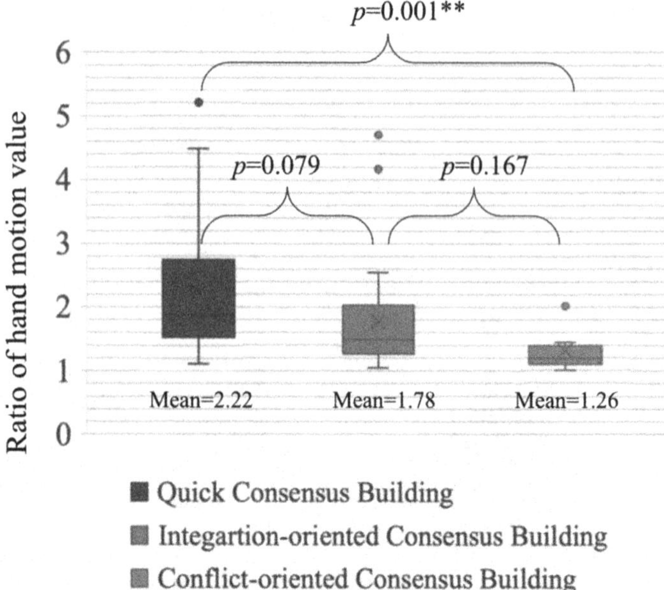

Fig. 4. Group-level hand motion ratio in different knowledge co-construction discourses.

hand movement during conflict-oriented consensus building discourses while in quick consensus building discourses, it was likely that one student employed much more hand movement than his/her collaborator.

5 Discussion

The findings expand the existing understanding of how students engage in knowledge co-construction by incorporating hand movement data. The significantly larger amount of hand movement during the conflict-oriented consensus building discourses than the other two discourses conditions can be explained by the high engagement level during the rejecting, modification, and negotiation of concepts during conflict-oriented consensus building [39]. Interestingly, students from the same group tended to use much more hand movement than their peers during quick consensus building, indicating that students not only engaged in the knowledge co-construction with limited cognitive contribution but also employed less body movement simultaneously. In comparison, students were found to employ an equal amount of hand movement during the conflict-oriented consensus building, illustrating that collaborating students were actively engaged in the idea negotiation via both verbal and non-verbal communication modes. This study highlights the important role of students' bodily engagement during the meaning-making and consensus building processes, adding value to the embodied cognition theories in the collaborative learning context [15].

This study has some limitations. A limited sample size of students was included in this study and a larger group of students are expected to strengthen the generalizability of the findings. Besides, only hand movement was analyzed without specifying the different hand gestures as well as collaborative hand gestures such as pointing, beating, and echoing to each other. In future studies, we will employ machine learning techniques to detect both hand movement and different hand gestures at larger scale data.

In summary, supported by the computer vision-based body recognition tool, this study demonstrates the feasibility and usability to conduct a moment-by-moment analysis of students' body movement during the different types of knowledge co-construction processes, opening a new window to study and refine collaborative learning from an embodied cognition perspective. The findings provide insights for future design and examination of collaborative learning in a novel and embodied approach.

References

1. OECD.: Reviews of National Policies for Education Education in Chile. OECD Publishing (2017)
2. Roschelle, J., Teasley, S.D.: The construction of shared knowledge in collaborative problem solving. In: Computer supported collaborative learning. Springer (1995)
3. Scardamalia, M., Bereiter, C.: Computer support for knowledge-building communities. J. Learn. Sci. **3**(3), 265–283 (1994)
4. Johnson, D.W., Johnson, F.P.: Joining Together: Group Theory and Group Skills. Prentice-Hall, Inc. (1991)
5. Hull, D.M., Saxon, T.F.: Negotiation of meaning and co-construction of knowledge: an experimental analysis of asynchronous online instruction. Comput. Educ. **52**(3), 624–639 (2009)
6. Stump, G.S., et al.: Collaborative learning in engineering students: gender and achievement. J. Eng. Educ. **100**(3), 475–497 (2011)
7. Marra, R.M., et al.: Beyond "group work": an integrated approach to support collaboration in engineering education. Int. J. STEM Educ. **3**(1), 1–15 (2016)
8. Du, X., et al.: Engineering students' conceptions of collaboration, group-based strategy use, and perceptions of assessment in PBL: a case study in Qatar. Int. J. Eng. Educ. **36**(1), 296–308 (2020)
9. Nguyen, A., et al.: Exploring socially shared regulation with an AI deep learning approach using multimodal data. In: Proceedings of International Conferences of Learning Sciences (ICLS) (2022)
10. Zhu, G., et al.: Exploring emotional and cognitive dynamics of knowledge building in grades 1 and 2. User Model. User-Adap. Inter. **29**(4), 789–820 (2019)
11. Lakoff, G., Núñez, R.: Where Mathematics Comes from, vol. 6. Basic Books, New York (2000)
12. Nathan, M.J., et al.: Teachers' attitudes about gesture for learning and instruction. Gesture **18**(1), 31–56 (2019)
13. Hidalgo, G., et al.: OpenPose library. CMU-Perceptual-Computing-Lab/https://github.com/CMU-Perceptual-Computing-Lab. (2017)
14. Güler, R.A., Neverova, N., Kokkinos, I.: Densepose: dense human pose estimation in the wild. In: Proceedings of the IEEE conference on computer vision and pattern recognition (2018)
15. Schneider, B., Worsley, M., Martinez-Maldonado, R.: Gesture and gaze: multimodal data in dyadic interactions. In: International Handbook of Computer-Supported Collaborative Learning, pp. 625–641. Springer (2021)
16. Weinberger, A., Fischer, F.: A framework to analyze argumentative knowledge construction in computer-supported collaborative learning. Comput. Educ. **46**(1), 71–95 (2006)

17. De Wever, B., et al.: Content analysis schemes to analyze transcripts of online asynchronous discussion groups: A review. Comput. Educ. **46**(1), 6–28 (2006)
18. Strijbos, J.-W., et al.: Content analysis: what are they talking about? Comput. Educ. **46**(1), 29–48 (2006)
19. Jewitt, C.: Technology, Literacy, Learning: A Multimodal Approach. Routledge (2012)
20. Mangaroska, K., et al.: Challenges and opportunities of multimodal data in human learning: the computer science students' perspective. J. Comput. Assist. Learn. **37**(4), 1030–1047 (2021)
21. Barsalou, L.W., et al.: Social embodiment. Psychol Learn. Motiv. **43**, 43–92 (2003)
22. Shapiro, L., Stolz, S.A.: Embodied cognition and its significance for education. Theory Res. Educ. **17**(1), 19–39 (2019)
23. Kiverstein, J.: The meaning of embodiment. Top. Cogn. Sci. **4**(4), 740–758 (2012)
24. Peng, Z., et al.: Recognizing teachers' hand gestures for effective non-verbal interaction. Appl. Sci. **12**(22), 11717 (2022)
25. Schwartz, D.L., Black, J.B.: Shuttling between depictive models and abstract rules: induction and fallback. Cogn. Sci. **20**(4), 457–497 (1996)
26. Yammiyavar, P., Clemmensen, T., Kumar, J.: Analyzing non-verbal cues in usability evaluation tests. In: International Conference on Usability and Internationalization Springer (2007)
27. Goldin-Meadow, S., Singer, M.A.: From children's hands to adults' ears: gesture's role in the learning process. Dev. Psychol. **39**(3), 509 (2003)
28. Cook, S.W., et al.: Hand gesture and mathematics learning: lessons from an Avatar. Cogn. Sci. **41**(2), 518–535 (2017)
29. Cash, P., Maier, A.: Prototyping with your hands: the many roles of gesture in the communication of design concepts. J. Eng. Des. **27**(1–3), 118–145 (2016)
30. Sakamoto, T., et al.: Radar-based hand gesture recognition using IQ echo plot and convolutional neural network. In: 2017 IEEE Conference on Antenna Measurements & Applications (CAMA), IEEE (2017)
31. Molchanov, P., et al.: Hand gesture recognition with 3D convolutional neural networks. In: Proceedings of the IEEE conference on computer vision and pattern recognition workshops (2015)
32. Pisharady, P.K., Saerbeck, M.: Recent methods and databases in vision-based hand gesture recognition: a review. Comput. Vis. Image Underst. **141**, 152–165 (2015)
33. Zhang, F., et al.: Mediapipe hands: on-device real-time hand tracking. arXiv preprint arXiv: 2006.10214 (2020)
34. Nguyen, H.: Let's teach Kibot: Discovering discussion patterns between student groups and two conversational agent designs. Brit. J. Educ. Technol. **53**, 1864–1884 (2022)
35. Tan, E., de Weerd, J.G., Stoyanov, S.: Supporting interdisciplinary collaborative concept mapping with individual preparation phase. Educ. Tech. Res. Dev. **69**(2), 607–626 (2021)

Scalable Educational Question Generation with Pre-trained Language Models

Sahan Bulathwela(✉), Hamze Muse, and Emine Yilmaz

Centre for Artificial Intelligence, University College London, London, UK
{m.bulathwela,hamze.muse.20,emine.yilmaz}@ucl.ac.uk

Abstract. The automatic generation of educational questions will play a key role in scaling online education, enabling self-assessment at scale when a global population is manoeuvring their personalised learning journeys. We develop *EduQG*, a novel educational question generation model built by adapting a large language model. Our extensive experiments demonstrate that *EduQG* can produce superior educational questions by further pre-training and fine-tuning a pre-trained language model on the scientific text and science question data.

1 Introduction

Digital learning resources such as Massively Open Online Courses (MOOC) and Open Educational Resources (OER) are abundant, but they often lack associated questions that enable self-testing and skill verification [3,5,7] once the learning resources are consumed. Generating scalable educational questions is crucial for democratising education [6]. While existing language models are used for question generation, their utility in education has only been explored recently. This work demonstrates how a large language model can be adapted for educational question generation. The experiments validate the improvement of questions through additional pre-training with educational text. The study also explores the impact of pre-training data size on question generation and investigates the enhancement of educational questions through fine-tuning with a science question dataset. The experimental results show that pre-training and fine-tuning with domain-specific scientific text can outperform a state-of-the-art baseline, providing significant evidence for building an effective educational question-generation model.

2 Related Work

This work focuses on developing AI systems capable of generating educational questions for technology-enhanced learning. It involves two main sub-tasks: Question Generation (QG), where a model generates a question based on given information, and Question Answering (QA), where a model generates a response to a question. QG is essential for QA and both tasks are part of reading comprehension tasks. This paper focuses on QG specifically.

N. Wang et al. (Eds.): AIED 2023, LNAI 13916, pp. 327–339, 2023.
https://doi.org/10.1007/978-3-031-36272-9_27

2.1 Automatic Question Generation (QG)

Automatic question generation involves creating valid and coherent questions based on given sentences and desired responses. Previous approaches have used rule-based and neural-based models, with neural models dominating in various applications [21]. Recent advancements in deep learning have led to the adoption of sequence-to-sequence models. By leveraging question-answering datasets, neural models can generate questions using both the context and expected response, ensuring high-quality questions. However, this approach often relies on an additional system to identify relevant responses [15], limiting its real-world applicability. The scarcity of public datasets also hinders the development of QG systems that generate both questions and answers. Alternatively, QG models can be trained to rely solely on the context, allowing the creation of questions that belong to a specific type [19] for the document, paragraph, or sentence level [8,9]. This work specifically focuses on the latter task setting, where only the context is used as input.

2.2 Pre-trained Language Models (PLMs) for Educational QG

In the field of educational neural question generation, state-of-the-art (SOTA) systems leverage pre-trained language models (PLMs) such as GPT-3 [2] and Google T5 [13]. These models, pre-trained on massive text corpora, enable zero-shot question generation without additional training. Recent research has demonstrated the potential for generating educational questions using GPT models [1,17].

Leaf, a cutting-edge question generation system, fine-tunes a large language model for the question and multiple-choice distracter generation [11]. It uses the SQuAD 1.1 dataset [14] to train its question generation component by fine-tuning a pre-trained T5 model [13]. This work diverges from SOTA approaches by employing pre-training to further enhance the PLM's handling of scientific language in the educational context [12], a technique that has shown promise in domain-specific applications like medicine [20].

Our hypothesis is that pre-training with scientific text can lead to better educational question generation even when models are fine-tuned for general-purpose tasks. To evaluate the quality of generated questions, various metrics are utilized, such as BLEU, ROUGE, METEOR, F1-Score, Human Ratings, Perplexity, and Diversity [1,11,17]. This study selects a representative subset of these metrics to measure success in terms of linguistic validity and fluency.

2.3 Related Datasets

S2ORC is a corpus comprising 81.1 million English scholarly publications across various academic fields [10]. For question generation (QG) and question-answering (QA) datasets, [21] offers a comprehensive review. The Leaf system, our baseline, is designed for educational purposes by fine-tuning the T5 model using the SQuAD 1.1 dataset, which focuses on reading comprehension [14]. However, this dataset is less suited for evaluating educational QG capabilities.

In contrast, SciQ [18] is a collection of 13,679 crowd-sourced scientific exam questions covering physics, chemistry, and other sciences. Although smaller than SQuAD, SciQ is more relevant for objectively evaluating educational QG models. Therefore, we use the SciQ dataset to assess the models developed in this work, aligning our evaluation with real-world scenarios.

3 Methodology

This study aims to study the effect of further pre-training and fine-tuning the Pre-trained Language Model (PLM) on Educational QG.

3.1 Research Questions

- **RQ1:** Can PLMs generate human-like educational questions?
- **RQ2:** Does pre-training PLMs with scientific text improve educational QG?
- **RQ3:** How does the training dataset size affect the pre-training?
- **RQ4:** Does fine-tuning the model with educational questions improve it?

3.2 Question Generations Models

Our experiments develop QG systems that utilise different PLMs trained using different task settings. It is important to note that we were not interested in training a neural model from scratch as this is impractical in real-world scenarios due to data scarcity and computational cost [2]. Instead, we used a PLM as the foundation of the different QG systems we developed for our experiments.

Baseline Leaf Model: Based on the relevant literature, we identified Leaf system [16] as the state-of-the-art educational question generation system to use as our baseline. In Leaf, the pre-trained language model, T5, a text-to-text transformer-based language model [13] (already trained on web-crawled data and Wikipedia articles) is fine-tuned for question generation using a reading comprehension dataset.

Proposed EduQG Models: The key differentiator between the baseline model and our proposal is that the EduQG model uses an additional pre-training step that further trains the PLM with scientific text documents before fine-tuning it for question generation. The expectation here is that the additional pre-training on scientific text is going to provide the PLM with more understanding of scientific language and knowledge that is relevant for generating good educational questions.

We also develop *Leaf+* and *EduQG+*, extending the Leaf model and the EduQG model that is further fine-tuned using an educational question dataset that is more specialised than a reading comprehension dataset that only contains general-purpose questions. We hypothesise that further pre-training harnesses the model's ability to generate educational questions.

3.3 Data

There are different types of datasets that are utilised in different stages of training the PLMs unto question generation models. These datasets allow us to:

1. Pre-train the PLM further with additional scientific language data
2. Fine-tune the PLM to carry out question generation, which is different from the initial task it was trained on
3. Objectively evaluate the performance of the question generation model

We incorporate a subset of datasets described in Sect. 2.3 in our experiments. While the pre-training step is skipped when building the baseline Leaf model, the S2ORC corpus [10] is used for pre-training the EduQG models. The resultant language model is fine-tuned for question generation using the SQuAD 1.1 dataset [14]. Finally, we use the test set data from the SciQ question dataset [18] for evaluation. This is because the SciQ dataset exclusively contains science questions from examinations making it suitable for objectively evaluating the model's suitability in *educational question generation.*

3.4 Evaluation Metrics

As identified in Sect. 2, two aspects of quality are considered when evaluating the QG models, i) the prediction accuracy and ii) the linguistic quality of the generated questions. To measure the predictive accuracy of the questions, we use the BLEU score and the F1 score that is used in prior work [1,11,14]. To measure how human-like the generated questions are (i.e. linguistic quality), we use perplexity and diversity [17]. A lower perplexity score indicates better coherence. The diversity score indicates how diverse the vocabulary of the generated questions is. Larger diversity values coupled with low perplexity, indicate the use of a richer vocabulary with grammatical precision.

3.5 Experimental Setup

Our experiments are designed to answer the research questions that are outlined in Sect. 3.1. To address RQ1, we calculate the linguistic quality-related metrics (specifically, perplexity and diversity) of the human-generated questions (the ground truth) in the SQuAD 1.1 and SciQ datasets. We hypothesise that the machine-generated questions are acceptable if they demonstrate superior or similar linguistic quality metrics in comparison to the metrics computed using the human-generated questions in the datasets (SQuAD and SciQ). The source code is available publicly[1].

[1] https://github.com/hmuus01/Educational_QG.

Figure 1 illustrates the experiments we set out to answer RQs 2–4. The foundational language model to all the developed models (baselines and proposals) is the *T5-small* language model (hereafter referred to as T5 model). Altogether 5 models are developed (coloured boxes in the figure), all of which are evaluated using the SciQ test data.

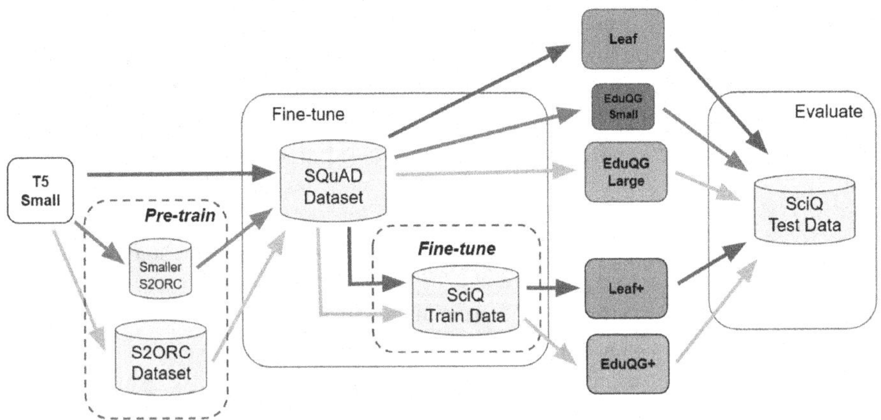

Fig. 1. Methodology for training and evaluating the baseline Leaf model (blue), novel EduQG Small (green) and EduQG Large (yellow) models (and their · + counterparts), introducing additional pre-training and fine-tuning steps (green dashed boxes) to address RQ 2,3 and 4. (Color figure online)

To address RQ2, we develop *Leaf* and *EduQG Large* models as per Fig. 1. As the baseline, we develop the *Leaf* model by fine-tuning the T5 model on the SQuAD 1.1 dataset (blue flow of arrows in Fig. 1 through the *Leaf* model). Our proposal, *EduQG Large*, additionally pre-trains the T5 model with a down-sampled version of the S2ORC dataset that contains approx. 23.2M scientific abstracts related to Chemistry, Biology and Physics research papers (yellow flow of in Fig. 1 through the *EduQG Large* model). To answer RQ3, we use two models, i) *EduQG Large* from the previous experiment, and ii) *EduQG Small* (green flow of arrows through the *EduQG Small* model) using a smaller number of training examples from 23.5M data points. To answer RQ4, we develop *Leaf+* and *EduQG+* (blue and yellow flows of arrows passing through the · + models), extensions of the Leaf and EduQG Large models (baselines for RQ4 experiment) that are further fine-tuned using the training data from the SciQ dataset. While the SQuAD dataset will help the PLM to learn question generation in general, the SciQ training data is expected to teach the model *educational question generation.* We hypothesise this change will lead to superior performance.

4 Results

As per Sect. 3.5, several experiments are executed. Table 1 shows the perplexity and diversity scores computed on the human-generated questions found in SQuAD 1.1 and SciQ datasets (RQ1). Table 2 presents the prediction accuracy and the linguistic quality metrics calculated for the models described in Sect. 3.2 (RQ 2 and 3). Figure 2 further elaborates the distribution of metric scores across the test data. Table 4 presents the improvement of predictive performance and the linguistic quality of the models *Leaf+* and *EduQG+* which are further fine-tuned using the SciQ training data (RQ4). Finally, Table 3 shows a handful of randomly selected test examples from the SciQ dataset where the baseline *Leaf* and the novel *EduQG Large* models have generated questions using the same context.

Table 1. Linguistic quality of the human-generated questions in the datasets.

Dataset	Perplexity ↓	Diversity ↑
SQuAD 1.1	84.16	0.779
SciQ	18.74	0.824

Table 2. Comparison of predictive performance and linguistic quality between Leaf (baseline) and EduQG (our proposals). The best and second best performance is indicated in **bold** and *italic* faces respectively. The proposed models that outperform the baseline counterpart ($p < 0.01$ in a one-tailed paired t-test) are marked with $\cdot^{(*)}$.

Model	Predictive Performance					Linguistic Quality	
	BLEU-1 ↑	BLEU-2 ↑	BLEU-3 ↑	BLEU-4 ↑	F1-Score ↑	Perplexity ↓	Diversity ↑
Leaf	27.07	20.22	17.17	*16.46*	30.90	**30.82**	0.735
EduQG Small	*29.07*$^{(*)}$	*21.52*$^{(*)}$	*17.49*$^{(*)}$	15.94	*33.12*$^{(*)}$	34.51	*0.736*
EduQG Large	**29.19**$^{(*)}$	**21.69**$^{(*)}$	**18.03**$^{(*)}$	**16.76**$^{(*)}$	**33.18**$^{(*)}$	*34.36*	**0.749**$^{(*)}$

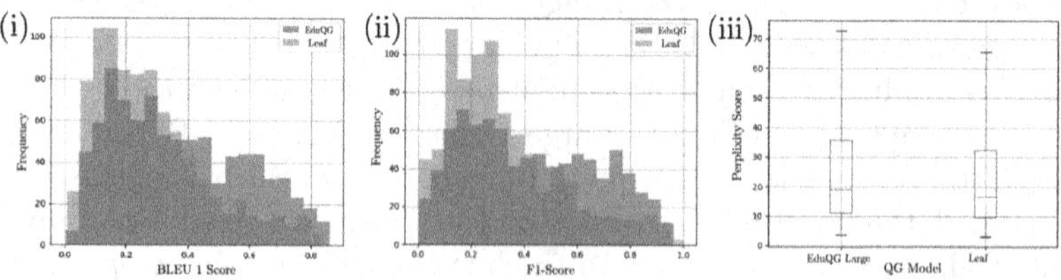

Fig. 2. The distribution of (i) BLEU 1, (ii) F1 and (iii) Perplexity Score between the Leaf and EduQG models.

Table 3. Randomly selected contexts From SciQ test data used to create questions using the Leaf and EduQG Large models.

Context	EduQG	Leaf
(1) Scientific models are useful tools for scientists. Most of Earth's systems are extremely complex. Models allow scientists to work with systems that are nearly impossible to study as a whole. Models help scientists to understand these systems. They can analyze and make predictions about them using the models. There are different types of models	What is used to analyze and make predictions about systems that are nearly impossible or easy to study as a whole?	What help scientists to understand the systems of Earth?
(2) Muscles That Move the Head The head, attached to the top of the vertebral column, is balanced, moved, and rotated by the neck muscles (Table 11.5). When these muscles act unilaterally, the head rotates. When they contract bilaterally, the head flexes or extends. The major muscle that laterally flexes and rotates the head is the sternocleidomastoid. In addition, both muscles working together are the flexors of the head. Place your fingers on both sides of the neck and turn your head to the left and to the right. You will feel the movement originate there. This muscle divides the neck into anterior and posterior triangles when viewed from the side (Fig. 11.14)	What is the major muscle that laterally rotates?	What is the major muscle that laterally flexes and rotates the head?
(3) Biodiversity refers to the variety of life and its processes, including the variety of living organisms, the genetic differences among them, and the communities and ecosystems in which they occur. Scientists have identified about 1.9 million species alive today. They are divided into the six kingdoms of life shown in the Figure below. Scientists are still discovering new species. Thus, they do not know for sure how many species really exist today. Most estimates range from 5 to 30 million species	What term refers to the variety of life and its processes?	How many species are identified today?
(4) Take-Home Experiment: The Pupil Look at the central transparent area of someone2019s eye, the pupil, in normal room light. Estimate the diameter of the pupil. Now turn off the lights and darken the room. After a few minutes turn on the lights and promptly estimate the diameter of the pupil. What happens to the pupil as the eye adjusts to the room light? Explain your observations. The eye can detect an impressive amount of detail, considering how small the image is on the retina. To get some idea of how small the image can be, consider the following example	What is the central transparent area of someone's eye?	What is the name of a take-home Experiment?
(5) In both eukaryotes and prokaryotes, ribosomes are the non-membrane bound organelles where proteins are made. Ribosomes are like the machines in the factory that produce the factory's main product. Proteins are the main product of the cell	What are the non-membrane bound organelles where proteins are made?	What is the main product of a cell?

Table 4. Comparison of predictive performance and linguistic quality between Leaf and EduQG models in Table 2 to the new proposals further fine-tuned on SciQ training data, *Leaf+* and *EduQG+*. The best and second best performance is indicated in **bold** and *italic* faces respectively. The new models that outperform the baseline counterparts ($p < 0.01$ in a one-tailed paired t-test) are marked with $\cdot^{(*)}$.

Model	Predictive Performance					Linguistic Quality	
	BLEU-1 ↑	BLEU-2 ↑	BLEU-3 ↑	BLEU-4 ↑	F1-Score ↑	Perplexity ↓	Diversity ↑
Leaf	27.07	20.22	17.17	16.46	30.90	*30.82*	0.735
EduQG	29.19	21.69	18.03	16.76	33.18	34.36	0.749
Leaf+	*36.67*$^{(*)}$	*31.45*$^{(*)}$	*28.17*$^{(*)}$	**24.26**$^{(*)}$	*41.65*$^{(*)}$	**26.43**$^{(*)}$	*0.801*$^{(*)}$
EduQG+	**37.20**$^{(*)}$	**33.86**$^{(*)}$	**28.49**$^{(*)}$	*22.35*$^{(*)}$	**43.04**$^{(*)}$	33.88$^{(*)}$	**0.812**$^{(*)}$

5 Discussion

The results presented in Sect. 4 provide sufficient information for us to answer the research questions pointed in Sect. 3.1.

5.1 Ability of PLMs to Generate Educational Questions (RQ1)

The results presented in Tables 1 and 2 together allow us to answer RQ1. It is seen from the linguistic quality metrics in Table 2 that the perplexity score obtained by all the trained models (both baseline and novel) is acceptable. That is, the perplexity scores obtained by the model-generated questions are much lower compared to the perplexity score of the SQuAD 1.1 questions that are human-generated. The language used in academic texts can be highly advanced and rich. This is reflected by the very low perplexity score and the high vocabulary diversity score of the SciQ questions in Table 1. While the proposed models haven't achieved a perplexity score close to the SciQ questions, having a superior perplexity in comparison to the SQuAD 1.1 question shows that the generated questions inherit coherent language and human readability. The random examples presented in Table 3 further reinforce this conclusion.

5.2 Effect of Pre-training with a Scientific Text Corpus (RQ2)

Table 2 demonstrates that the novel models, EduQG Small and EduQG Large, surpass the baseline Leaf model in nearly all evaluation metrics for predicting educational questions in the SciQ test dataset. This improvement highlights the impact of additional pre-training on scientific text for generating educational questions. With all models fine-tuned using the same question generation dataset (the SQuAD dataset), the only intervention in the proposed models is during the pre-training phase as per Fig. 1.

The T5 language model, the foundational PLM in this experiment, is trained primarily on web-crawled data and Wikipedia articles [13]. However, this training corpus lacks scientific texts, leading to a weaker understanding of scientific knowledge and language. The improvement in predicting educational questions signifies that additional pre-training enhances the model's grasp of scientific knowledge and language, even without specific training on educational questions during fine-tuning.

Table 2 shows higher mean perplexity scores for EduQG models, though the difference is not statistically significant. Figure 2(iii) indicates that the perplexity distribution between the two models is not statistically different. The observations in Table 3 further illustrate that the EduQG model generates more educational and pedagogically sound questions, as seen in rows 3 and 4.

5.3 Impact of the Training Size on the Question Quality (RQ3)

The results in Table 2 further points out the performance difference between models *EduQG Small* and *EduQG Large* where the only difference is the size of pre-training data (green vs. yellow arrows in Fig. 1). The *EduQG Large* model is superior in all evaluation metrics with the larger pre-training dataset of 23.2M data abstracts. The *EduQG Small* model outperforms the baseline *Leaf* model that uses fewer pre-training examples from the S2ORC dataset. This trend suggests that the increasing number of training examples used in the pre-training step leads to a better QG model. The increasing diversity values with the growing number of pre-training examples is also noticeable from Table 2. The improvement of BLEU and F1-Scores with diversity indicates that the validity of questions is not harmed by the diversity of the vocabulary used by the model.

5.4 Effect of Fine-Tuning Using Educational Questions (RQ4)

The experimental setups of RQ 2 and 3 use a *zero-shot* evaluation where no observations from the SciQ dataset are used during the training phase. On the contrary, the experiments relating to RQ4 ($\cdot$ + models in Fig. 1) use the training data from the SciQ dataset that allows the newly proposed models, *Leaf+* and *EduQG+* to learn from educational question examples. Table 4 indicates that the additional fine-tuning significantly improves the predictive accuracy. It is noteworthy that fine-tuning is also improving the perplexity score of the generated questions which was absent in the previous experiments. We can see that the new models are outperforming the baselines. This improvement attributes to the low perplexity score of SciQ questions as per Table 1 that are exposed to the model during training.

5.5 Opportunities

The examples in Table 3 with all the above results indicate that educational QG systems are very close to becoming part of human-facing technology-enhanced learning systems (Such as X5Learn that leverages Open Educational Resources [4]). Many works in the past have shown how zero-shot question generation is operationally feasible using very large language models gated behind an API from a large corporation (Model-as-a-Service architecture) [17]. However, our result contributes to this topic as we introduce methods to enhance openly-available PLMs (in our case, T5) to support educational QG. We intentionally use the *T5-Small* model that has 60M parameters in comparison models such as GPT-3 XL that has 1.3B parameters [2] to show that relatively small models can be trained with domestic hardware to create SOTA educational QG capabilities. Our method also gives the stakeholder full control and ownership, a critical feature for quality assurance of the downstream educational systems that rely on this model (contrary to having no control over a third party that can change their model over time). This work also informs the educational data mining community that domain-specific data can be used with language models to harness them to specific educational use cases (e.g. extend to other domains, different question types that support diverse pedagogy etc.). While the proposed systems are not perfect, the quality of AI-generated questions indicates that a teacher or an educator can re-purpose these questions with minimum effort and time. Human-in-the-loop systems can be built to support educators while their corrections will harvest more training data to improve the models over time. Educational questions can be generated at scale using the proposed model both for existing and newly created learning resources, adding more testing opportunities for learners/teachers to use when needed.

We see our work being foundational to building a series of tools that can support educators with scalable/personalised assessments. Ultimately, we have the opportunity to improve these models to the point where an intelligent tutor can rely on them to create on-demand questions to verify a learner's knowledge state with no human intervention.

5.6 Limitations

We need to be cautious to avoid the obvious pitfalls of such automatic systems. Intelligent QG models we build tend to exhibit the patterns in the data that we feed them. We need to be mindful that we take rigorous steps to validate the datasets to be ethically and pedagogically sound. Putting emphasis on quality assurance of the training data will help us to build ethical, unbiased QG models that can benefit all learners equally.

Many intelligent learning systems exploit learner engagement signals to determine what characteristics of the system should sharpen and weaken [3]. In the context of question generation, it is important to distinguish between *bad* questions vs. *difficult* questions as the latter, although demanding, may positively impact a learner while the former will only hinder and diminish learning gains. The AI-generated questions should allow users to improve their learning gains over time.

Another gap in this work is the lack of human evaluation of the AI-generated questions. While offline evaluation on labelled datasets is useful, having teachers and learners evaluate and contrast between human vs. AI-generated questions will provide much more insightful findings that can improve this line of research in the future. Our subsequent work will focus on this aspect.

6 Conclusion

This work demonstrates the operational feasibility of adapting pre-trained language models for educational question generation. Specifically, we argue that a relatively small language model manageable with domestic hardware can be further trained and harnessed with low computational costs and produce a humanly-acceptable educational question generation model. We validate that a PLM fine-tuned with question generation data can generate questions that are linguistically valid and humanl-like. We show that the quality of the educational questions generated can be significantly improved by pre-training using domain-specific corpora alone. We use a corpus of scientific abstracts to empirically demonstrate this while we point out the relationship between the prediction quality and the amount of data. Going further, we improve the model's question generation capabilities significantly by further fine-tuning it using a domain-specific question dataset, indicating fine-tuning can be used to further improve the model.

A few promising steps remain to take this work to the future. Validating the generalisability of our approach to other PLMs such as GPT [2] and extending evaluation to human experts [1,17] are the immediate next steps. Establishing methods to audit the ethical and pedagogical value of training datasets will improve the use of the downstream QG models. Identifying systematic approaches (e.g. using curriculum learning) to identify the most useful training examples would allow us to make QG models significantly better with less number of training examples leading to computational cost savings. Finally, formalising concepts such as question difficulty, and value for learning will allow us to evaluate the quality of generated questions much more pragmatically.

Acknowledgements. This work is also partially supported by the European Commission-funded project "Humane AI: Toward AI Systems That Augment and Empower Humans by Understanding Us, our Society and the World Around Us" (grant 820437), EU Erasmus+ project 621586-EPP-1-2020-1-NO-EPPKA2-KA and the EPSRC Fellowship "Task Based Information Retrieval" (grant EP/P024289/1). This research is conducted as part of the X5GON project (www.x5gon.org) funded by the EU's Horizon 2020 grant No 761758.

References

1. Bhat, S., Nguyen, H.A., Moore, S., Stamper, J., Sakr, M., Nyberg, E.: Towards automated generation and evaluation of questions in educational domains. In: Proceedings of the 15th International Conference on Educational Data Mining, 701, vol. 704 (2022)
2. Brown, T., Mann, B., Ryder, N., Subbiah, M., Kaplan, J.D., Dhariwal, P., Neelakantan, A., Shyam, P., Sastry, G., Askell, A., et al.: Language models are few-shot learners. Adv. Neural. Inf. Process. Syst. **33**, 1877–1901 (2020)
3. Bulathwela, S., Perez-Ortiz, M., Yilmaz, E., Shawe-Taylor, J.: Truelearn: a family of Bayesian algorithms to match lifelong learners to open educational resources. In: AAAI Conference on Artificial Intelligence (2020)

4. Bulathwela, S., Kreitmayer, S., Pérez-Ortiz, M.: What's in it for me? Augmenting recommended learning resources with navigable annotations. In: Proceedings of the International Conference on Intelligent User Interfaces Companion (2020)
5. Bulathwela, S., Pérez-Ortiz, M., Yilmaz, E., Shawe-Taylor, J.: Semantic TrueLearn: using semantic knowledge graphs in recommendation systems. In: Proceedings of First KGSWC International Workshop on Joint Use of Probabilistic Graphical Models and Ontology (PGMOnto) (2021). https://arxiv.org/abs/2112.04368
6. Bulathwela, S., Pérez-Ortiz, M., Holloway, C., Shawe-Taylor, J.: Could AI democratise education? Socio-technical imaginaries of an edtech revolution. In: Proceedings of NeurIPS Workshop on ML4D. arXiv (2021). https://arxiv.org/abs/2112.02034
7. Bulathwela, S., Pérez-Ortiz, M., Yilmaz, E., Shawe-Taylor, J.: Power to the learner: towards human-intuitive and integrative recommendations with open educational resources. Sustainability **14**(18), 11682 (2022)
8. Du, X., Shao, J., Cardie, C.: Learning to ask: neural question generation for reading comprehension. arXiv preprint arXiv:1705.00106 (2017)
9. Guo, H., Pasunuru, R., Bansal, M.: Soft layer-specific multi-task summarization with entailment and question generation. arXiv preprint arXiv:1805.11004 (2018)
10. Lo, K., Wang, L.L., Neumann, M., Kinney, R., Weld, D.: S2ORC: the semantic scholar open research corpus. In: Proceedings of the Annual Meeting of the ACL. Online (2020)
11. Lopez, L.E., Cruz, D.K., Cruz, J.C.B., Cheng, C.: Simplifying paragraph-level question generation via transformer language models. In: Pham, D.N., Theeramunkong, T., Governatori, G., Liu, F. (eds.) PRICAI 2021. LNCS (LNAI), vol. 13032, pp. 323–334. Springer, Cham (2021). https://doi.org/10.1007/978-3-030-89363-7_25
12. Muse, H., Bulathwela, S., Yilmaz, E.: Pre-training with scientific text improves educational question generation (student abstract). In: AAAI Conference on Artificial Intelligence (2023)
13. Raffel, C., et al.: Exploring the limits of transfer learning with a unified text-to-text transformer. J. Mach. Learn. Res. **21**(140), 1–67 (2020)
14. Rajpurkar, P., Zhang, J., Lopyrev, K., Liang, P.: Squad: 100, 000+ questions for machine comprehension of text. CoRR abs/1606.05250 (2016)
15. Tamang, L.J., Banjade, R., Chapagain, J., Rus, V.: Automatic question generation for scaffolding self-explanations for code comprehension. In: Rodrigo, M.M., Matsuda, N., Cristea, A.I., Dimitrova, V. (eds.) AIED 2022. LNCS, vol. 13355, pp. 743–748. Springer, Cham (2022). https://doi.org/10.1007/978-3-031-11644-5_77
16. Vachev, K., Hardalov, M., Karadzhov, G., Georgiev, G., Koychev, I., Nakov, P.: Leaf: multiple-choice question generation. In: Proceedings of the European Conference on Information Retrieval (2022)
17. Wang, Z., Valdez, J., Basu Mallick, D., Baraniuk, R.G.: Towards human-like educational question generation with large language models. In: Rodrigo, M.M., Matsuda, N., Cristea, A.I., Dimitrova, V. (eds.) AIED 2022. LNCS, vol. 13355, pp. 153–166. Springer, Cham (2022). https://doi.org/10.1007/978-3-031-11644-5_13
18. Welbl, J., Liu, N.F., Gardner, M.: Crowdsourcing multiple choice science questions. In: Proceedings of the 3rd Workshop on Noisy User-generated Text. ACL (2017). https://doi.org/10.18653/v1/W17-4413
19. Wu, X., Jiang, N., Wu, Y.: A question type driven and copy loss enhanced framework for answer-agnostic neural question generation. arXiv preprint arXiv:2005.11665 (2020)

20. Xu, H., Van Durme, B., Murray, K.: BERT, mBERT, or BiBERT? A study on contextualized embeddings for neural machine translation. In: Proceedings of Conference on Empirical Methods in Natural Language Processing (2021)
21. Zhang, R., Guo, J., Chen, L., Fan, Y., Cheng, X.: A review on question generation from natural language text. Trans. Inf. Syst. **40**(1), 1–43 (2021)

Involving Teachers in the Data-Driven Improvement of Intelligent Tutors: A Prototyping Study

Meng Xia[1(✉)], Xinyi Zhao[2], Dong Sun[3], Yun Huang[1], Jonathan Sewall[1], and Vincent Aleven[1]

[1] Carnegie Mellon University, Pittsburgh, USA
{mengxia,sewall,va0e}@andrew.cmu.edu, yunhuanghci@cmu.edu
[2] Fudan University, Shanghai, China
xinyizhao19@fudan.edu.cn
[3] Nio Inc., Shanghai, China

Abstract. Several studies show that log data analysis can lead to effective redesign of intelligent tutoring systems (ITSs). However, teachers are seldom included in the data-driven redesign of ITS, despite their pedagogical content knowledge. Examining teachers' possible contributions is valuable. To investigate what contributions teachers might make and whether (and how) data would be useful, we first built an interactive prototype tool for visualizing student log data, SolutionVis, based on needs identified in interviews with tutor authors. SolutionVis presents students' problem-solving processes with an intelligent tutor, including meta-cognitive aspects (e.g., hint requests). We then conducted a within-subjects user study with eight teachers to compare teachers' redesign suggestions obtained in three conditions: a baseline "no data" condition (where teachers examined just the tutor itself) and two "with data" conditions in which teachers worked with SolutionVis and with a list representation of student solutions, respectively. The results showed that teachers generated useful redesign ideas in all three conditions, that they viewed the availability of data (in both formats) as helpful and enabled them to generate a wider range of redesign suggestions, specifically with respect to hint design and feedback on gaming-the-system behaviors and struggle. The current work suggests potential benefits and ways of involving teachers in the data-driven improvement of ITSs.

Keywords: Intelligent tutoring system · Data-driven redesign · Students' log visualization

1 Introduction

Data-driven optimization of intelligent tutoring systems (ITSs) using design insights derived from student learning data and log data can substantially improve student learning outcomes [14]. Existing work has largely consisted of

N. Wang et al. (Eds.): AIED 2023, LNAI 13916, pp. 340–352, 2023.
https://doi.org/10.1007/978-3-031-36272-9_28

researchers' analyzing and refining the system's knowledge component model and then redesigning the system to optimize the learning of the revised knowledge components [9].

In this paper, we present a first exploration of whether and how teachers might be involved in the data-driven redesign of tutoring systems. Teachers have often been involved in the initial design and development of tutors [11,17] as well as analytics dashboards for use with ITSs [7,8]. However, we are not aware of any projects in which teachers were involved in data-driven redesign of tutors, although in one instance they successfully used data-driven methods to adapt the text of course materials for teaching English to non-native speakers [10]. In another project, teachers redesigned hints for an ITS, but without using data [25]. Beyond teachers' proven role on tutor design teams, there is reason to think that teachers could make valuable contributions to the data-driven refinement of tutoring systems. Teachers' rich pedagogical content knowledge and practical experience in giving effective feedback might put them in a great position to suggest improvements to an intelligent tutor. In particular, data from student problem solving with the given ITS might jog teachers' memory of what happened in the classroom, and reveal trends or events that they were not aware of. It is worth investigating what contributions teachers might make to the data-driven redesign process, given their unique sources of knowledge, and what tools would be useful in this regard, given their unique sources of knowledge.

Visual analytics have shown potential in summarizing and presenting problem-solving processes to enhance understanding of domain competencies or demands. The CTAT Behavior Graph visualizes pre-defined solutions within a given problem [19]. DataShop's widely used "learning curves" are line charts depicting students' mastery of targeted knowledge components over time. [13]. Later studies [16,20,23,24] visualized the problem-solving behaviors of groups of students to find common and distinct strategies. For example, Xia et al. [23] proposed QLens, a glyph-based Sankey diagram, to show how a group of students solve drag-and-drop problems step by step, allowing question designers to easily identify common difficult situations encountered. However, existing work misses detail in the visualization of student actions (e.g., hint requests) and has not investigated how to help teachers generate ideas for improving a tutoring system.

To this end, we first conducted a needs-finding study through semi-structured interviews with five intelligent tutor researchers and two school math teachers. From these interviews, we derived initial design requirements for how to present students' multi-step problem-solving processes to teachers in a way that might spur ideas for redesign. From these requirements, we then developed an interactive visualization interface prototype, SolutionVis. The tool presents students' problem-solving processes, including meta-cognitive aspects (e.g., hint requests). Finally, we conducted a within-subjects user study with another eight math teachers to explore the following research questions: RQ1. What kinds of redesign suggestions do teachers generate? RQ2. Does giving them a representation of students' interaction data help them generate redesign ideas? RQ3. To help teachers generate redesign suggestions, how does an interactive visualization of the data compare to presenting it in a standard list format?

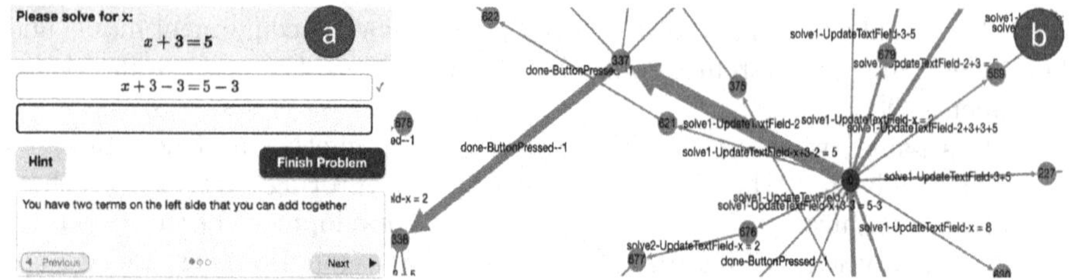

Fig. 1. (a) An example problem x + 3 = 5 of the intelligent tutor. (b) The node-link diagram of students' log data from the intelligent tutor with the problem: x + 3 = 5.

2 Needs-Finding Study

We conducted semi-structured interviews with five ITS researchers and two school math teachers, with more than 10 years of research or teaching experience on average. The purpose was to derive initial design requirements for how to visualize the data of a tutoring system in a way that would help teachers understand students' interactions with that tutoring system and generate ideas for improvements of that tutoring system. Having a mix of experienced authors and teachers gives us a wide variety of perspectives.

Data Used and Procedure. We selected Lynnette, a linear-equation-solving tutor [15] for our investigation because it had key features that are characteristic of an ITS [4,21] and a reasonable level of complexity. Specifically, it supports problems with multiple steps, multiple solution paths within each problem, and step-level guidance in the form of hints and feedback. Its design was based on instructional design research [12], but this unit had not been data-tuned. As shown in Fig. 1(a), we used the problem $x + 3 = 5$ as an example for the interview. We used 27 students' log data (records of individual student interactions with the tutoring software) from DataShop [13]. Each record includes the timestamp, the action (e.g., button pressed, text filled), and the evaluation results (e.g., correct or incorrect) from the intelligent tutor.

The interview lasted one hour for each participant. We presented a visualization of students' log data in the form of a lo-fi node-link diagram, derived from existing work [16], to learn how we could improve from that. See Fig. 1(b) above. The graph shows the solution paths (including errors and hint requrests of all 27 students. Each node in the graph represents a state in the problem-solving process; and each link represents a student's action. The number in the node shows the index of the transactions in the log data. Wider links indicate more students taking that action. The link label "solve1-updateText-2" means that the student answered "2" at the first step (indicated by 1 in solve1). We asked questions such as "*What information can you find from the visualization that can be used to improve the tutor design?*"

Initial Design Requirements. After the interviews, three authors analyzed the interview transcripts. The author who conducted the interviews extracted all suggestions for improving the visualization to gain insights for the tutor redesign from the transcripts, and all three authors worked together to derive the following initial design requirements using an affinity diagram [6].

- *R1: Make long paths easier to follow.* Participants mentioned that the initial radial layout makes it hard to identify long sequences and align different paths. They preferred a horizontal layout like a train station map.
- *R2: Make it easier to identify paths that capture similar problem-solving strategies by merging steps of paths.* Participants mentioned different ways to merge, such as, merging the same steps in different paths and collapsing steps from paths into stages (e.g., performing distribution of "()").
- *R3: Make it easier to identify common problematic steps using visual cues.* Participants did not find it easy to distinguish correct and incorrect steps. They mentioned it might be important to pay attention to common incorrect steps for diagnosing problems in the tutor and suggested using different colors to encode the edges and highlight the problematic steps.
- *R4: Make it easier to understand students' metacognition by showing hint request behaviors.* Participants emphasized that students' hint requests could reflect hint quality and students' metacognition.
- R5: *Make it easier to understand the problem-solving context by showing the tutor and a list of steps in a solution path.* Participants wanted to know what problems the students are solving and also a detailed list of steps, perhaps vertically, could contribute to a better understanding.

3 SolutionVis

To address all the requirements, we developed a visualization interface prototype, SolutionVis, to present students' log data for teachers to explore. We used Python to process the log data and save each problem's data as a JSON file. We then used Cytoscape.js [2] to visualize the node-link digram of the selected problem. SolutionVis consists of three views: Tutor View, Student Path View, and Sequence View.

The Tutor View, shown in Fig. 2(A), displays the "live" tutor so that teachers can interact with it and understand the problem-solving context (see requirement R5, above).

The Student Path View, shown in Fig. 2(B), is designed for teachers to understand how all the students (as a group) proceeded step by step. The steps in the paths are positioned from left to right for easy reading-of long paths especially (R1). This view supports zooming and panning. By default, it shows a fully-zoomed-out overview as shown in Fig. 3 below. The user zooms in to show details, as in Fig. 2(B). The nodes represent states on students' paths to a solution; the links represent students' actions. **Links:** The color of a link is determined by the evaluation of the student's action-green for correct input, red for incorrect input, and yellow for hint requests. We merged duplicate actions and used the thickness of the link to represent how many times students repeated the step (R2, R3). **Nodes:** The nodes in the graph represent the intermediate

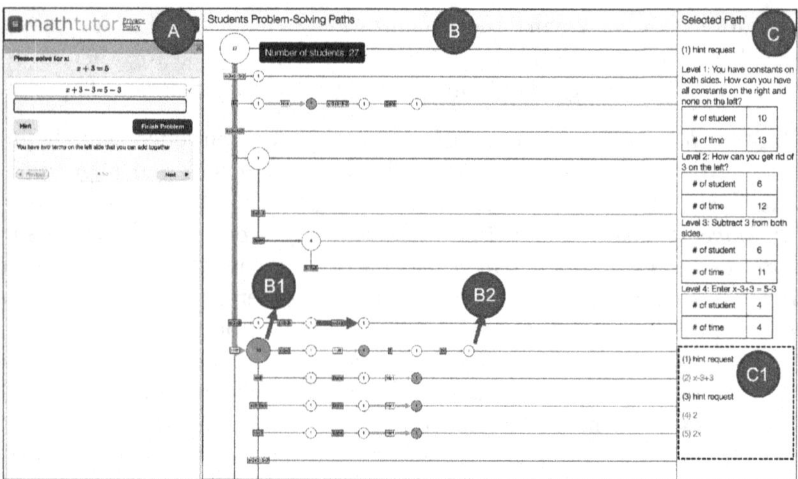

Fig. 2. SolutionVis: The Tutor View (A) shows the original intelligent tutor for teachers to interact with; The Student Path View (B) shows students' aggregated problem-solving paths, and the Sequence View (C) lists, at the user's request, a solution path or hints. Specifically, when the user clicks a step node is clicked (e.g., B2), the detailed steps will be listed (C1). When the user clicks a hint node (e.g., B1), apart from the steps, the details of the hint node will be shown (C).

states on students' solution paths. The number in the node represents how many students performed the same previous actions while the first node shows the total number of students. We use a pie chart to represent the ratio of students who asked for the bottom-out hint at that stage (R4). For example, as shown in Fig. 2(B1), the pie chart shows that 4 out of 10 students asked for the bottom-out hint at the first step. By clicking the pie chart, the user can see the text shown at each hint level and the number of students asking for that level. **Layout:** The x-axis position of the node represents which stage the student is in, in an effort to address R2. Solving basic linear equations includes four stages: removing parentheses, adding/subtracting terms, combining like terms, and dividing both sides of the equation by the coefficient of the variable term. As shown in Fig. 3, SolutionVis classified each node (state) along a solution path into different stages according to students' attempts. (The example omits the first stage, removing parentheses, since there are no parentheses in this problem.) Moreover, when students do not follow the given order of the four stages, SolutionVis shifts the node a fixed distance down along the y-axis (see the inset in Fig. 3).

The Sequence View shows detailed information about a selected path that one or more students took. When the user clicks on a node in the graph, the sequence view will display the steps from the start up to the selected node. For example, if the user clicks Fig. 2(B2), the path will be shown as Fig. 2(C1). In addition, if the node clicked is a hint request (e.g., Fig. 2(B1)), the levels of hints requested will be listed in the Sequence View (e.g., Fig. 2(C)).

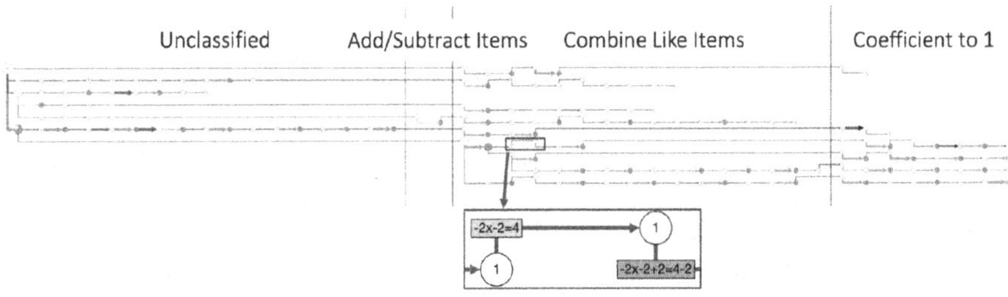

Fig. 3. The layout of the Student Path View in SolutionVis with the stages (at the top) for solving "−4x + 1 − 3 + 2x = 4". The enlarged part shows that the student input "−2x − 2 = 4" belongs to the stage "Combine Like Terms", followed by "−2x − 2 + 2 = 4 − 2", which does not belong to "Combine Like Terms" and is therefore shifted down.

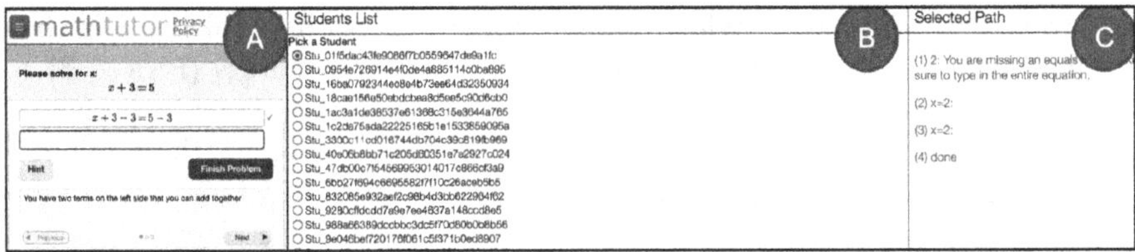

Fig. 4. List Interface: Tutor View (A), Students List (B), and Selected Path (C).

4 User Study

We conducted a user study to investigate the three research questions stated at the end of the Introduction.

Participants. Eight middle school math teachers (three males and five females, aged 30–59) participated in the user study, with an average of 12.9 years of experience in grades 5–12. They were from the US, Croatia, Taiwan, and Germany. While six had no ITS experience, one had experience with MATHia, and another used Khan Academy. Five had experience with visualization dashboards, four in creating data visualizations, two in reading interactive infographics, and one in reading self-tracking data charts.

Conditions and Datasets. To address our research questions, we compared three conditions: a baseline "no data" condition (i.e., working with just the tutor itself) and two "with data" conditions in which teachers worked with SolutionVis and with a list representation (List Interface, Fig. 4), respectively. The List Interface shows a list representation of students' IDs (Fig. 4(B)) and, at the user's request, a single student's log data (Fig. 4(C)). This kind of list representation is a common feature of commercial learning software such as Graspable Math [1]. Key differences between SolutionVis and the List Interface are that (a) SolutionVis has a graph of students' aggregated paths and (b) the List Interface lists a single student's solution path. We used a within-subjects study design in which all participants experienced all conditions. All teachers first worked with the Original Tutor and then with both the List Interface and SolutionVis. This way, we could ascertain whether data helps over and above working with just the tutor. To avoid a learning effect, we counterbalanced, across participating teachers, the order of the List Interface and SolutionVis. Participants were shown two equation-solving problems from the given intelligent tutor (Lynnette) with different difficulty levels: $x + 3 = 5$ and $-4x + 1 - 3 + 2x = 4$. We presented log data from past use of these problems by 27 and 10 sixth-grade students, respectively; no data from the teachers' own students was included.

Procedure. We conducted and recorded one 90-minute study session for each participant remotely through Zoom. The user study consisted of the following activities: (1) We asked participants to sign in the consent form to get their permission for Zoom recording. (2) We provided background information about the project. (3) We first provided links to these two tutor problems. We then provided links to the List Interface (Fig. 4) and the SolutionVis (Figs. 2, 3) to participants in a counterbalanced order. (Thus, each participant experienced all three conditions.) For each of the three conditions, we spent three minutes introducing the interface. We then asked participants to do a task: explore different interfaces and give suggestions on how to improve the tutor for each of the two tutor problems. Participants were given 15 to 20 min for each interface. (4) Lastly, for each of the three interfaces, we asked them to fill in a 5-point Likert scale questionnaire with answers ranging from 1 (Strongly disagree) to 5 (Strongly Agree) (Fig. 6) to rate the usability (Q1–Q5) and usefulness of the interface (Q6–Q10). We also asked which interface they preferred and why. The questionnaire was adapted from a standard usability test [5] and from questionnaires meant to evaluate the usefulness of teacher dashboards [22]. We also asked the participants to rate how easy it was to understand SolutionVis's interface. Each participant received an Amazon gift card ($40) for participating.

5 Results

We address RQ1 with an analysis of teachers' redesign suggestions. For RQs 2 and 3, we report both teachers' redesign suggestions and of their survey data, including comments regarding the design of SolutionVis and the List Interface.

Analysis of Teachers' Redesign Ideas. As shown in Fig. 5, we extracted all the suggestions made by teachers in step (3) for how to improve the tutor. In total, we got 52 suggestions and 22 unique ones. On average, each participant generated three redesign suggestions when working with the Original Tutor, one additional suggestion when subsequently working with the List Interface, and two additional suggestions with SolutionVis. Two authors, using an affinity diagram, then grouped these suggestions into three main categories, namely, (1) interface/logic, (2) hints, and (3) feedback on gaming-the-system behaviors and persistent struggle. Category 2 (hints), was subdivided into four subcategories: adaptivity, clarity/correctness, visibility, and composition.

	Original Tutor	List Interface	SolutionVis
Interface /logic design	Need to provide instruction that using "Enter" instead of "Finish problem" to go to next step. (5) The correct step with a different order of the items in the equation is not accepted by the tutor. (5) Cursor should go to the next line automatically.(1) Need to say explicitly to input the final answer or the intermediate step.(2)		
Hint Adaptivity	Address what the students did in the hint. (3)	If the answer is super close to the final answer (e.g., missing a negative sign), provide more concrete feedback (e.g., check your sign) (1)	Provide hints based on how many times the student asks for it. If the student asks a second time, showing a different hint. (1)
Hint Clarity/ Correctness	The hint "You can get the variable by itself by dividing both sides by the coefficient." is not correct for "-2x-4=2". Should "add 4 on both sides first".(3) Rewrite the hint "Your input in not valid algebra." as "You need to have 'x ='"(2) The hint "You have two terms on the left side that you can add together" is not well designed. Explain and give examples about "term".(4)	Provide the number line in the hint for students to understand positive and negative numbers for the steps where they need to move items from one side to the other side. (1)	Check the first hint of each step and make sure it is clear and easy to understand. (1)
Hint Visibility	Don't show the bottom hint.(1) Let the hints pop up automatically.(1) Show part of the hints to let students think more at each step; teaching them but not correcting them.(1)	Don't encourage students to ask for hints at beginning, but ask "what would be your first step?" (1)	
Hint Composition	–	–	Ask a question about the knowledge in the hint to let them think. (1) Show some exmaples in the hint. (1)
Feedback on Gaming the system behavior/ protracted struggle	–	When seeing students gaming the system or inputting random things, provide feedback like "show me your efforts", "show your work"(2) When seeing student submitting the same thing multiple times, let the tutor give the answer and move on.(1)	When seeing students gaming the system or inputting random things, provide feedback like "show me your efforts", "show your work" (4) When seeing student submitting the same thing multiple times, provide feedback like "stop clicking this, please try again" (2) Show cartons or funny animals to encourage them.(1)

Fig. 5. Suggestions made by teachers using different interfaces. For the List Interface and SolutionVis, the table lists only the suggestions that were not mentioned when working with the Original Tutor. "-" means no suggestion in the cell and empty means there are repetitive suggestions. Suggestions about hints are highlighted in yellow. The numbers in the parentheses indicate the number of participants. (Color figure online)

The main findings are: first, teachers generated many suggestions for how to improve the design of the tutor interface, the tutor's hints, and its feedback (RQ1). Second, with the data provided in the List Interface and by SolutionVis, teachers generated additional suggestions, compared to the ones they generated with the Original Tutor (RQ2). With the Original Tutor, they generated a mixture of ideas regarding the redesign of the interface/logic, content of hints, and pedagogical aspects of hints. With the List Interface and SolutionVis (the "with data" conditions), they showed a greater focus on hint design (highlighted in yellow) and on feedback on gaming and struggling behaviors. For example, for the category hint adaptivity, P7 noticed, using the List Interface, that one student repeatedly submitted a solution that just missed a negative sign and suggested: *"[I]f the answer is super close to the final answer (e.g., missing a negative sign), then provide more concrete feedback (e.g., check your sign)."* Also,

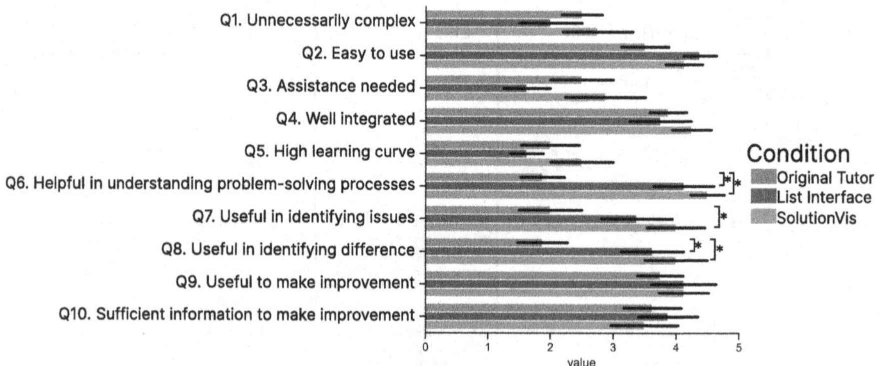

Fig. 6. Participants' rating results for usability and usefulness on a 5-point Likert scale under three conditions: Original Tutor, List Interface, and EqLens. ($* : p < .05$)

when teachers saw consecutive attempts like "x = 1, x = 2, ...", they suggested feedback like "show me your efforts." When they saw behaviors indicative of struggle (e.g., long sequences), they suggested letting the tutor give the answer or cartoons to encourage students. Second, when using SolutionVis, participants focused on the steps with thick lines, large nodes, and paths containing both; they gave more suggestions addressing students' common issues, e.g., gaming-the-system/struggling behaviors (RQ3).

Teachers' Ratings of the Interfaces. Figure 6 shows the ratings for the three interfaces in terms of usability (Q1-Q5) and usefulness (Q6-Q10). We ran a one-way ANOVA on each questionnaire item, with the three interfaces as the independent factor, followed by the Bonferroni posthoc test on measures with statistically significant differences [18]. Regarding usability, with SoutionVis the teachers felt they need slightly more assistance (Q3) and had a higher learning curve (Q5) than the Original tutor and the List interface, although the difference was not statistically significant. Participants could understand the visual representation in SolutionVis after the introduction. In terms of usefulness (see Q6–Q8), teachers found SolutionVis and the List Interface more helpful than the Original Tutor (without students' data) for understanding students' problem-solving behaviors, identifying common issues, and different problem-solving strategies (RQ2). There was no significant difference between the SolutionVis and the List Interface in this regard (RQ3).

Qualitative Feedback. We summarized participants' comments and suggestions regarding SolutionVis and the List Interface (RQ3) and found that four participants preferred the List Interface and four preferred SolutionVis. All would like to have an integrated version of both interfaces.

SolutionVis. Teachers' positive feedback about SolutionVis is that it helps them quickly understand students' data by providing aggregation and highlighting common errors. "*SolutionVis is nice because it does some aggregation*" (P1).

"*You can see, just by, like, the thickness of those arrows, you could see that the same hints were being asked many times or the same/similar answer was being entered over and over and over and to me, like. glaring red or those bold things would indicate that there's a problem there*" (P5). Teachers' suggestions for improving SolutionVis are as follows. First, they hope SolutionVis could let them trace back to the specific students who made particular mistakes so they could help these students directly. Second, it would be better to provide a compact overview of the paths that fits on a single page. Third, the tool could highlight gaming-the-system behaviors. Fourth, a replay function could be added to show students' actual interactions with the tutor interface [3].

List Interface. Teachers liked the List Interface because they thought they could use it to review individual students' data from their own class (the data used in the study was not from the teachers' own students), better interpret the data, and provide help to the students directly "*I want to connect the data to students' behaviors in class to understand them better*" (P2). They also mentioned several limitations of the List Interface. "*It is helpful for debugging but not helpful for the classroom*" (P5). "*There's no way to have 50 or 60 or 70 students every day and then 13, 14, 15 questions*" (P2).

6 Discussion

In general, for RQ1, teachers generated useful redesign ideas in both without data and with data (graph and list format) conditions. For RQ2, data about student learning (whether in graph or list format) was rated helpful (over and above working with just the tutor) and enabled teachers to generate a wider range of redesign suggestions. For RQ3 (comparison of graph v. list format), some teachers preferred SolutionVis because showing the aggregated data in a graph helps them find the tutor's problems efficiently, and others preferred the List Interface for tracing each individual student's solution. From these results we may conclude that SolutionVis enables teachers to be more active participants in the data-driven improvement of tutoring systems. Below, we further discuss design considerations, generality, and limitations of the current study.

Design Considerations. *(1) Combine access to aggregated and student-specific process information.* All participants would like to have an integrated version of both interfaces. and we suggest an integrated version with aggregated graphical representations of students' problem-solving processes and easy access to each individual student's problem-solving steps. *(2) Provide more context information to better support attribution analysis.* In the user study, we found that teachers wanted to utilize their understanding of a particular student's background (e.g., hard-working or not) to analyze the data. For example, they suggested that if a good student is experiencing struggle with the tutoring system, then maybe the tutoring system is to blame. We suggest providing a student's past data (e.g.,previous error rate) to aid teachers in attributing results to features of the tutors.*Integrate the tool into a teacher's workflow.* It is still open questions on how a tool like this might fit into a teacher's workflow. For example, education institutions could invite teachers to the data-driven intelligent tutors' redesign process periodically, or the tool could be connected with tutor authoring tools (e.g., CTAT) so teachers can customize the tutors themselves.

Generality. The design requirements and the visualization tool generated are applicable for data-driven optimization of step-level problem-solving (e.g., ITSs). ITSs have been built for many domains, where the method should apply by defining different problem-solving stages, although we have yet to deal fully with problems in which these stages do not have a fixed order.

Limitations. Our study has some limitations, including a limited number of participants, testing with data from a single tutor and a small problem set, and no evaluation of the impact on student learning. Despite this, we engaged more participants than previous studies [16,23] and gathered valuable quantitative data. Future work will enhance the prototype, address design considerations, and expand evaluations to include more teachers, tutors, and problems.

7 Conclusion

We investigated whether involving teachers in the data-driven of an ITS might be valuable, and what tools would be useful to harness teachers' unique sources of knowledge. The current study may well be the first in the AIED literature to address these questions. First, we derived five initial design requirements through a needs-finding study. We then built a visualization tool prototype, SolutionVis, that embodies the five requirements to present students' log data. Results from a within-subjects user study showed that teachers generated useful redesign ideas especially when supported by data, presented either as aggregated data from a group of students using a graph visualization (SolutionVis) or individual students' data in a list (List Interface). Teachers rated both interfaces as more helpful than working with just the tutoring system itself. The study shows two formats have complementary strengths and could be integrated in the next iteration of SolutionVis. These findings open up data-driven redesign for an audience that was not previously involved but may make unique contributions.

References

1. Grastable math (2022). https://activities.graspablemath.com/. Accessed 10 Sept 2022
2. Cytoscape.js (2023). https://js.cytoscape.org/. Accessed 29 Apr 2023
3. Aleven, V., Blankestijn, J., Lawrence, L., Nagashima, T., Taatgen, N.: A dashboard to support teachers during students' self-paced AI-supported problem-solving practice. In: Hilliger, I., Muñoz-Merino, P.J., De Laet, T., Ortega-Arranz, A., Farrell, T. (eds.) EC-TEL 2022. LNCS, vol. 13450, pp. 16–30. Springer, Cham (2022). https://doi.org/10.1007/978-3-031-16290-9_2

4. Aleven, V., Sewall, J.: The frequency of tutor behaviors: a case study. In: Micarelli, A., Stamper, J., Panourgia, K. (eds.) ITS 2016. LNCS, vol. 9684, pp. 396–401. Springer, Cham (2016). https://doi.org/10.1007/978-3-319-39583-8_47
5. Bangor, A., Kortum, P.T., Miller, J.T.: An empirical evaluation of the system usability scale. IJHCI **24**(6), 574–594 (2008)
6. Hartson, R., Pyla, P.S.: The UX Book: Process and Guidelines for Ensuring a Quality User Experience. Elsevier, Amsterdam (2012)
7. Heffernan, N.T., Heffernan, C.L.: The assistments ecosystem: building a platform that brings scientists and teachers together for minimally invasive research on human learning and teaching. IJAIED **24**(4), 470–497 (2014)
8. Holstein, K., McLaren, B.M., Aleven, V.: Co-designing a real-time classroom orchestration tool to support teacher-AI complementarity. J. Learn. Anal. **6**(2) (2019)
9. Huang, Y., et al.: A general multi-method approach to data-driven redesign of tutoring systems. In: LAK21, pp. 161–172 (2021)
10. Jin, T., Lu, X.: A data-driven approach to text adaptation in teaching material preparation: design, implementation, and teacher professional development. TESOL Q. **52**(2), 457–467 (2018)
11. Koedinger, K.R., Aleven, V.: An interview reflection on "intelligent tutoring goes to school in the big city". IJAIED **26**(1), 13–24 (2016)
12. Koedinger, K.R., Anderson, J.R.: Illustrating principled design: the early evolution of a cognitive tutor for algebra symbolization. Interact. Learn. Environ. **5**(1), 161–179 (1998)
13. Koedinger, K.R., Stamper, J.C., Leber, B., Skogsholm, A.: LearnLab's datashop: a data repository and analytics tool set for cognitive science. Top. Cogn. Sci. **5**(3), 668–669 (2013)
14. Koedinger, K.R., Stamper, J.C., McLaughlin, E.A., Nixon, T.: Using data-driven discovery of better student models to improve student learning. In: Lane, H.C., Yacef, K., Mostow, J., Pavlik, P. (eds.) AIED 2013. LNCS (LNAI), vol. 7926, pp. 421–430. Springer, Heidelberg (2013). https://doi.org/10.1007/978-3-642-39112-5_43
15. Long, Y., Aleven, V.: Enhancing learning outcomes through self-regulated learning support with an open learner model. User Model. User-Adap. Inter. **27**, 55–88 (2017)
16. McBroom, J., Yacef, K., Koprinska, I., Curran, J.R.: A data-driven method for helping teachers improve feedback in computer programming automated tutors. In: Penstein Rosé, C., et al. (eds.) AIED 2018. LNCS (LNAI), vol. 10947, pp. 324–337. Springer, Cham (2018). https://doi.org/10.1007/978-3-319-93843-1_24
17. Murray, T., Woolf, B.P.: Tools for teacher participation in ITS design. In: Frasson, C., Gauthier, G., McCalla, G.I. (eds.) ITS 1992. LNCS, vol. 608, pp. 593–600. Springer, Heidelberg (1992). https://doi.org/10.1007/3-540-55606-0_69
18. Norman, G.: Likert scales, levels of measurement and the "laws" of statistics. Adv. Health Sci. Educ. **15**(5), 625–632 (2010)
19. Simon, H.A., Newell, A.: Human problem solving: the state of the theory in 1970. Am. Psychol. **26**(2), 145 (1971)
20. Tsung, S., Wei, H., Li, H., Wang, Y., Xia, M., Qu, H.: Blocklens: visual analytics of student coding behaviors in block-based programming environments. In: Proceedings of the Ninth ACM Conference on Learning@ Scale, pp. 299–303 (2022)
21. VanLehn, K.: The behavior of tutoring systems. AIED **16**(3), 227–265 (2006)
22. Xia, M., et al.: Peerlens: peer-inspired interactive learning path planning in online question pool. In: Proceedings of the 2019 CHI, pp. 1–12 (2019)

23. Xia, M., Velumani, R.P., Wang, Y., Qu, H., Ma, X.: QLens: visual analytics of multi-step problem-solving behaviors for improving question design. IEEE Trans. Visual Comput. Graph. **27**(2), 870–880 (2020)
24. Xia, M., et al.: Visual analytics of student learning behaviors on k-12 mathematics e-learning platforms. arXiv preprint arXiv:1909.04749 (2019)
25. Yang, K.B., Nagashima, T., Yao, J., Williams, J.J., Holstein, K., Aleven, V.: Can crowds customize instructional materials with minimal expert guidance? Exploring teacher-guided crowdsourcing for improving hints in an ai-based tutor. Proc. ACM Hum.-Comput. Interact. **5**(CSCW1), 1–24 (2021)

Reflexive Expressions: Towards the Analysis of Reflexive Capability from Reflective Text

Andrew Gibson[(✉)], Lance De Vine, Miguel Canizares, and Jill Willis

Queensland University of Technology, Brisbane, Australia
andrew.gibson@qut.edu.au

Abstract. As a first step towards computational analysis of reflexive capability, this research established a process of identifying and classifying short groups of words prominent in reflective writing resulting in a corpus of reflexive expressions. The expressions were classified into theoretical categories of reflexivity, which were evaluated through a series of experiments that crowd-sourced human judgements of expression-category suitability. The purpose of the work was to ascertain: (a) the feasibility of computationally identifying expressions of reflexivity in reflective text, and (b) the extent to which computational classification of the expressions accord with human judgements. Success could advance the computational analysis of reflective text aiding the identification of reflexive capability at scale. The study involved (1) Social-technical derivation of theoretically informed categories, (2) Computational generation of a corpus, and (3) crowdsourced human judgements. We found that computational generation of English reflexive expressions was feasible, and that some categories accord well with human judgements drawn from a fluent English population. The work is expected to provide a foundation for ongoing inquiry, and a basis for more general use in identifying evidence of learner reflexivity.

Keywords: Reflexivity · Reflection · Reflective Writing Analytics

1 Introduction

Reflective writing analytics has been an active research topic in learning analytics for some time [4], however research has tended to focus mostly on identifying evidence of reflection quality within academic reflective writing (mostly within higher education contexts). This work aims to shift the focus from evidence of writing quality to evidence of reflexive capability - the extent to which a learner can *bend back* [1], synthesise their past experiences with current context and imagine a future direction. By attending to the reflexive capabilities of the learner, the research has wider potential than academic writing analytics, particularly in areas of professional and informal learning. For example, a medical practitioner's reflexive capability may help them adapt their practice to a local

N. Wang et al. (Eds.): AIED 2023, LNAI 13916, pp. 353–364, 2023.
https://doi.org/10.1007/978-3-031-36272-9_29

context, and a professional athlete's reflexive capability may help them improve psychological aspects of their performance. To the best of our knowledge, this is a new direction in reflective writing analytics research, and is a novel application of existing natural language processing technologies.

This study represents the beginning of a larger program of research directed toward developing analysis techniques and intelligent technologies for understanding and enhancing reflexive capability at scale. The work involved the computational identification of groups of words (linguistic *expressions*) expected to be reasonable indicators of *reflexive* interactions between the reflector and their experiences. For example, the expression "I hope to" can be an indicator of the learner considering some future action, and therefore may indicate a possible mental interaction between their current situation and an imagined future. Similarly, the lack of any expressions directed toward the future may indicate a lack of consideration of alternative paths or ways in which the learner may change.

The work was undertaken in 3 phases: **Phase T** - Theoretical development of reflexive expression categories; **Phase C** - computational identification of n-grams that occur frequently in reflective text but seldom in non-reflective texts, and computational classification of the n-grams in terms of the theoretical categories; and **Phase V** - verifying the extent to which these computationally derived n-grams can be recognised as appropriately categorised reflexive expressions by a more general population of fluent English language speakers.

2 From Reflective Properties, to Reflexive Interactions

The research involved a shift away from the typical computational analysis of *properties* of reflective writing, toward the more novel computational analysis for *interactions* between the author and their experiences. This shift was driven by a philosophical commitment to understand the learner in a more holistic way than just the source of their reflective writing. It involved viewing the learner, their experiences, their acts of reflection, and their writing as a whole complex adaptive system [9]. Due to their non-reductive nature, complex adaptive systems need to be analysed in terms of interactions between the components of the system and between the system and its environment, rather than properties of components [9]. Thus, rather than analysing properties of the reflective writing, this work sought to identify expressions with the potential to evidence interactions, such as an author's interactions with their past experiences or current capabilities. Reflective writing involves a person writing about themselves and their experiences. In a learning context, reflection can help establish meaningful connections between past experiences, what is currently being learnt, and future practice.

While the 'property' approach lends itself to identifying topical expressions that describe the content of the text, the 'interaction' approach identifies expressions that indicate a connection between the author and something else that may or may not be reflected in the text. By way of example, consider the difference between a 'property' focus and an 'interaction' focus in the following fictional reflective text:

property: I've been thinking carefully about my last primary prac, and I've decided I want to investigate changing to secondary teaching.
I love teaching, but I feel that I would be better with high school students than primary students which has been a bit of a struggle.
Making this change might also open up additional career possibilities.

interaction: I've been thinking carefully about my last primary prac, and I've decided I want to investigate changing to secondary teaching.
I love teaching, but I feel that I would be better with high school students than primary students which has been a bit of a struggle.
Making this change might also open up additional career possibilities.

For reflexive capability, it is not so much about *what* the learner is reflecting on (properties), but more *how* they are making connections between themselves and their experiential context (interactions). Thus, this work focused on identifying the kinds of expressions highlighted in the second 'interaction' example.

3 Reflexivity and Reflective Writing Analytics

Reflexive capability involves a recognition of multidimensional temporal and affective relationships - it is a 'bending back' of thought, feeling and experience [1]. It is an ongoing evaluation about what is best, now and in the future discerned from the many role options in social contexts. Reflexivity informs professional learning and writing as the individual can look back over tangible records of deliberations. They can identify the topics or 'concerns' of reflections, the commentaries relating to physical wellbeing, performative achievement and sense of identity and self-worth, and the attending emotions that provide 'shoving power' for action, and point to what is valued [1]. Reflexivity also informs personal learning. Through stories they tell themselves the story teller selects what is salient and makes meaning out of their experiences, and crafts those experiences towards a socially aware purpose [8]. Personal and social identities are crafted in the inner conversations [1] that occur as reflexive individuals make continuous streams of evaluations and decisions.

Previous approaches to reflective writing analytics have tended to focus on the less complex notion of reflection, and the more specific objective of identifying aspects of reflective writing quality. This work as tended to be domain focused where techniques are tuned for specific contexts such as health sciences higher education [2,3]. A range of technical approaches have been taken from more linguistic informed techniques [3] to more machine learning based techniques [6,7,10,12]. Regardless of the adopted approaches, most have tended to focus on identifying properties important to academic reflective writing. An aim of this study was to advance the state of the art of reflective writing analytics research through the shift of focus from reflective writing properties to reflexive interactions.

4 Methodology

Informed by theory and previous research (see Sect. 4.1), we expected that expressions indicating reflexive interactions would be visible in learners' reflective writing as: (1) explicit references to past and recent experiences, (2) the learner relating themselves to identified experiences, and (3) the learner projecting beyond their immediate situation, relating aspects of their personal story to an imagined future. The research was conducted in 3 phases. **Phase T** brought prior work [5,11] to the study, in order to inform the selection of theoretical categories. **Phase C** was a data-driven computational approach to creating a corpus of reflective ngrams classified into categories from phase T. **Phase V** involved human validation of the categorised ngrams through a series of experiments using crowd-sourced human judgements.

4.1 Phase T - Theoretical Categories

The theoretical development of reflexive categories grew from two of us (Gibson and Willis) engaging in a socio-technical process over nearly a decade of research collaboration. Inductive qualitative analyses traced patterns of connection and meaning for learners from their reflective writing. Identified patterns or potential lines of inquiry were explored through computational/technical analysis, which was then brought together with social analysis. Theorising occurred as part of the iterative meaning making process, drawing on Archer's concepts within reflexivity [1], as well as domain specific research fields.

An example of this socio-technical process can be seen in work identifying linguistic patterns associated with themes of assessment capability [11]. Prior successes with this socio-technical approach informed early discussions amongst us while evaluating the results from the early computational clustering of n-grams (see the work on Phase C in Sect. 4.2). Staying true to the socio-technical approach that has proven fruitful in previous work [5], we allowed theory and experience together with computational analysis to jointly inform the development of categories. This was an iterative process that spanned many weeks.

At the conclusion of Phase T, we had selected the 8 most promising categories, split into 2 groups of 4. See Table 1 for category descriptions and example expressions. **Positional:** purpose focused, locate the author. Includes categories: Egoreflexive (ER), Retroreflexive (RR), Anteroreflexive (AR), and Vertoreflexive (VR). **Expressive:** character focused, describe the author. Includes categories: Evaluative (EV), Epistemic (EP), Contending (CN), and Affective (AF). Other potential categories not included, but which may be investigated in future included Temporal (references to time), and Corporeal (references to concrete things).

4.2 Phase C - Computational Ngrams

Phase C of the study involved the creation of a reflexive ngram corpus classified into categories (from Phase T). It was conducted in 2 iterations, C1 leading to

Table 1. Category examples

	Cat	Category	Description	Examples
Positional	ER	Egoreflexive	Referring to myself, relating to myself, putting myself in the bigger picture	to say that i am * i feel confident that i * point in time i am * i am finding that i * i feel that i 'm * as i feel that i * because i am not
	RR	Retroreflexive	Referring to previous events, looking back, relating to past experience, thinking about the past	as we had to * i have done a lot * and we were all * i 've learnt a lot * i have n't had much * i have been reading * i have completed my
	AR	Anteroreflexive	Referring to possibilities, relating to what is coming next, thinking about the future	think we will be able * to move forward * to do in the future * looking forward to this * i know i can do * is going to go * and i hope to
	VR	Vertoreflexive	Making a change, making a resolution, changing direction, turning from something old to something new, committing to a new approach	to apply it to my * i will have to do * decided as a group to * to start working on the * will have to do the * where i want to be * we need to have
Expressive	AF	Affective	Feelings and emotions, emotionally responding, reacting with feeling, expressing passion	i am excited about * i am happy with what * i am a little nervous * a little nervous about the * was very happy with this * to feel this way * 'm a little bit worried
	CN	Contending	Struggling, striving, dealing with problems, overcoming issues, beating the odds	was unable to complete the * struggling to come up with * i struggle to * struggling a bit with the * i was able to understand * i was struggling to * a lot more confident in
	EP	Epistemic	Knowing, thinking, believing, imagining, wondering, knowledge and facts, ideas and thoughts	with the idea that the * so i think i * a better idea of how * but i think it could * of the idea * up with an idea that * i know that this is
	EV	Evaluative	Evaluating, assessing, or judging, making comparisons, weighing up options	think it is a good * that we will need to * and should be able to * was on the right track * time to think about * m on the right track * could have improved my performance

experiments 1 & 2, and C2 leading to experiment 3. Common Natural language processing (NLP) techniques (such as tokenising, lemmatising, word embedding, sentence embedding) were used to process a corpus of reflections to extract ngrams, and remove non-reflective ngrams. Ngrams are groups of tokens/words which are N length long. This work involved the creation of 3-grams, 4-grams, and 5-grams. The classification of the final ngrams in terms of the 8 reflexive categories developed in phase T also made use of common well-known strategies (such as k-means clustering) and technologies (like fasttext[1]) for classification. Our intention in this study was not to develop new technical processes, but to use common readily available technologies in service of a novel objective - to identify and classify expressions indicative of reflexive interactions.

All computational analysis used preexisting reflective writing data collected from undergraduate and postgraduate students using the *GoingOK*[2] web application over a period of 8 years. Data were comprised of 13,841 short written reflections about their personal experiences within 13 week learning contexts. These data were a subset of the larger corpus ($n > 17,500$), and spanned the years 2014 to 2022. Ethics approval was provided by the university for the computational analysis of the reflections without consent.

Phase C1 involved work that was used in experiments 1 and 2 of Phase V. NLP processing of the reflective writing corpus yielded 80,631 3-grams, 36,405 4-grams, and 10,585 5-grams. The top 10,000 most frequently occurring of each ngram group were processed further, with the remainder being discarded. The resulting ngrams included 1,822,750 tokens with a vocabulary of 27,801 words. The computational workflow consisted of: (1) Pre-processing of corpus to obtain clean text; (2) Sentence detection; (3) converting to lower case and removal of punctuation with the exception of apostrophes in contractions (e.g. *I 'm*); (4) Removal of duplicate sentences or lines; (5) Normalisation - transformation of words in text using POS tags, lemmas or dictionaries; (6) Identification of n-grams in the corpus. N-grams can be words, POS tags, lemmas, or dictionary words/codes. Ngrams (3–5) were collected, counted, and ranked according to their frequency. Only ngrams occurring a minimum of 3 times were retained; (7) After the n-grams were created, any identifying information was removed (e.g. people's names); (8) Comparison of sets of n-grams from different corpora using set operations (e.g. intersection and difference).

The guiding hypothesis was that some ngrams occurring frequently within the reflective writing corpus might be distinctive in reflective writing, and not likely to occur frequently in more general text. In order to test this hypothesis, ngrams were extracted from a general English language corpus, the British National Corpus (BNC)[3]. Processing the BNC resulted in 1,901,642 3-grams, 825,342 4-grams, and 254,108 5-grams. The resulting ngrams included 71,805,497 tokens with a vocabulary of 592,343 words. As with the reflective writing corpus ngrams, the 10,000 most frequently occurring ngrams for each ngram group were retained.

[1] https://fasttext.cc/.

[2] http://goingok.org.

[3] http://www.natcorp.ox.ac.uk.

The resulting BNC ngrams were used to filter out common English ngrams from the reflective writing corpus ngrams leaving only those ngrams that occur frequently in the reflective text, but not the BNC general text. This yielded a total of 7,157 3-grams, 8,389 4-grams, and 9,209 5-grams. The most frequent 5000 of each ngram group were retained. Filtering was finalised by removing any n-grams that formed a part of another ngram. The final reflective ngrams included 1,708 3-grams, 1,978 4-grams, and 5000 5-grams.

A classification process classified the reflective ngrams into categories. This involved: (1) Training word embeddings (high dimensional vector representations) on the original reflective writing corpus; (2) Constructing ngram and sentence embeddings from word embeddings; (3) Clustering embeddings using K-means clustering; (4) Performing nearest neighbour retrieval of ngrams using ngram embeddings; (5) Using final ngram clusters to inform category selection (see Sect. 4.1 for more detail); and (6) Constructing small sets of positive and negative examples for each category. Approximately 110 positive training ngram examples were used for each category. Computational processing used readily available NLP technologies written in Python and C++. Classification utilised the fasttext C++ command line application. In this phase C1 a multi-category classification approach was used.

In **Phase C2** the classified reflective ngrams corpus was modified for experiment 3 based on a preliminary analysis of experiment 1 and 2 results. To increase the likelihood of high quality n-grams, it was decided to reduce any potential ambiguity between reflective writing and BNC ngram sets by removing fiction and spoken text from the corpus. This was because these sub-corpora were relatively small, but could be likely to include reflective language. Another change was to move from one multi-category classifier operating across all categories to a single binary classifier for each category.

4.3 Phase V - Verification Judgements

Phase V of the research involved the use of a crowd-sourcing platform (Prolific) to engage people fluent in everyday English language to assess if a reflective ngram could be appropriately described by written description associated with a computer assigned category. Participants were paid for their time at a rate close to the average rate for crowd-sourcing which was £8 and £10 per hour. The crowd-sourcing platform included the facility to ensure that participants represented a balance of male and female and a range of ages. Note that disclosure of gender and age were NOT requirements of the study, but the platform setting to balance male and female representation in the participants was selected for this study. Participants were required to be fluent in English with a nominated nationality of an English language dominant country: Australia, Canada, the United Kingdom, and the United States. While we acknowledge that this introduces a significant bias into the data, this approach was necessary to ensure similar understandings of the English language expressions. We have not considered the use of this approach with other cultures and languages, as we do not know the extent to which other languages relate to the concept of reflexivity. In a series of experiments, crowd-sourced participants were shown randomly selected n-grams and asked if they fit a given category description (e.g. epistemic - Knowing, thinking, believing, imagining, ideas and thoughts). The repetition of small judgements with hundreds of participants for thousands of ngrams created response data which was analysed to determine the extent of agreement both at the category level and the ngram level. Experiments were developed in a serverless web application hosted by Amazon Web Services (AWS).

Experiment 1 had an approximate duration of 20 min for each participant. Participants evaluated 20 n-grams in each category and were presented with 4 categories. Each participant evaluated a total of 80 n-grams and each n-gram was evaluated by 4 participants. The n-grams selected to be evaluated were the ones with confidence > 0.7. The experiment consisted in showing the definition of the category and asking the question "Does the description above match the following expression?" The expression was the n-gram to be evaluated, and the participant had 3 options to answer the question: yes, it is a match; no, it is not a match and I can't decide. To maintain the attention of the participant the order of the agree and disagree options was random. Additionally, a visual cue was included in the experiment to warn the user when the category along with its description changed. An evaluation of the results of experiment 1 identified that the tasks were too cognitively demanding and time-consuming causing some issues with participants completing the task. Further, with only 4 participants per ngram judgement, the granularity of the judgements was possibly too course to be very effective at the ngram level, and if participants were eliminated due to issues with their responses, the total number of judgements would become ineffectual. Changes were made for experiment 2.

Experiment 2 had an approximate duration of 4 min for each participant. Participants evaluated 50 n-grams in a single category. Each n-gram was evaluated by 10 participants. The n-grams selected to be evaluated were the top 250 for each category. The experiment had a similar layout as experiment 1. An evaluation of the results of this experiment found that the task was too simple and resulted in participants completing the task inattentively, to get paid with the least effort. In the data cleaning process, many of these participants were identifiable as outliers, and were removed. However, this resulted in a significant loss of data. Thus, changes were made for experiment 3.

Experiment 3 targeted 5–6 min for each participant. The number of ngrams were the same as experiment 2, but closer attention was paid to the spread of high and low confidence ngrams. Approximately 20% of the 50 ngrams were drawn from a pool for 50 ngrams from the bottom confidence scores for each category. The remaining 80% were selected from the top confidence scores. Closer attention was paid participant selection with a requirement for higher Prolific approval ratings. Further, participants were advised in advance that they may not be paid if they failed attention checks - obvious negative cases. These changes together with the modification to the corpus in phase C2 resulted in significantly high quality data for this experiment.

5 Findings and Discussion

A very early finding of the study was that the process of identifying reflective ngrams (Phase C) was successful in producing expressions that had a reflective character, and which were valuable in finalising category selection. A computational clustering process also produced additional categories (like a temporal category) which were not used in this study, but may be fruitful in future. The early success provided confidence of the existence of expressions characteristic of reflective writing and not other genres, and that these were discoverable at scale. The resulting corpus of ngrams is unique, and the authors hope to gain ethical approval to release the corpus as an open dataset to seed future research.

A minimum number of crowd-sourced judgements were required to include an ngram in the final corpus. For experiment 1, this was 3 as most ngrams were only evaluated by 4 participants. For experiments 2 and 3, a minimum of 6 judgements were required. As there was variation on the number of judgements, a percentage of agree, disagree, uncertain was calculated. Then an ngram score which was calculated according to the following: $(a - d) \times (1 - u)$ where a, d, u are agree %, disagree %, and uncertain % respectively. The resulting scores ranged from -1 fully disagree, no agree or uncertain, to $+1$ fully agree, with no disagree or uncertain. The uncertain percentage moderates the score by reducing the degree of agreement by the level of uncertainty. For the analysis shown in Fig. 1, the *best* label was assigned to ngrams scoring at least $+0.5$, the *worst* label was assigned to ngrams scoring at most -0.5, and ngrams between -0.5 and $+0.5$ were given the label *mid*.

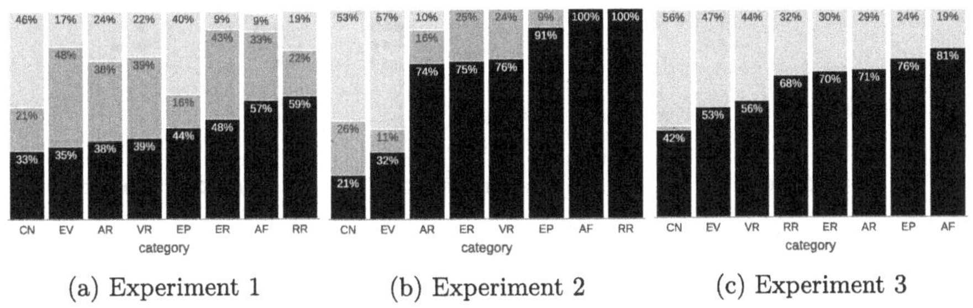

(a) Experiment 1 (b) Experiment 2 (c) Experiment 3

Fig. 1. Normalised charts showing the proportion of best (black), mid (dark), and worst (light) responses for each category for the 3 experiments.

Improvement from experiment 1 to experiment 3 can be seen in a larger margin between best and worst judgements for each category. This is obvious in the charts shown in Fig. 1 with Experiment 3 showing very good polarisation between responses. These charts also highlight the best and worst performing categories across the 3 experiments. Although experiment 3 provided the most reliable data in this study, *Contending* (CN) and *Evaluative* (EV) were clear poor performers in all experiments. *Affective* (AF) was a clear good performer in all studies, and *Epistemic* (EP) was a high performer in experiments 2 and 3. These are not surprising as statements about *how I feel* (AF) and *what I think* (EP) are common and readily understood. Expressions making an assessment (EV) or struggling (CN) tend to be more ambiguous. At the time of writing, the richness of the data has yet to be fully explored, and will be a key component of future work. However, it appears that corpora of Affective and Epistemic ngrams could be justifiably produced, and CN and EV ngrams should be re-assessed in terms of their prospects moving forward.

Another significant finding relates to the correlation between ngram scores which were calculated from the response data and ngram confidence where defined in advance computationally. While initial correlations were moderate at best (Exp3 Pearson = 0.409), analysis showed considerable disagreement where computational confidence and ngram score were oppositional in many categories. When this data were removed, there were marked improvements with all experiments exhibiting high correlation (Exp1 = 0.892, Exp2 = 0.798, Exp3 = 0.916). This can be seen visually in Fig. 2. This is a positive finding, as it will allow the production of a corpus of ngrams where human judgement strongly correlates with computer confidence at the ngram level. This can be useful for improving the training of classifiers. Further, such a corpus could also be useful for identifying and disambiguating those categories for reflective text where those ngrams are present.

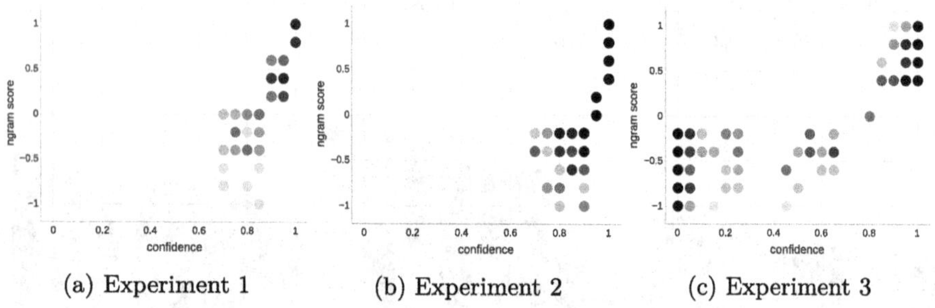

(a) Experiment 1 (b) Experiment 2 (c) Experiment 3

Fig. 2. An increase in correlation between ngram scores and confidence after removing ambiguous ngrams. More occurances are darker markers.

While the crowd-sourcing process was very rewarding in terms of the scale possible, it posed significant challenges that resulted in lengthy data cleaning processes, and multiple modifications between experiments. However, the improvements in our approach to the crowd-sourcing yielded a noticable improvement for experiment 3, reducing a loss of data from 49% to 12%. Given this change, as part of future work we intend to run a 4th experiment that keeps the improvements in crowd-sourcing as learnt in the first 3 experiments, and also includes feedback from best performing and least performing ngrams into the corpus creation and classification process. We would expect to see both a confirmation of experiment 3 results with small improvements in each category.

6 Limitations and Future Work

Analysis conducted thus far is too large for a full description in this paper, and undertaking this study has revealed many opportunities for improvement. As described in the methodology, a good portion of this project was dedicated to improving the crowd-source experiments. There is more analysis to do in order to identify further improvements which may allow for further experiments in the future. Due to the size of this study, incorporation of feedback for training the classifiers was minimal. However, the results from experiment 3 in particular are likely to provide a rich source of data which might be used to improve the classifiers and ultimately the model. This could be aided by further refinement of the categories including removing those which are too problematic. The process involving the creation of vector-space language models could be refined extensively using positive and negative feedback from phase V results. A final model could then be developed and tested for annotating unseen reflective text in terms of the training categories. This kind of model could then be hosted within an entity recognition service (such as AWS Comprehend) providing a mechanism for general use of the model similar to named entity recognition in general NLP services.

A significant limitation of the work was confining it to English language only, and to sourcing judgements from a limited number of western English speaking nations. This will have introduced biases in the data such that the results cannot be generalised to other languages and cultures without further research. Despite this limitation, we believe that the approach we have taken could be replicated in other languages (with available and comparable NLP technologies) in future. Another possible area for investigation is whether there might exist reflexive expressions that are universal, and thus useful for forming the foundation of a multi-lingual corpus. This possibility could also pave the way for research into minority indigenous languages, particularly where those languages are based on personal narratives similar to those identified in this research.

Another limitation was the deliberate ommition of other categories. One of these, temporality, holds theoretical significance and is likely to be a key feature of future work. Although clusters of temporal ngrams were readily identifiable in the phase C work, they were not easily disambiguated from other categories. A challenge for future work will be to identify how theoretically relevant categories might be included without adding unnecessary ambiguity in the classification. One approach may be to make a category like temporality a qualifier for the positional categories which tend to have an implied reference to the past, present or future.

7 Conclusion

This study has provided a solid basis from which to pursue new lines of inquiry for the analysis of reflexive capability which focus on interactions rather than properties of the writing. Improvements in the computational techniques to accommodate complex models of interactions are anticipated to catalyse further research on reflexivity in the areas of professional, personal and informal learning.

Acknowledgements. This research was supported by 'First Byte (2022)' funding from the QUT Centre for Data Science.

References

1. Archer, M.S.: Being Human: The Problem of Agency. Cambridge University Press, Cambridge (2000)
2. Cui, Y., Wise, A.F., Allen, K.L.: Developing reflection analytics for health professions education: a multi-dimensional framework to align critical concepts with data features. Comput. Hum. Behav. **100**, 305–324 (2019). https://doi.org/10.1016/j.chb.2019.02.019
3. Gibson, A., Aitken, A., Sándor, Á., Buckingham Shum, S., Tsingos-Lucas, C., Knight, S.: Reflective writing analytics for actionable feedback. In: Proceedings of the Seventh International Learning Analytics & Knowledge Conference, pp. 153–162. ACM, Vancouver (2017). https://doi.org/10.1145/3027385.3027436
4. Gibson, A., Kitto, K.: Analysing reflective text for learning analytics: an approach using anomaly recontextualisation. In: Proceedings of the Fifth International Conference on Learning Analytics And Knowledge - LAK 2015, pp. 275–279. ACM Press, Poughkeepsie (2015). https://doi.org/10.1145/2723576.2723635
5. Gibson, A., Willis, J.: Ethical challenges and guiding principles in facilitating personal digital reflection. In: Burr, C., Floridi, L. (eds.) Ethics of Digital Well-Being. PSS, vol. 140, pp. 151–173. Springer, Cham (2020). https://doi.org/10.1007/978-3-030-50585-1_8
6. Liu, M., Kitto, K., Buckingham Shum, S.: Combining factor analysis with writing analytics for the formative assessment of written reflection. Comput. Hum. Behav. **120**, 106733 (2021). https://doi.org/10.1016/j.chb.2021.106733
7. Magooda, A., Litman, D., Ashraf, A., Menekse, M.: Improving the quality of students' written reflections using natural language processing: model design and classroom evaluation. In: Rodrigo, M.M., Matsuda, N., Cristea, A.I., Dimitrova, V. (eds.) AIED 2022. LNCS, vol. 13355, pp. 519–525. Springer, Cham (2022). https://doi.org/10.1007/978-3-031-11644-5_43
8. Parr, G., Bulfin, S.: Professional learning and the unfinalizable: English educators writing and telling stories together. Chang. Engl. **22**(2), 157–175 (2015)
9. Silberstein, M., Chemero, A.: Complexity and extended phenomenological-cognitive systems. Top. Cogn. Sci. **4**(1), 35–50 (2012)
10. Ullmann, T.D.: Automated analysis of reflection in writing: validating machine learning approaches. Int. J. Artif. Intell. Educ. **29**(2), 217–257 (2019). https://doi.org/10.1007/s40593-019-00174-2
11. Willis, J., Gibson, A.: The emotional work of being an assessor: a reflective writing analytics inquiry into digital self-assessment. In: Fox, J., Alexander, C., Aspland, T. (eds.) Teacher Education in Globalised Times, pp. 93–113. Springer, Singapore (2020). https://doi.org/10.1007/978-981-15-4124-7_6
12. Wulff, P., Mientus, L., Nowak, A., Borowski, A.: Utilizing a pretrained language model (BERT) to classify preservice physics teachers' written reflections. Int. J. Artif. Intell. Educ. (2022). https://doi.org/10.1007/s40593-022-00290-6

Algebra Error Classification with Large Language Models

Hunter McNichols(✉), Mengxue Zhang, and Andrew Lan

University of Massachusetts Amherst, Amherst, USA
wmcnichols@umass.edu

Abstract. Automated feedback as students answer open-ended math questions has significant potential in improving learning outcomes at large scale. A key part of automated feedback systems is an error classification component, which identifies student errors and enables appropriate, predefined feedback to be deployed. Most existing approaches to error classification use a rule-based method, which has limited capacity to generalize. Existing data-driven methods avoid these limitations but specifically require mathematical expressions in student responses to be parsed into syntax trees. This requirement is itself a limitation, since student responses are not always syntactically valid and cannot be converted into trees. In this work, we introduce a flexible method for error classification using pre-trained large language models. We demonstrate that our method can outperform existing methods in algebra error classification, and is able to classify a larger set of student responses. Additionally, we analyze common classification errors made by our method and discuss limitations of automated error classification.

Keywords: Error Classification · Large Language Models · Math Education

1 Introduction

Quality math education, particularly at a young age, is of crucial importance for students growing into an increasingly technology-driven world. Intelligent tutoring systems (ITSs) have demonstrated their effectiveness in improving math learning outcomes [1,10,24]. Many ITSs have a component that provides automated feedback for students while they solve math questions. This component enables teachers to provide personalized feedback at scale, since they can write feedback once that applies to many students. Moreover, in some ITSs, this feedback can direct a student to the precise error in their problem solving process. This direct, timely feedback enables immediate learning and helps students refine their math skills at their own pace [21].

One crucial part of feedback generation is the task of error classification, i.e., detecting student errors and the class corresponding to the error type. Once detected, the error class then informs the feedback generator which feedback is

N. Wang et al. (Eds.): AIED 2023, LNAI 13916, pp. 365–376, 2023.
https://doi.org/10.1007/978-3-031-36272-9_30

appropriate to provide to the student. This is especially important for open-ended math questions, where students have to reason step-by-step and one error can lead a student away from the correct solution. A related line of work is automated grading of open-ended math responses [2,3,9,15,31], which is closely related to the automated short answer grading (ASAG) and automated essay scoring (AES) tasks [5,20,27,28]. However, student error classification has an important difference from automated scoring: the former operates at a finer-grained level, focusing on individual solution steps, while the latter operates at a higher level, focusing on one score for the entire response.

Traditionally, error classification systems require a domain expert to develop hand-crafted rules that outline the different patterns of errors possible in a question [4,11,13]. As a result, these systems are resource-intensive to create and do not generalize to responses that are not foreseen by the system designers. Data-driven methods have the potential to generalize to a wider range of responses, but few have been explored and most methods only process a final solution and not intermediate steps. The leading data-driven method to intermediate error classification relies on a tree embedding method [32]. This method is highly restrictive since it requires mathematical expressions in student responses to be converted to syntax trees [29]. Student solutions (particularly incorrect ones) are not always syntactically valid, which means they cannot be processed by this method. Additionally, this method only uses equations as input and cannot be easily extended to include other student response information, such as natural language text or ITS-recorded interactions.

Contributions. In this paper, we propose a method that overcomes the above limitations by using pre-trained Large Language Models (LLMs) for algebra error classification. Our contributions are:

- We outline a method towards using pre-trained LLMs for algebra error classification, which enables us to handle any student response, regardless of syntactic validity.
- We demonstrate the effectiveness of this method through experimental evaluation using an algebra problem-solving dataset. We compare various pre-trained LLMs to the existing data-driven baseline, and show that BERT (and no other LLM) outperforms it.
- We showcase the flexibility of our method in that we can easily add more information into our classification system, which results in further performance improvement. We show how one can augment our method by adding ITS-recorded interactions to the input, or through Domain Adapation (DA) that incorporates further domain-specific pre-training to the LLM.
- We perform qualitative analysis on the cause of classification errors in our method and discuss avenues of future work.

2 Methodology

In this section, we outline how we formulate the problem of algebra error classification. Then, we outline our method, which uses a LLM to classify errors.

2.1 The Algebra Error Classification Task

The setup of the algebra error classification problem in an open-ended math setting is as follows. We have a set of predefined error classes for a category of math questions $C = \{c_1 \dots c_m\}$. We consider a solution that a student provides to a math question as a series of steps $S = \{s_1, \dots, s_n\}$ (for simplicity of exposition, question statement can be included in the first step, s_1). A step encompasses a resulting intermediate equation and descriptions of the process by which a student comes to this result. This description includes, but is not limited to, the application of a theorem, transformation, or intermediate calculation. Assume that, during step s_t, a student makes an error. We aim to identify the type of error c_i made by a student at s_t given the step history $S_t = \{s_1 ... s_t\}$.

Concretely, we consider a question from an introduction to algebra class:

$$\text{Solve for } x: \; x + 4 = 8$$

With a student's solution:

$$\begin{aligned} &\text{Step 1: } x + 4 = 8 \\ &\text{Step 2: } x + 4 + 4 = 8 + 4 \\ &\text{Step 3: } x = 12 \end{aligned}$$

In this case, an error occurs at s_2, where a student adds to both sides instead of subtracting. While this is a valid algebra, it indicates the student is not progressing towards the solution. The task is to identify the class c_i which corresponds to a "wrong operation" error. As input, we are given $S_t = \{s_1, s_2\}$. A classification could also be done for the addition error at s_3 using $S_t = \{s_1, s_2, s_3\}$.

It is important to note that step information is not limited to the equation. It could include any additional information available about how the student arrived at that equation. For example, notes in the margin between equations or ITS-recorded interactions which indicate student intent.

2.2 Our Method

Our method to solve the algebra error classification task is to fine-tune a pre-trained LLM, which we detail below.

Model Architecture. We experiment with two main types of LLMs: those pre-trained on the masked language modeling (MLM) objective, such as BERT [8], and those pre-trained on next token prediction in an autoregressive way, such as (GPT-2) [22]. Each type requires a different setup to be used as an error classifier.

For MLM-type LLMs, the output contains the vector embedding of a [**CLS**] token. This token represents the contextual information of the entire input text sequence. As such, these models are designed for sequence classification. To form our classification head, we connect a linear layer from the [**CLS**] token to a probability space, which is the size of our number of error classes. For autoregressive

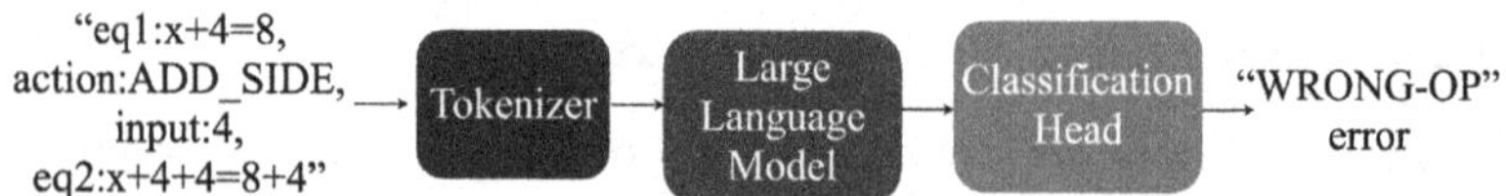

Fig. 1. Architecture diagram for our LLM-based method. Input is an arbitrary string for a set of steps. We depict the action information described in Sect. 3.1 as additional step information. Output is a class corresponding to error detected in student response.

models, such as GPT-2 [22], there is no **[CLS]** token. To form our classification head, we instead connect a linear layer from the final hidden state representation (i.e. representation after the entire input sequence is encoded) to a probability space, which is the size of our number of error classes (Fig. 1).

As input to the LLM, we tokenize a string that represents the step history. In our experiments, we try different methods of representing step history and compare the results. We note that input formulation is the most flexible part of our architecture and one of the main advantages of this method. First, it does not constrain the student responses to be a valid syntax tree. For example, response of "$x + 3 =$ unicorn" is still a valid string. Second, it can be easily adapted to include further information. In our experiments, we only represent information from two steps s_{t-1} and s_t. This representation is used to provide fair comparison to the leading data-driven method, which only considers the current and prior equation trees. However, this method could be easily extended to include the entire sequence history S_t or responses to prior questions.

Task-Specific Training. To train our model for the task of algebra error classification, we minimize the standard cross-entropy loss [19] for error class predictions. We backpropagate this loss to learn the linear classification head from scratch and fine-tune the LLM. We also experiment with performing DA prior to fine-tuning. In DA, we perform additional pre-training on the LLM to adapt it to the domain-specific vocabulary of our task. This is performed with a domain-related dataset. In our case, we use a subset of the AMPS Mathematica dataset [12] containing solutions to PEMDAS questions generated by Mathematica. This DA step adds additional flexibility to our method, since we can readily adapt the LLM to more complex math questions, such as calculus or probability theory. This pre-training may make the LLM more familiar with advanced mathematical notation and reasoning. We leave this exploration for future work.

3 Experimental Evaluation

In this section, we detail our experiments and analyses. First, we show our quantitative experiments that compare the performance of our method to the performance of the baseline method. Then, we show quantitative experiments on two variations of our method, which use different strategies to include additional contextual information. Finally, we perform a qualitative error analysis on our best performing method.

3.1 Dataset Details

For our experiments, we use the CogTutor dataset accessed via PSLC Datashop [14,23]. The dataset contains logs of students' step-by-step responses to questions in an Algebra I course. These logs were recorded during student interactions with an ITS named Cognitive Tutor. In Cognitive Tutor, students select from a set of predefined actions to manipulate an algebra equation. After selecting an action, the student inputs the expected resulting equation. Interactions are present in the dataset for each part of this process. We group the action interaction and the equation writing interactions into a single step.

The Dataset, in total, contains 130,823 interactions. These interactions span 9590 responses across 95 unique students. No demographic information is provided about the students. From the total interactions, we use a grouped subset of 5,744 steps that have a detected student error called *BUG* [32]. In these cases, the ITS showed an automated feedback message to the student indicating an error made during the equation-solving process. These messages are based on a set of 92 predefined rules in the ITS. The feedback messages vary in length from short hints, such as "Check your sign." to more descriptive feedback, such as "You are dividing a positive by a negative. The result should be negative".

Baseline Labeling. For our initial comparisons, we use the step labeling introduced in the leading data-driven method which has 24 distinct labels [32]. This labeling was present for 3,318 of the 5,744 *BUG* steps; the remaining 2,426 steps contained equations which were not compatible with the Tree-Embedding approach utilized in the baseline method.

New Labeling. In subsequent experiments, where we compare variations of our method, we utilize our own labeling of the *BUG* steps, which has 19 labels on 5,339 of the 5,744 *BUG* steps. The remaining 403 steps were dropped from our analysis since they were either i) system errors not related to math operations or ii) rare errors that were present less than 30 times across the entire dataset. We make this new dataset, called *CogTutorBugs*, publicly available[1].

3.2 Metrics and Baselines

For error classification, our primary metric is classification accuracy, i.e., percentage of correctly predicted classes in the test set. We compare the following:

- **TE+C** Best performing baseline method introduced in prior work [32]. This method converts intermediate equations in student response steps into tree-embeddings [29], which are used to learn math operation embeddings. The operation embeddings are then used in a supervised learning manner for error classification.

[1] https://github.com/umass-ml4ed/CogTutorBugs.

- **GRU+C** Additional method explored in the baseline paper. Concatenates the equations together and passes the combined sequence character-by-character into a GRU [6]. Final output state is used for error classification.
- **BERT** Bidirectional encoder representations from transformers (BERT) [8]. The predominant MLM-based LLM, with a format that is well-suited for classification tasks.
- **GPT-2** Generative Pre-trained Transformer 2 (GPT-2). An autoregressive LLM pre-trained for next sequence prediction instead of MLM. GPT-2 is well-suited for text generation tasks, but can be used for classification [22].
- **MathBERT** A BERT-based LLM pre-trained on a corpus of mathematics textbooks and course material [26].
- **XLM-RoBERTa** Multilingual variation of RoBERTa [17], a BERT-based LLM pre-trained on specific natural language processing benchmark tasks such as question answering, reading comprehension, and natural language understanding. XLM-RoBERTa often outperforms BERT on these tasks [7].

3.3 Implementation Details

We perform five-fold cross-validation on the *BUG* subsets of the CogTutor dataset. We use four folds for fine-tuning the LLMs and reserve the final fold for calculating test accuracy. For all experiments we train our models for 30 epochs. We report the mean and standard deviation in final-epoch test accuracy across the five folds.

All pre-trained LLMs are sourced from the HuggingFace [30] transformers library. For fine-tuning we use the AdamW optimizer [18], batch size of 300, and learning rate of $5 \cdot 10^{-5}$. We do not perform hyperparameter tuning and use fixed hyperparameters across all models. All models were fine-tuned for 30 epochs. We found this number of epochs was sufficient to obtain optimal performance. All models were fine-tuned on a single NVIDIA RTX 8000 GPU. For each fold, fine-tuning took between 4 and 9 min, depending on the selected LLM.

3.4 Results and Analysis

Table 1. Classification accuracy, using the error labeling in [32], for our method with selected LLMs and the baseline methods. **Bold** indicates the best result.

Method	Accuracy ↑
GRU+C	75.35 ± 1.41
TE+C	78.71 ± 1.74
BERT	$\mathbf{80.68 \pm 1.1}$
GPT-2	71.10 ± 2.58
MathBERT	72.66 ± 2.21
XLM-RoBERTa	72.75 ± 1.54

Comparison to Baseline Methods. We report the accuracy of our method with each LLM and the baseline methods in Table 1. We observe that BERT has the best performance across all LLMs and is a slight improvement over the leading baseline method (TE+C). Moreover, it achieves this performance without needing the tree-embeddings of the response equations used in TE+C. Perhaps surprisingly, other LLMs do not perform as well as either baseline method. This observation suggests that the choice of the LLM architecture and initial weights configuration is crucial for the success of LLMs in classification tasks. BERT appears to be more robust in fine-tuning for the specific task of algebra error classification. On the contrary, GPT-2 isn't well-suited for classification likely due to its autoregressive design. XLM-RoBERTa and MathBERT are pre-trained on other tasks and show limited ability to generalize to our task. Their less competitive performance suggests that pre-training an LLM for one task may degrade performance in another, completely different task. We find it surprising that MathBERT, which is specifically designed for mathematical content, does not perform well on our task. One possible explanation is that MathBERT was pre-trained on textbook content, which can be different from problem-solving content. It also is pre-trained on student work in other ITSs, which may differ in structure to Cognitive Tutor.

Table 2. Classification accuracy, using our new error labeling, for our method with selected LLMs. **Control** is the accuracy only with equation information in the input. **Action Inc.** is the accuracy with additional student action context provided by Cognitive Tutor ITS logs. **Bold** indicates the best result.

Language Model	Control ↑	Action Inc. ↑
BERT	82.02 ± 1.09	$\mathbf{85.90 \pm 1.86}$
GPT-2	76.46 ± 1.47	74.21 ± 2.08
MathBERT	76.08 ± 2.05	81.27 ± 0.32
XLM-RoBERTa	77.73 ± 2.59	84.23 ± 2.65

Use of Additional Context. An advantage in using a LLM-based method for the error classification task is the flexibility of the input. Therefore, we experiment with including information about the student-selected action in the input (in addition to the step equations). This action information is recorded by Cognitive Tutor and provides context on student intent. We report the classification accuracy for the selected LLMs in Table 2. For these experiments, we use our new feedback labeling. We observe that introducing action information improves the performance of all MLM models significantly. XLM-RoBERTa, in particular, has the largest improvement of 6.5%. We note that GPT-2 actually shows a small performance deterioration when action information is included. This deterioration is small and could be explained by random noise in the experiments. Another potential explanation is that GPT-2 is pre-trained on natural language text, but the action information is programmatic, system log text.

Table 3. Classification accuracy for BERT with various degrees of Domain Adaptation (DA). For example, BERT + 3 Epoch DA indicates 3 epochs of DA before classification fine-tuning and testing.

Model	Accuracy ↑
BERT	82.02 ± 1.09
BERT + 3 epoch DA	81.10 ± 2.40
BERT + 10 epoch DA	81.00 ± 1.72

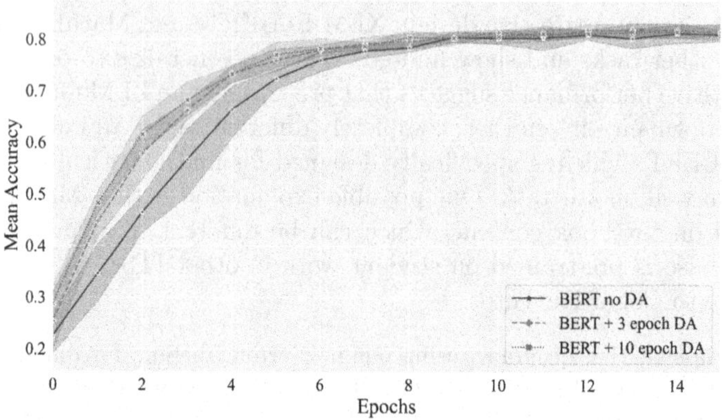

Fig. 2. Test accuracy comparison of DA methods. We observe only a slight improvement in initial epochs with DA and no significant difference in final performance (compared to no DA). The graph only shows first 15 epochs for clearer observation of early epochs. Shaded regions indicate range of accuracy across all folds. Best viewed in color.

Domain Adaptation. LLMs, such as BERT and GPT-2, are pre-trained on a variety of textual data but do not have domain-specific vocabulary for middle-school-level math. To adapt the LLM to better understand domain-specific vocabulary, we perform additional pre-training with in-domain but not task-related data. This step is done prior to task-specific fine-tuning and often called Domain Adaptation (DA). In Table 3, we detail our experiments on performing DA with the PEMDAS subset of the AMPS Mathmatica dataset [12]. We observe no significant change in overall accuracy with the introduction of either 3 or 10 epochs of DA. One possible explanation is the small vocabulary size of our domain: only numbers and fundamental math operations: $+, -, /, *, =, ()$. However, we do see a non-trivial performance improvement in initial epochs of training as depicted in Fig. 2. This observation suggests that DA does provide some context on basic math vocabulary and mathematical reasoning, but this context is sufficiently learned from task fine-tuning within the first few epochs.

Qualitative Error Analysis. We now analyze the cases when our best trained method (BERT + Action) mis-classifies student errors in a sample cross-validation fold. We show a mis-classification heatmap in Fig. 3. We observe

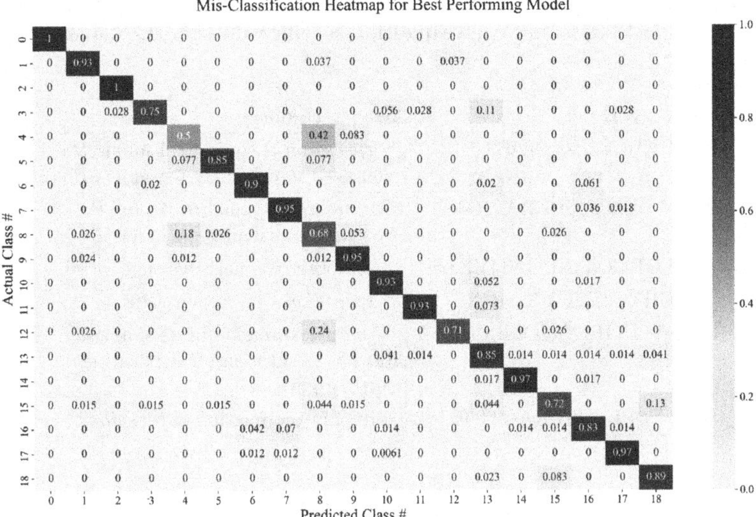

Fig. 3. Mis-classification heatmap for qualitative error analysis of our method. Value in each cell is the ratio of the predicted class count to total number of actual class instances in the test set (each row sums to 1). Best viewed in color.

two major causes of mis-classification: between two error classes where intent is ambiguous and between classes that are equivalent, but vary in level of specificity. Detailed information about each class mentioned is provided in Table 4.

The most frequent mis-classification is between error class #4 and #8 (REVERSED_SIDES and WRONG_OPERATION respectively). Class #4, 41% of the time, is incorrectly classified as class #8. 18% of the time the reverse mis-classification occurs. We observe that, in some scenarios, a REVERSED_SIDES error is equivalent to a WRONG_OPERATION error. Consider the question $y + 5 = -2$ and a student response $y = -3$. The student could have attempted to subtract 5 from both sides, but dropped the negative on -2. Equivalently, the student could have attempted to subtract -5 from both sides, erroneously thinking subtraction of -5 would cancel the 5 on the left hand side and confusing the rules of subtracting negatives. Both error classes can be appropriate, depending on the student's thought process. However, the true student intent is unknown, so it's ambiguous which class is correct.

Other frequent mis-classifications occur between classes that are almost the same but different in level of specificity. For example, class #12 (FORGOT_NEGATIVE) is 23.6% of the time mis-classified as class #8 (WRONG_OPERATION). In the case of addition and subtraction, a sign error can be explained by the selection of an incorrect operation. In that case, an operation error is a more specific error than the sign error. A similar type of mis-classification occurs 13.2% of the time from class #15 to class #18, and 11% of the time from class #3 to class #13.

Table 4. Example feedback templates for classes described in qualitative error analysis. The actual feedback students see have question-specific values for A, B, and *expression*.

#	Class Label	Sample Feedback
4	REVERSED_SIDES	"*expression* is equal to A minus B. You need to calculate $-A$ minus B"
8	WRONG_OPERATION	"*expression* is equal to A plus B. You need to calculate A minus B"
12	FORGOT_NEGATIVE	"You forgot the negative sign"
15	CHECK_YOUR_SIGN	"*expression* Check your sign"
3	MULTIPLY_TO_SIMPLIFY	"Put the expression in its simplest form by performing multiplication on the right side"
13	SIMPLIFY_FRACTION	"Simplify the fraction *expression*"

Through this qualitative analysis, we observe the most frequent misclassifications of our method are explainable. They often occur because of ambiguous or overlapping classes. These observations suggest our method may be performing better than the accuracy metrics indicate. The observations also suggest that there is an upper limit to the performance possible on our chosen dataset, depending on how error classes are defined. Furthermore, the feedback selected by the rules of the Cognitive Tutor ITS should not be seen as the absolute truth. Without a perfect knowledge of student intent for each step of a response, it is impossible to have certainty about the true cause of student error. This is a general limitation which underlies all error classification systems.

4 Discussion, Conclusion, and Future Work

In this paper, we detailed a large language model-based method for the task of math error classification in open-ended questions. We now discuss our key observations and outline avenues for future work. First, we observe that our method, when combined with BERT, performs better than the best-performing, data-driven method for math error classification. Additionally, since it does not rely on syntactically correct responses, the method can operate on a wider range of student responses than the baseline method. Second, we observe that incorporating additional information about student intent is easily achievable with our method and can provide a significant performance improvement. We hypothesize that providing further contextual information, such as student knowledge levels, will continue to improve performance, e.g., using models that can understand open-ended responses [16,25]. Third, we observe, through our qualitative analysis, that the errors made by our method are mainly due to ambiguous error class labels. This observation suggests an avenue for future exploration, where we extend our method to generate feedback, rather than classify errors.

Acknowledgements. The authors thank the NSF (grants 1917713, 2118706, 2202506, 2215193) for partially supporting this work.

References

1. Aleven, V., McLaughlin, E.A., Glenn, R.A., Koedinger, K.R.: Instruction based on adaptive learning technologies. In: Handbook of Research on Learning and Instruction, pp. 522–560 (2016)
2. Baral, S., Botelho, A.F., Erickson, J.A., Benachamardi, P., Heffernan, N.T.: Improving automated scoring of student open responses in mathematics. Int. Educ. Data Min. Soc. (2021)
3. Baral, S., Seetharaman, K., Botelho, A.F., Wang, A., Heineman, G., Heffernan, N.T.: Enhancing auto-scoring of student open responses in the presence of mathematical terms and expressions. In: Rodrigo, M.M., Matsuda, N., Cristea, A.I., Dimitrova, V. (eds.) AIED 2022. LNCS, vol. 13355, pp. 685–690. Springer, Cham (2022). https://doi.org/10.1007/978-3-031-11644-5_68
4. Brusilovsky, P., Peylo, C.: Adaptive and intelligent web-based educational systems. Int. J. Artif. Intell. Educ. **13**(2–4), 159–172 (2003)
5. Burstein, J.: The e-rater® scoring engine: automated essay scoring with natural language processing (2003)
6. Chung, J., Gulcehre, C., Cho, K., Bengio, Y.: Empirical evaluation of gated recurrent neural networks on sequence modeling. arXiv preprint arXiv:1412.3555 (2014)
7. Conneau, A., et al.: Unsupervised cross-lingual representation learning at scale. arXiv preprint arXiv:1911.02116 (2019)
8. Devlin, J., Chang, M.W., Lee, K., Toutanova, K.: BERT: pre-training of deep bidirectional transformers for language understanding (2018). https://arxiv.org/abs/1810.04805
9. Erickson, J.A., Botelho, A.F., McAteer, S., Varatharaj, A., Heffernan, N.T.: The automated grading of student open responses in mathematics. In: International Conference on Learning Analytics & Knowledge, pp. 615–624 (2020)
10. Fancsali, S.E., Ritter, S.: Context personalization, preferences, and performance in an intelligent tutoring system for middle school mathematics. In: International Conference on Learning Analytics and Knowledge, pp. 73–77 (2014)
11. Heeren, B., et al.: Fine-grained cognitive assessment based on free-form input for math story problems. In: Pammer-Schindler, V., Pérez-Sanagustín, M., Drachsler, H., Elferink, R., Scheffel, M. (eds.) EC-TEL 2018. LNCS, vol. 11082, pp. 262–276. Springer, Cham (2018). https://doi.org/10.1007/978-3-319-98572-5_20
12. Hendrycks, D., et al.: Measuring mathematical problem solving with the math dataset. In: NeurIPS (2021)
13. Koedinger, K.R., Anderson, J.R., Hadley, W.H., Mark, M.A.: Intelligent tutoring goes to school in the big city. Int. J. Artif. Intell. Educ. **8**, 30–43 (1997)
14. Koedinger, K.R., Baker, R.S., Cunningham, K., Skogsholm, A., Leber, B., Stamper, J.: A data repository for the EDM community: the PSLC datashop. Handb. Educ. Data Min. **43**, 43–56 (2010)
15. Lan, A.S., Vats, D., Waters, A.E., Baraniuk, R.G.: Mathematical language processing: automatic grading and feedback for open response mathematical questions. In: Proceedings of the ACM conference on learning@scale, pp. 167–176 (2015)
16. Liu, N., Wang, Z., Baraniuk, R., Lan, A.: Open-ended knowledge tracing for computer science education. In: Conference on Empirical Methods in Natural Language Processing, pp. 3849–3862 (2022)

17. Liu, Y., et al.: RoBERTa: a robustly optimized BERT pretraining approach (2019). https://doi.org/10.48550/ARXIV.1907.11692, https://arxiv.org/abs/1907.11692
18. Loshchilov, I., Hutter, F.: Decoupled weight decay regularization. arXiv preprint arXiv:1711.05101 (2017)
19. Murphy, K.P.: Machine Learning: A Probabilistic Perspective. MIT Press, Cambridge (2021)
20. Page, E.B.: The imminence of grading essays by computer. Phi Delta Kappan **47**(5), 238–243 (1966)
21. Pane, J.F., Griffin, B.A., McCaffrey, D.F., Karam, R.: Effectiveness of cognitive tutor algebra i at scale. Educ. Eval. Policy Anal. **36**(2), 127–144 (2014)
22. Radford, A., Wu, J., Child, R., Luan, D., Amodei, D., Sutskever, I., et al.: Language models are unsupervised multitask learners. OpenAI blog **1**(8), 9 (2019)
23. Ritter, S., Anderson, J.R., Koedinger, K.R., Corbett, A.: Cognitive tutor: applied research in mathematics education. Psychon. Bull. Rev. **14**(2), 249–255 (2007)
24. Roschelle, J., Feng, M., Murphy, R.F., Mason, C.A.: Online mathematics homework increases student achievement. AERA Open **2**(4), 2332858416673968 (2016)
25. Scarlatos, A., Lan, A.: Tree-based representation and generation of natural and mathematical language. In: Association for Computational Linguistics (ACL) (2023). https://arxiv.org/abs/230207974)
26. Shen, J.T., Yamashita, M., Prihar, E., Heffernan, N.T., Wu, X., Lee, D.: MathBERT: a pre-trained language model for general NLP tasks in mathematics education. CoRR abs/2106.07340 (2021). https://arxiv.org/abs/2106.07340
27. Uto, M., Uchida, Y.: Automated short-answer grading using deep neural networks and item response theory. In: Bittencourt, I.I., Cukurova, M., Muldner, K., Luckin, R., Millán, E. (eds.) AIED 2020. LNCS (LNAI), vol. 12164, pp. 334–339. Springer, Cham (2020). https://doi.org/10.1007/978-3-030-52240-7_61
28. Wang, Z., Lan, A., Waters, A., Grimaldi, P., Baraniuk, R.: A meta-learning augmented bidirectional transformer model for automatic short answer grading. In: Proceedings of the 12th International Conference on Education and Data Mining (EDM), pp. 1–4 (2019)
29. Wang, Z., Lan, A.S., Baraniuk, R.G.: Mathematical formula representation via tree embeddings. In: iTextbooks@ AIED, pp. 121–133 (2021)
30. Wolf, T., et al.: Huggingface's transformers: state-of-the-art natural language processing (2019). https://doi.org/10.48550/ARXIV.1910.03771, https://arxiv.org/abs/1910.03771
31. Zhang, M., Baral, S., Heffernan, N., Lan, A.: Automatic short math answer grading via in-context meta-learning. arXiv preprint arXiv:2205.15219 (2022)
32. Zhang, M., Wang, Z., Baraniuk, R.G., Lan, A.S.: Math operation embeddings for open-ended solution analysis and feedback. CoRR abs/2104.12047 (2021). https://arxiv.org/abs/2104.12047

Exploration of Annotation Strategies for Automatic Short Answer Grading

Aner Egaña, Itziar Aldabe(✉), and Oier Lopez de Lacalle

HiTZ Center - University of the Basque Country UPV/EHU, San Sebastian, Spain
{aegana028,itziar.aldabe,oier.lopezdelacalle}@ehu.eus

Abstract. Automatic Short Answer Grading aims to automatically grade short answers authored by students. Recent work has shown that this task can be effectively reformulated as a Natural Language Inference problem. State-of-the-art is defined by the use of large pretrained language models fine-tuned in the domain dataset. But how to quantify the effectiveness of the models in small data regimes still remains an open issue. In this work we present a set of experiments to analyse the impact of different annotation strategies when not enough training examples for fine-tuning the model are available. We find that when annotating few examples, it is preferable to have more question variability than more answers per question. With this annotation strategy, our model outperforms state-of-the-art systems utilizing only 10% of the full-training set. Finally, experiments show that the use of out-of-domain annotated question-answer examples can be harmful when fine-tuning the models.

Keywords: Automatic Short Answer Grading · Natural Language Processing · Natural Language Inference · Transfer Learning

1 Introduction

Automatic content scoring is an important application in the area of automatic educational assessment. In this context, the evaluation of short answers authored by students is referred to as Automatic Short Answer Grading (ASAG) and the available datasets usually consist of questions, reference answers and student answers. Current state-of-the-art in Natural Language Processing (NLP) has shown that task reformulation (e.g., transforming specific tasks as Natural Language Inference or Question Answering) is an effective way to transfer knowledge across tasks and improve results [15,16]. Similarly, recent work in ASAG has demonstrated that reformulating the ASAG as an entailment problem is an effective method to obtain strong results [3].

Methods that fine-tune large pretrained language models (LM) with large amounts of labelled data have established the state-of-the-art [10]. Nevertheless, due to differing languages, topic of questions, grading scale and the cost of human annotation, there is typically only a small number of labelled examples in real-world applications—and these models perform poorly. As an alternative,

N. Wang et al. (Eds.): AIED 2023, LNAI 13916, pp. 377–388, 2023.
https://doi.org/10.1007/978-3-031-36272-9_31

methods that require few (few-shot) or no (zero-shot) examples have emerged. Still, the way we should select training examples is an open question in ASAG. In this paper we focus on using entailment models to explore zero- and few-shot learning in student short answer grading. We define different scenarios where we assume there are no sufficient training examples for fine-tuning the model, and pose the following research questions in order to devise better strategies for data annotation:

RQ1 Having a task-agnostic generic entailment model, what would be the best way to annotate data and how much data would be needed to obtain state-of-the-art results?
RQ2 Can we effectively transfer task knowledge to new domains? And similarly, having a NLI-based fine-tuned model in one domain, how much data do we need to be annotated in a new one?

We attempt to answer the questions stated above empirically conducting experiments in the Semeval-2013 SRA dataset [5] and make the following contributions: 1) We show that annotation strategy can have a significant impact on results. This is because the annotation that increases the variability on the question side, at the cost of decreasing the amount of annotated answers per question, is preferable to having the same number of annotated examples with fewer questions and more answers. 2) Reformulating ASAG as an entailment problem and fine-tuning an entailment model allows us to obtain state-of-the-art results. 3) Related to this, we demonstrate that zero-shot entailment models can perform close to state-of-the-art results.4) We illustrate that the impact of the domain can be larger than the knowledge that can be acquired from the task. That is, using a generic entailment model is more effective than fine-tuning it with out-of-domain examples.

2 Related Work

Current approaches for ASAG can be categorized into three types [8]: 1) Hand-engineered feature-based machine learning (ML) approaches, which still get competitive results, 2) supervised deep learning approaches that fit parameters directly from training, and 3) large pretrained language models fine-tuned on the target task.

Hand-crafted features rely on the extraction of features from the questions and reference and student answers in order to find lexical, syntactic and semantic similarities between the student answer with the reference answer [9,12].

Deep-learning models contributed to significantly improving results in ASAG. They provide the opportunity to learn from different related tasks (e.g. transfer learning) and representations (e.g. feature types). For instance, [14] trained a bidirectional LSTM on the SNLI dataset [2] and adapted the feature extraction in combination with hand-crafted features for ASAG in the Semeval-2013 dataset [5].

Regarding transfer learning, today's state-of-the-art in ASAG is defined by the use of large pretrained language models fine-tuned in the specific ASAG dataset [3]. The main difference between these approaches comes with the selection of the pretrained language model and the strategies to fine-tune it. [10] explore the potential of using T5 and XLNET, among others, as pretrained models for ASAG, and [17] proposed new ways to enhance the performance of BERT by further pretraining it as a language model on domain specific data such as textbooks and use labeled automatic short answer grading data.

Textual Entailment as a Pivoting Task. The task of Textual Entailment, better known as Natural Language Inference (NLI), was first introduced by [4]. Given a textual premise and hypothesis, the task consists in classifying whether the premise entails or contradicts (or is neutral to) the hypothesis. The current state-of-the-art uses large pretrained LMs fine-tuned in NLI datasets [18].

Textual Entailment has been shown to be useful as a pivot task for few/zero-shot learning. For instance, entailment models are highly effective for text classification [16], and Information Extraction tasks [15], among others. The core idea relies on recasting the task at hand as an entailment task in which the original input is transformed into a classification problem of entailment, contradiction or neutral. Pretraining LMs in existing large entailment datasets and recasting the task into an entailment problem has demonstrated that it is possible to reduce annotation effort and still obtain state-of-the-art results [15].

Regarding ASAG, [3] explore the effect of transfer learning by fine-tuning a variety of pretrained LM models on the Semeval-2013 dataset. In a similar fashion to us, they also explored the impact of transfer learning with a model fine-tuned on the MNLI dataset [19]. They showed that models trained on this dataset are capable of transferring knowledge to the task of short answer grading, but did not quantify the effectiveness of the model in small data regimes. Other research lines are also appearing. To mention a few, [13] explore how to evaluate an automated grader in small-scale testing scenarios to help teachers and students in the use of such systems, [1] propose a similarity-based model, and [6] examine human-in-the-loop frameworks to guarantee grading quality.

3 Entailment Based Answer Grading

3.1 Problem Formulation

The Automatic Short Answer Grading task can be defined as follows. Given a triplet of *question*, *reference answers*, and *student answers* as input in our system, the system must assess the student answer by classifying it with a label that denotes the degree of correctness. We conduct our experiments in SciEntsBank and Beetle datasets, which were made available in the Semeval-2013 task 7 [5] and are one of the most used datasets. Note that Semeval-2013 datasets include three sets of labels that correspond to 2-, 3- and 5-way task problems, respectively. In this paper, we focus on the 3-way task, in which each answer is labeled as either correct, contradictory, or incorrect. Figure 1 depicts an example of a correct answer to a given question and the corresponding reference answer.

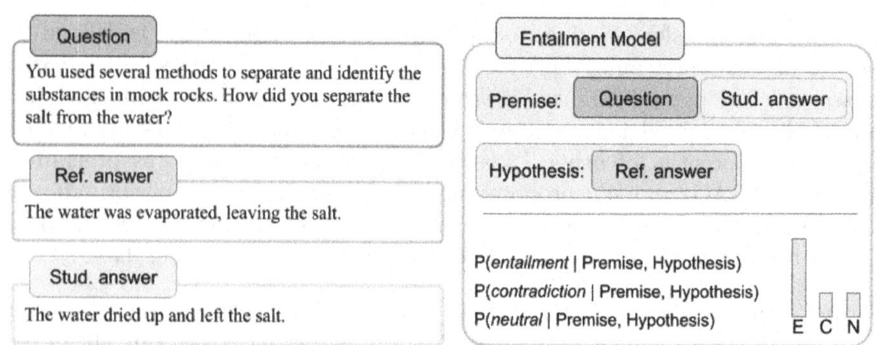

Fig. 1. Schema of the NLI-based ASAG model where the input of *question*, *reference answer*, and *student answer* are reformulated as an entailment model. Concatenation of the question and student answer form the *premise* of the NLI model, whereas the *hypothesis* is generated with the reference answer. Prediction of the entailment model is then mapped to the ASAG 3-way label.

Evaluation Scenarios. The SemEval-2013 challenge gives three different test scenarios in order to evaluate model generalization capabilities across problems and domains:

- **Unseen answers** (UA): A set containing held-out student answers from questions which are available for training the system and contain some other student answers.
- **Unseen questions** (UQ): A set containing held-out questions in order to assess the system in non-seen questions, but still laying in the same domain as the one used for training.
- **Unseen domains** (UD): Available only for SciEntsBank, a domain independent test set of responses to topics not seen in the training data.

3.2 Model Description

According to the standard definition of Textual Entailment, given two text fragments called Premise (P) and Hypothesis (H), P entails H if, typically, a human reading P would infer that H is most likely true [4]. In a typical answer assessment scenario, we expect that a correct student answer would entail the reference answer, while an incorrect answer would not. However, students often skip details that are mentioned in the question or may be inferred from it, while reference answers often repeat or make explicit information that appears in or is implied from the question [5]. Hence, a more precise formulation of the task in this context considers the entailing text P as consisting of both the original question and the student answer, while H is the reference answer.

Figure 1 shows the schema of our entailment-based ASAG model, where the input of *question*, *reference answer*, and *student answer* are reformulated as a textual entailment problem. Concatenation of the question and student answer form the premise (P) of the NLI model, whereas the hypothesis (H) is created using the reference answer.

In our experiments we focus on the 3-way classification task so the predictions of the entailment model are mapped to the 3-way set of labels in the Semeval-2013 dataset. That is, the predictions of *entailment*, *contradiction*, and *neutral* of the NLI model are mapped to `correct`, `contradictory` or `incorrect`, respectively.

3.3 Fine-Tuning the ASAG Model

We take advantage of NLI's ability to represent other NLP downstream tasks, ASAG in this case. Taking a large LM fine-tuned on MNLI as a base (RoBERTa-MNLI) [11], fine-tuning the ASAG model is carried out by reformulating the triplets (question, reference answer, student answer) provided in both SciEntsBank and Beetle datasets as traditional inference pairs (premise, hypothesis) as shown in Fig. 1.

4 Annotation Strategies

The paper's main contribution is to explore the effectiveness of different annotation strategies when there is a need to have new annotated examples. RQ1 not only deals with data quantity, but also selects new samples to effectively save time and effort. Similarly, RQ2 takes into account the importance of selecting unseen data wisely in order to take advantage of the annotation to the fullest extent possible.

In order to answer these research questions, we explore two strategies of data annotation using the SemEval-2013 dataset. As the dataset has multiple student answers for a given question, the sampling of labeled data can be done answer- or question-wise. Specifically, we define two ways for sampling the training set of our experiments:

- *One question per student (1Q1S)* This scenario annotates a unique question and student answer pair. That is, if we had 10 students, we would create 10 different questions and would have 10 different answers. The goal of this strategy is to increase the variability of the questions, losing the capacity to generalize over the answers. Note that having very few examples for a given question might necessarily be a better strategy. Note as well that in some cases it is not possible to sample the defined dataset as there are not enough questions in the dataset. In those cases, we tried to generate an approximated dataset as in the previous strategy.
- *One question for all students (1Q4A)* This scenario annotates multiple answers for a single question with the goal of having larger variability on the answers side. That is, if we had 10 students, we would create and ask a single question to all the students in order to get many answers for the question. Note that in most cases there are not enough answers for a single question, so we attempted to sample a dataset that approximated it as much as possible.

Table 1 shows the number of questions and student answers that each few-shot setting contains. As can be seen, the ideal 1Q1S and 1Q4A annotations are not always collected since there are limited questions and student answers per question. Even so, each annotation strategy aims to add more variability to either the questions or the student answers at hand.

Table 1. Number of questions (#Q) and student answers (#A) for each few-shot scenario according to the specific annotation strategy (Ann.) as well as the number of training examples (Total) for each few-shot setting and dataset. FT stands for full training.

Dataset	Ann.	1%			2%			5%			10%			FT
		#Q	#A	Total	#Q	#A	Total	#Q	#A	Total	#Q	#A	Total	Total
SciEntsBank	1Q1S	40	1	40	80	1	80	100	2	200	100	4	400	3966
	1Q4A	4	10		8	10		20	10		40	10		
Beetle	1Q1S	28	1	28	28	2	56	35	4	140	35	8	280	2833
	1Q4A	1	28		1	56		2	70		5	56		

5 Experimental Setting

We use the data provided in the SemEval-2013 shared task for our experiments. As explained above, the dataset consists of two distinct subsets: SciEntsBank and Beetle. The former is based on a corpus of student answers to assessment questions collected in 15 science domains, whereas the latter is based on transcripts of students interacting with the Beetle II dialogue system in the basic electricity and electronics domain. Although both subsets show similar structure, Beetle contains more than one reference answer for each question[1] while a single reference answer is given in SciEntsBank. SciEntsBank includes 150 assessment questions with 150 reference answers and 6242 student answers in total. By comparison, Beetle is a smaller subset, which includes 56 questions, 283 reference answers, and 5199 student answers in total.

Since there is no validation set in the SemEval-2013 dataset, we created one by separating some examples from the original training set. We obtained a specific validation set for each test scenario. For the UA scenario, the selection of validation examples was done answer-wise and we held out a set of student answers for questions existing in the training part. For UQ, the selection was carried out question-wise and we selected the same number of questions that were extracted for the test set. We sample 15 and 9 questions from the SciEntsBank and Beetle training datasets, respectively. As all the training data belongs to the same domain, it was not possible to create a validation set that met the conditions for the UD test scenario. In all the cases, we select the validation examples so that class distribution is kept as similar as possible to the training

[1] We use one reference answer in our experiments.

Table 2. Number of validation examples for unseen answers and unseen question scenarios.

	SciEntBank		Beetle	
	UA	UQ	UA	UQ
#Questions	120	15	38	9
#Reference answers	120	15	170	35
#Student answers	472	531	351	757

dataset. Table 2 displays the sizes of the validation sets in terms of number of questions, reference answers, and student answers for each dataset and test scenario.

In order to measure the effectiveness of the annotation strategies in different few-shot scenarios, we generate the same training sizes for 1Q1S and 1Q4A. We created samples of 1%, 2%, 5%, 10% of the remaining training set and we reduced the validation set according to the same ratio. Table 1 shows the number of training examples for each few-shot scenario in the *Total* columns. Although FT denotes full training, it actually contains 1003 and 1108 fewer examples than the original training set as a consequence of utilizing a certain amount of training examples as a validation set.

As in [3], we used the RoBERTa large [11] fine-tuned on the MNLI dataset as the base model for our zero- and few-shot experiments. The model is publicly available at Huggingface[2]. We performed the following hyperparameter exploration for each few-shot scenario: We ran our model for 25 epochs with a batch size of 4 and selected the best learning rate between 1e−5, 5e−5, and 4e−6, as well as the best gradient accumulation between 8 and 32. For the model selection we took the checkpoint with the lowest loss (cross-entropy) value in those 25 epochs.

6 Few-Shot Experiments

Table 3 shows the results of the effect of the annotation strategies in the few-shot scenario and seeks to answer the question the question concerning what would be the best strategy to annotate new data with a pretrained NLI model. It displays the macro F-score for the few-shot experiments in which we fine-tune an entailment model (NLI-roberta) using 1%, 2%, 5%, and 10% of training data and evaluated in unseen answers (UA), unseen questions (UQ), and unseen domains (UD).

The results indicate that, overall, increasing the number of annotated questions at the cost of reducing the number of different answers (1Q1S) seems to be the best strategy compared to increasing the variability of answers (at the cost of reducing the variability of seen questions) when annotating new question-answer pairs. This trend is confirmed in the bottom rows of the table, where

[2] https://huggingface.co/roberta-large-mnli.

Table 3. Results for the few-shot experiments. 1Q1S annotation correspond to training data where we annotate one question per student and 1Q4A correspond to the one question for all students annotation procedure.

Domain	Scenario	Annotation	0%	1%	2%	5%	10%	FT
SciEntsBank	UA	1Q1S	56.2	**59.5**	**63.2**	63.9	**67.0**	71.0
		1Q4A		58.3	60.0	**64.1**	59.6	
	UQ	1Q1S	65.8	**67.0**	**66.7**	64.4	64.2	68.6
		1Q4A		62.7	65.6	**65.9**	**66.8**	
	UD	1Q1S	59.0	57.9	58.7	**58.8**	**61.2**	67.6
		1Q4A		**58.2**	**59.2**	56.0	58.2	
Beetle	UA	1Q1S	51.0	50.0	**52.3**	**52.7**	**56.6**	73.8
		1Q4A		**50.1**	50.8	52.5	51.5	
	UQ	1Q1S	36.1	**37.0**	36.8	**38.0**	**43.1**	61.8
		1Q4A		34.8	**37.8**	36.5	37.1	
Overall F-score		1Q1S	–	**55.7**	56.5	**56.6**	**59.2**	–
		1Q4A	–	55.2	56.5	54.6	55.8	–

we report the macro-average of each few-shot setting. In addition, results suggest that 1Q1S annotation strategy yields better generalization properties as we increase the number of examples. For instance, when we annotate 400 examples in SciEntsBank and 280 examples in Beetle (10% few-shot setting), 1Q1S outperforms 1Q4A by almost 4 points and F-score increases steadily compared to the rest of the few-shot settings.

7 Cross-Domain Experiments

Table 4 shows the results on cross domain evaluation. Results of the top rows try to answer the questions posed in **RQ2**. First, we analyse if it is better to fine-tune in a related task but in a different domain or to simply apply a zero-shot model on the new dataset. The column with the 0% headline stands for this setting in which we have an entailment-based ASAG model fine-tuned in an out-of-domain dataset (e.g. Beetle) and evaluate it in the target domain dataset (i.e. SciEntsBank). We compare the fine-tuned (task-aware) model to the zero-shot entailment base model (task-agnostic, in parenthesis) in order to measure the effect of using out-of-domain task-related examples in learning.

Contrary to our expectation, task-aware fine-tuned models obtain significantly lower results compared to the task-agnostic model that is only pretrained in the MNLI dataset and not fine-tuned in the specific task. Results suggest that the impact of the domain is larger than the knowledge that can be acquired from the task. The drop is larger in the unseen questions scenario (UQ) in both Bettle and SciEntsBank datasets. This can be explained assuming that unseen question scenarios require a higher capacity of generalization to perform better.

Table 4. Results of cross-domain few-shot evaluation. BT stands for BEETLE and SB for SCIENTSBANK. In the top rows % indicates the amount of in-domain data included in the training set, whereas the bottom rows refer to the amount of out-of-domain data.

Train → test	Scenario	0%	5%	10%
BT+%SB → SB	UA	55.8 (↓56.2)	58.9 (↓63.9)	**63.3** (↓67.0)
	UQ	59.7 (↓65.8)	62.5 (↓65.9)	**62.8** (↓66.8)
	UD	53.9 (↓59.0)	56.0 (↓58.8)	**59.3** (↓61.2)
SB+%BT → BT	UA	50.4 (↓51.0)	51.0 (↓52.7)	**54.0** (↓56.6)
	UQ	33.8 (↓36.1)	34.6 (↓38.0)	**37.9** (↓43.1)
SB+%BT → SB	UA	**71.0**	70.6	68.5
	UQ	68.6	69.8	**74.3**
	UD	**67.6**	66.7	64.0
BT+%SB → BT	UA	**73.8**	72.7	71.0
	UQ	**61.8**	59.9	55.7

In that sense, results suggest that generalization can not achieve using related tasks for transfer learning. In order to effectively transfer task related nuances the domain must be related as well.

Similarly, the results of columns 5% and 10% in the top rows of Table 4 also try to answer the question posed in **RQ2**. In this scenario we assume that we already have an entailment-based ASAG model fine-tuned in an out-of-domain dataset (e.g SCIENTBANK) and we get some annotated examples of our target domain (i.e. BEETLE). We evaluate the performance of adding target domain examples into the out-of-domain model.

Results demonstrate that adding few in-domain examples improves the outcome compared to the model trained only in the out-of-domain scenario. However, they are significantly worse compared to in-domain few-shot models (figures in parenthesis). The results are in accordance with those obtained in Table 3 and suggest that the domain differences can affect negatively even if we are modeling the same task (which is something unexpected according to some recent work [15]). That is, we can conclude that, having an entailment model, it is better to start from scratch rather than learning an out-of-domain ASAG model and retraining with a few in-domain examples.

When we defined a new setting where we do have an in-domain ASAG model (NLI model fine-tuned with target domain examples) and added some out-of-domain examples, we observed the model behaved similarly as in the previous settings. Results are shown in the bottom rows of Table 4. In general, we can conclude that mixing in-domain examples with out-of-domain examples is not helpful (only the unseen questions scenario in SCIENTSBANK obtains any improvement).

Table 5. Comparison to SOTA F-Macro results. Underlined figures denote that current results outperform previous state-of-the-art models. * for results not directly comparable with ours. Bold for best among comparable results. In ours FT and 10% experiments, the validation examples are included in the training set.

Model	SciEntsBank			Beetle	
	UA	UQ	UD	UA	UQ
CoMeT [12]	64.0	38.0	40.4	71.5	46.6
ETS [9]	64.7	45.9	43.9	71.0	58.5
(Galhardi et al., 2018) [7]	70.2	49.3	53.7	67.7	58.8
(Saha et al., 2018) [14]	66.6	49.1	47.9	–	–
(Sung et al., 2019) [17]	72.0*	57.5*	57.9*	–	–
(Camus and Filighera, 2020) [3]	78.3*	65.7*	70.9*	–	–
Ours 10% (1Q1S)	67.1	67.3	62.5	58.9	48.2
Ours FT	**76.5**	**72.3**	**69.1**	**76.7**	**70.0**

8 Comparison to the State-of-the-Art

Table 5 details the comparison of our model with state-of-the-art systems in SciEntsBank and Beetle datasets and the corresponding evaluation scenarios: Unseen answers (UA), unseen questions (UQ), and unseen domain (UD). The table is organized into three groups: 1) top rows include the best systems that took part in the Semeval-2013 shared-task, which correspond to hand-engineered feature-based systems; 2) middle rows include systems that rely on fine-tuned language models; 3) bottom rows include our model, fine-tuned using 10% of the data annotated with the 1Q1S strategy and fine-tuned utilizing the whole set of the original training examples. It is worth noting that the best-performing systems in SciEntsBank [3,17] are not directly comparable with the rest of the models as it is not clear how model selection was carried out.

Regarding our few-shot model (10%-1Q1S), results show that annotating only 10% of examples (cf. Table 1) for training following the 1Q1S strategy is effective to outperform state-of-the-art systems in SciEntsBank dataset but not in the case of Beetle. It is also remarkable that Beetle appears more demanding, as recent state-of-the-art models [7]) are not able to surpass systems that participate in the SemEval-2013 shared task.

When we fine-tune our model employing all the data available in the training set, the model yields state-of-the-art results in both datasets and shows impressive generalization capabilities in those scenarios that presumably are more challenging. For example, our few-shot model improves in 18.0 F-score points in SciEntsBank compared to the best comparable model in the unseen questions (UQ) scenario (49.3 vs 67.3) and we increase the margin up to 23.0 points when we utilize the whole training set for fine-tuning the model (ours FT). In Beetle improvements rise to 11.2 F-score points with the full-training model.

9 Conclusion

In this study we reformulate Automatic Short Answer Grading as an entailment problem and explore the extent to which annotation strategies are effective in few-shot scenarios. Experiments show that increasing the variety of questions in the annotation is more effective than annotating more answers of the same question. Our method makes effective use of available labeled examples and, utilizing only 400 annotated examples, is able to perform on par with state-of-the-art approaches in SciEntBank. Moreover, when we use full-training, our model outperforms the rest of the models in the two datasets. Our analysis indicates that employing cross-domain annotated examples is not beneficial and it is more effective to use a task-agnostic general purpose entailment model. Actually, zero-shot obtains strong results, which indicates that the reformulation of ASAG into an entailment problem can be done naturally.

In the future we hope to explore methods to improve the limitation of grading answers into a more fine-grained level of entailment (subject to some arbitrary evaluation rubrics). In that sense, using generative large language models to learn reasoning on answers assessments seems to be a promising research avenue. On the other hand, adopting active learning to find refined ways of selecting questions would be a complementary approach to be explored. It would also help in measuring the variance of the sample selection and in obtaining robust findings.

Acknowledgements. Work supported by the Basque Government (excellence research group IT1570-22) and by the DeepKnowledge (PID2021-127777OB-C21) project funded by MCIN/AEI/10.13039/501100011033 and by "ERDF A way of making Europe".

References

1. Bexte, M., Horbach, A., Zesch, T.: Similarity-based content scoring - how to make S-BERT keep up with BERT. In: Proceedings of the 17th Workshop on Innovative Use of NLP for Building Educational Applications (BEA 2022), Seattle, Washington, pp. 118–123 (2022)
2. Bowman, S.R., Angeli, G., Potts, C., Manning, C.D.: A large annotated corpus for learning natural language inference. In: Proceedings of the 2015 Conference on Empirical Methods in Natural Language Processing, Lisbon, Portugal, pp. 632–642 (2015)
3. Camus, Leon, Filighera, Anna: Investigating transformers for automatic short answer grading. In: Bittencourt, Ig Ibert, Cukurova, Mutlu, Muldner, Kasia, Luckin, Rose, Millán, Eva (eds.) AIED 2020. LNCS (LNAI), vol. 12164, pp. 43–48. Springer, Cham (2020). https://doi.org/10.1007/978-3-030-52240-7_8
4. Dagan, I., Glickman, O., Magnini, B.: The PASCAL recognising textual entailment challenge. In: Quiñonero-Candela, J., Dagan, I., Magnini, B., d'Alché-Buc, F. (eds.) MLCW 2005. LNCS (LNAI), vol. 3944, pp. 177–190. Springer, Heidelberg (2006). https://doi.org/10.1007/11736790_9
5. Dzikovska, M., et al.: SemEval-2013 task 7: The joint student response analysis and 8th recognizing textual entailment challenge. In: Proceedings of the Seventh

International Workshop on Semantic Evaluation (SemEval 2013), Atlanta, Georgia, USA, pp. 263–274 (2013)
6. Funayama, H., Sato, T., Matsubayashi, Y., Mizumoto, T., Suzuki, J., Inui, K.: Balancing cost and quality: an exploration of human-in-the-loop frameworks for automated short answer scoring. In: Rodrigo, M.M., Matsuda, N., Cristea, A.I., Dimitrova, V. (eds.) AIED 2022. LNCS, vol. 13355, pp. 465–476. Springer, Cham (2022). https://doi.org/10.1007/978-3-031-11644-5_38
7. Galhardi, L.B., de Mattos Senefonte, H.C., de Souza, R.C.T., Brancher, J.D.: Exploring distinct features for automatic short answer grading. In: Anais do XV Encontro Nacional de Inteligência Artificial e Computacional, pp. 1–12. SBC (2018)
8. Haller, S., Aldea, A., Seifert, C., Strisciuglio, N.: Survey on automated short answer grading with deep learning: from word embeddings to transformers. arXiv preprint arXiv:2204.03503 (2022)
9. Heilman, M., Madnani, N.: ETS: domain adaptation and stacking for short answer scoring. In: Proceedings of the Seventh International Workshop on Semantic Evaluation (SemEval 2013), Atlanta, Georgia, USA, pp. 275–279 (2013)
10. Khayi, N., Rus, V., Tamang, L.: Towards improving open student answer assessment using pretrained transformers. In: The International FLAIRS Conference Proceedings, vol. 34 (2021)
11. Liu, Y., et al.: Roberta: A robustly optimized bert pretraining approach. arXiv preprint arXiv:1907.11692 (2019). https://arxiv.org/abs/1907.11692
12. Ott, N., Ziai, R., Hahn, M., Meurers, D.: CoMeT: integrating different levels of linguistic modeling for meaning assessment. In: Proceedings of the Seventh International Workshop on Semantic Evaluation (SemEval 2013), Atlanta, Georgia, USA, pp. 608–616 (2013)
13. Padó, U.: Assessing the practical benefit of automated short-answer graders. In: Rodrigo, M.M., Matsuda, N., Cristea, A.I., Dimitrova, V. (eds.) AIED 2022. LNCS, vol. 13356, pp. 555–559. Springer, Cham (2022). https://doi.org/10.1007/978-3-031-11647-6_114
14. Saha, S., Dhamecha, T.I., Marvaniya, S., Sindhgatta, R., Sengupta, B.: Sentence level or token level features for automatic short answer grading?: use both. In: Penstein Rosé, C., et al. (eds.) AIED 2018. LNCS (LNAI), vol. 10947, pp. 503–517. Springer, Cham (2018). https://doi.org/10.1007/978-3-319-93843-1_37
15. Sainz, O., Gonzalez-Dios, I., Lopez de Lacalle, O., Min, B., Eneko, A.: Textual entailment for event argument extraction: zero- and few-shot with multi-source learning. In: Findings of the Association for Computational Linguistics: NAACL-HLT 2022, Seattle, Washington (2022)
16. Schick, T., Schütze, H.: Exploiting cloze-questions for few-shot text classification and natural language inference. In: Proceedings of the 16th Conference of the European Chapter of the Association for Computational Linguistics: Main Volume, pp. 255–269. Association for Computational Linguistics (2021)
17. Sung, C., Dhamecha, T.I., Mukhi, N.: Improving short answer grading using transformer-based pre-training. In: Isotani, S., Millán, E., Ogan, A., Hastings, P., McLaren, B., Luckin, R. (eds.) AIED 2019. LNCS (LNAI), vol. 11625, pp. 469–481. Springer, Cham (2019). https://doi.org/10.1007/978-3-030-23204-7_39
18. Wang, S., Fang, H., Khabsa, M., Mao, H., Ma, H.: Entailment as few-shot learner (2021)
19. Williams, A., Nangia, N., Bowman, S.: A broad-coverage challenge corpus for sentence understanding through inference. In: Proceedings of the 2018 Conference of the North American Chapter of the Association for Computational Linguistics, pp. 1112–1122 (2018)

Impact of Learning a Subgoal-Directed Problem-Solving Strategy Within an Intelligent Logic Tutor

Preya Shabrina(✉), Behrooz Mostafavi, Min Chi, and Tiffany Barnes

North Carolina State University, Raleigh, NC 27695, USA
{pshabri,bzmostaf,mchi,tmbarnes}@ncsu.edu

Abstract. Humans adopt various problem-solving strategies depending on their mastery level, problem type, and complexity. Many of these problem-solving strategies have been integrated within intelligent problem-solvers to solve structured and complex problems efficiently. One such strategy is the means-ends analysis which involves comparing the goal and the givens of a problem and iteratively setting up subgoal(s) at each step until the subgoal(s) are straightforward to derive from the givens. However, little is known about the impact of explicitly teaching novices such a strategy for structured problem-solving with tutors. In this study, we teach novices a subgoal-directed problem-solving strategy inspired by means-ends analysis using a problem-based training intervention within an intelligent logic-proof tutor. As we analyzed students' performance and problem-solving approaches after training, we observed that the students who learned the strategy used it more when solving new problems, constructed optimal logic proofs, and outperformed those who did not learn the strategy.

Keywords: Means-ends Analysis · Subgoal · Problem Solving · Intelligent Tutor

1 Introduction

The existing literature frequently mentions three problem-solving strategies in the domain of structured problem-solving: 1) Forward strategy - starts from the givens of a problem and moves towards the goal by applying valid rules and actions in the problem domain [1], 2) Backward strategy - starts from the goal and at each step, the goal is refined to a new goal (also referred to as subgoal) until the givens are reached [1], and 3) Means-ends analysis [12] - carries out problem solving as a search for subgoals at each step to recursively reduce the distance between the goal and givens until the subgoal can be directly derived from the givens [2]. By definition, backward strategy and means-ends analysis involve subgoaling, where subgoaling can be referred to as refining the given goal to a new goal (or subgoal) to reduce the distance between the givens and the goal [8,11,19].

N. Wang et al. (Eds.): AIED 2023, LNAI 13916, pp. 389–400, 2023.
https://doi.org/10.1007/978-3-031-36272-9_32

Researchers have often stated means-ends analysis to be closely aligned with the natural human strategy for complex problem-solving where they work forwards while keeping backward subgoals (identified by comparing the goal and the givens) in mind [12,16]. The comparison carried out between the goal and givens reduces the search space of possible next steps and the type of comparison usually depends on the nature of the problem. Also, the comparison is set out to select the best action or subgoal to optimally reduce the distance between the goal and the givens [6,16]. This general problem-solving strategy rarely leads to dead-ends when there is a specific goal [16]. Thus, researchers have integrated Means-ends analysis in automated problem-solvers for efficient problem-solving (for example, in *General Problem Solver* [12]). Prior research suggests that although novices try to adopt means-ends analysis more due to their low prior knowledge (to reduce search space of possible next steps) [17], experts might be more able to use this strategy than novices [14]. However, methods to train and motivate novices to learn/use this strategy or the impact of learning this strategy on their problem-solving skills have been rarely explored.

Thus, in this study, we integrated problems that enforce the use of a mixed problem-solving strategy (MS) inspired by means-ends analysis within the training session of an intelligent logic-proof tutor. In the MS problems, the first few steps must be subgoaling steps carried out using the backward strategy (BS). After the subgoaling steps, the subgoals may be achieved using forward strategy (FS). To analyze the impact of learning MS, we implemented two training conditions within the tutor: 1) Control: this group was not taught MS, and 2) Treatment: this group was taught MS using examples and practice problems. We evaluated the impact of the mixed strategy training on the basis of the following research questions:

RQ1 (Students' Experience): How difficult is the mixed strategy training for students when integrated with an intelligent tutor? [Note: This question is important since novices often use intelligent tutors in the absence of human tutors. Thus, the difficulty level of a training intervention in a tutor should not be too high so that students can persist and learn successfully].
RQ2 (Impact on Performance): How does learning mixed strategy impact students' performance in new problems?
RQ3 (Impact on Problem-solving approach and skills): How does learning mixed strategy impact students' problem-solving and subgoaling approach and skills?

2 Related Work

The means-ends analysis (MEA) was first introduced by Newell and Simon [12] in an AI problem solver called *General Problem Solver* (GPS). They emphasized that MEA is actually a simulation of natural human thought processes. MEA involves calculating the distance between the goal and givens of a problem at each step using a function or method appropriate for the problem type. Then, the best action that will generate a subgoal to optimally reduce the goal-givens distance is executed. Since the 50s, this method has been used in AI problem solvers for efficient problem-solving. For example, GPS (general problem solver) [12], Prodigy (math problem solver) [20], Multilevel Flow Modeling or MFM (industrial process planner) [9] etc.

Although being a very well-known efficient problem-solving strategy, little research is found on the impact of explicitly MEA learning on students' problem-solving skills. In a recent study, Permatasari and Jauhariyah [13] found that incorporating MEA in problem-solving-based physics learning improved students' learning. Researchers have also recently found this strategy to be effective in improving students' critical thinking skills (skills to analyze and execute decisions) for mathematical [4,7,18] and geometrical [21] problem-solving. However, these results need to be further verified in other domains. Specifically, the integration of problems or processes to have students learn such strategy needs to be explored within problem-based intelligent tutors.

3 Method

3.1 Deep Thought (DT), the Intelligent Logic Tutor

We conducted this study using Deep Thought or DT, an intelligent logic tutor, that teaches students logic-proof construction [10]. Generally, each problem within DT is either a worked example (WE) of logic-proof construction constructed by the tutor or a problem-solving (PS) problem needed to be solved by the students. In each problem, the given premises and goal conclusion are shown as visual nodes Fig. 1a. Valid logic rules from a given rule palette need to be iteratively applied on the nodes to generate new propositions or nodes to complete the proof. DT is organized into one pretest level (level 1), 5 training levels (levels 2–6), and one posttest level (level 7). In the pretest level, students are first shown two sample logic-proof problems (1.1–1.2) to acquaint them with the different features of DT. Then, they solve two pretest problems (1.3–1.4). After the pretest, students go through 5 training levels each containing 4 logic-proof problems (x.1–x.4). In the first three problems in each training level (x.1–x.3), the tutor offers on-demand step-level hints if the problem is of type PS. The last problem in each training level (x.4) is a PS that students need to solve independently without any tutor support. These are called training-level tests. After training, students enter a posttest level containing 6 posttest problems (7.1–7.6). Each student receives a score for each of the pretest, training-level test, and posttest problems. The score is a function of problem-solving time, step count (count of derived nodes), and logic-rule application accuracy and is scaled between 0 and 100. Solutions constructed in less time with fewer steps and fewer incorrect rule applications get higher scores.

3.2 Experiment Design

Problem Types: To facilitate our study, we used 4 types of logic-proof problems implemented within training levels of DT. The problem types are:

1. *Problem Solving (PS):* These are the default logic-proof construction problems in DT where students need to construct proofs themselves using forward/FW (derivations that move from the given premises towards the goal Fig. 1a), backward/BW (derivations that move from the goal towards given premises Fig. 1b), or both strategies without the tutor requiring them to use a specific strategy.
2. *Worked Examples(WE):* These are default worked examples available in DT where the tutor demonstrates logic-proof construction using only forward (FW) derivations.
3. *Mixed-strategy Problem Solving(MPS):* MPSs require the student to construct logic proofs using mixed strategy. Most DT problems require 5–10 steps to solve. If the subgoals derived in the first 2–3 BW steps are correct, the rest of the proof becomes straightforward. Thus, in MPS, students must derive the first three steps backward Fig. 1b and during this time, forward derivations are disabled. Then, they can use the forward strategy if they like Fig. 1c. The purpose of having the students start with backward steps is to explicitly motivate and involve them in subgoaling as in means-ends analysis.
4. *Mixed-strategy Worked Examples(MWE):* These are worked examples showing the mixed problem-solving strategy. The tutor carries out the first 2–3 steps backward refining the goal into subgoals which are followed by forward steps to derive the subgoal(s).

PS and MPS are visually depicted in Fig. 1. The interface for WE and MWE is the same as PS/MPS. The only difference is that WEs and MWEs are constructed by the tutor step by step as the students click on a 'Next Step' button.

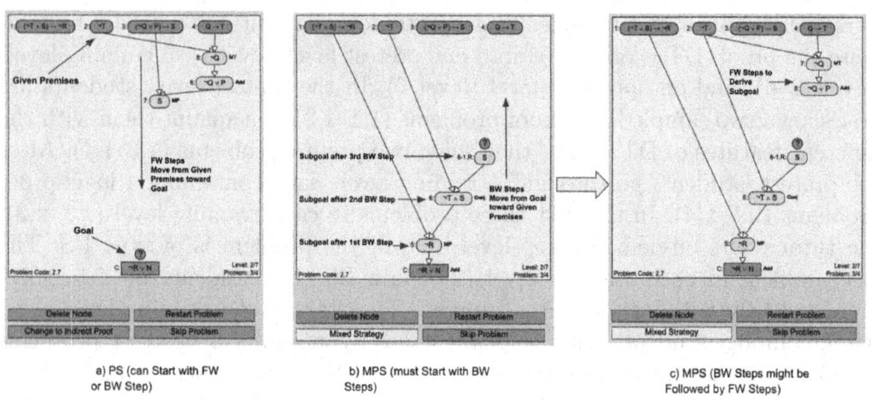

Fig. 1. PS (a) and MPS (b and c) Interface in DT.

Training Conditions: Using the 4 types of problems, we implemented two training conditions as described below:

1. *Control(C):* Students assigned to this condition received only PS and WE.
2. *Treatment(T):* Students assigned to this condition received MWE and MPS in addition to PS and WE.

The problem organization within the two training conditions is shown in Fig. 2. Notice that the treatment group students receive MWEs in each training level (up to level 5) to remind them of the mixed problem-solving strategy [blue squares in Fig. 2]. However, they receive MPSs only in the first half of the training [green squares in Fig. 2]. This was done to ensure that the students were given an opportunity to explore the strategy independently in the second half of the training. In a prior study [15], such organization of training treatment was found to be effective in having students learn a strictly BW strategy. Also, note that our problem organization ensures that students from both training conditions get an equal amount of examples (WE/MWE) and practice (PS/MPS) during training.

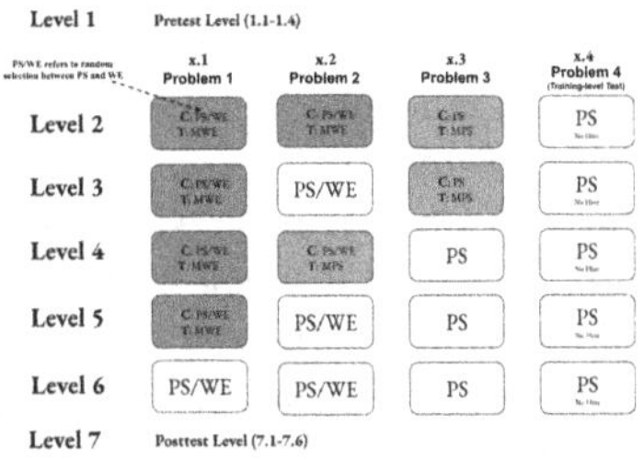

Fig. 2. Problem Organization in the DT Training Levels for the 2 Training groups. (Color figure online)

Data Collection: We deployed DT with the two training conditions in an undergraduate logic course offered at a public research university in the USA. The students taking the course were required to submit a DT assignment that involved completing all levels of DT to receive full credit. Each student after completing the pretest was automatically assigned to one of the training conditions. Our assignment algorithm ensures equal distribution of students between the conditions. It also ensures that the pretest scores of the training conditions come from a similar distribution. At the end of the experiment, 50 students in the C group and 45 students in the T group completed the tutor. We collected students' pretest, training-level test, and posttest scores. Additionally, we collected the time-series log data from DT that detail each step (derivation/deletion of nodes, direction, and time of derivation) of students while they construct proofs within DT. We used this data to carry out statistical and graph-mining-based proof-construction approach analysis to answer our research questions.

4 RQ1 (Students' Experience): Difficulties Across MPS and PS Problems

To understand students' experience with mixed strategy learning training, we compared the difficulty level of training MPS problems against training PS problems in terms of the time students took (i.e. the problem-solving time) to solve each type of problem. The reason for considering PS as a baseline is that PSs are simple proof-construction problems where students are not restricted to using a specific strategy unlike in MPSs. In a prior study [15], we observed that being restrictive about using a strategy could lead to significant difficulties for students. However, in MPS, the strict restriction is applied for only the first 3 steps where students must carry out subgoaling using BW steps which should limit the difficulties associated with MPS. Also, note that we did not consider MWEs and WEs in this analysis, since those are solved by the tutor.

To compare the difficulty level of MPS and PS, we carried out a mixed-effect regression analysis with problem-type (all MPS against all PS problem-solving instances) as fixed-effect, problem ID as random-effect (to eliminate the impact of differences in problems), and problem-solving time as the dependent variable. The analysis gave a p-value of < 0.001 [n(PS) = 410, n(MPS) = 135, problem-solving time(PS) = 14.4 (7.4) min, problem-solving time(MPS) = 18.0 (20.4) min]. The p-value indicates a significant difference in the difficulty levels of MPS and PS problems where MPS seemed to be more difficult in terms of problem-solving time. However, further investigation showed that this difference was introduced by only 25% (above 75th percentile) MPS problem-solving instances (n = 34, mean problem-solving time for these instances = 41.8 min). The rest of the 75% (up to 75th percentile) MPS instances (n = 101, mean problem-solving time for these instances = 10.0 min) had no significant differences from PS problems in terms of problem-solving time as per a Mann-Whitney U test. Recall that during training, the T group students solved three training MPS problems. As we further analyzed each of these problems separately, we observed that 95% of the 34 MPS instances with significantly higher problem-solving time than PS occurred in the first MPS problem that the students received in training level 2. Problem-solving time for the MPS problems in training levels 3 and 4 did not have any significant differences from the problem-solving time of the PS instances. Additionally, we identified three training problems for which we found both PS (n = 67) and MPS (n = 71) problem-solving instances in the collected data. However, interestingly, MPS instances had marginally lower step counts than the PS instances according to Mann-Whitney U test [Step Count(PS, MPS) = 10.1, 8.9, p = 0.08] which indicated adopting mixed-strategy possibly led students to more efficient proof construction.

Overall, our analysis results indicate that in most cases MPS problems are as easy as PS problems. However, students might require more time to solve the first MPS they receive during training, possibly to figure out how to address the strict requirement of using the mixed strategy. However, solving a problem using mixed strategy may help students to solve logic-proof construction problems with higher efficiency.

5 RQ2: Students' Performance After Training

To understand the impact of learning mixed problem-solving strategy on students' performance, we compared students' test scores across the control (C) and treatment (T) groups. We performed two mixed-effect regression analyses: one for the training-level test (4th problem in each training level: 2.4–6.4) and one for the posttest (problems in level 7: 7.1–7.6) problems to understand improvement over the period of training and after training respectively. Note that these problems are solved by students without any tutor help and thus, are good candidates for performance analysis. In both of the analyses, we incorporated training conditions (control (C)/treatment (T)) as the fixed effect, problem IDs as the random effect (to exclude the impact of differences in the problems), and problem scores as the dependent variable. The analysis for the training-level test problems did not show any significant differences across the two training groups: n(50,45), mean(std)[C, T] = 65.11(24.08), 65.14(23.36), p = 0.99. However, the analysis for the posttest problems showed that overall the treatment group significantly outperformed the control group: n(50, 45), mean(std) [C, T] = 70.42(22.88), 74.10(21.51), p = 0.04 [$p < 0.05$ indicates significant difference]. As we analyzed students' scores for each of the problems separately using Mann-Whitney U tests, we observed for most of the problems T group students had higher scores than the C group. This trend can be visualized in Fig. 3a, where the curve for the T group (solid orange line) mostly lies above the curve for the C group (dashed blue line) starting from problem 5.4. The p-values in the figure show the problems with significantly higher averages for the T group. Also, note that we did not find any problem where the T group received significantly lower scores than the C group.

To further investigate the reason for the difference in problem scores across the training groups, we analyzed their step counts, problem-solving time, and logic rule application accuracy in the training-level test and posttest problems. We did not find any significant differences in the rule application accuracies and problem-solving time (Fig. 3c) across the training groups. However, in a mixed-effect regression analysis similar to the one for problem scores, we found significant differences in step counts across C and T [Step Count (C, T) = 13.4, 10.1, p = 0.031]. Also, the T groups had lower step counts than the C group in most of the problems [notice the orange curve for the T group in Fig. 3b starting from problem 5.4].

Overall, the results indicate that learning mixed problem-solving strategy helped students to construct logic-proofs with fewer steps which led to higher scores. In RQ3, we analyzed students' proof construction approach to identify how learning mixed strategy helped to achieve this efficiency.

6 RQ3: Proof-Construction and Subgoaling Approach/Skills After Training

To investigate students' proof-construction and subgoaling approach and skills across the training conditions, we modeled students' proof-construction attempts using approach maps [3] for each of the training-level test and posttest problems that they solved independently without the tutor requiring them to use a specific strategy. Approach maps are high-level graphical representations of students'

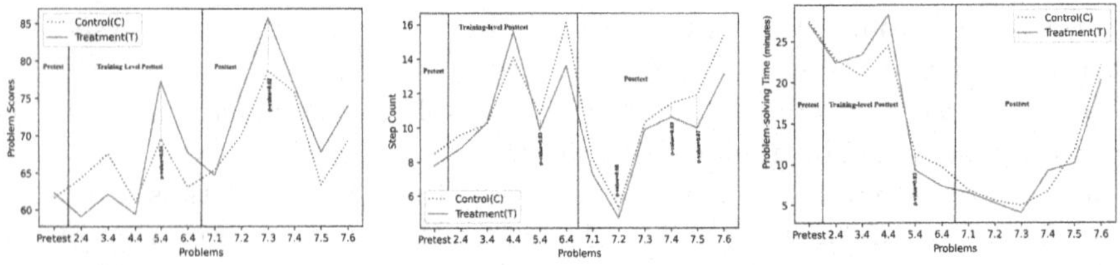

Fig. 3. Problem Score, Step Count, and Problem-solving Time across the Control (C) and Treatment (T) groups over the period of pretest, training, and posttest [p-values < 0.05 (obtained from Mann-Whitney U tests) indicates a significant difference].

proof construction attempts for a given problem that show the propositions they commonly derived during proof construction [Fig. 4]. To generate the approach map for a problem, students' state transitions during the construction of a proof for that problem are presented using a graph called an interaction network. Here, a state contains all nodes or propositions (lexicographically ordered) present on DT interface at a moment during proof construction. Students move from state to state during proof construction by deriving/deleting nodes, i.e. via steps. The Girvan-Newman clustering algorithm [5] is applied on this interaction network to identify clusters of closely connected states. Later, from those clusters, we extracted effective propositions that contributed to students' final solutions. As shown in Fig. 4, finally, approach maps become a graph where the start state (containing only given premises) is connected to the goal via clusters of effective propositions. Each path from start to goal via the clusters is a student approach (A1, A2, etc.). The propositions written in purple were derived using BW derivation (i.e. subgoaling) by students who adopted mixed strategy. Note that for simplicity, we do not detail all effective or unnecessary propositions[1] derived by all students in the approach maps. We only show the commonly derived effective nodes that are sufficient to describe differences in student approaches. However, we recorded counts of all effective/unnecessary propositions derived by each student since both increase step counts. In the next sections, we explain differences in student approaches across the training conditions using approach maps and statistically analyze proof derivation efficiency (using Mann-Whitney U tests). Lower time, fewer unnecessary, and effective propositions indicate higher efficiency. Note that this method to analyze student approaches can be adopted for structured problems from any domain where states and transitions during problem solving can be defined definitely. Effective or unnecessary steps can be identified using the differences between the final problem state and earlier states.

6.1 Student Approaches in Training-Level Test Problems

In the training-level test problems (2.4–6.4), up to 49% of the T group students used mixed strategy, whereas this percentage for C group students is only 24%.

[1] Propositions derived by the students but were not part of their final solutions.

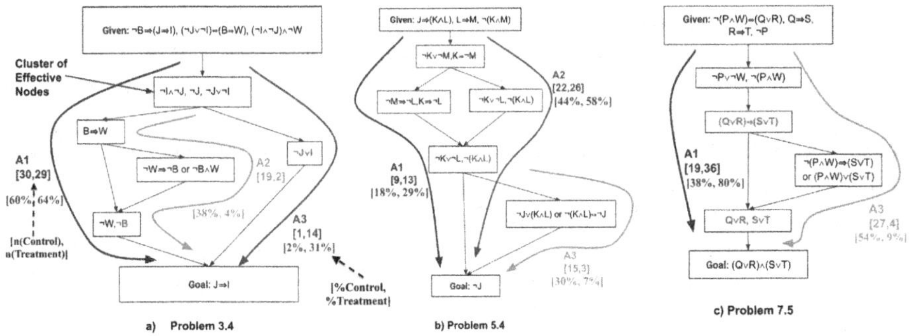

Fig. 4. Approach Maps for Training-level Test Problems a) 3.4, b) 5.4, and c) Posttest Problem 7.5.

The student counts adopting mixed strategy for each of the problems are shown in Fig. 5a. We also observed that, in these problems, T group showed sporadic signs of improved skills (marginally shorter solutions/fewer unnecessary propositions). For example, in problem 2.4, the proof length was marginally lower for the T group than the C group [mean(std) = C: 5.8(5.5), T:4.5(5.4), test statistic = U(50, 45) = 1213.0, p-value = 0.09]. In problem 3.4, T group students had significantly fewer unnecessary propositions in their solution attempt than C students [C: 7.3(10.2), T:5.2(4.7), U(50, 45) = 1398.5, 0.003]. Additionally, T students who adopted mixed strategy or T_{MS} (n(T_{MS}) = 17 in 3.4) had significantly shorter solutions than those of C and marginally shorter solutions than those of T students who did not adopt the strategy [$C = 8.1(2.0)$, ${T_{FS}}^2 = 7.2(3.4)$, $T_{MS} = 5.3(2.8)$, $U_{T_{MS}<C}(50, 17) = 534.0, 0.01$; $U_{T_{MS}<T_{FS}}(28, 17) = 281.5, 0.08$]. The approach map for problem 3.4 is shown in Fig. 4a. As shown in the figure, A3 is the shortest solution for this problem. 13 out of 17 of the T_{MS} students derived optimal subgoal $\neg J \vee I$ using BW derivation which led them to the shortest solution A3. We observed a similar pattern in problem 4.4 as well. From these statistics, we concluded that by the first three training levels, not all T group students achieved an equal level of mastery in using mixed strategy. However, those who *were* able to learn and adopt the strategy (i.e. the T_{MS} students) constructed proofs more efficiently.

In problem 5.4, only 4.4% of T group students adopted mixed strategy. However, overall T group students had significantly smaller solution length than C group students [C: 8.02(1.18), T:7.0(2.3), U(49,45) = 1224.5, 0.007]. The approach map for 5.4 is shown in Fig. 4b. As shown in the figure, A1 and A2 are the optimal solutions for this problem where $\neg(K \wedge L)$ is a subgoal. Although only 2 students from the T group explicitly used mixed strategy and derived $\neg(K \wedge L)$ as a subgoal using BW derivation at the beginning of their proof construction attempt, almost all T students were observed to follow one of the 2 optimal approaches (A1/A2). On the other hand, 15 C group students adopted a lengthier approach, A3, while solving this problem. Moreover, C-group students derived

[2] Treatment group students who only used forward strategy.

significantly more unnecessary propositions [C: 4.1(2.2), T:2.7(1.3), U(49,45) = 1126.5, 0.01]. In problem 6.4, overall, T-group students showed a similar efficiency by identifying subgoals that led to shorter proofs. These results indicate that by the time the students finished the training levels (i.e. level 6), T students were more skilled in proof construction and identifying better subgoals that will lead to shorter proofs. However, we did not observe all T students explicitly use the mixed strategy. Possibly, the T group students identified subgoals using mixed strategy but carried out the steps in forward direction in the tutor (i.e. BW strategy in mind for *implicit subgoaling*, which was not reflected by their derivations in the tutor).

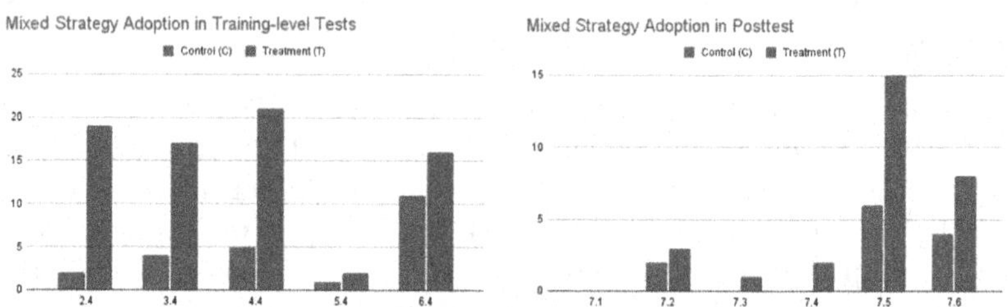

Fig. 5. Students across the control (C) and Treatment (T) groups Adopting Mixed Strategy in a) training level posttest problems, and b) posttest problems.

6.2 Student Approaches in Posttest Problems

In the first 4 posttest problems (7.1–7.4), we observed only a few students from the T and C groups demonstrated the adoption of mixed strategy. However, in problems 7.5–7.6, more students explicitly adopted the mixed strategy (17–33% T students and 8–12% C students). Note that the posttest problems are organized in increasing order of difficulty and Fig. 5b indicates that more students adopted subgoaling-based mixed strategy explicitly in the last two, harder problems.

Problem 7.1 is a trivial problem with a 3-step shortest solution and we did not find much difference for this problem across the training groups. However, T students constructed significantly shorter proofs for problems 7.2–7.5 [For 7.2, C:3.9(1.6), T:3.3(1.3), U(50, 45) = 1315.5, 0.04. For 7.3, C: 8.0(1.4), T:7.4(1.1), U(50,45) = 1183.5, 0.03. For 7.4, C: 8.7(2.0), T:7.5(1.9), U(50, 45) = 1415.0, 0.001. For 7.5, C: 6.4(2.7), T:5.6(2.6), U(50,45) = 1355.5, 0.04]. As a sample, we showed the approach map of Problem 7.5 in Fig. 4c. For this problem, the optimal solution is A1 as shown in the figure. T students who used mixed strategy derived $Q \vee R$ and $S \vee T$ as the first set of subgoals (resulted from their first BW derivation). They derived $(Q \vee R) \Rightarrow (S \vee T)$ as the next subgoal. These subgoaling steps led them to optimal solution A1. The T students who did not demonstrate explicit subgoaling also constructed the same proof using only FW derivations. However, above 50% C group students constructed a longer proof (approach A3 in Fig. 4c). Additionally, in problem 7.6, the T group derived significantly fewer unnecessary propositions than the C group [C: 8.6(5.9), T:5.9(3.7),

U(50,45) = 1167.5, 0.005]. The results from our approach map analysis showed that the T-group students who learned mixed strategy with MWE+MPS during training achieved higher efficiency in proof construction and subgoaling. They possibly learned to compare the goal and givens effectively to derive better subgoals leading to shorter proofs, and fewer unnecessary propositions. However, note that our statistical tests did not detect a similar efficiency in the few C students who demonstrated mixed strategy but were not trained beforehand unlike T students.

7 Discussion

Overall, the results of our study showed that learning the mixed strategy with MWE+MPS during training posed a similar level of difficulty in most cases as learning only the forward strategy. Additionally, learning and using this strategy to construct proofs via subgoaling could be beneficial for students to help improve their proof-construction efficiency (i.e. shorter proofs with subgoaling). However, our observations indicate that explicit adoption of mixed strategy can depend on the complexity of the problem or the mastery level of students in using mixed strategy. Some T-group students were observed to explicitly use the strategy in the first few training-level tests (2.4–3.4) and showed improved performance early. On the contrary, some of the students did not use mixed strategy explicitly at all, or only in harder problems, but demonstrated improved performance over time (possibly by using an implicit subgoaling strategy).

8 Conclusion and Future Work

The contributions of this paper are 1) an efficient training strategy for mixed strategy learning within an intelligent tutor using problems that involves subgoaling first and then deriving the subgoal as in means-ends analysis [this training strategy can be adopted within any tutor containing structured problems with specific goals that can be refined into subgoals using actions from a finite set], and 2) an evaluation of the impact of learning this subgoal-directed strategy that showed that it helped to improve students' subgoaling and proof construction skills. Students who learned the strategy achieved higher scores by deriving better subgoals and constructing shorter proofs. This efficiency can be compared to what automated problem-solvers achieve by exploiting means-ends analysis to reduce the search space of possible next steps and select the most efficient one. One limitation of this study is that we observed many treatment group students (trained with MWE+MPS) demonstrated improved performance without explicitly using the strategy which we identified as potential cases of implicit use. Thus, in future studies, students learning the mixed strategy should be interviewed to confirm how they used mixed strategy to identify better subgoals and construct better proofs.

Acknowledgments. This research was supported by the NSF Grant 2013502.

References

1. Al-Ajlan, A.: The comparison between forward and backward chaining. Int. J. Mach. Learn. Comput. **5**(2), 106 (2015)
2. Andrade, W.: Cognitive Psychology. LibreTexts (2021)
3. Eagle, M., Barnes, T.: Exploring differences in problem solving with data-driven approach maps. In: Educational Data Mining 2014 (2014)
4. Fitrianna, A.Y., Dahlan, J.A.: Means end analysis (MEA) learning model in developing algebraic reasoning ability: a literature study. J. Anal. **8**(1), 1–10 (2022)
5. Girvan, M., Newman, M.E.: Community structure in social and biological networks. Proc. Natl. Acad. Sci. **99**(12), 7821–7826 (2002)
6. Heyworth, R.M.: Procedural and conceptual knowledge of expert and novice students for the solving of a basic problem in chemistry. Int. J. Sci. Educ. **21**(2), 195–211 (1999)
7. Hidayat, A.W., Surya, A., Sukarno, J.I.S.P.: The effect of means ends analysis model on reasoning skills of planes. In: 3rd International Conference on Learning Innovation and Quality Education (ICLIQE 2019), pp. 629–633. Atlantis Press (2020)
8. Laird, J., Rosenbloom, P., Newell, A.: Universal Subgoaling and Chunking: The Automatic Generation and Learning of Goal Hierarchies, vol. 11. Springer, Heidelberg (2012)
9. Morten, L.: An introduction to multilevel flow modeling. J. Nucl. Saf. Simul. **2**, 22–32 (2011)
10. Mostafavi, B., Barnes, T.: Evolution of an intelligent deductive logic tutor using data-driven elements. Int. J. Artif. Intell. Educ. **27**(1), 5–36 (2017)
11. Newell, A.: Unified Theories of Cognition. Harvard University Press (1994)
12. Newell, A., Simon, H.A.: GPS, a program that simulates human thought. Technical report, Rand Corp Santa Monica Calif (1961)
13. Permatasari, N.Y., Jauhariyah, M.N.R.: The students' problem-solving skills improvement by using integration of means-ends analysis on problem-based learning model. Inovasi Pendidikan Fisika **9**(2) (2020)
14. Priest, A., Lindsay, R.: New light on novice-expert differences in physics problem solving. Br. J. Psychol. **83**(3), 389–405 (1992)
15. Shabrina, P., Mostafavi, B., Abdelshiheed, M., Chi, M., Barnes, T.: Investigating the impact of backward strategy learning in a logic tutor: aiding subgoal learning towards improved problem solving. Int. J. AI Educ. (IJAIED) (2023)
16. Sweller, J.: Cognitive load during problem solving: effects on learning. Cogn. Sci. **12**(2), 257–285 (1988)
17. Sweller, J., Mawer, R.F., Ward, M.R.: Development of expertise in mathematical problem solving. J. Exp. Psychol. Gen. **112**(4), 639 (1983)
18. Umar, W.: Constructing means ends analysis instruction to improve students' critical thinking ability and mathematical habits of mind dispositions. Int. J. Educ. Res. **5**(2), 261–272 (2017)
19. VanLehn, K.: Problem solving and cognitive skill acquisition. Technical report, Carnegie-Mellon Univ. Pittsburgh PA Artificial Intelligence and Psychology ... (1988)
20. Veloso, M., Carbonell, J., Perez, A., Borrajo, D., Fink, E., Blythe, J.: Integrating planning and learning: the prodigy architecture. J. Exp. Theor. Artif. Intell. **7**(1), 81–120 (1995)
21. Yoradyastuti, R.: The effect of using means ends analysis (MEA) model in learning geometry. In: Proceeding International Seminar of Primary Education, vol. 2 (2019)

Matching Exemplar as Next Sentence Prediction (MeNSP): Zero-Shot Prompt Learning for Automatic Scoring in Science Education

Xuansheng Wu[1], Xinyu He[2], Tianming Liu[1](✉), Ninghao Liu[1](✉), and Xiaoming Zhai[2](✉)

[1] School of Computing, University of Georgia, Athens, GA, USA
{xuansheng.wu,tliu,ninghao.liu}@uga.edu
[2] AI4STEM Education Center, University of Georgia, Athens, GA, USA
{xinyu.he1,xiaoming.zhai}@uga.edu

Abstract. Developing natural language processing (NLP) models to automatically score students' written responses to science problems is critical for science education. However, collecting sufficient student responses and labeling them for training or fine-tuning NLP models is time and cost-consuming. Recent studies suggest that large-scale pre-trained language models (PLMs) can be adapted to downstream tasks without fine-tuning by using prompts. However, no research has employed such a prompt approach in science education. As students' written responses are presented with natural language, aligning the scoring procedure as the next sentence prediction task using prompts can skip the costly fine-tuning stage. In this study, we developed a zero-shot approach to automatically score student responses via **M**atching **e**xemplars as **N**ext **S**entence **P**rediction (MeNSP). This approach employs no training samples. We first apply MeNSP in scoring three assessment tasks of scientific argumentation and found machine-human scoring agreements, Cohen's Kappa ranges from 0.30 to 0.57, and F1 score ranges from 0.54 to 0.81. To improve scoring performance, we extend our research to the few-shots setting, either randomly selecting labeled student responses at each grading level or manually constructing responses to fine-tune the models. We find that one task's performance is improved with more samples, Cohen's Kappa from 0.30 to 0.38, and F1 score from 0.54 to 0.59; for the two other tasks, scoring performance is not improved. We also find that randomly selected few-shots perform better than the human expert-crafted approach. This study suggests that MeNSP can yield referable automatic scoring for student-written responses while significantly reducing the cost of model training. This method can benefit low-stakes classroom assessment practices in science education. Future research should further explore the applicability of the MeNSP in different types of assessment tasks in science education and further improve the model performance. Our code is available at https://github.com/JacksonWuxs/MeNSP.

N. Wang et al. (Eds.): AIED 2023, LNAI 13916, pp. 401–413, 2023.
https://doi.org/10.1007/978-3-031-36272-9_33

Keywords: Prompt Learning · Pre-trained Language Model · written Response · Automatic Scoring · Natural Language Processing

1 Introduction

To engage students in meaningful science learning, it is crucial that science assessments can elicit student knowledge-in-use so that instructors can better understand and support student learning [23]. The need for assessing student competence in using scientific knowledge to solve problems or design solutions has been identified in the actualization of 21st-century science education [4]. While such science and engineering practices are desirable for students to enact, it has been challenging to assess with multiple-choice questions. More complicated written response assessments are needed to assess such scientific practices.

Although the written response assessment allows for the freedom of expression elicited in responses, it accounts for a number of challenges in the process of scoring. Since scoring is expected to be fair and reliable, it requires, among other efforts, that educators are trained using pre-designed rubrics to enhance reliability. Despite the expertise and rubric familiarity of the raters, it demands a lot of time to score the learners' responses appropriately.

In contrast to some of the issues associated with human scoring, machine learning, such as Natural Language Processing (NLP), offers quicker, less expensive, and more consistent scoring [35]. However, most NLP scoring models need to be tested with human scoring as benchmarks before applying in classroom settings. Their validity and performance are a function of the former [33,37]. The majority of the models needed to achieve this require a large sample of data for training, validation, and testing purposes [39]. Given that the conditions necessary to collect a reasonable size of data are a huge effort in the field of education, satisfying the needs of these models is cumbersome [38].

To overcome these challenges, this study employs prompt learning with pre-trained language models (PLMs) to develop algorithmic models for automatic scoring. Prompt learning utilizes little or limited labeled data (i.e., zero-shot) in developing a model and has shown great potential in accomplishing NLP tasks with significant efficiency [41]. To verify the efficiency of this new approach, we develop NLP scoring models to automatically score students' written arguments to science phenomena when engaging in scientific argumentation practices.

This study will contribute to the efficiency of developing innovative assessments in education broadly and science education specifically. Our approach will significantly save the time and cost of developing NLP scoring algorithmic models for automatic scoring, which is a bottleneck that has prevented the broad use of machine learning-based assessment practices in classrooms [39]. Implementing this approach could benefit millions of students using automatically scored written response assessments that have been developed [8].

2 Related Work

2.1 Natural Language Processing for Automatic Scoring

Natural language processing (NLP) is a field of computer science using computational techniques to learn, understand, and produce human language content. There has been a long time to apply NLP in language education, such as to correct students' writing errors, conduct semantic analysis, or assess language skills [12]. With the improvement of using NLP to evaluate students' writing skills, research has begun to explore using NLP to evaluate the content of students' writing and study the quality of writing at the level of domain-specifics such as science learning. Written response assessment in science education is one of the specific domains needing NLP for automatic scoring [39].

To help teachers better understand students' scientific thinking, researchers have explored using NLP technologies to score student-written responses a decade ago automatically. Haudek et al. [9] used SPSS Text Analytics for Surveys to score students' biology understanding automatically. The program can extract key linguistic features of student writing to classify student responses according to scoring rubrics. Researchers [20] at Carnegie Mellon University developed SIDE (current version named LightSIDE) that integrates various algorithmic functions. Nehm et al. [20] employed his package to develop a portal EvoGrader to examine students' understanding of biology concepts. Educational Testing Service developed a C-rater [13] for the automatic scoring of GRE essays and short written-response answers. Later, they incorporated and upgraded the tool to C-rater-ML [7,11] and employed it to automatically score students' scientific argumentation and explanations. While prior tools employed individual algorithms, researchers [18,31] also developed tools that ensemble multiple algorithms to score student-written responses simultaneously. Most recently, researchers [1,25] also employed BERT [5] for automatic scoring of student-written responses for scientific practices.

Among these developments, recent surveys [38,39] suggested that developing these models requires a large number of human-scored written responses with varying accuracy. However, collecting and validating the student responses and rigorous rubrics for scoring takes much effort. Since training human experts to use the scoring rubric to assign scores to student responses reliably is challenging, obtaining datasets to develop these models is costly.

2.2 Prompt Learning

Prompt learning [3,30,32,34] leads to a new paradigm in NLP as it can achieve comparable performance to full-parameter fine-tuning with fewer training samples and parameters. "Prompt" typically is a short piece of text that include instructions for the task (zero-shot learning) or a few samples of the task (few-shots learning) [19]. By selecting appropriate prompts, it is possible to directly predict the target label using the pre-trained language models. Compared with the previous "pre-train and fine-tune" paradigm, prompt learning has the advantage of reducing the training cost and being applicable to multiple tasks by changing the prompts [14].

The new paradigm brings its advantage as well as a new challenge, that is, how to find the most appropriate prompt. The most natural way is to create intuitive prompts manually [15,27]. However, some researchers [14,29] found that the optimal prompt may not be readable by humans. Thus, many methods [15, 16,28,29] are proposed to learn a better prompt automatically. Besides learning prompts, providing and ordering a few additional answers in the prompts can also result in satisfied model performance [6,17].

To the best of our knowledge, the application of prompt learning in the education field is still at the beginning. Hart-Davidson et al. (2021) [21] applied prompt learning in the qualitative coding research task, which could be used to provide written feedback for student writing. Zhang et al. (2022) [40] applied prompt learning in the fine-tuning process to boost the performance for automatic scoring of short Math answers. How to give full play to The advantage of using prompt learning hasn't been fully discussed.

In summary, there is a lack of research exploring PLMs' zero-shot performance for automatic scoring in education, not even in science education. Therefore, this study applies prompt learning to automatically score student-written responses to science assessments. The approach can be applied to more tasks in education to save time and cost.

3 Approach

This section proposes a method to develop the system for scoring student-written arguments without fully labeled training datasets. Inherently, student responses are presented with natural language, which is readable by PLMs. Since PLMs are built under some pre-trained tasks over natural language corpus, we can reformat the scoring task as one of the pre-trained tasks so that the textual responses of students can be graded without further fine-tuning of the models on labeled responses. Reformatting a downstream task as a pre-trained task is also known as *prompt learning*, which has been widely applied to handle NLP tasks under the zero-shot and few-shots settings where labeled data are unavailable or limited. However, letting PLMs understand (1) the meaning of the "scoring" procedure, and (2) the scoring rubrics of each assessment item are two non-trivial challenges to achieving the goal. Technically, we need to find a pre-trained task to reformat the scoring procedure and combine student responses and rubrics as new inputs to feed for PLMs. In the rest of this section, we present a two-stage pipeline to score student responses based on the rubrics of the assessments.

3.1 Matching Exemplars

Most researchers [15,16,29] reformat their downstream tasks as the masked language modeling task [5], where PLMs are asked to fill the blank *[MASK]* that

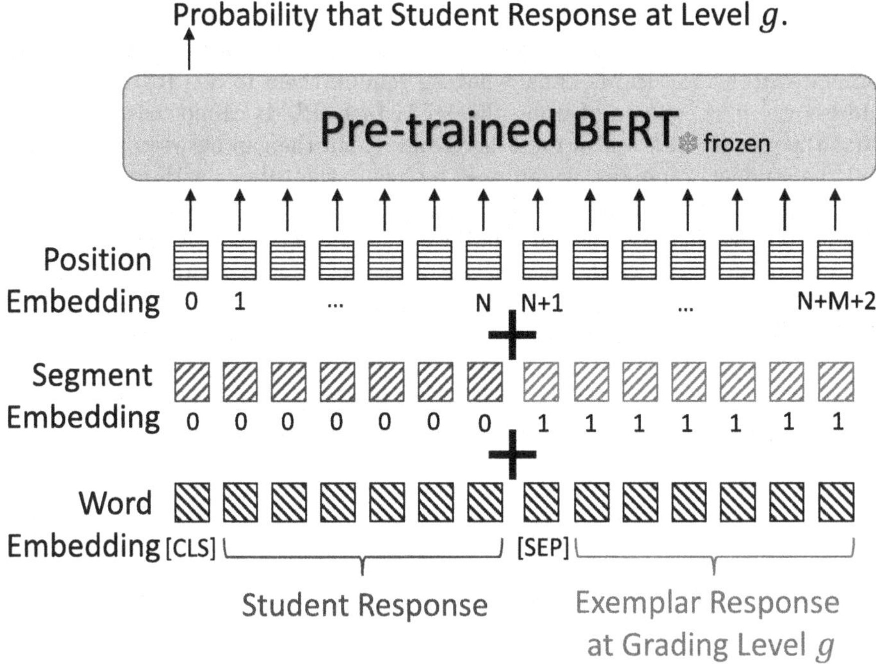

Fig. 1. Next sentence prediction as matching exemplars.

is shown in a context with a proper word. For example, by combining a student response and the corresponding rubric, we can construct a new input as *"The student responses are that [Response]. The rubric is that [Rubric]. Overall, the response can be grad as [MASK] point."*, where each underlined word is filled with the actual input, and the rest parts of this example are called *template*. Ideally, we expect the PLMs to fill the blank *[MASK]* with the grade of the response according to the given scoring rubric. However, our piloting data show that PLMs almost randomly fill in the blank without considering the context. We assert that this strategy is failed for two reasons. Firstly, both the responses and rubrics are so long that PLMs pay less attention to the task definition provided by the template. Secondly, PLM has no hints to bridge the gap between the rubrics and the final grading points.

To some extent, scoring student responses according to a given rubric can be considered a matching process if the rubrics can be refined to each grading level. Following this idea, we propose reforming the scoring procedure as the Next Sentence Prediction (NSP) task [5], which requires PLMs to determine whether the given two texts come from the same context. Specifically, assuming that the rubrics on each grading level are available, we first concatenate the student responses with each rubric's overall grading levels independently. After that, PLMs judge whether each response-rubric pair shows the same context by using the outputs of the NSP task. Finally, each score of a student response will be the grading level indicated by the rubrics that best match it. Since the objective of the NSP task has been clear to PLMs, we no more need to worry about designing templates to let the PLMs know what we require them to do. To this end, we address the first concern of using the MLM task: PLMs cannot easily identify what the scoring task is. On the other hand, to fill the gap between the rubrics and the student responses, we propose replacing the rubrics with exemplars as standards to score the responses. This strategy is necessary because the language style of the student responses dramatically differs from that of the assessment rubrics, which leads PLMs to constantly predict the rubric-response pair as different (Here is a sample case[1]). Figure 1 demonstrates the proposed method.

Formally, given a pre-defined vocabulary set $\mathcal{V}$, a student response $\mathcal{R} = [w^{(1)}, ..., w^{(n)}, ..., w^{(N)}]$ consists of N words $w^{(n)} \in \mathcal{V}$, and the g-level grade exemplar response $\mathcal{E}_g = [w_g^{(1)}, ..., w_g^{(m)}, ..., w_g^{(M)}]$ has M words $w^{(m)} \in \mathcal{V}$. We adopt a PLM $f_{\text{nsp}} : \mathcal{V}^L \rightarrow [0, 1]$ with a pre-trained NSP head to automatically score the student response as:

$$score(\mathcal{R}) = \arg\max_{g} f_{\text{nsp}}([w_{\text{cls}}; \mathcal{R}; w_{\text{sep}}; \mathcal{E}_g]) \,, \tag{1}$$

where w_{cls},$w_{\text{sep}} \in \mathcal{V}$ are special words, $[\cdot;\cdot]$ indicates the concatenating operation over the input words, and $L = M + N + 2$ in this case. According to Eq. 1, we achieve automatic scoring under the zero-shot scenarios (without training) by aligning the scoring process into an exemplar matching process and releasing the PLMs' ability learned from the pre-training stage.

3.2 Zero Grade Identifier

However, the above strategy raises a new challenge in obtaining fine-grained exemplars for each grading level. We simultaneously cope with this challenge by decoupling the rubric and the perfect response of an assessment item. Particularly, the perfect rubric can be separated into several points, where each point reflects a grading level. Therefore, it is easy to write down the fine-grained rubrics of each grading level. Similarly, we first develop an optimal student response that fits all grading points of the perfect (full score) rubric. Once the optimal response is given, we remove each part of the perfect response to generate exemplar responses at different levels gradually.

Although this strategy generates high-quality responses for non-zero grading levels, it cannot and is also impossible to enumerate all zero-point exemplars since the reasons that the responses receive high scores are limited (exhaustive). In contrast, the zero-point responses can be various (in-exhaustive). To reduce invalid scoring because of missing zero-point exemplars, we introduce a pre-stage

[1] Rubric: Student can specify Sam's claim and corresponding evidence, and explain the relationship between the claim and evidence properly. Response: Sam's claim is that gas particles float on the top in the box. The evidence is that bubbles in soda water float to the top. As gas particles and bubbles, all refer to air. Sam can infer that air in the box can also float on the top just as air in soda water does.

before the above method so that we find out those zero-point responses in an early stage and let Eq. 1 focus on how many points the responses reward. Recall that PLMs represent inputs as vectors in their interior, carrying rich semantic information. Since the zero-point response is very different from the perfect response, the distance between the vectors of them should be far away. Thus, PLMs generate vectors to represent both the perfect exemplar response and the given student response and measure the distance between these two vectors with the Cosine similarity. If the cosine similarity between them is smaller than a threshold, we directly grade them with zero point. The threshold is calculated by averaging the cosine similarities of the zero-score and the one-score exemplar response to the perfect exemplar response.

Theoretically, given a PLM $f_{\text{emb}} : \mathcal{V}^L \rightarrow \mathbb{R}^d$ that maps a piece of L-length text into a d-dimensional space, we collect the representations of the student response and the g-level exemplar $\mathbf{z}_{\text{R}} = f_{\text{emb}}(\mathcal{R})$ and $\mathbf{z}_g = f_{\text{emb}}(\mathcal{E}_g)$, respectively. We determine whether the response $\mathcal{R}$ belongs to 0-level grade by:

$$zero\_score(\mathcal{R}) = \begin{cases} Yes, & cos(\mathbf{z}_{\text{R}}, \mathbf{z}_{g=2}) < \theta, \\ No, & otherwise, \end{cases} \tag{2}$$

where

$$\theta = \frac{\cos(\mathbf{z}_{g=0}, \mathbf{z}_{g=2}) + \cos(\mathbf{z}_{g=1}, \mathbf{z}_{g=2})}{2}, \tag{3}$$

$\cos(\mathbf{x}, \mathbf{y}) = \frac{\mathbf{x}\mathbf{y}^{\mathsf{T}}}{||\mathbf{x}||_2||\mathbf{y}||_2}$ is the cosine distance among vectors, and $\mathbf{z}_{g=i} \in \mathbb{R}^d$ is the embedding of the i-th grading level exemplar. The combination of Eq. 2 and 3 identifies most zero-score responses without enumerating all possible zero-level exemplars. Since Eq. 2 and 3 are non-parametric, we can identify zero-score responses under the zero-shot setting.

4 Experiment

This section aims to quantitatively justify the following three research questions (**RQ**): (1) How accurate is the proposed method in scoring student responses? (2) Can the performance of the proposed method be further improved if a few sample responses are available? (3) How does the quality of sample responses affect the method performance under the few-shots setting?

4.1 Setup

Dataset. We choose a subset of an existing dataset of argumentation items [36] for our experiments. The dataset originally consists of eight written response items sharing the same context regarding gases. The items require varied levels of cognitive demands aligned with a learning progression of argumentation in science [22]. Students' responses were scored based on their performance on claim stating, evidence clarifying, and warrants using. According to the different cognitive demands of each item, the scoring rubrics varied between 2 to 4 levels.

Meanwhile, each item is designed with different levels of complexity, diversity (refers to the three-dimensional science learning requirement), and structure (refers to the learning progression of scientific argumentation [10]. In this study, we select three items (G4, G5, and G6) sharing the same diversity, structures, and rubrics level but distributing in two complexity levels as our downstream tasks. Overall, the dataset contains 2081 labeled responses (770 for item G4, 669 for item G5, and 642 for item G6) from 931 students of grades 5 to 8.

Dataset Splitting To mimic a few-shot setting, we only leave three samples at each grading level to answer **RQ2** and **RQ3**. Here, no valid set remains because tuning the hyper-parameters for each item is not encouraged. Since almost the entire dataset is considered the test set, the K-folds setting is unnecessary.

Metric. We calculate the Cohen's kappa (Kappa) and F1 scores to measure the performance of the machine learning models for auto-scoring. Cohen's kappa is one of the standard rater agreement indices to quantify levels of agreement between computer scoring and human expert scoring [2]. Kappa value ranges from -1 to 1, where 1 indicates a strong agreement between the human score and the machine score and 0 refers to an opposite agreement. Typically, a machine learning model can be accepted if it reaches around 0.4 Kappa score [39]. We also report the F1 score, the weighted harmonic mean of Recall and Precision, to evaluate the model performance [24].

Baseline. We compare MeNSP with some baseline methods to measure the effectiveness of MeNSP. Under the zero-shot setting, we perform **Random** strategy as our baseline, which randomly scores the student responses from 0 to 2 (three grading levels). Under the few-shots setting, we follow the previous studies [36] that performed machine learning in the automatic scoring task. Specifically, we choose some popular ensemble models as the baselines, including Gradient-Gased Decision Tree (**GBDT**), Random Forest with Decision Tree (**RFDT**), and the simple Voting strategy (**Vote**) over five basic models (e.g., Naive Bayes, Decision Tree, Logistic Regression, Multilayers Perceptron, and Support Vector Machine). All baselines make decisions based on the TF-IDF [26] scores of words presented in the student responses.

Exemplar Design. We first manually develop exemplars for each task according to the rubric of the highest level (level 2) [10]. Human experts are involved to ensure that the level 2 exemplars contain all elements of a perfect argument to the greatest extent. Then, we delete elements level-by-level and adapt sentences to meet the rubrics of level 1 and level 0. These exemplars are the prompts used in the zero-shot experiment.

Sample Design. To examine the model performance training with few samples, we use two strategies to generate samples for the few-shots tuning: (1)

Table 1. Machine automatically scoring performance.

Shot	Sample	Model	G4		G5		G6	
			Kappa (%)	F1 (%)	Kappa (%)	F1 (%)	Kappa (%)	F1 (%)
0	–	Random	$-0.2_{\pm 3.4}$	$32.9_{\pm 1.9}$	$-1.1_{\pm 2.2}$	$35.8_{\pm 1.5}$	$-3.3_{\pm 3.6}$	$30.6_{\pm 2.0}$
		MeNSP	$30.3_{\pm 0.3}$	$54.2_{\pm 0.2}$	$57.2_{\pm 1.1}$	$81.1_{\pm 0.5}$	$34.5_{\pm 0.3}$	$57.0_{\pm 0.3}$
1	Random	RFDT	$-1.1_{\pm 2.9}$	$23.8_{\pm 5.5}$	$6.5_{\pm 9.4}$	$50.0_{\pm 14.8}$	$-1.1_{\pm 4.5}$	$24.1_{\pm 11.3}$
		GBDT	$5.6_{\pm 7.2}$	$34.7_{\pm 7.3}$	$6.1_{\pm 16.2}$	$43.6_{\pm 17.9}$	$3.3_{\pm 9.9}$	$34.6_{\pm 11.7}$
		Vote	$3.1_{\pm 11.2}$	$33.6_{\pm 9.1}$	$20.9_{\pm 12.2}$	$62.0_{\pm 8.9}$	$7.3_{\pm 9.5}$	$36.5_{\pm 14.1}$
		MeNSP	$35.9_{\pm 3.2}$	$57.9_{\pm 2.2}$	$43.0_{\pm 10.1}$	$74.3_{\pm 6.1}$	$25.7_{\pm 4.3}$	$53.2_{\pm 3.6}$
	Manual	RFDT	$0.1_{\pm 0.1}$	$26.3_{\pm 0.1}$	$0.3_{\pm 0.3}$	$51.5_{\pm 0.1}$	$0.0_{\pm 0.0}$	$34.2_{\pm 0.0}$
		GBDT	$0.0_{\pm 0.0}$	$26.2_{\pm 0.0}$	$1.5_{\pm 1.5}$	$52.5_{\pm 1.0}$	$0.7_{\pm 0.3}$	$34.8_{\pm 0.3}$
		Vote	$7.0_{\pm 1.7}$	$34.0_{\pm 1.4}$	$5.6_{\pm 0.7}$	$52.2_{\pm 0.4}$	$-0.14_{\pm 0.0}$	$34.2_{\pm 0.0}$
		MeNSP	$35.5_{\pm 7.1}$	$59.1_{\pm 1.8}$	$27.0_{\pm 4.5}$	$56.5_{\pm 0.5}$	$30.9_{\pm 1.8}$	$56.8_{\pm 0.6}$
3	Random	RFDT	$5.5_{\pm 6.2}$	$33.3_{\pm 5.9}$	$13.2_{\pm 14.4}$	$49.8_{\pm 19.8}$	$3.6_{\pm 2.2}$	$33.1_{\pm 11.2}$
		GBDT	$4.0_{\pm 7.2}$	$33.2_{\pm 7.4}$	$14.0_{\pm 16.0}$	$49.6_{\pm 19.9}$	$13.6_{\pm 3.6}$	$46.2_{\pm 1.7}$
		Vote	$15.6_{\pm 3.4}$	$44.8_{\pm 2.0}$	$20.9_{\pm 16.7}$	$57.4_{\pm 10.8}$	$13.1_{\pm 4.9}$	$44.1_{\pm 6.9}$
		MeNSP	$37.5_{\pm 3.3}$	$59.0_{\pm 2.0}$	$51.6_{\pm 8.2}$	$78.7_{\pm 3.6}$	$27.8_{\pm 6.7}$	$53.4_{\pm 4.6}$
	Manual	RFDT	$8.1_{\pm 3.9}$	$34.2_{\pm 3.7}$	$0.7_{\pm 0.4}$	$51.6_{\pm 0.1}$	$0.0_{\pm 0.0}$	$34.2_{\pm 0.0}$
		GBDT	$17.1_{\pm 0.4}$	$45.3_{\pm 0.3}$	$7.6_{\pm 0.2}$	$53.2_{\pm 0.1}$	$1.03_{\pm 0.0}$	$35.64_{\pm 0.0}$
		Vote	$14.8_{\pm 1.0}$	$43.9_{\pm 1.0}$	$8.1_{\pm 2.1}$	$52.5_{\pm 1.1}$	$7.83_{\pm 0.6}$	$40.07_{\pm 0.4}$
		MeNSP	$33.8_{\pm 1.2}$	$56.2_{\pm 0.9}$	$24.8_{\pm 1.1}$	$58.2_{\pm 1.4}$	$30.1_{\pm 0.5}$	$56.9_{\pm 0.3}$

We randomly select student responses for each level from the dataset. (2) We use ChatGPT[2] to generate new responses based on the exemplar at each grading level and then conduct manual inspection and adjustment to ensure the machine-generated responses meet the rubrics.

4.2 Results

To reduce the uncertainty of experiments, we run through our experiment codes over five random seeds, including data splitting and few-shots tuning, and report both the mean and standard deviation of each metric on each item in Table 1.

RQ1: MeNSP Effectively Scores Responses Based on Context and the Exemplars Without Training (Zero-Shot). Under the zero-shot setting, all the Kappa values of the three items increase from negative to more than 0.30, and the F1 scores also increase among the three items, which indicates the effectiveness of MeNSP in the zero-shot condition (compared with Random baseline). One of the items (G5) has its Kappa exceed 0.50 with MeNSP, which reaches an acceptable benchmark of the trained model. Although MeNSP surpluses a 0.4 Kappa score only on G5, it still achieves at least a 0.3 Kappa score on the other two items, showing the great potential of MeNSP.

[2] ChatGPT is available at https://chat.openai.com/chat.

RQ2: MeNSP Can be Improved by Training on a Few Sample Responses on Easier Items (Few-Shots). To discuss with the few-shots setting, MeNSP performs better than itself on item G4, with the Kappa of random MeNSP increasing from 0.30 (zero-shot) to 0.36 (1-shot), and to 0.38 (3-shots). However, increasing the number of sample responses does not improve MeNSP further on G5 and G6. Although MeNSP's performance fails on two items, it still has a higher average Kappa and lower standard deviation than all baselines under the 1-shot and 3-shots settings on all three items. This indicates that MeNSP performs better and is more stable than other models.

Considering the lower complexity level of G4 (level 2) than G5 and G6 (level 3), we first conclude that MeNSP's performance can be improved by training on some responses on the low complexity items. However, for items with high complexity, we argue that the performance of MeNSP heavily relies on the characteristics of the sample responses provided for training. According to our experiments, we observe an improvement of Kappa on G5 from 0.59 to 0.61 with random seed 55301 and from 0.57 to 0.60 with random seed 9, and a gain of Kappa on G6 from 0.34 to 0.35 with random seed 45983. A possible speculation of the successful improvement with these random seeds is that the sample responses selected for training are diverse enough[3] to help MeNSP to learn complete and unbiased knowledge about the general scoring rubrics. This speculation also aligns with the finding of RQ3.

RQ3: The Quality of the Provided Samples Affects the Few-Shots Learning Performance. We compare MeNSP on item G4 with different sample-gathering strategies (random or manual). The results show that with the 1-shot setting, both strategies lead MeNSP to a similar performance, with the average Kappa close to 0.36. However, with the 3-shot setting, the random strategy performs better than the manual strategies, with the average Kappa increase from 0.34 (manual) to 0.38 (random). As the random strategy extracts samples from the real student responses, the manual strategy provides augmented student responses similar to the exemplars[4]. Therefore, combined with the observation of improvements with three random seeds on G5 and G6 discussed earlier, we summarize that the diversity of the given samples is a potential factor that impacts the efficiency of few-shots learning.

[3] G5 random samples at grade 2: (1) if the gas bubbles in soda go to the top, then so should the gas particles in the box. (2) Sam could use the Gas bubble idea to support his claim. Gas bubbles are made up of particles. The gas bubbles float to the top of the can. So, the particles in the box will rise to the top also. (3) The gas particles go to the top, like in soda.

[4] G4 manual samples at grade 1: (1) The evidence suggests that the air in the balloon has spread throughout. (2) Charlie's claim is right because the air in the balloon spreads everywhere. (3) That air in the balloon spreads everywhere supports claim.

5 Conclusion and Discussion

In this study, we develop a zero-shot approach (MeNSP) to score student responses automatically. We propose three research questions and examine MeNSP's performance with 1-shot and 3-shot scenarios. Through experiments, we prove the effectiveness of MeNSP on automatically scoring student responses based on exemplars without training. We also find that increasing training samples can improve MeNSP's performance on items with lower complexity. However, the quality of the improvement is related to the characteristics of the training samples, for example, the diversity of the sample responses. Given that our goal of this study is to preliminary demonstrate the feasibility of MeNSP and the prompt learning methods in automatic scoring science assessment, the machine scoring accuracy may be used for low-stake formative assessment practices. Future research should improve the performance of MeNSP and use varying datasets to verify the approach.

Acknowledgement. The study was funded by National Science Foundation (NSF) (Award # 2101104, 2138854, PI: Zhai). Any opinions, findings, conclusions, or recommendations expressed in this material are those of the author(s) and do not necessarily reflect the views of the NSF.

References

1. Amerman, H., et al.: Does transformer deep learning yield more accurate sores on student written explanations than traditional machine learning? In: AERA (2023)
2. Bejar, I.I.: A methodology for scoring open-ended architectural design problems. J. Appl. Psychol. **76**, 522 (1991)
3. Brown, T., et al.: Language models are few-shot learners. Adv. Neural Inf. Process. Syst. **33**, 1877–1901 (2020)
4. Council, N.R., et al.: A Framework for K-12 Science Education: Practices, Crosscutting Concepts, and Core Ideas. National Academies Press (2012)
5. Devlin, J., Chang, M.W., Lee, K., Toutanova, K.: BERT: pre-training of deep bidirectional transformers for language understanding. In: ACL (2019)
6. Gao, T., Fisch, A., Chen, D.: Making pre-trained language models better few-shot learners. arXiv preprint arXiv:2012.15723 (2020)
7. Gerard, L., Kidron, A., Linn, M.C.: Guiding collaborative revision of science explanations. Int. J. Comput.-Support. Collab. Learn. **14**(3), 291–324 (2019). https://doi.org/10.1007/s11412-019-09298-y
8. Harris, C.J., et al.: Designing knowledge-in-use assessments to promote deeper learning. Educ. Meas. Issues Pract. **38**, 53–67 (2019)
9. Haudek, K.C., et al.: What are they thinking? Automated analysis of student writing about acid-base chemistry in introductory biology. Life Sci. Educ. **11**, 283–293 (2012)
10. Haudek, K.C., Zhai, X.: Exploring the effect of assessment construct complexity on machine learning scoring of argumentation (2021)
11. Lee, H.S., et al.: Automated text scoring and real-time adjustable feedback: supporting revision of scientific arguments involving uncertainty. Sci. Educ. **103**, 590–622 (2019)

12. Litman, D.: Natural language processing for enhancing teaching and learning. In: Thirtieth AAAI Conference on Artificial Intelligence (2016)
13. Liu, O.L., et al.: Automated scoring of constructed-response science items: prospects and obstacles. Educ. Meas. Issues Pract. **33**, 19–28 (2014)
14. Liu, P., et al.: Pre-train, prompt, and predict: a systematic survey of prompting methods in natural language processing. arXiv preprint arXiv:2107.13586 (2021)
15. Liu, X., et al.: GPT understands, too. arXiv preprint arXiv:2103.10385 (2021)
16. Liu, X., et al.: P-tuning v2: prompt tuning can be comparable to fine-tuning universally across scales and tasks. arXiv preprint arXiv:2110.07602 (2021)
17. Lu, Y., et al.: Fantastically ordered prompts and where to find them: overcoming few-shot prompt order sensitivity. arXiv preprint arXiv:2104.08786 (2021)
18. Maestrales, S., et al.: Using machine learning to score multi-dimensional assessments of chemistry and physics. J. Sci. Educ. Technol. **30**, 239–254 (2021)
19. Mayer, C.W., Ludwig, S., Brandt, S.: Prompt text classifications with transformer models! An exemplary introduction to prompt-based learning with large language models. J. Res. Technol. Educ. **55**, 125–141 (2022)
20. Nehm, R.H., Ha, M., Mayfield, E.: Transforming biology assessment with machine learning: automated scoring of written evolutionary explanations. J. Sci. Educ. Technol. **21**, 183–196 (2012)
21. Omizo, R., Meeks, M., Hart-Davidson, W.: Detecting high-quality comments in written feedback with a zero shot classifier. In: ACM ICDC (2021)
22. Osborne, J.F., et al.: The development and validation of a learning progression for argumentation in science. J. Res. Sci. Teach. **53**, 821–846 (2016)
23. Pellegrino, J.W.: Proficiency in science: assessment challenges and opportunities. Science **340**, 320–323 (2013)
24. Powers, D.M.: What the F-measure doesn't measure: features, flaws, fallacies and fixes. arXiv preprint arXiv:1503.06410 (2015)
25. Riordan, B., et al.: An empirical investigation of neural methods for content scoring of science explanations. In: Proceedings of the Fifteenth Workshop on Innovative Use of NLP for Building Educational Applications (2020)
26. Salton, G., Buckley, C.: Term-weighting approaches in automatic text retrieval. Inf. Process. Manage. **24**, 513–523 (1988)
27. Schick, T., Schütze, H.: It's not just size that matters: small language models are also few-shot learners. arXiv preprint arXiv:2009.07118 (2020)
28. Schick, T., Schütze, H.: Exploiting cloze-questions for few-shot text classification and natural language inference. In: Proceedings of the 16th Conference of the European Chapter of the Association for Computational Linguistics: Main Volume, pp. 255–269 (2021)
29. Shin, T., et al.: AutoPrompt: eliciting knowledge from language models with automatically generated prompts. arXiv preprint arXiv:2010.15980 (2020)
30. Su, Y., et al.: On transferability of prompt tuning for natural language processing. In: NACL, pp. 3949–3969 (2022)
31. Uhl, J.D., et al.: Introductory biology undergraduate students' mixed ideas about genetic information flow. Biochem. Mol. Biol. Educ. **49**, 372–382 (2021)
32. Vu, T., et al.: SPoT: better frozen model adaptation through soft prompt transfer. arXiv preprint arXiv:2110.07904 (2021)
33. Wolfe, E.W., Wendler, C.L.W.: Why should we care about human raters? Appl. Meas. Educ. **33**(3), 89–190 (2020)
34. Wu, X., et al.: A survey of graph prompting methods: techniques, applications, and challenges. arXiv preprint arXiv:2303.07275 (2023)

35. Zhai, X.: Practices and theories: how can machine learning assist in innovative assessment practices in science education. J. Sci. Educ. Technol. **30**, 139–149 (2021)
36. Zhai, X., Haudek, K.C., Ma, W.: Assessing argumentation using machine learning and cognitive diagnostic modeling. Res. Sci. Educ. **53**, 405–424 (2022)
37. Zhai, X., Krajcik, J., Pellegrino, J.W.: On the validity of machine learning-based next generation science assessments: a validity inferential network. J. Sci. Educ. Technol. **30**, 298–312 (2021)
38. Zhai, X., Shi, L., Nehm, R.H.: A meta-analysis of machine learning-based science assessments: factors impacting machine-human score agreements. J. Sci. Educ. Technol. **30**, 361–379 (2021)
39. Zhai, X., Yin, Y., Pellegrino, J.W., Haudek, K.C., Shi, L.: Applying machine learning in science assessment: a systematic review. Stud. Sci. Educ. **56**, 111–151 (2020)
40. Zhang, M., et al.: Automatic short math answer grading via in-context meta-learning. arXiv preprint arXiv:2205.15219 (2022)
41. Zhong, R., Lee, K., Zhang, Z., Klein, D.: Adapting language models for zero-shot learning by meta-tuning on dataset and prompt collections. In: EMNLP (2021)

Learning When to Defer to Humans for Short Answer Grading

Zhaohui Li[1(✉)], Chengning Zhang[1], Yumi Jin[1], Xuesong Cang[2], Sadhana Puntambekar[2], and Rebecca J. Passonneau[1]

[1] Department of Computer Science and Engineering, Pennsylvania State University, State College, USA
{zjl5282,cbz5118,yfj5089,rjp49}@psu.edu

[2] Department of Educational Psychology, University of Wisconsin-Madison, Madison, USA
xcang@wisc.edu, puntambekar@education.wisc.edu

Abstract. To assess student knowledge, educators face a tradeoff between open-ended versus fixed-response questions. Open-ended questions are easier to formulate, and provide greater insight into student learning, but are burdensome. Machine learning methods that could reduce the assessment burden also have a cost, given that large datasets of reliably assessed examples (labeled data) are required for training and testing. We address the human costs of assessment and data labeling using selective prediction, where the output of a machine learned model is used when the model makes a confident decision, but otherwise the model defers to a human decision-maker. The goal is to defer less often while maintaining human assessment quality on the total output. We refer to the deferral criteria as a *deferral policy*, and we show it is possible to learn when to defer. We first trained an autograder on a combination of historical data and a small amount of newly labeled data, achieving moderate performance. We then used the autograder output as input to a logistic regression to learn when to defer. The learned logistic regression equation constitutes a deferral policy. Tests of the selective prediction method on a held out test set showed that human-level assessment quality can be achieved with a major reduction of human effort.

Keywords: Selective prediction · Automated short answer grading

1 Introduction

Assessment is important both for student learning and for instructors' implementation of a curriculum. Asking students to show their reasoning through writing is widely believed to facilitate better understanding [9], which would argue for open-ended questions instead of fixed response ones, such as multiple choice or true/false. But the choice between the two is heavily weighted towards fixed response types because they can be graded automatically in little time, thus minimizing assessment effort and maximizing timeliness. Deep learning methods that could reduce the assessment burden also have a cost, since they depend

N. Wang et al. (Eds.): AIED 2023, LNAI 13916, pp. 414–425, 2023.
https://doi.org/10.1007/978-3-031-36272-9_34

on large datasets of reliably assessed examples (labeled data) for training and testing autograder models. To offset the costs of human assessment of student work and of providing labeled data for machine learning, we present a human-in-the-loop [15] method that also leverages transfer learning from historical data.

The dataset used in this study was from middle school students participating in a unit on physics. In this unit, students learned about the basic concepts in mechanics while designing a roller coaster using a simulation. The curriculum was designed for students to build on science knowledge, adding more concepts as students iteratively engaged in the roller coaster design process. After each design cycle, students wrote short answers to open-ended questions based on the content they just learned. The questions helped teachers and researchers assess students' current understanding, as well as how they progressively learned science ideas throughout the unit. The data was collected from a single school from students in classes taught by three teachers. The school was situated in a rural-suburban area in Midwestern United States. Students from rural as well as suburban areas attended the school. Students in this district came from diverse backgrounds representing 22 different languages. Approximately twenty two percent of students in the school district identified as economically disadvantaged.

Human-in-the-loop approaches include selective prediction, which uses confident decisions from a machine learned model, but otherwise defers to a human. Recent work indicates that softmax probabilities from machine learned models can serve as a measure of relative confidence [11,23], and that the human decision makers perform better when they are given no information about the model prediction [2], with the result that human effort can be offline. We refer to the deferral criteria as a *deferral policy*. Using an existing high-performing automated short answer grading (ASAG) model [13], we first train this model on a combination of historical data and a small amount of newly labeled data, achieving moderate accuracy. Then we compare a very simple, and therefore elegant, manually derived heuristic deferral policy with machine learned policies. On a held out test set, the learned policies generalize better than the heuristic one. Further, the learned policy has a tuning parameter to control the tradeoff between accuracy and human effort. To our knowledge, our work is the first to learn controllable selective prediction deferral policies.

2 Related Work

Recent machine learning for automatic short answer grading (ASAG) relies primarily on deep learning, especially transformer models such as BERT [5]. Most work trains a pipeline consisting of an encoder for the student answer and reference answer followed by a classifier layer to determine answer correctness. Two often-used benchmark datasets are SemEval [6] and ASAP [21]. SemEval has 2-way, 3-way and 5-way correctness labels in three increasingly challenging settings: unseen answers, unseen questions, and unseen domain. Using a biLSTM, Riordan et al. [17] achieved quadratic weighted kappa of 0.77 on ASAP. On the same dataset with a model that encodes rubric elements, Wang et al. [21] achieved similar results. On a large, proprietary dataset from psychology, Liu et al. [14] trained separate encoders for student and reference answers, followed by another transformer layer to merge them, followed by a multilayer perceptron (MLP) classifier. They achieved 0.89 accuracy against a simple logistic regression baseline that achieved 0.83. Also on a proprietary dataset, Sung et al. [19] achieved accuracies between 0.78 and 0.81 using BERT to separately encode the question, and the concatenation of the student and reference answer. Saha et al. [18] separately encoded the question, the reference answer and the student answer using InferSent [3], and combined the three vectors with manually engineered features as input to a simple classifier. On the nine SemEval tasks, their accuracies ranged from 0.51 to 0.79. Using only the student answer and reference answer as input, Ghavidel et al. [8] compared BERT (cased and uncased) with XLNET on the SciEntsBank subset of SemEval, achieving SOTA accuracies. On the full SemEval, Li et al. [13]'s work SFRN used BERT to separately encode the question, the reference answer, and the student answer, and learned a relation vector over the three encodings, followed by an MLP classifier. SFRN achieved accuracies on SemEval from 0.40 to 0.91, and either beat the state-of-the-art or was competitive. In particular, SFRN did far better than other approaches on the 5-way classification. Given that SFRN is superior on public datasets to the other models examined here, we use SFRN in our selective prediction system. Multiple approaches to human-machine teams exist, including human-in-the-loop, where humans intervene on items that are difficult for a given algorithm [15], and algorithm-in-the-loop, which privileges human decision-makers but incorporates algorithms for efficiency or improvements in total accuracy [10]. Factors that influence whether a human-machine team works better than either agent alone include the type of decision-making algorithm, the level of risk of the decision, the transparency of the machine decisions, and human attitudes towards AI (e.g., automation bias) or biases (e.g., racial) [4]. Evidence suggests that for the supervised learning classification task of assigning a correctness label, overall accuracy can be maximized by a selective prediction approach to human-in-the-loop AI [22]. In a selective prediction experiment on classification of images of landscapes to detect animals to benefit wildlife in the Serengeti, Bondi et al. [2] found that the total accuracy of a human-in-the-loop system was best with defer-only, meaning humans are told only that the algorithm defers, rather than showing the algorithm's decision or confidence. That it is critical to present the right information to humans in a human-machine team is consistent with initial experiments on human-in-the-loop essay grading [1]. Based on defer-only, selective prediction can be developed by acquiring human decisions prior to developing the selective prediction system, which is what we do here. Hendrycks and Gimpel [11] introduced a simple selective prediction method using the maximum softmax probability as the confidence estimator. Other work trained a calibrator model on the softmax output as a selective prediction confidence [7,12,20]. Our work is most similar to calibration methods, but relies on a weight hyperparameter to control accuracy and deferral rate.

3 Description of the Data

Our dataset consists of short answer responses to open-ended questions from a middle school physics curriculum about roller coasters collected from classrooms of seven teachers. Responses were coded on a 3-point scale as incorrect, partially correct, or correct. Here we describe the nine open-ended questions and their relative difficulty for students, the coding reliability, the size of the dataset and the labeled subset, and our historical data used in transfer learning with SFRN.

Students participated in a three to four week unit to learn about the physics of energy transfer by designing a roller coaster in simulation. There were five labs that in turn addressed the following conceptual relationships: initial drop height and energy; the effect of hills on energy; mass and energy; height and speed; mass and speed. At the end of each lab, students responded to two to five open-ended questions along with some multiple-choice questions. There were thirteen open-ended questions and six multiple choice questions across labs. The open-ended questions covered nine relations among physics concepts (e.g., greater height corresponds to more potential energy), including the law of conservation of energy. The questions thus assessed students' grasp of the interconnectedness among science concepts. Our experiments used the nine open-ended questions that assessed understanding of physics relations.

Four researchers working in two groups coded the answers as correct, partially correct or incorrect. One group coded labs one and three; and the other group coded the rest. Each group coded a randomly selected set of responses for each lab. Inter-rater reliability (IRR) was assessed using two-way random, consistency average-measures of intra-class correlations (ICCs). A 95% confidence interval with two-tailed tests was performed. The cutoff for qualitative ratings of agreement based on ICC values was 0.90, and for each question was 0.80. The agreement measures for each question are displayed in the ICC column of Table 1. To achieve reliability, the coders went through two to four rounds of coding. Disagreements were resolved in team discussions. The coding scheme was refined or extended in the process of making decisions for the disagreements.

The full dataset consists of 4,703 items, shown broken down by question in Table 1. Table 1 also shows the breakdown by question for two phases of coding: the first phase where reliability was developed (Coding 1), and the second phase after reliability was achieved (Coding 2). Counts per question, ICC, human accuracy and average student score are shown for Coding 1, while Coding 2 and Combined (Coding 1 + Coding 2) show the counts for each question and average student score. We split Coding 1 into a train set of 408 examples to train SFRN, and a dev set of 447 examples to evaluate SFRN and to develop deferral policies. SFRN was trained on a combination of this small subset of the spring 2022 data, and historical data that is described below. We use the Coding 2 data as a test set to evaluate the deferral policies.

As shown by the average score column in Table 1, the number of questions per lab decreased in later labs. With the exception of Lab 2, questions within a lab generally increased in difficulty, with a trend towards greater difficulty in later labs. The questions that were easiest for students (Lab1.Q1, Lab3.Q3 and

Table 1. A breakdown by question is shown for the full data set, the first and second coding phases (Cod.1, Cod. 2), and the combination of the two codings (Comb.). For Cod. 1, the table shows ICC, average human accuracy, and average student scores. Cod. 2 and Comb. have the counts and average student scores. The last *Summary* row has column totals of counts (Full, Cod.1, Cod.2), average of averages for Hum. Acc. and Avg. Sc. columns, and the overall ICC score.

Qid	Full	Cod. 1	ICC	Hum. Acc.	Avg. Sc.	Cod. 2	Avg. Sc.	Comb.	Avg. Sc.
Lab1.Q1	539	92	0.93	0.97	1.88	56	1.68	148	1.80
Lab1.Q2	534	92	0.92	0.95	1.22	54	0.76	146	1.05
Lab1.Q4	513	92	0.92	0.92	1.12	54	0.70	146	0.97
Lab2.Q3	528	94	0.92	0.91	1.07	49	0.73	143	0.96
Lab2.Q4	514	94	0.84	0.89	1.43	50	1.00	144	1.28
Lab3.Q3	522	85	0.95	0.96	1.85	52	1.50	137	1.72
Lab3.Q4	507	85	0.94	0.94	1.72	50	1.60	135	1.67
Lab4.Q2	514	85	0.91	0.93	1.56	52	1.62	137	1.58
Lab5.Q2	532	136	0.87	0.91	1.31	50	1.46	186	1.35
Summary	4703	855	0.93	0.93	1.46	467	1.23	1322	1.37

Lab3.Q4) tended to have the highest ICC and human accuracy, but otherwise these rankings do not correlate well. In the results section, we show differences in selective prediction performance broken down by question.

The historical data consists of 6,956 student responses to 33 questions drawn from a decade of pre- and post-tests for assessment of middle school students' understanding of relationships among physics concepts, such as how the height of an inclined plane affects the amount of force needed to lift a load. Validity of questions was ensured through consultation with physics experts, teachers, and statisticians. As in our current data, questions were assessed on a 3-pt scale, with 10–25% of each test coded independently by two researchers who achieved at least 85% agreement, a standard metric at the time of the original studies. As the original reliability coding is no longer available, we cannot apply an agreement coefficient post-hoc. However, chance-adjusted agreement scores reduce to percent agreement as chance agreement approaches zero [16], which would hold for this data where the proportions of correct, partially correct and incorrect were only mildly skewed (respectively 25%, 42% and 33%).

4 Methods

To produce a selective prediction system, we first trained a classifier to predict correctness of each student response. Here we describe how we used historical data combined with a small amount of labeled data to train the classifier (N = 408; from Coding 1 in Table 1), testing it on our dev data (N = 447; also drawn from Coding 1). Using this classifier, the SFRN output on the dev data was then used to develop deferral policies, which we evaluated on the test data

(N = 467; Coding 2). Here we describe the classifier, a heuristic deferral policy, and our use of linear machine learning to learn deferral policies.

For the classifier, we chose the SFRN algorithm for its superior performance, as reviewed above in Sect. 2. In the SemEval dataset that SFRN was developed on, there are often multiple reference answers per question prompt. For each student response, three text strings are first encoded using BERT [5]: the question prompt, a reference answer, and the student answer. This triple of embeddings is the input to a multilayer linear perceptron (MLP) that learns a single relation vector for each reference answer for a given student answer. A fusion relation is then learned that merges all the relation vectors for a given student answer, which is then classified for correctness by a final learned MLP. As well as creating the question prompts for the historical and spring 2022 data, we also created reference answers for each correctness class. We first trained SFRN on the historical data alone (6,956 student answers, 33 questions), then in combination with the training subset of Coding 1. When trained only on the historical data, SFRN had 45.86% accuracy on dev. By adding in the training subset of the spring 2022 data, SFRN performance improved to 62.41%. This is a substantial increase that shows the power of transfer learning, but it is still well below average human accuracy of 93%.

Given that the softmax probability from a neural model's output layer performs well as a confidence estimator for selective prediction [11,23], we develop a heuristic deferral policy D based on a probability threshold θ:

$$D(\mathcal{P}(x);\theta) = \begin{cases} 1, & \text{if } \max \mathcal{P}(x) \le \theta \\ 0, & \text{otherwise} \end{cases} \tag{1}$$

where $P(x)$ is the probability distribution over the three classes, 1 represents the decision to defer, and 0 triggers use of the model decision. We test deferral policies by retrieving the correct label when the policy says to defer.

Figure 1 is a confusion matrix of the SFRN predictions on dev, where each cell has the count and whisker plots for the probabilities of the three classes. The cells on the diagonal show that when SFRN predicts correct, it is both accurate (89.58%) and quite certain (mean probability of correct above 0.90), but has lower accuracy and confidence on its other predictions. The first column shows that a single threshold of around 0.85 or so when SFRN predicts correct would include most of the true correct, and exclude the predictions of correct that are best to defer on. However, a single θ is not obvious from this figure.

Again using SFRN output on dev, Fig. 2 shows a separate curve for each correctness class with values of θ on the x-axis and class accuracy on the y-axis. The incorrect and partially correct curves intersect at $\theta = 0.68$ with accuracy above 90%, but with lower class accuracy for correct. A heuristic policy with $\theta = 0.68$ applied to dev yields a total accuracy of 80%, and a low deferral rate of 30%. We report performance of this policy on the test set in the next section.

Based on the small size of the dev data for policy training, we use linear classifiers to learn deferral policies from SFRN's output on the dev data. We selected logistic regression and random forest for their interpretability and good

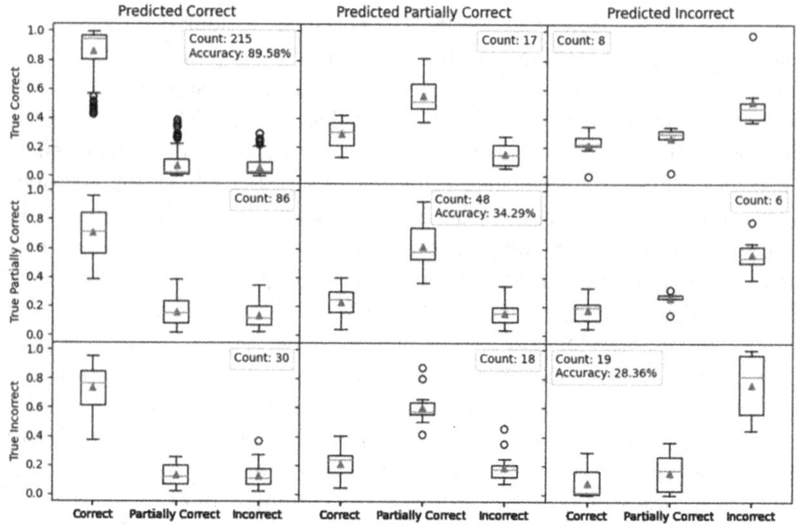

Fig. 1. Confusion matrix of SFRN on dev (N = 447). Each cell contains the count and three whisker plots for the probability distribution over the three classes.

performance. We tested a variety of feature representations, drawing from a set of 11: SFRN probabilities for each class (P0, P1, P2), SFRN prediction (Max), class with the next highest SFRN probability (Mid), SFRN accuracy on the given question (SFRN-Qdiff), average student score per question (stud-Qdiff), ICC score per question (coder-Qdiff). We also tested a feature for noise injection, sampling numbers from a Gaussian distribution with mean and standard deviation of 0.5. Max is the inverse of the proportion of each class in SFRN output. This captures the class influence, and puts the values on the same scale as the other features (0 to 1) for better interpretability of the learned weights.

As a proxy for a ground truth label of whether to defer, we use the SFRN correctness to label each example as +defer if the model was incorrect and -defer otherwise. We use a weighted cross entropy loss, where a weight hyperparameter λ on the positive class allows us to increase the weight on the decision to defer, resulting in a higher deferral rate but also higher accuracy. In the next section we show how accuracy and deferral rate on the test set varies with values of λ.

5 Results

Here we report performance of the deferral policies on the test data (N = 467). Because logistic regression (LR) models performed somewhat better than random forest, we only discuss the LR policies. Below we present the LR training parameters, followed by a discussion of the features that performed best. For the best performing set of features, we then discuss how the λ weight in our loss function affects the tradeoff between accuracy and deferral rate. Then we

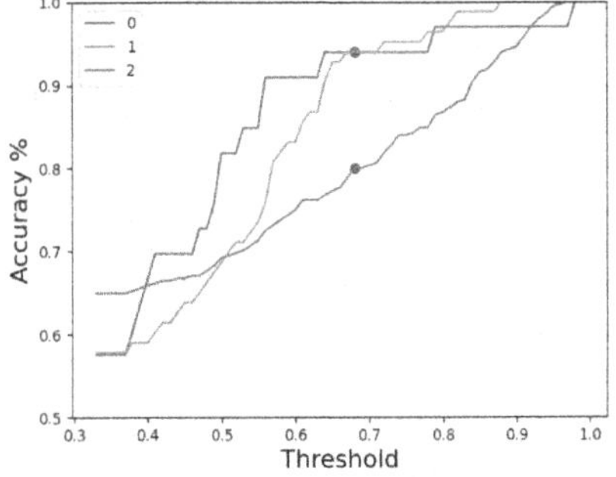

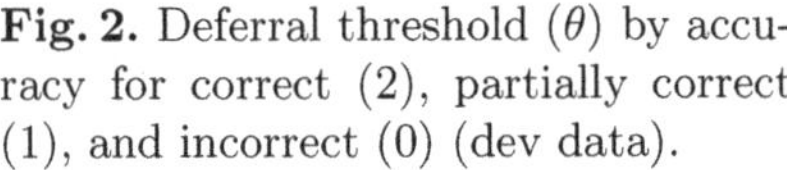
Fig. 2. Deferral threshold (θ) by accuracy for correct (2), partially correct (1), and incorrect (0) (dev data).

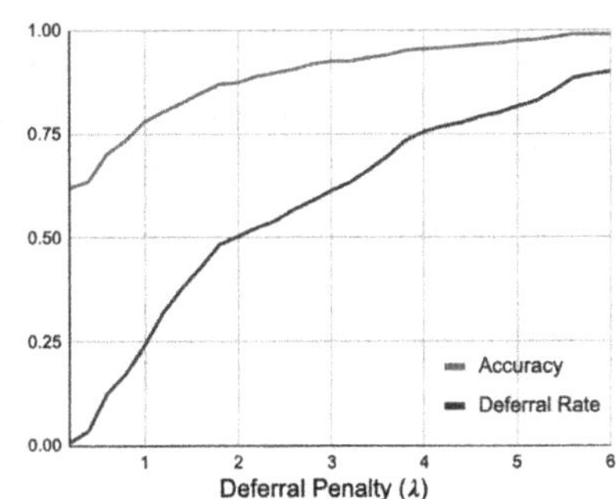

Fig. 3. Sensitivity plot for the loss weight (λ) by accuracy (blue) and deferral rate (red) (test data). (Color figure online)

Table 2. Feature sets (N = 3 to 5), weights and performance for three LR models.

Learned Weights					Intercept	Performance	
SFRN-QDiff	Max	P0	P1	P2	Intercept	Accuracy	Deferral rate
NA	*NA*	−0.0963	0.8442	−0.8959	5.4066	71.30	17.98
NA	−13.2667	2.6016	2.4180	−1.9343	5.3704	75.18	21.61
−3.4324	−13.6363	3.4137	3.5853	−1.5924	5.9448	**77.97**	**24.17**

compare the performance of SFRN alone, the heuristic policy, and several of the LR policies, showing that the heuristic policy does not generalize well. An LR policy that has the same accuracy as humans has less reduction of human effort, however reasonable accuracy is achieved with a significant reduction of effort. Finally, we report differences in performance across the nine questions prompts.

Using the logistic regression from the scikit learn python library, we found good performance with the default hyperparameters, apart from assigning no regularization penalty (versus the default L2 norm), and a high inverse regularization strength C = 20 versus the default of 1. Of the 11 features we experimented with, we found that six did not contribute to model performance. We compared multiple feature sets from the remaining five.

The five features that had an impact on performance were the SFRN prediction (Max), the SFRN training accuracy on the question (SFRN-QDiff), and the softmax probabilities (P0, P1, P2). A logistic regression deferral policy $\mathcal{D}_{\mathcal{LR}}$ is the learned linear equation for the log odds of the positive class (+defer):

$$\mathcal{D}_{\mathcal{LR}} = \alpha + (\sum_i W_i \times F_i) + \epsilon \tag{2}$$

where α is the intercept, F_i represents the value of feature i for a given input, W_i represents the learned weight for that feature, and ϵ is the residual error. The

Table 3. Accuracies and deferral rates of the heuristic and LR deferral policies on the Dev and Test sets, compared to SFRN alone (None). For the learned policies, we show 95% confidence intervals (CIs) on the test set, and also the ICC score with the ground truth labels to compare with human ICC.

Deferral Policy	Dev		Test		ICC
	Accuracy	Def. Rate	Accuracy (95% CI)	Def. Rate (95% CI)	
None	63.08	*NA*	61.75 (60.71, 62.85)	*NA*	0.64
Heuristic	80.08	30.42	76.33 (75.47, 77.38)	27.09 (26.18, 28,09)	0.80
LR ($\lambda = 1$)	77.79	25.75	77.97 (76.66, 79.28)	24.17 (22.85, 25.24)	0.80
LR ($\lambda = 0.9$)	75.32	20.38	75.82 (74.52, 77.14)	**20.48** (19.04, 21.66)	0.78
LR ($\lambda = 1.1$)	79.36	28.88	79.49 (78.33, 80.71)	27.19 (25.70,28.33)	0.82
LR ($\lambda = 3.3$)	95.29	66.72	**92.94** (92.38, 93.57)	65.03 (63.56, 66.43)	**0.94**

sign of W_i indicates whether F_i increases or decreases the log odds of +defer, and higher magnitudes of W_i indicate F_i has greater influence.

Table 2 shows the learned weights and the performance of the LR policies for the softmax probability features alone or in combination with other features; NA indicates the feature was not used. The 5-feature model has the highest accuracy, but also the highest (worst) deferral rate. Because we prioritize accuracy first, our remaining results pertain to this LR. Two of the three negative features (Max, SQRN-Qdiff) have high weights, meaning high values of these features reduce the log odds more: the policy is increasingly less likely to defer for each next class ordered as incorrect, partially correct, correct, and the policy is less likely to defer when it is more accurate on the given question. It has a somewhat lower negative weight on the probability of the correct class. The high positive weights on P0 and P1 mean that the higher the probabilities of incorrect or partially correct, the more often the policy will defer, which is consistent with what we saw in Fig. 1 of lower SFRN accuracies on incorrect and partially correct.

Figure 3 shows the sensitivity of accuracy and deferral rate to values of λ for the 5-feature LR policy. For $\lambda = 1$, accuracy is 77.97% with a low deferral rate of 24.17%. As λ increases, the deferral rate increases faster than accuracy, meaning that in our setting reducing human effort without reducing accuracy is very challenging. Reducing λ below 1 does not increase the accuracy.

Table 3 compares SFRN with no selective prediction, the heuristic policy, and the 5-feature LR policy with 4 different values of λ. To assess variance, we computed confidence intervals (CIs) using 200 iterations of bootstrapped samples (with replacement) using 90% of the test data. All policies are clearly better than the classifier on its own, showing it is possible to greatly increase the accuracy through selective prediction at less than 100% human effort. The heuristic policy, which had higher dev accuracy than LR $\lambda = 1$, does not generalize as well to the test set, and while the LR $\lambda = 1$ policy does not have confidently better accuracy and has equal ICC, it has a confidently lower deferral rate. Compared to LR $\lambda = 1$ policy, LR $\lambda = 1.1$ is about as accurate (overlapping CIs), but has a confidently higher deferral rate, further evidence that our setting is challenging. For $\lambda = 3.3$, the policy is as accurate as expert humans with somewhat higher ICC (0.94 versus 0.93), at two thirds the human effort.

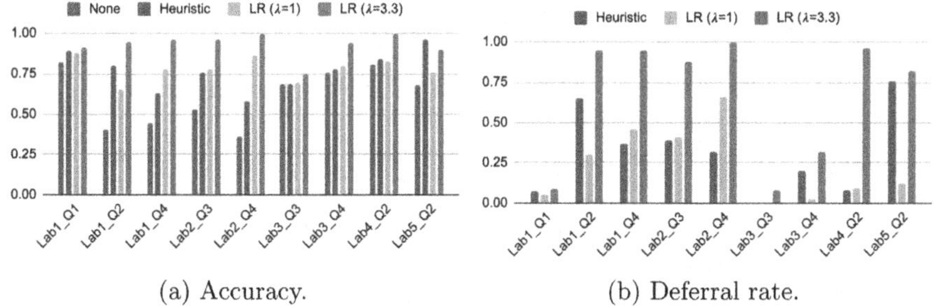

(a) Accuracy.

(b) Deferral rate.

Fig. 4. Performance broken down by question.

Because human reliability varied by question, we examined the performance of the deferral policies broken down by question. Figure 4a) shows that SFRN accuracy varies a lot by question. On questions that are harder for SFRN, there are also greater accuracy differences across policies, especially for Lab2_Q4. The heuristic policy, with an overall deferral rate of 27.09%, defers most for questions Lab1_Q2 and Lab5_Q2, and otherwise much less often. In contrast, LR $\lambda = 3.3$, which has a high overall deferral rate of 65.03%, has low deferral rates only on questions Lab1_Q1, Lab3_Q3 and Lab3_Q4.

6 Discussion and Conclusion

In this paper, we have shown it is possible to learn a selective prediction policy that can achieve expert human accuracy and ICC, while reducing human effort. Our immediate goal is to produce a reliable coding of our dataset of 4,703 items. By labeling 1,322 items, we have learned a selective prediction policy LR $\lambda = 3.3$ that can be expected to achieve expert human performance on the remaining unlabeled data (N = 3,381), where the expert humans would need to label 65.03%, or 2,199 more items. In sum, experts would have labeled 3,521 items, for a 25% reduction in human effort and expert human accuracy. In practice, we plan to learn new LR policies where we split our combined set of 1,322 labels into 50% for training and 50% for development. Instead of a reserved test set, we could verify the quality by re-coding random samples after we use the new LR policy to code the entire dataset using online hu. By retraining SFRN using the historical data combined with 611 labeled examples instead of 408, SFRN should improve. An LR policy learned on the remaining 611 examples should perform better than our current LR $\lambda = 3.3$ policy, given improved performance of SFRN and a larger dataset for policy training of 611 instead of 447 (an increase of 28%).

The main limitation of our work is that it would be preferable to find a learning method for selective prediction that can jointly optimize accuracy and the deferral rate. Another limitation is that both classifier performance and policy performance might have been higher if we had been able to code relatively more of the questions that turned out to be difficult for the classifier. Future work could investigate these two issues, as well as whether selective prediction methods could work in the classroom, which could potentially increase timeliness and accuracy of grading, and avoid the subjectivity of human graders.

References

1. Baikadi, A., et al.: An apprenticeship model for human and AI collaborative essay grading. In: Trattner, C., Parra, D., Riche, N. (eds.) Joint Proceedings of the ACM IUI 2019 Workshops Co-located with the 24th ACM Conference on Intelligent User Interfaces (ACM IUI 2019), vol. 2327 (2019). http://ceur-ws.org/Vol-2327/IUI19WS-UIBK-2.pdf
2. Bondi, E., et al.: Role of human-AI interaction in selective prediction. In: Proceedings of the AAAI Conference on Artificial Intelligence, vol. 36 (5), pp. 5286–5294 (2022). https://doi.org/10.1609/aaai.v36i5.20465
3. Conneau, A., Kiela, D., Schwenk, H., Barrault, L., Bordes, A.: Supervised learning of universal sentence representations from natural language inference data. In: Proceedings of the 2017 Conference on Empirical Methods in Natural Language Processing (EMNLP), pp. 670–680. ACL, Copenhagen (2017). https://doi.org/10.18653/v1/D17-1070
4. De-Arteaga, M., Fogliato, R., Chouldechova, A.: A Case for humans-in-the-loop: decisions in the presence of erroneous algorithmic scores. In: Proceedings of the 2020 CHI Conference on Human Factors in Computing Systems, pp. 1–12. ACM, Honolulu HI USA (2020). https://doi.org/10.1145/3313831.3376638
5. Devlin, J., Chang, M.W., Lee, K., Toutanova, K.: BERT: pre-training of deep bidirectional transformers for language understanding. In: Proceedings of the 2019 Conference of the North American Chapter of the ACLm, pp. 4171–4186. ACL (2019). https://doi.org/10.18653/v1/N19-1423
6. Dzikovska, M., et al.: SemEval-2013 task 7: the joint student response analysis and 8th recognizing textual entailment challenge. In: Proceedings of the Seventh International Workshop on Semantic Evaluation (SemEval 2013), pp. 263–274. ACL (2013). https://aclanthology.org/S13-2045
7. Garg, S., Moschitti, A.: Will this question be answered? Question filtering via answer model distillation for efficient question answering. In: Proceedings of the 2021 Conference on Empirical Methods in Natural Language Processing, pp. 7329–7346. Association for Computational Linguistics (2021). https://doi.org/10.18653/v1/2021.emnlp-main.583, https://aclanthology.org/2021.emnlp-main.583
8. Ghavidel, H., Zouaq, A., Desmarais, M.: Using BERT and XLNET for the automatic short answer grading task. In: Proceedings of the 12th International Conference on Computer Supported Education, pp. 58–67. SCITEPRESS - Science and Technology Publications, Prague, Czech Republic (2020). https://doi.org/10.5220/0009422400580067
9. Graham, S., Kiuhara, S.A., MacKay, M.: The effects of writing on learning in science, social studies, and mathematics: a meta-analysis. Rev. Educ. Res. **90**(2), 179–226 (2020). https://doi.org/10.3102/0034654320914744
10. Green, B., Chen, Y.: Algorithm-in-the-loop decision making. In: Proceedings of the AAAI Conference on Artificial Intelligence, vol. 34(9), pp. 13663–13664 (2020). https://doi.org/10.1609/aaai.v34i09.7115
11. Hendrycks, D., Gimpel, K.: A baseline for detecting misclassified and out-of-distribution examples in neural networks. http://arxiv.org/abs/1610.02136, number: arXiv:1610.02136

12. Kamath, A., Jia, R., Liang, P.: Selective question answering under domain shift. In: Proceedings of the 58th Annual Meeting of the Association for Computational Linguistics, pp. 5684–5696. Association for Computational Linguistics (2020). https://doi.org/10.18653/v1/2020.acl-main.503, https://aclanthology.org/2020.acl-main.503
13. Li, Z., Tomar, Y., Passonneau, R.J.: A Semantic feature-wise transformation relation network for automatic short answer grading. In: Proceedings of the 2021 Conference on Empirical Methods in Natural Language Processing (EMNLP), pp. 6030–6040. ACL (2021). https://doi.org/10.18653/v1/2021.emnlp-main.487
14. Liu, T., Ding, W., Wang, Z., Tang, J., Huang, G.Y., Liu, Z.: Automatic short answer grading via multiway attention networks. In: Isotani, S., Millán, E., Ogan, A., Hastings, P., McLaren, B., Luckin, R. (eds.) Artificial Intelligence in Education, pp. 169–173. Springer International Publishing, Cham (2019). https://doi.org/10.1007/978-3-030-23207-8_32
15. Mosqueira-Rey, E., Hernández-Pereira, E., Alonso-Ríos, D., Bobes-Bascarán, J., Fernández-Leal, A.: Human-in-the-loop machine learning: a state of the art. Artif. Intell. Rev. (2022). https://doi.org/10.1007/s10462-022-10246-w
16. Passonneau, R.J., Carpenter, B.: The benefits of a model of annotation. Trans. ACL **2**, 311–326 (2014). https://doi.org/10.1162/tacl_a_00185
17. Riordan, B., Horbach, A., Cahill, A., Zesch, T., Lee, C.M.: Investigating neural architectures for short answer scoring. In: Proceedings of the 12th Workshop on Innovative Use of NLP for Building Educational Applications, pp. 159–168. ACL, Copenhagen (2017). https://doi.org/10.18653/v1/W17-5017
18. Saha, S., Dhamecha, T.I., Marvaniya, S., Sindhgatta, R., Sengupta, B.: Sentence level or token level features for automatic short answer grading?: Use both. In: Penstein Rosé, C., et al. (eds.) Artificial Intelligence in Education, pp. 503–517. Springer International Publishing, Cham (2018). https://doi.org/10.1007/978-3-319-93843-1_37
19. Sung, C., Dhamecha, T., Saha, S., Ma, T., Reddy, V., Arora, R.: Pre-training BERT on domain resources for short answer grading. In: Proceedings of the 2019 Conference on Empirical Methods in Natural Language Processing and the 9th International Joint Conference on Natural Language Processing (EMNLP-IJCNLP), pp. 6071–6075. ACL, Hong Kong (2019). https://doi.org/10.18653/v1/D19-1628
20. Varshney, N., Mishra, S., Baral, C.: Investigating selective prediction approaches across several tasks in IID, OOD, and adversarial settings. In: Findings of the Association for Computational Linguistics: ACL 2022, pp. 1995–2002. Association for Computational Linguistics, Dublin, Ireland (2022). https://doi.org/10.18653/v1/2022.findings-acl.158
21. Wang, T., Inoue, N., Ouchi, H., Mizumoto, T., Inui, K.: Inject rubrics into short answer grading system. In: Proceedings of the 2nd Workshop on Deep Learning Approaches for Low-Resource NLP (DeepLo 2019), pp. 175–182. ACL, Hong Kong (2019). https://doi.org/10.18653/v1/D19-6119
22. Wiener, Y.: Theoretical foundations of selective prediction. Ph.D. thesis, Technion - Israel Institute of Technology, Israel (2013)
23. Xin, J., Tang, R., Yu, Y., Lin, J.: The art of abstention: selective prediction and error regularization for natural language processing. In: Proceedings of the 59th Annual Meeting of the Association for Computational Linguistics (ACL) and the 11th International Joint Conference on Natural Language Processing (IJCNLP), pp. 1040–1051. Association for Computational Linguistics (2021). https://doi.org/10.18653/v1/2021.acl-long.84

Contrastive Learning for Reading Behavior Embedding in E-book System

Tsubasa Minematsu[1(✉)], Yuta Taniguchi[2], and Atsushi Shimada[3]

[1] Data-Driven Innovation Initiative, Kyushu University, Fukuoka, Japan
minematsu.tsubasa.659@m.kyushu-u.ac.jp
[2] Research Institute for Information Technology, Kyushu University, Fukuoka, Japan
taniguchi.yuta.941@m.kyushu-u.ac.jp
[3] Faculty of Information Science and Electrical Engineering, Kyushu University, Fukuoka, Japan
atsushi@ait.kyushu-u.ac.jp

Abstract. When students use e-learning systems such as learning management systems and e-book systems, the operation logs are stored and analyzed to understand student learning behaviors. For implementing some applications, such as dashboard systems and at-risk student detection, the operation logs are mainly transformed into features designed by researchers. Such hand-crafted features, like the number of operations, are easily interpretable. However, the power of the hand-craft features may be limited for the recent large-scale educational dataset. In machine learning research, data-driven features are demonstrated to be a better representation than hand-crafted features. However, there are few discussions in the educational data due to a need for many operation logs. In this study, we collect reading logs of an e-book system. We propose a representation learning method for the reading logs based on contrastive learning. Our proposed method transforms time-series reading logs into reading behavior feature vectors directly without hand-crafted features. In our experiments, we demonstrate that the power of our feature representation is better than a traditional count-based hand-crafted feature representation in the at-risk student detection task. In addition, we investigate the characteristics of the feature space learned by our proposed method.

Keywords: Representation learning · Contrastive learning · Click-stream log · Reading behavior analysis

1 Introduction

Learning support systems and learning platforms with mechanisms for recording student behavior are becoming more widespread. Pattern recognition and machine learning are becoming fundamental technologies in educational data analysis. In other words, the growth of these technologies is significant because it can improve the accuracy of educational data analysis and the performance of

N. Wang et al. (Eds.): AIED 2023, LNAI 13916, pp. 426–437, 2023.
https://doi.org/10.1007/978-3-031-36272-9_35

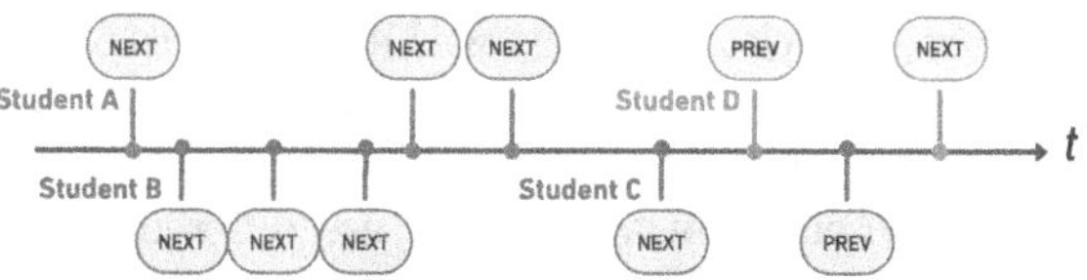

Fig. 1. Time-series operation log examples

learning support systems. It is critical to obtain discriminative feature representations of learning behaviors for achieving acceptable performance in learning support systems using data analysis methods, such as performance prediction and change detection.

In feature engineering, student features designed by researchers have been mainly proposed in various e-learning systems [11,12,14,15]. However, such hand-crafted features have limitations when comprehensively considering complex behavior patterns for handling recent large-scale educational data [2]. Figure 1 shows an example of click event data used in this study. E-learning systems record student operations and timestamps, such as operations to read the next or previous page (NEXT/PREV). An example of feature representation based on this event data is a count-based method aggregating the number of operations and operation times. Student A's and B's behaviors can be distinguished in a simple count-based feature representation. However, it is challenging to distinguish student C's logs from student D. It is necessary to redesign feature representations for various e-learning system case by case.

To overcome the above limitations of the power of feature representation and system dependence, we need feature representations that can be discriminative for large-scale educational data and can be applied flexibly to various educational data. Unsupervised representation learning methods have recently been proposed to obtain representative image/word features on large datasets [1,7]. Inspired by such representation learning, we propose a feature representation learning method for educational data in a data-driven manner. Our method can automatically learn feature representations that account for operation types, operation sequences, and time intervals. We acquire a feature representation that can distinguish between an operation sequence based on a contrastive learning manner. Our contributions are as follows:

- We propose a framework for automatically acquiring representative features from a large-scale educational dataset.
- The proposed method can be applied to various e-learning systems recording click event data.
- We acquire feature representations from reading logs of an e-book system and show its effectiveness in at-risk student detection.

2 Related Work

As basic feature representation, some researchers design features such as the number of interactions with the e-learning system and operation time. Park et al.

used the total number of click event logs within a specific period in Canvas to detect learning behavior change [15]. Northcut et al. use time differences in quiz responses from the same student across multiple accounts for cheat detection [12]. Other feature representation approaches use different e-learning systems. Okubo et al. used scores computed from lecture attendance rates and the usage of three e-learning systems for final grade prediction [14]. Marras et al. proposed features to assess student knowledge and alignment with course schedules from quiz responses and event logs of video views in MOOCs [11]. In addition, their experiments were conducted to compare with various existing feature representations, such as video viewing time and the number of views. Such hand-crafted features are highly interpretable. However, the approaches may reduce the power of the feature representation because they focus mainly on event log aggregation and limited learning activities.

On the other hand, there have been attempts to automatically acquire more refined feature representations from hand-crafted features. Ding et al. used count-based features from video and navigation activity logs, such as a play-video operation. They converted them into refined feature representations by the long short-term memory autoencoder [4]. Lopez et al. obtained effective feature representations from count-based features of the e-book reading logs using a variational autoencoder and demonstrated their effectiveness in at-risk student detection [8]. These studies showed that hand-crafted features might not be the optimal representation for the target downstream task. However, even in the studies, hand-crafted features are still used, and we consider that the feature representation can improve more.

In computer vision and natural language processing, as the dataset size has increased, it has become possible to obtain more effective feature representations based on unsupervised representation learning rather than hand-crafted features [1,7]. Large-scale educational datasets with click event logs, such as EdNet [2], are recently available, and it is expected that large-scale data can be utilized. However, more research on educational feature representation learning still needs to be done to leverage such large-scale data. Therefore, to tackle the problem, we propose a feature learning method to directly and automatically convert operation names and time information in event logs into feature vectors and explore the possibility of expression learning in educational data.

3 Contrastive Learning for Reading Behavior Embedding

Figure 2 illustrates the overview of our proposed method. Our objective is to acquire feature representations in a data-driven manner using raw data from clickstream event logs recorded in e-learning systems as time-series data without using hand-crafted feature transformations. Although we assume the reading logs of an e-book system as actual data, our method can also be applied to other event logs since it is not specific to reading logs.

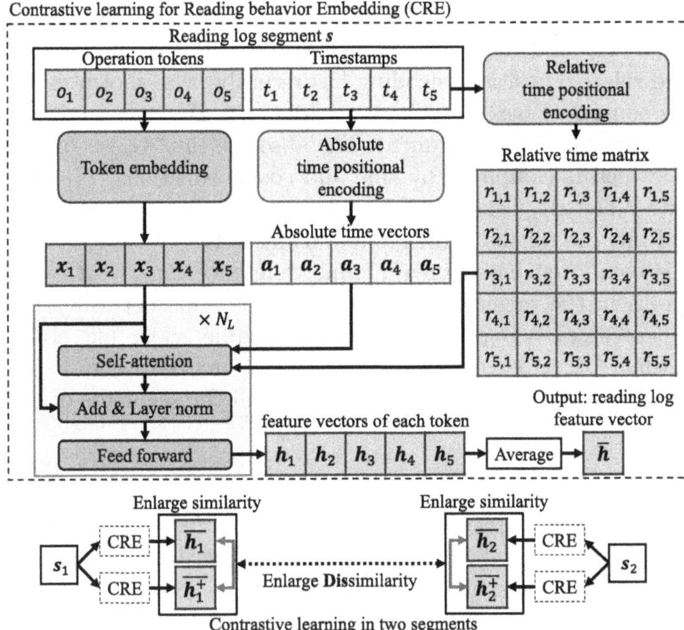

Fig. 2. Overview of contrastive learning for reading behavior embedding

3.1 Reading Log Segmentation

The reading log used in this study is similar to a typical click-stream event log, including user ID, operation name, and timestamp. The entire sequence length of student operations is different for each student, and the number of operations is numerous because the operation logs are collected throughout the lecture period. On the other hand, a short-term operation sequence exists, such as reading a page after moving to another page. Therefore, it is possible to characterize learning behavior during the entire lecture period by combining the features of short sequences of operations. In this study, we extracted the short-term reading logs as segments and learned their feature representations.

A reading log segment of each student is extracted as one segment when the time difference Δt between operations exceeds T_s. Each operation name within the segment corresponds to a unique ID as an operation token o. Furthermore, the timestamp of each operation is recorded as an integer value of UNIX time t. Therefore, each segment $\boldsymbol{s}$ comprises operation tokens $[o_1, \ldots, o_n]$ and their timestamps $[t_1, \ldots, t_n]$. In this study, T_s is the value of the 95th percentile of Δt in our dataset.

3.2 Absolute and Relative Time-Positional Encoding

Unlike representation learning in natural language processing, reading logs contain time information. To represent the time of the operation sequence, we perform time position embedding. We develop absolute and relative time positional embedding (ATPE and RTPE) in this study because the effectiveness of both absolute and relative position encoding has recently been shown in [6,16,17].

ATPE is computed based on the time difference $\Delta t_i = t_i - t_1$ from the first operation's timestamp of the segment. The absolute time vector of i-th position denotes $\boldsymbol{a}_i$, which is computed by sine and cosine functions [17] based on the position Δt_i. When $\Delta t_{i,1}$ exceeds T_a, we use $\boldsymbol{a}_{\mathrm{T}_a}$. we implement T_a as the 95th percentile of $\Delta t_{i,1}$ in all the segments of our dataset.

RTPE is computed based on the time difference $\Delta t_{i,j} = t_i - t_j$ between two operations in the segment. The relative time value between i-th and j-th positions denotes $r_{i,j} = r_{\Delta t_{i,j}}$, which is a learnable scalar like [16]. Similarly to ATPE, when $|\Delta t_{i,j}|$ exceeds T_r, we clip $\Delta t_{i,j}$ to the range $[-\mathrm{T}_r, \mathrm{T}_r]$. We implement T_r as the 95th percentile of $\Delta t_{i,j}$ in all the segments of our dataset.

3.3 Network Architecture

The upper part of Fig. 2 illustrates our network architecture. Our network comprises five parts: token embedding, absolute/relative time-potential encoding, transformer-based encoder, and feature aggregation.

In the token embedding, we apply the word embedding such as BERT [7] to the operation tokens. After the embedding process, we scale the token vectors by the square root of the dimensionality and apply dropout to the scaled vectors. We denote the token vector as $\boldsymbol{x}$. In addition, we perform the absolute time position embedding and relative time position embedding described in the previous section to obtain the absolute time vectors $\boldsymbol{a}$ and the relative time matrix $r_{i,j}$. Note that the dimension of $\boldsymbol{x}$ and $\boldsymbol{a}$ is the same value. The model dimension denotes d.

The encoder consists of Transfomer-based self-attention layers and feed-forward networks [17]. In this study, we use TUPE [6] to obtain attention considering both absolute and relative time-position embeddings. When computing attention, TUPE is designed to eliminate unnatural correlation calculations between position embeddings and token features. In this study, we do not use the [CLS] symbol; thus, we use the simpler TUPE. In TUPE, the feature $\boldsymbol{z}_i$ of the token at the i-th position is calculated as follows.

$$\begin{aligned} \boldsymbol{z}_i &= \sum_j \frac{e^{\alpha_{i,j}}}{\sum_k e^{\alpha_{i,k}}}(\boldsymbol{x}_j W^V) \\ \text{, where } \alpha_{i,j} &= \frac{1}{\sqrt{2d}}(\boldsymbol{x}_i W^Q)(\boldsymbol{x}_j W^K)^T + \frac{1}{\sqrt{2d}}(\boldsymbol{a}_i U^Q)(\boldsymbol{a}_j U^K)^T + r_{i,j} \end{aligned} \tag{1}$$

U^Q, U^K, W^Q, W^K, and $W^V \in \mathbb{R}^{d \times d}$ are the projection matrice for the positional embedding. Q, K, and V represent the self-attention layer's query, key, and value.

We modify the original TUPE because the original TUPE cannot distinguish between quick NEXT and slow NEXT operations like student A and B logs in Fig. 1. In the original TUPE, the word token vectors do not add to the positional embedding vectors. When the operation tokens are the complete same, and the timestamps are different in the two segments, $\boldsymbol{x}_i = \boldsymbol{x}_j \Rightarrow \boldsymbol{x}_i W^V = \boldsymbol{x}_j W^V$. Therefore, $\boldsymbol{z}_i = \boldsymbol{x}_i W^V (\sum_j \frac{e^{\alpha_{i,j}}}{\sum_k e^{\alpha_{i,k}}}) = \boldsymbol{x}_i W^V \times 1$. We use $(\boldsymbol{x}_j + \boldsymbol{a}_j)W^V$ instead of $\boldsymbol{x}_j W^V$ only in the first self-attention layer to avoid the problem.

We obtain the segment's feature vector $\bar{\boldsymbol{h}}$ as the network output in the feature aggregation. As mentioned above, the encoder can compute each token's feature vectors $\boldsymbol{h}$. We average $\boldsymbol{h}$ to convert the feature vectors $\boldsymbol{h}$ to $\bar{\boldsymbol{h}}$.

3.4 Network Training

We train our network by contrastive learning based on SimCSE [5]. In SimCSE, the network is given the same sentence twice as input. However, a different dropout mask is used for each input sentence. Therefore, different features can be obtained for the same sentence. The different dropout masks correspond to the transformations by data augmentation in contrastive learning [1]. An example of contrastive learning is shown in the lower part of Fig. 2. The training process increases the similarity between feature vectors computed from the same segment while decreasing the similarity between ones from different segments. The cost function for contrastive learning is as follows.

$$E = \frac{1}{N} \sum_{i=1}^{N} - \log \frac{e^{s_{\cos}(\bar{\boldsymbol{h}}_i, \bar{\boldsymbol{h}}_i^+)/\tau}}{\sum_{j=1}^{N} e^{s_{\cos}(\bar{\boldsymbol{h}}_i, \bar{\boldsymbol{h}}_j^+)/\tau}}, \tag{2}$$

where the function $s_{\cos}(\cdot, \cdot)$ is the cosine similarity, τ is a temperature hyperparameter, and N is the number of pairs in the mini-batch. $\bar{\boldsymbol{h}}_i$ is the output vector of $\boldsymbol{s}_i$, and $\bar{\boldsymbol{h}}_i^+$ is the positive instance. Therefore, a pair $\{\bar{\boldsymbol{h}}_i, \bar{\boldsymbol{h}}_j^+\}, i \neq j$ is the negative instance pair.

We use AdamW [9] as an optimizer to train our networks. Our training stage has a warm-up stage. The learning rate can increase from zero to the initial learning rate in the warm-up stage. After the warm-up stage, the learning rate decays linearly to $0.5 \times$ initial learning rate. In addition, we set the gradient clip norm to 1.0.

4 Experimental Settings

4.1 At-Risk Student Detection Settings

We set up an at-risk student detection task as a downstream task for evaluating the feature representation of the reading logs. In this study, at-risk students were defined as ones with C, D, and F in the grade point (5-point scale of A, B, C, D, and F). To detect at-risk students in each lecture course, we extract all the segments of each student who has taken the lecture. The segments are converted to feature vectors by CRE, and we aggregate the feature vectors to compute student features. Figure 3 illustrates the flow of at-risk student detection with CRE.

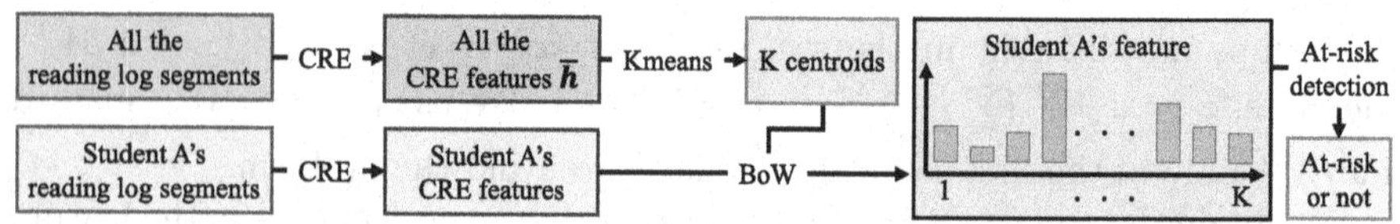

Fig. 3. Flow of at-risk student detection with CRE

Table 1. Parameter settings.FF: feed-forward.

Log preparation & AdamW		Training		Network	
T_s	97	Batchsize N	1024	#Layers (N_L)	6
T_a	79	Warm-up step	100	#Heads	8
T_r	79	Learning rate	0.001	d	128
AdamW β_1	0.9	Weight decay	0.1	d in FF	512
AdamW β_2	0.999	Training step	400000	τ	0.2
AdamW ϵ	1.00e-07	Maximum token length	394	dropout rate	0.1

First, all the segments in our dataset are encoded by CRE. After the encoding process, Kmeans clustering is applied to these CRE feature vectors to obtain the centroid of each cluster. Next, for each student who has taken a course, the student's CRE feature vectors are computed from all the segments of the student. And then, we select the centroids of the nearest neighbors of each student's CRE feature vector and count the number of times each centroid is selected. The histogram is used as each student feature vector for at-risk student detection. The above procedure is similar to bag-of-visual-words [3], which uses image features as the vocabulary of bag-of-words (BoW). Therefore, the dimension of the student's feature vector is the number of clusters K.

We used support vector machine (SVM), random forest (RF), and multi-layer perceptron (MLP) as at-risk student detectors. The detectors were trained using training datasets containing the student's feature vectors and the ground truth data showing whether the student was at-risk. The detail of the at-risk dataset is described in the following subsection. In addition, we examine the effect of K on the performance of at-risk student detection. As the baseline of each student's feature representation, we used a histogram of the number of times each student performed each operation. The dimension of the baseline vector is the number of types of operations, which is 32 in the experiments.

4.2 Dataset and Parameter Settings

In this study, we used BookRoll [13] as an e-book system to collect reading behaviors of textbooks used in lecture courses of our institution. Students can access e-textbooks anytime and leave notes, highlights, and bookmarks. BookRoll can record student operations on each page, including the user name, operation name, timestamp, and other information such as highlight positions. As our contrastive learning dataset, we used reading logs from 2019/Feb./19 to 2022/Oct./01. The operation logs included 6331 students and a total of

Table 2. At-risk datasets. I: information science, D: digital signal processing, F: first semester, S: second semester, and four digits: years

(a) The number of at-risk students

Course	Non-risk	At-risk
I2021F	62	45
I2021S	103	75
D2020F	68	19
D2021F	63	21

(b) The details of the at-risk datasets

Dataset	Train course	Test course
I→D	I2021F, I2021S	D2020F, D2021F
D→I	D2020F, D2021F	I2021F, I2021S
I2021F	The other courses	I2021F
I2021S	The other courses	I2021S
D2020F	The other courses	D2020F
D2021F	The other courses	D2021F

14,327,510 operations. Additionally, by decomposing the reading logs of the dataset into segments, we could extract 717,033 segments. Table 1 shows the hyperparameters used in CRE.

For the at-risk student detection dataset, we used grade point data from four courses for which we received permission to use from the instructors. Two of the four courses were information science lectures by Teacher A, and the others were digital signal processing lectures by Teacher B. These four courses were divided into six datasets. Of the six datasets, four datasets contain three training courses and the remaining one test course for evaluating the detectors. The remaining two datasets contain two training courses and two test courses divided by the course names (information science and digital signal processing). The details of the datasets are shown in Table 2. Note that because of the small number of students with C, D, and F in D2020F and D2021F, we set up the detection task as binary classification instead of the five-label classification.

Hyperparameter tuning of the at-risk student detectors was performed using 5-fold cross-validation for each dataset. We conducted a grid search method to find the best parameters for each dataset. The grid search settings were as follows: For SVM, we used the RBF kernel with a C value of [0.1, 1, 10, 100, 1000] and a gamma value of [1, 0.1, 0.01, 0.001, 0.0001]. For MLP, we set the maximum number of iterations to 3000, the hidden layer sizes to [(128), (256, 128), (512, 256, 128)], the initial learning rate to [0.001, 0.0001], and the L2 regularization parameter to [0.1, 0.01]. For the random forest, we set the number of trees to [10, 20, 30, 50, 100, 300] and the number of features for the best-split decision to $[\sqrt{n_f}, \log_2(n_f), n_f]$, where n_f is the number of features. Note that the input features for all three methods were standardized.

5 Experimental Results

5.1 Evaluation of At-Risk Student Detection

Table 3 shows the F1 scores for at-risk student detection. According to Table 3, the F1 scores of the detectors using CRE features were, on average higher than those of count-based features. In addition, cre-MLP-K1024 demonstrated the best averaged F1 score. The results indicate that CRE-based features are more

Table 3. F1 scores for at-risk student detection. K values represent the number of clusters in Kmeans clustering. The best three F1 scores are represented in bold style.

	I→D	D→I	I2021F	I2021S	D2020F	D2021F	Average
count-MLP	0.49	0.34	0.55	0.58	0.67	0.72	0.56
count-RF	0.46	0.26	0.52	0.58	0.64	0.73	0.53
count-SVM	0.50	0.34	**0.58**	0.56	0.63	0.71	0.55
cre-MLP-K32	0.47	0.34	0.45	0.68	0.69	0.59	0.54
cre-RF-K32	0.54	0.46	0.42	**0.73**	0.67	0.63	0.57
cre-SVM-K32	**0.73**	0.32	0.42	0.66	**0.77**	0.77	0.61
cre-MLP-K64	0.42	0.41	0.54	0.70	0.75	0.74	0.59
cre-RF-K64	0.58	0.54	0.40	0.69	0.69	0.73	0.61
cre-SVM-K64	0.41	0.48	0.39	**0.73**	0.68	**0.79**	0.58
cre-MLP-K128	0.54	0.49	**0.56**	**0.73**	0.72	**0.80**	**0.64**
cre-RF-K128	0.57	**0.60**	0.40	0.69	0.72	0.75	0.62
cre-SVM-K128	0.49	0.45	0.44	0.73	0.72	**0.81**	0.61
cre-MLP-K256	0.50	0.48	0.56	0.69	0.73	0.76	0.62
cre-RF-K256	0.59	0.49	0.43	0.64	0.59	0.73	0.58
cre-SVM-K256	0.44	**0.55**	0.44	0.72	0.74	0.72	0.60
cre-MLP-K512	0.54	**0.56**	0.55	0.67	0.74	0.68	0.62
cre-RF-K512	0.61	0.52	0.41	0.71	0.52	0.68	0.58
cre-SVM-K512	0.66	0.54	0.49	0.65	0.72	0.71	0.63
cre-MLP-K1024	**0.68**	0.50	**0.57**	0.72	**0.80**	0.68	**0.66**
cre-RF-K1024	0.55	0.50	0.48	0.66	0.73	0.75	0.61
cre-SVM-K1024	**0.71**	0.52	0.43	0.71	**0.76**	0.70	**0.64**

effective than count-based features in obtaining discriminative feature representations in at-risk student detection.

For the number of clusters K, we observed that the F1 scores were likely higher for larger K. This result implies that a good variety of vocabulary (cluster centers) of reading behavior patterns is needed for representing learning behaviors in the various segments. In the count-based feature, the feature dimension is limited to the number of types of operations. To improve the power of the feature representation, we need to design new hand-crafted features while tuning K can freely adjust CRE-based feature dimension.

In comparisons between datasets, the CRE-based feature was particularly effective on the I→D and D-I datasets. This result indicates that CRE-based features can acquire more useful across courses. In addition, the count-based feature was influenced by the domain shift across courses caused by differences in instructors and lecture content across courses. Figure 4 shows that the feature spaces of the count-based and CRE-based features are visualized by t-SNE [10]. In the figure, the red points mean at-risk students, and the blue points mean non-at-risk students. The shape of each point corresponds to the course name, with I2021F represented by an x, I2021S represented by a star, D2020F represented by a circle, and D2021F represented by a square. According to Fig. 4, the count-based features are biased across the courses. The relationship between the uniformity of features and performance in downstream tasks has been dis-

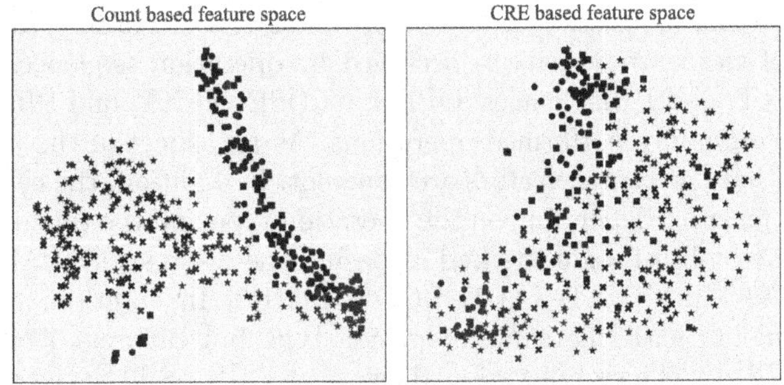

Fig. 4. Visualization of count-based and CRE-based feature space.

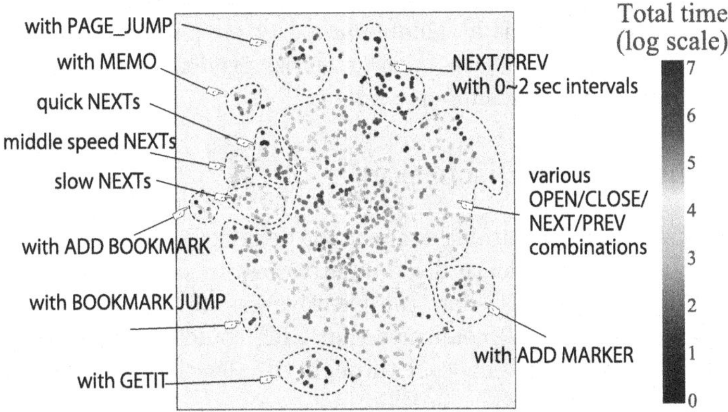

Fig. 5. Visualization of cluster centers of CRE feature vectors

cussed in contrastive learning [18]. We consider that the CRE-based feature was practical for downstream tasks thanks to their uniformity, although the feature was used for at-risk student detection. Hence, we concluded that CRE could automatically learn discriminative features in at-risk student detection.

5.2 CRE Feature Space Analysis

We conducted a qualitative analysis of the CRE features $\bar{\boldsymbol{h}}$ computed from the segments in our dataset. Since the number of segments in our dataset was numerous, we focused on the cluster center vectors of $\bar{\boldsymbol{h}}$ by Kmeans clustering. In addition to the cluster center vectors being representative vectors of all the segment features, we can indirectly evaluate the properties of the cluster center vectors by observing the segment closest to each cluster center vector. Considering the results of the evaluation of at-risk student detection, the following analysis uses Kmeans clustering results with $K = 1024$.

The feature space of cluster center vectors was visualized by t-SNE, as shown in Fig. 5. The colors indicate the time length of the segments in a log scale. We annotated the characteristics of each blob manually. According to Fig. 5, the major part of the feature space is occupied by operation sequences of OPEN, CLOSE, NEXT, and PREV pages. OPEN, CLOSE, NEXT, and PREV are students' most commonly performed operations. As an aspect of the time length, we observed that the center vectors are coherent. In addition, the center vectors had different features depending on the operation type and the time interval. For example, some blobs of segments had a specific operation such as ADD BOOKMARK of "NEXT→NEXT→ADD BOOKMARK". In addition, we observed blobs of segments with the same operation type but different time intervals, such as NEXT × 4. This observation shows that CRE could automatically separate features of quick NEXT from slow NEXT and represent such essential differences. In other words, our method CRE can automatically learn discriminative features by providing a series of operation names and their timestamps. The discriminative feature is a valuable property for pattern recognition tasks based on learning behavior data, such as at-risk student detection, which can help develop better learning support systems.

6 Conclusion and Discussion

We proposed contrastive learning for reading behavior embedding (CRE). Our method can automatically learn discriminative features from reading logs of e-books. Experimental results show the effectiveness of CRE-based features in at-risk student detection. We confirmed that CRE could obtain effective feature representations at the segment level through visualization of the CRE feature space.

This study has demonstrated the effectiveness of contrastive learning feature representation in clickstream event logs such as reading logs. In addition, regardless of the type of learning support system logs, feature vectors with arbitrary dimensionality can be applied to learning support system applications. High-dimensional feature representations are highly compatible with deep neural networks and are expected to improve performance in downstream tasks. However, our method could not utilize the context, such as the learning content on the page and when students study. To overcome this limitation, we need to add such context to the primitive operation features computed by CRE. In the future, we will consider a framework for learning various contexts and adding the context feature representation to the CRE features.

Acknowledgements. This work was supported by JSPS KAKENHI Grant Number JP21K17864.

References

1. Chen, T., Kornblith, S., Norouzi, M., Hinton, G.: A simple framework for contrastive learning of visual representations. In: Proceedings of the 37th International Conference on Machine Learning. Proceedings of Machine Learning Research, vol. 119, pp. 1597–1607 (2020)

2. Choi, Y., et al.: EdNet: a large-scale hierarchical dataset in education. In: Bittencourt, I.I., Cukurova, M., Muldner, K., Luckin, R., Millán, E. (eds.) AIED 2020. LNCS (LNAI), vol. 12164, pp. 69–73. Springer, Cham (2020). https://doi.org/10.1007/978-3-030-52240-7_13
3. Csurka, G., Dance, C., Fan, L., Willamowski, J., Bray, C.: Visual categorization with bags of keypoints. In: Workshop on Statistical Learning in Computer Vision, ECCV, pp. 59–74. Prague (2004)
4. Ding, M., Yang, K., Yeung, D.Y., Pong, T.C.: Effective feature learning with unsupervised learning for improving the predictive models in massive open online courses. In: Proceedings of the 9th International Conference on Learning Analytics & Knowledge, pp. 135–144 (2019)
5. Gao, T., Yao, X., Chen, D.: Simcse: simple contrastive learning of sentence embeddings. In: Proceedings of the 2021 Conference on Empirical Methods in Natural Language Processing, pp. 6894–6910 (2021)
6. Ke, G., He, D., Liu, T.Y.: Rethinking positional encoding in language pre-training. In: International Conference on Learning Representations (2020)
7. Kenton, J.D.M.W.C., Toutanova, L.K.: Bert: pre-training of deep bidirectional transformers for language understanding. In: Proceedings of NAACL-HLT, pp. 4171–4186 (2019)
8. Lopez Zapata, E., Minematsu, T., Taniguchi, Y., Okubo, F., Shimada, A.: Encoding students reading characteristics to improve low academic performance predictive models. In: Companion Proceedings 12th International Conference on Learning Analytics & Knowledge (LAK22), pp. 36–38, March 2022
9. Loshchilov, I., Hutter, F.: Decoupled weight decay regularization. In: International Conference on Learning Representations (2018)
10. Van der Maaten, L., Hinton, G.: Visualizing data using t-SNE. J. Mach. Learn. Res. **9**(86), 2579–2605 (2008)
11. Marras, M., Vignoud, J.T.T., Kaser, T.: Can feature predictive power generalize? Benchmarking early predictors of student success across flipped and online courses. In: 14th International Conference on Educational Data Mining, pp. 150–160 (2021)
12. Northcutt, C.G., Ho, A.D., Chuang, I.L.: Detecting and preventing "multiple-account" cheating in massive open online courses. Comput. Educ. **100**, 71–80 (2016)
13. Ogata, H., et al.: Learning analytics for E-Book-based educational big data in higher education. In: Yasuura, H., Kyung, C.-M., Liu, Y., Lin, Y.-L. (eds.) Smart Sensors at the IoT Frontier, pp. 327–350. Springer, Cham (2017). https://doi.org/10.1007/978-3-319-55345-0_13
14. Okubo, F., Yamashita, T., Shimada, A., Ogata, H.: A neural network approach for students' performance prediction. In: Proceedings of the Seventh International Learning Analytics & Knowledge Conference, pp. 598–599 (2017)
15. Park, J., Denaro, K., Rodriguez, F., Smyth, P., Warschauer, M.: Detecting changes in student behavior from clickstream data. In: Proceedings of the Seventh International Learning Analytics & Knowledge Conference, pp. 21–30 (2017)
16. Raffel, C., et al.: Exploring the limits of transfer learning with a unified text-to-text transformer. J. Mach. Learn. Res. **21**(140), 1–67 (2020)
17. Vaswani, A., et al.: Attention is all you need. Adv. Neural Inf. Process. Syst. **30**, 6000–6010 (2017)
18. Wang, F., Liu, H.: Understanding the behaviour of contrastive loss. In: Proceedings of the IEEE/CVF Conference on Computer Vision and Pattern Recognition, pp. 2495–2504 (2021)

Generalizable Automatic Short Answer Scoring via Prototypical Neural Network

Zijie Zeng[1], Lin Li[1], Quanlong Guan[2], Dragan Gašević[1], and Guanliang Chen[1](✉)

[1] Centre for Learning Analytics, Monash University, Melbourne, Australia
{zijie.zeng,lin.li,dragan.gasevic,guanliang.chen}@monash.edu
[2] Jinan University, Guangzhou, China
gql@jnu.edu.cn

Abstract. We investigated the challenging task of generalizable automatic short answer scoring (ASAS), where a scoring model is tasked with generalizing to target domains (provided only with limited labeled data) that have no overlap with the auxiliary domains on which the model is trained. To address this, we introduced a framework based on Prototypical Neural Network (PNN). Specifically, for a target short answer instance whose score needs to be determined, the framework first calculates the distance between this target instance and each cluster of support instances (support instances are a set of labeled short answer instances that are grouped to different clusters according to their labels, i.e., the ground-truth scores). Then, it rates the target instance using the ground-truth score of the cluster that has the closest distance to the target instance. Through extensive empirical studies on an open-source ASAS dataset consisting of 10 different question prompts, we observed that the proposed approach consistently outperformed other baselines across settings concerning different numbers of support instances. We further observed that the proposed approach performed better when with wider training data sources than when with restricted data sources for training, showing that including more data sources for training may add to the generalizability of the proposed framework.

Keywords: Generalizability · Automatic Short Answer Scoring · Few-shot Learning · Prototypical Neural Network

1 Introduction

Short answer scoring aims at accurately evaluating the short textual responses written by learners based on certain grading criteria [1]. The completion of this task used to rely on the manual efforts of human instructors. On the one hand, manual grading can be rather accurate and reliable when performed by human experts (with enough time for scoring each assignment). On the other hand, the lack of qualified human experts may hinder the timely and accurate assessment of short text answers when facing a large number of students (e.g., the

N. Wang et al. (Eds.): AIED 2023, LNAI 13916, pp. 438–449, 2023.
https://doi.org/10.1007/978-3-031-36272-9_36

student-teacher ratio in MOOC can be up to 10,000:1 [18]). Moreover, it has been documented that human graders can hardly keep their scoring precision and consistency at high levels throughout the whole grading process, i.e., grading performance can be of high quality at the beginning and then gradually becomes poor, which can result in unfairness to some students [9,28]. Therefore, researchers have been exploring the feasibility of applying computational techniques to score short written text automatically, i.e., automatic short answer scoring (ASAS). ASAS approaches vary from the early rule-based methods [15] that rate a textual short answer based on the extent the answer matches the specific rules, to the traditional machine learning approaches [17,23] based on hand-crafted features, then to the deep learning-based approaches [24,25,29] that can automatically engineer features from raw input text.

Certain ASAS approaches have demonstrated impressive scoring performance [24,25,29], e.g., the ASAS model proposed in [25] achieved predictive performance up to human-agreement levels on the task of scoring short textual responses from psychology domain. However, there remain issues that may prevent educators from adopting such models, i.e., their inabilities to generalize to unseen domains [5,33] and the significant amount of manual efforts on labeling data required by machine learning or deep learning-based ASAS approaches [17,25,29] to train reliable models. To address this challenge, existing studies attempt to develop generalizable ASAS systems [5,33], which are capable of generalizing to unseen domains (e.g., unseen questions and unseen prompts) when equipped with very limited labeled data (from the unseen domains) for reference. In this line of research, Condor et al. [5] investigated how the combinations of different input content (e.g., responses, question text, scoring rubric) affect the generalizability of the deep learning-based ASAS model. META [33] goes further to include a set of support instances (i.e., labeled short answer responses from the target domain) as an additional source of input content. Specifically, the support instances were appended to the end of the original input, i.e., the target response to be scored as well as other contextual contents. This practice of explicitly appending support instances to the input is expected to reduce the scoring task to the relatively easier task of finding similar instances [33].

Although META [33] has demonstrated state-of-the-art performance for generalizable ASAS, it suffers from the following limitations: (i) It requires that the scoring range of the training domains must be consistent with the scoring range of the testing domains, limiting its ability to generalize to real-world data sources of various scoring ranges; (ii) The maximum number of support instances is limited by the structure of its input format, i.e., it packs all relevant elements including the target response and the set of support instances to form a single input string. Note that some modern large language models have limits with regard to their input length (e.g., the input length limit is 512 tokens for BERT [6]). The above limitations might hamper META's ability to apply to real-world short answer scoring scenarios commonly seen where scoring ranges and subject areas might vary from task to task. As an example, the short answer scoring datasets used in this study (see Table 1) consisted of 10 prompts across various subjects and various scoring ranges, where the average length of responses could be up to 50 tokens for some prompts. To better adapt to such real-world scenarios, it is necessary to develop generalizable ASAS approaches that are free from the above limitations.

Therefore, in this paper, we intend to add to the existing generalizable ASAS studies by proposing a framework based on Prototypical Neural Network (PNN) [22], considering its success in generalizable image classification problem in computer vision. Specifically, for a target short answer instance whose score needs to be determined, the proposed framework scores it following two steps: (i) Calculate the distance between the target instance and each cluster of support instances (support instances are a set of labeled short answer instances that are grouped to different clusters according to their labels, i.e., the ground-truth scores); and (ii) Rate the target instance as score S, where S is the ground-truth score of the cluster that has the closest distance to the target instance. With the proposed generalizable ASAS framework, in this paper, we intended to investigate the following research question:

- To what extent can the model trained with the proposed ASAS framework based on PNN generalize to new domains (prompts or questions unseen during training)?

To answer this question, we conducted extensive empirical experiments on an open-source short answer scoring dataset and summarized the main findings as follows: (i) The proposed approach consistently outperformed other baselines across settings concerning different numbers of support instances; (ii) The proposed approach performed better when with wider training data sources than when with restricted data sources for training, showing that the generalizability of the proposed framework could be boosted by including more data sources for training.

2 Related Work

Our generalizable ASAS study can be included in the broader automatic text scoring (ATS) studies that aim at learning reliable models with limited annotation efforts on specific domians [5,8,12,13,21,32,33]. These studies are mainly driven by the fact that in real-world settings, scoring tasks oftentimes involve unseen domains about which little knowledge is known and the cost of learning domain-specific models from scratch can be fairly high. Jiang et al. [12] investigated automated essay scoring in the one-shot setting where they proposed to augment the one-shot data with pseudo-labeled data to guarantee effective training. Similar attempts can also be found in [8,13,32], where authors tried to reduce annotation efforts by only annotating the centroids of clusters that consisted of similar answers [32] or sampling a subset of the most informative essays to annotate [8]. Jin et al. [13] proposed a two-stage method where a set of pseudo-labeled essays were extracted based on the prediction of a model trained on non-target domains, which would be then used for training a target domain-specific model. Ridley et al. [21] developed a zero-shot neural-based model that concatenates the part-of-speech embedding and a set of prompt-irrelevant features to form representations of the textual data, which were then used as the input to a linear layer for regression to predict the essay scores. Condor et al. [5] examined the generalizability of a series of supervised models when coupled with different types of text representation approaches (e.g., *Sentence-Bert, Word2Vec, Bag-of-words*) and various sources of input content (e.g., *question text, question context, rubric text, and question bundle identifier*). They showed that generalizability could hardly be achieved even by using the state-of-the-art sentence-BERT for text representation or by including context information such as question text. META [33] introduced labeled instances as additional input information to augment the query instances whose scores are to be predicted. Then, META employed a BERT-based classifier to predict over the augmented query instances. Our generalizable ASAS framework distinguishes itself from the existing studies for the following: (i) We focus on the task of scoring short textual answers, which is different from existing essay scoring studies [8,12,13,21]; (ii) Our framework requires no further training over the previously unseen target domains and can make predictions with merely a few support instances from the target domain (i.e., different from existing related works [12,13]); (iii) As a generalizable ASAS approach, our framework is similar to META [33] but with more flexibility w.r.t. the scoring range of train/test data. Moreover, our framework generates a representation for each query/support instance instead of concatenating query instances and support instances to form a single and long input string, which might exceed the length limit of some text encoders (e.g., BERT has an input limit of 512 tokens).

3 Method

In the setting of ASAS, generalizability refers to the capacity of a model to predict over previously unseen domains [33] (e.g., questions, prompts, or subject areas). We identified the few-shot learning based on PNN [22] as a promising approach to address the generalizable ASAS problem, considering its success in the generalizable image classification problem in computer vision [22]. Specifically, for a query instance Q to be scored, as well as the set of labeled support instances of different scores, denoted as S (note that both Q and S are from the same domain, whether it be the training domain or the testing domain), the PNN-based model scores Q following four steps: (i) Compute a low-dimensional representation Q' for Q with a certain text encoder f_θ; (ii) Compute a low-dimensional representation for each of the support instances from S and then group them into different clusters according to their ground-truth scores; (iii) Average the representations within each cluster to form a representation p for each score, i.e., prototype; (iv) Calculate the distances between Q' and each prototype p and then predict the score of Q as s, where s is the ground-truth score of the prototype that has the shortest distance from Q'. Note that when

Input: Text encoder f_θ with learnable parameters θ; Training data $\mathcal{D} = \{\mathcal{D}_1, \mathcal{D}_2, ..., \mathcal{D}_r\}$ where $\mathcal{D}_i = \{(\mathbf{x}_1, y_1), (\mathbf{x}_2, y_2), ...\}$ denotes the dataset of the ith domain; Test data (i.e., target domain) $\mathcal{D}_{test}$ and note that $\mathcal{D}_{test} \notin \mathcal{D}$; Number of training episodes N; Distance function d where $d(v_1, v_2)$ meansures the distance between v_1 and v_2; Number of support instances per class (score) N_S; Number of query instances per class (score) N_Q.

Output: Trained text encoder f_θ for short answer scoring on **unseen** domains

for *episode* $\in \{1, 2, ..., N\}$ **do**
  $\mathcal{D}_j \leftarrow$ RandomSample($\mathcal{D}$, 1) ▷ Select a domain for episode
  $\mathcal{C}_j \leftarrow$ The set of distinct scores in $\mathcal{D}_j$ ▷ Obtain the scoring range
  for score $i \in \mathcal{C}_j$ **do**
    $\mathcal{D}_{j,i} \leftarrow$ The set of instances in $\mathcal{D}_j$ with score i
    $S_i \leftarrow$ RandomSample($\mathcal{D}_{j,i}$, N_S) ▷ Select support instances
    $Q_i \leftarrow$ RandomSample($\mathcal{D}_{j,i} \backslash S_i$, N_Q) ▷ Select query instances
    $p_i \leftarrow \frac{1}{|S_i|} \sum_{(\mathbf{x},y) \in S_i} f_\theta(\mathbf{x})$ ▷ Compute prototype for score i
  end for
  $Q \leftarrow \bigcup_{i \in \mathcal{C}_j} Q_i$ ▷ Obtain all query instances
  for $(\mathbf{x}, y) \in Q$ **do**
    $p_y \leftarrow$ The prototype with score y
    for $p \in \{p_i | i \in \mathcal{C}_j \text{ and } p_i \neq p_y\}$ **do**
      $\Delta_{Dist} \leftarrow d(f_\theta(x), p) - d(f_\theta(x), p_y)$
      $\mathcal{L}_\theta \leftarrow \frac{1}{N_Q|\mathcal{C}_j|}[-\log(\sigma(\Delta_{Dist}))]$ ▷ Obtain the loss
      Compute the gradients for θ with respect to $\mathcal{L}_\theta$ ▷ Backward pass
    end for
  end for
  Update θ based on the gradients accumulated in this episode ▷ Update
end for
Predict over unseen data $\mathcal{D}_{test}$ (i.e., target domain) using f_θ ▷ Testing

Fig. 1. The training algorithm for generalizable ASAS. Note that RandomSample(S, N) returns a subset of N elements randomly selected from set S.

the distance metric d is fixed, accurate scoring requires that the text encoder f_θ with learnable parameters θ should be able to encode the instances so that of all the prototypes $\mathcal{P} = \{p_1, p_2, ...\}$, Q' should have the minimum distance from prototype p_q ($p_q \in \mathcal{P}$), where p_q shares the same score with Q'. This can be achieved by training the text encoder f_θ following the algorithm in Fig. 1. The algorithm follows an episodic training paradigm widely adopted in few-shot learning studies [10,11,22,27] and formalizes the scoring task as the C-way-N_S-shot-N_Q-query problem, i.e., at each episode, it mimics the few-shot task by predicting over a batch of query instances (i.e., N_Q query instances per score) using a few labeled support instances (i.e., N_S support instances per score) and minimizes the loss accordingly. Note that C is the number of distinct scores.

Generally, for a query instance x, the PNN-based model [11,16,22] employs the softmax over x's distances to the prototypes to generate a predicted distribution over all classes (i.e., distinct scores). The probability of x being predicted as score y is as follows [22]:

$$p_\theta(score = y|x) = \frac{exp(-d(f_\theta(x), p_y))}{\sum_{k'} exp(-d(f_\theta(x), p_{k'}))}, \tag{1}$$

where p_y is the prototype with score y. Then the negative log-probability loss for true class (let y be the ground-truth score here) is formalized as [22]:

$$\mathcal{L}_\theta' = -\log p_\theta(score = y|x), \tag{2}$$

which will be minimized via certain optimization algorithms (e.g., SGD). However, the cost of computing score probabilities with softmax grows linearly with the number of classes (the sum term in the denominator of Eq. (1), making it too expensive in real-world applications when the number of classes (i.e., distinct scores in ASAS setting) is large [3]. Moreover, in the backward passing stage, the optimization algorithm needs to compute the gradients with respect to the loss, which also requires a RAM linearly correlated with the number of classes to store parameters' gradients. Considering that $\mathcal{L}_\theta'$ in Eq. (2) is a function of the learnable parameters θ, of which the size can be up to millions for large language models (e.g., BERT-base [6] model has 110M parameters), we argue that minimizing the negative log loss in Eq. (2) can be memory-demanding, particularly with domains of wide scoring range, i.e., containing a large set of distinct scores.

To manage the memory required for model training to a controllable magnitude, we alternatively formalize the loss in a pairwise manner as follows [20,31,34]: (i) Given the query instance x (encoded as $f_\theta(x)$) and a pair of prototypes $< p_y, p >$, where p_y shares the same ground-truth score with x while p is another prototype with a different ground-truth score, we denote the distance between $f_\theta(x)$ and prototype p_y as $d(f_\theta(x), p_y)$ and $d(f_\theta(x), p)$ is defined similarly; (ii) Then, we subtract $d(f_\theta(x), p_y)$ from $d(f_\theta(x), p)$ and denote the difference as Δ_{Dist} (Line 14 in Fig. 1); (iii) We follow [20,31] to employ $\sigma(\Delta_{Dist})$ as the proxy for the probability of $d(f_\theta(x), p) > d(f_\theta(x), p_y)$, where $\sigma(x) = \frac{1}{1+e^{-x}}$ is the sigmoid function. For example, when $d(f_\theta(x), p) - d(f_\theta(x), p_y) = 0.2$, we have $\sigma(0.2) = 0.55$, meaning that we have a 55% probability of $d(f_\theta(x), p) > d(f_\theta(x), p_y)$, i.e., a 55% chance of successfully rating x as the ground-truth score against the incorrect score; (iv) Finally, we maximize the probability $\sigma(\Delta_{Dist})$ by minimizing the following loss over the pair of distances $< d(f_\theta(x), p), d(f_\theta(x), p_y) >$ [20,31]:

$$\mathcal{L}_\theta = -log(\sigma(\Delta_{Dist})), \tag{3}$$

note that $\mathcal{L}_\theta$ always involves two prototypes, i.e., the pair of $< p_y, p >$, making the required memory (RAM) needed irrelevant to the number of classes (scores) when computing the gradients of θ w.r.t $\mathcal{L}_\theta$ when performing backward passing (Line 16 in Fig. 1). This enables the proposed framework to be applied to domains with wide scoring ranges when merely modest RAM is available[1].

[1] All the experiments were completed on NVIDIA Tesla T4 GPU with 16GB RAM.

4 Experiment

4.1 Data

We based our generalizable ASAS study on the Short Answer Scoring datasets[2], which contain about 17,000 short textual answers from US students of Grade 10. The datasets consist of 10 prompts across four subject areas. The adopted datasets distinguish themselves from those used in existing generalizable ASAS studies [5,33] in that their scoring ranges and subject areas vary from prompt to prompt. Consequently, the existing studies [26,29,30] based on these datasets tend to build one prompt-specific model for each prompt without considering the model generalizability[3] across domains.

Table 1. Statistics of the ASAS dataset. ELA is short for English Language Arts.

Prompt ID	Subject	#Answers	Score Range	Average Length	Prompt ID	Subject	#Answers	Score Range	Average Length
1	Science	1672	0–3	47 words	6	Biology	1797	0–3	23 words
2	Science	1278	0–3	59 words	7	English	1799	0–2	41 words
3	ELA	1808	0–2	48 words	8	English	1799	0–2	53 words
4	ELA	1657	0–2	40 words	9	English	1798	0–2	50 words
5	Biology	1795	0–3	25 words	10	Science	1640	0–2	41 words

4.2 Baseline Methods

- **PNN-BERT**: PNN [22] is widely adopted as a few-shot learning framework for image classification [11,22]. To adapt to the generalizable ASAS setting, we equip it with BERT [6] as the encoder to deal with textual input, considering that BERT has been the core component for many ASAS studies [24,26,30,33]. Details of this approach can be found in Fig. 1. Specifically, PNN-BERT was further broken down into three versions of different ranges of training sources:
  - *All.* PNN-BERT (*All*) was trained on all prompts except for the target being tested, e.g., when prompt 1 was the target prompt that our model needed to generalize to, we trained PNN-BERT (*All*) over prompt 2–10.
  - *Restricted.* PNN-BERT (*Restricted*) was trained only over the prompts that shared the same scoring range with the target prompt, e.g., when prompt 1 was the target prompt, we exerted restrictions on PNN-BERT so that it could only be trained over prompts 2, 5 and 6 because prompts 1, 2, 5 and 6 shared the same scoring range of [0, 3]. This restricted version was meant to match META w.r.t. its training sources[4].

[2] https://www.kaggle.com/competitions/asap-sas/data.

[3] We follow [33] to define generalizability as the capacity of a model to predict over previously unseen domains.

[4] This restriction also applies to META due to its own limitation (see Sect. 1).

  - *NoTuned.* PNN-BERT (*NoTuned*) means that we directly adopted the pre-trained BERT as the encoder without further fine-tuning it.
- **META** [33]: META is one of the strongest baseline approaches for generalizable ASAS. For a query instance q from the target prompt U, META appends a series of labeled support instances (from the same prompt, i.e., U) to q to form the input string (usually very long), which is then put to the BERT-based classifier to predict the score of q. Note that for this method, only the first 512 tokens of the input string could be encoded and the exceeding part will be truncated due to the input limit of BERT.
- **SBERT-C** [2]: SBERT-C encodes both support instances and query instances with SBERT [19]. Then it computes the Canberra distance [14] between the query instance q and each of the support instances. Then it predicts the query instance q using M's ground-truth score, where M has the minimum distance from q among all support instances.

4.3 Training, Evaluation, Metrics and Implementation Details

Following Jin et al. [13], we adopted a prompt-wise cross validation for evaluation, i.e., ten-fold cross validation. In each fold, student answers of the target prompt were reserved for testing while answers from the remaining prompts were used for model training. In addition, 30% of the training data were reserved for validation and early-stopping would be triggered once no significant loss over the validation data was observed. Following existing few-shot learning studies [4,11,22], we adopted accuracy as the evaluation metric. We also reported quadratic weighted kappa (QWK) following existing ASAS studies [26,30,33].

For the text encoder, we adopted the pre-trained BERT-base-cased implemented in the Python package Transformers[5]. We followed its default settings to set dropout rate as 0.1 and activation function as GELU and adopt AdamW as the optimizer. For the episodic training, we adopt 5-shot, and 5-query setting as informed by previous studies [7,11,22]. Note that all methods were tested under different n-shot settings (i.e., 5-shot, 10-shot and 15-shot) to better understand the effects of n on their generalizability. We followed [22] to adopt Euclidean distance as the distance metric for PNN-BERT and searched the learning rate from $\{1e-5, 2e-5, 5e-5\}$ for all methods. As informed by previous studies [5,33], we also adopted the prompt text as additional input content, which was by default appended to each query q for all methods in our implementations. It is noteworthy that none of the methods were fine-tuned on test (target) prompts due to the nature of our study being to investigate generalizability, which is also consistent with existing few-shot learning studies [11,22]. All experiments were run on NVIDIA Tesla T4 GPU with 16 GB RAM and the codes are available via https://github.com/douglashiwo/GeneralizableASAS.

[5] https://github.com/huggingface/transformers.

5 Results

We summarized the results in Table 2 and organized our analysis as follows.

Overall Performance of the Proposed ASAS Framework. We observed that PNN-BERT (*All*) consistently outperformed all other approaches across different n-shot settings. Specifically, for different versions of the proposed PNN-BERT, we always had the observations w.r.t. the performance comparisons that *All* > *Restricted* > *NoTuned* (across different n-shot settings). Firstly, it is not surprising to see *NoTuned* being the worse because it had not been trained on any training data sources. Secondly, the observation of *All* > *Restricted* might suggest that the generalizability of the proposed PNN-BERT could benefit from including more data sources for training, even when the data sources have different scoring ranges with the target domain that the model is expected to generalize to. It should be pointed out that this observation could be of great significance to our proposed framework. Because one of the advantages of our proposed framework against META (the state-of-the-art baseline in generalizable ASAS) is the flexibility that our proposed framework requires no consistency between the scoring ranges of the train data and that of the test data. The observation of *All* > *Restricted* might suggest that the generalizability of the proposed framework can benefit from such flexibility.

Table 2. Experiment results (testing) of different generalizable ASAS approaches under the n-shot settings. Note that each entry of the table is the average over all the 10 prompts while the result for each prompt is the average over 500 episodes.

Method	Test Settings (n-shot)					
	Accuracy			QWK		
	5-shot	10-shot	15-shot	5-shot	10-shot	15-shot
PNN-BERT (*All*)	0.453	0.462	0.463	0.387	0.410	0.415
PNN-BERT (*Restricted*)	0.425	0.448	0.446	0.332	0.370	0.379
PNN-BERT (*NoTuned*)	0.365	0.376	0.380	0.183	0.206	0.216
META	0.430	0.432	0.430	0.364	0.362	0.363
SBERT-C	0.422	0.448	0.455	0.311	0.361	0.376

Effects of the Number of Shots. We observed that all methods, except META, performed better as the number of shots grew. For PNN-BERT, this could be explained as follows: the prototypes of the classes (i.e., distinct scores) could be more accurately represented as more support instances (shots) were provided, leading to better classification (scoring) performance. This is consistent with existing few-shot learning studies [10,22]. Similar to PNN-BERT, SBERT-C also adopts a distance-based idea for classification but it directly assigns a query instance q to the support instance s that has the shortest distance to q, i.e., rating q with the score of s. When the number of support instances (i.e., n) increased, the query q became more likely to be correctly assigned to a support instance that shares the same score with q. Finally, we noticed that META performed quite well under the 5-shot setting (the second best and only worse than PNN-BERT (*All*)). However, its performance got no better (stayed stable) as the number of shots (i.e., n) increased. This can be explained by its limitation mentioned in Sect. 1, i.e., it augments the query instance by appending to it all support instances to form a single input string, which would probably exceed the length limit of the adopted text encoder (they adopt BERT as the encoder and it accepted input of 512 tokens at most). Such practice is acceptable when the average length of the instances is short, which was actually the case for the ASAS study on mathematics domain [33] where META was first proposed. However, for datasets with relatively long average instance lengths, the disadvantage caused by this limitation becomes manifest. For example, when we were testing META on prompt 2 in the 5-shot setting, we would have a string of length l (estimated) when we packed all the support instances together to form a single input, where $l = 59 * (4 * 5 + 1) + c > 512$. Note that 59 is the average length of instances from prompt 2 (see Table 1), 4 is the number of distinct scores for prompt 2, and 5 is the number of shots while 1 means that we have 1 query and c denotes the length of other input contents (e.g., prompt text was used in our case). Note that only the first 512 tokens could be encoded and the exceeding part will be truncated, i.e., the additional shots beyond the limit were actually invalid and could unlikely bring any improvement. To summarize, due to the input length limit of the adopted text encoder (i.e., BERT) and the way it constructed its input, META eventually failed to benefit from increasing the number of shots. We could see the performance of META dropped from the second best (out of five methods) in the 5-shot setting to the second worse in the 15-shot settings.

6 Conclusion and Future Work

In this paper, we proposed a framework based on PNN [22] for the generalizable ASAS problem. Through extensive empirical studies on the open-source ASAS datasets, we summarized the following findings: (i) Across different n-shot settings (n denotes the number of support instances for each class, i.e., score), we observed that the proposed method consistently outperformed other baselines, including the state-of-the-art baseline META [33]; (ii) The generalizability of the proposed framework could benefit from including more data sources for training; (iii) The performance of the proposed approach improved as more and more support instances from the target domains were provided. For future work, we would like to extend our empirical experiments by including other text encoders (e.g., Sentence-BERT [19]). Another potential direction would be to evaluate the effectiveness of the proposed framework on automatic essay scoring (AES), which is considered a more challenging task.

References

1. Alikaniotis, D., Yannakoudakis, H., Rei, M.: Automatic text scoring using neural networks. In: Proceedings of the 54th Annual Meeting of the Association for Computational Linguistics (Volume 1: Long Papers), pp. 715–725 (2016)

2. Baral, S., Botelho, A.F., Erickson, J.A., Benachamardi, P., Heffernan, N.T.: Improving automated scoring of student open responses in mathematics. Int. Educ. Data Min. Soc. (2021)
3. Blanc, G., Rendle, S.: Adaptive sampled softmax with kernel based sampling. In: International Conference on Machine Learning, pp. 590–599. PMLR (2018)
4. Boney, R., Ilin, A., et al.: Active one-shot learning with prototypical networks. In: ESANN (2019)
5. Condor, A., Litster, M., Pardos, Z.: Automatic short answer grading with sbert on out-of-sample questions. In: Proceedings of the 14th International Conference on Educational Data Mining (2021)
6. Devlin, J., Chang, M.W., Lee, K., Toutanova, K.: Bert: pre-training of deep bidirectional transformers for language understanding. In: Proceedings of the 2019 Conference of the North American Chapter of the Association for Computational Linguistics: Human Language Technologies, Volume 1 (Long and Short Papers), pp. 4171–4186 (2019)
7. Dong, N., Xing, E.P.: Few-shot semantic segmentation with prototype learning. In: BMVC, vol. 3 (2018)
8. Dronen, N., Foltz, P.W., Habermehl, K.: Effective sampling for large-scale automated writing evaluation systems. In: Proceedings of the Second (2015) ACM Conference on Learning@ Scale, pp. 3–10 (2015)
9. Fazal, A., Dillon, T., Chang, E.: Noise reduction in essay datasets for automated essay grading. In: Meersman, R., Dillon, T., Herrero, P. (eds.) OTM 2011. LNCS, vol. 7046, pp. 484–493. Springer, Heidelberg (2011). https://doi.org/10.1007/978-3-642-25126-9_60
10. Geng, R., Li, B., Li, Y., Zhu, X., Jian, P., Sun, J.: Induction networks for few-shot text classification. In: Proceedings of the 2019 Conference on Empirical Methods in Natural Language Processing and the 9th International Joint Conference on Natural Language Processing (EMNLP-IJCNLP), pp. 3904–3913 (2019)
11. Jakubik, J., Blumenstiel, B., Voessing, M., Hemmer, P.: Instance selection mechanisms for human-in-the-loop systems in few-shot learning. **6** (2022)
12. Jiang, Z., Liu, M., Yin, Y., Yu, H., Cheng, Z., Gu, Q.: Learning from graph propagation via ordinal distillation for one-shot automated essay scoring. In: Proceedings of the Web Conference 2021, pp. 2347–2356 (2021)
13. Jin, C., He, B., Hui, K., Sun, L.: TDNN: a two-stage deep neural network for prompt-independent automated essay scoring. In: Proceedings of the 56th Annual Meeting of the Association for Computational Linguistics (Volume 1: Long Papers), pp. 1088–1097 (2018)
14. Jurman, G., Riccadonna, S., Visintainer, R., Furlanello, C.: Canberra distance on ranked lists. In: Advances in Ranking NIPS 09 Workshop (2009)
15. Leacock, C., Chodorow, M.: C-rater: automated scoring of short-answer questions. Comput. Humanit. **37**(4), 389–405 (2003)
16. Li, O., Liu, H., Chen, C., Rudin, C.: Deep learning for case-based reasoning through prototypes: a neural network that explains its predictions. In: Proceedings of the AAAI Conference on Artificial Intelligence, vol. 32 (2018)
17. Nau, J., Haendchen Filho, A., Passero, G.: Evaluating semantic analysis methods for short answer grading using linear regression. Sciences **3**(2), 437–450 (2017)
18. Pappano, L.: The year of the MOOC. N. Y. Times **2**(12), 2012 (2012)
19. Reimers, N., Gurevych, I.: Sentence-bert: sentence embeddings using siamese bert-networks. In: Proceedings of the 2019 Conference on Empirical Methods in Natural Language Processing and the 9th International Joint Conference on Natural Language Processing (EMNLP-IJCNLP), pp. 3982–3992 (2019)

20. Rendle, S., Freudenthaler, C., Gantner, Z., Schmidt-Thieme, L.: BPR: Bayesian personalized ranking from implicit feedback. In: Proceedings of the Twenty-Fifth Conference on Uncertainty in Artificial Intelligence, pp. 452–461 (2009)
21. Ridley, R., He, L., Dai, X., Huang, S., Chen, J.: Prompt agnostic essay scorer: a domain generalization approach to cross-prompt automated essay scoring. arXiv preprint arXiv:2008.01441 (2020)
22. Snell, J., Swersky, K., Zemel, R.: Prototypical networks for few-shot learning. Adv. Neural Inf. Process. Syst. **30** (2017)
23. Sultan, M.A., Salazar, C., Sumner, T.: Fast and easy short answer grading with high accuracy. In: Proceedings of the 2016 Conference of the North American Chapter of the Association for Computational Linguistics: Human Language Technologies, pp. 1070–1075 (2016)
24. Sung, C., Dhamecha, T., Saha, S., Ma, T., Reddy, V., Arora, R.: Pre-training bert on domain resources for short answer grading. In: Proceedings of the 2019 Conference on Empirical Methods in Natural Language Processing and the 9th International Joint Conference on Natural Language Processing (EMNLP-IJCNLP), pp. 6071–6075 (2019)
25. Sung, C., Dhamecha, T.I., Mukhi, N.: Improving short answer grading using transformer-based pre-training. In: Isotani, S., Millán, E., Ogan, A., Hastings, P., McLaren, B., Luckin, R. (eds.) AIED 2019. LNCS (LNAI), vol. 11625, pp. 469–481. Springer, Cham (2019). https://doi.org/10.1007/978-3-030-23204-7_39
26. Surya, K., Gayakwad, E., Nallakaruppan, M.: Deep learning for short answer scoring. Int. J. Recent Technol. Eng. **7**(6), 1712–1715 (2019)
27. Vinyals, O., Blundell, C., Lillicrap, T., Wierstra, D., et al.: Matching networks for one shot learning. Adv. Neural Inf. Process. Syst. **29** (2016)
28. Wind, S.A., Peterson, M.E.: A systematic review of methods for evaluating rating quality in language assessment. Lang. Test. **35**(2), 161–192 (2018)
29. Xia, L., Guan, M., Liu, J., Cao, X., Luo, D.: Attention-based bidirectional long short-term memory neural network for short answer scoring. In: Guan, M., Na, Z. (eds.) MLICOM 2020. LNICST, vol. 342, pp. 104–112. Springer, Cham (2021). https://doi.org/10.1007/978-3-030-66785-6_12
30. Zeng, Z., Li, X., Gasevic, D., Chen, G.: Do deep neural nets display human-like attention in short answer scoring? In: Proceedings of the 2022 Conference of the North American Chapter of the Association for Computational Linguistics: Human Language Technologies, pp. 191–205 (2022)
31. Zeng, Z., Lin, J., Li, L., Pan, W., Ming, Z.: Next-item recommendation via collaborative filtering with bidirectional item similarity. ACM Trans. Inf. Syst. (TOIS) **38**(1), 1–22 (2019)
32. Zesch, T., Heilman, M., Cahill, A.: Reducing annotation efforts in supervised short answer scoring. In: Proceedings of the Tenth Workshop on Innovative Use of NLP for Building Educational Applications, pp. 124–132 (2015)
33. Zhang, M., Baral, S., Heffernan, N., Lan, A.: Automatic short math answer grading via in-context meta-learning. In: Proceedings of the 15th International Conference on Educational Data Mining (2022)
34. Zhu, Z., Wang, J., Caverlee, J.: Measuring and mitigating item under-recommendation bias in personalized ranking systems. In: Proceedings of the 43rd International ACM SIGIR Conference on Research and Development in Information Retrieval, pp. 449–458 (2020)

A Spatiotemporal Analysis of Teacher Practices in Supporting Student Learning and Engagement in an AI-Enabled Classroom

Shamya Karumbaiah[1(✉)], Conrad Borchers[2], Tianze Shou[2], Ann-Christin Falhs[3], Pinyang Liu[2], Tomohiro Nagashima[4], Nikol Rummel[3], and Vincent Aleven[2]

[1] University of Wisconsin-Madison, Madison, USA
shamya.karumbaiah@wisc.edu
[2] Carnegie Mellon University, Pittsburgh, USA
{cborcher,vincent.aleven}@cs.cmu.edu, tshou@andrew.cmu.edu, pinyangl@stanford.edu
[3] Ruhr University Bochum, Bochum, Germany
{ann-christin.falhs,nikol.rummel}@rub.de
[4] University of Saarland, Saarbrücken, Germany
nagashima@cs.uni-saarland.de

Abstract. Research indicates that teachers play an active and important role in classrooms with AI tutors. Yet, our scientific understanding of the way teacher practices around AI tutors mediate student learning is far from complete. In this paper, we investigate spatiotemporal factors of student-teacher interactions by analyzing student engagement and learning with an AI tutor ahead of teacher visits (defined as episodes of a teacher being in close physical proximity to a student) and immediately following teacher visits. To conduct such integrated, temporal analysis around the moments when teachers visit students, we collect fine-grained, time-synchronized data on teacher positions in the physical classroom and student interactions with the AI tutor. Our case study in a K12 math classroom with a veteran math teacher provides some indications on factors that might affect a teacher's decision to allocate their limited classroom time to their students and what effects these interactions have on students. For instance, teacher visits were associated more with students' in-the-moment behavioral indicators (e.g., idleness) than a broader, static measure of student needs such as low prior knowledge. While teacher visits were often associated with positive changes in student behavior afterward (e.g., decreased idleness), there could be a potential mismatch between students visited by the teacher and who may have needed it more at that time (e.g., students who were disengaged for much longer). Overall, our findings indicate that teacher visits may yield immediate benefits for students but also that it is challenging for teachers to meet all needs – suggesting the need for better tool support.

Keywords: Spatial analytics · Temporality · Teaching · Student Engagement · Human-AI Partnership · Multimodality

N. Wang et al. (Eds.): AIED 2023, LNAI 13916, pp. 450–462, 2023.
https://doi.org/10.1007/978-3-031-36272-9_37

1 Introduction

Previous studies suggest that teachers play an active role in supporting student learning with AI tutors [2, 10]. With adaptive instruction and immediate feedback, it is often argued that AI tutors are designed to free up teachers' time so they can focus on students who need their time the most [9]. Accordingly, teachers are observed to be highly engaged in such AI-enabled classrooms, moving from one student to another providing individualized help [2] and socio-emotional support [11], and helping students get out of an unproductive rut [9]. Yet, our scientific understanding of teacher practices around AI tutors in classrooms and the ways in which they mediate student learning is limited [cf. 9, 17]. Besides being interesting in its own right, this kind of understanding could be helpful in designing better tools that support teachers as they help students during AI-enabled classroom sessions, a focus within AIED research in recent years [9, 13].

1.1 Spatiotemporal Factors in Teacher Practices with AI Tutors

To better understand teacher practices in classrooms with AI tutors, we need to go beyond student log data (an often-used source of data in AIED research) and explore the physical classroom where the student-teacher interaction is happening while students are learning with AI tutors [7]. In a previous study, Holstein and colleagues [10] highlighted the importance of analyzing spatial factors of student and teacher behaviors to understand the role of "out-of-software events" such as teachers' help-giving in AI-enabled classrooms. More broadly, research in education highlights the importance of studying teachers' and students' physical location and interaction in a classroom to understand teachers' pedagogical practices and their impact on students [5, 16]. More recently, spatial pedagogy [14] – a framework on teacher positioning and movement in classrooms – has been used in learning analytics to better understand teachers' spatial pedagogical approaches [15]. However, it is hard to borrow these insights as-is for AI-enabled classrooms since the presence of the AI tutor drastically transforms the classroom environment and its social structures [2]. For example, the teacher's role shifts from that of a lecturer at the front of the class to a coach working with students individually in their close proximity [2, 9].

Due to the often one-to-one nature of teachers' individualized attention to students in AI-enabled classrooms [2], we focus our spatial analysis on teachers' close physical proximity to individual students. We use the term *teacher visits* to denote shorter or longer episodes where the teacher stays close to the same student (see illustration in Fig. 1; *left*). Teacher visits to students in AI-enabled classrooms could serve several purposes, such as monitoring, help-giving, motivational support, or simple reassurance. Teachers may visit an individual student or a small group of students after observing them from afar or while monitoring them routinely. Often, these visits help them gather additional information to improve their sensemaking and interpretation of noteworthy events (aka shaping [17]). Some teacher visits may involve specific intervention (e.g., conceptual guidance, socio-emotional support), either proactively based on what they notice or find out through talking to students directly, or after being prompted by a student's request for help (e.g., by raising their hand). While teacher visits have been reported to be positively related to learning and engagement in traditional classroom settings [6], they have not been studied thoroughly in human-AI hybrid teaching. One notable exception is a study by Holstein et al. [10] which collected manual observations of teacher visits and found a positive effect of teacher monitoring on student learning.

While exploring the spatial dimension of student-teacher interactions in classrooms with AI tutors seems promising, we argue that it is necessary to further contextualize the role of teacher visits by juxtaposing it with student learning and engagement happening in the AI tutor just prior to and just after each visit. That is, instead of aggregating measures of students' learning or engagement, we look at change (e.g., in idleness or struggle [12]) from before to after a teacher visit. Doing so enables us to explore factors that may correspond to a teacher's choice to visit a given student at a given point in time, and what change each teacher visit may bring for the visited student. This quantitative spatiotemporal lens complements earlier qualitative explorations of teacher practices using a replay tool [10].

1.2 Research Questions and Hypotheses

Prior interviews of teachers in AI-enabled classrooms (e.g., [10]) help us identify two key factors that may help determine a teacher's choice of the student to visit: 1) teachers' prior knowledge about student abilities and behavioral tendency, and 2) their perception of student needs while observing the class and monitoring student work on the students' computer screens. For our analysis, we translate these into two types of proxies: 1) a broader, static measure of student need for teacher support such as low prior knowledge, and 2) students' in-the-moment struggle and disengagement while learning with an AI tutor. Determining students' in-the-moment needs by direct observation may be harder for teachers in classrooms with AI tutors since there is a gap between what the teacher can perceive in the physical classroom and what their students do behind the screens with the AI tutor [9]. Certain student disengagement behaviors may be more overt (e.g., being idle) than others (e.g., misusing the tutor) for teachers' direct observation. A previous qualitative analysis showed that despite wanting to pay attention to students who are off-task, teachers overlooked students who spent greater time off-task [10]. Also, more so than in traditional classrooms, students are likely to work on divergent activities in the AI tutor and have different needs for teacher conceptual guidance while struggling, which makes us ask if there are differences in the effectiveness of teacher intervention based on student needs. Lastly, since teacher time is limited, we also expect teachers' decisions on who they will visit to depend on what they perceive a student's need is relative to other students' needs at a given time. These motivations translate to the following research questions and hypotheses:

RQ1. What factors about student learning and engagement with AI tutors relate to a teacher's choice of students they visit?
H1: Teacher visits are related to broader, static measures of student need for teacher support (i.e., low prior knowledge).
H2a: Teacher visits are related to students' in-the-moment needs while learning with the AI tutor, that is, students who are currently struggling or disengaged (i.e., being idle, misusing the tutor).
H2b: Teacher prioritizes students who have been struggling or disengaged the longest.
RQ2. How do teacher visits relate to student learning and engagement with AI tutors?

H3: Teacher visits are associated with less struggle or disengagement after the visit has taken place, compared to before.
H4: Teacher visits positively relate to learning, as measured by in-tutor performance and out-of-tutor knowledge tests.

Answering these questions could generate insights on analytics that could aid teachers while co-orchestrating their classrooms with AI tutors and generate scientific knowledge about effective teaching practices that, in turn, can support teacher learning and reflections. To enable such investigations, we need time-synchronized data about teachers' visits and student interaction with the AI tutor. In this work, we use position sensors to automatically record teachers' positions in the classroom and algorithmically infer visits (with reasonable albeit imperfect accuracy; [8]). We detail our methodology in the next section. Then, we present the findings from a case study in an authentic setting (i.e., a K12 math classroom using an AI tutor), addressing the research questions above. Lastly, we discuss the methodological and empirical contributions of this work on understanding teacher practices in supporting student learning with AI tutors.

2 Methods

2.1 Case Study Context

For the duration of three days in Summer 2022, we collected teacher position and student tutor log data in a public school in the United States. The participants were eighty-five 7th graders across five different classes (aka Periods 1–5), all taught by the same mathematics teacher. At the school, in 2022, 45.9% of all students were categorized as "Below Basic" for their performance on the end-of-course test on Algebra 1 [3]. The teacher who participated in the study was already familiar with the capabilities of an AI tutor due to prior experience participating in similar studies. They had been teaching mathematics at the same school for 16 years at the time of this study.

During their regular math class time (for approximately 20 min each day), all students used *Lynnette,* an AI tutor designed for middle school algebra (Fig. 1; *center*). In this tutor, students solve equation problems using interactive scaffolding including step-by-step feedback and next-step hints. All students in the study were assigned the same set and sequence of four problems across 12 problem levels (48 problems in total), which started from basic equations and then gradually increased in complexity. Before using the tutor, students worked on a web-based pretest on conceptual and procedural knowledge of algebra. Then, as a post-test, students worked on an isomorphic version of the test after three days of using the tutor. We used two forms, counterbalanced across pre- and post-test so that half the students got the one form as pre-test and the other as post-test, whereas for the other students it was the other way around. The tests contained conceptual items that tested students' conceptual reasoning on problem-solving in algebra [1] and procedural items that asked students to solve algebra problems similar to the ones they practiced in the tutor.

We used Pozyx's UWB (ultrawide-band)-based position sensors to collect the real-time X-Y coordinates of the teacher in the classroom. The positioning system estimates a person's real-time position based on the signal transmitted by UWB tags in a lanyard worn around their neck (Fig. 1; *right*) to six anchors (mounted on tripods) along the periphery of the classroom. The positioning system treats the entire classroom as a 2-D coordinate system. Once activated, it samples teacher positions at each second. Next, we measured static coordinates for each student's desk which were mapped to their student IDs in the AI tutor. Any changes to the student seating (a very rare occurrence in this case study) were recorded, so as to allow for accurate tracking of teacher positions in relation to students in the classroom. After time-synchronizing position data with tutor log data for each student, we were able to identify what the student was doing when the teacher was around them. We also measured the static coordinates of all the major objects in the classroom, including the teacher's desk, blackboard, window, and door. These data were used to create visualizations for follow-up interviews with the teacher participating in this study.

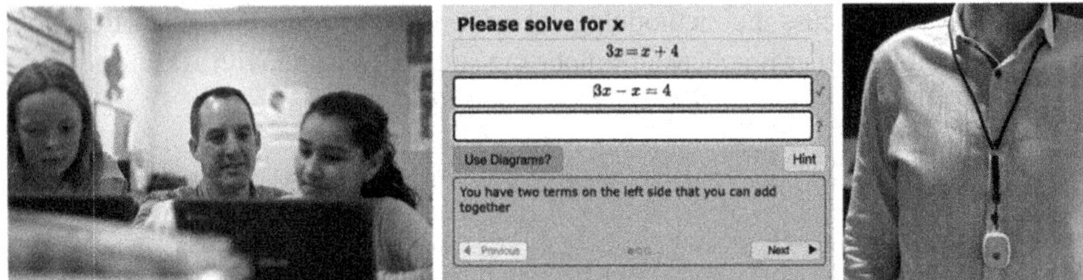

Fig. 1. An illustrative instance of a teacher visit (*left*; image credit: Mathia); An example problem in *Lynnette*, the AI tutor used in the study (*center*); Pozyx tag in a lanyard (*right*; image credit: pozyx.io)

2.2 Teacher Visits in the Temporal Context of Student Learning

Using the teacher's position data and students' seating coordinates, we infer teacher visits to a particular student through a stop detection algorithm. We extended the quantitative definition of *stopping* [15] proposed by adjusting the algorithm parameters to the spatial context of a K-12 classroom (smaller and more densely populated than the open learning spaces from previous studies) and the teaching context of interest (teacher stopping close to a student for individualized attention). The algorithm marks a teacher visit when the teacher's X-Y coordinates are within a small area for a chosen duration of time, d or longer. The small area is defined by a circular moving window with the teacher coordinates' centroid as the center and a chosen radius, r. Unlike previous studies that set these two parameters (r and d) ad hoc or using a heuristic, we chose to compute a more accurate combination of parameters with respect to human-coded training data of teacher visits collected in this study. To determine the final set of parameters, we maximized visit recall to 0.17 while constraining the parameter search space during grid search to a precision of at least 0.2, which was more generalizable than unbounded maximization of precision and recall based on cross-validation (see [8] for more details). We also added new logic to detect the student that the teacher is visiting. We define another parameter called range (*rng*). Students with a Euclidean distance smaller than *rng* to the centroid of the detected stop are marked as visited by the teacher. The final set of optimal parameters chosen from this procedure are $d = 21$ s, $r = 600$ mm (approx. 2 ft), and $rng = 700$ mm (approx. 2.3 ft). On average, the teacher visited a student once (SD = 2.1) for an average duration of 36 s (SD = 77.6 s).

Next, we computed the frequency of machine-predicted disengagement measures (i.e., idleness, misuse) and student struggle from AI tutor log data based on models developed in [9]. The models classify the presence of idleness at a threshold of 2 min, struggle at 25 s (operationalized as slow skill mastery), and misuse at 25 s (operationalized as either hint abuse or rapid sequences of attempts). For each student, each tutor interaction is annotated with indicator variables representing the presence or absence of idleness, tutor misuse, and struggle. These variables can then be aggregated within time windows to compute the percentage of interactions with each behavior. For example, one student may show, on average, idleness at every 10th interaction (0.1) while another student at every 5th (0.2). We similarly aggregate in-system performance measures of learning (i.e., correctness and the average number of errors per problem step) for each student. To investigate the association of teacher visit timing with in-the-moment student needs (i.e., disengagement and struggle), we compare the frequency of these behaviors recorded closer to an upcoming visit (pre-visit) to those closer to a past visit (post-visit). We also created a more granular classification, further splitting up pre- and post-visit interactions. To avoid post-visit interactions overlapping with pre-visit interactions of the next upcoming visit, the more granular classification first assigns each classification a pre- or post-visit class based on whether it is closer to an upcoming or past visit and then classifies its proximity to the closest visit. Specifically, we sample the 50% and 25% closest tutor interactions ahead of teacher visits (closer than 265 s and 123 s, respectively) and the 50% furthest pre-visit interactions (further away than 265 s). We do the same for post-visit interactions with the 25% and 50% closest interactions being classified at 117 s and 284 s, respectively. We compare these frequencies to a baseline subset of students the teacher never visited (see [3] for code and data).

3 Results

Our analysis was driven by the two research questions presented earlier (see Sect. 1.2). First (RQ1), we explore factors that may be related to the teacher's choice of students to visit at a given time. Second (RQ2), we investigate the associations between teacher visits and student learning and engagement with the AI tutor.

3.1 RQ1: Factors Associated with Teacher's Choice of Students to Visit

We hypothesized that the teacher visits are related to both a broader, static measures of student needs such as low prior knowledge (H1) and in-the-moment behavioral indicators from their interaction with the AI tutor (i.e., struggle or disengagement; H2a). We also hypothesized that teachers would additionally prioritize students that struggled or were disengaged the longest relative to other students (H2b).

We find H1 to be not supported by our correlation analysis. Averaging students' scores across procedural and conceptual items on the pre-test, we find that students' prior knowledge was uncorrelated to whether the teacher visited them ($r(48) = -0.22$ [−0.47, 0.06], $p = .118$), the number of visits they received ($r(48) = -0.03$ [−0.31, 0.25], $p = .838$) and the total length of the teacher visits ($r(48) = -0.01$ [−0.28, 0.28], $p = .990$). These findings are based on 50 students instead of 68, as 18 students in the sample were missing test scores. However, we find differences based on the timing of teacher visits concerning students' prior knowledge, as we will elaborate on in our results for H4.

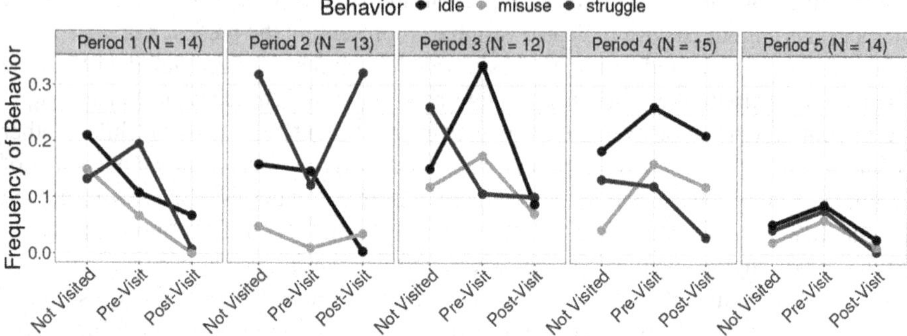

Fig. 2. Frequency of the presence of disengagement and struggle at any given tutor interaction before and after teacher visits per period with students who were never visited as a baseline. Standard error bars are excluded as they are too small to be meaningfully visible.

We find partial support for H2a as there were differential associations with our three behavioral indicators. In the 20-min sessions, on average, students were idle for 50.5 s (SD = 164.3 s), struggled for 29.4 s (SD = 123.4 s), and misused the system for 22.1 s (SD = 74.2 s). The data are shown in Fig. 2, separately for each of the five classes. We compare the struggle and disengagement rates of students the teacher never visited with students' behavior before a visit. The hypothesis would be confirmed if the pre-visit value is higher than the not-visited value. In three out of five classrooms, the teacher tended to allocate visits to students with higher idleness and tutor misuse. By contrast, in three out of five classrooms, the struggle rate was higher in students whom the teacher never visited compared to those who were visited. We note that these comparisons are between individual students, with students that were never visited making up a minority of students (N = 25; 29.07%). Therefore, as an additional way of testing H2a, we conduct additional, one-sided binomial tests. We compare whether students were more likely to show struggle and disengagement right before a teacher visit (recorded as their last tutor interaction before a teacher visit) than the behavioral rates across the whole learning session. We find that, only for idleness, the behavioral frequency was significantly high right before visits (p = 0.23, CI95% = [0.17, 0.29] compared to a baseline of p = 0.14]. We learned from a follow-up interview with the teacher that different periods have varying characteristics, which might explain the period-level difference shown in Fig. 2. According to the teacher, students in Period 1 are not highly-motivated needing them to "stand beside [students] to keep them motivated and working"; Period 3 has most IEP (Individualized Education Program) students that they "quickly figured out the hint strategy and misused the system"; Period 5 is the honors class that is "intrinsically motivated."

Moving on to H2b, we find no indications that the teacher preferentially visited students who exhibited struggle and disengagement earliest in scenarios where multiple students exhibited struggle and disengagement simultaneously. We find that only for 14% of visits, the teacher visited a student who showed the longest (i.e., least recent) period of struggle or disengagement ($p = 0.14$, CI95% = [0.09, 0.20]), including behavioral sequences up to 30 min ahead of a visit. There were no significant differences regarding this finding among the three behavioral dimensions across classrooms. Thus, H2b is not confirmed in our data.

3.2 RQ2: Teacher Visit Associations with Student Engagement and Learning

To test H3, we compare the struggle and disengagement rates of students before teacher visits to after (Fig. 2). H3 would be confirmed if the struggle/disengagement is lower post-visit than pre-visit. In all five periods, idleness was significantly lower after a teacher visit. Similarly, tutor misuse, except for Period 2, was significantly lower after a teacher visit. Notably, struggle was lower after teacher visits in all classes except Period 2. Taken together with findings for H2a and H2b, we find that tutor idleness was most robustly associated with the timing of teacher visits (increasing before and decreasing after visits). In line with teachers visits being related to student idlenss, we estimate that per additional SD (i.e., 3.96 min) of time students spend in the AI tutor without being re-visited by the teacher, the odds of showing idleness in the tutor approximately halves ($\beta = -0.04$, OR = 0.55, CI95% = [0.47, 0.64], $p < .001$).

Given heterogeneous associations of our struggle and disengagement measures with teacher visits, we ask which indicators had the strongest association with teacher visits. To answer this question, we calculate the aggregated mean difference between struggle and disengagement metrics before and after teacher visits broken out by period. Given that differences may be larger or smaller depending on how often the behavior occurs generally, we standardize this difference by dividing it by the behavioral frequency in interactions ahead of visits. We find that idleness had the largest standardized change, with idleness being between 37.4% to 97.3% less frequent after than before visits. Averaged by period, this reduction was 59.6%. Struggle had the second largest change; it was reduced by an average of 21.5% (although for period 2, the struggle was increased by 164.0%). Misuse showed an average reduction of 0.6%, ranging from a reduction of 96.3% to an increase of 261.0%. Thus, overall H3 is confirmed for idleness and struggle but not for misuse.

We test associations of teacher visits with student learning in the AI tutor (H4). We do this by associating teacher visits with in-system performance and learning gain on out-of-tutor tests. We begin by reporting relations between teacher visits and the two measures of in-system performance, correctness of individual student responses in the tutor and the average number of attempts per step. We compare logistic regression models via likelihood-ratio tests to investigate whether the correctness of students' first attempt at each step is significantly different pre- and post-visits (as operationalized via our time binning described in Sect. 2.2). We compare a model featuring an indicator variable representing whether the given attempt is the student's first attempt at the step with a model additionally featuring teacher visit timing represented through the time binning variable. We find that adding teacher visit timing significantly improved model fit ($\chi 2(6) = 41.58$, $p < .001$). In addition, adding prior knowledge (measured as pre-test score) as an additive effect ($\chi 2(1) = 541.21$, $p < .001$) and in interaction with teacher visit timing ($\chi 2(6) = 11.29$, $p = .080$) further improved model fit. We follow a similar procedure using Poisson count models inferring the number of attempts at a given step. Here, a model with just an intercept served as a baseline. Adding teacher visit timing ($\chi 2(1) = 2541.90$, $p < .001$), prior knowledge ($\chi 2(1) = 5781.50$, $p < .001$), and the interaction of both ($\chi 2(1) = 263.20$, $p < .001$), significantly improved model fit. Like before, we omitted 18 cases due to missing test scores in this part of the analysis. Taken together, these findings highlight the utility of the relative timing of teacher visits in student in-tutor correctness, although the association was not uniformly positive, as assumed by H4, visualized in Fig. 3.

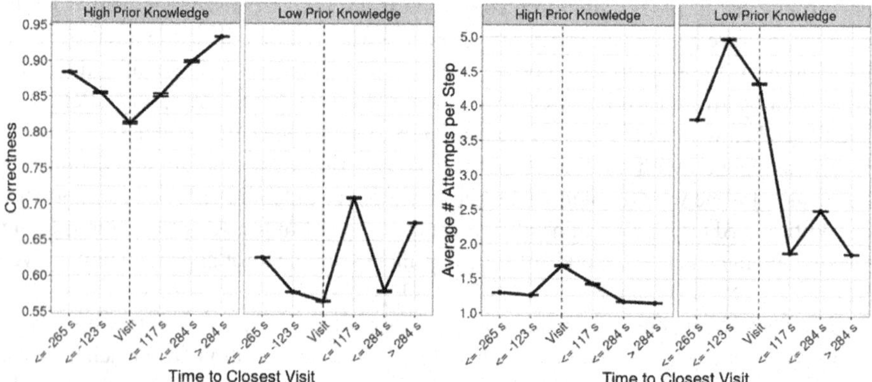

Fig. 3. Average correctness and the number of attempts per problem step broken out by the relative timing of tutor interactions with respect to teacher visits and whether students exhibited an above-median prior knowledge in the pre-test including 2 SE bars. Observations to the right of the dashed line designate aggregations of data points closer to past than upcoming visits.

Based on Fig. 3, we find two notable patterns. First, while all visited students exhibited decreasing correctness before teacher visits, students with high prior knowledge continuously increased their correctness at steps after a teacher visit. On the other hand, low prior knowledge students only had correctness improvements after teacher visits for around 2 min before falling back in correctness, except after around 5 min if the teacher did not revisit them. Second, high prior knowledge students only exhibited an increased number of attempts per step 2 min ahead of a teacher visit. In comparison, students with low prior knowledge already exhibited that behavior more than 4 min before a visit. In other words, students with high prior knowledge were visited by the teacher earlier after their number of attempts per step started to rise.

Finally, to continue our test of H4, we investigate the association of teacher visits, that is, their number and length, with learning as measured by pre/post tests. Including the length of visit in this analysis was motivated by observing considerable variance in the duration of teacher visits ($M = 36.85$ s, $SD = 18.53$ s). Given heterogeneous associations between teacher visits and in-tutor performance as displayed in Fig. 3, we e select a set of control variables through an *AIC*-based backward search for Gaussian regression models that predict pre/post procedural and conceptual learning gains, to isolate the effects of the number and length of visits. Positive marginal effects of the number and length of visits given these control variables would confirm H4. The procedure removes the least significant feature from a model including all features until there is no further improvement in *AIC*. Our feature search space includes features representing student struggle and disengagement (e.g., the average number and length of idleness sequences), student prior knowledge, and whether students were missing on the three days of data collection. We find that the number of teacher visits was significantly negatively associated with procedural ($\beta = -0.31$, CI95% = [−0.56, −0.06], p = .018) and conceptual ($\beta = -0.95$, CI95% = [−1.78, −0.12], p = .045) learning gain (see Sect. 2.1 for definitions of these measures). In addition, the total time spent with a student was significantly positively associated with conceptual learning gain ($\beta = 1.02$, CI95% = [0.06, 1.99], p = .038). Overall, this analysis reveals that teacher visits correlate with student leaning gains, though not in a straightforward manner. We unpack these estimated effects. First, increasing the number of visits while keeping the total length of visits constant (i.e., students experiencing more, shorter visits) was associated with lower procedural and conceptual learning gain. Second, increasing the total length of visits while keeping the number of visits constant (i.e., students experiencing fewer, longer visits) was associated with higher conceptual learning gains but not higher procedural learning gains. In a follow-up interview, the teacher noted that the duration of their visits varied and that some students needed short but frequent attention for reassurance, while others needed longer assistance on conceptual understanding. Taken together, while teacher visits exhibited significant associations with student learning, as measured by in-tutor performance and out-of-tutor knowledge tests, our hypothesis that this association would be uniformly positive was not supported.

4 Discussion and Conclusion

In recent years, the AIED community has increasingly focused on developing teacher tools and dashboards for AI-enabled classrooms. As we progress in such work on augmenting teachers' practices around AI tutors, it could be highly beneficial to have a deeper theoretical and empirical understanding of decisions teachers make in the physical classroom and their impact on student learning and experience with AI tutors. In this paper, we investigate spatiotemporal aspects of student-teacher interactions to better understand how teachers support student learning with AI tutors. Our analysis reveals that teacher visits were associated more with students' in-the-moment behavioral indicators (e.g., idleness) than with a broader, static measure of student needs such as low prior knowledge. While teacher visits were often associated with positive changes in student behavior afterward (e.g., decreased idleness), it was not clear that the teacher always selected a student who may have needed a visit the most (e.g., sometimes there were students who had been disengaged for much longer than the visited student; students with high prior knowledge were often visited earlier than low prior knowledge students). Furthermore, we find that *longer* teacher visits were associated with higher conceptual learning gain, while *frequent* visits were associated with lower conceptual and procedural learning gain. The latter might represent a selection effect, not an indication that the teacher's help was detrimental. Students who have difficulty with the material may attract many teacher visits. Therefore, there may be qualitative differences between teacher visits that may relate to conceptual and procedural support that calls for further scrutiny in future research. Lastly, we also observe differences in after-visit patterns between different student groups that potentially are not desirable. Understanding student-level differences in the antecedents and impact of teacher visits remains an important topic for future research.

The empirical contributions of this study have several implications for the development of teacher support for orchestration and reflection. First, the insights derived suggest that it may be useful if future teacher support tools were to present to teachers information about how their spatial pedagogical practices in the classroom (and visits, specifically) relate to student learning and engagement. Second, our findings also suggest that teacher behavior may not be ideal in some cases (e.g., teachers may have missed visiting struggling students), further motivating the need for well-designed tools to improve teachers' awareness and sensemaking [9]. Third, our study provides some new forms of analytics that, embedded in teacher tools, could help teachers better prioritize their limited classroom time between conflicting student needs (e.g., analytics that help identify students who were disengaged for longer). Fourth, the quantitative conceptualization of teacher practices opens up new possibilities for teaching analytics focused on helping teachers reflect on their own practices to make pedagogical improvements. As a preliminary illustration of that idea, while reflecting on a visualization of their visits, the teacher from the current case study said in one of the follow-up interviews, "Look at [student's name], I barely stopped by him, and he's the kid who is struggling, but I was hardly there. So, I want to check in on him a little bit more often." They were referring to a student who was seated at the back of the classroom. Such reflections led the teacher to imagine concrete classroom enactments, such as potential changes to the seating arrangement to have better visibility to students who may misuse the system, to have a peer tutor in close proximity to struggling students, and to group multiple struggling students to provide conceptual intervention.

In addition, there are several methodological implications of the current study for AIED research. First, combining data sources from the physical and virtual spaces enables an integrated, temporal analysis of student behaviors in the AI tutor and student-teacher behaviors in the classroom. Such analyses are necessary if we aim to improve our understanding of teachers' roles and practices in supporting students' learning with AI tutors. Second, the current study demonstrates the feasibility of collecting and analyzing such data in an automated way. To the best of our knowledge, the current study is the first to collect fine-grained, time-synchronized data on teacher position and log data of student-tutor interaction in an AI-enabled classroom in an automated way. Unlike other automated approaches like videos used in previous studies to quantify student-teacher interactions [17], position sensing is less intrusive, preserves student privacy, has a lower risk of unintended surveillance [4], and requires lower post-processing [18] to generate position coordinates. Third, we demonstrate how an existing algorithm (stop detection;

[8]) could be extended to quantitatively conceptualize a teaching construct of interest (visit) in a K12 classroom.

The study has several limitations that point to interesting opportunities for future work. First, our analysis revealed differences in the nature of visits (e.g., infrequent long visits versus frequent short visits). While we explored visit length in this study, future research needs to further contextualize teacher visits, for example, by encoding their content (e.g., help-giving, socioemotional support, reassurance). Second, the quantitative definition of teacher visits in this paper does not capture student-teacher interactions happening without physical proximity and may misattribute instances when a teacher stands near one student but talks to another, or observes the class, or does nothing. Third, our exploratory analysis doesn't consider other factors in an ecological setting like a classroom that may play a role (e.g., students seeking help by raising their hand, students getting help from a peer). Lastly, this is a case study with one teacher and limited time. The findings need to be tested broadly for generalizability.

Our findings confirm that teachers play an important role in AI-enabled classrooms, but also that they are limited in their abilities to perceive and prioritize *all* student needs in real-time. We have seen that tools designed to improve teachers' awareness and sensemaking of students' learning with AI tutors lead to better student learning [9]. Our results inform future design of teacher support tools that intentionally bring out the best of teacher abilities and overcome some of their limitations.

References

1. Crooks, N.M., Alibali, M.W.: Defining and measuring conceptual knowledge in mathematics. Dev. Rev. **34**(4), 344–377 (2014)
2. Schofield, J.W., Eurich-Fulcer, R., Britt, C.L.: Teachers, computer tutors, and teaching: the artificially intelligent tutor as an agent for classroom change. Am. Educ. Res. J. **31**(3), 579–607 (1994)
3. https://pslcdatashop.web.cmu.edu/Project?id=879. Last accessed 7 May 2023
4. Derry, S.J., et al.: Conducting video research in the learning sciences: guidance on selection, analysis, technology, and ethics. J. Learn. Sci. **19**(1), 3–53 (2010)
5. Fernandes, A.C., Huang, J., Rinaldo, V.: Does where a student sits really matter?-The impact of seating locations on student classroom learning. IJAES **10**, 1 (2011)
6. Stang, J.B., Roll, I.: Interactions between teaching assistants and students boost engagement in physics labs. Phys. Rev. Special Topics-Phys. Educ. Res. **10**(2) (2014)
7. Dessus, P., Cosnefroy, O., Luengo, V.: Keep your eyes on'em all!: a mobile eye-tracking analysis of teachers' sensitivity to students. In: EC-TEL, pp. 72–84 (2016)
8. Shou, T., Borchers, C., Karumbaiah, S., Aleven, V: Optimizing parameters for accurate position data mining in diverse classrooms layouts. In: EDM (2023)
9. Holstein, K., McLaren, B.M., Aleven, V.: Student learning benefits of a mixed-reality teacher awareness tool in AI-enhanced classrooms. In: Penstein Rosé, C., et al. (eds.) AIED 2018. LNCS (LNAI), vol. 10947, pp. 154–168. Springer, Cham (2018). https://doi.org/10.1007/978-3-319-93843-1_12
10. Holstein, K., McLaren, B.M., Aleven, V.: SPACLE: investigating learning across virtual and physical spaces using spatial replays. In: Proceedings of the LAK, pp. 358–67 (2017)
11. Kessler, A., Boston, M., Stein, M.K.: Exploring how teachers support students' mathematical learning in computer directed learning environments. ILS **121**(1/2), 52–78 (2019)

12. Knight, S., Wise, A.F., Chen, B.: Time for change: why learning analytics needs temporal analysis. J. Learn. Anal. **4**(3), 7–17 (2017)
13. Knoop-van Campen, C.A.N., Wise, A., Molenaar, I.: The equalizing effect of teacher dashboards on feedback in K-12 classrooms. Interact. Learn. Env. 1–17 (2021)
14. Lim, F.V., O'Halloran, K.L., Podlasov, A.: Spatial pedagogy: mapping meanings in the use of classroom space. Camb. J. Educ. **42**(2), 235–251 (2012)
15. Martinez-Maldonado, R., et al.: Moodoo the tracker: Spatial analytics for characterising classroom pedagogies (2020)
16. Mei, C.C.Y., Chin, H.B., Taib, F.: Instructional proxemics and its impact on classroom teaching and learning. IJMAL **1**(1), 69–85 (2017)
17. van Es, E.A., Sherin, M.G.: Expanding on prior conceptualizations of teacher noticing. ZDM – Math. Educ. **53**(1), 17–27 (2021)
18. Prieto, L.P., Sharma, K., Kidzinski, Ł, Rodríguez-Triana, M.J., Dillenbourg, P.: Multimodal teaching analytics: automated extraction of orchestration graphs from wearable sensor data. J. Comput. Assist. Learn. **34**(2), 193–203 (2018)

Trustworthy Academic Risk Prediction with Explainable Boosting Machines

Vegenshanti Dsilva, Johannes Schleiss(✉), and Sebastian Stober

Otto-von-Guericke University, Magdeburg, Germany
{vegenshanti.dsilva,johannes.schleiss,stober}@ovgu.de

Abstract. The use of predictive models in education promises individual support and personalization for students. To develop trustworthy models, we need to understand what factors and causes contribute to a prediction. Thus, it is necessary to develop models that are not only accurate but also explainable. Moreover, we need to conduct holistic model evaluations that also quantify explainability or other metrics next to established performance metrics. This paper explores the use of Explainable Boosting Machines (EBMs) for the task of academic risk prediction. EBMs are an extension of Generative Additive Models and promise a state-of-the-art performance on tabular datasets while being inherently interpretable. We demonstrate the benefits of using EBMs in the context of academic risk prediction trained on online learning behavior data and show the explainability of the model. Our study shows that EBMs are equally accurate as other state-of-the-art approaches while being competitive on relevant metrics for trustworthy academic risk prediction such as earliness, stability, fairness, and faithfulness of explanations. The results encourage the broader use of EBMs for other Artificial Intelligence in education tasks.

Keywords: Explainable AI in Education · Responsible AI · Trustworthy Machine Learning · Academic Risk Prediction · Virtual Learning

1 Introduction

Predictive models in education are used for a range of tasks, such as predicting enrollment numbers or student performance, engagement and satisfaction of classroom activities, or identifying at-risk students [22,25,29]. In this context, academic risk prediction focuses on predicting if a student fails or drops out of a course or class based on either past academic data or learning behaviors [1,3,22,25]. Thus, the use of predictive models in education should support learners in their learning journey and provide enriched information for educators or staff to give additional support in the classroom or on a system level [37].

Like in other AI domains, we face the trade-off between interpretable models and accurate models [9,17]. Interpretability in this context refers to a situation where humans can understand or even predict a model's output [21].

N. Wang et al. (Eds.): AIED 2023, LNAI 13916, pp. 463–475, 2023.
https://doi.org/10.1007/978-3-031-36272-9_38

To bring AI to the high-stake environment of education, it is core to address the need for stakeholders to understand the models' predictions from a user and a regulatory perspective [14]. The European General Data Protection Regulation (EU GDPR) [8], for example, states that "the existence of automated decision-making should carry meaningful information about the logic involved, as well as the significance and the envisaged consequences of such processing for the data subject." Thus, users require additional information on the underlying logic of automated decision-making. Similarly, trustworthy AI [37], ethics of AI [14], and the importance of explainability [9,17] are discussed in the context of AI in education. In this context, Fiok et al. [9] mention transparency, causality, privacy, fairness, trust, usability, and reliability as goals of explainable AI (XAI) that build the basis for the responsible and trustworthy use of AI in education. This is especially relevant when aiming to understand the most important causes and underlying factors in academic risk prediction [3,12,25,36].

When interpreting model results, explanations can be categorized as either *ante-hoc* or *post-hoc*. *Post-hoc* explanations, such as LIME [26] or SHAP [20], attempt to explain the predictions of a black-box model after it has been trained. However, these methods are limited by the assumption that a model can be approximated and require a lot of computing time. Additionally, Swamy et al. [35,36] found that *post-hoc* XAI methods often disagree and that human experts are not good validators of explainability. Therefore, using inherently interpretable models, or *ante-hoc* explanations, is recommended.

Recently, Explainable Boosting Machines (EBMs) [24] were introduced as an extension to Generative Additive Models [13]. Through its additive nature, the model is inherently interpretable while achieving state-of-the-art performance on tabular data on a range of tasks [4,23,24,38]. This paper explores the use of EBMs as an interpretable model for the task of academic risk prediction. The main contribution of the paper is to demonstrate the benefits of EBMs trained with online learning behaviors. We show how the interpretability of EBMs aligns with the causality a teacher might use to provide support to students in real life. Moreover, we compare EBMs to other state-of-the-art models on performance and relevant metrics in trustworthy academic risk prediction, such as earliness, stability, fairness, and faithfulness of explanations.

The paper is structured as follows. Section 2 gives the background on academic risk prediction, explainable AI in education, and EBMs. Next, Sect. 3 introduces the analysis approach including the data, training, and the proposed model assessment. Section 4 presents the experimental results of performance and other metrics and discusses the implications of the results. Last, Sect. 5 concludes the study and gives an outlook on future research.

2 Background

2.1 Academic Risk Prediction

Academic risk prediction is a part of predictive learning analytics and describes the task to identify students that might fail or drop a class or a course as a classification task [22,25,29]. Before the Covid-19 pandemic, demographic data and prior academic data were typically used as features for academic risk prediction [25,27]. Recently, more online learning behavioral data, such as interactions with Learning Management Systems, is available and used as input data for predictive models [25,29]. Common approaches include Artificial Neural Networks, Naive Bayes, K-Nearest-Neighbor as well as tree-based models like Random Forest, Decision Tree, or Gradient Boosting [25,29]. Typically, classification models are evaluated using performance measures such as the accuracy, F-measure, precision, or area under the curve [22,25,29]. In this context, most papers for academic risk prediction only assess a one-dimensional performance rubric (e.g. accuracy or precision), missing out on key considerations of trustworthy and explainable AI, such as earliness, stability, or quantitative explainability metrics [2,31].

2.2 Explainable AI in Education

Explanations are needed to gain the user's trust, improve the design, support the user's understanding of the recommendation and prediction, set the context of the prediction, and justify and rationalize an action, as means to communicate results [33]. Thus, explanations should focus on stakeholders' needs, and improve the perception, trust, and acceptance of users. Shin et al. [30] show that users evaluate explanations based on existing beliefs and partly on their understanding of the model. Moreover, they find that fairness, accountability, and transparency influence causability and trust in AI systems.

Similar to XAI in general, XAI in education focuses on global and local approaches for explanation [9,17,36]. However, state-of-the-art explainable post-hoc methods are not yet applied to most models of student performance predictions [2,6]. Moreover, the review of Almari et al. [2] shows that none of the analyzed studies has quantified the explainability of the proposed models, making it difficult to compare them on these metrics. They conclude that explainability metrics should be included next to the accuracy metric in the analysis and development of models for AI in education.

2.3 Explainable Boosting Machines

EBM [24] is an improved additive machine learning model based on Generalized Additive Models [13]. It can compete with state-of-the-art machine learning models on tabular data while being inherently interpretable [4,24,38]. EBMs address the need for local explanations through their additive model design, where each feature contributes individually to the prediction [24]. This means it has a feature importance build-in, which can simplify the understanding of the model's predictions. We demonstrate the explainability of EBMs in the context of academic risk prediction in Sect. 4.1. Moreover, Nori et al. [23] also have added Differential Privacy to EBM, which addresses the need to protect privacy, for example in educational data [28].

An EBM consists of feature functions $f_j(x_j)$ that are learned individually for each feature x_j, and pairwise feature interactions $f_{ij}(x_i, x_j)$ for the most important feature pairs. In the training phase, individual trees are trained on each feature separately in a round-robin fashion with a low learning rate (to diminish the effects of feature sequence dependence) for several iterations using boosting and bagging methods. The learned trees are averaged to a feature function that provides the feature score for different feature value bins. Based on this, the pairwise feature interactions are trained using the FAST algorithm [19]. The final prediction $E[y]$ can be expressed as,

$$g(E[y]) = \beta_0 + \sum_j f_j(x_j) + \sum f_{ij}(x_i, x_j) \tag{1}$$

with x_i and x_j expressing features, f_j the feature function, and f_{ij} the pairwise interaction function for the most important feature pairs. $g(.)$ refers to the link function for different learning tasks and β_0 describes the intercept.

Related to AI in education, Jayasundara et al. [16] conducted performance prediction of students with an EBM using socio-economic features and pre-course academic performance and compared the model accuracy to other approaches. In terms of interpretability, they analyzed the correlation between features and labels and compared them with global and local feature importance. Building up on this, we propose training EBMs on data of online learning behaviors. Furthermore, we explore the use of EBMs as trustworthy models for academic risk prediction by demonstrating their explainability and evaluating them on various metrics relevant for trustworthy academic risk prediction.

3 Methods

3.1 Study Area and Data

Task. In our investigation, we evaluate EBM for two academic risk prediction tasks: failure prediction and dropout prediction.[1] We implement the task as binary classification so that the feature functions generated by EBM are easy to interpret.

Dataset. The study uses the Open University Learning Analytics Dataset (OULAD) [18]. It contains data from 32,593 students for seven course modules offered in two terms, each course lasting for approximately nine months. The dataset includes demographic information (e.g. gender and age), pre-course information (e.g. studied credits and whether a course was previously attempted by the student), and clickstream data from the virtual learning environment (VLE) for 20 types of VLE resources. Moreover, it contains four target classes: Distinction, Pass, Fail, and Withdrawn.

[1] The code is available under https://gitlab.com/vegeedsilva/trustworthy-academic-risk-prediction-with-explainable-boosting-machine.git.

Online Learning Behaviors as Features. The clickstream data available through the interaction with VLE can reveal learning behaviors of students. Existing research explores how learning behaviors like engagement, regularity, and curiosity influence students' academic performance. Using learning behaviors as features can allow predictions to be actionable when provided with a relevant explanation [11,32]. Here, we created learning behaviors as features based on the interactions of a learner with the VLE. Table 1 describes the learning behaviors used as features for this dataset.

Table 1. Online learning behaviors used as features for academic risk prediction with the count indicating the number of a type of feature

Learning Behavior	Description	Count
Engagement	Total number of interactions as sum of clicks that a learner has with each VLE resource	20
Session Count	Total number of study sessions spent by a learner for each VLE resource	20
Regularity	Count of blocks (continuous weeks of learning interactions with the VLE) as maximum, minimum, average, and variance	5
Lateness	Percentage of assessment deadlines missed by a learner	1
Curiosity	Percentage of VLE resource covered by the learner	20
Assessment Coverage	Percentage of assessments completed by a learner	1

After generating all features, we have 67 features in total. The target labels are encoded as 1 for *at-risk* and 0 for *non-risk* learners. We perform feature scaling and select 30 features using the MinMaxScaler and SelectK functions from the scikit-learn package. After filtering student records with multiple attempts for a course and only retaining the latest records in such cases, we have 31,284 student records. The dataset is split randomly with a 70-10-20 split for training, validation, and test set.

3.2 Training

To evaluate EBMs against other state-of-the-art models, we train a Logistic Regression (LR), Decision Tree (DT), Random Forest (RF), and Feed-forward Neural Network (FNN) for failure prediction and dropout prediction. The length of the longest course in OULAD is 39 weeks. Hence, we compute the appropriate feature values based on the course week and train the models with the final result as the target. The models have been optimized for their respective hyperparameters. As a separate model was trained for each week for the two tasks, we have only mentioned important parameter details of the selected models. The selected EBM models were trained for 5000 rounds with 10 feature pairs and a quantile method for binning. The LR models were trained for 200 iterations. The DT and RF models used Gini impurity as their splitting criteria and the selected FNN models were composed of three layers, with ReLU activation, Adam optimizer, and binary cross-entropy loss.

3.3 Model Assessment

Accuracy. As the OULAD is a balanced dataset, we use accuracy as a performance measure and compare the accuracy development through the 39 weeks of the course period for different models.

Earliness and Stability. Especially in the context of education, early prediction of risk can facilitate scaffolding and intervention at the right time and enable learners to improve on their learning path [1,5]. Hence, Soussia et al. [31] introduced earliness and stability measures to evaluate academic risk prediction systems. *Earliness* is computed as the average time period when the first correct predictions for each record are made. One drawback here is that earliness does not reflect the accuracy of the model. Hence, the authors propose using the harmonic mean (HM) of earliness and accuracy to have a well-rounded assessment of prediction systems [31]. We use HM to evaluate the earliness of different models as follows,

$$HM = \frac{2 * (1 - earliness) * accuracy}{(1 - earliness) + accuracy} \tag{2}$$

In this context, a higher HM value is desirable for early and accurate predictions.

(Temporal) stability helps to gain user trust in the model's ability to make correct predictions over time. It describes the average of the highest number of weeks the model can deliver successive correct predictions over the course period. A higher stability score indicates a stable predictor. Thus, earliness and stability are significant parameters for evaluation when investigating the suitability of academic risk predictors.

Fairness. Bias can slide into a machine learning pipeline at several stages, including imbalanced datasets or missing data, bias during model training, and user perception during data collection. Here, we assess the algorithmic bias of the dataset and the mentioned models. For this purpose, we compute group fairness metrics such as the *statistical parity difference (SPD)* [7] and *equal opportunity difference (EOD)* [10]. *SPD* is defined as the difference between the probabilities of the protected and the majority groups in obtaining a favorable decision. It is described as

$$SPD = P(Y_{pred} = 1 \mid F = min) - P(Y_{pred} = 1 \mid F = maj) \tag{3}$$

where Y_{pred} are the model predictions and F is the group of the sensitive attribute. A SPD value of zero indicates fairness.

EOD also takes into account the class label and measures the deviation from equal opportunity. In this context, equal opportunity implies that both the privileged and unprivileged group have the same probability of obtaining a favorable class. Thus, an EOD value of zero indicates fairness. EOD can be expressed as

$$EOD = P(Y_{pred} = 1 \mid F = min, Y = 1) - P(Y_{pred} = 1 \mid F = maj, Y = 1) \tag{4}$$

where Y_{pred} are the model predictions, F is the sensitive attribute and Y are the true labels.

Table 2. Mean feature importance score assigned by EBM for predicting dropouts in week 20 for the top five features.

Feature	Mean Importance Score
Lateness	0.667
Curiosity: quiz	0.422
Assessment Coverage	0.414
Regularity: Block count	0.366
Curiosity: subpage	0.234

Faithfulness of Local Explanations. A faithful explanation should provide insight into the rationale of the model to arrive at its prediction. Although there is no standard method to quantitatively evaluate the faithfulness of explanations, different measures have been introduced in the field of XAI [15,34]. Here, we measure the faithfulness of a model by computing the average of recall obtained after re-training the model over the top important features that contribute at least 50% towards the final outcome in the local explanations of the test set. This measure is called *recall on important features (ROIF)* [26]. It allows evaluating the consistency between the feature importance in the local explanations and the true importance given by the model.

4 Experimental Results

4.1 Explainability of EBMs

Before providing the results of the model assessments, we want to demonstrate the interpretability of EBMs in the context of educational data. The first step is to understand how EBMs work on inference time. To provide a prediction outcome, EBMs use feature functions and pairwise feature interaction heatmaps as lookup tables. Thus, EBMs provide the possibility to access the feature importance by default.

For example, Table 2 shows the overall mean importance of the features in EBM for dropout prediction in week 20 for the five most important features. Furthermore, we can also plot the feature shape function to understand the influence of the feature, as demonstrated in Fig. 1 for the feature "lateness". We can observe that when the lateness is very low (below 0.2), the model gives a score of -1, pushing toward the *non-risk* class (i.e. label 0). Correspondingly, when lateness is high, its score is positive, pushing the overall prediction towards *at-risk* (i.e. label 1). In terms of model understanding, this corresponds to our didactic understanding. For example, assume a teacher observes that a student has been submitting most assignments late, we can expect that the teacher will give special attention to the student, as there might be a risk of dropout or failure. Thus, using EBMs with online learning behavior data allows interpretable and reasonable predictions that are easy to access and provide support for students and teachers.

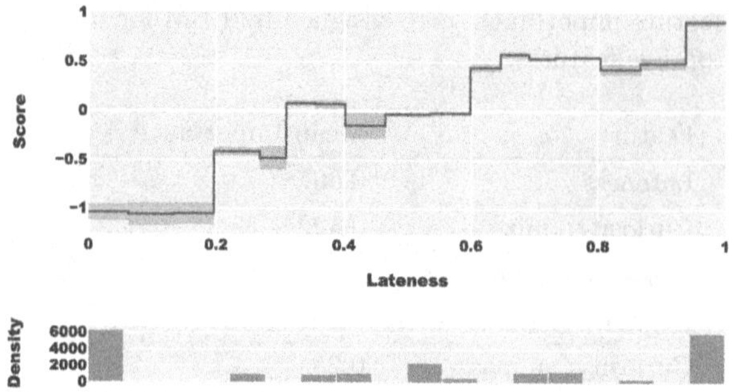

Fig. 1. Feature function for the feature "lateness" generated by EBM for predicting learners at risk of dropout in week 20 with the gray boxes representing variance around a score. The lower graph represents the density in the different feature value bins.

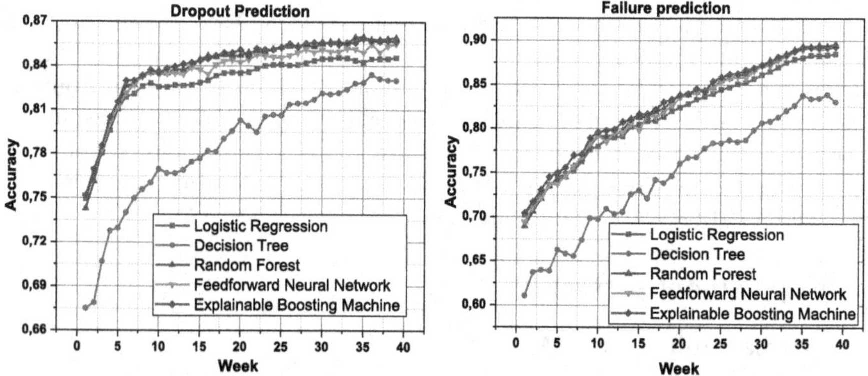

Fig. 2. Comparison of model accuracy evolution through course period for dropout prediction (left) and failure prediction (right).

4.2 Accuracy

The accuracy of the models per course week for the dropout and failure prediction is shown in Fig. 2. We can observe that the accuracy to predict the events increases over time, with all models except DT performing in a similar accuracy range. This demonstrates that EBMs perform equally well compared to established methods in the context of academic risk prediction with online learning behaviors.

4.3 Earliness and Stability

The earliness and stability results are presented in Table 3. When evaluating models for earliness the group of learners *at-risk* is of special focus to identify such learners as early as possible. For the task of predicting dropouts, we observe similar HM earliness of around 85% for the *at-risk* group for EBM along with DT,

RF, and FNN. For the failure prediction, DT has the highest HM earliness score. Here, the earliness of EBM, although not the best, is at par with the remaining models (except DT). Concerning the stability of models, we can observe that LR provides stable correct predictions for the longest period. At the same time, the EBM is comparably stable in both tasks.

Table 3. Harmonic mean earliness and stability measures for task of dropout prediction and failure prediction for *at-risk* (AR) and *non-risk* (NR) students.

Model	HM Earliness (Percentage)				Stability (No. of weeks)			
	Dropout		Failure		Dropout		Failure	
	(AR)	(NR)	(AR)	(NR)	(AR)	(NR)	(AR)	(NR)
EBM	85.53	**74.64**	78.79	70.02	22.86	33.97	25.67	**36.47**
LR	84.15	73.94	77.40	**71.42**	**23.94**	**34.35**	**26.24**	36.28
DT	85.19	60.64	**82.77**	54.62	22.54	24.59	20.51	26.94
RF	85.61	71.55	79.26	65.95	**23.93**	33.12	24.31	36.10
FNN	**85.88**	74.61	79.01	67.46	22.98	33.57	24.33	36.09

Table 4. Algorithmic fairness for gender and disability expressed through statistical parity difference (SPD) and equal opportunity difference (EOD) for both tasks of dropout and failure prediction.

Model	Gender				Disability			
	Dropout		Failure		Dropout		Failure	
	SPD	EOD	SPD	EOD	SPD	EOD	SPD	EOD
EBM	−0.031	−0.024	−0.024	**0.003**	0.053	0.028	0.054	0.012
LR	**−0.023**	**−0.010**	**−0.012**	0.011	0.048	0.028	0.056	0.018
DT	−0.028	−0.014	−0.023	−0.006	**0.036**	0.026	**0.050**	**−0.009**
RF	−0.039	−0.025	−0.018	0.008	0.050	**0.017**	0.063	0.028
FNN	−0.032	−0.017	−0.022	0.006	0.056	0.034	**0.050**	0.015

4.4 Fairness

Before looking at the model behavior, we need to assess the balance of the dataset for the gender and disability attributes. The SPD of the OULAD for the attribute gender is 0.002 for dropout and −0.019 for failure prediction with females as the privileged group for both cases. Respectively, the SPD for the attribute disability is 0.076 for dropout and 0.065 for failure prediction with non-disabled learners as the privileged group. Thus, the dataset is balanced for both attributes. Table 4 illustrates the results for fairness metrics SPD and EOD in week 39 for all trained models. With this balanced dataset in place, we can observe that all models perform mostly similarly concerning the evaluated fairness metrics.

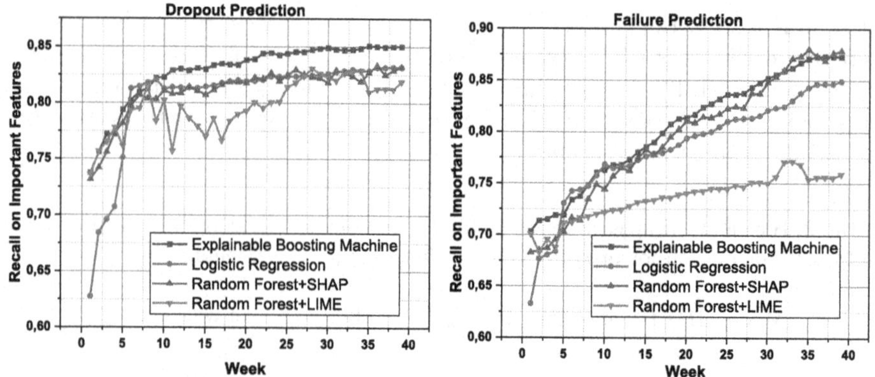

Fig. 3. Average ROIF (for top five features) for dropout prediction (left) and failure prediction (right).

4.5 Faithfulness of Explanations

We compare the faithfulness of local explanations given by EBM to LR, another inherently interpretable model, and to the application of post-hoc explainability methods such as LIME [26] and SHAP [20] on a RF model. Figure 3 shows the ROIF score of the methods through the course period. In the dataset, the initial period is critical for retaining students as 60% of the total dropouts occur in the first ten weeks of the course period. From our results, we can observe that for dropout and failure prediction the ROIF score for EBM is generally higher than all other explainability methods.

4.6 Discussion

The experiments evaluated EBM in comparison to other state-of-the-art models for academic risk prediction on online learning behavior data. Furthermore, we showed how EBMs are inherently interpretable and how this can benefit the trustworthiness of their use in an academic risk prediction task. The experiments on one dataset indicate that EBMs are equally accurate as other models and perform very stable. All models perform in a similar range on other metrics such as earliness and fairness. Concerning the faithfulness of local explanations, EBMs show slightly better ROIF scores compared to LR as another interpretable model or the use of model-agnostic post-hoc explainability methods, such as SHAP and LIME used in combination with RF. At the same time, more analysis is needed to generalize these results.

Using EBMs in production is well supported through the InterpretML package [24]. They are especially useful for datasets that are well understood and features that have no complex interdependence. However, EBMs currently only support tree structures as base learners. Furthermore, the application of EBMs for multi-class prediction settings, for example identifying the risk of failure and dropouts in the same setting, is still under ongoing research [24]. This also supports the argument for improving and using model-agnostic post-hoc interpretability methods. In any case, to develop trustworthy academic risk prediction, we need to make it a standard to include quantified explainability metrics. Regarding EBMs, we can conclude that it is a useful model for tabular datasets in education in tasks that require interpretability of the model.

5 Conclusion and Outlook

To address the need for trustworthiness, we demonstrated the use of EBMs for the task of academic risk prediction using data from online learning behaviors. Through their additive nature, EBMs are inherently interpretable as the contribution of each feature is visible. In our experiments, we demonstrated the interpretability of EBMs and quantified the trustworthiness of the model in comparison to other state-of-the-art models using metrics such as faithfulness of explanation, fairness, earliness, and stability. We observe that EBMs are able to capture the influence of different learning behaviors on the risk of dropout and failure in accordance with the causality a teacher might use to provide support to students in real life. Moreover, we find that EBMs perform similarly to other models for the metrics discussed in the academic risk context. Thus, we can conclude that its explainability could be a significant advantage that can give insights and create trust for all stakeholders involved [35]. The results encourage broader use of EBM for other tasks in AI in education that use tabular data.

Future work will test the model on other open access datasets, compare it to further developments of GAMs, and explore the suggested application of Differential Privacy on EBM [23] with educational data. Moreover, a user study is planned to analyze the impact of explainability in a real scenario.

References

1. Adnan, M., et al.: Predicting at-risk students at different percentages of course length for early intervention using machine learning models. IEEE Access **9**, 7519–7539 (2021)
2. Alamri, R., Alharbi, B.: Explainable student performance prediction models: a systematic review. IEEE Access **9**, 33132–33143 (2021)
3. Baranyi, M., Nagy, M., Molontay, R.: Interpretable deep learning for university dropout prediction. In: Proceedings of the 21st Annual Conference on Information Technology Education, pp. 13–19 (2020)
4. Bussmann, N., Giudici, P., Marinelli, D., Papenbrock, J.: Explainable machine learning in credit risk management. Comput. Econ. **57**(1), 203–216 (2021)
5. Chen, F., Cui, Y.: Utilizing student time series behaviour in learning management systems for early prediction of course performance. J. Learn. Anal. **7**(2), 1–17 (2020)
6. Cohausz, L.: Towards real interpretability of student success prediction combining methods of XAI and social science. In: Proceedings of the 15th International Conference on Educational Data Mining, pp. 361–367 (2022)

7. Dwork, C., Hardt, M., Pitassi, T., Reingold, O., Zemel, R.: Fairness through awareness. In: Proceedings of the Innovations in Theoretical CS Conference, pp. 214–226 (2012)
8. EU: Regulation EU 2016/679 of the European Parliament and of the Council of 27 April 2016. Official Journal of the European Union (2016)
9. Fiok, K., Farahani, F.V., Karwowski, W., Ahram, T.: Explainable artificial intelligence for education and training. J. Defense Model. Simul. **19**(2), 133–144 (2022)
10. Hardt, M., Price, E., Srebro, N.: Equality of opportunity in supervised learning. Adv. Neural Inf. Process. Syst. **29**, 3315–3323 (2016)
11. Hasan, R., Fritz, M.: Understanding utility and privacy of demographic data in education technology by causal analysis and adversarial-censoring. Proc. Priv. Enhanc. Technol. **2022**(2), 245–262 (2022)
12. Hasib, K.M., Rahman, F., Hasnat, R., Alam, M.G.R.: A machine learning and explainable AI approach for predicting secondary school student performance. In: IEEE 12th Annual Computing and Communication Workshop and Conference, pp. 0399–0405. IEEE (2022)
13. Hastie, T., Tibshirani, R.: Generalized additive models: some applications. J. Am. Stat. Assoc. **82**(398), 371–386 (1987)
14. Holmes, W., et al.: Ethics of AI in education: towards a community-wide framework. Int. J. Artif. Intell. Educ. **32**(3), 504–526 (2022)
15. Hooker, S., Erhan, D., Kindermans, P.J., Kim, B.: A benchmark for interpretability methods in deep neural networks. Adv. Neural Inf. Process. Syst. **32**, 9737–9748 (2019)
16. Jayasundara, S., Indika, A., Herath, D.: Interpretable student performance prediction using explainable boosting machine for multi-class classification. In: 2022 2nd International Conference on Advanced Research in Computing (ICARC), pp. 391–396. IEEE (2022)
17. Khosravi, H., et al.: Explainable artificial intelligence in education. Comput. Educ. Artif. Intell. **3**, 100074 (2022)
18. Kuzilek, J., Hlosta, M., Zdrahal, Z.: Open university learning analytics dataset. Sci. Data **4**(1), 1–8 (2017)
19. Lou, Y., Caruana, R., Gehrke, J., Hooker, G.: Accurate intelligible models with pairwise interactions. In: Proceedings of the 19th ACM SIGKDD International Conference on Knowledge Discovery and Data Mining, pp. 623–631 (2013)
20. Lundberg, S.M., Lee, S.I.: A unified approach to interpreting model predictions. Adv. Neural Inf. Process. Syst. **30**, 4765–4774 (2017)
21. Miller, T.: Explanation in artificial intelligence: insights from the social sciences. Artif. Intell. **267**, 1–38 (2019)
22. Namoun, A., Alshanqiti, A.: Predicting student performance using data mining and learning analytics techniques: a systematic literature review. Appl. Sci. **11**(1), 237 (2020)
23. Nori, H., Caruana, R., Bu, Z., Shen, J.H., Kulkarni, J.: Accuracy, interpretability, and differential privacy via explainable boosting. In: International Conference on Machine Learning, pp. 8227–8237 (2021)
24. Nori, H., Jenkins, S., Koch, P., Caruana, R.: Interpretml: a unified framework for machine learning interpretability. arXiv preprint arXiv:1909.09223 (2019)
25. de Oliveira, C.F., Sobral, S.R., Ferreira, M.J., Moreira, F.: How does learning analytics contribute to prevent students' dropout in higher education: a systematic literature review. Big Data Cogn. Comput. **5**(4), 64 (2021)

26. Ribeiro, M.T., Singh, S., Guestrin, C.: Why should i trust you? Explaining the predictions of any classifier. In: Proceedings of the 22nd ACM SIGKDD International Conference on Knowledge Discovery and Data Mining, pp. 1135–1144 (2016)
27. Rubiano, S.M.M., Garcia, J.A.D.: Formulation of a predictive model for academic performance based on students' academic and demographic data. In: 2015 IEEE Frontiers in Education Conference (FIE), pp. 1–7. IEEE (2015)
28. Schleiss, J., Günther, K., Stober, S.: Protecting student data in ML pipelines: an overview of privacy-preserving ML. In: Rodrigo, M.M., Matsuda, N., Cristea, A.I., Dimitrova, V. (eds.) Artificial Intelligence in Education. Posters and Late Breaking Results, Workshops and Tutorials, Industry and Innovation Tracks, Practitioners' and Doctoral Consortium. AIED 2022. LNCS, vol. 13356, pp. 532–536. Springer, Cham (2022). https://doi.org/10.1007/978-3-031-11647-6_109
29. Sghir, N., Adadi, A., Lahmer, M.: Recent advances in predictive learning analytics: a decade systematic review (2012–2022). Educ. Inf. Technol. 1–35 (2022)
30. Shin, D.: The effects of explainability and causability on perception, trust, and acceptance: Implications for explainable AI. Int. J. Hum. Comput. Stud. **146**, 102551 (2021)
31. Soussia, A.B., Labba, C., Roussanaly, A., Boyer, A.: Assess performance prediction systems: Beyond precision indicators. In: Proceedings of the 14th International Conference on Computer Supported Education, pp. 489–496 (2022)
32. Soussia, A.B., Treuillier, C., Roussanaly, A., Boyer, A.: Learning profiles to assess educational prediction systems. In: Rodrigo, M.M., Matsuda, N., Cristea, A.I., Dimitrova, V. (eds.) Artificial Intelligence in Education. AIED 2022. LNCS, vol. 13355, pp. 41–52. Springer, Cham (2022). https://doi.org/10.1007/978-3-031-11644-5_4
33. Srinivasan, R., Chander, A.: Explanation perspectives from the cognitive sciences-a survey. In: Proceedings of the Twenty-Ninth International Conference on International Joint Conferences on Artificial Intelligence, pp. 4812–4818 (2021)
34. Sundararajan, M., Taly, A., Yan, Q.: Axiomatic attribution for deep networks. In: Proceedings of the 34th International Conference on Machine Learning, pp. 3319–3328 (2017)
35. Swamy, V., Du, S., Marras, M., Kaser, T.: Trusting the explainers: teacher validation of explainable artificial intelligence for course design. In: LAK23: 13th International Learning Analytics and Knowledge Conference, pp. 345–356 (2023)
36. Swamy, V., Radmehr, B., Krco, N., Marras, M., Käser, T.: Evaluating the explainers: black-box explainable machine learning for student success prediction in MOOCS. In: Proceedings of the International Conference on Educational Data Mining (2022)
37. Vincent-Lancrin, S., van der Vlies, R.: Trustworthy artificial intelligence (AI) in education. OECD Educ. Work. Pap. **218** (2020)
38. Wang, C., Han, B., Patel, B., Rudin, C.: In pursuit of interpretable, fair and accurate machine learning for criminal recidivism prediction. J. Quant. Criminol. **39**, 519–581 (2023)

Automatic Educational Question Generation with Difficulty Level Controls

Ying Jiao(✉), Kumar Shridhar, Peng Cui, Wangchunshu Zhou, and Mrinmaya Sachan(✉)

ETH Zurich, Zurich, Switzerland
yijiao@ethz.ch, {kumar.shridhar,peng.cui,wangchunshu.zhou, mrinmaya.sachan}@inf.ethz.ch

Abstract. We consider the task of automatically generating math word problems (MWPs) of various difficulties that meet the needs of teachers in teaching and testing students in corresponding educational stages. Existing methods fail to produce high-quality problems while allowing the teacher control over the problem difficulty level. In this work, we introduce a controllable MWP generation pipeline that samples from an energy language model with various expert model components for realizing the target attributes. We control the difficulty of the resulting MWPs from mathematical and linguistic aspects by imposing constraints on equations, vocabulary, and topics. We also use other control attributes including fluency and distance to the conditioning sequence to manage language quality and creativity. Experiments and evaluation results demonstrate our approach improves upon the baselines in generating solvable, well-formed, and diverse MWPs of controlled difficulty levels. Lastly, we solicit feedback from various math educators who approve the effectiveness of our system for their MWP design processes. They suggest our outputs align with the expectations of problem designers showing a possibility of using such problem generators in real-life educational scenarios. Our code and data are available on request.

Keywords: Math Word Problem generation · Automatic Educational Question generation · Controllable text generation

1 Introduction

Math Word Problem (MWP) is an important pedagogical tool in teaching arithmetic and situational reasoning skills to students through real-world situations and stories [11,18]. As MWPs are widely used in Education [9,17] and public standardized tests at different learning phases of students, these problems of varying difficulty levels are in high demand. However, MWPs are often created by human experts, and manual constructing and updating the question banks with diverse difficulties is costly [11]. This has led to a growing focus on automatically generating MWPs with controllable difficulty [8,20].

N. Wang et al. (Eds.): AIED 2023, LNAI 13916, pp. 476–488, 2023.
https://doi.org/10.1007/978-3-031-36272-9_39

Previous template-based studies have explored imposing difficulty controls in MWP generations [8, 20]. However, their results are limited in quality and diversity due to the dependence on fixed templates and domain knowledge. Recent works [21, 24] increasingly rely upon general-purpose neural-network-based text generation models. While these models produce outputs with high language quality and valid mathematical relations, they offer limited control over the problem difficulty; thus they tend to only output simple problems. Moreover, due to their lack of controllability, it is hard to incorporate new user constraints into these approaches, leading to poor adoption of these methods in Education.

In this work, we propose an MWP generation framework that allows for the generation of solvable, well-formed, and innovative problems of prespecified difficulty levels. Instead of treating "difficulty" as a black box, we follow [4, 18] from the learning science domain and disassemble this concept into mathematical and linguistic difficulty. We allow constraints on the solution equations to manage the mathematical difficulty and constraints on the topics and vocabulary to control for the linguistic difficulty.

Our system adopts the Mix and Match Language Model (M&M LM) [16] which samples from an energy LM using the Gibbs-Metropolis-Hastings sampler [6]. Our energy value is composed of scores from expert models including a topic discriminator, an equation generator, a large pre-trained Masked LM (MLM) assessing fluency, as well as Hamming distance and BertScore [23] measuring the distance to the conditioning problem. In contrast to prior neural network-based methods, this energy-based model allows easy integration of various constraints and modification of existing ones.

We evaluate the quality and difficulty controllability of our results using automatic and human evaluations. The automatic evaluation shows improved perplexity, coherence, and distance to the conditioning problem over previous approaches. Then, we asked several human evaluators to assess 5 facets of the generated problem: well-formedness, solvability, creativity, topic matching, and difficulty matching. The evaluation results show that our system improves upon the baselines in terms of language quality, solvability, innovation as well as controllability of topic and difficulty. We further examine the usability of our system through a user study with real math teachers. The teachers in this study suggest our results are almost indistinguishable from human-authored ones in our dataset. We find this very encouraging as it brings automatic question-generation methods one step closer to deployment in real educational settings.

2 Related Work

MWP Generation: Earlier explorations on MWP generation use template-based approaches [8, 17, 20]. Their generation results based on hand-crafted templates are limited in diversity, which could lead to rote memorization. Wang et al. [20] impose difficulty control on equation complexity and amount of redundancy. Their results are of a limited scope and language quality. Khodeir et al. [8] explore the difficulty-controllable generation of questions described with rhetorical schemes.

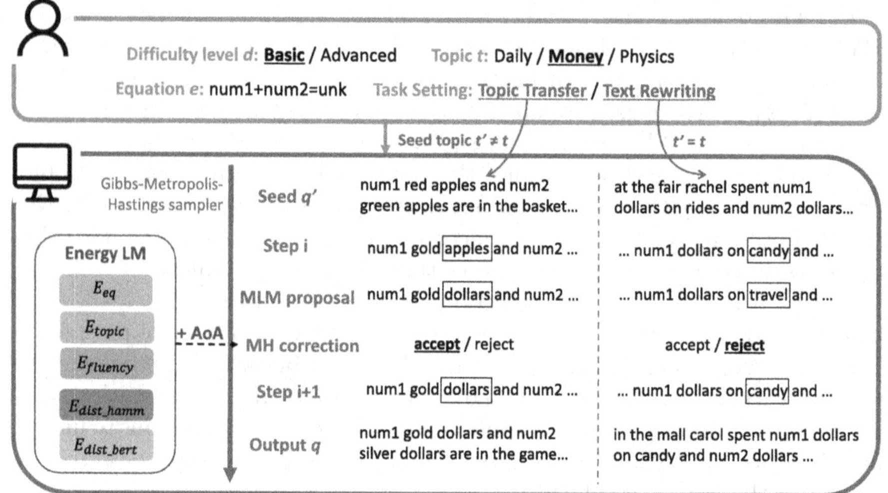

Fig. 1. A schematic illustrating one step of the M&M sampling procedure.

The majority of recent works are generative models built upon neural networks [13,21,24]. Koncel-Kedziorski et al. [9] propose a replacement-based approach. They change the topics of seed problems by replacing content words while maintaining syntactic and semantic properties. The above works do not allow for difficulty control. Most of the generated problems are elementary level.

Controllable Text Generation: Our task is closely related to the emerging body of research on generating text that satisfies certain control attributes and constraints [5,12,25]. In this work, we adopt M&M LM [16], which treats controllable generation as sampling from an energy-based model. It allows encoding our constraints as energy components and mixing them conveniently.

3 Problem Formulation

Our goal is to generate MWPs that satisfy certain constraints. Figure 1 shows an overview of our framework. We assume that we are provided a set of target attributes for the desired problem namely, problem difficulty level d, topic t, and equation e. The objective of our system is to output a novel well-formed question q with topic t and solution equation e. q should also satisfy the vocabulary constraint of difficulty d i.e. it should not use words that are more complex than the requisite difficulty level d.

We produce q with two task settings: **topic transfer** and **text rewriting**. Topic transfer starts from a seed problem q' of difficulty d, topic t', and solution e, where t' and t are not the same. The goal is to produce a problem with the same characteristics but on a different topic. For text rewriting, t' equals t.

The goal here is to produce diverse problems of the target difficulty and math concepts, which help students to practice and avoid rote memorization.

4 Approach

Our work uses a general-purpose controllable text generation model, Mix and Match (M&M) LM [16] as a backbone. With this approach, we generate $q = (\bar{q}_1 ... \bar{q}_m)$ by sampling from an energy-based model $p(q;\theta) = \frac{e^{-E(q;\theta)}}{\sum_{q' \in \mathcal{Q}} e^{-E(q';\theta)}}$, where $\mathcal{Q}$ represents the space of possible problem candidates and $E(q;\theta)$ parameterized by θ refers to the energy of q. We describe the details of our vocabulary constraint and energy function designed for difficulty-controllable MWP generation in Sect. 4.1 and 4.2.

4.1 Age of Acquisition Based Sampling

We incorporate the vocabulary constraint imposed by the target difficulty directly in the sampling procedure. In particular, we constrain the generated samples by their *Age of Acquisition* (AoA) [10]. AoA($\bar{q}_n$) refers to the age at which a child typically acquires a new word.[1] We require the sampled question to only contain words with AoA below a certain difficulty-specified threshold Thres$_d$. This constraint ensures the words in the output question q will be within the knowledge of the target students. At Markov chain Monte Carlo (MCMC) step i, a random position n of the current problem q^i is masked out, and a new problem q^{i+1} is proposed. Mathematically, we accept q^{i+1} with a probability $p(q^{i+1}, q^i) = \min\left(1, \frac{e^{-E(q^{i+1};\theta)} p_{\text{mlm}}\left(\bar{q}_n^i | \bar{q}_{\setminus n}^i\right)}{e^{-E(q^i;\theta)} p_{\text{mlm}}\left(\bar{q}_n^{i+1} | \bar{q}_{\setminus n}^{i+1}\right)} \cdot \mathbb{1}(\text{AoA}(\bar{q}_n^{i+1}) < \text{Thres}_d)\right)$. Here, $\mathbb{1}(c) = 1$ if the condition c is True else 0. p_{mlm} represents the conditional softmax of MLM at the n-th position.

4.2 Energy Components

We model the energy $E(q|q', t, e; \theta)$ of a new problem q using a number of ways to achieve the desired attributes in the generated question. We define:

$$\begin{aligned} E(q|q', t, e; \theta) = &\alpha E_{\text{eq}}(q|e) + \beta E_{\text{topic}}(q|t) + \gamma E_{\text{fluency}}(q) \\ &+ \delta E_{\text{dist\_hamm}}(q|q') + \phi E_{\text{dist\_bert}}(q|q') \end{aligned} \quad (1)$$

The various components of the energy function are as follows:

- $\mathbf{E}_{\text{eq}}(q|e)$: The equation energy measures the Levenshtein distance between the solution equation of q and the target equation e. This promotes producing questions whose solutions are close to the target equation and thus controls the solvability and mathematical difficulty of generated questions. We obtain the predicted solutions from a pretrained model EqGen (Sect. 5.3) and then define $E_{\text{eq}}(q|e) = \text{Levenshtein}(\text{EqGen}(q), e)$.

[1] http://crr.ugent.be/archives/806.

- $\mathbf{E}_{\text{topic}}(q|t)$: This energy component is defined as the logit of the topic classification model (Sect. 5.3) for the target topic t given a problem q.
- $\mathbf{E}_{\text{fluency}}(q)$: This component models the fluency of q. In particular, we use an MLM (Sect. 5.3) to judge the question's well-formedness [23]. For a problem q of length N, $E_{\text{fluency}}(q) = -\sum_{n=1}^{N} f\left(q_n, h\left(q_{\backslash n}\right)\right)/N$, where q_n is the n-th token and $h\left(s_{\backslash n}\right)$ is a representation learned by MLM which contains the context of both directions around q_n. $f\left(q_n, h\left(q_{\backslash n}\right)\right)$ is the raw logit at position n. It is computed from MLM's forward pass while masking the n-th position [6].
- $\mathbf{E}_{\text{dist\_hamm}}(q|q')$ and $\mathbf{E}_{\text{dist\_bert}}(q|q')$: Given a seed question q', these two quantities are defined as Hamming distance (normalized by the length of q') and BertScore [23] between q and q'. They can be used to encourage changes at the token and semantic levels and lead to creative outputs.

5 Experiments

5.1 Data Preparation

We build a basic seed set from ASDiv [15] and GSM8K [3] as well as an advanced one from MathQA [1]. The basic seed set contains 7K questions for elementary school students, which can be solved within 4 steps (one operator counts one step) using operators +, −, *, and /. The 6.7K GRE/GMAT questions in the advanced seed set require 5 to 10 solving steps with the same four operators. The problems solved by more sophisticated operators are left for future explorations. Questions and equations of the two seed sets are pre-processed by replacing numbers in problems with special tokens num1, num2, etc. Sub-results in equations are replaced with sub1, sub2, etc., and the final answer is represented by the token unk. For the problem topic, we perform K-means clustering using TF-IDF vectors and assign a topic label for each cluster. We have 6 topic categories: basic daily, money, physics, and advanced daily, money, physics. From the statistics in Table 1 and the example seed problems in Table 2, we can see most basic problems are close to primary-school student's daily life while advanced ones are about more abstract concepts like profit and speed.

Table 1. Statistics of seed sets.

Seed Set	#MWPs	Avg #tokens per MWP	Avg #symbols per eq	Topic proportion		
				Daily	Money	Physics
Basic	7,000	38.90	12.02	59.49%	24.81%	15.70%
Advanced	6,698	45.74	29.72	23.52 %	40.80%	35.68%

Table 2. Examples of generated MWPs. BTT, BTR and ATR refer to basic-level topic transfer, basic-level and advanced-level text rewriting. D, M and P mean topic daily, money and physics. Subscript B and A are for basic and advanced level.

Task: BTT; **Target topic:** M_B; **Seed topic:** D_B
Seed: mrs. hilt saw num1 bees in the hive. the next day she saw num2 times that many. how many bees did she see on the second day?
Output: mrs. hilt spent num1 dollars on a cake and the following day she spent num2 times the amount. how many dollars did she spend exactly the following day?
Task: BTR; **Target topic:** P_B; **Seed topic:** P_B
Seed: mrs. hilt measured her bookcase. it was num1 inches long. how many feet long was the bookcase?
Output: mrs. foster enlarged her bookcase until it was num1 inches wide. how many feet wide is her bookcase?
Task: ATR; **Target topic:** P_A; **Seed topic:** P_A
Seed: a cyclist rides a bicycle num1 km at an average speed of num2 km/hr and again travels num3 km at an average speed of num4 km/hr. what is the average speed for the entire trip?
Output: a student began his journey num1 km at an average speed of num2 km/hr. he continued num3 km at another average speed of num4 km/hr. what was his average speed for the whole journey?

5.2 Experiment Settings

Our experiments cover (1) transferring 15 randomly sampled basic daily problems to basic money and basic physics outputs denoted as Ours_{BTT}; and (2) rewriting 10 randomly sampled seeds from each of the basic and advanced topics, denoted as Ours_{BTR} and Ours_{ATR}. Table 2 demonstrates examples of the generation results. The topic transfer task is only performed for the basic problem generation since its preliminary test on the advanced level produces less satisfying results. The reason could be the advanced questions have more specialized storylines for each topic, which costs more token replacements to transfer to another topic and is thus more challenging. For Ours_{BTT}, we set the hyperparameters α, β, δ, and ϕ in Eq. 1 as 7000, 1000, 0, and 0. γ is tuned between 10, 20, 30, 50, 100, and 200. For Ours_{BTR} and Ours_{ATR}, α, β, γ is set to 1000, 1000, 100. δ and ϕ are assigned the same value and tuned between 500, 1000, and 2000. Our system runs a separate MCMC chain for 300 epochs for each of the 10 parallel outputs. An epoch refers to one masking cycle of all positions in random order.

5.3 Expert Models

We utilize fine-tuned `T5-large` and `BertForSequenceClassification` on our seed sets (Sect. 5.1) for E_{eq} and E_{topic}. `bert-base-uncased` and `bert-large-`

`uncased` fine-tuned on the seed sets are used as the MLM for computing $E_{fluency}$ and offering proposal distributions in the sampling process. We apply `bert-base-uncased` in basic-level problem generation for its diverse proposals benefit topic transfer and innovative problem production. `bert-large-uncased` which produces sophisticated storylines of higher consistency is used in advanced-level experiments. All the above-mentioned models are from Huggingface [22]. `T5-large`, `BertForSequenceClassification`, `bert-base-uncased`, and `bert-large-uncased` are fine-tuned for 8, 5, 5, and 5 epochs with a batch size of 8, 32, 16, and 16. We use AdamW optimizer [14] with a learning rate of 1×10^{-4}, 2×10^{-5}, 5×10^{-5}, and 5×10^{-5} respectively.

5.4 Baselines

We compare our results with the following baselines:

- **GPT2** [19] is a vanilla GPT2-based MWP generation model. Its input is the concatenation of difficulty level, topic, equation, and math problem. GPT2 with basic-level and advanced-level inputs are denoted as GPT2_B and GPT2_A.
- **MAGNET** [24] is a seq2seq model based on (bi-directional) Gate Recurrent Unit (GRU) [2] that fuses information from topic and equation and utilizes an entity-enforced loss to enhance the relevance to the given equation. We use MAGNET_B and MAGNET_A to represent the model fed with topics and equations of the basic and advanced levels.
- **Theme Rewriter** [9] is a rewriting-base method optimized for text coherence measures. It changes the theme of seed questions while maintaining their storylines. Experiments on our model and Theme Rewriter share the same task settings and seed problems. Basic-level topic transfer, basic-level text rewriting, and advanced-level text rewriting tasks using Theme Rewriter are denoted as TR_BTT, TR_BTR and TR_ATR respectively.

The GPT2 baseline is initialized with `GPT2-small` from Huggingface [22]. It is fine-tuned on our seed sets (Sect. 5.1) for 7 epochs using AdamW optimizer [14] with a learning rate of 5×10^{-5} and a batch size of 16. We use the released code for the rest two baselines for experiments.

5.5 Question Quality Evaluations and Observations

Our evaluations are conducted on the 30 top1 results from each of Ours_BTT, Ours_BTR, Ours_ATR, TR_BTT, TR_BTR and TR_ATR, as well as 30 randomly sampled outputs from each of MAGNET_B, MAGNET_A, GPT2_A and GPT2_B.

Automatic Evaluations. Automatic evaluation metrics are utilized to assess the language quality and creativity of generated MWPs. We use perplexity (PPL) values computed by `GPT2-large` [22] to indicate the fluency of our

Table 3. Automatic evaluation results of question quality.

Diffculty Level	Method	PPL↓	Coherence↑	Hamming↑	BertScore ↓
Basic	$GPT2_B$	**67.806**	-0.838	–	–
	$MAGNET_B$	96.142	-0.874	–	–
	TR_{BTT}	152.238	-0.883	0.090	0.948
	TR_{BTR}	101.147	-0.833	**0.112**	0.965
	$Ours_{BTT}$	104.750	**-0.759**	0.098	**0.928**
	$Ours_{BTR}$	98.461	-0.790	0.100	0.935
Advanced	$GPT2_A$	79.856	-0.722	–	–
	$MAGNET_A$	80.820	-0.716	–	–
	TR_{ATR}	93.405	-0.719	0.045	0.969
	$Ours_{ATR}$	**70.646**	**-0.674**	**0.089**	**0.943**

Table 4. Human annotations of well-formedness (Well-formed.), topic match (T.M.), solvability (Solv.), and creativity. Accept. refers to acceptable.

Difficulty Level	Method	Well-formed.%			T.M.%		Solv.%		Creativity%		
		Yes	Accept.	No	Yes	No	Yes	No	Yes	Accept.	No
Basic	$GPT2_B$	53.3	40.0	6.7	76.7	23.3	36.7	63.6	–	–	–
	$MAGNET_B$	20.0	50.0	30.0	80.0	20.0	16.7	83.3	–	–	–
	TR_{BTT}	13.3	46.7	40.0	0.0	100.0	63.3	36.7	36.7	33.3	30.0
	TR_{BTR}	40.0	30.0	30.0	**100.0**	0.0	70.0	30.0	3.3	40.0	56.7
	$Ours_{BTT}$	63.3	36.7	0.0	76.7	23.3	90.0	10.0	**80.0**	20.0	0.0
	$Ours_{BTR}$	**73.3**	26.7	0.0	**100.0**	0.0	**96.7**	3.3	60.0	36.7	3.3
Advanced	$GPT2_A$	40.0	33.3	26.7	76.7	23.3	26.7	73.3	–	–	–
	$MAGNET_A$	30.0	40.0	30.0	60.0	40.0	13.3	86.7	–	–	–
	TR_{ATR}	40.0	40.0	20.0	93.3	6.7	53.3	46.7	3.3	30.0	66.7
	$Ours_{ATR}$	**76.7**	23.3	0.0	**96.7**	3.3	**93.3**	6.7	**33.3**	46.7	20.0

results. `facebook/bart-large` [22] is applied for coherence evaluation following the method from CTRLEval [7]. For creativity, the Hamming distance and BertScore [23] between output MWPs and their seed questions are reported to measure lexical and semantic differences.

The automatic evaluation results are listed in Table 3. For the **basic level**, we can see $Ours_{BTT}$ and $Ours_{BTR}$ outperform the others in coherence and BertScore. Though GPT2, MAGNET, and $Ours_{BTR}$ show comparable PPL scores, the better coherence scores of our model suggest **our results are more logical and more consistent between sentences**, which is necessary for well-formed and solvable MWPs. Also, the Hamming distance and BertScore show **our model leads to more semantically diverse outputs** compared to their seeds while making fewer lexical changes. The performance of $Ours_{BTT}$ and $Ours_{BTR}$ demonstrate a compromise between language quality and creativity, $Ours_{BTT}$ provides innovative results at the sacrifice of fluency and $Ours_{BTR}$ vice versa. For the **advanced level**, $Ours_{ATR}$ performs **better than the baselines for all the metrics**.

Table 5. Human annotations of difficulty-level match (D.M.), topic-difficulty-level match (T.D.M.) and equation-difficulty-level match (E.D.M.).

Ground Truth Difficulty Level	Method	D.M.%			T.D.M.%		E.D.M.%		
		Yes	No	Uncertain	Yes	No	Yes	No	Uncertain
Basic	$GPT2_B$	33.3	10.0	56.7	90.0	10.0	33.3	3.3	63.3
	$MAGNET_B$	16.7	3.3	80.0	96.7	3.3	20.0	0.0	80.0
	TR_{BTT}	63.3	0.0	36.7	**100.0**	0.0	63.3	0.0	36.7
	TR_{BTR}	70.0	0.0	30.0	**100.0**	0.0	70.0	0.0	30.0
	$Ours_{BTT}$	90.0	0.0	10.0	**100.0**	0.0	90.0	0.0	10.0
	$Ours_{BTR}$	**96.7**	0.0	3.3	**100.0**	0.0	**96.7**	0.0	3.3
Advanced	$GPT2_A$	3.3	36.7	60.0	76.7	23.3	6.7	20.0	73.3
	$MAGNET_A$	16.7	20.0	63.3	80.0	20.0	16.7	6.7	76.7
	TR_{ATR}	53.3	6.7	40.0	93.3	6.7	60.0	0.0	40.0
	$Ours_{ATR}$	**90.0**	3.3	6.7	**96.7**	3.3	**90.0**	3.3	6.7

Human Evaluation. We invited 10 graduate students majoring in Computer Science, Mechanical Engineering, and Economics as our evaluators.

Each of them answered an evaluation questionnaire of multiple choice (MC) questions on the following facets of question quality:

- **Well-formedness:** This examines the semantic correctness of a question. The annotation scale is `yes, acceptable` and `not acceptable`. Our results suggest that despite some grammar errors, the meaning of the generated questions is still largely inferrable.
- **Solvability:** This checks whether the question is mathematically solvable. The annotation scale is `yes` or `no`.
- **Topic match:** This assesses whether the question belongs to the target topic. The annotation scale is `yes` and `no`. We provide the evaluators with 3 example seed questions for each topic category for their reference.
- **Creativity:** This checks whether the generated problem is innovative compared to its seed question. The annotation scale is `yes, acceptable` and `not acceptable`. The latter is chosen when only background information (not related to computation) is changed, but the generated question is not too close to the seed, enabling rote learning. This is assessed for rewriting-based approaches (our model and Theme Rewriter) by comparing with the seed questions.

From the human evaluation results in Table 4, we observe that **our method provides better performance across all dimensions for both difficulty levels**. The topic matching ratio of $Ours_{BTT}$ shows that with the topic discriminator, our model can successfully transfer the basic daily problems to the ones of other basic topics in most cases. In comparison, TR_{BTT} fails to do the transfer since tf-idf scores lead to similar topic-related content words for the three basic topics. The Theme Rewriter approach needs to be applied to more separated topics. For solvability, Ours_{ATR} maintains a high solvable rate for the advanced level, which shows the effectiveness of our equation constraint. In addition, though Ours_{ATR} outperforms TR_{ATR} in creativity, the percentage of innovative results from Ours_{ATR} is only 33.3%. More investigations are required to increase the rate of novel results conditioned on advanced MWPs with sophisticated storylines.

5.6 Difficulty Controllability Analysis

For the resulting MWPs mentioned in Sect. 5.5, the 10 graduate students also evaluated whether their difficulty levels match the targets from three dimensions: the overall questions, the topics, and the equations. The topic difficulty match is marked as `yes` or `no`. The annotation scale for equation and overall difficulty match is `yes`, `no` and `uncertain`. As described in Sect. 5.1, the basic-level equations have no more than 4 solving steps while the advanced-level ones have 5 to 10 steps. The equation difficulty is uncertain if the question is unsolvable. The overall difficulty is uncertain if the topic difficulty is matched but the equation difficulty is uncertain. According to the results in Table 5, **our system can successfully control the difficulty level of all three dimensions** for the most outputs. Though the Theme Rewriter method has a higher matching rate of topic difficulty level than the other baselines and is even comparable with our approach, a large proportion of its generated questions is unsolvable, which makes it impossible to decide their equation difficulty and the overall difficulty.

6 User Study

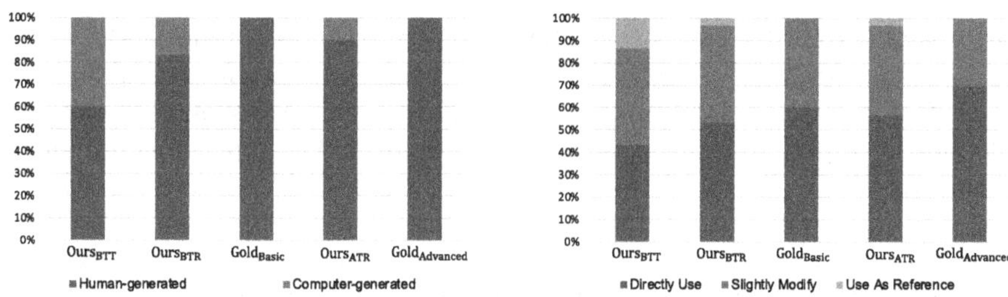

Fig. 2. User study results from human experts. **Left:** Percentages of MWPs recognized as human-generated or computer-generated. **Right:** Percentages of the MWPs' utilities in real-life educational activities.

To verify the usability of our difficulty-controllable MWP generation system in real classes of different grades, we invited three math educators from Upwork[2] to evaluate the quality and difficulty of our resulting questions. The teachers'

[2] https://www.upwork.com/.

experience of math teaching and math content development varies from primary school to college.

We presented 2 sets of math problems to the teachers. One set contained 30 top1 outputs from Ours_{BTT}, 30 top1 outputs from Ours_{BTR} and 10 randomly sampled questions from the basic seed set ($\text{Gold}_{\text{Basic}}$). The other set was composed of 30 top1 results from Ours_{ATR} and 10 randomly sampled questions from the advanced seed set ($\text{Gold}_{\text{Adanced}}$). For each question in these two sets, the teachers were asked to 1) check whether it is suitable for primary school students or not; 2) distinguish whether it is produced by a human math expert or a computer; 3) assess whether it could be used in their real-life classes, assignments, and exams.

The feedback from the math teachers in Fig. 2 shows the majority of our generated problems of both difficulties are recognized as human-authored. **Over 90% of our results in basic and advanced difficulty levels can be directly used in the primary- and high-school educational activities or require minor modifications to remove any potential confusion from students**. The minor modifications they could accept mean no more than three changes of words or rearranging the word order of only one sentence. The rest results can be used as references and help the teachers to produce new questions faster than starting from scratch. **The teachers suggest our framework can effectively assist their MWP creation and reduce their efforts for writing novel storylines**. Overall, the teachers encouraged the development of our automatic educational question-generation system. They were interested in the future update of our framework that generates MWPs for each grade since each teacher is normally responsible for one grade.

7 Conclusion

We present an energy-based controllable MWP generation framework, which allows us to restrain the outputs problem's difficulty with mathematical and linguistic factors such as equations, vocabulary, and topics. We show that our outputs match the desired problem difficulty. Their solvability, topic, language quality, and creativity also satisfy human expectations. Finally, our user study with math educators validates the utility of our framework in teaching and testing students of different stages. Our future work includes constructing the required resources and building systems with finer-grained difficulty control and more diverse math content.

References

1. Amini, A., Gabriel, S., Lin, P., Koncel-Kedziorski, R., Choi, Y., Hajishirzi, H.: Mathqa: towards interpretable math word problem solving with operation-based formalisms. arXiv preprint arXiv:1905.13319 (2019)
2. Cho, K., et al.: Learning phrase representations using RNN encoder-decoder for statistical machine translation. arXiv preprint arXiv:1406.1078 (2014)

3. Cobbe, K., et al.: Training verifiers to solve math word problems. arXiv preprint arXiv:2110.14168 (2021)
4. Daroczy, G., Wolska, M., Meurers, W.D., Nuerk, H.C.: Word problems: a review of linguistic and numerical factors contributing to their difficulty. Front. Psychol. **6**, 348 (2015)
5. Dathathri, S., et al.: Plug and play language models: a simple approach to controlled text generation. arXiv preprint arXiv:1912.02164 (2019)
6. Goyal, K., Dyer, C., Berg-Kirkpatrick, T.: Exposing the implicit energy networks behind masked language models via metropolis-hastings. arXiv preprint arXiv:2106.02736 (2021)
7. Ke, P., et al.: CTRLEval: an unsupervised reference-free metric for evaluating controlled text generation. arXiv preprint arXiv:2204.00862 (2022)
8. Khodeir, N.A., Elazhary, H., Wanas, N.: Generating story problems via controlled parameters in a web-based intelligent tutoring system. Int. J. Inf. Learn. Technol. **35**(3), 199–216 (2018)
9. Koncel-Kedziorski, R., Konstas, I., Zettlemoyer, L., Hajishirzi, H.: A theme-rewriting approach for generating algebra word problems. arXiv preprint arXiv:1610.06210 (2016)
10. Kuperman, V., Stadthagen-Gonzalez, H., Brysbaert, M.: Age-of-acquisition ratings for 30,000 English words. Behav. Res. Methods **44**(4), 978–990 (2012)
11. Kurdi, G., Leo, J., Parsia, B., Sattler, U., Al-Emari, S.: A systematic review of automatic question generation for educational purposes. Int. J. Artif. Intell. Educ. **30**(1), 121–204 (2020)
12. Liu, A., et al.: Dexperts: decoding-time controlled text generation with experts and anti-experts. arXiv preprint arXiv:2105.03023 (2021)
13. Liu, T., Fang, Q., Ding, W., Liu, Z.: Mathematical word problem generation from commonsense knowledge graph and equations. arXiv preprint arXiv:2010.06196 (2020)
14. Loshchilov, I., Hutter, F.: Fixing weight decay regularization in adam (2018)
15. Miao, S.Y., Liang, C.C., Su, K.Y.: A diverse corpus for evaluating and developing English math word problem solvers. arXiv preprint arXiv:2106.15772 (2021)
16. Mireshghallah, F., Goyal, K., Berg-Kirkpatrick, T.: Mix and match: learning-free controllable text generation using energy language models. arXiv preprint arXiv:2203.13299 (2022)
17. Polozov, O., O'Rourke, E., Smith, A.M., Zettlemoyer, L., Gulwani, S., Popović, Z.: Personalized mathematical word problem generation. In: Twenty-Fourth International Joint Conference on Artificial Intelligence (2015)
18. Pongsakdi, N., Kajamies, A., Veermans, K., Lertola, K., Vauras, M., Lehtinen, E.: What makes mathematical word problem solving challenging? Exploring the roles of word problem characteristics, text comprehension, and arithmetic skills. ZDM **52**(1), 33–44 (2020)
19. Radford, A., Wu, J., Child, R., Luan, D., Amodei, D., Sutskever, I., et al.: Language models are unsupervised multitask learners. OpenAI blog **1**(8), 9 (2019)
20. Wang, K., Su, Z.: Dimensionally guided synthesis of mathematical word problems. In: IJCAI, pp. 2661–2668 (2016)
21. Wang, Z., Lan, A.S., Baraniuk, R.G.: Math word problem generation with mathematical consistency and problem context constraints. arXiv preprint arXiv:2109.04546 (2021)
22. Wolf, T., et al.: Transformers: state-of-the-art natural language processing. In: Proceedings of the 2020 Conference on Empirical Methods in Natural Language Processing: System Demonstrations, pp. 38–45 (2020)

23. Zhang, T., Kishore, V., Wu, F., Weinberger, K.Q., Artzi, Y.: BERTScore: evaluating text generation with BERT. arXiv preprint arXiv:1904.09675 (2019)
24. Zhou, Q., Huang, D.: Towards generating math word problems from equations and topics. In: Proceedings of the 12th International Conference on Natural Language Generation, pp. 494–503 (2019)
25. Zhou, W., Jiang, Y.E., Wilcox, E., Cotterell, R., Sachan, M.: Controlled text generation with natural language instructions (2023)

Getting the Wiggles Out: Movement Between Tasks Predicts Future Mind Wandering During Learning Activities

Rosy Southwell(✉), Candace E. Peacock, and Sidney K. D'Mello

University of Colorado Boulder, Boulder, CO 80301, USA
rosy.southwell@colorado.edu

Abstract. Mind wandering ("zoning out") is a frequent occurrence and is negatively related to learning outcomes, which suggests it would be beneficial to measure and mitigate it. To this end, we investigated whether movement from a wrist-worn accelerometer between tasks could predict mind wandering as 125 learners read long, connected, informative texts. We examined random forest models using both basic statistical and more novel nonlinear dynamics movement features, finding that the former were more predictive of future (i.e., about 5 min later) reports of mind wandering. Models generalized across students with AUROCS up to 0.62. Importantly, vertical movement as measured by the Z-axis accelerometer channel, e.g. flexion or extension of the elbow in stretching, was the most predictive signal, whereas horizontal arm movements (measured by X- and Y-axis channels) and rotational movement were not predictive. We discuss implications for theories of mind wandering and applications for intelligent learning interfaces that can prospectively detect mind wandering.

Keywords: mind wandering · fidgeting · reading comprehension · movement

1 Introduction

Mind wandering (MW) is defined as 'zoning out', or where attention disengages from the primary task towards task-unrelated thoughts [10, 18]. Students are estimated to mind wander approximately 30% of the time during learning [33], varying from 20% to 40% depending on the learning activity, with greater occurrence in passive forms of learning such as lecture viewing (35%, [33]) than when engaging with interactive learning technology (23% [19]). Although MW has benefits such as enabling future planning and creativity [23], meta-analyses indicate that it is negatively related to performance for activities involving focused attention and concentration [10, 14]. Accordingly, researchers have investigated various interventions to alleviate the effects of MW in education. Proactive interventions are (usually non-individualized) modifications to task parameters made beforehand, such as changing text properties to reduce MW [10, 13, 17] whereas reactive interventions aim to detect and mitigate effects of MW after its onset, for example, by repeating content or asking a probing question to reengage attention and mitigate any learning deficits [8, 22].

R. Southwell and C.E. Peacock—These authors contributed equally

N. Wang et al. (Eds.): AIED 2023, LNAI 13916, pp. 489–501, 2023.
https://doi.org/10.1007/978-3-031-36272-9_40

This raises the question of whether there are patterns of behavior that can prospectively predict that future MW will occur. If so, detecting such patterns could be used in learning technologies to enable proactive interventions or to improve prediction of MW when it occurs. These possibilities depend on the ability to identify behaviors that can prospectively predict MW compared to existing research that focuses on detecting it when it occurs [3, 8, 14, 22, 28]. As a step in this direction, the present study examines whether patterns of movement during pre-defined intra-task intervals can prospectively predict self-reports of MW as probed roughly 5 min later during learning from complex texts. We model movement because well-established theoretical and empirical research links it to engagement and attention [3, 5, 9, 14, 24, 28] and movement can be measured using low-cost wearable accelerometers scalable for use in classroom settings. In addition to the practical applications, the present work also advances empirical knowledge of the behaviors that underlie MW.

1.1 Background and Related Work

Movement and Student Engagement. Mind wandering is an attentional subcomponent of (dis)engagement [6], and one way that movement has been studied in relation to engagement is through monitoring posture. In a classic study, Mota and Picard (2003) measured how posture (as measured by pressure sensors on a seat and back of a chair) related to a child's level of interest [24]. D'Mello et al. extended this approach to test how a student's posture, also measured via a pressure-sensitive chair, was related to their level of engagement and found that boredom was associated with leaning back and changes in seat pressure and that flow was associated with leaning forward [5, 9]. Fidgeting, which indicates restlessness, has also been associated with MW in classrooms using self-report measures [3]. Seli et al. (2014) expanded beyond self-reports of fidgeting and found that increased MW was associated with more fidgeting as indexed by participants sitting on a Wii balance board [28]. Farley et al. (2013) demonstrated that fidgeting (as coded from video) and self-reports of MW increased as a function of time during lecture viewing and that fidgeting was negatively predictive of the retention of lecture material beyond the role of MW [14].

Whereas the above research suggests a link between body movement and engagement, it is unknown whether there are patterns of movement that are indicative of future MW. Nevertheless, the literature on aerobic breaks (e.g., running on the spot, jumping exercises, stretching) can shed some light. Indeed, when aerobic exercise was added to breaks, participants reported decreases in fatigue and increases in vigor compared to unstructured breaks in which students could do what they wanted as long as they remained in their seats [1]. Furthermore, when exercise breaks (calisthenic exercises), non-exercise breaks (computer game), and no breaks were introduced into computerized lectures, it was found that exercise breaks promoted attention (lower MW) and better retention throughout the lecture relative to the non-exercise breaks and no breaks condition [15]. These studies suggest that moving more during breaks in learning tasks can have some positive benefits in terms of reducing MW, but have yet to yield predictive models of MW from behavior measured during breaks.

Prospectively Predicting Mind Wandering. There is a small body of Work that focuses on whether it is possible to prospectively predict MW, although without considering movement. In a simulation study Kopp et al. (2014) explored whether individual attributes (e.g., working memory capacity) could be used to identify reading situations (defined as text difficulty and utility value) that resulted in the lowest MW rates for an individual [20], suggesting that a learning system that places learners in conditions to reduce MW is feasible (though they did not test this possibility). Another study found that eye movements were more dispersed (spread out) before people later reported MW [34], suggesting that gaze could prospectively predict MW. However, cross-validation was not done so it is unknown whether their models would generalize across learners. In another recent study using electroencephalography, causality analysis of timeseries of oscillatory neural activity indicated that it predicted MW several minutes later [25]. However, taking such neural measurements would be impractical for a learning environment.

1.2 Current Work: Contributions and Novelty

We tested the intriguing possibility as to whether machine-learning models of movement during task breaks can prospectively predict MW. Specifically, we measured movement (via accelerometer data) during short intra-task intervals between reading each of five expository texts and collected probe-caught MW responses as students read each text. Random forest models were trained to predict MW from movement measured prior to reading (i.e., during breaks) in a learner-independent fashion.

Our study is novel in multiple respects. To our knowledge, this is the first attempt to prospectively predict MW in a student-generalizable manner, which has several applications if successful (see Discussion). Second, prior work has employed basic statistical movement features (e.g., mean movement, variance). As an alternative we employed a technique from nonlinear dynamical analyses called recurrent quantification analysis (RQA [32]) to characterize various patterns of movement, including fidgeting, which has been associated with MW [3, 5, 9, 14, 24, 28].

We examined three research questions. First, we evaluated how the prospective MW models performed relative to several baselines including a context model that captured situational factors independent of movement, a "gaze dispersion" model inspired by [34], and a model of movement during the MW episode. Second, we examined whether models consisting of statistical movement features or RQA were better predictors of MW. A final question examined which channels and features of movement were most predictive of mind wandering.

2 Methods

2.1 Data Collection

Data were collected as part of a larger study investigating neurophysiology during learning from complex texts. Only aspects germane to the present study are presented here. The data analyzed here have not been previously published. Participants (N = 156, age

23 ± 6 years, 67% female, 32% male, 1% other) were students from a large public University in the Western US (77% White, 9% Asian, 7% Hispanic, 5% Other, 2% prefer not to say). Participants were paid $20 per hour plus $10 for a follow-up survey via Amazon gift cards. All procedures were approved by the Institution's internal review board (IRB00000191 protocol #19-0396) and all participants provided informed consent after being given the opportunity to read the consent form prior to the study and upon arrival, and to ask the researcher any questions. We analyzed data from the 125 participants with complete accelerometer data.

Participants each read five expository texts of around 1000 words each split into 10 pages (screens of text) on the topic of behavioral research methods: Bias, Hypothesis, Casual Claims, Validity, and Variables. The texts had a mean Flesch-Kincaid grade level of 13.2 indicating an advanced reading level [16] suitable for college students. Reading was self-paced in that participants pressed a key to advance to the next page but could not return to a previous page. On average, participants read each text for 5.5 min (SD = 1.8), for a total of 27.6 min (SD = 9.2).

As participants read the texts, they were probed for MW at two pseudorandom points roughly corresponding to the first and second half of the text. Specifically, upon advancing to the next page of text, they were presented with the following question: *Were you zoning out when you read the previous page?* Participants responded with "yes" or "no" using mouse clicks. Such self-reports of MW are a standard and validated method for measuring conscious phenomena [10, 29]. Participants were trained on how to respond to MW probes prior to the main reading session. Here, they were given a definition of MW (i.e., abbreviated as thinking about something other than reading or "zoning out") and were informed that they would be asked about these zone outs as they read. Participants were asked to be honest about these zone outs, because they are perfectly natural and were reassured that responding "yes" would not affect their progress in the study. As a knowledge check, they were asked to identify an example of zoning out in a multiple-choice question.

Participants wore a Shimmer3 (sampling rate = 51.2 Hz) on their non-dominant wrist. The sensors include electrodermal activity, photoplethysmography, two accelerometers, a gyroscope, and a magnetometer, but models focused on the latter three sensors. Participants also underwent gaze tracking using the Eyelink 1000+ desktop mounted eye-tracker. Chin and forehead rests were used to minimize head movement during reading, but participants were encouraged to stretch and reposition prior to reading each text: participants were presented with a prompt ("*Before proceeding, if you need a short break to stretch or close your eyes, you can do so now.*") and to press a button to proceed when ready. Next the eyetracker was recalibrated by collecting fixation positions from nine known target positions (presented in random order) to map raw eye data to gaze position on-screen. Following calibration, participants went through a validation procedure which measured the difference between the computed fixation position and the fixation position for the target during calibration. The error reflects the accuracy of the calibration. Analyses focused on movement (accelerometer, gyroscope, magnetometer) between texts, i.e. the intra-task interval comprising the break and eyetracker recalibration. Eye-tracker calibration error was also analyzed as this may increase if the participant moves or fidgets during the calibration. The time stamps corresponding to the initial eyetracker calibration (preceding the first text read) were not recorded so we focus on the recalibration periods prior to the final four texts.

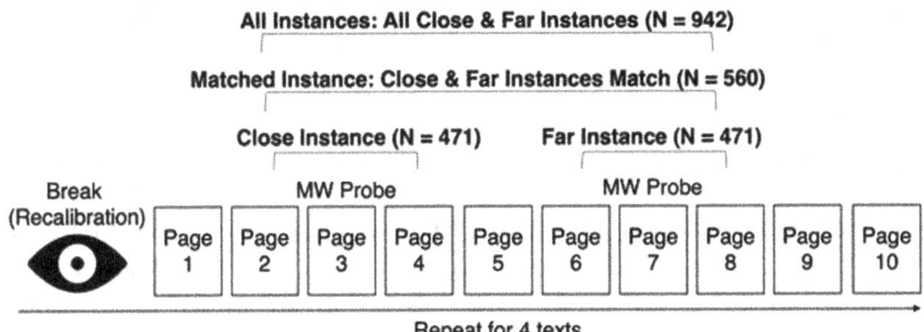

Fig. 1. Example event sequence. First, participants were recalibrated on the eyetracker during a break and then proceeded to read 10 pages of an expository text. They were probed for MW on two pseudo randomly selected pages. N refers to the number of instances.

2.2 Machine Learning Models

We chose Random Forest classifiers since they incorporate nonlinearity and interactivity among features and have good generalization properties. The random forest classifier was implemented in *sklearn*, with 100 estimators, minimum of 15 samples per leaf, and the maximum number of features set to the square root of the total number of features. The class weights of the models were balanced by setting the weights to be inversely proportional to the number of samples in each class. Note that no resampling was done on either the training or testing sets: setting the class weights to 'balanced' simply penalizes incorrect predictions made on the minority class.

We tested several different models. One model was trained on only the first MW probe per text (Fig. 1) to test whether the features were more sensitive to future MW that occurred closer in time to calibration ("closest"). A second model was trained on only the second MW probe per text (Fig. 1) to test the endurance of the model ("farthest"). A third model was only tested on cases where MW responses for the nearest and farthest probes were the same (i.e., positive or negative MW) ("matched"). The matched model provided the purest test of whether movement was predictive of future MW since people consistently mind wandered (or not) throughout the duration of the text. A final model was tested on all MW probes administered during each text ("all").

We used four-fold cross-validation at the participant-level to ensure generalizability to new participants [7]. Here, the dataset was randomly split into four folds, with the data from a given participant only being in a single fold. Predictions were pooled across the four folds prior to computing AUROCs. The process was repeated 101 times with a different random partitioning of the folds for each run. The same fold assignments were used to train all the models (including baseline models) per run. Performance was evaluated using the area under the receiver operator characteristic (AUROC) curve, which ranges from 0 to 1 with an AUROC of 0.5 indicating baseline chance-level performance.

Here, we focus on the run with the median-performing model for each model specification as the results were highly consistent across runs (95% confidence intervals at most differed by 0.01 AUROC across models). We used the roc.test function from the pROC package in R [27] (bootstrap test with 10,000 iterations) to statistically compare the median run's AUROC between models. The false discovery rate correction was used to correct for six multiple comparisons.

2.3 Sensors, Data Processing, and Feature Extraction

Sensors and Signals. The Shimmer3 has a low noise accelerometer with a lower range of values (range: ± 2.0 g), a wide range accelerometer that has a wider range of possible values (range: ± 2.0 g; ± 4.0 g; ± 8.0 g; ± 16.0 g), a gyroscope, and a magnetometer. All sensors were sampled at 51.2 Hz. Accelerometers and gyroscopes measure linear and rotational motion in X, Y, and Z coordinates, respectively, whereas magnetometers indicate direction. From both accelerometer's *acceleration* time series, we additionally derived *velocity* [acceleration * change in time] and *distance* [acceleration * (change in time)^2]. This resulted in 6 accelerometer time series (3 measures [acceleration, velocity, distance] × 2 sensors [low noise vs. wide range]) for each axis (X,Y,Z). From the *angular velocity* (i.e. gyroscope), we computed *degrees rotated* [angular velocity * change in time] totaling 2 timeseries for each axis. Combining these accelerometer and gyroscope signals with the magnetometer *direction* time series this totaled 9 movement measures * 3 axes, yielding a timeseries for each of 27 "channels" from which statistical features were derived (see below). We smoothed each time series using the denoise_wavelet function in Scikit Image (*wavelet* = 'sym4', *mode* = 'soft', *wavelet_levels* = 3) then removed values that exceeded 3 SD from the mean to account for noise and linearly interpolated over removed data. Figure 2 depicts example accelerometer time series for intra-task intervals corresponding to positive and negative MW responses.

Statistical Movement Features. The following eight statistical features were computed from each of the 27 time series: mean, median, maximum, minimum, standard deviation, kurtosis, skew, and range [11, 26]. Statistical features were z-scored within-participant. Because the question of interest was whether movement intra-task intervals was prospectively predictive of MW, we computed statistical features from the time series between reading blocks, which included a self-paced break followed by eyetracker recalibration (total $M = 142.8$ s; $SD = 90.2$ s). As the intra-task interval contained both a break and the recalibration, we also extracted features from just the break ($M = 13.4$ s, $SD = 16.5$ s), and just the calibration ($M = 101.8$ s, $SD = 73.6$ s), and fit models using the features from these intervals alone. For comparison with the pre-reading movement models, we also extracted features during reading the page corresponding to the MW probe (as a baseline check; $M = 34.0$ s; $SD = 10.4$ s).

Calibration Error Features. During each validation period, the eye-tracker outputs the degrees, horizontal (X pixels), and vertical Error (Y pixels) based on how far each

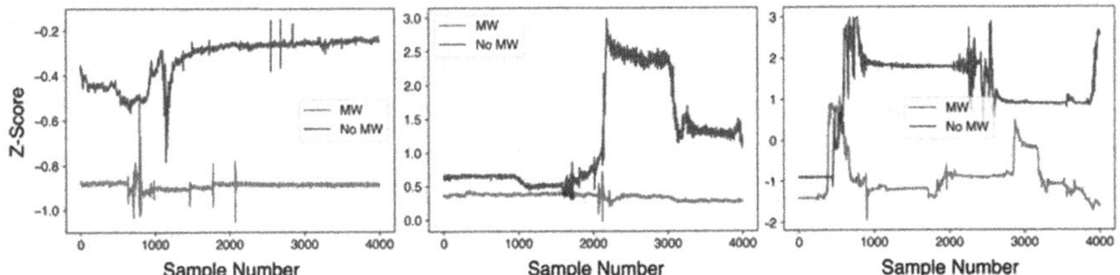

Fig. 2. Example time series from the z component of the low noise accelerometer for breaks corresponding to positive (red lines) and negative (blue lines) responses to MW probes. (Color figure online)

fixation was from each of the nine calibration points. We computed the same eight statistical features over the 9 calibration errors, resulting in 24 calibration error features.

Context Features. Context features capture situational factors independent of movement and were used as a baseline measure to gauge the added value of movement features. They comprised calibration duration and text order.

Model Comparisons. To evaluate which channels would provide the most information, we first trained separate models for the 9 movement measures, each on 24 statistical movement features: 8 statistical features * 3 axes, plus the two context features. The angular velocity, degrees rotated, and direction channels were not predictive of MW (all AUROCs at chance). We found that the low noise and wide range accelerometer features yielded Very similar (AUROCS within 0.01) results, so we focused on the low-noise accelerometer channels. The most strongly predictive accelerometer-derived channel was distance. For this low noise distance channel, we then trained separate models using the 8 statistical features of the X, Y, and Z components. Overall, the Z component which indexed the direction Perpendicular to the wrist surface (i.e., flexion and extension of the elbow/arm, such as may occur during stretching, or lifting the hand from the keyboard and mouse) [30] contained the most information (M AUROC = 0.62, 95% CI = [0.62, 0.62]) relative to X and Y (which indexed horizontal, planar movements such as moving the arm across the desk [30]) which were at chance, so we focused models on the Z component statistical features from the low noise distance channel.

Recurrent Quantification Analysis Features. We conducted recurrent quantification analysis (RQA) to complement the statistical movement features. RQA is an analytic suite used to investigate the dynamics of complex timeseries, and is a powerful tool that avoids several of the assumptions of traditional techniques (e.g., assumptions of stationary, homoscedasticity) [32]. We computed the auto recurrence (the amount that a time series recurs with itself at all time lags) of the Z component of the low noise distance channel during recalibration and during reading of each page with a MW probe. This entailed transforming the time series into a distance matrix where each cell captures the pairwise distance between the elements of the time series with itself (Fig. 3). The diagonal is the line of identity (LOI) or the distance between elements that occur at the same point (0 for auto-recurrence). Diagonal lines parallel to the LOI represent points that occur at a particular time lag with lines further from the LOI representing greater time lags. The distance matrix is then transformed to a recurrence matrix with a radius which captures whether two points recur. If the distance of a given cell is less than the radius, then the cell is defined as a recurrent point and a value of 1 is assigned to the cell (Fig. 3). Diagonal lines represent patterned behaviors (e.g., fidgeting) and vertical lines represent smooth or slow changes to movement.

We used the "crqa" package in R [4], set the delay to 0, the embedding dimension to 1 (i.e., no phase space embedding), standardized each time series, and used Euclidean distance metric. We tested radius values between 0.03 and 0.15 with a step size of 0.005 on a random 10% of participants and selected the value corresponding to approximately 5% mean recurrence. We repeated this procedure with another random 10% of participants, to confirm values. We proceeded with radii of 0.04 and .095 for time series associated with recalibration and MW pages, respectively. We computed the following RQA features from the recurrence matrices: recurrence rate (the fraction of recurrent time points over the total number of points), and features related to vertical and diagonal lines (determinism, laminarity, averaged diagonal line length, trapping time, longest diagonal line, longest vertical line, ratio, entropy) [4].

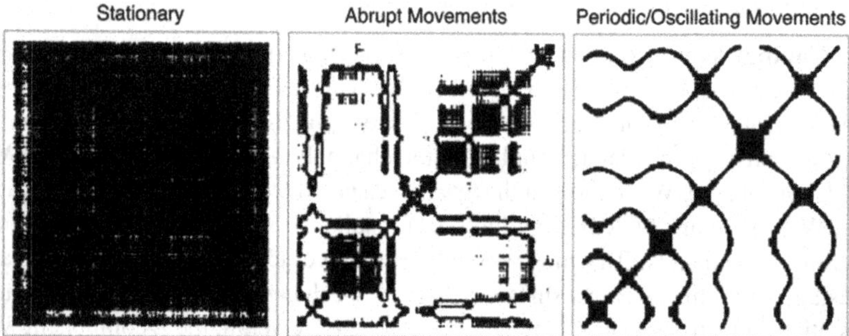

Fig. 3. Example recurrence plots. The first panel represents a stationary system (e.g., no movement), the second panel represents a system with abrupt movements (e.g., stretching), and the third panel represents periodic movements (e.g., fidgeting).

3 Results

3.1 Model Comparisons

All four statistical movement models (all", "matched", "closest", "farthest") computed during intra-task intervals outperformed chance (i.e., AUROCs > 0.5, CIs do not overlap 0.5), whereas the RQA and calibration error models did not (Table 1). The context-only models also performed at chance except for the "farthest" model, but a pairwise ROC test indicated that its accuracy was lower than the "farthest" statistical model ($p = .04$). Pairwise ROC tests between the statistical feature models revealed no significant differences ($ps > 0.05$). Quantitatively, the statistical movement "matched" and "farthest" models yielded the highest AUROCs of 0.62. The best AUROCs for the separate break models (all: 0.53 [0.50, 0.58]) and calibration (matched: 0.59 [0.56, 0.65]) models were significant but lower than for the main model using the entire intra-task interval. Interestingly, the statistical features model during the probed reading page performed at chance, while there was a weak signal for the RQA movement model.

Table 1. AUROCs and 95% bootstrapped CIs from the median performing model. Bolded values indicate models that outperformed chance.

Model Type	Movement Models		Baseline Models	
	Statistical	RQA	Context	Calibration Error
Matched	**0.62 [0.58, 0.67]**	0.46 [0.41, 0.50]	0.48 [0.44, 0.53]	0.49 [0.44, 0.54]
Closest	**0.54 [0.50, 0.60]**	0.46 [0.41, 0.51]	0.44 [0.39, 0.49]	0.54 [0.49, 0.58]
Farthest	**0.62 [0.56, 0.66]**	0.50 [0.45, 0.56]	**0.55 [0.50, 0.60]**	0.48 [0.42, 0.53]
All	**0.59 [0.55, 0.62]**	0.48 [0.44, 0.51]	0.49 [0.46, 0.53]	0.49 [0.45, 0.52]
Page Of	0.49 [0.49, 0.49]	**0.52 [0.51, 0.52]**	–	–

3.2 Predictive Features

We used Shapley Additive exPlanations (SHAP) [21] on the median-performing "matched" model as it provides the cleanest mapping of movement during breaks and MW. SHAP gives feature importance and direction of influence. The top three features were kurtosis, median, and minimum of the Z component of distance, capturing flexion and extension of the elbow/arm (e.g., stretching) (Fig. 4). Positive kurtosis (i.e. the peak of the distribution is sharp and the tails are short) was associated with future MW. This might suggest that there were fewer outliers during periods that were followed by MW. Future MW was also associated with a lower median distance. This suggests that when people produced smaller movements during the recalibration break, they mind wandered more later. Finally, the minimum distance was smaller before texts where MW was reported, again suggesting that smaller movements during breaks predict more MW later. Together, these findings suggest that when people produce bigger movements during intra-task intervals via flexion and extension of the arms, later MW is reduced.

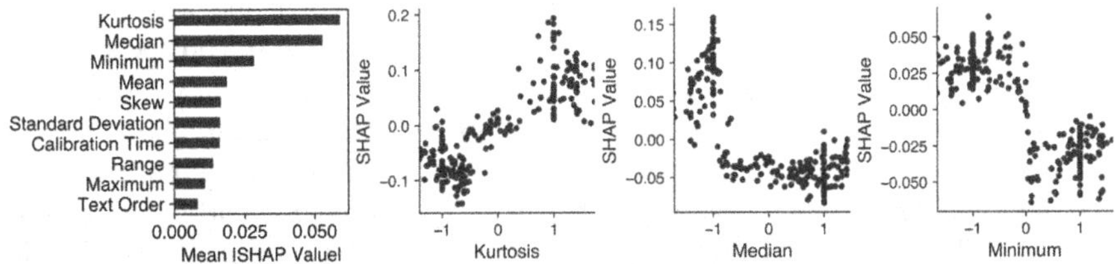

Fig. 4. Feature importances (Panel 1) and directions (Panels 2–4) for the top 3 features. X-axes represent the value of each feature and the Y-axis represents the SHAP value in predicting the probability of MW (values > 0 indicate MW). Each dot represents an instance from the test set.

4 Discussion

The purpose of the present work was to identify whether there are patterns of behavior between task intervals that can prospectively predict MW during later learning. Overall, our results showed a consistent pattern across participants, namely that statistical features computed from an accelerometer time series between task blocks predicted future MW better than the page of MW, a context model, and a model that considered calibration error. Such models were predictive of MW both early during reading (on average 1.7 min into each text) and later (after 5.6 min of reading) with a stronger (albeit nonsignificant) effect for the latter. These patterns generalized across learners. We achieved a highest AUROC of 0.62, while admittedly modest, is consistent with AUROCs of 0.60 obtained by both computer vision algorithms and majority voting by nine human observers in a study detecting MW from visual features during reading [2]. It is lower than AUROCs obtained from eye gaze [12], but this is unsurprising because eye gaze provides a more direct measure of visual attention.

We also found that learners who moved more during intra-task intervals, as indicated by greater flexion and extension of the arms, experienced less MW later, suggesting that how students behave during breaks matters. Indeed, prior work has found that exercise breaks reduce MW during lecture viewing relative to non-exercise breaks [15]. Our findings provide converging evidence and suggest that even in a controlled lab setting, students who moved more tended to MW less during later reading. As participants were seated during the entire task, including the break, these results might be akin to breaks that occur during standardized testing. A surprising finding was that there were no consistent temporal patterns of movement that predicted MW as people read, as prior work has identified a link between movement and concurrent engagement [3, 5, 9, 14, 24, 28]. Although RQA should detect fidgeting behaviors, it might be the case that due to the study constraints and instructions (i.e., participants wore sensors and were asked to minimize movements as they read), participants might have moved less than they would have in more naturalistic reading conditions. Therefore, the present study is likely underestimating the influence of movement during MW.

Applications. The knowledge that there are consistent patterns in movement that can prospectively predict MW has the potential to be incorporated into intelligent learning interfaces. If such a model detected that MW might occur, intelligent technology could suggest to a learner to stretch or walk before completing the next learning activity. If applied to multiple learners, and the prospect of future MW is high, the instructor could be notified that a calisthenic break might be appropriate. Furthermore, a prospective MW detector could be paired with existing proactive and reactive interventions to reduce the negative effects of MW. Specifically, a system might be able to increase the frequency of proactive interventions to decrease MW, such as changing the textual properties [13, 17] or interspersing more test questions [31] if MW is predicted. Similarly, reactive MW detectors could be "seeded" with prospective MW predictions to potentially increase the accuracy of these detectors.

Limitations. As described above, the lab setting might have influenced how learners moved during reading relative to naturalistic reading settings, however participants were encouraged to move at the start of the recalibration, which is the focus of the analyses.

Relatedly, because participants were seated the present models might not generalize to all task contexts. For example, the finding that the Z channel, which indexed flexion and extension of the arms, contained the most information (contrary to X and Y which captured horizontal, planar arm movements which might correspond to less naturalistic movements when seated) may only be relevant where learners remain seated during breaks (e.g., standardized testing). However, it is likely the X and Y channels would be more relevant for active breaks, such as walking. It will be critical for future work to test whether the present model generalizes to other break contexts. The outcome measure we predict in the present study is self-reported mind-wandering, probed after reading on (up to) ten selected pages, and at a coarse, binary level. However, MW is a multifaceted phenomenon [29] and future work should consider using a wider range of MW measures. Other relevant measures of individual differences, e.g. working memory, were not considered here but controlling for these may nuance the findings. In addition, the sample demographics are not representative of the US as a whole [35], with a narrow age range, no Black respondents, and Hispanic participants underrepresented, again highlighting the need to check whether these results generalize.

Conclusion. Although MW has some benefits, multiple studies indicate that it is negatively correlated with learning outcomes (see Introduction), leaving the question open as to how we might mitigate it. Our results suggest that how people move between task blocks is related to how much they will mind wander later. In addition to its theoretical relevance, the ability to prospectively model mind wandering has exciting implications for intelligent learning systems that aim to reduce MW during learning.

References

1. Blasche, G., et al.: Comparison of rest-break interventions during a mentally demanding task. Stress. Health **34**(5), 629–638 (2018). https://doi.org/10.1002/smi.2830
2. Bosch, N., D'Mello, S.K.: Can computers outperform humans in detecting user zone-outs? Implications for intelligent interfaces. ACM Trans. Comput.-Hum. Interact. **29**, 2, 1–33 (2022). https://doi.org/10.1145/3481889
3. Carriere, J.S.A., et al.: Wandering in both mind and body: individual differences in mind wandering and inattention predict fidgeting. Can. J. Exp. Psychol. **67**(1), 19–31 (2013). https://doi.org/10.1037/a0031438
4. Coco, M.I., Dale, R.: Cross-recurrence quantification analysis of categorical and continuous time series: an R package. Front. Psychol. **5** (2014)
5. D'Mello, S., et al.: Posture as a predictor of learner's affective engagement. In: Proceedings of the Annual Meeting of the Cognitive Science Society, vol. 29, no. 29, p. 7, Merced, CA (2007)
6. D'Mello, S.K.: Giving eyesight to the blind: towards attention-aware AIED. Int. J. Artif. Intell. Educ. **26**(2), 645–659 (2016). https://doi.org/10.1007/s40593-016-0104-1
7. D'Mello, S.K., et al.: Machine-learned computational models can enhance the study of text and discourse: a case study using eye tracking to model reading comprehension. Discourse Process. **57**(5–6), 420–440 (2020). https://doi.org/10.1080/0163853X.2020.1739600
8. D'Mello, S.K., et al.: Zone out no more: mitigating mind wandering during computerized reading. In: Proceedings of the 10th International Conference on Educational Data Mining. International Educational Data Mining Society, Wuhan, China (2017)

9. D'Mello, S.K., Graesser, A.C.: Mining bodily patterns of affective experience during learning. In: Proceedings of the Third International Conference on Data Mining (2010)
10. D'Mello, S.K., Mills, C.S.: Mind wandering during reading: an interdisciplinary and integrative review of psychological, computing, and intervention research and theory. Lang. Linguist. Compass. **15**(4), e12412 (2021). https://doi.org/10.1111/lnc3.12412
11. Erdaş, Ç.B., et al.: Integrating features for accelerometer-based activity recognition. Proc. Comp. Sci. **98**, 522–527 (2016). https://doi.org/10.1016/j.procs.2016.09.070
12. Faber, M., Bixler, R., D'Mello, S.K.: An automated behavioral measure of mind wandering during computerized reading. Behav. Res. Methods **50**(1), 134–150 (2017). https://doi.org/10.3758/s13428-017-0857-y
13. Faber, M., Mills, C., Kopp, K., D'Mello, S.: The effect of disfluency on mind wandering during text comprehension. Psychon. Bull. Rev. **24**(3), 914–919 (2016). https://doi.org/10.3758/s13423-016-1153-z
14. Farley, J., et al.: Everyday attention and lecture retention: the effects of time, fidgeting, and mind wandering. Front. Psychol. **4**, 619 (2013). https://doi.org/10.3389/fpsyg.2013.00619
15. Fenesi, B., et al.: Sweat so you don't forget: exercise breaks during a university lecture increase on-task attention and learning. J. Appl. Res. Mem. Cogn. **7**(2), 261–269 (2018). https://doi.org/10.1016/j.jarmac.2018.01.012
16. Flesch, R.: A new readability yardstick. J. Appl. Psychol. **32**(3), 221–233 (1948)
17. Forrin, N.D., Risko, E.F., Smilek, D.: On the relation between reading difficulty and mind-wandering: a section-length account. Psychol. Res. **83**(3), 485–497 (2017). https://doi.org/10.1007/s00426-017-0936-9
18. Fox, K.C.R., Christoff, K.: The Oxford Handbook of Spontaneous Thought: Mind-Wandering, Creativity, and Dreaming. Oxford University Press (2018)
19. Hutt, S., et al.: Automated gaze-based mind wandering detection during computerized learning in classrooms. User Model. User-Adap. Inter. **29**(4), 821–867 (2019). https://doi.org/10.1007/s11257-019-09228-5
20. Kopp, K., Bixler, R., D'Mello, S.: Identifying learning conditions that minimize mind wandering by modeling individual attributes. In: Trausan-Matu, S., Boyer, K.E., Crosby, M., Panourgia, K. (eds.) ITS 2014. LNCS, vol. 8474, pp. 94–103. Springer, Cham (2014). https://doi.org/10.1007/978-3-319-07221-0_12
21. Lundberg, S.M., Lee, S.-I.: A unified approach to interpreting model predictions. In: Advances in Neural Information Processing Systems. Curran Associates, Inc. (2017)
22. Mills, C., et al.: Eye-Mind reader: an intelligent reading interface that promotes long-term comprehension by detecting and responding to mind wandering. Hum.-Comp. Interact. **36**(4), 306–332 (2021). https://doi.org/10.1080/07370024.2020.1716762
23. Mooneyham, B.W., Schooler, J.W.: The costs and benefits of mind-wandering: a review. Can. J. Exp. Psychol. **67**(1), 11–18 (2013)
24. Mota, S., Picard, R.W.: Automated posture analysis for detecting learner's interest level. In: 2003 Conference on Computer Vision and Pattern Recognition Workshop, p. 49 IEEE, Madison, Wisconsin, USA (2003)
25. Nakatani, C., et al.: Context-dependent neural effects predict mind wandering minutes in advance. http://biorxiv.org/lookup/doi/https://doi.org/10.1101/2021.09.04.458977, (2021). https://doi.org/10.1101/2021.09.04.458977
26. Ravi, N., et al.: Activity recognition from accelerometer data. In: American Association for Artificial Intelligence, 6 (2005)
27. Robin, X., et al.: pROC: an open-source package for R and S+ to analyze and compare ROC curves. BMC Bioinformatics **12**(1), 77 (2011)
28. Seli, P., et al.: Restless mind, restless body. J. Exp. Psychol. Learn. Mem. Cogn. **40**(3), 660–668 (2014). https://doi.org/10.1037/a0035260

29. Smallwood, J., Schooler, J.W.: The Science of Mind Wandering: Empirically Navigating the Stream of Consciousness. Annu. Rev. Psychol. **66**(1), 487–518 (2015). https://doi.org/10.1146/annurev-psych-010814-015331
30. Straczkiewicz, M., Glynn, N.W., Harezlak, J.: On placement, location and orientation of wrist-worn tri-axial accelerometers during free-living measurements. Sensors **19**(9), 2095 (2019). https://doi.org/10.3390/s19092095
31. Szpunar, K.K., et al.: Interpolated memory tests reduce mind wandering and improve learning of online lectures. PNAS **110**(16), 6313–6317 (2013)
32. Webber, C., Zbilut, J.: Recurrence quantification analysis of nonlinear dynamical systems. Tutorials in Contemporary Nonlinear Methods for the Behavioral Sciences (2005)
33. Wong, A.Y., et al.: Task-unrelated thought during educational activities: a meta-analysis of its occurrence and relationship with learning. Contemp. Educ. Psychol. **71**, 102098 (2022). https://doi.org/10.1016/j.cedpsych.2022.102098
34. Zhang, H., Jonides, J.: Pre-trial gaze stability predicts momentary slips of attention. https://psyarxiv.com/bv2uc/ (2021). https://doi.org/10.31234/osf.io/bv2uc
35. U.S. Census Bureau QuickFacts: United States. https://www.census.gov/quickfacts/fact/table/US/PST045221. Last accessed 14 Jan 2023

Efficient Feedback and Partial Credit Grading for Proof Blocks Problems

Seth Poulsen[(✉)], Shubhang Kulkarni, Geoffrey Herman, and Matthew West

University of Illinois Urbana-Champaign, Urbana, IL 61801, USA
sethp3@illinois.edu

Abstract. Proof Blocks is a software tool that allows students to practice writing mathematical proofs by dragging and dropping lines instead of writing proofs from scratch. Proof Blocks offers the capability of assigning partial credit and providing solution quality feedback to students. This is done by computing the *edit distance* from a student's submission to some predefined set of *solutions*. In this work, we propose an algorithm for the edit distance problem that significantly outperforms the baseline procedure of exhaustively enumerating over the entire search space. Our algorithm relies on a reduction to the minimum vertex cover problem. We benchmark our algorithm on thousands of student submissions from multiple courses, showing that the baseline algorithm is intractable, and that our proposed algorithm is critical to enable classroom deployment. Our new algorithm has also been used for problems in many other domains where the solution space can be modeled as a DAG, including but not limited to Parsons Problems for writing code, helping students understand packet ordering in networking protocols, and helping students sketch solution steps for physics problems. Integrated into multiple learning management systems, the algorithm serves thousands of students each year.

Keywords: Mathematical proofs · Automated feedback · Scaffolding

1 Introduction

Traditionally, classes that cover mathematical proofs expect students to read proofs in a book, watch their instructor write proofs, and then write proofs on their own. Students often find it difficult to jump to writing proofs on their own, even when they have the required content knowledge [25]. Additionally, because proofs need to be graded manually, it often takes a while for students to receive feedback on their work.

Proof Blocks is a software tool that allows students to construct a mathematical proof by dragging and dropping instructor-provided lines of a proof instead of writing from scratch (similar to Parsons Problems [19] for writing code—see Fig. 1 for an example of the Proof Blocks user interface). This tool scaffolds students' learning as they transition from reading proofs to writing proofs while

N. Wang et al. (Eds.): AIED 2023, LNAI 13916, pp. 502–514, 2023.
https://doi.org/10.1007/978-3-031-36272-9_41

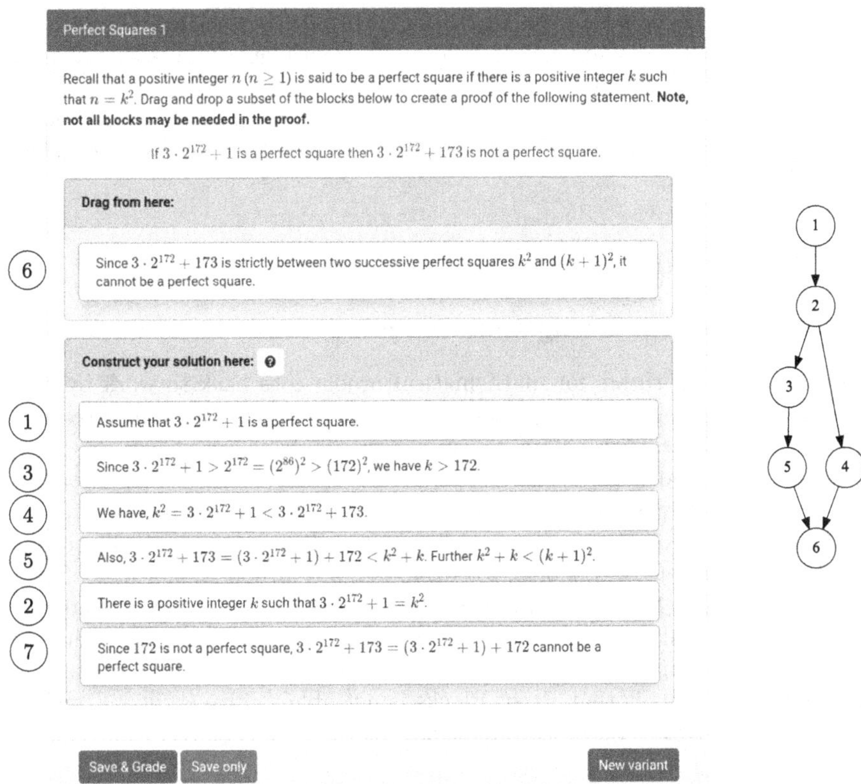

Fig. 1. An example of the Proof Blocks student-user interface. The instructor wrote the problem with 1, 2, 3, 4, 5, 6 as the intended solution, but the Proof Blocks autograder will also accept any other correct solution as determined by the dependency graph shown. For example, both 1, 2, 4, 3, 5, 6 and 1, 2, 3, 5, 4, 6 would also be accepted as correct solutions. Line 7 is a distractor that does not occur in any correct solution.

also providing instant machine-graded feedback. To write a Proof Blocks problem, an instructor specifies lines of a proof and their logical dependencies. The autograder accepts any ordering of the lines that satisfies the dependencies.

Calculating the least edit distance from a student submission to some correct solution solves two problems: (1) assigning students partial credit (a key concern for students taking exams in a computerized environment [5,8]) and (2) giving students instant feedback on their work, which can be a huge help for students [3]. The baseline algorithm performance is not sufficient to scale to provide immediate feedback to students in large classrooms, necessitating a more efficient solution. This paper makes the following contributions:

- An efficient algorithm for calculating the minimum edit distance from a sequence to some topological ordering of a directed acyclic graph (DAG)
- Application of this algorithm to grading and feedback for Proof Blocks problems

- Mathematical proofs that the algorithm is correct, and has asymptotic complexity that allows it to scale to grading many student submissions at once, even for large problems
- Benchmarking results on thousands of student submissions showing that the efficient algorithm is needed to enabling the performance necessary for classroom deployment

2 Related Work

2.1 Software for Learning Mathematical Proofs

Work in intelligent tutors for mathematical proofs goes back to work by John Anderson and his colleagues on The Geometry Tutor [3,4,12]. More recently, researchers have created tutors for propositional logic, most notably Deep Thought [16,17] and LogEx [13,14]. The authors' prior work reviews other software tools that provide visual user interfaces for constructing proofs [23].

Most of these tools cover only a small subset of the material typically covered in a discrete mathematics course, for example, only propositional logic. Those tools that are more flexible require learning complex theorem prover languages. In contrast, Proof Blocks enables instructors to easily provide students with proof questions on any topic by abstracting the content from the grading mechanism. The downside of this is that Proof Blocks is not able to give students content-specific hints as some of the other tools are.

2.2 Edit Distance Based Grading and Feedback

To our knowledge, no one has ever used edit distance based grading as a way of providing feedback for mathematical proofs, but edit distance based grading algorithms have been used in other contexts.

Chandra et al. [6] use edit distance to assign partial credit to incorrect SQL queries submitted by students, using reference solutions provided by the instructor. Edit distance based methods, often backed by a database of known correct solutions, have also been used to give feedback to students learning to program in general purpose programming languages [9,18] One difference between these and our method is that in programming contexts, the solution space is very large, and so the methods work based on edit distance to some known correct solution (manually provided by the instructor or other students). Because we model mathematical proofs as DAGs, we are able to constrain the solution space to be small enough that our algorithm can feasibly check the shortest edit to *any* correct solution.

Alur et al. [2] provide a framework for automatically grading problems where students must construct a deterministic finite automata (DFA). They use edit distance for multiple purposes in their multi-modal grading scheme.

3 Proof Blocks

Prior work has shown that Proof Blocks are effective test questions, providing about as much information about student knowledge as written proofs do [22], and also show promise in helping students save time when learning to write proofs [21]. To write a Proof Blocks problem, an instructor provides the proof lines and the logical dependencies between the lines of the proof. These logical dependencies form a DAG. The autograder gives the student points if their submission is a topological sort of the dependency graph. On exams or during in-class activities, students are often given multiple tries to solve a problem, so it is critical that they receive their feedback quickly. Additional details about the instructor and student user interfaces, as well as best practices for using Proof Blocks questions are given in a tool paper [23].

Proof Blocks is currently integrated into both PrairieLearn [26] and Runestone Interactive [15]. In PrairieLearn, students who submit their work are shown their score and told the first line of their proof that isn't logically supported by previous lines. The Runestone implementation highlights certain lines of the student submission that should be moved in order to get to a correct solution. Research and discussion about which types of feedback are most helpful for student learning are of crucial importance, but are beyond the scope of this paper, which will focus on the technical details of the edit distance algorithm which enables the construction of feedback.

Our algorithm assumes that each block is of equal weight for assigning partial credit. The benefit of this is that the algorithm can assign partial credit for any Proof Blocks problem without needing to know anything about the content of the blocks, making it quicker and easier for instructors to write questions. We have also had instructors using Proof Blocks express an interest in having the ability to have some blocks weighted more than others in the grading. We leave this to future work.

4 The Edit Distance Algorithm

Before defining our grading algorithms rigorously, it will first help to set up some formalism about Proof Blocks problems. We then give the baseline version of the algorithm, followed by an optimized version which is necessary for production deployment. For simplicity, our focus in communicating the algorithms will be on calculating the edit distance, but the proof of the correctness of the algorithm also explicitly constructs the edit sequence used to give students feedback on how to edit their submission into a correct solution.

4.1 Mathematical Preliminaries

Graph Theory. Let $G = (V, E)$ be a DAG. Then a subset of vertices $C \subseteq V$ is a *vertex cover* if every edge in E is incident to some vertex in C. The *minimum vertex cover* (MVC) problem is the task of finding a vertex cover of minimum cardinality. In defining our algorithms, we will assume the availability of a few classical algorithms for graphs: AllTopologicalOrderings(G) to return a set containing all possible topological orderings of a graph G [11], ExistsPath(G, u, v) returns a boolean value to denote if there is a path from the node u to the node v in the graph G, and MinimumVertexCover(G) to return an MVC of a graph G by exhaustive search.

Edit Distance. For our purposes, we use the *Longest Common Subsequence* (LCS) edit distance, which only allows deletion or addition of items in the sequence (it does not allow substitution or transposition). This edit distance is a good fit for our problem because it mimics the affordances of the user interface of Proof Blocks. Throughout the rest of the paper, we will simply use "edit distance" to refer to the LCS edit distance. We denote the edit distance between two sequences S_1 and S_2 as $d(S_1, S_2)$. Formally defined, given two sequences S_1, S_2, the edit distance is the length of the shortest possible sequence of operations that transforms S_1 into S_2, where the operations are: (1) **Deletion** of element s_i: changes the sequence $s_1, s_2, ...s_{i-1}, s_i, s_{i+1}, ...s_n$ to the sequence $s_1, s_2, ...s_{i-1}, s_{i+1}, ...s_n$. (2) **Insertion** of element t after location i: changes the sequence $s_1, s_2, ...s_i, s_{i+1}, ...s_n$ to the sequence $s_1, s_2, ...s_i, t, s_{i+1}, ...s_n$. We assume the ability to compute the edit distance between two sequences in quadratic time using the traditional dynamic programming method [24]. We also identify a topological ordering O of a graph G with a sequence of nodes so that we can discuss the edit distance between a topological ordering and another sequence $d(S, O)$.

4.2 Problem Definition

A *Proof Blocks problem* $P = (C, G)$ is a set of *blocks* C together with a DAG $G = (V, E)$, which defines the logical structure of the proof. Both the blocks and the graph are provided by the instructor who writes the question (see [23] for more details on question authoring). The set of vertices V of the graph G is a subset of the set of blocks C. Blocks which are in C but *not* in V are blocks which are not in any correct solution, and we call these *distractors*, a term which we borrow from the literature on multiple-choice questions. A *submission* $S = s_1, s_2, ...s_n$ is a sequence of distinct blocks, usually constructed by a student who is attempting to solve a Proof Blocks problem. If a submission S is a topological ordering of the graph G, we say that S is a *solution* to the Proof Blocks problem P.

If a student submits a submission S to a Proof Blocks problem $P = (C, G)$, we want to assign partial credit with the following properties: (1) students get 100% only if the submission is a solution (2) partial credit declines with the number of edits needed to convert the submission into a solution (3) the credit received is guaranteed to be in the range 0–100. To satisfy these desirable properties, we assign partial credit as follows: score $= 100 \times \frac{\max(0, |V| - d^*)}{|V|}$, where d^* is the minimum edit distance from the student submission to some correct solution of P, that is: $d^* = \min\{d(S, O) \mid O \in \text{AllTopologicalOrderings}(G)\}$. This means, for example, that if a student's solution is 2 deletions and 1 insertion (3

Algorithm 1. Baseline Algorithm

Input
S The student submission being graded
P The Proof Blocks problem written by the instructor
Output
The minimum number of edits needed to transform S into a solution
procedure GetMinimumEditDistance($S = s_1, s_2, ...s_\ell$, $P = (C, G)$)
Brute force calculation of d^*:
return $\min\{d(S, O) \mid O \in \text{AllTopologicalOrderings}(G)\}$
end procedure

edits) away from a correct solution, and the correct solution is 10 lines long, the student will receive 70%. If the edit distance is greater than the length of the solution, we simply assign 0.

4.3 Baseline Algorithm

The most straightforward approach to calculating partial credit as we define it is to iterate over all topological orderings of G and for each one, calculate the edit distance to the student submission S. We formalize this approach as Algorithm 1. While this is effective, this algorithm is computationally expensive.

Theorem 1. *The time complexity of Algorithm 1 is $\mathcal{O}(m \cdot n \cdot n!)$ in the worst case, where n is the size of G and m is the length of the student submission after distractors are discarded.*

Proof. The algorithm explicitly enumerates all $\mathcal{O}(n!)$ topological orderings of G. For each ordering, the algorithm forms the associated block sequence, and computes the edit distance to the student submission, requiring $\mathcal{O}(m \cdot n)$ time. □

4.4 Optimized (MVC-Based) Implementation of Edit Distance Algorithm

We now present a faster algorithm for calculating the Proof Blocks partial credit, which reduces the problem to the *minimum vertex cover* (MVC) problem over a subset of the student's submission. Rather than iterate over all topological orderings, this algorithm works by manipulating the student's submission until it becomes a correct solution. In order to do this, we define a few more terms. We call a pair of blocks (s_i, s_j) in a submission a *problematic pair* if line s_j comes before line s_i in the student submission, but there is a path from s_i to s_j in G, meaning that s_j must come *after* s_i in any correct solution.

We define the *problematic graph* to be the graph where the nodes are the set of all blocks in a student submission that appear in some problematic pair, and the edges are the problematic pairs. We can then use the problematic graph to

Algorithm 2. Novel Algorithm using the MVC

1: **Input**
2: S The student submission being graded
3: P The Proof Blocks problem written by the instructor
4: **Output**
5: The minimum number of edits needed to transform S into a solution
6: **procedure** GetMinimumEditDistance($S = s_1, s_2, ...s_\ell$, $P = (C, G)$)
Construct the problematic graph:
7: $E_O \leftarrow \{(s_i, s_j) \mid i > j \text{ and } \textsc{ExistsPath}(G, s_i, s_j)\}$
8: $V_O \leftarrow \{s_i \mid \text{there exists } j \text{ such that } (s_i, s_j) \in E_O \text{ or } (s_j, s_i) \in E_O\}$
9: problematicGraph $\leftarrow (V_O, E_O)$

Find number of insertions and deletions needed:
10: mvcSize $\leftarrow$ |MinimumVertexCover(problematicGraph)|
11: numDistractors $\leftarrow |\{s_i \in S \mid s_i \notin V\}|$
12: deletionsNeeded $\leftarrow$ numDistractors + mvcSize
13: insertionsNeeded $\leftarrow |V| - (|S| -$ deletionsNeeded)
14: **return** deletionsNeeded + insertionsNeeded
15: **end procedure**

guide which blocks need to be deleted from the student submission, and then we know that a simple series of insertions will give us a topological ordering. The full approach is shown in Algorithm 2, and the proof of Theorem 2 proves that this algorithm is correct.

4.5 Worked Example of Algorithm 2

For further clarity, we will now walk through a full example of executing Algorithm 2 on a student submission. Take, for example, the submission shown in Fig. 1. In terms of the block labels, this submission is $S = 1, 3, 4, 5, 2, 7$. In this case, block 2 occurs after blocks 3, 4, and 5, but because of the structure of the DAG, we know that it must come before all of those lines in any correct solution. Therefore, the problematic graph in this case is problematicGraph $= (\{2, 3, 4, 5\}, \{(2, 3), (2, 4), (2, 5)\})$. The minimum vertex cover here is $\{2\}$, because that is the smallest set which contains at least one endpoint of each edge in the graph. Now we know that the number of deletions needed is 1 + 1 = 2 (vertex cover of size one, plus one distractor line picked, see Algorithm 2 line 12), and the number of insertions needed is 2 (line 2 must be reinserted in the correct position after being deleted, and line 6 must be inserted). This gives us a least edit distance (d^*) of 4, and so the partial credit assigned would be score $= 100 \times \frac{\max(0, |V| - d^*)}{|V|} = 100 \times \frac{6-4}{6} \approx 33\%$.

4.6 Proving the Correctness of Algorithm 2

First we will show that the Algorithm constructs a feasible solution, and then we will show that it is minimal.

Lemma 1 (Feasability). *Given a submission $S = s_1, s_2, ...s_\ell$, there exists a sequence of edits $\mathcal{E}$ from S to some solution of P such that $|\mathcal{E}|$ is equal to* GETMINIMUMEDITDISTANCE(S, P) *as computed by Algorithm 2.*

Proof. Given the MVC computed on line 10 of Algorithm 2, delete all blocks in the MVC from S, as well as all distractors in S, and call this new submission S'. Now S' is a submission such that it contains no distractors, and its problematic graph is empty.

Now, for all i where $1 \leq i < \ell$, add the edge (s_i, s_{i+1}) to the graph G, and call this new graph G'. Because there are no problematic pairs in S', we know that adding these new edges does not introduce any new cycles, so G' is a DAG. Now, a topological ordering O of the graph G' will be a topological ordering of G with the added constraint that all blocks which appeared in the submission S' are still in the same order with respect to one another. Then, since there are no distractors in S', S' will be a subsequence of O. Thus, we can construct O simply by adding blocks to S. The length of this sequence $\mathcal{E}$ is exactly what Algorithm 2 computes. □

Lemma 2 (Minimality). *Let E' be any edit from the submission S to some correct solution of P. Then the length of E' is greater than or equal to the output of Algorithm 2.*

Proof. Let E be the edit sequence constructed in Lemma 1. We will show that the number of deletions and the number of insertions in E' is greater than or equal to the number of deletions and insertions in E.

If there is any problematic pair (s_i, s_j) in the student submission, one of s_i and s_j must be deleted from the submission to reach a solution. Because there is no substitution or transposition allowed, and because each block may only occur once in a sequence, there is no other way for the student submission to be transformed into some correct solution unless s_i or s_j is deleted and then re-inserted in a different position.

Therefore, the set of blocks deleted in the edit sequence E' must be a vertex cover of the problematic graph related to S in E and we delete only the blocks in the minimum vertex cover of the problematic graph. In both cases, all distractors must be deleted. So, the number of deletions in E' is greater than or equal to the number of deletions in E.

The number of insertions in any edit sequence must be the number of deletions, plus the difference between the length of the submission and the size of the graph, so that the final solution will be the correct length. Then since the number of deletions in E' is at least as many as there are in E, the number of insertions in E' is also at least as many as the number of insertions in E.

Combining what we have shown about insertions and deletions, we have that E' is at least as long of an edit sequence as E. □

Theorem 2. *Algorithm 2 computes d^*—the minimum edit distance from the submission S to some topological ordering of G—in $\mathcal{O}(m^2 \cdot 2^m)$ time, where m is the length of the student submission after distractors are discarded.*

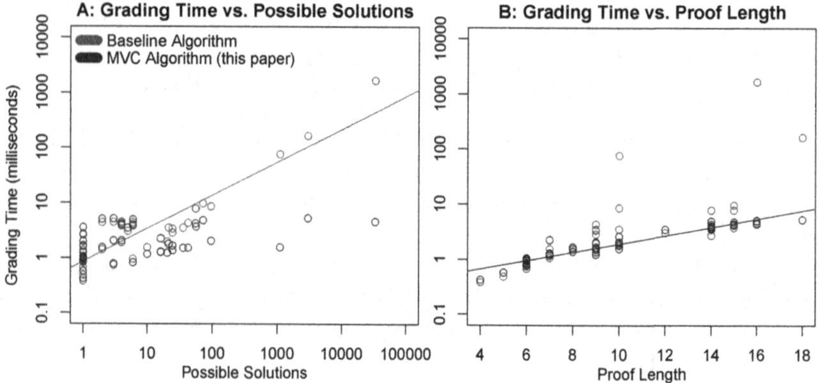

Fig. 2. Comparison of grading time for the two grading algorithms. Subplot (A) is a log-log plot showing that the baseline algorithm scales with the number of possible solutions, Subplot (B) is a log-linear plot showing that the MVC Algorithm runtime scales with the length of the proof. This is a critical difference, because the number of topological orderings of a DAG can be $n!$ for a graph with n nodes.

Proof. The correctness of the algorithm is given by combining Lemma 1 and Lemma 2. To see the time complexity, consider that constructing the problematic graph requires using a breadth first search to check for the existence of a path between each block and all of the blocks which precede it in the submission S, which can be completed in polynomial time. Naïvely computing the MVC of the problematic graph has time complexity $\mathcal{O}(m^2 \cdot 2^m)$. Asymptotically, the calculation of the MVC will dominate the calculation of the problematic graph, giving an overall time complexity of $\mathcal{O}(m^2 \cdot 2^m)$. □

Remark 1. The complexity of Algorithm 2 is due to the brute force computation of the MVC, however, there exists a a $\mathcal{O}(1.2738^k + kn)$-time *fixed parameter tractable* (FPT) algorithm where k is the size of the minimum vertex cover [7]. While we focus in this paper on the brute force MVC method since it is sufficient for our needs, using the FPT method may give further speedup, especially considering the often small size of k in real use cases (see Table 1).

Proof Blocks also supports a feature known as *block groups* which enables Proof Blocks to handle proof by cases using an extended version of Algorithm 2. Full details of the extended version of Algorithm 2 and proofs of its correctness can be seen in the first author's dissertation [20]. In the benchmarking results shown below in Table 1, problems 2, 4, 6, 9, and 10 make use of this extended version.

5 Benchmarking Algorithms on Student Data

5.1 Data Collection

We collected data from homework, practice tests, and exams from Discrete Mathematics courses taught in the computer science departments at the University of Illinois Urbana-Champaign and the University of Chicago. Problems covered topics including number theory, cardinality, functions, graphs, algorithm analysis, and probability theory. Some questions only appeared on optional practice exams, while others appeared on exams. Also, more difficult questions received more incorrect submissions as students were often given multiple tries. This explains the large discrepancy between the number of submissions to certain questions seen in Table 1. In total, our benchmark set includes 7,427 submissions to 42 different Proof Blocks problems.

5.2 Benchmarking Details

All benchmarking was done in Python. For Algorithm 1, we used the NetworkX library [10] to generate all topological orderings of G and used the standard dynamic programming algorithm for LCS edit distance to calculate the edit distance between each submission and each topological ordering. Our implementation of Algorithm 2 also used NetworkX to store the graph, and then found the MVC using the naïve method of iterating over all subsets of the graph, starting from the smallest to the largest, until finding one which is a vertex cover. Benchmarks were run on an Intel i5-8530U CPU with 16GB of RAM. The implementation of Algorithm 2 used for benchmarking is the implementation used in production in PrairieLearn [1]. Runestone Academy uses an alternate implementation in JavaScript.

5.3 Results

Fig. 2 shows the algorithm run time of our novel MVC-based algorithm (Algorithm 2) and the baseline algorithm (Algorithm 1) for all of the problems in our benchmark set, compared both to the number of possible solutions (A) and the Proof Length (B). This demonstrates that the theoretical algorithm runtimes hold empirically: the run time of Algorithm 1 scales exponentially with the number of topological orderings of the proof graph (A), and the run time of Algorithm 2 scales exponentially with the length of the proof (B). This is a critical difference, because the number of topological orderings of a DAG can be $n!$ for a graph with n nodes. Thus, a relatively short Proof Blocks problem could have a very long grading time with Algorithm 1, while with Algorithm 2, we can guarantee a tight bound on grading time given the problem size.

Algorithm 1 performed about twice as fast as Algorithm 2 when there was one solution—more trivial cases when both Algorithms took less than a millisecond. Performance was comparable for problems with between 2 and 10 possible solutions Algorithm 2 performed significantly faster for all problems with over 10 possible solutions ($p < 0.001$).

Table 1 gives further benchmarking details for the 10 questions from the data set with the greatest number of possible solutions (the others are omitted due to space constraints). These results show that Algorithm 2 is far superior in performance. The mean time of 1.7 s for the most complex Proof Blocks problem under Algorithm 1 may not seem computationally expensive, but it does not scale

Table 1. Performance of baseline vs. MVC algorithm for 10 problems with the most topological sorts. For all problems shown, speedup was statistically significant at $p < 0.001$. Numbers in parentheses are standard errors.

Question Number	Proof Length	Possible Solutions	Distractors	Submissions	Submission size (mean)	Prob. Graph Size (mean)	MVC Size (mean)	Baseline Alg. Time (mean ms)	MVC Alg. Time (mean ms)	Speedup Factor
1	9	24	0	529	8.9	3.3	1.6	3.36 (0.90)	1.66 (0.50)	2.0
2	9	35	5	376	8.5	1.2	0.4	3.50 (0.63)	1.51 (0.28)	2.3
3	9	42	0	13	8.8	1.5	0.7	4.29 (0.40)	1.52 (0.15)	2.8
4	15	55	0	29	6.9	2.2	0.4	7.83 (1.39)	4.21 (0.75)	1.9
5	14	56	0	324	6.4	1.8	0.6	7.82 (1.32)	3.68 (0.98)	2.1
6	15	72	0	260	7.4	1.6	0.5	9.69 (2.20)	4.80 (1.27)	2.0
7	10	96	0	145	8.3	2.5	1.0	8.56 (1.27)	2.02 (0.48)	4.2
8	10	1100	0	616	9.4	3.1	1.3	76.05 (8.78)	1.56 (0.21)	48.7
9	18	3003	0	253	8.2	1.6	0.5	166.4 (15.9)	5.30 (1.05)	31.4
10	16	33264	0	97	4.9	1.2	0.4	1676.0 (235.)	4.60 (1.43)	364.8

to having hundreds of students working on an active learning activity, or taking exams at the same time, all needing to receive rapid feedback on their Proof Blocks problem submissions. Furthermore, this grading time could easily be 10 or even 100 times longer per question if the DAG for the question was altered by even a single edge.

6 Conclusions and Future Work

In this paper, we have presented a novel algorithm for calculating the edit distance from a student submission to a correct solution to a Proof Blocks problem. This information can then be used to give students feedback and calculate grades. This algorithm can also be used with Parsons Problems, task planning problems, or any other type of problem where the solution space can be modeled as a DAG. We showed with student data that our algorithm far outperforms the baseline algorithm, allowing us to give students immediate feedback as they work through exams and homework. Now deployed in dozens of classrooms across multiple universities, this algorithm benefits thousands of students per semester.

Acknowledgements. We would like to thank Mahesh Viswanathan, Benjamin Cosman, Patrick Lin, and Tim Ng for using Proof Blocks in their Discrete Math courses and allowing us to use anonymized data from their courses for the included benchmarks.

References

1. pl-order-blocks documentation (2021). https://prairielearn.readthedocs.io/en/latest/elements/#pl-order-blocks-element. Accessed June 2021
2. Alur, R., D'Antoni, L., Gulwani, S., Kini, D., Viswanathan, M.: Automated grading of DFA constructions. In: Twenty-Third International Joint Conference on Artificial Intelligence (2013)
3. Anderson, J.R., Corbett, A.T., Koedinger, K.R., Pelletier, R.: Cognitive tutors: lessons learned. J. Learn. Sci. **4**(2), 167–207 (1995)

4. Anderson, J., Boyle, C., Yost, G.: The geometry tutor. In: Proceedings of the 9th International Joint Conference on Artificial Intelligence (1985)
5. Apostolou, B., Blue, M.A., Daigle, R.J.: Student perceptions about computerized testing in introductory managerial accounting. J. Account. Educ. **27**(2), 59–70 (2009). https://doi.org/10.1016/j.jaccedu.2010.02.003
6. Chandra, B., Banerjee, A., Hazra, U., Joseph, M., Sudarshan, S.: Automated grading of SQL queries. In: 2019 IEEE 35th International Conference on Data Engineering (ICDE), pp. 1630–1633, April 2019. https://doi.org/10.1109/ICDE.2019.00159, iSSN: 2375-026X
7. Chen, J., Kanj, I.A., Xia, G.: Improved upper bounds for vertex cover. Theor. Comput. Sci. **411**(40), 3736–3756 (2010). https://doi.org/10.1016/j.tcs.2010.06.026
8. Darrah, M., Fuller, E., Miller, D.: A comparative study of partial credit assessment and computer-based testing for mathematics. J. Comput. Math. Sci. Teach. **29**(4), 373–398 (2010)
9. Gulwani, S., Radiček, I., Zuleger, F.: Automated Clustering and Program Repair for Introductory Programming Assignments, p. 16 (2018)
10. Hagberg, A.A., Schult, D.A., Swart, P.J.: Exploring network structure, dynamics, and function using NetworkX. In: Varoquaux, G., Vaught, T., Millman, J. (eds.) Proceedings of the 7th Python in Science Conference, pp. 11–15. Pasadena, CA USA (2008)
11. Knuth, D.E., Szwarcfiter, J.L.: A structured program to generate all topological sorting arrangements. Inf. Process. Lett. **2**(6), 153–157 (1974)
12. Koedinger, K.R., Anderson, J.R.: Abstract planning and perceptual chunks: elements of expertise in geometry. Cogn. Sci. **14**(4), 511–550 (1990)
13. Lodder, J., Heeren, B., Jeuring, J.: A comparison of elaborated and restricted feedback in LogEx, a tool for teaching rewriting logical formulae. J. Comput. Assist. Learn. **35**(5), 620–632 (2019). https://doi.org/10.1111/jcal.12365
14. Lodder, J., Heeren, B., Jeuring, J.: Providing hints, next steps and feedback in a tutoring system for structural induction. Electron. Proc. Theor. Comput. Sci. **313**, 17–34 (2020). https://doi.org/10.4204/EPTCS.313.2
15. Miller, B.N., Ranum, D.L.: Beyond pdf and ePub: toward an interactive textbook. In: Proceedings of the 17th ACM Annual Conference on Innovation and Technology in Computer Science Education, pp. 150–155. ITiCSE 2012, ACM, New York, NY, USA (2012). https://doi.org/10.1145/2325296.2325335
16. Mostafavi, B., Barnes, T.: Evolution of an intelligent deductive logic tutor using data-driven elements. Int. J. Artif. Intell. Educ. **27**(1), 5–36 (2016). https://doi.org/10.1007/s40593-016-0112-1
17. Mostafavi, B., Zhou, G., Lynch, C., Chi, M., Barnes, T.: Data-driven worked examples improve retention and completion in a logic tutor. In: Conati, C., Heffernan, N., Mitrovic, A., Verdejo, M.F. (eds.) AIED 2015. LNCS (LNAI), vol. 9112, pp. 726–729. Springer, Cham (2015). https://doi.org/10.1007/978-3-319-19773-9_102
18. Paassen, B., Hammer, B., Price, T.W., Barnes, T., Gross, S., Pinkwart, N.: The Continuous hint factory - providing hints in vast and sparsely populated edit distance spaces. J. Educ. Data Min. **10**(1), 1–35 (2018)
19. Parsons, D., Haden, P.: Parson's programming puzzles: a fun and effective learning tool for first programming courses. In: Proceedings of the 8th Australasian Conference on Computing Education, vol. 52, pp. 157–163. ACE '06, Australian Computer Society Inc., AUS (2006)
20. Poulsen, S.: Proof blocks: autogradable scaffolding activities for learning to write proofs. Ph.D. thesis (2023)

21. Poulsen, S., Gertner, Y., Cosman, B., West, M., Herman, G.L.: Efficiency of learning from proof blocks versus writing proofs. In: Proceedings of the 54th ACM Technical Symposium on Computer Science Education. Association for Computing Machinery, New York, NY, USA (2023)
22. Poulsen, S., Viswanathan, M., Herman, G.L., West, M.: Evaluating proof blocks problems as exam questions. In: Proceedings of the 17th ACM Conference on International Computing Education Research, pp. 157–168 (2021)
23. Poulsen, S., Viswanathan, M., Herman, G.L., West, M.: Proof blocks: autogradable scaffolding activities for learning to write proofs. In: Proceedings of the 27th ACM Conference on Innovation and Technology in Computer Science Education, vol. 1. Association for Computing Machinery, New York, NY, USA (2022). https://doi.org/10.1145/3502718.3524774
24. Wagner, R.A., Fischer, M.J.: The string-to-string correction problem. J. ACM (JACM) **21**(1), 168–173 (1974)
25. Weber, K.: Student difficulty in constructing proofs: the need for strategic knowledge. Educ. Stud. Math. **48**(1), 101–119 (2001). https://doi.org/10.1023/A:1015535614355
26. West, M., Herman, G.L., Zilles, C.: PrairieLearn: mastery-based online problem solving with adaptive scoring and recommendations driven by machine learning. In: 2015 ASEE Annual Conference & Exposition, pp. 26.1238.1-26.1238.14. No. 10.18260/p.24575, ASEE Conferences, Seattle, Washington, June 2015

Dropout Prediction in a Web Environment Based on Universal Design for Learning

Marvin Roski[1(✉)], Ratan Sebastian[2], Ralph Ewerth[2,3], Anett Hoppe[2,3], and Andreas Nehring[1]

[1] Institute for Science Education, Leibniz University Hannover, Hannover, Germany
{roski,nehring}@idn.uni-hannover.de
[2] TIB – Leibniz Information Centre for Science and Technology, Hannover, Germany
{ratan.sebastian,ralph.ewerth,anett.hoppe}@tib.eu
[3] L3S Research Center, Leibniz University Hannover, Hannover, Germany

Abstract. Dropout prediction is an essential task in educational Web platforms to identify at-risk learners, enable individualized support, and eventually prevent students from quitting a course. Most existing studies on dropout prediction focus on improving machine learning methods based on a limited set of features to model students. In this paper, we contribute to the field by evaluating and optimizing dropout prediction using features based on personal information and interaction data. Multiple granularities of interaction and additional unique features, such as data on reading ability and learners' cognitive abilities, are tested. Using the Universal Design for Learning (UDL), our Web-based learning platform called I_3Learn aims at advancing inclusive science learning by focusing on the support of all learners. A total of 580 learners from different school types have used the learning platform. We predict dropout at different points in the learning process and compare how well various types of features perform. The effectiveness of predictions benefits from the higher granularity of interaction data that describe intermediate steps in learning activities. The cold start problem can be addressed using assessment data, such as a cognitive abilities assessment from the pre-test of the learning platform. We discuss the experimental results and conclude that the suggested feature sets may be able to reduce dropout in remote learning (e.g., during a pandemic) or blended learning settings in school.

Keywords: Dropout prediction · Science Education · Inclusion

1 Introduction

With the Covid-19 pandemic, Web-based learning platforms are experiencing a resurgence. With remote learning by necessity, learners and educators worldwide were forced to rely on digital alternatives to the traditional classroom setting [10].

N. Wang et al. (Eds.): AIED 2023, LNAI 13916, pp. 515–527, 2023.
https://doi.org/10.1007/978-3-031-36272-9_42

By providing wide-ranged access to a multitude of learners, online learning platforms are popular alternatives to face-to-face instruction (a) in times of a pandemic, (b) for rural or less affluent areas without the necessary infrastructure, or (c) for blended learning opportunities [2,20,28]. Furthermore, online learning environments can contribute to achieving UNESCO's fourth Substantial Development Goal (SDG) [25,27]. The goal is to achieve "Education for All" by 2030, providing learners worldwide with access to sustainable and high-quality education regardless of the prerequisites of the learners [26]. Online learning also has many advantages: Learners can organize their learning, learn at their own pace, and evaluate their learning process [12]. However, this independence can be challenging: It may lead to frustration and loss of motivation due to problems with technology, missing immediate feedback, or the lack of suitable learning spaces at home [12,21,28]. These and other reasons can lead to increased dropout from Web-based learning platforms, whether the learning is aimed at children or adults, formal or informal education [6,11,21,28]. Dedicated artificial intelligence and learning analytics systems promise to identify at-risk learners, providing the opportunity for interventions to prevent dropout [1,3]. Vulnerable groups might benefit from such a system.

In this paper, we analyze dropout behavior in a learning platform suitable for use in remote or blended learning settings. We investigate various features and their impact on performance prediction. While common approaches rely on publicly available datasets [22], we collected our dataset through an learning platform called I$_3$Learn for chemistry education at school level in Germany. I$_3$Learn is characterized by being a smaller educational offering that runs over a shorter period of time and is suitable for remote or blended learning sessions. Several features differ from platforms formerly presented in the literature, such as the availability of assessment data (generated from assessments for reading ability, cognitive ability, and topic knowledge) and interaction data with enhanced granularity [11,22]. The learning platform has been designed to advance inclusive science learning.

In the next section, related work on dropout prediction is described. The resulting research questions follow in Sect. 3. Subsequently, Sect. 4 and 5 present our methods and dataset respectively, followed by a discussion of the results in Sect. 6. The paper closes with conclusion in Sect. 7.

2 Related Work

Dropout prediction is a broadly defined task that seeks to identify which students will disengage from a learning activity so that corrective measures can be taken [22]. Depending on the context, the data available for dropout prediction differs. De Oliveira et al. [19] propose a categorization that divides popular dropout prediction features into personal and academic features. The academic features further subdivide into features related to the current course of study and results of prior studies. Personal features include demographic features as well as various ability measures, self-reported psychological factors, and self-concepts.

Other surveys [11,16] classify available features into background information and online (current study) behavior at the top level.

Based on these, we classify our features into personal (as in [19]), and interaction data corresponding to current study as in [11] (see Table 1). We further divide interaction data into clickstream - after the schema defined in [22] - and learning-process related. The latter is a newly introduced here and refers to clickstream-derived features that follow from recommendations of educational scientists, the need for which is highlighted in [14]. A recent survey on dropout prediction [11] highlights a limitation of existing works: Rich personal and interaction features are only rarely available in the same study. This follows from the contexts in which dropout prediction is deployed. When data is collected from institutional student records, the personal information axis of the student model is more affluent, and when the data is collected from Learning Management System or Massiv Open Online Course (MOOC) interactions, more information along the interactional axis is available.

This study aims to address this shortcoming since data along both axes are collected in I3Learn allowing us to compare them. Manrique et al. [16] examine dropout prediction in a context where both kinds of features are available and find that academic features are good predictors when interactional data are scarce as at the beginning of a course (the cold-start problem). The data collected in I3Learn gives us (a) interaction data that are a lot more detailed (fine-grained interactions with learning materials), allowing us to see if this addresses the cold-start problem, and (b) personal features that measure student ability and self-concept, which might be better predictors than prior knowledge (which was the only feature in this category employed in the above study).

3 Research Questions

Dropout prediction forms an intersection of computer science and (science) education. Numerous efforts have been made to improve predictive power in this multidisciplinary research field. However, most dropout predictions focus on between-semester dropouts or week-to-week MOOC dropouts [13,29]. MOOCs represent an entire course and extend over several weeks with limited interaction capabilities. Dropout from institutional settings, such as university or school, is long-term and does not take advantage of the characteristics of student interactions. A smaller educational offering, such as I3Learn, teaches one topic, runs over a few weeks, and is suitable for remote learning or blended learning sessions. Consequently, it creates a unique set of conditions for learning and combines the typically available data from institutional and MOOC dropouts.

RQ1 Transfer of methods for dropout prediction: Can the results of predicting dropout from institutional settings and MOOCs be applied to educational resources with shorter duration and smaller scope, such as the I3Learn learning platform?

I3Learn produces more detailed interaction data than MOOCs. A more fine-grained and detailed dataset that contains more interaction data is promising to improve dropout prediction.

RQ2 Dropout prediction with data of varied granularity: Does dropout prediction benefit from finer-grained interaction recording?

Interaction data are weak predictors of dropout when not much interaction between student and system has been collected [16]. For this reason, we explore currently less considered features based on personal information that we created using different types of assessments (cognitive ability, reading ability, and topic knowledge). These are collected as part of a pre-test at the beginning of using the learning platform I3Learn and thus available prior to most interaction data.

RQ3 Assessments for predicting dropout: What assessment data best predict dropout from an internet-based learning platform?

4 Methodology

4.1 The Learning Platform I3Learn and Dropout Level

The Web-based learning platform I3Learn provides learning opportunities for the topic of ion bonding, a crucial topic in K12 chemistry education in German schools [4,9]. The platform design is based on Universal Design for Learning (UDL) guidelines aiming at providing access, regardless of the prerequisites and characteristics of the learners [5]. With I3Learn, we intend to reduce barriers to learning and thus improve inclusivity for everyone, not just learners with special needs or disabilities. The architecture of the learning platform is designed to allow learners to follow an individual learning path while achieving a common learning goal. Learners can voluntarily repeat important, and prerequisite contents such as ion formation, noble gas configuration, and electron transfer [9]. The individual chapters allow the definition of dropout on several levels. When learners have reached the post-test and answered at least one question, it is considered a successful completion of the learning platform. An overview of the individual chapters and the dropout definitions can be found in Fig. 1.

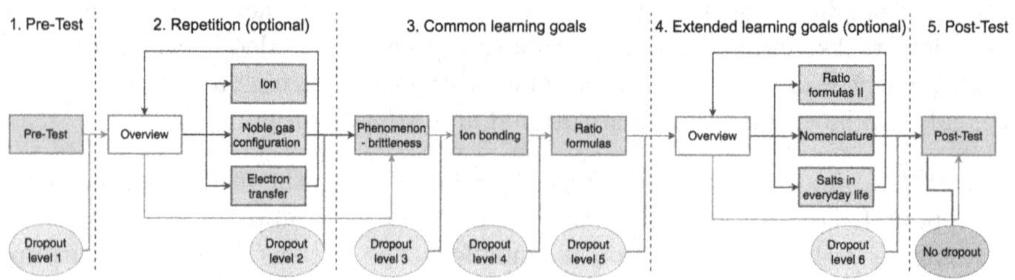

Fig. 1. An overview of the architecture of the I3Learn learning platform. It shows the individual chapters and the six dropout levels.

4.2 Data Collection and Features

Educators were contacted by email and informed about I3Learn. If they were interested in using the learning platform in their lessons, the teachers received a personal introduction through an online meeting. The general classroom use was entirely up to the teachers, who did not receive any remuneration. Learners create data by interacting with the learning platform. We distinguish between personal and interaction data. Personal data is generated by the learners' input, for example, when they take assessments, or can be found in the metadata of the accounts they use. Most of the personal information features were self-reported as part of the pre-test, which all learners had to pass before the learning opportunities. However, any information provided by the learners are voluntary. This includes gender, language spoken at home, socio-economic background [24], school grade, type of school, last chemistry grade, self-concept chemistry [7], topic knowledge assessment, reading ability [17], and cognitive ability (only the part figures analogy) [8]. The topical knowledge assessment is an adapted variant of the Bonding Representations Inventory (BRI) test by Luxford and Bretz [15]. Features like class size, school, and school type, were available from metadata linked to user ids. Furthermore, learners can evaluate their learning success after each learning opportunity. Learning opportunities include videos, texts, and tasks [7,8,15,17,24].

Interaction data were derived from two sources: (a) clickstream data and (b) learning process-related data. The former category consists of comprehensive features that quantify various aspects of learners' interactions, like the kind of interaction (e.g., viewing a video), the length of the interaction, and the relative position of the interaction in the clickstream. The latter category consists of engineered features that, from an educational perspective, can be assumed to reflect relevant interaction with the learning platform. These go beyond plain click data. All data and corresponding feature categories can be found in Table 1.

Table 1. Overview of features derived from I3Learn data.

Personal Information		Interaction data	
Demographic	Learning-related	Clickstream	Learning Process-related
Gender	Last grade	Count of interactions	Attempts Repetion
Language	Self-concept	Duration of interactions	Participates in Assessment
Socio-economics	Prior knowledge	Position of interactions	Revises content
School grade	Reading ability	Activity per week	Technical issues
School type	Cognitive ability	Activity per day	Worked in classroom/at home
School	Assessment answers		
Class size			

4.3 Data Aggregation

We consider three interaction quantifications (counts, position, and duration) and two kinds of aggregation levels (time-delimited and structure-delimited):

As shown in Fig. 1, there are five modules/chapters in I_3Learn. Each module comprises multiple activities, like watching a video, reading text, or working on learning tasks. Each of these activities can lead to numerous log events corresponding to steps in a learning task, or navigation on video and text content. The structure-delimited aggregations are made at the level of the module, activity, and activity step. For instance, the structure-delimited count features would be the count of the number of interactions in each module, activity, activity step, and overall. This also applies to aggregations of position and duration, except for an overall position aggregation that would make no sense since it would be the same for every student. Unlike structure-delimited aggregations, time-delimited ones aggregate quantities within time windows like week and day. For instance, a time-delimited count aggregation would be the number of interactions each week or day. Time-delimited position aggregations are not considered since it does not make sense to compare when one kind of interaction occurred relative to another in a week or day. At the weekly level, count and duration features are considered. Means of the count of interactions per week and total active time per week are taken since different students participate in different weeks and for other numbers of weeks. We also include a standard deviation aggregation to quantify the distribution of the interaction count across weeks. Further, we include a count of weekly sessions following indications in the MOOC dropout literature. The quantities being aggregated - here called the interaction quantifications are:

- Count: the number of interactions in a time window or structural component.
- Position: the normalized median of the indexes in the clickstream that a given structural component is interacted with.
- Duration: how long the student interacts with a specific activity. Duration is calculated based on the time between subsequent interactions with the cut-off that interactions more than 30 min apart fall in separate sessions.

For all the features derived from the user logs, we consider only logs created before a given dropout target is reached for prediction. Since there is a significant class imbalance for most dropout levels, the weighted F1-score is used to compare the performance of various models and across feature sets. All model training is done using pycaret[1] with 3-fold cross-validation. The best-performing model (of the 20 kinds that PyCaret tries to fit) across all splits is chosen, and its performance measures are reported in Table 4. The best performing model type is cited in the results tables according to the following legend: GBC: Gradient Boosting Classfier, KNN: K-Nearest Neighbours, LR: Logistic Regression, LightGBM: Tree-based gradient boosting[2] and RF: Random Forests Splits are constructed so that a proportional representation of the true and false classes are included in each split.

[1] https://pycaret.org.

[2] https://lightgbm.readthedocs.io/en/v3.3.2/index.html.

5 Dataset

Between January and June 2022, 580 learners in grades 9 and 10 in Germany interacted with I3Learn. The 27 classes were from integrative comprehensive schools (IGS), cooperative comprehensive schools (KGS), and high schools from Lower Saxony, Germany. The data show a balanced distribution in terms of gender and school class. The distribution of schools corresponds to a similar scope as the actual distribution of learners among the school types in Lower Saxony [18]. The sample characteristics can be taken from Table 2. Since learners always have the option of explicitly not answering, the number of responses to the different characteristics varies. An overview of the distribution of learners who can be assigned to the individual dropout levels can be seen in Table 3.

Table 2. Sample characteristics of the data collected with the digital learning platform I3Learn.

Sample characteristic		N	Percentage
Gender	Male	247	45.7%
	Female	246	45.5%
	Divers	23	4.3%
	Not specified	25	4.5%
School	IGS	169	31.6%
	KGS	77	14.4%
	High school	279	52.2%
	Other	9	1.7%
Grade	9	317	59.5%
	10	216	40.5%
Language at home	German	468	87.6%
	Other	66	12.4%

Table 3. Dropout of learners in the process of learning with I3Learn.

Dropout level	Definition	% dropped	# datapoints
Level 1	Pre-test	25.19%	532
Level 2	Repetition	9.80%	398
Level 3	Chapter 1	12.81%	359
Level 4	Chapter 2	8.31%	313
Level 5	Chapter 3	14.28%	287
Level 6	Post Test	21.14%	246

6 Results

6.1 RQ 1: Transfer of Methods for Dropout Prediction

The results in Table 4 show that both personal and interaction data can predict the dropout of learners in I$_3$Learn. Personal features perform better in predicting early-stage dropout. This confirms previous results [16] where a similar effect appears since these features (i.e., school among the demographic features and cognitive ability among the learning-related features) have more information than behavioral features early on in the learning process (the cold-start problem).

Table 4. Comparison of the individual feature categories achieved weighted F1 scores for each dropout level. The best category is highlighted.

Feature Categories	Dropout Level					
	Level1	Level 2	Level 3	Level 4	Level 5	Level 6
Personal Information						
Demographic	0.733	**0.618** (GBC)	0.485	0.674	0.531	**0.745** (GBC)
Learning-related	**0.868** (KNN)	0.395	0.457	0.517	0.452	0.615
Interaction Data						
Clickstream	0.663	0.551	0.612	0.681	0.528	0.646
Learning process-related	0.780	0.560	**0.710** (LR)	**0.695** (LightGBM)	**0.652** (RF)	0.614

Once the cold-start problem is overcome with enough interaction data, we start to see that learning-process-related features outperform the more indiscriminate clickstream features at earlier stages of dropout. This shows that explainable natural features perform better only when less information is available. Their performance is comparable at later dropout stages, meaning the more detailed features add little in predictiveness while reducing explainability for this data. Concerning dropout level 6, the demographic features have a higher predictive power again, with the "school" feature standing out. The results suggest that dropout prediction can be implemented for a classroom-based usage scenario of I$_3$Learn, which differs in the already discussed characteristics (Sect. 3) from prediction of institutional or MOOC dropout.

6.2 RQ 2: Dropout Prediction with Data of Diverse Granularity

The results in Table 5 show that finer-grained features can predict learner dropout more effectively. It should be noted that these results refer to the dropout in dropout level 3 - Chapter 1, which was selected because it was the first target for which interaction data such as these performed adequately. Furthermore, structure-delimited and time-delimited features can be compared. The structure-delimited perform better than the time-delimited features (Table 6).

Table 5. Weighted F1 scores of the features which are captured at multiple grain sizes for dropout level 3.

Grain	Features				
	Count of interactions	Duration of interactions	Position of interactions	Revision Count	Technical Issues
Activity-Step	0.478	**0.567** (RF)	**0.515** (RF)	**0.619** (GBC)	0.641 0.641
Activity	**0.707** (LR)	0.559	0.373	0.407	**0.695** (LR)
Module	0.417	0.395	0.378	0.360	0.473
Overall	0.445	0.295	N/a	0.356	0.200

Table 6. Weighted F1 scores of the structure- and time-delimited features.

Feature categories	Dropout level					
	Level 1	Level 2	Level 3	Level 4	Level 5	Level 6
Structure-delimited						
Count of interactions	0.637	0.524	0.612	0.567	0.554	0.578
Duration of interactions	**0.671** (RF)	**0.551** (GBC)	0.557	0.681	0.480	**0.646** (RF)
Position of interactions	0.600	0.465	0.509	0.515	0.456	0.581
Revision count	0.660	0.446	**0.710** (LR)	**0.695** (LightGBM)	**0.652** (RF)	0.614
Technical issues	0.667	0.541	0.637	0.619	0.598	0.600
Time-delimited						
Activity per week	0.662	0.367	0.335	0.465	0.445	0.485
Activity per day	0.621	0.449	0.372	0.328	0.331	0.468

6.3 RQ 3: Assessments for Predicting Dropout

The results from research questions 1 and 2 have shown that features based on personal information are best suited to predict the first two dropout levels. A breakdown of the F1 scores of the individual features in the personal category shows that the cognitive ability test used is the most effective for level 1 dropouts, and the school is the best predictor for level 2 (Table 7). However, the baseline performance for level 2 is low due to a highly skewed class distribution (see Table 3). Thus, focusing on only Level 1 dropout, cognitive ability tests may be a good predictor of dropout when behavioral data does not yet exist.

Table 7. Comparison of F1 scores attained with each feature for each dropout level.

Features	Dropout level 1	Dropout level 2
Answers to evaluation question	0.488	0.200
Class size	0.676	0.453
Cognitive ability	**0.868**	0.367
Gender	0.438	0.381
Grade	0.333	0.347
Knowledge assessment	0.686	0.395
Language at home	0.325	0.516
Last chemistry grade	0.333	0.347
Reading ability	0.401	0.276
School	0.733	**0.618**
School type	0.617	0.398
Self-concept chemistry	0.503	0.399
Socio-economic background	0.581	0.372

7 Conclusion

The Covid-19 pandemic and the resulting emergency remote learning reveal the need for flexible Web-based learning platforms, such as I$_3$Learn. The results show that for a learning platform designed for use in schools, and thus significantly different from the widely used MOOCs, the dropout of learners at different levels can be defined, and features determined that have the strongest predictive power for dropout at different levels. Furthermore, it is shown that with an increase in information about the learning process, both through a higher degree of granularity and longer learning time, the effectiveness of interaction data for predicting dropout increases. Lack of information at the beginning of the learning process can be mediated by data from assessments, in our case, mainly a cognitive assessment. Comparing the structured- and time-delimited features showed that the structured-delimited performed better to predict dropout. Intuitively, this could be due to the embedding in a classroom setting. Students are compelled by the teacher to access the platform and to show some level of activity. Information on how students interact with specific resources (i.e. length of interaction, revision activities), seem to be more indicative of learning progress and hence, lead to more accurate predictions of dropout.

Web-based learning platforms such as I$_3$Learn can enhance future classrooms. To date, they typically do not collect personal information of students. Our analysis has shown, however, that this information can be helpful to detect potential dropout in the beginning of a technology-based learning process, when individual interaction data are not yet available. This aspect should be considered when balancing data economy and data quality. Previous work [23] pointed out a research gap with respect to the behaviour of learners in UDL-based learning platforms specifically. I_3Learn and the data collected is a first step to bridge this gap, and to gain insights in how UDL may influence learner behavior. Moreover, adapted dropout prediction algorithms may enable us to detect barriers to learning at a data-based level: Currently, identifying such barriers is a labour-intensive process for educators and researchers. An approach based on behavioral data and learning analytics might facilitate (semi-)automatic methods, leading to scalable solutions to universal learning support and equality.

Limitations of this work lie in the dependencies of the analyzed features to other, not captured ones. The "school" features has, for instance, a high F1 score compared to the captured assessment features. It is, however, possible that the influencing factor is not the school itself, but the teacher that guides the use of I_3Learn in the classroom. Furthermore, we still investigate the possibility of other, unintentionally surveyed factors in the assessments: We used the figure analogy scale as a cognitive ability test. Since the test is challenging, there is a possibility that learners who complete it will also have a higher tendency to sucessfully complete the other learning opportunities in I_3Learn.

Acknowledgements. This work has been supported by the PhD training program LernMINT funded by the Ministry of Science and Culture, Lower Saxony, Germany.

References

1. Arnold, K.E., Pistilli, M.D.: Course signals at Purdue: using learning analytics to increase student success. In: ACM International Conference Proceeding Series, pp. 267–270 (2012). https://doi.org/10.1145/2330601.2330666
2. Azhar, N., Ahmad, W.F.W., Ahmad, R., Bakar, Z.A.: Factors affecting the acceptance of online learning among the urban poor: a case study of Malaysia. Sustainability (Switzerland) **13**(18) (2021). https://doi.org/10.3390/su131810359
3. Baker, R.: Using learning analytics in personalized learning. In: Murphy, M., Redding, S., Twyman, J. (eds.) Handbook on Personalized Learning for States, Districts, and Schools, pp. 165–174 (2016)
4. Barke, H.D., Pieper, C.: Der ionenbegriff - historischer spätzünder und gegenwärtiger außenseiter. Chemkon **15**(3), 119–124 (2008). https://doi.org/10.1002/ckon.200810075
5. CAST: Universal design for learning guidelines version 2.2 (2018)
6. Dalipi, F., Imran, A.S., Kastrati, Z.: MOOC dropout prediction using machine learning techniques: review and research challenges. In: IEEE Global Engineering Education Conference, EDUCON, vol. 2018-April, pp. 1007–1014. IEEE Computer Society, May 2018. https://doi.org/10.1109/EDUCON.2018.8363340
7. Elert, T.: Course Success in the Undergraduate General Chemistry Lab, Studien zum Physik- und Chemielernen, vol. 184 (2019)
8. Heller, K.A., Perleth, C.: Kognitiver Fähigkeitstest für 4. bis 12. Klassen, Revision. Beltz Test (2000)

9. Hilbing, C., Barke, H.D.: Ionen und ionenbindung: Fehlvorstellungen hausgemacht! ergebnisse empirischer erhebungen und unterrichtliche konsequenzen. CHEMKON **11**(3), 115–120 (2004). https://doi.org/10.1002/ckon.200410009
10. Hodges, C., Moore, S., Lockee, B., Trust, T., Bond, A.: The difference between emergency remote teaching and online learning (2020). https://er.educause.edu/articles/2020/3/the-difference-between-emergency-remote-teaching-and-online-learning
11. Ifenthaler, D., Yau, J.Y.-K.: Utilising learning analytics to support study success in higher education: a systematic review. Educ. Tech. Res. Dev. **68**(4), 1961–1990 (2020). https://doi.org/10.1007/s11423-020-09788-z
12. Kauffman, H.: A review of predictive factors of student success in and satisfaction with online learning. Res. Learn. Technol. **23** (2015). https://doi.org/10.3402/rlt.v23.26507
13. Kloft, M., Stiehler, F., Zheng, Z., Pinkwart, N.: Predicting MOOC dropout over weeks using machine learning methods. In: Proceedings of the EMNLP 2014 Workshop on Analysis of Large Scale Social Interaction in MOOCs, pp. 60–65. Association for Computational Linguistics, Doha, Qatar, October 2014. https://doi.org/10.3115/v1/W14-4111
14. Lodge, J.M., Corrin, L.: What data and analytics can and do say about effective learning. NPJ Sci. Learn. **2**(1) (2017). https://doi.org/10.1038/s41539-017-0006-5
15. Luxford, C.J., Bretz, S.L.: Development of the bonding representations inventory to identify student misconceptions about covalent and ionic bonding representations. J. Chem. Educ. **91**(3), 312–320 (2014). https://doi.org/10.1021/ed400700q
16. Manrique, R., Nunes, B.P., Marino, O., Casanova, M.A., Nurmikko-Fuller, T.: An analysis of student representation, representative features and classification algorithms to predict degree dropout. In: ACM International Conference Proceeding Series, pp. 401–410. Association for Computing Machinery, March 2019. https://doi.org/10.1145/3303772.3303800
17. Mayringer, H., Wimmer, H.: Salzburger Lese-Screening für die Schulstufen 2–9 (SLS 2–9). hogrefe (2014)
18. Niedersächsisches Kultusministerium: Die niedersächsischen allgemein bildenden Schulen Zahlen und Grafiken (2021). https://www.mk.niedersachsen.de/startseite/service/statistik/die-niedersaechsischen-allgemein-bildenden-schulen-in-zahlen-6505.html
19. de Oliveira, C.F., Sobral, S.R., Ferreira, M.J., Moreira, F.: How does learning analytics contribute to prevent students' dropout in higher education: a systematic literature review. Big Data Cogn. Comput. **5**(4) (2021). https://doi.org/10.3390/bdcc5040064
20. Parkes, M., Gregory, S., Fletcher, P., Adlington, R., Gromik, N.: Bringing people together while learning apart: creating online learning environments to support the needs of rural and remote students. Australian Int. J. Rural Educ. **25**(1), 66–78 (2015). https://doi.org/10.3316/aeipt.215238
21. Patricia Aguilera-Hermida, A.: College students' use and acceptance of emergency online learning due to COVID-19. Int. J. Educ. Res. Open **1** (2020). https://doi.org/10.1016/j.ijedro.2020.100011
22. Prenkaj, B., Velardi, P., Stilo, G., Distante, D., Faralli, S.: A survey of machine learning approaches for student dropout prediction in online courses. ACM Comput. Surv. **53**(3) (2020). https://doi.org/10.1145/3388792
23. Roski, M., Walkowiak, M., Nehring, A.: Universal design for learning: the more, the better? Educ. Sci. **11**, 164 (2021). https://doi.org/10.3390/educsci11040164, https://www.mdpi.com/2227-7102/11/4/164

24. Torsheim, T., et al.: Psychometric validation of the revised family affluence scale: a latent variable approach. Child Indic. Res. **9**(3), 771–784 (2015). https://doi.org/10.1007/s12187-015-9339-x
25. Tovar, E., et al.: Do MOOCS sustain the UNESCOs quality education goal? In: 2019 IEEE Global Engineering Education Conference (EDUCON), pp. 1499–1503 (2019)
26. UNESCO: Guidelines for Inclusion: Ensuring Access to Education for All (2005)
27. UNESCO-UNEVOC: Medium-Term Strategy for 2021–2023 (2021)
28. de la Varre, C., Irvin, M.J., Jordan, A.W., Hannum, W.H., Farmer, T.W.: Reasons for student dropout in an online course in a rural k-12 setting. Distance Educ. **35**(3), 324–344 (2014). https://doi.org/10.1080/01587919.2015.955259
29. Xenos, M., Pierrakeas, C., Pintelas, P.: A survey on student dropout rates and dropout causes concerning the students in the course of informatics of the Hellenic open university. Comput. Educ. **39**(4), 361–377 (2002). https://doi.org/10.1016/S0360-1315(02)00072-6

Designing for Student Understanding of Learning Analytics Algorithms

Catherine Yeh, Noah Cowit, and Iris Howley(✉)

Williams College, Williamstown, MA, USA
{cy3,nqc1,ikh1}@williams.edu

Abstract. Students use learning analytics systems to make day-to-day learning decisions, but may not understand their potential flaws. This work delves into student understanding of an example learning analytics algorithm, Bayesian Knowledge Tracing (BKT), using Cognitive Task Analysis (CTA) to identify knowledge components (KCs) comprising expert student understanding. We built an interactive explanation to target these KCs and performed a controlled experiment examining how varying the transparency of limitations of BKT impacts understanding and trust. Our results show that, counterintuitively, providing some information on the algorithm's limitations is not always better than providing no information. The success of the methods from our BKT study suggests avenues for the use of CTA in systematically building evidence-based explanations to increase end user understanding of other complex AI algorithms in learning analytics as well as other domains.

Keywords: Explainable artificial intelligence (XAI) · Learning analytics · Bayesian Knowledge Tracing

1 Introduction

Artificial Intelligence (AI) enhanced learning systems are increasingly relied upon in the classroom. Due to the opacity of most AI algorithms, whether from protecting commercial interests or from complexity inaccessible to the public, there is a growing number of decisions made by students and teachers with such systems who are not aware of the algorithms' potential biases and flaws. Existing learning science research suggests that a lack of understanding of learning analytics algorithms may lead to lowered trust, and perhaps, lower use of complex learning analytics algorithms [26]. However, in different educational contexts, research has evidenced that increased transparency in grading can lead to student dissatisfaction and distrust [15], so more information is not guaranteed to be better. Furthermore, opening and explaining algorithms introduces different issues, as cognitive overwhelm can lead to over-relying on the algorithm while failing to think critically about the flawed input data [24]. As a first step toward realizing this relationship between algorithm, user understanding, and outcomes, this work answers the following questions about BKT:

N. Wang et al. (Eds.): AIED 2023, LNAI 13916, pp. 528–540, 2023.
https://doi.org/10.1007/978-3-031-36272-9_43

- What are the knowledge components of algorithmic understanding for BKT?
- What factors impact successful learning with our interactive explanation?
- How does our explanation impact user attitudes toward BKT and AI?

To answer these questions, we first systematically identify the knowledge components (KCs) of BKT, then design assessments for those KCs. Next, we implement a post-hoc, interactive explanation of BKT using these KCs, evaluating our explanation in light of the assessments and user perspectives of the system. Finally, we run an additional experiment varying the amount of information participants are shown about BKT and measuring how this reduction in transparency impacts understanding and perceptions.

2 Prior Work

A student may use a learning analytics system displaying which skills they have mastered, and which they have not. This information can assist the student in determining what content to review, learning activities to pursue, and questions to ask. This information can also inform the learning analytics system of which practice problems to select, creating a personalized, intelligent tutoring system. Underneath the system display, there may be a Bayesian Knowledge Tracing algorithm predicting mastery based on student responses in addition to various parameters such as the likelihood of learning or guessing [2]. As an AI algorithm, BKT is inherently prone to biases and flaws [9]. Without a proper mental model for BKT, students may make decisions based on their own observations of system outputs, which may not be accurate. A sufficient understanding of the underlying algorithm may influence trust in the system as well as decision-making [14].

Researchers studying fairness, accountability, and transparency of machine learning (ML) view this topic from two angles: 1) where the ML algorithm can automatically produce explanations of its internal working, and 2) post-hoc explanations constructed after the ML model is built [19]. To achieve this second goal, researchers create **post-hoc explanations** to teach the concepts of particular algorithms [19]. Often, these explanations require the reader to have extensive prior knowledge of machine learning, despite a large proportion of algorithmic decision systems being used by non-AI/ML experts, such as healthcare workers and criminal justice officials. Additional ML work suggests using the basic units of cognitive chunks to measure algorithmic understanding [11]. A means to identify what knowledge experts rely on to understand complex algorithms is necessary to bridge this gap between post-hoc explanations for ML researchers and typical users, such as students with learning analytics systems.

Previous work has involved measuring and evaluating what it means to know a concept from within the learning sciences, but this question manifests itself somewhat differently in explainable AI (XAI) research. One approach is to adopt definitions from the philosophy of science and psychology to develop a generalizable framework for assessing the "goodness" of ML explanations. For example, in decision-making research, the impact of different factors on people's understanding, usage intent, and trust of AI systems is assessed via hypothetical scenario decisions [21]. Using pre-/post-tests to measure learning about AI algorithms is one possibility for bridging the ML and learning science approaches [27].

While increasing research explores different ways of evaluating post-hoc explanations in the ML community, it is not clear how XAI designers identify the concepts to explain. **Cognitive Task Analysis (CTA)** provides a rigorous conceptual map of what should be taught to users of ML systems by identifying the important components of algorithms according to existing expert knowledge [7].

Intelligent tutoring system design uses CTA to decompose content into the knowledge and sub-skills that must be learned as part of a curriculum [20]. Lovett breaks CTA down into 2×2 dimensions, the theoretical/empirical and the prescriptive/descriptive. Our study focuses on the *empirical/prescriptive* dimension of CTA, where a think aloud protocol is used as experts solve problems pertaining to the domain of interest (e.g., a particular algorithm). We chose to leverage a form of expert CTA, as studying expertise elucidates what the results of "successful learning" look like and what kinds of thinking patterns are most effective and meaningful for problem-solving [20]. Ultimately, these results from CTA can be used to design more effective forms of instruction for novices, such as explanations of learning analytics algorithms.

The knowledge and skills revealed by CTA are called **knowledge components** or **KCs**. KCs are defined as "an acquired unit of cognitive function or structure that can be inferred from performance on a set of related tasks" [16]. In this paper, we use CTA to systematically identify the different knowledge components that comprise the AI algorithm, Bayesian Knowledge Tracing, which are ultimately evaluated through observable assessment events.

Bayesian Knowledge Tracing (BKT) models students' knowledge as a latent variable and appears in Technology Enhanced Learning systems such as the Open Analytics Research Service [5]. BKT predicts whether a student has mastered a skill or not (either due to lack of data or low performance) using four parameters: `P(init)`, `P(transit)`, `P(guess)`, and `P(slip)`. In practice, these parameters are fit through a variety of methods [2] and may be shared across an entire class of students [9]. Additionally, `P(transit)`, `P(guess)`, and `P(slip)` are often not updated, remaining at their preset initial values [2]. BKT updates its estimates of mastery, `P(init)`, as a student proceeds through a lesson [2].

As a probabilistic algorithm, BKT falls subject to certain biases and limitations. For example, model degeneracy occurs when BKT does not work as expected due to its initial parameter values being outside an acceptable range [9]. BKT's parameters also do not account for certain events, such as forgetting [8] or the time it takes a student to answer a question, which are relevant and important to consider when assessing learning and mastery.

BKT is a sufficiently complex algorithm as to not be easily understood, but also sufficiently explainable as the parameters and how they interact are all known. While we use BKT as our algorithm of interest for this study, it is possible to apply these same methods of examination to other learning analytics algorithms that students and teachers may find difficult to understand.

3 Knowledge Components of BKT

We conducted a CTA to gain knowledge about expert understanding with respect to our selected AI algorithm, BKT. We examine BKT from the student perspective, as they represent one of BKT's target user groups. Our CTA protocol involves interviewing student experts of BKT and having them think aloud and step-through various scenarios that may be encountered when using a BKT system. This is analogous to the approach described in [20], which uses CTA for the design of intelligent tutoring systems for mathematics.

The participants in this study were seven undergraduate students at a rural private college who previously studied BKT as part of past research experiences and in some cases had implemented small-scale BKT systems. Interviews were semi-structured with a focus on responding to ten problems and lasted 30–60 minutes in duration. By recording comprehensive, qualitative information about user performance during these interviews [7], we were able to identify the knowledge components of student BKT expertise.

We developed our own BKT scenarios, as identifying problems for experts to solve is less straightforward than identifying problems for statistics experts to solve as in [20]. We adapted our approach from Vignette Survey design [1] to generate BKT problems within the context that experts ordinarily encounter. Vignette Surveys use short scenario descriptions to obtain individual feedback [1]. Additionally, the numerous social indicators of BKT parameters and weighing of subjective factors necessary for model evaluation make vignettes an optimal tool for this study. For instance, a lack of studying, sleep, or prior knowledge can all lead to a low starting value of `P(init)`.

For each scenario, we constructed a vignette describing background information about a hypothetical student followed by one or more questions regarding BKT (e.g., "Amari loves debating. They are very well spoken in high school debate club. Although Amari's vocabulary is impressive, they often have difficulty translating their knowledge into their grades. For example, Amari gets flustered in their high school vocab tests and often mixes up words they would get correct in debate. These tests are structured in a word bank model, with definitions of words given the user must match to a 10-question word bank. (1) What do you think are reasonable parameters for BKT at the beginning of one of these vocab tests? Please talk me through your reasoning..."). We were not only interested in comprehension of BKT's parameters and equations, but also the context in which BKT systems are used. Thus, our CTA protocol includes additional details such as test anxiety and other potential student differences that may create edge cases for interpretations of BKT output.

Our data was compiled after interviews were completed. First, the initial and final states (i.e., the given information and goal) for each scenario were identified. Questions with similar goals were grouped together, forming broader knowledge areas (e.g., "Identifying Priors"). Next, each participant's responses were coded to identify the steps taken to achieve the goal from the initial state. Then, we identified common steps used in each scenario. Final knowledge components were created by matching similar or identical processes from questions in the same knowledge area. If a certain step was taken by the majority of participants but not all, we denote it as an "optional" KC by using *italics.*

We ultimately divided our analysis of BKT into four discrete but related knowledge areas: (1) **Identifying Priors**, (2) **Identifying Changed Parameters**, (3) **Evaluating P(init)**, and (4) **Limitations of BKT**. Each knowledge area consisted of 4–5 knowledge components, resulting in a total of 19 KCs:

Identifying Priors concerns the processing of subjective vignettes into reasonable numerical values for the four initial parameters of BKT.

1. Recall range of "normal values" and/or definitions for the parameter in question. This may involve recognizing (implicitly or explicitly) what `P(init / transit / guess / slip)` is and how it is calculated.
2. Synthesize (summarize or process) information from vignette, identifying specific evidence that is connected to the parameter in question.
3. *Consider BKT's limitations & how this could impact this parameter's value.*
4. Make an assessment about the parameter in question based on this qualitative evidence (or lack thereof).
5. Choose a parameter value by converting to a probability between 0 and 1.

Identifying Changed Parameters focuses on the direction of change in parameter values (if any).

1. Consider the prior parameter level of `P(init / transit / guess / slip)`.
2. Synthesize new information given, identifying specific evidence that suggests a change in parameter value (or a lack thereof).
3. Make an assessment about the parameter in question based on this qualitative evidence (or lack thereof).
4. Decide direction of change (increase, decrease, or stays the same).
5. If prompted, choose a new parameter value by converting assessment to a probability between 0 and 1.

Evaluating P(init) addresses how the parameter `P(init)` is essential for evaluating practical applications of BKT. Sometimes experts arrived at different answers for these more open-ended problems, but our participants typically followed similar paths to arrive at their respective conclusions.

1. Synthesize information from vignette, considering parameter level of `P(init)`.
2. Make a judgment as to the magnitude of `P(init)` (e.g., low, moderate, high, moderately high, etc.).
3. Consider magnitude with respect to the situation and BKT's definition of mastery. Some situations call for a very high level of knowledge-and thus a very high `P(init)` (e.g., space travel), while in other situations, a moderate level of knowledge is acceptable (e.g., a high school course).
4. Take a stance on the question. Often: "With this value of `P(init)`, has X achieved mastery?" or "...is 0.4 a reasonable value for `P(init)`?"
5. *Explain why BKT's predictions might not be accurate in this case due to its limitations, probabilistic nature, etc.*

Limitations of BKT covers three limitations of BKT within this protocol: model degeneracy [9], additional non-BKT parameters such as time taken and forgetfulness between tests, and the probabilistic nature of BKT. In many cases, these problems also related to the "Evaluating P(Init)" knowledge area.

1. Synthesize information from vignette, identifying any "irregular" pieces of information (e.g., anything that's relevant to learning/mastery but not encompassed by the standard 4 BKT parameters, like whether a student is being tested before or after their summer vacation).
2. *If relevant, consider previous parameter values.*
3. Experiment with irregular information and consider limitations of BKT. This often involved asking open ended questions about learning/mastery.
4. Make a statement about BKT's analysis (correct or not correct, sensible/intuitive or not, etc.), or answer the posed question(s) accordingly, after determining that BKT does not account for this irregular information.

4 The BKT Interactive Explanation

With KCs established, we designed our BKT explanation using principles from user-centered design[1]. Following this iterative design process, we went through several cycles of brainstorming, prototyping, testing, and revising. Our final explanation is an interactive web application that uses American Sign Language to motivate and illustrate the behavior of BKT systems (Fig. 1).

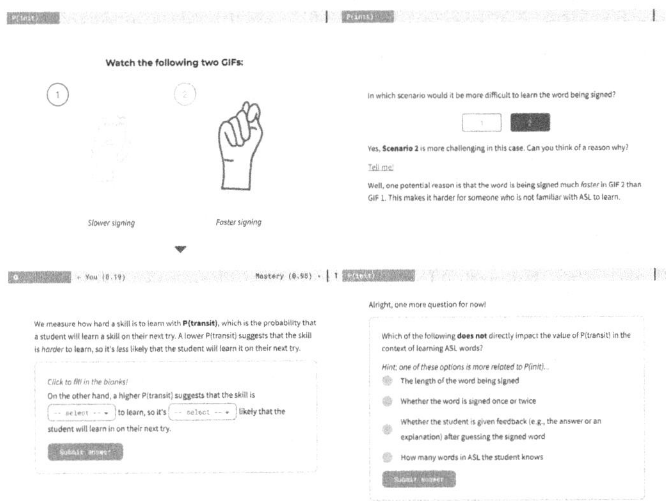

Fig. 1. P(transit) Module from BKT Explanation, artwork by Darlene Albert (https://darlenealbert.myportfolio.com/)

[1] https://dschool.stanford.edu/resources/design-thinking-bootleg.

Along with the pedagogical principles of Backward Design [25], we made additional design decisions following best practices in learning and instruction, active learning, and self-explanation in particular. "Learning by doing" is more effective than passively reading or watching videos [17], and so our explanation is interactive with immediate feedback which research shows leads to increased learning. To ensure that we targeted the BKT KCs identified previously, we mapped each activity in our explanation to its corresponding KCs.

After learning about all four parameters, we bring the concepts together for a culminating mini game in which participants practice their ASL skills by identifying different finger-spelled words until they achieve mastery. Following the BKT mini game are four modules to teach BKT's flaws and limitations: (1) "When do you lose mastery?", (2) "What Causes Unexpected Model Behavior?", (3) "How do Incorrect Answers Impact Mastery?", and (4) "What is the Role of Speed in BKT?". This is essential to our goal of encouraging deeper exploration of the algorithm and helping users develop realistic trust in BKT systems. For example, in our "When Do You Lose Mastery?" module, participants are asked to assess the magnitude of `P(init)` at different points in time. This module demonstrates how BKT does not account for forgetting, which may bias its estimates of mastery.

To implement our interactive explanation, we created a web application coded in JavaScript, HTML, and CSS. We iteratively tested and revised our implementation with participants, until we reached a point of diminishing returns in which no new major functionality issues arose. The final design is a dynamic, interactive, publicly accessible explanation: https://catherinesyeh.github.io/bkt-asl/.

After the implementation phase, we assessed the effectiveness of our BKT explanation with a formal user study. We designed pre- and post-tests to accompany our BKT explanation in a remote format. Our pre-test consisted mainly of questions capturing participant demographics and math/computer science (CS) background. Prior work suggests that education level impacts how users learn from post-hoc explanations [27], so we included items to assess participant educational background and confidence. Math/CS experience questions were adapted from [13] using Bandura's guide for constructing self-efficacy scales [3]. We also collected self-reported familiarity with Bayesian statistics/BKT systems and general attitudes toward AI; these questions were based on prior work [10].

Our post-test questions were inspired by [22], which outlines evaluation methods for XAI systems. To evaluate user *mental models* of BKT, our post-test includes questions specifically targeting our BKT KCs, such as Likert questions like: "BKT provides accurate estimates of skill mastery." Many of our other questions were similar to the vignette-style problems included in our CTA protocol, mirroring the scenarios we present to participants throughout the explanation. To measure *usability and user satisfaction*, we adapted questions from the System Usability Scale [4] and similar scales [3,12,23]. We also measured *user trust* with a modified version of the 6-construct scale from [6]. Finally, we compared *user attitudes* toward AI algorithms more generally before and after completing our explanation using the same set of questions from our pretest [10].

User study participants were nine undergraduate students from the same rural private college as above. Each participant completed a pre-test, stepped-through the explanation, and ended with a post-test. We do not go in-depth into this user study here, as our follow-up experiment uses the same measures with a larger sample size. Preliminary results suggest that any participant can learn from our BKT explanation regardless of their math/CS background, and that users received our BKT explanation positively. Satisfied with this initial evaluation, we moved to the next stage of this work: examining how the information in the interactive explanation impacts user understanding and other outcomes.

5 Impact of Algorithmic Transparency on User Understanding and Perceptions

With an effective post-hoc explanation of BKT, we can answer the following research questions: **RQ1:** How does decreasing the transparency of BKT's limitations affect algorithmic understanding? **RQ2:** How does decreasing the transparency of BKT's limitations affect perceptions of algorithmic fairness and trust? As these questions focus on trust and other user outcomes in decision-making situations, the most impactful factor of this is likely related to the user's understanding of the algorithm's limitations or flaws. And so, we designed a controlled experiment examining how three levels of information about BKT's limitations impact user perceptions of BKT vs. Humans as the decider of a hypothetical student's mastery in high-stakes and low-stakes evaluation circumstances.

To vary the amount of information provided about BKT's limitations, we included three explanation conditions. The *Long Limitations* condition included the original four limitations modules at the end of the BKT explanation. The *Short Limitations* condition reduced the multiple pages and interactive activities with a text summary and images or animations illustrating the same concepts. The *No Limitations* condition had none of the limitations modules. Participants were randomly assigned to an explanation condition.

Similar to prior work on user perceptions of fairness of algorithmic decisions [18], we developed scenarios to examine algorithmic understanding's impact on user outcomes. In our case, we were interested in low-stakes and high-stakes situations involving BKT as the decision-maker, as compared to humans. Each scenario had a general context (i.e., "At a medical school, first-year applicants must complete an entrance exam. To be recommended for admission, applicants must score highly on this exam. An AI algorithm assesses their performance on the entrance exam.") and a specific instance (i.e., "Clay applies to the medical school. The AI algorithm evaluates their performance on the entrance exam."). These decision scenarios were added to our post-test previously described, along with measures from prior work asking about the fairness of each decision [18].

We recruited 197 undergraduate students from across the United States, 74 of which completed the pre- and post-tests satisfactorily. Of these, 50% identified as female, 43% male, 5% other, and 2% did not respond. 82% reported being from the USA, with the remainder representing most other inhabited continents.

47% reported a major in math or engineering, and the rest were a mix of social & natural science, humanities, business, and communication. We later dropped 10 of these respondents due to outlying survey completion times or re-taking the survey after failing an attention check. 24 participants were assigned to the Long Limitations condition, 21 to Short, and 19 to No Limitations.

How Does Decreasing the Transparency of BKT's Limitations Affect Algorithmic Understanding? There was no statistically significant relationship of time participants spent on the study and their post-test scores, nor was there a statistically significant difference of explanation conditions on time spent. Participants in the Long Limitations condition ($\mu = 0.94$, $\sigma = 0.09$) had a higher average score on the **Limitations of BKT** knowledge area than the Short Limitations group ($\mu = 0.92$, $\sigma = 0.08$), which in turn had a higher score than the No Limitations group ($\mu = 0.88$, $\sigma = 0.1$). As understanding of BKT's limitations will depend on a more general understanding of BKT, we conducted a one-way ANCOVA to identify a statistically significant difference between explanation condition on learning in the Limitations of BKT knowledge area, controlling for performance in the three other knowledge areas, $F(3, 64) = 3.85$, $p < 0.05$. A Student's t-test shows that the No Limitations condition performed significantly worse on the Limitations of BKT knowledge area as compared to the other two explanation conditions. This suggests that our manipulation was mostly effective at impacting participant understanding of BKT's limitations.

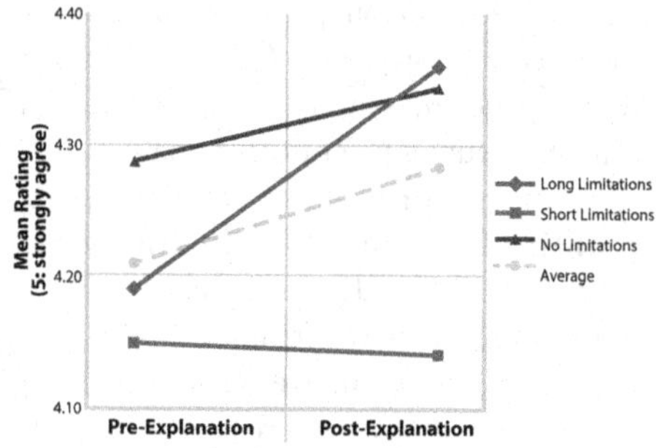

Fig. 2. Changes in Average Participant Attitudes Toward AI

How Does Decreasing the Transparency of BKT's Limitations Affect Perceptions of Algorithmic Fairness and Trust? There was a significant interaction between experimental condition and initial attitudes about AI on final attitudes about AI, $F(2, 64) = 3.21$, $p < 0.05$, as shown in Fig. 2.

All of our additional self-reported perceptions of the explanation design ($F(2, 64) = 10.77$, $p < 0.0001$), explanation effectiveness ($F(2, 64) = 4.89$, $p < 0.05$), and trust in the BKT algorithm ($F(2, 64) = 3.96$, $p < 0.05$) show a statistically significant effect of explanation condition. In all three of these cases, a Student's

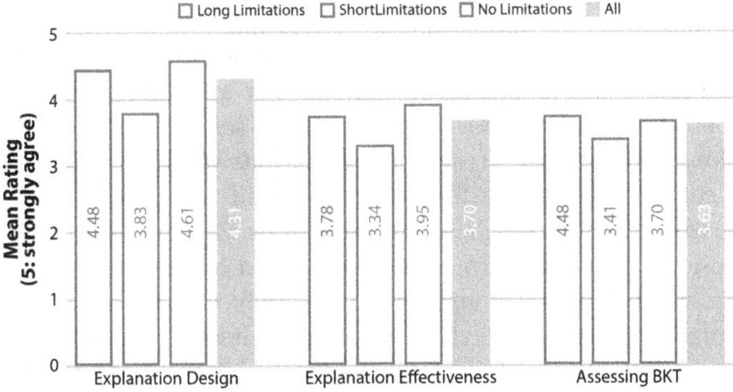

Fig. 3. Perceptions of Explanation Design, Explanation Effectiveness, and BKT Trust

t-test shows that the Short Explanation condition has a significantly lower mean than the other two conditions, as shown in Fig. 3.

We did not find significant results for the low/high-stakes X human/AI as the decision-maker questions. We likely need more than one question per category, or possibly longer exposure to BKT to measurably impact decision-making. However, in all cases, the Short Limitations condition had lower means than the other two conditions. This aligns with our prior results.

These results show that less information about an algorithm is not always worse. Participants in the Short Limitations group did not experience a positive increase in general attitudes about AI, unlike the Long and No Limitations groups, and the Short Limitations condition also perceived the BKT algorithm significantly less positively than the other two conditions. Despite the fact that the Short Limitations participants learned significantly more about BKT's limitations than the No Limitations condition, students in our middle-level information condition appear to have a significantly less positive perception of both our specific AI and AI more generally. For designers of interactive AI explanations, understanding how the design of an explanation impacts user perceptions is critical, and our work on student understanding of the learning algorithms they use provides a method for investigating that relationship.

6 Conclusion

This work shows that CTA can identify the necessary components of understanding a learning analytics algorithm and therefore, the necessary learning activities of an interactive explanation of the algorithm. Understanding the algorithm underlying learning analytics systems supports users in making informed decisions in light of the algorithm's limitations. Our CTA results identify four main knowledge areas to consider when explaining BKT: (1) **Identifying Priors**, (2) **Identifying Changed Parameters**, (3) **Evaluating P(init)**, and (4) **Limitations of BKT**. We then varied the length of the limitations module in the implementation of an interactive BKT explanation. Results revealed that using a limitations module with reduced information can have surprising effects, mostly, a less than statistically expected impact on general perceptions of AI, as well as on perceptions of the learning algorithm itself.

Limitations of this work arise from limitations of the methods. The think aloud protocol for CTA shares limitations with all think aloud protocols: as a method for indirectly observing cognitive processes which are not directly observable, it is possible that some processes were missed by the think aloud protocol. Additionally, while these KCs apply to BKT, they may not generalize directly to another algorithm, although the CTA method itself certainly does extend to other contexts. The Short Limitations condition did not learn significantly less on the BKT limitations post-test as compared to the Long Limitations section, and so our results looking at explanation condition are likely capturing an effect based on more than just algorithmic understanding. Our posttest measures of high/low stakes human/AI decision makers only tested one decision scenario of each type, and needs to be expanded to be more generalizable. Furthermore, students are not the only users of learning analytics. Teachers are also important stakeholders, so next steps include repeating the process for instructor users of BKT. This information can be used to decide whether different explanations should be constructed for different stakeholders, or if a more general AI explanation could suffice, given that user goals are sufficiently aligned.

Our findings also inform future work involving other complex algorithms, with the larger goal of measuring how user understanding affects system trust and AI-aided decision-making processes. This process of applying CTA methods to identify important expert concepts that novices should learn about an algorithm, designing explanatory activities to target each KC, and then evaluating knowledge acquisition and shifts in decision-making patterns connected to each KC provides a generalizable framework for building evidence-based post-hoc AI explanations that are accessible even to non-AI/ML experts.

References

1. Atzmüller, C., Steiner, P.: Experimental vignette studies in survey research. Methodology **6**(3), 128–138 (2010)
2. Baker, R.S.J., Corbett, A.T., Aleven, V.: More accurate student modeling through contextual estimation of slip and guess probabilities in Bayesian knowledge tracing. In: Woolf, B.P., Aïmeur, E., Nkambou, R., Lajoie, S. (eds.) ITS 2008. LNCS, vol. 5091, pp. 406–415. Springer, Heidelberg (2008). https://doi.org/10.1007/978-3-540-69132-7_44
3. Bandura, A.: Guide for constructing self-efficacy scales. In: Self-Efficacy Beliefs of Adolescents, pp. 307–337. Information Age, USA (2006)
4. Bangor, A., Kortum, P., Miller, J.: An empirical evaluation of the system usability scale. Int. J. Hum.-Comput. Interact. **24**(6), 574–594 (2008)
5. Bassen, J., Howley, I., Fast, E., Mitchell, J., Thille, C.: Oars: exploring instructor analytics for online learning. In: Proceedings of the ACM Conference on L@S, p. 55. ACM (2018)

6. Berkovsky, S., Taib, R., Conway, D.: How to recommend? User trust factors in movie recommender systems. In: Proceedings of the International Conference on Intelligent User Interfaces. pp. 287–300. ACM (2017)
7. Clark, R., Feldon, D., van Merriënboer, J., Yates, K., Early, S.: Cognitive task analysis. In: Handbook of Research on Educational Communications & Technology, pp. 577–593. Routledge, USA (2008)
8. Doroudi, S., Brunskill, E.: The misidentified identifiability problem of Bayesian knowledge tracing. In: Proceedings of the International Conference on Educational Data Mining, pp. 143–149. International Educational Data Mining Society (2017)
9. Doroudi, S., Brunskill, E.: Fairer but not fair enough on the equitability of knowledge tracing. In: Proceedings of the International Conference on LAK, pp. 335–339. ACM (2019)
10. Dos Santos, D.P., et al.: Medical students' attitude towards artificial intelligence: a multicentre survey. Eur. Radiol. **29**(4), 1640–1646 (2019)
11. Doshi-Velez, F., Kim, B.: Towards a rigorous science of interpretable machine learning (2017). https://doi.org/10.48550/ARXIV.1702.08608
12. Fu, F.L., Su, R.C., Yu, S.C.: EGameFlow: a scale to measure learners' enjoyment of e-learning games. Comput. Educ. **52**(1), 101–112 (2009)
13. Hutchison, M., Follman, D., Sumpter, M., Bodner, G.: Factors influencing the self-efficacy beliefs of first-year engineering students. J. Eng. Educ. **95**(1), 39–47 (2006)
14. Khosravi, H., et al.: Explainable artificial intelligence in education. Comput. Educ. Artif. Intell. **3**, 100074 (2022)
15. Kizilcec, R.: How much information? Effects of transparency on trust in an algorithmic interface. In: Proceedings of the SIGCHI Conference on Human Factors in Computing Systems, pp. 2390–2395. ACM (2016)
16. Koedinger, K., Corbett, A., Perfetti, C.: The knowledge-learning-instruction framework: bridging the science-practice chasm to enhance robust student learning. Cogn. Sci. **36**(5), 757–798 (2012)
17. Koedinger, K., Kim, J., Jia, J.Z., McLaughlin, E., Bier, N.: Learning is not a spectator sport: doing is better than watching for learning from a MOOC. In: Proceedings of the ACM Conference on Learning at Scale, pp. 111–120. ACM (2015)
18. Lee, M.K.: Understanding perception of algorithmic decisions: fairness, trust, & emotion in response to algorithmic management. Big Data Soc. **5**(1) (2018)
19. Lipton, Z.: The mythos of model interpretability. ACM Queue **16**(3), 31–57 (2018)
20. Lovett, M.C.: Cognitive task analysis in service of intelligent tutoring system design: a case study in statistics. In: Goettl, B.P., Halff, H.M., Redfield, C.L., Shute, V.J. (eds.) ITS 1998. LNCS, vol. 1452, pp. 234–243. Springer, Heidelberg (1998). https://doi.org/10.1007/3-540-68716-5_29
21. Lu, J., Lee, D., Kim, T.W., Danks, D.: Good explanation for algorithmic transparency. In: Proceedings of the AAAI/ACM Conference on AIES, p. 93. ACM (2020)
22. Mohseni, S., Zarei, N., Ragan, E.: A multidisciplinary survey & framework for design & evaluation of explainable AI systems. ACM TiiS **11**(3–4), 1–45 (2021)
23. Phan, M., Keebler, J., Chaparro, B.: The development & validation of the game user experience satisfaction scale. Hum. Factors **58**(8), 1217–1247 (2016)
24. Poursabzi-Sangdeh, F., Goldstein, D., Hofman, J., Wortman Vaughan, J., Wallach, H.: Manipulating & measuring model interpretability. In: Proceedings of the SIGCHI Conference on Human Factors in Computing Systems, pp. 1–52. ACM (2021)

25. Wiggins, G., Wiggins, G., McTighe, J.: Understanding by Design. Association for Supervision & Curriculum Development, USA (2005)
26. Williamson, K., Kizilcec, R.: Effects of algorithmic transparency in BKT on trust & perceived accuracy. International Educational Data Mining Society (2021)
27. Zhou, T., Sheng, H., Howley, I.: Assessing post-hoc explainability of the BKT algorithm. In: Proceedings of the AAAI/ACM Conference on AIES, pp. 407–413. ACM (2020)

Feedback and Open Learner Models in Popular Commercial VR Games: A Systematic Review

YingAn Chen[1(✉)], Judy Kay[1], and Soojeong Yoo[2]

[1] The University of Sydney, Sydney, NSW 2006, Australia
yche3373@uni.sydney.edu.au

[2] University College London, London WC1E 6BT, UK

Abstract. Virtual reality (VR) educational games are engaging as VR can enable high levels of immersion and presence. However, for effective learning, educational literature highlights the benefits if such games have effective feedback. We aimed to understand the nature of feedback provided in popular commercial VR educational games. To discover this, we systematically reviewed 260 commercially available educational games from VIVEPORT, Oculus and Steam, the key platforms for VR games. We assessed if they offered key forms of feedback we identified from the literature, score, levels, competition, self, self-comparison, till correct, accuracy, process, and Open Learner Models (OLMs). We found that just 67 games (26%) had any of these forms of feedback and just four had OLMs. Our key contributions are: (1) the first systematic review of feedback in commercial VR games; (2) literature-informed definition of key forms of feedback for VR games; (3) understanding about OLMs in commercial VR games.

Keywords: Feedback · Open Learner Model · Virtual Reality Educational Games · Systematic Review

1 Introduction

Virtual Reality (VR) technology may provide important benefits for learning by allowing users to explore a virtual environment [4,7,23,30]. Learners can visit and explore arbitrary places and work through tasks in simulated scenarios. VR's immersiveness can provide a very engaging learning environment. It can enable learning in simulations of dangerous and inaccessible places and events, such as an earthquake [8,19]. With the dropping costs of VR technology, there has been a growth in the available commercial educational VR games. This means that they have the potential to play an important role in education if they are well-designed for learning.

An important potential benefit of VR games over real environments is that virtual environments can give learners detailed, timely, personalised feedback.

N. Wang et al. (Eds.): AIED 2023, LNAI 13916, pp. 541–552, 2023.
https://doi.org/10.1007/978-3-031-36272-9_44

VR games can incorporate "stealth assessment" [24] which can provide data for feedback. Feedback is important for learning in general [10]. Although feedback can sometimes have negative effects on learning motivation, it is valuable because it ensures that a learner understands their actual learning progress and has information on what they need to improve [27]. Feedback has also been demonstrated to improve learning in games [28]. Some is simply from the immediate interaction in the VR environment, but VR can also provide other rich forms of feedback based on the combination of suitable game design and the collection of interaction data through the game [3]. Some simple forms of feedback are confirmation of a correct answer or whether a task is complete. Frequently, games also have feedback in terms of game levels, accuracy, scores, and comparisons with other players.

Feedback on learning is particularly important for commercial games, as they can be widely used outside classrooms; so they should provide feedback for effective learning in standalone use. By contrast, in classroom use, a teacher can create activities that provide feedback to complement the VR game experience. We wanted to study the forms of feedback in popular commercial games for several reasons. Firstly, the most popular games are likely to be the most influential and so the effectiveness of the learning is important. Secondly, we were keen to discover the forms of feedback they provide and whether we could draw lessons from them. Finally, given the substantial body of AIED research on Open Learner Models including evidence of the benefits for learning [1,11,16,18], we wondered how whether they are used in popular commercial games, and if so how extensively they are used and the forms they take.

There have been many reviews of the benefits of feedback for learning in general [10,14,22,27], and in educational VR games [5,9,13,15,17,20,21,23,25,29,31]. However, we have not found a review of feedback in commercial educational VR games. Therefore, our work aimed to fill that gap, based on the following research questions:

- RQ1: What forms of feedback are provided in popular commercial VR educational games?
- RQ2: Are there OLMs in popular commercial VR educational games? And if so, what forms do they take?

This paper is organised as follows: the next section reviews key literature on educational feedback and previous reviews of educational VR games. We then describe the study method and results, followed by a discussion of the implications of the findings for educational VR games.

2 Background

This section has background on two main bodies of research that informed our work. First, and most fundamental, is work on the nature of feedback and the key forms that it takes. The second is the background from key systematic reviews on educational VR games.

Feedback in Educational Games

At the core of VR games, there is continuous feedback for players as they interact with the environment of the game and have feedback that encourages them to continue to play the game. Educational games should be more effective if they go beyond this and give feedback on learning progress [22,24,26]. Table 1 shows key forms of such feedback we identified from comprehensive meta-analyses of feedback in education [10,27] and guidance on feedback in educational games [14]. We use these categories in the rest of the paper.

The first block is typical *game-progress* feedback based on a *Score* (ID-1), *Level* (ID-2) and *Competition* that compares these with other players (ID-3). These can provide extrinsic motivation, encouraging a learner to continue to play (and potentially learn) more and giving them unscheduled rewards. However, *Competition* feedback is a form of social comparison which has educational risks, especially for learners who are not near the top and become discouraged. Even so, it is very common and learners value it for making sense of their own progress. It can be selectively used, for example with a leader-board showing just people close to the level of the individual learner.

Table 1. Key forms of feedback to facilitate learning in VR games. These are in three blocks separated by horizontal lines: the first block is mainly on game progress; the second is mainly related to game progress; the last block had feedback on learning. In the final column, examples in italics show what a learner actually sees and the others are descriptions.

ID-Form	Description	Example
1. Score	Progress on game	*Score is 99*
2. Level	Simple progression measure	*You just finished Level 3!*
3. Competition	Rank compared to others	*You are in the top 10% of players.*
4. Self	Broad performance	*You are doing well*
5. Self-Comparison	Current against previous	*You are 10% ahead of your PB*
6. Till Correct	Indication when wrong	Need to continue until correct
7. Accuracy	Report on correct performance	*You got 75% correct!*
8. Process	Hints, clues or information on how to complete tasks	*Use the scalpel to dissect the frog.*
9. OLMs	Progress on learning objectives	Skill-meter showing mastery of four maths skills

The second block is also primarily feedback about *game-progress*, both broad encouraging comments about individual progress (*Self*, ID-4) and progress that is relative to previous performance (*Self-Comparison*, ID-5). These may be overall or at the task level, for example, informing the learner that they did well to get the last task right. These forms of feedback may build self-confidence and self-efficacy.

The third block is more tightly linked to communicating *learning progress*, as opposed to *game progress.* These forms enable a learner to see their progress on the learning objectives of the game. *Till correct* (I-6) can help players to learn from mistakes. *Accuracy* (ID-7) can also inform players about learning progress. *Process* feedback (ID-8) is any information that helps a learner proceed - it is timely and, may give valuable feedback on success in learning. The final form, OLMs, as they appear in a substantial body of AIED research [1,2]. This can make use of data collected during play to create a learner model, which can drive personalisation of the game, for example, by creating personalised hints. A suitable form of the learner model can be "opened" to the learner, in various forms to help them better self-monitor their learning progress. We now present some examples of these forms of feedback in educational VR games.

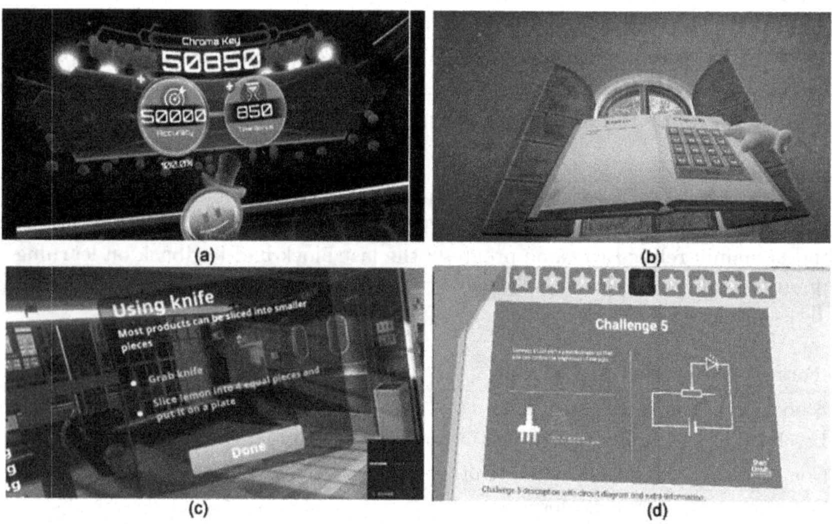

Fig. 1. Examples of the forms of feedback: screenshots from (a) HoloLAB Champions - VG31 (This is the ID in the appendix (https://bit.ly/44zs2bd.) which describes each game) (b) CodingKnights - VG13 (c) Cooking Simulator - VG6 (d) Short Circuit VR - S02

Figure 1 shows screenshots illustrating each of these forms of feedback. The first screenshot (a) is for *HoloLAB Champions*, which teaches chemistry; it shows the score (50,850) and player accuracy rating (100%) in doing the set tasks. In (b) for *CodingKnights*, the player needs to instruct the avatar to overcome obstacles to reach an endpoint. The feedback is in the form of progress through 4 chapters and their the levels. This figure shows progress for Chapter 1 in the book – the four levels in the top row are unlocked where the 12 other levels are locked. Image (c) is for *Cooking Simulator* and show green text for steps a play has done correctly (incorrect steps would be red). Part (d) is for *Short Circuit VR*, where the player learns how to build electronic circuits. The screenshot shows how it a circuit (at the right) provided as a hint for the initial task described at the top left using the components at the lower left.

Systematic Reviews of Educational VR Games

There have also been several reviews of educational VR games [5,9,13,15,17, 20,21,23,25,29,31]. Notably, several provide meta-analyses have reported learning benefits, for example these recent reviews [5,20,25]. Similarly, papers that provide guidance on game design, such as Johnson et al., [14] review considerable seminal work demonstrating the importance of feedback both broadly and in-game contexts.

As would be expected, reviews of educational VR games are based on searches over research literature. Typically, this has a focus on particular educational contexts. For example, a 2014 meta-analysis of 67 studies, in K-12 and higher education, concluded that VR-based instruction was quite effective [21]. More recently, a 2019 systematic review of the effectiveness of VR for health professions education analysed 31 studies and conducted a meta-analysis of eight of them, reporting benefits for knowledge and skills outcomes compared to various other forms of education [17]. In the context of higher education, a 2020 paper [23] discussed six previous reviews and analysed 38 articles in terms of the technology used, approaches to research design, data collection and analysis, learning theories, learning outcomes evaluation, domains, and content. Another recent meta-analysis of the effectiveness of immersive VR analysed 25 studies and showed some benefits, particularly for K-12 learners, for science education and to developing specific abilities [29]. A 2021 review of 29 publications analysed the reported learning benefits of immersive VR [9]. This highlighted limitations in the methodology for measuring learning benefits both in the papers they reviewed and in earlier systematic reviews.

In summary, there has been considerable work analysing various aspects of VR educational games. In spite of the importance of feedback for learning, we did not find work that studied the nature of the *forms of feedback* in educational VR games. Nor did we find a study of *commercial* VR games and their feedback. Our work aims to fill this gap in the literature. This is particularly timely due to 3 trends: (1) the recent prominence of commercial AI in education such as Khanmigo[1] and Duolingo Max[2] (2) the dropping costs for VR hardware, and (3) increasing numbers of widely available commercial educational VR games.

3 Methods: Game Selection and Coding

This section first explains the process we followed to identify relevant commercial educational VR games. Then it explains how we coded their feedback. Fig. 2 is

[1] https://www.khanacademy.org/khan-labs.
[2] https://blog.duolingo.com/duolingo-max/.

an overview of the process using the Preferred Reporting Items for Systematic Reviews and Meta-Analyses (PRISMA) model for systematic reviews [12].

The first step was to identify the popular platforms for commercial VR games: VIVEPORT[3]; Oculus[4]; and Steam[5]. The top-left box in the figure summarises the game identification search process for each platform. The screening phase is outlined in the second box at the left using exclusion criteria in the top right box. We now describe the identification and screening for each platform.

- VIVEPORT-GAMES: the search on 'Education' gave 148 games. The screening retained only (1) games available to us as members the platform or because they were free and (2) they operate on the HTC-Vive (a popular VR game platform).
- VIVEPORT-APPs: the same education search retrieved 263 games. To make the analysis more manageable, we included only the 69 games with a 4+ star ranking.
- Steam: the keyword 'Education' with tags 'Education', 'VR only', and 'Free' gave 66 games (58 after removing duplicates).
- Oculus: As it does not have a similar search to the above, we used the comprehensive list of Oculus educational games[6] We did not have the device but wanted to study this important platform, so used platform videos when available and for others, we searched and analysed gameplay videos on YouTube.

We acknowledge that our approach has limitations. (1) We only had a HTC-Vive. While this is a widely used device for playing VR games, we had to use the strategy above to include Oculus games. (2) We studied only games that were either free or available for VIVEPORT members; this may have missed paid games may have more advanced game mechanisms, including more complex feedback systems. (3) For VIVEPORT-APPs, we restricted the analysis to games with a 4+ star ranking. As this still provided 69 games, we felt that this was a substantial and important set of popular commercial games.

From February 1st, 2022 to April 20th, 2022, the first author played each game found on VIVEPORT and STEAM to determine the nature of the learning feedback it provided and to capture images of the forms of feedback. Typically, this took 30 min for the more sophisticated games. The first author captured the video footage of the gameplay for use in discussions between the authors. Then, to analyse the videos of games on Oculus, from 1 May 2022 to 10 May 2022, the first author spent approximately 10 min on each video, focusing on the gameplay and checking for feedback, and capturing images of that feedback. The authors discussed the coding in terms of the forms of learning feedback in Table 1.

[3] https://www.viveport.com/game.html.
[4] https://www.oculus.com/experiences/quest/.
[5] https://store.steampowered.com/.
[6] Futuclass: https://bit.ly/423odJD.

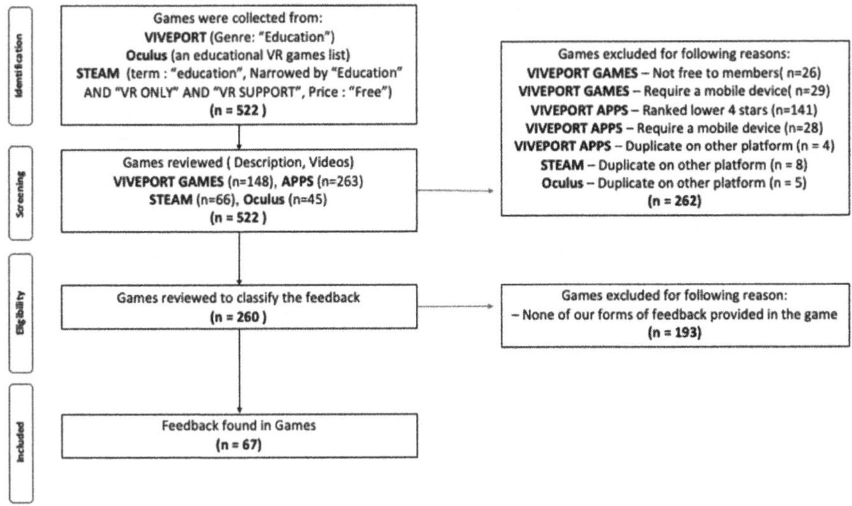

Fig. 2. Summary of the games identified, screened, eligible, and included in the analysis

Table 2 summarises the number of games we analysed from each platform. The first column shows the numbers after screening as described above. The next column shows the numbers of these excluded at eligibility stage shown in Fig. 2. Based on analysis of the games, those excluded had none of our types of feedback (in Table 1. It may seem surprising that so many games failed this eligibility criteria. In fact, this was because so many of the games were designed a free experiences of three main types:

- 72 games were interactive tours of places like a museum, art gallery or space station with free exploration and where interactivity could reveal information to learn about the place;
- 41 were purely 3-D videos on topics such as famous attractions around the world, ocean pollution or historic events
- 17 were social games for events like parties, conferences and classes.

None of the above types of games had any of our forms of feedback. In addition, there were 130 simulation games. This genre of game is particularly well suited to provide valuable learning feedback and, indeed, 67 of these games did provide at least one of our forms of feedback. (Details of these are in the appendix[7])

[7] https://bit.ly/44zs2bd.

Table 2. Counts of VR games on the four platforms after screening.

Platform	Items screening	Excluded on Eligibility	Included
VIVEPORT-Games	93	51	42 (45%)
VIVEPORT-Apps	69	66	3 (4%)
STEAM	58	47	11 (19%)
Oculus	40	29	11 (28%)
Total	**260**	**193**	**67** (26%)

4 Results

This section summarises the results in terms of our two research questions for the 67 games that met the eligibility criteria.

RQ1: What forms of feedback are provided in popular commercial VR educational games?

Table 3 shows the frequencies of each form of feedback. Scores were the most common form of feedback, appearing in almost half of the games; they were the sole form of feedback in 8 of these, meaning that for most games, there was also additional feedback. As we had expected, *Score* and *Level* feedback were particularly common forms of games progress feedback. It was pleasing to discover that the next most common forms are in hte last four rows for learning feedback. Here *Process* feedback was in close to half the games and the sole form in 8 of them.

Table 3. Frequency of each form of feedback in the commercial VR games. The earlier rows are mainly game feedback. The last four italicised rows are learning feedback.

Feedback Type	N (%)	Only this form
Score	33 (49%)	8
Levels	25 (37%)	0
Competition	3 (4%)	0
Self	5 (7%)	0
Self-Comparison	4 (6%)	0
Till Correct	25 (37%)	3
Accuracy	15 (22%)	0
Process	30 (45%)	8
OLM	4 (6%)	0
Total		**19**

RQ2: Are there OLMs in popular commercial VR educational games? And if so, what forms do they take?

The first part of the answer to this question is that there were indeed 4 games with OLMs: *VirtualSpeech*[8], *ENHANCE*[9], *Virtual Indus Discovery*[10] and *VR Waste Classification Experience System*[11]. Figure 3 shows a screenshot of the OLM in each of these. We now describe them briefly to answer the second part of this research question.

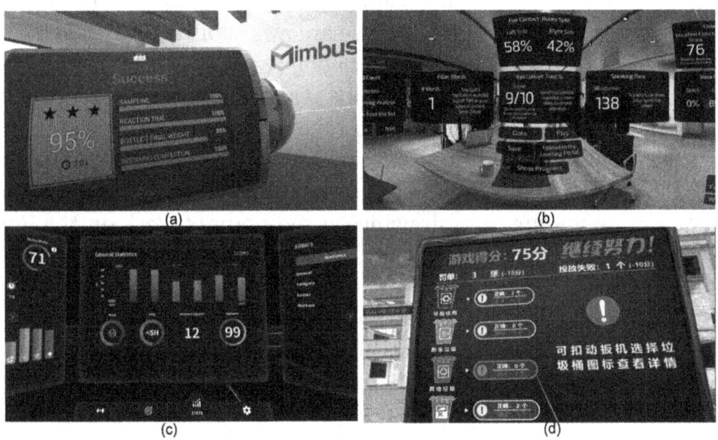

Fig. 3. Screenshots of the OLMs. (a) Virtual Indus Discovery, (b) VirtualSpeech, (c) ENHANCE and (d) VR Waste Classification Experience System

Virtual Indus Discovery, from VIVEPORT APPS, where players follow steps on a virtual watch to learn to control industrial equipment. Figure 3 (a) shows feedback for one task, with an overall accuracy score and bar charts and scores on four aspects of task performance, aspects modeled from game interaction.

In *VirtualSpeech*, available on STEAM, VIVEPORT APPS, and Oculus, players practice giving a speech, building skills like eye-contact, as in Fig. 3(b). The game models eye-contact, filler-word usage, and speaking pace. There is also feedback associated with some scores; for example, the speaking pace information may give advice to slow down. There are also personalised hints.

The OLM for *ENHANCE*, from Oculus, in Fig. 3(c) uses data from players' self-reports on aspects such as mood and sleep. They can play various games to develop thinking ability and sensitivity. The results are logged and used to identify areas of improvement. This game also provides daily, monthly and yearly statistics so players can see their progress as in Fig. 3(c).

[8] https://www.oculus.com/experiences/quest/3973230756042512/.
[9] https://www.oculus.com/experiences/quest/3696764747036091/.
[10] https://www.viveport.com/8cd8d9f0-fe9c-4d9f-9c20-82c553e5658f.
[11] https://www.viveport.com/6076d21f-c4e4-4bc5-ac1c-79d7e4128bd1.

The OLM in *VR Waste Classification Experience System* from VIVEPORT GAMES, in Fig. 3(d) shows a learner's success in putting items into the correct bins within the 60 s time limit. It shows an overall knowledge score, the number of items classified and just how it calculated the score from correctly or incorrectly classifying trash types.

5 Discussion and Conclusions

There have been many systematic reviews and meta-analyses [5,6,9,13,17,20,21, 23,25,29,31]. Our work is the first that we are aware of to study *widely available commercial games* to understand the *forms of feedback* they provide. As AIED becomes increasingly core to commercially available learning resources, there is a growing value for academic research to take account of it, for the potential to both learn from and to contribute to it. In particular, we see it as important to study popular commercial VR games through the lenses of established learning sciences knowledge, in our case, drawing on the body of work on the important roles and forms of feedback that can enhance learning.

We analysed a comprehensive body of popular VR educational games: 260 games on four popular VR games platforms. One striking finding is that just 67 (26%) games had any of the forms of feedback with 52% of simulation games doing so. This points to the potential learning benefits if many more educational VR games incorporate feedback. For example, an interactive tour through a museum could readily provide valuable feedback for learning. Our work highlights that many commercial games may well be far more effective in a classroom setting, where the teacher can design complimentary activities to give feedback on learning progress. At the outset, we are unsure whether we would find any OLMs; we found four games had OLMs and this adds to the picture of their use [1].

We are aware of limitations in our work. We only played the games available for the HCT-Vive. We relied on videos for the other important platform, Oculus. On VIVEPORT GAMES, we only analysed games that were free or available to members. Other games may have had richer feedback. On VIVEPORT APPS, we analysed only games with four+ star popularity rating. Notably, of these 69 analysed games, only 3 had one of our forms of feedback. Our analysis was in terms of the nine forms of feedback that we identified from the academic literature on feedback in general and in games. A final limitation is about the goals since commercial games are typically blackbox software. It is not feasible to change them. In spite of this, studying them has the potential to provide lessons for our research and for us to be more knowledgeable when we deal with the commercial AIED community.

Our work makes three main contributions. The first is the identification of a set of forms of feedback that could be made available in many VR games. This could provide a short checklist for designers to consider; the games we identified with each of these forms of feedback might serve as examples to inform design. The second key contribution is the first characterisation of the nature of feedback available across popular commercial VR games; future studies could track improvements. We also were unable to find systematic reviews of academic literature that analysed the nature of feedback available in the games reviewed; this points to a new criterion should be considered for future reviews. The third contribution was to gain understanding about OLMs in these games. Overall, our work can contribute both to the design of future games and for future analysis of educational VR games.

References

1. Bull, S.: There are open learner models about! IEEE Trans. Learn. Technol. **13**(2), 425–448 (2020)
2. Bull, S., Kay, J.: Student models that invite the learner in: the SMILI:() open learner modelling framework. Int. J. Artif. Intell. Educ. **17**(2), 89–120 (2007)
3. Checa, D., Bustillo, A.: A review of immersive virtual reality serious games to enhance learning and training. Multimed. Tools Appl. **79**(9), 5501–5527 (2020)
4. Christopoulos, A., Pellas, N., Laakso, M.J.: A learning analytics theoretical framework for stem education virtual reality applications. Educ. Sci. **10**(11), 317 (2020)
5. Coban, M., Bolat, Y.I., Goksu, I.: The potential of immersive virtual reality to enhance learning: a meta-analysis. Educ. Res. Rev., 100452 (2022)
6. Di Natale, A.F., Repetto, C., Riva, G., Villani, D.: Immersive virtual reality in k-12 and higher education: a 10-year systematic review of empirical research. Br. J. Educ. Technol. **51**(6), 2006–2033 (2020)
7. Elmqaddem, N.: Augmented reality and virtual reality in education. Myth or reality? Int. J. Emerg. Technol. Learn. **14**(3), 37 (2019)
8. Feng, Z., et al.: Towards a customizable immersive virtual reality serious game for earthquake emergency training. Adv. Eng. Inf. **46**, 101134 (2020)
9. Hamilton, D., McKechnie, J., Edgerton, E., Wilson, C.: Immersive virtual reality as a pedagogical tool in education: a systematic literature review of quantitative learning outcomes and experimental design. J. Comput. Educ. **8**(1), 1–32 (2021)
10. Hattie, J., Timperley, H.: The power of feedback. Rev. Educ. Res. **77**(1), 81–112 (2007)
11. Hou, X., Nguyen, H.A., Richey, J.E., Harpstead, E., Hammer, J., McLaren, B.M.: Assessing the effects of open models of learning and enjoyment in a digital learning game. Int. J. Artif. Intell. Educ. **32**(1), 120–150 (2022)
12. Hutton, B., et al.: The prisma extension statement for reporting of systematic reviews incorporating network meta-analyses of health care interventions: checklist and explanations. Ann. Intern. Med. **162**(11), 777–784 (2015)
13. Jiawei, W., Mokmin, N.A.M.: Virtual reality technology in art education with visual communication design in higher education: a systematic literature review. Educ. Inf. Technol., 1–19 (2023)
14. Johnson, C.I., Bailey, S.K., Buskirk, W.L.V.: Designing effective feedback messages in serious games and simulations: a research review. Instr. Tech. Facilitate Learn. Motiv. Serious Games **3**(7), 119–140 (2017)
15. Kavanagh, S., Luxton-Reilly, A., Wuensche, B., Plimmer, B.: A systematic review of virtual reality in education. Themes Sci. Technol. Educ. **10**(2), 85–119 (2017)
16. Kay, J., Rus, V., Zapata-Rivera, D., Durlach, P.: Open learner model visualizations for contexts where learners think fast or slow. Des. Recommendations Intell. Tutoring Syst. **6**, 125 (2020)

17. Kyaw, B.M., et al.: Virtual reality for health professions education: systematic review and meta-analysis by the digital health education collaboration. J. Med. Internet Res. **21**(1), e12959 (2019)
18. Long, Y., Aleven, V.: Enhancing learning outcomes through self-regulated learning support with an open learner model. User Model. User-Adap. Interact. **27**(1), 55–88 (2017)
19. Lovreglio, R., et al.: Prototyping virtual reality serious games for building earthquake preparedness: the Auckland city hospital case study. Adv. Eng. Inf. **38**, 670–682 (2018)
20. Luo, H., Li, G., Feng, Q., Yang, Y., Zuo, M.: Virtual reality in k-12 and higher education: a systematic review of the literature from 2000 to 2019. J. Comput. Assist. Learn. **37**(3), 887–901 (2021)
21. Merchant, Z., Goetz, E.T., Cifuentes, L., Keeney-Kennicutt, W., Davis, T.J.: Effectiveness of virtual reality-based instruction on students' learning outcomes in k-12 and higher education: a meta-analysis. Comput. Educ. **70**, 29–40 (2014)
22. Panadero, E., Lipnevich, A.A.: A review of feedback models and typologies: towards an integrative model of feedback elements. Educ. Res. Rev. **35**, 100416 (2022)
23. Radianti, J., Majchrzak, T.A., Fromm, J., Wohlgenannt, I.: A systematic review of immersive virtual reality applications for higher education: design elements, lessons learned, and research agenda. Comput. Educ. **147**, 103778 (2020)
24. Shute, V.J.: Focus on formative feedback. Rev. Educ. Res. **78**(1), 153–189 (2008)
25. Villena-Taranilla, R., Tirado-Olivares, S., Cozar-Gutierrez, R., González-Calero, J.A.: Effects of virtual reality on learning outcomes in k-6 education: a meta-analysis. Educ. Res. Rev., 100434 (2022)
26. Westera, W.: Why and how serious games can become far more effective: accommodating productive learning experiences, learner motivation and the monitoring of learning gains. J. Educ. Technol. Soc. **22**(1), 59–69 (2019)
27. Wisniewski, B., Zierer, K., Hattie, J.: The power of feedback revisited: a meta-analysis of educational feedback research. Front. Psychol. **10**, 3087 (2020)
28. Wouters, P., Van Oostendorp, H.: A meta-analytic review of the role of instructional support in game-based learning. Comput. Educ. **60**(1), 412–425 (2013)
29. Wu, B., Yu, X., Gu, X.: Effectiveness of immersive virtual reality using head-mounted displays on learning performance: a meta-analysis. Br. J. Educ. Technol. **51**(6), 1991–2005 (2020)
30. Zhang, L., Bowman, D.A., Jones, C.N.: Exploring effects of interactivity on learning with interactive storytelling in immersive virtual reality. In: 2019 11th International Conference on Virtual Worlds and Games for Serious Applications (VS-Games), pp. 1–8. IEEE (2019)
31. Zhonggen, Y.: A meta-analysis of use of serious games in education over a decade. Int. J. Comput. Games Technol. **2019**, 100 (2019)

Gender Differences in Learning Game Preferences: Results Using a Multi-dimensional Gender Framework

Huy A. Nguyen[1(✉)], Nicole Else-Quest[2], J. Elizabeth Richey[3], Jessica Hammer[1], Sarah Di[1], and Bruce M. McLaren[1]

[1] Carnegie Mellon University, Pittsburgh, PA 15213, USA
hn1@cs.cmu.edu
[2] University of North Carolina at Chapel Hill, Chapel Hill, NC 27599, USA
[3] University of Pittsburgh, Pittsburgh, PA 15260, USA

Abstract. Prompted by findings of gender differences in learning game preferences and outcomes, education researchers have proposed adapting games by gender to foster learning and engagement. However, such recommendations typically rely on intuition, rather than empirical data, and are rooted in a binary representation of gender. On the other hand, recent evidence from several disciplines indicates that gender is best understood through multiple dimensions, including gender-typed occupational interests, activities, and traits. Our research seeks to provide learning game designers with empirical guidance incorporating this framework in developing digital learning games that are inclusive, equitable, and effective for all students. To this end, we conducted a survey study among 333 5th and 6th grade students in five urban and suburban schools in a mid-sized U.S. city, with the goal of investigating how game preferences differ by gender identity or gender-typed measures. Our findings uncovered consistent differences in game preferences from both a binary and multi-dimensional gender perspective, with gender-typed measures being more predictive of game preferences than binary gender identity. We also report on preference trends for different game genres and discuss their implications on learning game design. Ultimately, this work supports using multiple dimensions of gender to inform the customization of learning games that meet individual students' interests and preferences, instead of relying on vague gender stereotypes.

Keywords: digital learning games · gender studies · game preferences · survey study

1 Introduction

While digital learning games have been shown to be a promising form of instruction thanks to their motivational and learning benefits [22], designing effective games requires a clear understanding of the preferences of different player populations. For instance, there are consistent gender differences in game preferences, such that boys tend to prefer

N. Wang et al. (Eds.): AIED 2023, LNAI 13916, pp. 553–564, 2023.
https://doi.org/10.1007/978-3-031-36272-9_45

faster paced, action-style games, while girls tend to prefer games with puzzle and social interaction elements [2, 5]. With digital learning games specifically, girls tend to rank goal clarity and social interaction as more important than boys, while boys tend to prefer challenge, progress feedback, and visual appeal [11]. These gendered preferences can produce meaningful differences in learning behaviors and outcomes. For example, girls have sometimes been shown to enjoy learning games more [1] and have greater learning outcomes [24, 30] than boys. Different features of learning games have also been shown to induce gendered effects, such as girls benefiting more from a digital learning companion [4].

Prompted by these findings, researchers have proposed adapting digital learning games based on gender to create more inclusive and equitable learning experiences [25, 35]. However, such recommendations for gender-based adaptation typically rely on game designers' intuitions, stereotypes, or preferences observed through playtesting and focus groups [13, 33], rather than experimental studies. Moreover, such efforts are limited by a grounding in the gender binary, which views gender as one of two discrete categories, male and female, framed as biologically-based, apparent at birth, and stable over time [23]. Conflating gender with binary, birth-assigned sex is not only imprecise, but may also contribute to gender-stereotyped interests and gender disparities in academic achievement [6, 15]. In contrast, evidence from multiple disciplines has demonstrated that gender is complex, fluid and dynamic, comprising multiple interrelated but separate dimensions [23].

Thus, a more sophisticated and nuanced approach to understanding gender differences in learning game preferences should take into account not only self-reported gender identity (e.g., male, female, non-binary, trans), but also other gender dimensions – such as gender-typed occupational interests, activities, and traits [26] – which are continuous and more fine-grained than gender identity. Such an approach is consistent with best practices in gender studies research, including with late elementary and middle school youth [14, 23], among whom these dimensions of gender are only modestly correlated with one another [8, 31]. Notably, prior work has shown that middle school children can be differentiated along these dimensions of gender and can reliably report on their gender-typed behaviors [8, 26, 27].

Motivated by this multi-dimensional approach, our work seeks to better understand students' digital game preferences and how they relate to different dimensions of gender, through a survey deployed to 333 young students. The first half of the survey asked students to rank their preferred game genres (e.g., *action*, *strategy*, *sandbox*) and game narratives (the overarching game world and story – e.g., fighting pirates, hunting treasures), while the second half queried about dimensions of gender identity and gender-typed occupational interests, activities, and traits. With this survey design, our primary research questions are as follows.

RQ1: Are there significant gender differences – based on gender identity or gender-typed interests, activities, and traits – in game genre preferences?

RQ2: Are there significant gender differences – based on gender identity or gender-typed interests, activities, and traits – in game narrative preferences?

By addressing these questions, our work contributes to research on young students' preferences in digital games. We also demonstrate, through statistical testing and qualitative analyses, how additional dimensions of gender can better reflect individual preferences than binary gender identity. In turn, this knowledge can enable the design and development of more inclusive and effective learning games.

2 Methods

2.1 Participants

Our sample comprises $n = 333$ students who participated in a classroom study in 5th ($n = 100$) and 6th grades ($n = 233$) across five urban and suburban public schools in a mid-sized U.S. city. Students ranged in age from 10 to 13 years ($M = 11.06$, $SD = .69$). In terms of self-reported gender identity, 52.0% ($n = 173$) described themselves as male, 47.0% ($n = 156$) as female, 0.3% ($n = 1$) as trans or nonbinary, and 0.6% ($n = 2$) preferred not to disclose their gender. Because the subsample of students in the last two categories was small, these 3 students were excluded from statistical analyses of gender identity.

2.2 Materials and Procedures

Surveys were administered as part of a classroom study of a digital learning game in mathematics. Students first filled out a demographic questionnaire that queried about their age, grade level, and an open-ended item to self-identify their gender. Next, they completed two sets of surveys related to gender dimensions and game preferences.

Gender-Typed Behaviors. This survey is based on the Children's Occupational Interests, Activities, and Traits-Personal Measure (COAT-PM - [26]), which assesses youth's interests, activities, and traits in relation to gender-stereotyped norms. The COAT-PM has two scale scores comprising masculine- and feminine-typed occupational interests, activities, and traits. For brevity, we adapted the measure by removing 9 gender-neutral items, 6 masculine-typed items, and 6 feminine-typed items, such that it consisted of 54 items in total (out of the original 75), rated on Likert scales from 1 to 4. The *occupational interests* domain indicates how much one wants to pursue certain jobs, and includes 18 items rated from 1 (not at all) to 4 (very much), such as "hairstylist" or "nurse" (feminine) and "construction worker" or "engineer" (masculine). The *activity* domain indicates how often one performs certain activities, and includes 18 items rated from 1 (never) to 4 (often or very often), such as "make jewelry" or "take dance lessons" (feminine) and "play basketball" or "go fishing" (masculine). The *traits* domain indicates how much one would describe themselves, and includes 18 items rated from 1 (not at all like me) to 4 (very much like me), such as "gentle" or "neat" (feminine) and "adventurous" or "confident" (masculine). Note that these items do not reflect actual gender differences (e.g., we do not assume "confidence" is a male-only trait); rather, they only portray *stereotypical* perceptions of gender differences. Our goal in analyzing students' ratings of these items is to quantify how much they shape their behaviors around traditional gender norms and stereotypes [26]. Each domain also included two gender-neutral items for filler which are excluded from analysis (e.g., "YouTuber," "practice an instrument," "friendly"). The internal consistency (Cronbach's α) was good for both the masculine-typed occupational interests, activities, and traits scale ($\alpha = .80$) and the feminine-typed occupational interests, activities, and traits scale ($\alpha = .85$).

Game Genre and Narrative Preferences. We examined digital learning game preferences in terms of game genres and narratives. First, students were asked to indicate their top three preferred game genres from a list of seven options, selected to capture the most popular genres among young players [16, 37], and then rank them from most to least liked. The game genres (and example games) listed were: *action* (e.g., Fortnite, Splatoon); *sports & racing* (e.g., Rocket League, FIFA); *strategy* (e.g., Age of Empires, Civilization); *sandbox* (e.g., Roblox, Minecraft); *music & party* (e.g., Pianista, Just Dance); *role-playing* (e.g., Stardew Valley, Legend of Zelda); and *casual* (e.g., Bejeweled, Animal Crossing).

Next, students were given the descriptions of four game narratives: *Amusement Park* (lead new alien friends around an amusement park), *Treasure Hunt* (search for hidden treasure among ocean landmarks, racing against an arch-nemesis), *Helping a Sea Friend* (help a sea creature save an underwater city that has lost power before it's too late), and *War at Sea* (fight criminal naval masterminds and disable their secret weapon to save the world). The narratives were brainstormed by the research team to vary along multiple dimensions – such as world building, goal focus and presence of competition or cooperation – that were shown to be differentially preferred by boys and girls in prior work [4, 11]. Based on the provided descriptions, students were asked to rank the four game narratives from most to least interesting.

3 Results

3.1 Game Genre Preferences

We focused our analyses on students' first and second choices among game genres. A series of chi-square analyses of gender identity (boy or girl) by preference for each game genre revealed significant binary gender differences in most game preferences, such that boys tended to prefer the genres of *action* and *sports & racing*, whereas girls tended to prefer *sandbox*, *music & party*, *role-playing*, and *casual* games. Boys and girls were similarly likely to prefer *strategy* games. Table 1 displays the percentage of boys and girls who ranked each game genre as their 1st or 2nd ranked options. Following Cohen's guidelines [7], the reported significant gender identity differences on game preferences all have small to medium effect sizes ($.10 < \phi < .50$).

To test for significant differences in digital game genre preferences based on masculine- and feminine-typing, we ran a series of one-way between-subject ANOVAs comparing the masculine- and feminine-typed behaviors of students who ranked specific game genres as their 1st or 2nd options. We found that students who preferred the genres of *action* and *sports & racing* reported significantly higher masculine-typed

Table 1. Gender differences in game genre preferences according to gender identity. Percentages sum to 200 to represent first and second rank choices. (***) $p < .001$, (**) $p < .01$, (*) $p < .05$.

Game Genre	% Boys	% Girls	χ^2 (1)	ϕ
Action	70.5	39.1	32.80***	−.32
Sports & Racing	45.1	14.1	37.22***	−.34
Strategy	11.6	5.8	3.42	−.10
Sandbox	46.8	59.0	4.86*	.12
Music & Party	5.2	37.8	53.23***	.40
Role-Playing	12.7	21.2	4.19*	.11
Casual	8.1	23.1	14.29***	.21

behaviors, whereas students who preferred the genres of *sandbox*, *music & party*, and *casual* games reported significantly lower masculine-typed behaviors. By contrast, students who preferred the genres of *action* and *sports & racing* reported significantly lower feminine-typed behaviors, and students who preferred the *music & party* genre reported significantly higher feminine-typed behaviors. See Table 2 for the descriptive statistics, effect sizes and F-statistics; here a game genre is considered not preferred (Not P.) if it is not within the top two ranked options. Following Cohen [7], most of the significant gender-typing differences in game preferences can be interpreted as medium ($d \cong .50$), with the exception of the large gender difference in preferring *music & party* games ($d > .80$).

Next, we computed the correlations between gender dimensions. Binary gender identity (with "female" coded as 1 and "male" coded as 0) was positively correlated with feminine-typed scales ($r = .62$) and negatively correlated with masculine-typed scales ($r = -.33$). In addition, feminine-typed scales were positively correlated with masculine-typed scales ($r = .19$). Thus, these three gender dimensions were moderately correlated but not redundant. Because there is no evidence of multicollinearity, our next analysis involves a series of logistic regressions predicting each game genre preference (i.e., whether it is included in the top two choices) with binary gender, masculine-typed scale, and feminine-typed scale (Table 3). This allowed us to assess how much each variable predicted preferences when controlling for the other variables. For the *action, sports & racing*, *sandbox*, and *music & party* genres, masculine- and feminine-type scales were significant predictors of preference while binary gender was not. For only one genre - *casual* games - was binary gender a significant predictor while masculine- and feminine-types were not. Models predicting *role-playing* and *strategy* genres were not significant and are excluded from Table 3.

3.2 Game Narrative Preferences

A series of chi-square analyses of gender identity (boy or girl) by preference for each game narrative revealed significant gender differences in two of the four narrative preferences, such that the *War at Sea* narrative was preferred by boys and *Help a Sea Friend*

Table 2. Gender differences in game genre preferences according to feminine- and masculine-typed occupational interests, activities, and traits.

Game Genre	Feminine-typed				Masculine-typed			
	Pref. *M* (*SD*)	Not P. *M* (*SD*)	*F*	*d*	Pref. *M* (*SD*)	Not P. *M (SD)*	*F*	*d*
Action	2.17 (.47)	2.32 (.48)	8.04**	-.31	2.42 (.47)	2..23 (.42)	14.64***	.41
Sports & Racing	2.09 (.42)	2.30 (.49)	14.40***	-.44	2.52 (.39)	2.26 (.46)	24.14***	.57
Strategy	2.08 (.42)	2.25 (.48)	3.34	-.35	2.38 (.45)	2.33 (.46)	.29	.10
Sandbox	2.24 (.49)	2.23 (.47)	.01	.01	2.25 (.44)	2.44 (.46)	15.40***	-.42
Music & Party	2.56 (.43)	2.15 (.45)	45.62***	.86	2.22 (.49)	2.37 (.44)	6.41*	-.34
Role-Playing	2.31 (.42)	2.22 (.49)	1.775	.19	2.26 (.43)	2.36 (.46)	2.16	-.21
Casual	2.32 (.51)	2.22 (.47)	2.03	.22	2.22 (.44)	2.36 (.46)	3.96*	-.30

Table 3. Logistic regressions predicting genre preferences according to gender identity (female = 0), masculine-typed scale, and feminine-typed scale. *Exp(B)* represents the change in odds corresponding to a unit change in the coefficient.

Game Genre	Constant	Binary gender	Feminine-typed scale	Masculine-typed scale
Action	$B = -2.10$*, $Exp(B) = .12$	$B = .27$, $Exp(B) = 1.31$	$B = -.68$*, $Exp(B) = .51$	$B = 1.07$***, $Exp(B) = 2.91$
Sports & Racing	$B = -1.87$, $Exp(B) = .15$	$B = .60$, $Exp(B) = 1.82$	$B = -1.48$***, $Exp(B) = .23$	$B = 1.22$**, $Exp(B) = 3.37$
Sandbox	$B = .044$, $Exp(B) = 1.05$	$B = -.13$, $Exp(B) = .88$	$B = .72$*, $Exp(B) = .2.06$	$B = -1.03$***, $Exp(B) = .36$
Music & Party	$B = -2.69$*, $Exp(B) = .07$	$B = -1.22$, $Exp(B) = .30$	$B = 1.54$*, $Exp(B) = .4.67$	$B = -1.33$*, $Exp(B) = .27$
Casual	$B = -1.73$, $Exp(B) = .18$	$B = -2.75$**, $Exp(B) = .064$	$B = -.74$, $Exp(B) = .48$	$B = -.58$, $Exp(B) = 1.78$

narrative was preferred by girls (Table 4). Following Cohen [7], these effect sizes are medium. On the other hand, boys and girls were similarly likely to prefer the *Amusement Park* and *Treasure Hunt* narratives.

Table 4. Gender differences in game narrative preferences according to gender identity.

Game Narrative	% Boys	% Girls	χ^2 (1)	ϕ
Amusement Park	42.8	50.0	1.72	.07
Treasure Hunt	50.3	56.4	1.23	.06
Help a Sea Friend	32.9	53.2	13.77***	.21
War at Sea	74.0	40.4	38.04***	− .34

Table 5. Gender differences in game narrative preferences according to feminine- and masculine-typed occupational interests, activities, and traits.

Game Narrative	Feminine-typed				Masculine-typed			
	Pref. *M (SD)*	Not P. *M (SD)*	*F*	*d*	Pref. *M (SD)*	Not P. *M (SD)*	*F*	*d*
Amusement Park	2.25 (.48)	2.23 (.48)	.18	.05	2.33 (.45)	2.35 (.47)	.13	− .04
Treasure Hunt	2.25 (.46)	2.22 (.50)	.21	.05	2.34 (.43)	2.34 (.50)	.01	.01
Help a Sea Friend	2.36 (.52)	2.13 (.42)	18.07***	.46	2.28 (.50)	2.38 (.42)	4.59*	− .24
War at Sea	2.12 (.44)	2.39 (.49)	27.44***	− .56	2.39 (.46)	2.27 (.45)	5.80*	.27

To test for significant differences in game narrative preferences based on masculine- and feminine-typing, we ran four one-way between-subjects ANOVAs comparing the masculine- and feminine-typed behaviors of students who ranked specific game narratives as their 1st or 2nd option. With regards to feminine-typed occupational interests, activities, and traits, there were medium effects of students who preferred the *Help a Sea Friend* narrative reporting significantly higher feminine-typed behaviors, and those who preferred the *War at Sea* narrative reporting significantly lower feminine-typed behaviors. In addition, there were small effects of students who preferred the *War at Sea* narrative reporting significantly higher masculine-typed behaviors, and those who preferred the *Help a Sea Friend* narrative reporting significantly lower masculine-typed behaviors. See Table 5 for descriptive statistics, effect sizes (Cohen's *d*), and *F*-statistics.

Next, we conducted a series of logistic regressions predicting each game narrative preference with binary gender, masculine-typed scale, and feminine-typed scale in a single model (Table 6). For the *Help a Sea Friend* narrative, masculine- and feminine-type scales were significant predictors of preference, while binary gender was not. For the *War at Sea* narrative, both binary gender and the feminine-type scale were significant predictors. Models predicting the *Amusement Park* and *Treasure Hunt* narratives were not significant and are excluded.

Table 6. Logistic regressions predicting game preferences according to gender identity (female = 0), masculine-typed scale, and feminine-typed scale. (*) $p < .05$, (**) $p < .01$, (***) $p < .001$.

Game Narrative	Constant	Binary gender	Feminine-typed scale	Masculine-typed scale
Help a Sea Friend	$B = -.82$, $Exp(B) = .44$	$B = -.24$, $Exp(B) = .79$	$B = .98$**, $Exp(B) = 2.67$	$B = -.68$*, $Exp(B) = .51$
War at Sea	$B = .52$, $Exp(B) = 1.69$	$B = .84$**, $Exp(B) = 2.33$	$B = -.82$*, $Exp(B) = .44$	$B = .53$, $Exp(B) = 1.70$

3.3 Post-hoc Analyses

As a follow-up, we analyzed whether the additional gender dimensions, reflected by the masculine- and feminine-typed scores, would reveal more nuances about students' game preferences. We identified 13 boys who had stronger feminine-typed ($M = 2.18$, $SD = 0.29$) than masculine-typed ($M = 1.95$, $SD = 0.30$) behaviors and 31 girls who had stronger masculine-typed ($M = 2.52$, $SD = 0.47$) than feminine-typed ($M = 2.31$, $SD = 0.41$) behaviors. In each of these groups, the *action* and *sandbox* genres were most popular. Additionally, *Treasure Hunt* was the most preferred narrative among the 13 boys, and *War at Sea* was the most preferred narrative among the 31 girls.

4 Discussion and Conclusion

In this work, we examined young students' game preferences through the lens of a multi-dimensional gender framework. Our research was motivated by the need to characterize gender differences in game preferences while considering additional gender dimensions beyond binary gender identity, including gender-typed occupational interests, activities and traits [23, 26]. This topic is especially relevant to the design of inclusive learning games that match students' gaming interests without relying upon vague or outdated gender stereotypes. Beyond reporting students' preferred game genres and narratives, our work also supports the analysis of multiple gender dimensions in educational research. We discuss the insights from these results below.

Overall, we observed consistent gender differences in game preferences when representing gender as binary categories and as continuous gender-typed scores. In particular, boys and those with strong masculine-typed behaviors tended to prefer the *action* and *sports & racing* genres, as well as the battle-oriented game narrative, *War at Sea*. Meanwhile, girls and those with strong feminine-typed behaviors reported more interest in the *casual* and *music & party* genres, and the co-operative game narrative, *Help a Sea Friend*. These patterns are consistent with those reported in past surveys [2, 5], indicating that gender-based game preferences have remained stable over the years.

However, our results suggest two significant advantages of using a multi-dimensional gender representation over the traditional binary one. First, based on Cohen's interpretation guidelines [7], we observed gender differences with medium to large effect sizes using gender-typing measures, but only small to medium effect sizes using gender identity, suggesting that the former approach better reflects the influence of gender on children's game interests. This observation is further supported by logistic regression analyses showing that gender-typed scales predicted game preferences better than binary gender identity, when both were included in the same models. In short, a more nuanced assessment of gender is a more powerful predictor of game preferences. Second, our post-hoc analysis revealed distinct preferences from boys with stronger feminine-typed behaviors and from girls with stronger masculine-typed behaviors, compared to other students in their respective gender identity groups. Thus, when the students' gender identity and gender-typed behaviors do not completely align, their gaming interests can be better distinguished by measures of gender-typed interests, activities, and traits. That is, the multi-dimensional assessment of gender is more precise. Based on these findings, we encourage AIED researchers to adopt multidimensional gender measures in their future studies to more closely examine how students' background factors may influence their learning interest and experience.

The findings from our survey also have implications for learning game design. First, we observed that the *strategy*, *role-playing* and *casual* genres were preferred by only 5–25% of students. Notably, these three genres often feature game elements – such as slow pace, reflection and study of the environment – that are frequently used in learning games [3, 29] because they can foster a playful experience without inducing high cognitive load and interfering with learning. At the same time, they may cause the identified genres to be less favored by students, for several possible reasons. First, many modern games, regardless of genre, tend to incorporate elements that require fast and accurate decision making from the players [10]. If children are used to this style of game play, games that feature slower and more methodical play may be less engaging. Alternatively, prior work has shown that middle-school students may omit reporting that they enjoy games which do not fit with the gamer identity they are attempting to project [18]. Further work is needed to disambiguate these conjectures, but in both cases, increasing children's exposure to slow and reflective entertainment games is likely to improve their reaction to those features in digital learning games.

The remaining four game genres – *action*, *sports and racing*, *sandbox*, *music and party* – were all highly favored by both boys and girls. Among them, the *sandbox* genre was the most universally popular, being included in the top two genres of close to half of the boys and girls in the study. This attribute may be explained by the rapid rise in popularity of sandbox games, most notably *Minecraft* and *Roblox*, during the COVID-19 pandemic [9, 19]. These platforms allow players to not only play games created by others, but also to freely design their own activities. In this way, they are effective at fostering player connections [9] and have also been used for educational purposes [12]. Digital learning game researchers can take advantage of this trend by introducing sandbox elements, such as player agency [32], into their games to promote engagement. Incorporating player agency has also been shown to result in better learning efficiency [17] and outcomes [36] in learning games. Additionally, there are successful examples of full-fledged sandbox digital learning games, such as *Physics Playground*, which utilize stealth assessment techniques to measure student learning without disrupting the sandbox experience [34]. In turn, this prior work provides a solid foundation for investigating how digital learning games can take advantage of the sandbox genre's increasing popularity among young students.

One limitation of this study is that students may not have a clear picture of all the game genres and narratives included in the survey, given their brief text descriptions. Providing graphical illustrations of the example games in each genre and the storylines in each narrative would likely help students make more informed selections. At the same time, the findings thus far suggest several novel research directions. First, we could evaluate the role of additional gender dimensions, such as gender typicality (i.e., perceived similarity to other children of one's own and other gender [27]), in explaining students' game preferences. In addition, future research should investigate the connection between gender differences in game preferences and in learning outcomes, for example by manipulating a learning game's narrative to be either masculine- or feminine-oriented while retaining the game mechanics and instructional materials. This comparison would be particularly interesting for learning games that have been shown to produce gender differences in learning outcomes [4, 24, 30]. For instance, the math game *Decimal Point* [28], which features an amusement park metaphor, has yielded consistent learning benefits for girls over boys across several studies [20, 21, 30]. To follow up, one could investigate the outcomes of an alternate version of the game with a more masculine-oriented narrative, such as *War at Sea* – in this case, would gender differences in learning favoring girls still emerge, or would boys have an advantage instead? More generally, the study of how gender-based preferences interact with the game features and instructional materials is an important step towards understanding the nuanced role of gender in shaping students' playing and learning experience.

In conclusion, our results indicate that a nuanced approach to gender is more predictive of game preferences than binary gender. That is, assessing multiple dimensions of gender is more precise, particularly for students whose activities, interests, and traits are less stereotypical of their binary gender identity. To our knowledge, this is the first research that utilizes a multi-dimensional gender framework for examining children's game preferences, and it suggests that this approach is more appropriate for game designers seeking to customize learning games to better suit individual students' interests and preferences. Ultimately, we envision that these customizations may also be driven by AI techniques which construct accurate and inclusive models of students' learning and preferences, using our identified multi-dimensional gender features as a starting point.

Acknowledgements. This work was supported by NSF Award #DRL-2201796. The opinions expressed are those of the authors and do not represent the views of NSF.

References

1. Adamo-Villani, N., Wilbur, R., Wasburn, M.: Gender differences in usability and enjoyment of VR educational games: a study of SMILE™. In: 2008 International Conference Visualisation, pp. 114–119. IEEE (2008)
2. Aleksić, V., Ivanović, M.: Early adolescent gender and multiple intelligences profiles as predictors of digital gameplay preferences. Croatian J. Educ.: Hrvatski časopis za odgoj i obrazovanje **19**(3), 697–727 (2017)

3. Amory, A.: Building an educational adventure game: theory, design, and lessons. J. Interact. Learn. Res. **12**(2), 249–263 (2001)
4. Arroyo, I., Burleson, W., Tai, M., Muldner, K., Woolf, B.P.: Gender differences in the use and benefit of advanced learning technologies for mathematics. J Educ. Psychol. **105**(4), 957 (2013)
5. Chou, C., Tsai, M.-J.: Gender differences in Taiwan high school students' computer game playing. Comput. Hum. Behav. **23**(1), 812–824 (2007)
6. Chung, B.G., Ehrhart, M.G., Holcombe Ehrhart, K., Hattrup, K., Solamon, J.: Stereotype threat, state anxiety, and specific self-efficacy as predictors of promotion exam performance. Group Org. Manag. **35**(1), 77–107 (2010)
7. Cohen, J.: Statistical Power Analysis for the Behavioral Sciences. Routledge (2013)
8. Cook, R.E., Nielson, M.G., Martin, C.L., DeLay, D.: Early adolescent gender development: the differential effects of felt pressure from parents, peers, and the self. J. Youth Adolesc. **48**(10), 1912–1923 (2019)
9. Cowan, K., et al.: Children's digital play during the covid-19 pandemic: insights from the play observatory. Je-LKS: J. e-Learning Knowl. Soc. **17**(3), 8–17 (2021)
10. Dale, G., Kattner, F., Bavelier, D., Green, C.S.: Cognitive abilities of action video game and role-playing video game players: Data from a massive open online course. Psychol. Popular Media **9**(3), 347 (2020)
11. Dele-Ajayi, O., Strachan, R., Pickard, A., Sanderson, J.: Designing for all: exploring gender diversity and engagement with digital educational games by young people. In: 2018 IEEE Frontiers in Education Conference (FIE), pp. 1–9. IEEE (2018)
12. Dundon, R.: Teaching Social Skills to Children with Autism using Minecraft®: A Step by Step Guide. Jessica Kingsley Publishers (2019)
13. Farrell, D., Moffat, D.C.: Adapting cognitive walkthrough to support game based learning design. Int. J. Game-Based Learn. (IJGBL) **4**(3), 23–34 (2014)
14. Fast, A.A., Olson, K.R.: Gender development in transgender preschool children. Child Dev. **89**(2), 620–637 (2018)
15. Galdi, S., Cadinu, M., Tomasetto, C.: The roots of stereotype threat: when automatic associations disrupt girls' math performance. Child Dev. **85**(1), 250–263 (2014)
16. GameTree Team: Industry Results: Genre and Platform Preferences (Age & Gender), https://gametree.me/blog/global-gamer-insights-report/. Last accessed 01 Nov 2023
17. Harpstead, E., Richey, J.E., Nguyen, H., McLaren, B.M.: Exploring the subtleties of agency and indirect control in digital learning games. In: Proceedings of the 9th International Conference on Learning Analytics & Knowledge, pp. 121–129 (2019)
18. Higashi, R., Harpstead, E., Solyst, J., Kemper, J., Odili Uchidiuno, J., Hammer, J.: The design of co-robotic games for computer science education. In: Extended Abstracts of the 2021 Annual Symposium on Computer-Human Interaction in Play, pp. 111–116 (2021)
19. Hjorth, L., Richardson, I., Davies, H., Balmford, W.: Playing during covid-19. In: Exploring Minecraft, pp. 167–182. Springer (2020)
20. Hou, X., Nguyen, H.A., Richey, J.E., Harpstead, E., Hammer, J., McLaren, B.M.: Assessing the effects of open models of learning and enjoyment in a digital learning game. Int. J. Artif. Intell. Educ. **32**, 1–31 (2021)
21. Hou, X., Nguyen, H.A., Richey, J.E., McLaren, B.M.: Exploring how gender and enjoyment impact learning in a digital learning game. In: International Conference on Artificial Intelligence in Education, pp. 255–268. Springer (2020)
22. Hussein, M.H., Ow, S.H., Elaish, M.M., Jensen, E.O.: Digital game-based learning in K-12 mathematics education: a systematic literature review. Educ. Inf. Technol. **27**, 1–33 (2021)
23. Hyde, J.S., Bigler, R.S., Joel, D., Tate, C.C., van Anders, S.M.: The future of sex and gender in psychology: five challenges to the gender binary. Am. Psychol. **74**(2), 171 (2019)

24. Khan, A., Ahmad, F.H., Malik, M.M.: Use of digital game based learning and gamification in secondary school science: the effect on student engagement, learning and gender difference. Educ. Inf. Technol. **22**(6), 2767–2804 (2017)
25. Kinzie, M.B., Joseph, D.R.: Gender differences in game activity preferences of middle school children: implications for educational game design. Educ. Tech. Research Dev. **56**(5–6), 643–663 (2008)
26. Liben, L.S., Bigler, R.S.: The Development Course of Gender Differentiation. Blackwell publishing (2002)
27. Martin, C.L., Andrews, N.C., England, D.E., Zosuls, K., Ruble, D.N.: A dual identity approach for conceptualizing and measuring children's gender identity. Child Dev. **88**(1), 167–182 (2017)
28. McLaren, B.M., Adams, D.M., Mayer, R.E., Forlizzi, J.: A computer-based game that promotes mathematics learning more than a conventional approach. Int. J. Game-Based Learn. (IJGBL) **7**(1), 36–56 (2017)
29. McLaren, B.M., Nguyen, H.A.: Digital learning games in artificial intelligence in education (AIED): a review. In: Handbook of Artificial Intelligence in Education, p. 440 (2023). https://doi.org/10.4337/9781800375413
30. Nguyen, H.A., Hou, X., Richey, J.E., McLaren, B.M.: The impact of gender in learning with games: a consistent effect in a math learning game. Int. J. Game-Based Learn. (IJGBL) **12**(1), 1–29 (2022)
31. Perry, D.G., Pauletti, R.E., Cooper, P.J.: Gender identity in childhood: a review of the literature. Int. J. Behav. Dev. **43**(4), 289–304 (2019)
32. Ryan, R.M., Rigby, C.S., Przybylski, A.: The motivational pull of video games: a self-determination theory approach. Motiv. Emot. **30**(4), 344–360 (2006)
33. Seaborn, K., Fels, D.I.: Gamification in theory and action: a survey. Int. J. Hum Comput Stud. **74**, 14–31 (2015)
34. Shute, V., et al.: Maximizing learning without sacrificing the fun: stealth assessment, adaptivity and learning supports in educational games. J. Comput. Assist. Learn. **37**(1), 127–141 (2021)
35. Steiner, C.M., Kickmeier-Rust, M.D., Albert, D.: Little big difference: gender aspects and gender-based adaptation in educational games. In: International Conference on Technologies for E-Learning and Digital Entertainment, pp. 150–161. Springer (2009)
36. Taub, M., Sawyer, R., Smith, A., Rowe, J., Azevedo, R., Lester, J.: The agency effect: the impact of student agency on learning, emotions, and problem-solving behaviors in a game-based learning environment. Comput. Educ. **147**, 103781 (2020)
37. Yee, N.: Beyond 50/50: Breaking Down The Percentage of Female Gamers by Genre, https://quanticfoundry.com/2017/01/19/female-gamers-by-genre. Last accessed 01 Nov 2023

Can You Solve This on the First Try? – Understanding Exercise Field Performance in an Intelligent Tutoring System

Hannah Deininger[1(✉)], Rosa Lavelle-Hill[1,2], Cora Parrisius[1,3], Ines Pieronczyk[1], Leona Colling[1], Detmar Meurers[1], Ulrich Trautwein[1], Benjamin Nagengast[1], and Gjergji Kasneci[4]

[1] University of Tübingen, Tübingen, Germany
hannah.deininger@uni-tuebingen.de
[2] University of Copenhagen, Copenhagen, Denmark
[3] Karlsruhe University of Education, Karlsruhe, Germany
[4] Technical University Munich, Munich, Germany

Abstract. The analysis of fine-grained data on students' learning behavior in intelligent tutoring systems using machine learning is a starting point for personalized learning. However, findings from such analyses are commonly viewed as uninterpretable and, hence, not helpful for understanding learning processes. The explainable AI method SHAP, which generates detailed explanations, is a promising approach here. It can help to better understand how different learning behaviors influence students' learning outcomes and potentially produce new insights for adaptable intelligent tutoring systems in the future. Based on K-12 data ($N = 472$ students), we trained several ML models on student characteristics and behavioral trace data to predict whether a student will answer an exercise field correctly or not—a low-level approximation of academic performance. The optimized machine learning models (lasso regression, random forest, XGBoost) performed well ($AUC = 0.68$; $F_1 = [0.63; 0.66; 0.69]$), outperforming logistic regression ($AUC = 0.57$; $F_1 = 0.52$). SHAP importance values for the best model (XGBoost) indicated that, besides prior language achievement, several behavioral variables (e.g., performance on the previous attempts) were informative predictors. Thus, specific learning behaviors can help explain exercise field performance, independent of the student's basic abilities and demographics—providing insights into areas of potential intervention. Moreover, the SHAP values highlight heterogeneity in the effects, supporting the notion of personalized learning.

Keywords: Explainable AI · Learning Analytics · Academic Performance Prediction · Intelligent Tutoring Systems

N. Wang et al. (Eds.): AIED 2023, LNAI 13916, pp. 565–576, 2023.
https://doi.org/10.1007/978-3-031-36272-9_46

1 Introduction

Achieving an ideal learning experience for each student has recently been regarded as a major educational objective [18]. To ensure the best possible learning experience and learning outcomes, it is crucial to teach in line with each student's unique learning rhythm [10]. This is described by the term *personalized learning*, which refers to a learning environment that is optimized to the individual learner's needs [18]. In the traditional school context, personalized learning is resource-heavy and time-consuming for teachers. With the support of digital learning tools such as intelligent tutoring systems (ITS), personalized learning is more achievable in practice in the near future [18]. By using ITS in schools, monitoring students' learning progress could become easier if the system itself provides data about students' learning behaviors. Bringing personalized learning to the next level, *precision education* pursues two goals: (1) identifying a student's learning status and (2) providing timely interventions for those students who are struggling [32]. An important step towards including the promising concept of precision education in schools is the ability to infer students' current performance (i.e., learning status).

Although many studies have looked at academic performance prediction (APP) [6], few have taken it a step further and investigated what kind of student characteristics (e.g., demographics, previous performance) and behavioral variables influence APP [9,16]. However, understanding which variables influence APP is crucial for precision education because the knowledge about predictive behavioral features is a prerequisite for targeted behavioral interventions. To date, the increased use of data-driven machine learning (ML) algorithms for APP has led to an increase in predictive power compared with traditional regression approaches [11,13,28]. Such ML algorithms are able to process high-dimensional, unstructured data at different scales and can implicitly model complex, non-linear relationships and interactions in the data [4]. However, highly predictive ML models are often too complex to be fully understandable for humans. Thus, extracting information on influential variables can be challenging. Therefore, previous studies examining which variables influence APP have focused on easily interpretable, traditional regression approaches [8,15], forgoing the potential advantages of ML models. Yet, [9] argue that knowledge of helpful or harmful learning behavior gained from highly predictive models is critical when trying to improve the learning process with targeted interventions.

To contribute to the two overarching goals of precision education (accurate APP and implementation of targeted interventions), we suggest combining an ML model with an explainable AI (XAI) component. Adding to the literature on APP, we examined which exact indicators are relevant for APP in our model. To this end, we first selected the best model to predict whether a student can solve an exercise field in an ITS, which serves as a low-level approximation of academic performance, and expanded it with an XAI component providing high-quality explanations for the model and enabling insights into the models' internals [12]. That is, we aim to get a first impression of which learning behaviors promote or prevent learning success. Our contributions are as follows:

1. We show that XAI can be a useful tool to gain insights from ML models when predicting exercise performance.
2. We present which (behavioral) variables influence the ML model's prediction of exercise performance most, and how.

2 Related Work

2.1 Academic Performance Prediction

Since the introduction of ITS has made the collection of fine-granular learning process data easier, the application of ML models for APP has increased—paving the way for achieving precision education's first goal, namely predicting students' current academic performance [28,32]. In the Learning Analytics (LA) community, APP based on learner data is particularly prevalent with a focus on overall academic performance (e.g., grades or overall course performance scores) as the prediction outcome [6]. Predicting the overall academic performance is relevant to monitor students' learning progress, for example, by a teacher, and detecting those at risk of failure. At a lower level, accurate prediction of a student's performance in certain exercises within an ITS is a good way to model a student's current ability, and hence to adjust the learning process within the ITS accordingly [14]. Instead of trying to predict a student's learning status of a whole domain, the model determines how likely the student is to solve a specific exercise correctly, or even a single gap in a gap-filling exercise [2]. This so-called *exercise field performance* can serve as a low-level approximation of the current learning status [5] and enables the adaptive presentation of exercises during practice sessions that best fit the current learning status of the student in question [10]. Employing low-level APP allows for a more direct determination of which learning behaviors are most conducive to performance [2]. Furthermore, enabling adaptivity at the lowest possible level might be the most efficient mechanism for adaptive systems to present the next best exercise (i.e., faster adaption to the learning status). Yet, exercise performance as a prediction outcome is less common than high-level APP measures [2,14].

2.2 Investigating Factors that Influence Academic Performance

For precision education to be implemented in practice, understanding students' learning processes and discovering possible areas for interventions is important. So far, only a few approaches have tried to understand the relationship between input data and APP, namely, how a prediction model has arrived at its output [9]. One such approach, proposed by [15] examined the relationship between multiple practicing variables derived from an ITS (e.g., time on task) and learning gains. They demonstrated the predictive capabilities of behavioral ITS data for APP in a similar context to the current investigation. However, they focused on simple linear regression models only; they report an R^2 of .37, which is relatively low compared to the predictive potential of more complex ML methods such as tree-based methods [13]. Furthermore, using linear regression requires the pre-specification and selection of model features, which leads to reduced granularity in the insights produced.

2.3 Explainable AI for Academic Performance Prediction

To overcome some of the limitations of traditional statistical models, it is reasonable to evaluate the utility of more complex ML methods in APP research. However, a potential drawback of using ML lies in the complexity of highly predictive models, which can make the direct assessment of variable effects difficult to disentangle [12]. Applying XAI techniques can make APP mechanisms visible. Particularly, insights into the impact of actionable variables, namely fundamentally changeable predictors (such as learning behavior), can serve as a decision basis for adaptive systems [16]. Various XAI methods have been developed in the past years that can roughly be classified based on their application options [31]: XAI techniques that are independent of the ML model–so-called *model-agnostic* methods–can illuminate the black box of any ML classifier; *model-specific* methods rely on information specific to a class of ML algorithms (e.g., a variable importance measure based on the weights in a neural network). Another distinction criterion for XAI methods is the level at which they provide explanations [31]: *Global* methods provide sample-wide explanations that give insights into the overall importance of features, whereas *local* techniques focus on individually relevant variables on the person-level.

To be applicable in the context of APP, the selected XAI method needs to be model-agnostic because it is not clear beforehand which model will perform best, and it should provide local as well as global explanations to enable individual-students as well as whole-sample insights. Particularly local explanations are relevant for personalized learning because prediction mechanisms for individual students can serve as starting points for targeted interventions. All these requirements are fulfilled by the explanation method SHAP (SHapley Additive exPlanations) [19], which is widely used across different research domains [21]. In addition, SHAP can take high correlations among the input features for the explanations into account by splitting up their shared impact among them [19]. However, it has to be mentioned that SHAP also has its limitations (which hold for other XAI methods as well): (1) it explains the correlational structure defined by the model, which does not imply causal relationships by default; and (2) it explains the relevance of a variable in the model–depending on the model quality, this might differ from reality [17,22].

Even though the community agrees on the importance of XAI for APP, most APP studies focus on developing highly predictive models but do not address the key question of *how* a model reached a prediction [1,9]. However, this is necessary to foster a deeper understanding of the complex relationships between students' individual prerequisites, learning processes, and learning outcomes [15,16]. Among the APP studies applying XAI, only a few focus on XAI results and mostly cover global interpretability [1]. However, as stated above, local explanations are relevant for truly personalized learning.

3 Methodology

The current study is based on data collected within the *Interact4School study* [24], a cluster-randomized controlled field trial that investigates second-language learning in secondary school through individualized practice phases with the ITS *FeedBook* (see [29] for a technical description of the system).

3.1 Dataset

The FeedBook is a web-based workbook for English as a second language targeting seventh-grade students in Germany. Within the Interact4School study, 24 classes used the FeedBook during the school year 2021/22 in four learning units for three weeks each. Within each unit, students could individually practice specific grammar topics (e.g., Simple Past) and vocabulary with exercises in the FeedBook. The exercises come in different formats (e.g., fill-in-the-blanks and categorize exercises). Every unit was preceded and followed by a language test. For this investigation, we used data collected by the FeedBook on the specific practicing behavior during the first unit only to avoid data reduction due to dropout ($N = 472$ students; 52.9% female, $M_{age} = 12.5$ years, $SD_{age} = 0.41$), as well as the students' corresponding language pre- and posttest data and demographics collected at the beginning of the school year.

Our data extraction process was guided by literature, interviews we conducted with the students regarding their usage of the ITS, and exploration of the data collected by the system. Each exercise in the FeedBook consists of several items including a gap that students had to fill—the exercise field. The students could reenter an exercise field unlimited times. Within this investigation, only the students' first attempt for each exercise field was used, which could be either correct or incorrect ($N = 73,471; 77.6\%$ correct). To predict whether an exercise field is going to be solved on the first try or not, we used two different data sources as input features: student characteristics and behavioral data (see Table 1). Students' demographics were assessed via surveys or provided by the schools. Students' previous academic performance was measured by their previous school year's grades, a general English proficiency test score (i.e., a C-test score [27]), and a specific proficiency test score on grammar topics covered in the following unit. The behavioral data consists of students' practicing behavior in the system. It includes metadata of the specific exercise field a student attempted (e.g., corresponding exercise's grammar topic and type), as well as practicing information such as the student's performance in previous attempts (i.e., across exercises and learning sessions).

3.2 Data Analysis

Following [20] for the C- and pretest scores, we computed a sum score for each participant. The C-test score [27] refers to students' general English proficiency at the beginning of the school year. The pretest score reflects students' previous knowledge of the grammar topics covered within a learning unit. For both tests,

Table 1. Overview of all raw variables used as the basis for the ML models. All demographic and performance test variables were split into class-level aggregation variables ($Mclass, SDclass$) and group mean-centered individual variables during data preparation.

Data Source	Variable Name	Description	Type
Student Characteristics			
Demographics	sex	Student's sex	binary
	age	Student's age in years	numeric
	edu	Student's parents' education (at least one parent has a high school diploma)	binary
	migra	Student's migration background (at least one parent or the student was not born in Germany)	binary
Study Info	cond_class	Study condition of a class (three different system versions focusing on motivational aspects)	category
	cond_individ	Student's study condition (receiving corrective feedback or not)	binary
Previous Performance	gl6e	English grade year six	numeric
	gl6g	German grade year six	numeric
	gl6m	Math grade year six	numeric
	ctestscoreT1	Student's C-Test score	numeric
	pretestscoreT1	Student's pretest score	numeric
Behavioral Data			
Exercise Field Metadata	task type	Type of corresponding exercise	category
	grammar topic	Grammar topic of the corresponding exercise	category
	more practice	Whether the exercise is a "more practice" exercise (i.e., voluntary exercise for further practicing)	binary
	capstone exercise	Whether the exercise is a capstone exercise (i.e., human ranked most complex exercise per grammar topic)	binary
Practicing Data	only attempted	Whether the exercise field was attempted, but not submitted	binary
	is weekend	Whether the attempt was made at weekends	binary
	time of day	Time of the day the attempt was made	category
	prev. attempt's correctness	Whether the previous attempt was correct or incorrect	binary
	average previous performance	Average performance across all previous first attempts	numeric

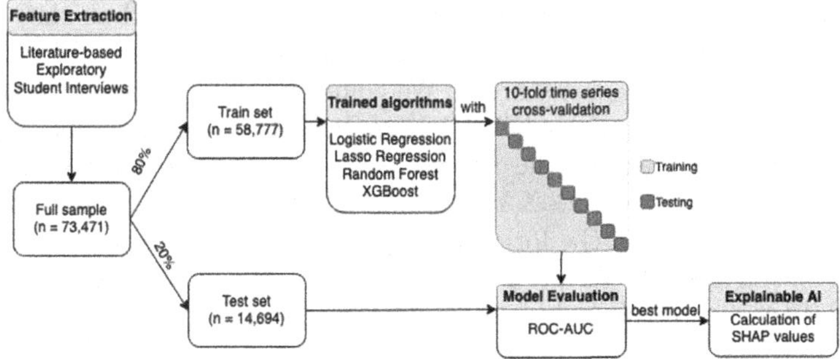

Fig. 1. Overview of the analysis procedure followed within this study.

higher scores indicate higher proficiency. To facilitate interpretation for a non-German audience, we re-coded the grades (1 = worst to 6 = best). For all student characteristics, M and SD per class were introduced as features to capture the class-related dependencies in the dataset. In addition, the students' raw values were group-mean centered. From a total of 472 students, 245 (52%) did not provide their parents' educational background, 28 (6%) lacked their migration background, and for 17 students (4%), their German, English, and math grades from the previous school year were missing. Given that the used implementations of the ML algorithms except for XGB were not able to handle missing data, KNN imputation was utilized to complete missing values [25]. To ensure comparability between all algorithms, we built two XGB models, one on the imputed dataset and one on the original dataset including missing values [30]. Information on the daytime and the weekday of practice was extracted from the logging timestamps of each attempt. Two features were calculated based on the chronological order of the attempts: information on whether the previous first attempt was solved, and the average performance across all previous first attempts. Multi-categorical features were one-hot encoded, expanding the dataset to 50 features.

For the comparison of the ML algorithm's predictive performance, we drew on approaches for tabular data varying in complexity that recent studies used to predict learning outcomes [3,6], namely logistic regression (LR), lasso regression (LassoR; [33]), random forest (RF; [4]), and eXtreme gradient boosting (XGB; [7]). The extracted dataset was split into train and test set based on an 80:20 partition using the most recent 20% of all attempts as a test set to account for the time dependency in the data (see Fig. 1). To avoid information leakage based on temporal dependencies between attempts during training, we employed 10-fold time-series cross-validation. To measure the models' performance, we assessed accuracy, F_1-score, and AUC [26]. F_1-score and AUC are especially well-suited to describe the predictive quality of imbalanced datasets. In addition, for the AUC, no manual prediction threshold needs to be set [26]. These two measures are therefore used as selection criteria to identify the best-performing model. As a first step, we trained all models on student characteristics only to assess whether the extracted behavioral data contained information relevant to the overall prediction. Of these, the best performing was selected as the "student data only" baseline model. Second, all models mentioned above were trained with all features using a grid search[1]. We compared the predictive power of all models to the "student data only" XGB baseline. Third, we identified the most predictive model trained on the full dataset to generate explanations for the test set predictions using the XAI method SHAP [19]. SHAP generates post-hoc, local explanations in the form of Shapley values. It quantifies the influence per variable on the outcome's difference to the base value (i.e., the average prediction), with positive values indicating an increase and negative values a decrease in the model's prediction [19].

4 Results

Of all models trained on student characteristics only, the XGB model performed best with an AUC of 0.56 and was therefore selected as the "student data only" XGB baseline. When including the full set of variables (i.e., student characteristics and behavioral data), the predictive power increased to AUC = 0.68 (LassoR, RF, XGB; 0.5 = chance level) except for LR, for which no increase was found. Additionally, all ML models outperformed LR on the full variable set (i.e., an absolute difference of 11% points in terms of AUC (corresponding to an increase of 22 Gini points); see Fig. 2). Among the ML models, we could not identify any differences in performance regarding the AUC. There was also no difference in performance between the XGB models trained on the imputed or the original dataset. Concerning the F_1 score, XGB showed the best performance, followed by RF (see Fig. 2).

Among the ten most influential features for predicting exercise performance given the XGB model were four variables related to previous academic performance and six related to the student's learning behavior (see Fig. 3). Generally, the most influential feature was the behavioral trace variable *average correctness of previous attempts* (a proximal performance indicator): A higher previous performance tended to be associated with a higher exercise field performance. However, there are some students for which lower previous performance was associated with higher exercise field performance as well.

5 Discussion and Conclusion

The study demonstrates how predictive ML models can be enhanced using XAI methods to understand learning behavior and lays the groundwork for precision education in the future. All tested ML models showed meaningfully better

[1] Final hyper-parameter grid and optimal values in bold: LR: tol: [0.00001, **0.0001**], solver: [**saga**], class weight: [none, **balanced**]; LassoR: penalty: [**L1**, L2], C: [**0.01**, 0.5], tol: [0.00001, **0.0001**], solver: [**saga**], class weight: [none, **balanced**]; RF: n estimators [50, **200**], max depth: [2, **6**, 20], max features: [**0.2**, 15, auto]; XGB: n estimators [50, **100**, 150], max depth: [**3**, 4, 5], colsample bytree: [0.1, **1**], learning rate: [0.01, **0.05**] Note: For computational reasons, grids were adjusted iteratively.

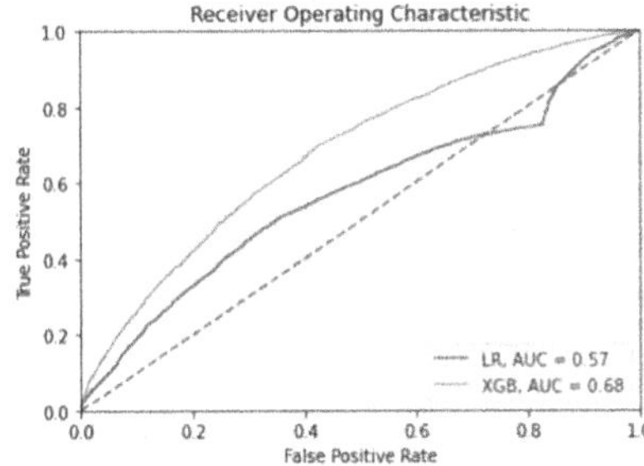

Classifier	Accuracy	F_1
"student data only" XGB baseline	0.76	0.66
LR	0.49	0.52
LassoR	0.60	0.63
RF	0.76	0.66
XGB	0.77	**0.69**

Fig. 2. *Right*: ROC-Curves and AUC scores for the LR and XGB model. *Left*: Overview of evaluation measures for all trained models. Note: For a better presentation, we have omitted the curves for the other models, which were very similar to that of XGB and resulted in the same AUC of 0.68. LR = Logistic Regression, XGB = eXtreme Gradient Boosting.

performance than the XGB baseline model trained on student characteristics only, supporting the notion that behavioral ITS data can improve the predictive power of APP models [8,15,20]. In general, all ML models performed better than the LR suggesting that more insights can potentially be gained using ML compared to regression models traditionally used when interpreting prediction models [16]. The findings indicate that there may be irrelevant features within the data that the LR cannot eliminate during training as can LassoR. Both regression approaches do not perform better than the "student data only" XGB baseline indicating that there are non-linearities and complex interactions in the data that the tree-based models could capture. Indeed, it is often argued that psychological phenomena are highly complex and multi-causal, which is why they cannot be explained solely by uncorrelated input variables and linear relationships [35]. As XGB can model conditional dependencies among variables best [7] and achieved the best F_1-score, we argue that in our investigation, this approach is preferred over the other approaches.

In line with previous findings, we could see previous performance measures and behavioral variables among the ten most influential features within the XGB model [5,6,8,13,15,34]. Notably, the average C-test score per class seems to have a non-linear effect on the exercise field performance, which can be derived from the heterogeneity of high SHAP values corresponding to high and low C-test scores. This non-linear relationship could only be captured in a complex ML model and might reflect learning as it takes place. Interestingly, variables related to the time of practicing (e.g., weekend vs. weekday) or demographics were not among the top ten most influential features, in contrast to what has been shown in the literature [13]. This suggests that our best-performing XGB model can identify the direct relationship between behavior and learning outcome, regardless of the demographic confounder variables that might affect both behavior and learning outcome [13,23]. Furthermore, demographics appearing as not being relevant in our model questions the common practice of including

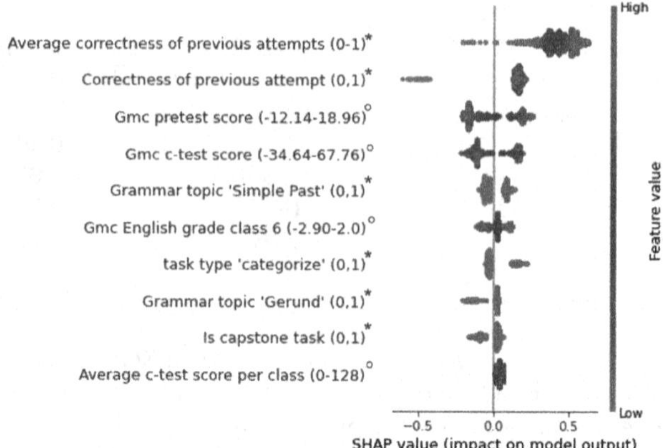

Fig. 3. SHAP summary plot for the test set given the XGB model trained on all features. Each point corresponds to one student. The y-axis displays the 10 most influential variables. The x-axis corresponds to the SHAP values (negative values indicate lower, positive values higher predicted exercise performance). The color represents the variable value from low (=blue) to high (=red). Values in parentheses depict the value range. Note: Gmc=group mean-centered, *: behavioral variables; ∘: prev. performance. (Color figure online)

them in APP models, especially as these variables mainly reflect socio-economic factors and biases, and are not fundamentally changeable through interventions [23]. On a final note, SHAP has not only the advantage of disentangling interaction effects between input variables by sharing these effects among them [19]; but additionally, the combination of XGB with SHAP enabled individual-level analysis beyond group-level analytics, making it a promising technology for realizing truly personalized learning. However, it has to be stated that prediction does not imply causation. Therefore, the results have to be validated carefully before implementing corresponding interventions in practice.

In summary, we showed how predictive ML models can be enriched by XAI methods, revealing which behavioral variables are related to low-level academic performance and expanding upon previous research from the LA community. Such basic research on understanding learning behavior paves the ground for implementing precision education in the near future.

Acknowledgements. Hannah Deininger and Leona Colling are doctoral students at the LEAD Graduate School & Research Network, which is funded by the Ministry of Science, Research and the Arts of the state of Baden-Württemberg. Furthermore, this research was funded by the German Federal Ministry of Research and Education (BMBF; 01JD1905A and 01JD1905B). It was additionally supported by Dr. Kou Murayama's Humboldt Professorship (Alexander von Humboldt Foundation) for the employment of Rosa Lavelle-Hill. Ines Pieronczyk is now affiliated with the DLR Projektträger Bonn.

References

1. Alamri, R., Alharbi, B.: Explainable student performance prediction models: a systematic review. IEEE Access **9**, 33132–33143 (2021)
2. Beck, J.E., Woolf, B.P.: High-level student modeling with machine learning. In: Gauthier, G., Frasson, C., VanLehn, K. (eds.) ITS 2000. LNCS, vol. 1839, pp. 584–593. Springer, Heidelberg (2000). https://doi.org/10.1007/3-540-45108-0_62
3. Borisov, V., Leemann, T., Seßler, K., Haug, J., Pawelczyk, M., Kasneci, G.: Deep neural networks and tabular data: a survey. IEEE Trans. Neural Netw. Learn. Syst. (2022)
4. Breiman, L.: Random forests. Mach. Learn. **45**(1), 5–32 (2001)
5. Cetintas, S., Si, L., Xin, Y.P., Hord, C.: Predicting correctness of problem solving from low- level log data in intelligent tutoring systems. Educ. Data Min. 230–239 (2009)
6. Chakrapani, P., Chitradevi, D.: Academic performance prediction using machine learning: A comprehensive & systematic review. In: 2022 International Conference on Electronic Systems and Intelligent Computing (ICESIC), pp. 335–340. IEEE (2022)
7. Chen, T., Guestrin, C.: Xgboost: a scalable tree boosting system. In: Proceedings of the 22nd ACM SIGKDD International Conference on Knowledge Discovery and Data Mining (KDD 2016), pp. 785–794 (2016)
8. Chen, X., Breslow, L., DeBoer, J.: Analyzing productive learning behaviors for students using immediate corrective feedback in a blended learning environment. Comput. Educ. **117**, 59–74 (2018)
9. Chitti, M., Chitti, P., Jayabalan, M.: Need for interpretable student performance prediction. In: 2020 13th International Conference on Developments in eSystems Engineering (DeSE), pp. 269–272. IEEE, Liverpool (2020)
10. Corno, L., Snow, R.E., et al.: Adapting teaching to individual differences among learners. Handb. Res. Teach. **3**, 605–629 (1986)
11. Costa-Mendes, R., Oliveira, T., Castelli, M., Cruz-Jesus, F.: A machine learning approximation of the 2015 Portuguese high school student grades: a hybrid approach. Educ. Inf. Technol. **26**(2), 1527–1547 (2021)
12. Das, A., Rad, P.: Opportunities and challenges in explainable artificial intelligence (xai): A survey, pp. 1–24. arXiv preprint arXiv:2006.11371 (2020)
13. Daza, A., Guerra, C., Cervera, N., Burgos, E.: Predicting academic performance through data mining: a systematic literature. TEM J., 939–949 (2022)
14. Ding, X., Larson, E.C., Doyle, A., Donahoo, K., Rajgopal, R., Bing, E.: EduAware: using tablet-based navigation gestures to predict learning module performance. Interact. Learn. Environ. **29**(5), 720–732 (2021)
15. Hui, B., Rudzewitz, B., Meurers, D.: Learning processes in interactive CALL systems: linking automatic feedback, system logs, and learning outcomes. Lang. Learn. Technol., 1–22 (2022)
16. Khosravi, H., et al.: Explainable Artificial intelligence in education. Comput. Educ.: Artif. Intell. 1–34 (2022)
17. Kumar, I.E., Venkatasubramanian, S., Scheidegger, C., Friedler, S.: Problems with shapley-value-based explanations as feature importance measures. In: International Conference on Machine Learning, pp. 5491–5500. PMLR (2020)
18. Luan, H., Tsai, C.C.: A review of using machine learning approaches for precision education. Educ. Technol. Soc. **24**, 250–266 (2021)

19. Lundberg, S.M., Lee, S.I.: A unified approach to interpreting model predictions. Adv. Neural. Inf. Process. Syst. **30**, 1–10 (2017)
20. Meurers, D., De Kuthy, K., Nuxoll, F., Rudzewitz, B., Ziai, R.: Scaling up intervention studies to investigate real-life foreign language learning in school. Ann. Rev. Appl. Linguist. **39**, 161–188 (2019)
21. Molnar, C.: Interpretable Machine Learning. Lulu. com (2020)
22. Molnar, C., et al.: General pitfalls of model-agnostic interpretation methods for machine learning models. In: Holzinger, A., et al. (eds.) xxAI-Beyond Explainable AI: International Workshop, Held in Conjunction with ICML 2020, July 18, 2020, Vienna, Austria, Revised and Extended Papers, vol. 13200, pp. 39–68. Springer, Cham (2022). https://doi.org/10.1007/978-3-031-04083-2_4
23. Paquette, L., Ocumpaugh, J., Li, Z., Andres, A., Baker, R.: Who's learning? using demographics in EDM research. J. Educ. Data Min. **12**(3), 1–30 (2020)
24. Parrisius, C., et al.: Using an intelligent tutoring system within a task-based learning approach in English as a foreign language classes to foster motivation and learning outcome (interact4school): Pre-registration of the study design, pp. 1–74 (2022)
25. Pedregosa, F., et al.: Scikit-learn: machine learning in python. J. Mach. Learn. Res. **12**, 2825–2830 (2011)
26. Provost, F., Fawcett, T.: Analysis and visualization of classifier performance: comparison under imprecise class and cost distributions. In: Proceedings of the 3rd International Conference on Knowledge Discovery and Data Mining, pp. 1–6 (1997)
27. Raatz, U., Klein-Braley, C.: The C-test–A modification of the cloze procedure (1981)
28. Romero, C., Ventura, S.: Educational data mining and learning analytics: an updated survey. WIREs Data Min. Knowl. Discovery **10**(3), 1–21 (2020)
29. Rudzewitz, B., Ziai, R., De Kuthy, K., Meurers, D.: Developing a web-based workbook for English supporting the interaction of students and teachers. In: Proceedings of the Joint Workshop on NLP for Computer Assisted Language Learning and NLP for Language Acquisition, pp. 36–46 (2017)
30. Sheetal, A., Jiang, Z., Di Milia, L.: Using machine learning to analyze longitudinal data: a tutorial guide and best-practice recommendations for social science researchers. Appl. Psychol. 1–26 (2022)
31. Stiglic, G., Kocbek, P., Fijacko, N., Zitnik, M., Verbert, K., Cilar, L.: Interpretability of machine learning-based prediction models in healthcare. Wiley Interdisc. Rev.: Data Min. Knowl. Discovery **10**(5), 1–13 (2020)
32. Tempelaar, D., Rienties, B., Nguyen, Q.: The contribution of dispositional learning analytics to precision education. Educ. Technol. Soc. **24**(1), 109–121 (2021)
33. Tibshirani, R.: Regression shrinkage and selection via the lasso. J. R. Stat. Soc.: Ser. B (Methodol.) **58**(1), 267–288 (1996)
34. Villagrá-Arnedo, C.J., Gallego-Durán, F.J., Compañ, P., Llorens Largo, F., Molina-Carmona, R., et al.: Predicting academic performance from behavioural and learning data. Int. J. Des. Nat. Ecodyn. **11**(3), 239–249 (2016)
35. Yarkoni, T., Westfall, J.: Choosing prediction over explanation in psychology: lessons from machine learning. Perspect. Psychol. Sci. **12**(6), 1100–1122 (2017)

A Multi-theoretic Analysis of Collaborative Discourse: A Step Towards AI-Facilitated Student Collaborations

Jason G. Reitman(✉), Charis Clevenger, Quinton Beck-White, Amanda Howard, Sierra Rose, Jacob Elick, Julianna Harris, Peter Foltz, and Sidney K. D'Mello

University of Colorado Boulder, Boulder, CO 80309, USA
jason.reitman@colorado.edu

Abstract. Collaboration analytics are a necessary step toward implementing intelligent systems that can provide feedback for teaching and supporting collaborative skills. However, the wide variety of theoretical perspectives on collaboration emphasize assessment of different behaviors toward different goals. Our work demonstrates rigorous measurement of collaboration in small group discourse that combines coding schemes from three different theoretical backgrounds: Collaborative Problem Solving, Academically Productive Talk, and Team Cognition. Each scheme measured occurrence of unique collaborative behaviors. Correlations between schemes were low to moderate, indicating both some convergence and unique information surfaced by each approach. Factor analysis drives discussion of the dimensions of collaboration informed by all three. The two factors that explain the most variance point to how participants stay on task and ask for relevant information to find common ground. These results demonstrate that combining analytical tools from different perspectives offers researchers and intelligent systems a more complete understanding of the collaborative skills assessable in verbal communication.

Keywords: Collaboration Analytics · Collaborative Problem Solving · Team Communication · Academically Productive Talk · Talk Moves

1 Introduction

Collaboration as a skill is increasingly emphasized at work and in school as both technology- and pandemic-related changes highlight the importance of developing collaboration skills. Researchers have studied collaboration – and how to measure it – for decades [1], but efforts to develop intelligent systems to support collaboration in human teams have only emerged more recently [2]. Crucially, much of that work has focused on the collaborative practices and AI supports

N. Wang et al. (Eds.): AIED 2023, LNAI 13916, pp. 577–589, 2023.
https://doi.org/10.1007/978-3-031-36272-9_47

that predict content learning and task performance outcomes. There is comparatively little work on intelligent systems that emphasize the development of collaboration skills as valued outcomes in their own right. Accordingly, our long-term focus is on developing AIED systems that monitor and facilitate effective small group classroom collaboration where outcomes of interest are both collaboration skills and learning measures.

Measurement is essential for any intelligent system that aims to provide feedback support for collaborative processes. However, a major challenge of collaboration assessment is the complex nature of the construct. Researchers working from different perspectives have developed a variety of approaches to assess collaboration based on different assumptions, foci, and analytic approaches [3]. Whereas this multidisciplinary approach affords an expansive view on collaboration it risks the jingle-jangle problem, where the same term can refer to different items and different terms can refer to the same item. Conversely, the use of a single framework to investigate collaboration can result in the opposite problem of construct deficiency, where the approach is too narrow to encompass all relevant phenomena. Researchers, therefore resort to defining their own frameworks to analyze collaboration as it unfolds in a particular context, which limits generalizability [4] – a major goal for developing intelligent collaborative measures and supports.

Taking a somewhat different approach, the present study focuses on analyzing classroom collaboration data using a multi-theoretical approach spanning three different perspectives towards measuring and modeling collaborative processes in small groups: Collaborative Problem Solving (CPS) [5], Academically Productive Talk (APT) [6], and communication in Team Cognition (TC) [7]. Our goal is to examine incidence of collaborative behaviors across these different frameworks, examine shared variance across frameworks, and identify underlying latent dimensions of collaboration common to these three schemes. Taken together, our results suggest certain communication practices that can be measured and encouraged to assess and support collaborative learning. They provide a stepping stone toward implementing intelligent systems that can give monitor and facilitate collaborations in classrooms.

1.1 Background and Related Work

We focus on collaboration frameworks which emphasize communication, a central component of how groups learn and work together. Whereas communication can also occur non-verbally, we limit our scope to verbal communications.

Collaborative Problem Solving. CPS describes a set of skills and tasks related to problem solving [8]. It addresses with both the cognitive abilities required to problem-solve as a team and the social dynamics between group members as they coordinate their knowledge and actions [8,9]. Over the last decade, researchers have proposed several frameworks to measure CPS [8]. For example, Sun et al.'s CPS framework identified three core facets of CPS measurable in verbal discourse: "constructing shared knowledge, negotiation/coordination, and maintaining team function" [10]. This framework blends both the cognitive and social dimensions of collaboration whereas other frameworks, such as Andrews-Todd's CPS ontology delineates among these dimensions [11].

Accountable Talk Theory & Academically Productive Talk. With a focus on how essential certain kinds of talk are to many students' learning, education research has focused on how teachers and students use and react to certain kinds of verbal communication. Talk moves, or accountable talk theory, are families of utterances that are derived from ethnographic studies of teachers who were successful in promoting rich, student-driven, academically productive talk in their classrooms. Those talk moves serve three broad instructional goals of ensuring (1) accountability to the learning community (LC), (2) accountability to content knowledge (CK), and (3) accountability to rigorous thinking (RT) [12]. Research has consistently documented how implementing such talk moves promotes student learning [13] and contributes to educational equity [14]. By using these talk moves, students not only contribute their ideas but also attend to and build on their classmates' ideas, helping to ensure they are collaboratively engaging in challenging academic work.

Team Cognition & Communication. Team cognition describes how teams gather, retain, and use information through interactions and communication to work toward their goals [15]. Marlow, Lacerenza, and Salas's [7] framework describes the role of communication in team cognition. They note that how much a team communicates can have little to do with how well they function together, but communication quality impacts how teams perform cognitive work to a greater degree [7]. Communication quality is broad, though, and qualities of communication with relationships to team performance range from clarity of speech to timeliness of information [16]. The specific qualities that are most relevant to the present work are informed by peer mentoring styles identified by Leidenfrost et al. [17], who identified key aspects of communication styles in student mentor-mentee relationships that related to mentee performance: motivating, informative, and timely mentor communication. These are also aspects of communication quality emphasized in Marlow, Lacerenza, and Salas's [7] framework of team communication.

Related work. Whereas most studies analyze collaboration from a unitary perspective, some researchers have utilized multiple perspectives. Jordan and Henderson's [18] foundational work emphasized that the complex nature of interaction data benefits from analysis by theoretically diverse working groups, as it "reveals and challenges idiosyncratic biases" (p. 43). Suthers, Lund, Rosé, and Teplovs' [19] productive multivocality is a guiding example of how to effectively integrate theory and practice from multiple disciplines. Their work presenting shared data sets to groups of analysts from diverse traditions yielded insights into how challenging "individual operationalizations of complex constructs" (p. 588) can reveal the limited scope of how different literatures deal with broad constructs. They argue that analyzing the same data from different perspectives can offer deeper understanding. Analysts from diverse theoretical backgrounds participated in the Productive Multivocality Project over the course of five years, analyzing shared data corpora using their own methods and juxtaposing their analyses with those of the other participants, producing insights into the difficulties and advantages of combining their disparate perspectives. One of these data sets consisted of video data student groups working on chemistry problems in an undergraduate chemistry class. The methods brought to these data included ethnographic analysis , coding and counting analysis , and social network analysis . The researchers found that although they had all studied leadership in the student groups, their interpretations of the construct differed in ways that made clear that each perspective on its own lacked the depth of understanding of the complexities of leadership that they were able to achieve with a comparative analysis of all three approaches [20]. For example, the qualitative and social network analyses both identified a difference between groups in terms of whether they primarily moved procedurally through the steps of a problem or spent more time discussing concepts relevant to the problem at hand. The analysts inferred based on this difference that the procedural approach lost out on meaningful group interactions that could support collaborative learning. The code and count analysis yielded the same difference between groups, but focusing on transactivity yielded findings that throughout the step-by-step procedure, group members were reasoning together and solving steps collaboratively [20]. The ability of disparate analytical approaches to challenge each other in this way argues for inclusion of diverse perspectives when measuring complex constructs. For our work, comparing results from multiple analyses is essential for rigorous measurement so that each perspective can inform our understanding of collaboration.

Current study. We applied three different measures for analyzing collaboration data to investigate how they relate to each other and what can be learned from a multi-theoretical approach. We have three guiding research questions: (RQ1) how often do the behaviors of interest for each measure occur, (RQ2) to what extent do these measures align, and (RQ3) what are the underlying factors of collaboration common to these measures?

Our work contributes to rigorous measurement of collaboration as a step toward developing AIED systems that can assess and support collaborative learning in classrooms. We provide insights into collaboration measurement from diverse theoretical backgrounds and detail how the resultant frameworks can work together to provide a more complete understanding of collaboration in classrooms. We focus on the process of collaboration, as opposed to content outcomes, because the development of collaboration skills is the outcome of interest. This research, which aims to identify underlying dimensions of collaborative discourse that are common across diverse theoretical frameworks, is an important step towards the development of generalizable computational models of collaborative processes for feedback and intervention.

2 Methods

Our data are 31 transcribed videos from dyads and triads of middle school students working on collaborative tasks as part of their normal classroom instruc-

Table 1. Excerpt of student dialog during sensor programming

Speaker	Utterance
S1	So, so far we have the "forever" button and it's if the sound intensity is over 1000 then it will show the icon and play this sound for four beats [TC: Inf, PresentFuture, Plural; APT: Relating, Providing; CPS: MonitoringExecution]
S2	So like four times that length. [TC: Inf, PresentFuture; APT: Relating, Making; CPS: MonitoringExecution]
S1	And if it's not over that, then it'll plot a bar graph. [TC: Inf, PresentFuture; APT: Relating, Providing; CPS: SharingUnderstanding]
S2	Yeah. So the "forever" makes it sense the sound intensity, and if it goes over a thousand then it's gonna show this icon and then play this [TC: Inf, PresentFuture; APT: Providing; CPS: Responding, MonitoringExecution]

tion. Video and audio were collected with iPads and Yeti Blue microphones set up at tables with participating groups during normal class time. Researchers transcribed and analyzed five minutes of audio from each group. These five-minute samples were randomly selected from longer videos with the constraints that the audio and video were of good quality and included the midpoint of the period of small-group work.

Participants. Participants were 34 middle school students from four classes taught by one teacher in a school district with 48.8% female students and a ethnic demography of 62.3% White, 30.0% Hispanic, 3.3% Asian, 3.0% two or more races, 0.9% Black, 0.3% American Indian or Alaska Native, and 0.1% Hawaiian/Pacific Islander students in the 2020–2021 school year. 22.0% of students in this district are receiving free or reduced price lunch. These participants each gave assent and provided consent forms signed by their parents, and the data collection procedure was approved by the appropriate Institutional Review Board.

Collaborative Task. The tasks are part of a Sensor Immersion (SI) curriculum that focuses on developing scientific models and computational thinking. Through wiring and programming sensors and data displays using Micro:Bit hardware and MakeCode block programming software, students practice computational thinking to design and describe their systems. Tasks build from using specific sensors to answer specific questions to students postulating and answering their own questions about their environment by wiring and programming a multi-sensor system.

Table 1 contains a sample of collaborative conversation between two students during a task in the SI curriculum annotated with the three coding schemes (see below). The students in this group are working to program the data display that they wired to a sound sensor. Their goal is to have the setup perform certain actions when the sound sensor registers specific volume levels.

Measures. We analyzed this data set with coding schemes based on each CPS, APT, and TC. Separate teams of annotators worked simultaneously to code the data with their respective scheme. The CPS team was comprised of four coders including one master coder who had previously worked with the CPS scheme. This team worked to reach consensus with the master coder, who checked all annotated transcripts. The APT and TC teams were each comprised of two coders who worked to consensus.

Collaborative Problem Solving - CPS. The CPS framework and coding scheme, as detailed by Sun et al. [10] is built around three facets of collaborative problem solving: constructing shared knowledge, negotiation/coordination, and maintaining team function. Each of these facets is comprised of sub-facets (see Table 2), which are in turn comprised of verbal and non-verbal indicators. These indicators were originally defined in terms specific to a different collaborative task. The CPS coding team, therefore, adapted them to SI data before analyzing this corpus. A team of three coders and one master coder annotated SI transcripts, recorded discrepancies they found in applying the coding scheme to the new data, supplemented the coding scheme with notes and examples from coding the SI data, and discussed utterances of high disagreement until they reached consensus. To ensure the reliability of the adapted coding scheme's application to the SI data, each annotated transcript was reviewed by the master coder until the original coder and the master coder reached consensus. The data can be analyzed at all three levels(facet, sub-facet, and indicator), and we chose to focus on the intermediate sub-facet level here. For this, the indicators were aggregated to the sub-facet level so that if an utterance was coded with any of the indicators that comprise a sub-facet, then it was coded with that sub-facet.

Academically Productive Talk - APT. The APT coding scheme draws on a set of student talk moves detailed by Suresh et al. [21], originally defined to analyze student speech during math lessons. Four talk moves were identified based on suggestions from experts and required relatively minor adaption for the SI curriculum (see Table 2).

Team Communication - TC. To measure communication quality and style as defined by Marlow et al.'s [7] team communication framework and Leidenfrost et al.'s [17] styles of peer mentoring, the TC coding scheme includes indicators of motivational, informative, timely, and collective speech (see Table 2). This coding scheme was originally developed to measure communication styles in competitive video game teams with the explicit purpose of measuring communication quality with a domain-agnostic method [22]. Because it is quite general, it did not require much adaptation to SI.

3 Results and Discussion

RQ1: Occurrence of Collaborative Behaviors Within Schemes. There was considerable variability in the different codes as noted in Table 2, which provides insights into their collaborative processes. For example, APT coding

Table 2. CPS, APT, and TC coding scheme definitions, examples, and mean frequencies

CPS Facet	Sub-facet	Occurrence Mean (SD)	Definition
Constructing shared knowledge	Sharing understanding of problems/solutions	0.26 (0.13)	Contributing expertise and ideas regarding particular problems and toward specific solutions
	Establishing common ground	0.06 (0.05)	Acknowledging ideas, confirming understanding, and clarifying misunderstanding
Negotiation/ coordination	Responding to others' questions/ideas	0.14 (0.06)	Providing feedback, offering reasons for/against claims, and implementing agreed on solutions
	Monitoring execution	0.09 (0.08)	Talking about strategy, progress, and results
Maintaining team function	Taking initiative to advance the collaboration process	0.06 (0.05)	Asking questions, acknowledging others' contributions, and helping to maintain team organization
	Fulfilling individual roles on the team	0.06 (0.05)	Performing own roles and responsibilities
	Participating in off-topic conversation	0.11 (0.16)	Initiating or joining off-topic conversation
APT Facet	**Code**	**Occurrence Mean (SD)**	**Definition**
LC	Relating to another student	0.52 (0.21)	Using, commenting on, or asking questions about a classmate's ideas
LC	Asking for more information	0.05 (0.04)	Requests more information or says they are confused or need help
CK	Making a claim	0.38 (0.15)	Makes a math claim, factual statement, or lists a step in their answer
RT	Providing evidence/ reasoning	0.02 (0.03)	Explains their thinking, provides evidence, or talks about their reasoning
TC Facet	**Code**	**Occurrence Mean (SD)**	**Definition**
Mot	Positive	0.04 (0.05)	Positive emotional valence
	Negative	0.07 (0.08)	Negative emotional valence
Inf	Informative	0.51 (0.20)	Contains or requests useful info
	Uninformative		Lacks substance
Tim	Present/Future	0.85 (0.11)	Present or future tense or referring to present or future
	Past		Past tense or referring to past
Col	Plural	0.11 (0.07)	Collective references to group and group members
	Singular	0.25 (0.10)	Singular references to group members
	No Reference		No reference to group members

indicated that students make a lot of claims (m = 0.38), but do not appear to explain their reasoning behind those claims (m = 0.02). Their CPS discourse which mainly consists of shared knowledge construction (m = 0.33), followed by negotiation/coordination (m = 0.23) and lastly maintaining team function (m = 0.12 – not counting off-topic conversations), mirrors distributions on other tasks [10]. Some indicators require more context to understand what their occurrence rates suggest about collaboration. Are students referring to individual team members (TC: Singular, m = 0.25) more than twice as much as they refer to their team as a collective (TC: Plural, m = 0.11), because they are not working as a team or because they are working together to assign aspects of the task to specific teammates? Are they encouraging off-topic conversations (m = 0.11, SD = 0.16) as much as almost every other CPS behavior, because they are not working together on the task or because they want to relate to their teammates to promote team cohesion?

RQ2: Relationships Between Schemes. To understand relationships across schemes, we aggregated the data by observations and computed Pearson correlation coefficients between CPS sub-facets and APT indicators, CPS sub-facets and TC indicators, and APT indicators and TC indicators. To aggregate from the coded utterances to the observation level, we computed the frequency of each code applied in relation to the total number of utterances in each observation. We found low to moderate mean correlations across all three schemes. CPS and APT yielded a mean r(29) of 0.29 (SD = 0.22, median = 0.30, range = <0.01–0.77). CPS and TC yielded a mean r(29) of 0.20 (SD = 0.18, median=0.18, range = <0.01-0.80). APT and TC yielded a mean r(29) of 0.28 (SD = 0.26, median = 0.20, range = <0.01–0.83). Overall, this suggests a modicum of overlap across the three different coding approaches, confirmation their different foci according to their respective theoretical foundations, while also demonstrating shared variance across schemes.

RQ3: Factors of Collaboration. We conducted a principle components factor analysis with a varimax rotation to identify latent factors comprised of the three different coding schemes. Measures that implied these data were factorable included at least moderate correlation between codes, an overall Kaiser-Meyer-Olkin measure of sampling adequacy of 0.59 (which approaches the recommended threshold of 0.60), and a significant Bartlett's test of sphericity ($\chi^2(78) = 475$, $p<0.001$). We selected six factors with eigenvalues > 1 that explained a cumulative 82% of the variance in our data (see Table 3).

We interpret the first factor as a measure of task-related communication. It includes codes from each scheme that indicate on-topic, informative discourse. The excerpt in Table 1 exemplifies this dimension. The students stay on task, share information, relate to each other, and make claims about their work with almost every utterance. This type of discourse is important for collaboration.

The second factor pertains to finding common ground. APT's "Asking for more information" and CPS's "Establishing common ground" load onto this factor strongly. This dimension is evident when group members prompt each other for relevant information that the group can orient their thinking around. For example, when a triad was struggling to get their data display to respond to sensor input, one student started asking the rest of the group for more information on how to proceed: "Okay, show strings? What should we tell it to do?" Thus, When a group is having difficulty, they center the collaborative effort around prompts for specific information.

The third factor is most strongly informed by CPS's "Monitoring execution" and is most visible in utterances where a group member is assessing the current state of the group's work by making a claim about how their sensor setup is functioning. For example, consider the following excerpt when a team was trying to make their data display react at a certain noise level and were struggling to find the right threshold: "So even my click was louder than this." The implied assessment the student is making is that they currently have the threshold set too low, because even their clicking the computer mouse was louder than their current threshold. CPS's "Sharing understanding of problems/solutions" loads negatively onto this dimension, meaning that the problem has been established, and now the focus is on finding a solution and debugging faulty solutions (i.e., monitoring execution).

Groups display the fourth factor when they focus on what is happening in the present or coming in the future, without communicating negative emotions about it. If something goes wrong in a collaborative task, it can be easy for a group to harp on that past event. Utterances that plan, strategize, or consider possible group achievements can be a vital part of what keeps a collaboration productive, as long as they are not presented with demotivating emotions. Many examples of this dimension occur when participants offer ideas to guide their work toward exciting designs, like a student suggesting fun ways to program the data display, like "We could make it look like it's exploding," and "Let's graph ASMR." These utterances spur on collaborative work toward a planned goal.

The fifth factor is where CPS's "Taking initiative to advance the collaboration process" and TC's "Positive" converge. This dimension assesses motivational discourse and expressly positive emotional valence. These can be simple feedback remarks, like "That is very cool," or personal compliments, like "We're pure geniuses." Discourse in this dimension encourages collaborators to work together and rewards them when they do.

The final factor is informed by only one code. CPS's "Fulfilling individual roles on the team" measures talk that indicates individual group members are owning their responsibilities and doing their part for the collaboration as a whole. This includes offering instructional support when the group is stuck (e.g. "Alright, you gotta do 'on button B press', so add another one of these.") and owning up to mistakes by apologizing for them (e.g. "No, no no no. Sorry. My bad. Scroll over to where you can see the whole thing.").

4 General Discussion

Measuring a complex construct like collaboration benefits from the nuance gained by combining different analytical perspectives [20]. Each coding scheme

Table 3. Factors and loadings yielded by factor analysis

Factor	Qualitative Description	Proportion of Variance	Codes	Loadings
Factor 1	Task content	0.26	TC:Informative	0.84
			APT:RelatingOther	0.81
			APT:MakingClaim	0.82
			CPS:SharingUnderstanding	0.74
			CPS:OffTopic	-0.82
Factor 2	Using information to find common ground	0.15	APT:AskingInfo	0.92
			CPS:EstablishesCG	0.88
Factor 3	Monitoring teamwork	0.11	CPS:MonitorsExecution	0.91
			APT:MakingClaim	0.33
			CPS:SharingUnderstanding	-0.47
			TC:Positive	-0.35
Factor 4	Strategizing and planning	0.11	TC:PresentFuture	0.83
			TC:Negative	-0.80
Factor 5	Motivating	0.10	TC:Positive	0.72
			CPS:Initiative	0.80
Factor 6	Fulfilling role	0.09	CPS:FulfillRole	0.92

used in these analyses focuses on different aspects of collaboration. They each have their own strengths and weaknesses for use as a foundation on which to develop computational models for AIED interventions to measure and support effective collaboration.

Regarding our first research question about occurrence of behaviors of interest, student groups in this sample show similar distribution of the CPS facets in the SI tasks as they do in previously researched tasks. They participate most in shared knowledge construction and least in maintaining team function. APT analysis revealed that these groups almost never provide their reasoning and infrequently ask for more information, despite frequently relating to each other and making claims about their group work. TC points to mostly informative talk, without strong emotional valence, that does not linger on events that have already passed. A system equipped with these assessments could, for example, elicit explicit reasoning and encourage maintenance of team function for the average group in these data.

For our second research question asking how related these measures are, correlations across schemes indicate that they are identifying behaviors that do relate to each other, but only weakly to moderately. We infer that these measures are in fact approaching the same construct from different perspectives and therefore contain some overlap with each other while also focusing on unique aspects of collaboration. Simply put, they are complimentary and not redundant.

The underlying dimensions of collaboration informed by these schemes surfaced by factor analysis answer our third research question. The measures, when taken together, do in fact point to underlying factors of collaboration that none of the schemes can fully assess on its own. Four of the six factors include codes from multiple schemes (factors 1, 2, 3, & 5), indicating dimensions of collaboration that no one scheme accounts for on its own. Two of these include input from all three schemes (factors 1 & 3). In other words, these factor analysis results support the inference that these schemes are complementary, yet distinct, approaches to measuring collaborative processes in student groups.

The overall implication of these findings is that it might be beneficial to incorporate discourse from multiple coding schemes and theoretical frameworks to obtain a more nuanced, contextual analysis of complex classroom collaboration. An initial hypothesis is that computational models might focus on predicting the factors which are largely composed of multiple schemes rather than scheme-specific indicators. An open question is also whether this will result in more accurate and generalizable models because they are focused on more abstract dimensions of collaboration.

Limitations and Future Work. While the data were collected in a real-world classroom setting, the sample was small and occurred in a single teacher's classes in a single school district. We are continuing to collect data in multiple classrooms in multiple schools for subsequent analysis with larger, more diverse samples. The small sample size precluded an analysis of the effects of varying curricula and group composition on the unfolding collaboration dynamics, so replication and generalizability across curricula, domains, and group compositions is an important item for future work.

Conclusion. In order to develop systems that can provide feedback and support collaboration, we require rigorous and reliable tools for measuring collaboration in the first place. This work is a necessary step toward implementing an AIED system that puts these collaboration analytics to work in live measurement and feedback for small groups in classrooms. Effective interventions will also need to consider how, when, and what metrics are presented to groups as they learn collaboratively. Work developing, implementing, and studying such a system is underway with the intent to inform design of interventions in collaborative learning situations and collaborative systems more broadly.

Acknowledgements. This research was supported by the NSF National AI Institute for Student-AI Teaming (iSAT) under grant DRL 2019805. The opinions expressed are those of the authors and do not represent views of the NSF.

References

1. Gedney, J.: Development of an instrument to measure collaboration and satisfaction about care decisions. J. Adv. Nurs. **20**(1), 176–182 (1994)
2. Stewart, A., D'Mello, S.K.: Connecting the dots towards collaborative AIED: linking group makeup to process to learning. In: Penstein Rosé, C., et al. (eds.) AIED 2018. LNCS (LNAI), vol. 10947, pp. 545–556. Springer, Cham (2018). https://doi.org/10.1007/978-3-319-93843-1_40

3. Forsell, J., Forslund Frykedal, K., Hammar Chiriac, E.: Group work assessment: assessing social skills at group level. Small Group Res. **51**(2), 87–124 (2020)
4. Walters, S.J., Stern, C., Robertson-Malt, S.: The measurement of collaboration within healthcare settings: a systematic review of measurement properties of instruments. JBI Evid. Synth. **14**(4), 128–197 (2016)
5. Fiore, S.M., Graesser, A., Greiff, S.: Collaborative problem-solving education for the twenty-first-century workforce. Nat. Hum. Behav. **2**(6), 367–369 (2018)
6. O'Connor, C., Michaels, S., Chapin, S., Harbaugh, A.G.: The silent and the vocal: participation and learning in whole-class discussion. Learn. Instr. **48**, 5–13 (2017)
7. Marlow, S.L., Lacerenza, C.N., Salas, E.: Communication in virtual teams: a conceptual framework and research agenda. Hum. Res. Manag. Rev. **27**(4), 575–589 (2017)
8. Graesser, A.C., Fiore, S.M., Greiff, S., Andrews-Todd, J., Foltz, P.W., Hesse, F.W.: Advancing the science of collaborative problem solving. Psychol. Sci. Public Interest **19**(2), 59–92 (2018)
9. Fiore, S.M., Rosen, M.A., Smith-Jentsch, K.A., Salas, E., Letsky, M., Warner, N.: Toward an understanding of macrocognition in teams: predicting processes in complex collaborative contexts. Hum. Factors **52**(2), 203–224 (2010)
10. Sun, C., Shute, V.J., Stewart, A., Yonehiro, J., Duran, N., D'Mello, S.: Towards a generalized competency model of collaborative problem solving. Comput. Educ. **143**, 103672 (2020)
11. Andrews-Todd, J., Kerr, D.: Application of ontologies for assessing collaborative problem solving skills. Int. J. Test. **19**(2), 172–187 (2019)
12. Resnick, L.B., Asterhan, C.S.C., Clarke, S.N.: Accountable talk: instructional dialogue that builds the mind. In: Marope, M., Vosniadou, S. (eds.) Educational Practices Series 29, pp. 14–32. The International Academy of Education (IAE) and the International Bureau of Education (IBE) of the United Nations Educational, Scientific and Cultural Organization (UNESCO), Geneva (2018)
13. Webb, N.M., Franke, M.L., Ing, M., Turrou, A.C., Johnson, N.C., Zimmerman, J.: Teacher practices that promote productive dialogue and learning in mathematics classrooms. Int. J. Educ. Res. **97**, 176–186 (2019)
14. O'Connor, C., Michaels, S.: Supporting teachers in taking up productive talk moves: the long road to professional learning at scale. Int. J. Educ. Res. **97**, 166–175 (2019)
15. Cooke, N.J., Gorman, J.C., Myers, C.W., Duran, J.L.: Interactive team cognition. Cogn. Sci. **37**(2), 255–285 (2013)
16. González-Romá, V., Hernández, A.: Climate uniformity: its influence on team communication quality, task conflict, and team performance. J. Appl. Psychol. **99**(6), 1042 (2014)
17. Leidenfrost, B., Strassnig, B., Schabmann, A., Spiel, C., Carbon, C.-C.: Peer mentoring styles and their contribution to academic success among mentees: a person-oriented study in higher education. Mentoring Tutoring: Partnership Learn. **19**(3), 347–364 (2011)
18. Jordan, B., Henderson, A.: Interaction analysis: foundations and practice. J. Learn. Sci. **4**(1), 39–103 (1995)
19. Suthers, D.D., Lund, K., Rosé, C.P., Teplovs, C.: Achieving productive multivocality in the analysis of group interactions. In: Suthers, D.D., Lund, K., Rosé, C.P., Teplovs, C., Law, N. (eds.) Productive Multivocality in the Analysis of Group Interactions. CCLS, vol. 15, pp. 577–612. Springer, Boston, MA (2013). https://doi.org/10.1007/978-1-4614-8960-3_31

20. Rosé, C.P.: A Multivocal Analysis of the Emergence of Leadership in Chemistry Study Groups. In: Suthers, D.D., Lund, K., Rosé, C.P., Teplovs, C., Law, N. (eds.) Productive Multivocality in the Analysis of Group Interactions, pp. 243–254. Springer, New York. (2013). https://doi.org/10.1007/978-1-4614-8960-3_13
21. Suresh, A., et al.: Using transformers to provide teachers with personalized feedback on their classroom discourse: The TalkMoves application. Paper presented to the 2021 AAAI Conference on Artificial Intelligence in K-12 Education (2021)
22. Reitman, J. G.: Generalizable Communication Styles in Novice and Expert Team Performance. Doctoral Dissertation. UC Irvine, (2022)

Development and Experiment of Classroom Engagement Evaluation Mechanism During Real-Time Online Courses

Yanyi Peng(✉), Masato Kikuchi, and Tadachika Ozono

Nagoya Institute of Technology, Nagoya, Aichi 4668555, Japan
pyy@toralab.org, {kikuchi,ozono}@nitech.ac.jp

Abstract. Student engagement is an essential indicator of the quality of teaching. However, during real-time online classes, teachers are required to balance course content and observe students' reactions simultaneously. Here, we aim to develop a web-based classroom evaluation system for teachers that evaluates student participation automatically. In this study, we present a novel mechanism that evaluates student participation based on multi-reaction. The system estimates students' head poses and facial expressions through the camera and uses the estimation results as criteria for assessing student participation. Additionally, we conducted two evaluation experiments to demonstrate the system's effectiveness in automatically evaluating student participation in a real-time online classroom environment with nine students. The instruction experiment was divided into head pose estimation, facial expression recognition, and multi-reaction estimation; the accuracy rates were 70.0%, 60.0%, and 80.0%, respectively. Although participants did not show many heads poses in the simulation classroom experiment, the system evaluated the classroom by assessing expression recognition, and 70% of the questions showed the same results as those of a questionnaire.

Keywords: Classroom engagement evaluation · Student behavior detection · Online classroom

1 Introduction

Student behavior is independent; even in a traditional face-to-face classroom, it is difficult for teachers to observe each student's classroom reaction. Moreover, students' reactions in the classroom are not monolithic; and are a combination of head poses, body movements, and facial expressions. This problem is prominent in online classrooms, where the teacher must observe individual student performance from a camera to determine real-time feedback of the students. This in turn causes the teachers to lose focus on the class content, which is further aggravated by the multiple reactions of students.

Here, we develop an engagement evaluation mechanism that automatically evaluates student participation based on students' multi-reaction in real-time

N. Wang et al. (Eds.): AIED 2023, LNAI 13916, pp. 590–601, 2023.
https://doi.org/10.1007/978-3-031-36272-9_48

classes. A novel platform to conduct online education activities and evaluate the overall classroom quality based on head pose estimation and facial expression recognition of all students without affecting the attention of students and teachers has been proposed [13]. In this paper, we discuss the system proposed in [13], and evaluate the system's performance through two evaluation experiments with nine participants, including an instruction experiment and a simulation classroom experiment. The first experiment evaluates students' system performance under the specified reactions (head posture response and expression response). The second experiment tests the relationship between students' autonomous reactions and classroom quality evaluated by the system.

2 Related Work

2.1 Behavior Estimation

Head orientation is used to communicate information during interactions, such as understanding or disagreement. Guo et al. [5] used a CNN model to detect classroom head-raising rates to measure participation. Xu et al. [16] proposed a method to analyze students' attention based on head pose estimation. Detecting students' head poses can help infer current class participation effectively. Thus, this study focused on detecting nodding and shaking head poses via a camera. Because only one student is visible in each camera, the multi-target situation does not apply, and estimation errors are reduced.

Human expressions also convey significant information and emotions during communication. Ekman et al. [4] identified universal facial expressions across cultures: anger, disgust, fear, happiness, sadness, and surprise. Currently, some well-known CNN structures are being applied to expression recognition, for example, VGGNet [14] is used to extract image features leveraging a simple structure and excellent versatility. We reference the basic expression and make adjustments based on student emotions in the teaching environment. The study selects happiness and sadness from the basic expressions and adds expressions of focused, confused, and sleepy.

2.2 Class Evaluation System

For the class evaluation system, Linson et al. [9] proposed a method to detect student engagement by detecting the direction of students' eyes while watching classroom lectures. Liu et al. [10] proposed a system for measuring teacher-student interaction in the classroom to determine student response to teacher instructions. However, their instruction library only included five common instructions. Recently, Chen et al. [2] presented a method for evaluating teaching that involves recognizing students' expressions and behaviors in offline classroom videos. Based on this research and the rise in online classes, we propose a classroom evaluation system using the results of detecting students' head poses and expressions, which can be applied to online classrooms.

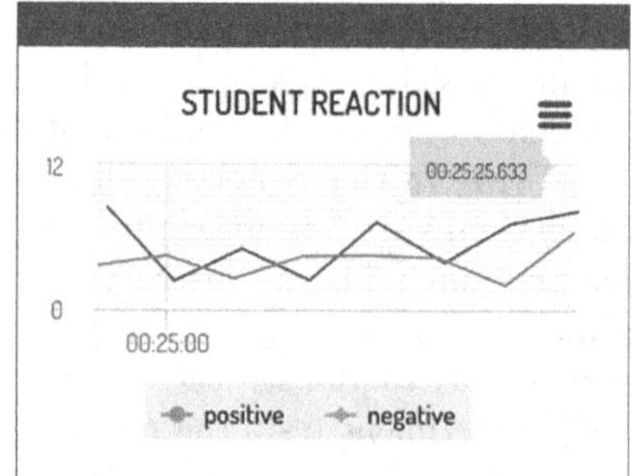

Fig. 1. Student reaction page for teachers

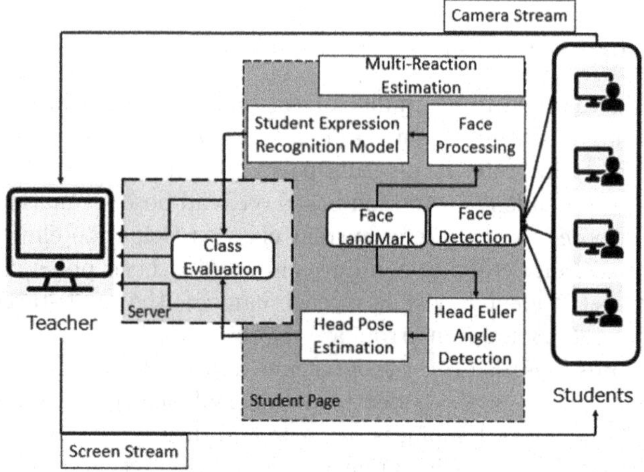

Fig. 2. Structure of online classroom system

3 Online Education Platform

We develop a simple online classroom system to provide a platform for teachers and students to conduct online learning activities. The system implements the various needs of teachers in multiple classrooms, and teachers create new classrooms through classroom creation panel function. Teachers can view the changes in the number of positive and negative reactions of students, as shown in Fig. 1.

Figure 2 shows the system's structure, which builds a communication channel between the student page and teacher page using WebRTC and WebSocket. The system uses the multi-reaction estimation module to identify the student's reactions in the camera through head pose estimation and expression recognition and determines the student's learning status and participation state (positive, negative, or neutral). To improve students' expression accuracy and real-time detection, the system conducts reaction detection on the student page.

4 Multi-reaction Estimation

Both teachers and students are affected by class performance in a classroom. The teacher adjusts the class content based on students' reactions, including head pose and expression feedback. The system evaluates students' reactions from multiple reactions, categorizing them as positive, negative, or neutral.

4.1 Student Head Reaction

Based on the behavior of East Asians, our system assumes that students nod their heads to indicate that they understand the teacher's instructions or content and shake their heads to indicate that they do not. The system predicts the student's head pose by face landmarks. Therefore, to improve the accuracy of head pose estimation, the students should place their heads in the middle of the camera area.

To reduce the pressure on the server in detecting multiple students' reactions simultaneously, we implement the head pose estimation on the student page. The video stream data from the student camera detect the current head pose of each student. We used the Tiny Face Detector [6], which leverages face-api.js[1] to detect students' faces in the video stream. The student page conducts face detection approximately every 150 ms, and the system calculates the Euler angles from the acquired face landmarks. The values of pitch, yaw, and roll are calculated from the rotation vector, which is obtained from the Euler angle. Then, the system obtains the student's head pose using a simple tracking method, which determines the change in head pose by the angle difference between adjacent frames.

4.2 Student Expression Reaction

Active behaviors require students' cooperation, such as head nods and shakes. Therefore, we propose a method to predict students' passive responses by detecting their expressions.

The classification rules in the conventional classification of expressions described in Sect. 2.1 are inappropriate for student expressions in the teaching environment. Therefore, we propose a classification method for student expressions based on the classroom environment. In this study, we primarily classify classroom expressions as happy, focused, confused, sad, and sleepy based on students' emotions in the classroom. Additionally, we consider students who exhibit no expression as neutral. The dataset of student expressions in the classroom we created is shown in Table 1.

In this study, we construct a facial expression recognition model based on "our frame" [7]. After training the expression recognition model, the Keras model is converted to TensorFlow.js Layers format via the TensorFlow.js API [1]. The

[1] https://github.com/justadudewhohacks/face-api.js/.

Table 1. Dataset of student expression for classroom expression recognition, containing 1,102 training data and 290 testing data from JAFFE [12], CK+ [11], SFEW 2.0 [3], and Google images.

Category	Samples	Train	Test
Happiness	216	170	46
Focused	85	68	17
Confused	134	106	28
Sad	152	114	38
Sleepy	96	76	20
Neutral	709	568	141
Total	1,392	1,102	290

model is loaded on the student page, and the converted model predicts the student's expression reactions directly on the student page. Thus, the stability of detection and accuracy of results can be ensured.

5 Online Classroom Evaluation

In our system, class evaluation is based on the students' reaction estimation. When the system estimates the students' head poses and expression reactions on each student page, the system assesses the current classroom participation by collecting the results of individual student responses. We believe that teaching content before and after the time is interconnected. Taking these factors into consideration, we increase the influence of positive responses on subsequent classroom evaluations. Therefore, we present a weighted method to calculate the overall student participation considering interference in the above mentioned cases.

As described in Sect. 4, we consider students' reactions as positive, negative, and neutral, and evaluate the quality of the class r_{class} using the following equation:

$$r_{class} = \sum_{i=1}^{n} w_i r_i, \tag{1}$$

where r_i and w_i are the reaction and weight of the i-th student s_i which considers the student's past reactions, respectively. r_{class} shows the value of a class reaction considering $r_1, r_2, r_3, ..., r_n$. When most students are positive in the classroom, the evaluation is positive, and when most students are negative, the evaluation is negative. Therefore, the current classroom status is considered positive when r_{class} is greater than 0.2, negative when r_{class} is less than -0.2, and neutral when r_{class} is -0.2 to 0.2. This method considers students' past and current reactions to reduce the influence of students who have not participated in the past on current evaluations.

To define the value of r_i for each student's reaction, we set 1 for positive, -1.2 for negative, and 0 for neutral. We present the positive and the negative reactions with unequal values, because teachers should conduct classroom activities that enable students to understand the content of the lesson in the classroom. Therefore, we assign a higher negative value to negative reactions. For the weight of each student, we define the value of w_i as

$$w_i = \frac{f_i}{\sum_{j=1}^{n} f_j}. \tag{2}$$

This weight calculation algorithm reduces the weight coefficient proportion f_i for student s_i as the number of students increases. When the number of students in the classroom is sufficiently large, the weight coefficient proportion f_i of student s_i approaches 0. Therefore, when using this method to assess classroom levels, we designate classrooms for approximately ten students.

Specifically, w_i gives the number of negative reactions n_i and the number of positive reactions p_i of the student s_i from the beginning to the current moment. Additionally, we consider the effect of the increasing number of student responses on the classroom evaluation. We implement a logarithmic function to reduce the effect of students' response times.

The weight coefficient f_i of student s_i is expressed as

$$f_i = \log(2p_i + n_i). \tag{3}$$

We implement the logarithmic function to reduce the effects of students' past reaction times on the weight coefficient, where p_i is twice n_i. This is because we hypothesize that students who responded more positively in the past would better understand the content taught by the teacher, and therefore, can easily understand the current classroom content.

6 Experiments

To assess the performance of our online classroom support system and class evaluation method, we conducted two experiments to evaluate head pose estimation, facial expression recognition, and classroom evaluation in the online real-time classroom environment. The participants were nine Asians; four females, and five males. All participants used MacBooks, and participated in the experiments on Google Chrome.

6.1 Experiment I: Instruction Experiment

We evaluated students' reaction estimation with three experiments, head pose estimation experiment, facial expression recognition experiment, and multi-reaction estimation experiment, each lasting 5 min. For each experiment, we divided the intervals in 10-s increments, and specified the reaction actions of each participant within the interval, as shown in Table 2. In these examples, the student was required to make negative reactions in 11–20 s, and positive

Table 2. Instruction example for a participant in 1 min

	0–10 s	11–20 s	21–30 s	31–40 s	41–50 s	51–60 s
Student	neutral	negative	neutral	neutral	positive	positive

reactions in 41–50 s and 51–60 s, according to the instructions. In each reaction interval, we asked participants to look directly into the camera and repeat the reaction three times.

In each interval, we obtained 65 detection points, among which we consider that the time is neutral when the sum of detected positive and negative reactions is less than ten detection points or the difference between positive and negative reactions is less than five. Otherwise, the reaction of the interval is taken as the greater of positive or negative.

Evaluation of Head Poses. In this experiment, we specified each participant's reaction as a nod, head shaking, or neutral. Within five minutes, we obtained 270 reaction intervals from nine participants. In this part, we considered nod as a positive reaction and shake head as a negative reaction. Within the 66 nod action interval, the system accuracy was 98.5%; however, in the case of the shaking and neutral actions, the accuracy rates were 86.0% and 83.7%, respectively.

Evaluation of Facial Expression Recognition. We obtained the participants' expressions (happy, focused, sad, confused, sleepy, and neutral) in ten seconds. The experiment comprised a total of 270 test intervals for the nine participants. In this part, the detection results of the system were positive (happy, focused), negative (sad, sleepy, confused), and neutral. The accuracy rates of positive, negative, and neutral expression recognition were 45.1%, 76.2%, and 96.3%, respectively. For focused expression (positive), we obtained an unsatisfactory accuracy in expression recognition. Within the 32 focused expression intervals, the system detected 8 negative and 19 neutral intervals. In addition, the expression recognition accuracy of some participants in the system was approximately 100%, while some participants had very low accuracy rates.

Evaluation of Multi-reaction Estimation. In this part, we estimated the participants' head pose and expression simultaneously. Because students may make head poses and facial expressions simultaneously, the system prioritizes head poses over facial expressions. After one minute of the experiment, one participant began to doze off. In order to eliminate the effect of this participant's other behaviors on the evaluation experiment, we did not consider the data for this participant. As a result, eight effective participants were considered in 240 reaction intervals. Finally, we obtained accuracy rates of 71.0%, 89.3%, and 95.9% for positive, negative, and neutral reactions, respectively.

6.2 Experiment II: Simulation Classroom Experiment

Classroom Evaluation in Instruction Experiment. To evaluate the evaluation method proposed in Sect. 5, we collected all students' reactions described in Sect. 6.1, and calculated the predicted r_{class}, and compared with the r_{class}

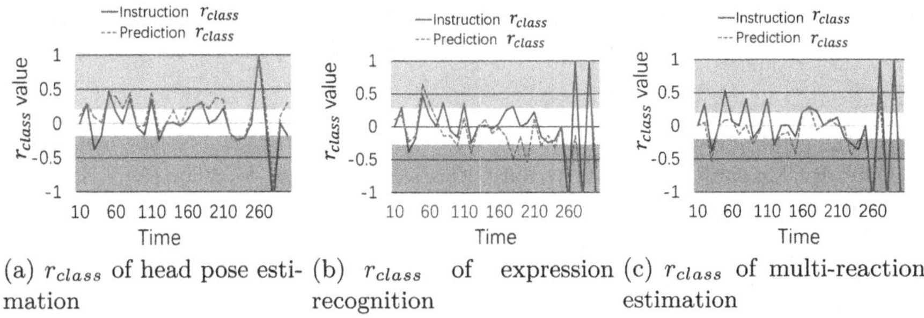

(a) r_{class} of head pose estimation (b) r_{class} of expression recognition (c) r_{class} of multi-reaction estimation

Fig. 3. Classroom evaluation result r_{class}

of instruction. The prediction r_{class} is plotted based on the system's predictions of student reactions. The instruction r_{class} is plotted based on the instructions to the students during the experiment. The r_{class} of each experiment is shown in Fig. 3. In Fig. 3, we used the blue section to mark positive r_{class}, and red to mark negative r_{class}. In this instruction experiment, the classroom evaluation function of the system could assess the current level of student participation in the classroom. The accuracy rates of student participation level was 70.0%, 60.0%, and 80.0% in head pose estimation, facial expression recognition, and multi-reaction estimation, respectively.

Classroom Evaluation in Simulation Classroom Experiment. We conducted a simulation classroom experiment to determine whether the system can accurately evaluate student participation in a teaching environment. Participants were asked to answer ten questions; the difficulty of each question is shown in Table 3. Then, reactions to each question were recorded in a questionnaire. During the experiment, the system evaluated the participants' reactions and assessed the overall quality of the class. For each question, we allowed 40 s for the participants to answer and 30 s for them to complete the questionnaire. We divided this interval into 5-s intervals to collect the students' reactions. For each interval, we obtained 32 detection points. We considered the time neutral if the number of positive or negative reactions was less than five and the difference between positive and negative reactions was less than three. Otherwise, we considered the interval's reaction to be positive or negative based on which was greater. For the survey questionnaire, the main contents included the following:

- Overall feeling of the question.
- How did you feel when you saw the question?
- How did you feel when you saw the answer?
- If there is hint: Do you think the hint is useful?

Figure 4 shows the change in the system's classroom evaluation results in r_{class} at various times for each question in the simulation classroom experiment. By observing r_{class}, the teacher can infer how the students' psychology changes

Table 3. Difficulty of each question; ^ means English question, * means wrong answer to the question. *Italic* type indicates easy questions, **bold** type indicates hard question.

	Q1	*Q2*	Q3	**Q4^**	**Q5***	Q6	*Q7*	Q8	**Q9^**	**Q10***
Difficulty	Medium	Easy	Medium	Hard	Hard	Medium	Easy	Medium	Hard	Hard

Table 4. The results of the questionnaire and system detection. Pos is positive, and Neg is negative; (num) denotes the difference between positive and negative reactions for all student detection intervals. For example, during Q1, the system detected five more negative student reactions than positive reactions.

Reaction	Question									
	Q1	*Q2*	Q3	**Q4^**	**Q5***	Q6	*Q7*	Q8	**Q9^**	**Q10***
Questionnaire	Neg	Neutral	Neg	Neg	Neg	Pos	Neutral	Neutral	Neg	Neg
Detection	Neg(5)	Neg(3)	Neg(12)	Neg(13)	Neg(16)	Neg(6)	Pos(3)	Neg(4)	Neg(13)	Neg(5)

for each question. In Q3, **Q4^**, **Q5***, and **Q9^**, r_{class} are mostly negative. However, in *Q2*, Q6, *Q7*, and **Q10***, the system's prediction from the classroom evaluation shows that the students are in a neutral state. For **Q10***, which has a mismatched question and answer, the evaluation results r_{class} of the students' reactions eventually improve, probably because the participants were happy that the evaluation experiment was over. In most questions, r_{class} improves after the students see the answers, reflecting positive changes in their behavioral reactions.

Based on the results of the questionnaires, participants found the hints for Q3 and Q6 to be helpful and the hint for **Q10*** to be the least useful. By comparing the evaluated results of the participants with the questionnaires, we confirmed that the system could correctly evaluate the overall quality of the classroom in most cases. The hint scores from the questionnaire also showed that the classroom evaluation results in r_{class} increased for 25–35 s in the more useful Q3 and Q6, but there was no significant change in r_{class} between 25–35 s in **Q10***, where the hint was not as helpful. For the easy *Q2* and *Q7*, r_{class} had more detection points within the neutral value segment compared to other questions. In addition, during the two difficult questions **Q4^** and **Q9^**, the results of the class evaluation from the system showed that the student's responses, although the r_{class} values increased when a hint appeared between 25–35 s, they decreased when the answers appeared between 35–40 s, probably due to the inconsistency between the answers to the questions and the student's ideas. In addition, although **Q10*** was a hard question, the r_{class} values assessed by the system were closer to neutral compared to the other three hard questions, which may be because after this the whole evaluation experiment would be over, making the participants happy. Through the analysis of r_{class} and the questionnaires, we believe that the system can make valid assessments of the classroom evaluation in small classrooms, even in teaching environments (Table 4).

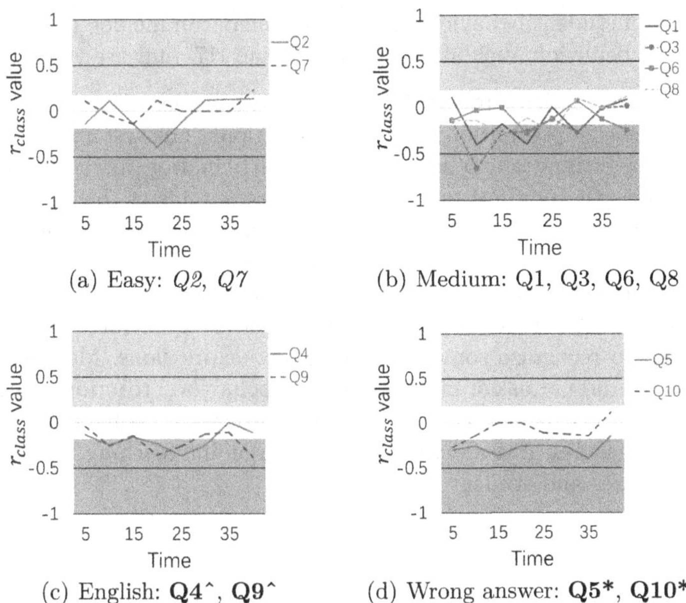

Fig. 4. Classroom evaluation result r_{class} for each question

7 Discussion

The experimental results showed that the system could effectively evaluate the quality of online classes in real-time. In the instruction experiment, the system demonstrated accuracy in detecting head pose, recognizing facial expressions, and recognizing multiple student responses. In the simulation classroom experiment, the system showed that its classroom evaluation function could effectively assess the overall student listening status, which can help reduce the teacher's burden of monitoring student reactions during online classes. However, there are some limitations to the system.

The Dataset for Classroom Experiment: Regarding student expression for classroom expression recognition, some samples were collected from laboratory-controlled samples, which caused the expressions of these samples to be intentional and exaggerated. As a result, the prediction results of the facial expression recognition model for some students' stable expressions were relatively low in the evaluation experiment. Li et al. [8] presented a novel large-scale spontaneous micro-expression dataset, which can effectively solve the above-mentioned problem.

Multi-reaction Estimation: We combined head pose and expression to accurately classify student feedback in the classroom. However, the system requires students to position their heads in the middle of the camera, and the system cannot easily detect extremely stable behavior reactions. Moreover, considering the diversity of student behavior, future researchers can include the detection of other student behavior, such as mouse interaction [17] and eye tracking [15].

Online Classroom Evaluation Method: In the proposed system, class evaluation is based on the estimation of students' reactions. The system uses head pose and expression estimation to assess student participation in the classroom. In the instruction classroom experiment, the accuracy rates of the classroom evaluation method were 70.0%, 60.0%, and 80.0%, in head pose estimation, facial expression recognition, and multi-reaction estimation, respectively. In the simulation classroom experiment, the assessment results of classroom evaluation were the same as those of the questionnaire in 70% of the questions, even when the system failed to recognize some students' micro-expressions. Moreover, in this experiment, we did not ask students to present behavioral reactions such as head pose. Future studies can consider evaluating the quality of the class by incorporating other measures of student engagement on online teaching platforms, such as in-class quizzes and student self-reports.

8 Conclusion

This paper presented a new online education support system that evaluates student participation based on the classroom evaluation method. In this method, we collected students' reactions, including head poses and expressions, and categorized these reactions as positive, negative, and neutral. Then, we conducted two experiments to evaluate the automatically evaluated classroom quality result from a real-time online classroom with nine students. During the instruction experiment, the system assessed the classroom evaluation with accuracy rates of 70.0%, 60.0%, and 80.0%, in head pose estimation, facial expression recognition, and multi-reaction estimation, respectively. When the system evaluated the classroom evaluation in the simulation classroom experiment, 70% of the questions showed the same response as those of the questionnaires, even when the system failed to recognize some students' micro-expressions. The experiment shows that the system effectively detects Asian students' reactions and evaluates the classroom in small online classes.

Acknowledgements. This work was supported in part by JSPS KAKENHI Grant Numbers JP19K12266, JP22K18006.

References

1. Abadi, M., Agarwal, A., Barham, P., et al.: TensorFlow: large-scale machine learning on heterogeneous systems (2015). http://arxiv.org/abs/1603.04467. Software available from http://tensorflow.org
2. Chen, Z., Liang, M., Yu, W., Huang, Y., Wang, X.: Intelligent teaching evaluation system integrating facial expression and behavior recognition in teaching video. In: Proceedings of the 2021 IEEE International Conference on Big Data and Smart Computing (BigComp), pp. 52–59 (2021)

3. Dhall, A., Murthy, O.R., Goecke, R., Joshi, J., Gedeon, T.: Video and image based emotion recognition challenges in the wild: EmotiW 2015. In: Proceedings of the 2015 ACM on International Conference on Multimodal Interaction (ICMI), pp. 423–426 (2015)
4. Ekman, P., Friesen, W.V.: Constants across cultures in the face and emotion. J. Pers. Soc. Psychol. **17**(2), 124–129 (1971)
5. Guo, Q.: Detection of head raising rate of students in classroom based on head posture recognition. Traitement du Signal **37**(5), 823–830 (2020)
6. Hu, P., Ramanan, D.: Finding tiny faces. In: Proceedings of the 2017 IEEE Conference on Computer Vision and Pattern Recognition (CVPR), pp. 951–959 (2017)
7. Kuo, C.M., Lai, S.H., Sarkis, M.: A compact deep learning model for robust facial expression recognition. In: Proceedings of the 2018 IEEE/CVF Conference on Computer Vision and Pattern Recognition Workshops (CVPRW), pp. 2202–2210 (2018)
8. Li, J., et al.: CAS(ME)^3: a third generation facial spontaneous micro-expression database with depth information and high ecological validity. IEEE Trans. Pattern Anal. Mach. Intell. **45**(3), 2782–2800 (2023)
9. Linson, A., Xu, Y., English, A.R., Fisher, R.B.: Identifying student struggle by analyzing facial movement during asynchronous video lecture viewing: towards an automated tool to support instructors. In: Proceedings of the 23rd International Conference on Artificial Intelligence in Education (AIED), pp. 53–65 (2022)
10. Liu, L., Liu, W., Fan, Z., Xu, J., Cheng, W.: A measuring system for teacher-student interaction in classroom. In: Proceedings of the 2019 International Conference on Modern Educational Technology (ICMET), pp. 71–76 (2019)
11. Lucey, P., Cohn, J.F., Kanade, T., Saragih, J., Ambadar, Z.: The extended Cohn-Kanade dataset (CK+): a complete dataset for action unit and emotion-specified expression. In: Proceedings of the 2010 IEEE Computer Society Conference on Computer Vision and Pattern Recognition Workshops (CVPRW), pp. 94–101 (2010)
12. Lyons, M.J., Akamatsu, S., Kamachi, M., Gyoba, J.: Coding facial expressions with Gabor wavelets. In: Proceedings of the 3rd IEEE International Conference on Automatic Face and Gesture Recognition (FG), pp. 200–205 (1998)
13. Peng, Y., Kikuchi, M., Ozono, T.: Online classroom evaluation system based on multi-reaction estimation. In: Proceedings of the 20th IEEE/WIC/ACM International Conference on Web Intelligence and Intelligent Agent Technology (WI-IAT), pp. 500–505 (2021)
14. Simonyan, K., Zisserman, A.: Very deep convolutional networks for large-scale image recognition. In: Proceedings of the 3rd International Conference on Learning Representations (ICLR), pp. 1–14 (2015)
15. Strohmaier, A.R., MacKay, K.J., Obersteiner, A., Reiss, K.M.: Eye-tracking methodology in mathematics education research: a systematic literature review. Educ. Stud. Math. **104**(2), 147–200 (2020)
16. Xu, X., Teng, X.: Classroom attention analysis based on multiple Euler angles constraint and head pose estimation. In: Proceedings of the 26th International Conference on Multimedia Modeling (MMM), pp. 329–340 (2020)
17. Zhang, Z., Li, Z., Liu, H., Cao, T., Liu, S.: Data-driven online learning engagement detection via facial expression and mouse behavior recognition technology. J. Educ. Comput. Res. **58**(1), 63–86 (2020)

Physiological Synchrony and Arousal as Indicators of Stress and Learning Performance in Embodied Collaborative Learning

Lixiang Yan(✉), Roberto Martinez-Maldonado, Linxuan Zhao, Xinyu Li, and Dragan Gašević

Monash University, Clayton, Victoria 3108, Australia
lixiang.yan@monash.edu

Abstract. Advancements in sensing technologies, artificial intelligence (AI) and multimodal learning analytics (MMLA) are making it possible to model learners' affective and physiological states. Physiological synchrony and arousal have been increasingly used to unpack students' affective and cognitive states (e.g., stress), which can ultimately affect their learning performance and satisfaction in collaborative learning settings. Yet, whether these physiological features can be meaningful indicators of students' stress and learning performance during highly dynamic, embodied collaborative learning (ECL) remains unclear. This paper explores the role of physiological synchrony and arousal as indicators of stress and learning performance in ECL. We developed two linear mixed models with the heart rate and survey data of 172 students in high-fidelity clinical simulations. The findings suggest that physiological synchrony measures are significant indicators of students' perceived stress and collaboration performance, and physiological arousal measures are significant indicators of task performance, even after accounting for the variance explained by individual and group differences. These findings could contribute empirical evidence to support the development of analytic tools for supporting collaborative learning using AI and MMLA.

Keywords: CSCL · Learning Analytics · Educational Data Mining · Physiological Synchrony · Arousal · Stress · Performance

1 Introduction

Embodied collaborative learning (ECL) provides unique opportunities for students to practice key procedural and collaboration skills in co-located, physical learning spaces where they need to interact with others and utilise physical and digital resources to achieve a shared goal [14]. This learning format is highly valued in high-stakes disciplines, such as healthcare education, where learning experiences can remain ecologically valid without exposing students to unnecessary risks [20]. Based on the embodied cognition theory, students' affective and

N. Wang et al. (Eds.): AIED 2023, LNAI 13916, pp. 602–614, 2023.
https://doi.org/10.1007/978-3-031-36272-9_49

cognitive states (e.g., stress) contain vital information for understanding their learning experience during ECL as such experiences are shaped by how their body responds to various stimuli [8]. Thus, capturing students' physiological data is critical to advance collaborative learning research and unpack individual students' collaborative learning experience in co-located physical settings [21].

Recent innovations in sensing technologies, artificial intelligence (AI) and multimodal learning analytics (MMLA) are enabling unprecedented opportunities to study ECL, and in general, collaborative learning, from the cognitive and affective perspectives [10,16]. In particular, wearable biometric sensors have made it possible to unobtrusively capture students' physiological data during collaboration, beyond laboratory-controlled settings, within authentic learning contexts, improving the ecological validity of educational research [23,28,31].

Physiological synchrony (PS) and physiological arousal (PA; defined in the Background), modelled from students' physiological signals, have been increasingly used in collaborative learning research. Specifically, both of these physiological features have shown evident relevance for unpacking the complexity of the cognitive [23,26], metacognitive [6,22], and affective processes [5,11,31] that unfold during collaborative learning. Modelling students' PS and PA could also capture essential latent variables linked with critical educational constructs (e.g., cognitive load [15], socially shared regulation [18], and self-efficacy [19]) that could provide meaningful insights to inform the pedagogical and instructional design of learning activities.

Emerging research has provided empirical evidence to support the value of analysing physiological signals captured by wearable biometric sensors in advancing collaborative learning research [5,23,28]. Yet, most of these studies have focused on PS and PA in collaborative situations that involved a relatively low level of physical activities (e.g., collaborative learning around a table [23]). To the best of our knowledge, none of the existing studies has investigated the potential of PS and PA as indicators of stress and learning performance in highly dynamic ECL situations where students can freely move around [33], which could have unknown effects on individuals' physiological signals. Additionally, prior studies rarely considered whether the physiological indicators of collaborative learning would remain significant after accounting for the variance explained by individual and group differences (e.g., self-efficacy and group allocation), especially since the outcomes and experiences of collaborative learning vary across different group compositions [27].

The study presented in this paper investigates whether PS and PA, modelled from students' heart rates using linear mixed-effects models (LMMs), are significant indicators of their perceived stress and learning performance in ECL after accounting for individual and group differences. The findings of the current study contribute empirical evidence regarding the potential of PS and PA as indicators of stress and learning performance in ECL. Such evidence could provide the empirical basis for future AI in education and LA studies that aim at developing tools for supporting formative assessment and reflective practices in ECL and other learning situations where students are physically active [5,33].

2 Background and Related Work

PS and PA, a pair of common indicators used in psychophysiology research for understanding cognition and affect, have shown increased relevance in collaborative learning research as measures of higher-order and latent constructs such as stress and regulation [17]. On the one hand, PS occurs when the physiological process of two or more individuals become interdependent or associated with each other [7]. Specifically, group-level PS has been related to students' learning success and their social regulation process in collaborative learning [12,21,28]. On the other hand, PA often occurs when individuals' sympathetic response activates after exposure to certain stimuli, resulting in increased heart rate and skin conductivity, and measurable changes in other physiological signals [4]. Existing findings have established the association between students' individual-level PA and their cognitive load and self-efficacy in collaborative learning [15,19].

Prior studies modelling PS and PA in collaborative learning settings have mainly relied on individuals' electrodermal activity (EDA) and heart rate data captured with wearable biometric sensors (e.g., Empatica E4) [12,30]. While EDA sensors can provide fine-grained temporal data [6,12], the reliability of EDA data captured by wearable sensors is questionable in contexts with constant hand movements (such as the case of ECL), whereas the heart rate data has proven reliable compared to other medical-grade devices [25]. For example, Sobocinski et al. [30] have used heart rate data to model students' physiological synchrony in advanced physics lessons that involved hands-on experiments. Consequently, heart rate data captured by wearable sensors can be a suitable and reliable source for capturing and modelling students' PS and PA in ECL.

Although the associations of PS and PA with individuals' stress and experiences have been extensively studied in individual psychophysiology studies (e.g., see [4] and [17], respectively), such relationships still need to be clarified in collaborative learning research, especially in embodied and highly dynamic learning contexts. Martinez-Maldonado et al. [24] presented the only study that visualised the sequential pattern of PA and students' learning actions in ECL through an end-user dashboard but did not explore further into the association between physiological indicators and students' stress and learning performance.

We address the aforementioned literature gaps by investigating the role of PS and PA as indicators of students' stress and learning performance in ECL. Two research questions were studied. **RQ1** – To what extent do *individual-level* PS and PA indicate students' perceived stress and learning performance after accounting for individual and group differences? **RQ2** – To what extent do *group-level* PS and PA indicate students' perceived stress and learning performance after accounting for individual and group differences?

3 Methods

3.1 Educational Context

The study was conducted in the context of an authentic undergraduate classes that involves high-fidelity healthcare simulations in a (7 by 9.5 m) specialised classroom. This is a physically simulated patient room with four beds, medical equipment, and four high-fidelity patient manikins with adjustable and measurable vital signs (e.g., heart rates and pulse). In each class, two simulation sessions are conducted. Each simulation consisted of a group of four students, with two (graduate nurses) entering the simulation classroom first and the other two (ward nurses) waiting outside to be called in. Students were randomly assigned to two different scenarios without prior knowledge of the scenarios. Scenario A involved a *covertly* deteriorating patient, among the four patients in the room, that can only be identified through conducting a systematic vital assessment. Scenario B involved an *overtly* deteriorating patient and her daughter (role-played by teaching staff) who actively approached the graduate nurses for assistance. In other words, in Scenario A students had to identify the patient that needed urgent help and in Scenario B they already knew which patient needed close attention. In both scenarios, the students were required to work collaboratively to identify and resolve the medical emergency of the deteriorating patient after being assigned several other secondary tasks. Each simulation can be divided into four different phases, including initial handover (Phase 1), clinical deterioration (Phase 2), ward nurses being called in (Phase 3), and medical emergency call (Phase 4). Our analyses focus on Phase 3 as this was the only phase where all four students were inside the simulation classroom without the presence of a physician (role-played by teaching staff) who enters and provides instructions in the final phase. Table 2 shows the average duration of Phase 3 in each scenario.

3.2 Apparatus and Data Collection

Empatica E4 wristbands were used to the collect physiological data of 224 students (56 simulations) who consented to this study, including heart rate, electrodermal activity (EDA), and blood pressure pulse (BVP). The current study focused on analysing heart rate, calculated automatically by Empatica from BVP, due to the poor quality of the EDA data (only eight sessions with valid data for each student), which may have resulted from the disconnection between the electrodes and the skin during constant physical activities. The heart rate data of 52 students (13 simulations) were not captured due to sensor malfunctions, resulting in a valid dataset of 172 students (43 simulations). Students' task and collaboration self-efficacy (Items S1 and S2) and performance (Items S4 and S5) were captured using a pre- and post-survey (Table 1). Their perceived stress was also captured before and after the simulation (Items S3 and S6). Each item was measured using a seven-point bipolar Likert scale, ranging from *strongly disagree* (1) to *strongly agree* (7). Ethical approval was obtained from [Anonymised] University (Project ID: [Anonymised]).

Table 1. Mean (M) and standard deviation (SD) of the student survey items.

Item	Details	M	SD
S1	I usually feel confident about participating in nursing simulations	4.59	1.43
S2	I usually feel confident about collaborating with other students	5.94	1.02
S3	I usually feel high levels of stress during simulations	5.83	1.08
S4	I am satisfied with my task performance during the simulation	4.49	1.37
S5	I am satisfied with the collaboration performance of my group	5.67	1.20
S6	I felt high levels of stress during the simulation	5.97	1.15

3.3 Feature Engineering

Four physiological features were computed to investigate the research questions. For RQ1, Pearson's correlation was calculated to derive the individual-level PS of each student [5]. This calculation was done by computing the correlation between the heart rate of each possible pair of students in a simulation and calculating the mean for a given student [11]. For example, with four students (A, B, C, and D), the individual-level PS of A was computed as the average correlation between A–B, A–C, and A–D. The individual-level PA was calculated by summing up the total duration of which a student's heart rate is above the average heart rate of that student during a session.

Regarding RQ2, the Multidimensional Recurrence Quantification Analysis (MdRQA) was used to compute the group-level PS [32]. This recurrence-based analysis technique has been frequently used in educational and psychological research to explore the synchronisation patterns of multiple individuals' physiological signals over time [6,11]. The parameters of MdRQA were calculated following Dindar et al.'s approach [6] using an entropy R package [13], which involved using the *mutual* and *false.nearest* functions for computing DEL and DIM, respectively, and iterating the radius to achieve an recurrence rate percentage (REC) between 1% to 5% for most groups [32]. The parameters used were EMB = 10, DEL = 1, DIM = 4, norm = 'euc', and radius = 0.10. The metrics of MdRQA, specifically the recurrence rate percentage (REC), the average diagonal length (ADL), and the maximum diagonal length (MDL), were normalised and averaged to calculate a single measure of group-level PS as these metrics were highly correlated (e.g., a Pearson's correlation of 0.79 between REC and MDL) [11]. Group-level PA was calculated as the sum of the individual-level PA of each student in a given simulation. Table 2 shows the mean and standard deviation of each feature across scenarios.

3.4 Analysis

We used LMMs to model whether individual and group-level PS and PA are significant indicators of students' perceived stress, task, and collaboration performance. The *lme4* package in R was used to conduct the analyses [1]. The models were analysed separately for each scenario due to the differences in the

Table 2. Descriptive statistics of physiological synchrony (PS) and arousal (PA).

	All		Scenario A		Scenario B	
	Mean	*SD*	*Mean*	*SD*	*Mean*	*SD*
Phase 3 duration (sec)	297	193	394	212	186	77
Individual-level PS (r)	0.04	0.29	0.02	0.28	0.06	0.30
Individual-level PA (sec)	140	90	186	96	87	39
Group-level PS	0.01	0.34	0.07	0.33	0.08	0.34
Group-level PA (sec)	559	340	744	359	346	128

learning design (as elaborated in Sect. 3.1). The assumption of linearity and homoscedasticity were met, whereas the normality assumption of residual was violated in both scenarios. However, this modelling technique remains appropriate as LMMs have shown remarkable robustness to violations of distributional assumptions [29]. The correlation between PS and PA were low at both individual (r = −0.10) and group-level (r = 0.07), indicating no violation of collinearity.

Regarding the first research question (RQ1), we built a model (1) to investigate whether individual-level PS and PA are significant indicators of students' post-survey items after considering the variance explained by the corresponding pre-survey items. For example, we investigated students' post-simulation stress using individual-level PS and PA after considering their pre-simulation stress. We also considered the potential group differences as students' stress and performance could depend on their group differences. For each post-survey item (j) of a given student (i), we included a fixed intercept (γ_{00}), a random intercept for students' pre-survey item (μ_{pre}), and a random intercept for students' simulation group (μ_{group}). $\text{PS}_{individual}$ and $\text{PA}_{individual}$ are continues variables with the corresponding coefficient γ_{10} and γ_{20}, respectively.

$$\text{Post}_{ji} = \gamma_{00} + \gamma_{10}\text{PS}_{individual} + \gamma_{20}\text{PA}_{individual} + \mu_{pre} + \mu_{group} \quad (1)$$

For the second research question (RQ2), a similar model (2) was constructed by replacing individual-level with group-level PS and PA. The fixed (γ_{01}) and random intercepts (μ_{pre} and μ_{group}) were similar to those in the first model as these individual and group differences could also be relevant for group-level PS and PA. Group-level variables PS_{group} and PA_{group} had the corresponding coefficient γ_{11} and γ_{21}, respectively. For both models, we dropped the interaction term of the independent variables (e.g., $\text{PS}_{individual} \times \text{PA}_{individual}$ for (1) and $\text{PS}_{group} \times \text{PA}_{group}$ for (2)) as they were non-significant in both scenarios [2].

$$\text{Post}_{ji} = \gamma_{01} + \gamma_{11}\text{PS}_{group} + \gamma_{21}\text{PA}_{group} + \mu_{pre} + \mu_{group} \quad (2)$$

4 Results

RQ1. The effect of $\text{PS}_{individual}$ (γ_{10}) and $\text{PA}_{individual}$ (γ_{20}) were non-significant across students' post-stress, task, and collaboration performance in Scenario A (Table 3). Students' task efficacy only explained 10% of the variance in their task performance and their pre-stress only explained 11% of their post-stress. The group differences explained much of the variances in students' collaboration performance ($\tau_{group} = 0.77$). On the other hand, in Scenario B (Table 4), students with higher $PS_{individual}$ were significantly less stressed post-simulation ($\gamma_{10} = -0.26$, $p = 0.046$) after accounting for the variance explained by their pre-stress level ($\tau_{stress} = 0.41$). Likewise, students with higher $PA_{individual}$ were significantly more satisfied with their task performance during the simulation ($\gamma_{20} = 0.62$, $p = 0.018$) after accounting for the variance explained by their task self-efficacy ($\tau_{task} = 0.25$). The random effect of students' simulation group (τ_{group}) was suppressed for both task and stress due to a zero value, indicating that students' post-stress and task performance were unaffected by group differences. This random effect explained some variances in students' collaboration performance ($\tau_{group} = 0.33$), and students' collaboration self-efficacy was suppressed due to a zero value. Neither $PS_{individual}$ nor $PA_{individual}$ significantly affected students' collaboration performance.

These differences between Scenario A and B can be explained by the learning design, where students may actively engage in collaboration to reduce their stress when dealing with the medical emergency of the overtly deteriorating patient in Scenario B. As students physically co-located around the same patient bed, they were more likely to work together on the same task, potentially leading to higher $PS_{individual}$. Such co-located situation could also spark more conversation and communication among the group members, potentially reducing their perceived stress level as they may have felt more supported by others. Likewise, $PA_{individual}$ in this scenario could potentially represent a higher level of cognitive load (e.g., thinking about the appropriate medical procedures) and physical engagement (e.g., putting the oxygen mask on the patient), leading to higher task performance. Whereas in Scenario A, the covertly deteriorating patient required students to actively identify her medical situation among the four patients, and thus, students could be occupied with several other tasks, and their physiological responses became less related to their stress and performance.

Table 3. Individual-level physiological synchrony and arousal (Scenario A).

	Task			Stress			Collaboration		
Predictors	*Estimates*	*CI*	*p*	*Estimates*	*CI*	*p*	*Estimates*	*CI*	*p*
(Intercept)	4.03	3.61–4.46	**<.001**	5.91	5.53–6.29	**<.001**	5.60	5.15–6.06	**<.001**
$PS_{individual}$	−0.02	−0.35–0.30	0.892	−0.06	−0.27–0.15	0.571	−0.07	−0.37–0.23	0.651
$PA_{individual}$	0.03	−0.26–0.32	0.832	0.12	−0.08–0.31	0.237	−0.29	−0.63–0.04	0.088
Random Effects									
σ_2	2.12			0.91			1.03		
τ_{00}	0.10_{task}			0.11_{stress}			0.77_{group}		
ICC	0.05			0.10			0.43		
N	7_{task}			6_{stress}			23_{group}		
Observations	92			92			92		
Marginal/Conditional R^2	0.001/0.046			0.019/0.121			0.052/0.458		

RQ2. Similar to the findings of RQ1, neither PS_{group} (γ_{11}) nor PA_{group} (γ_{21}) had significant effects on students' post-stress, task, and collaboration performance in Scenario A. Students' task efficacy and pre-stress only explained a

Table 4. Individual-level physiological synchrony and arousal (Scenario B).

	Task			Stress			Collaboration		
Predictors	*Estimates*	*CI*	*p*	*Estimates*	*CI*	*p*	*Estimates*	*CI*	*p*
(Intercept)	5.20	4.62–5.77	**<.001**	5.63	4.85–6.41	**<.001**	6.07	5.59–6.54	**<.001**
$PS_{individual}$	0.04	−0.17–0.26	0.684	-0.26	−0.51–−0.01	**0.046**	−0.14	−0.37–0.09	0.232
$PA_{individual}$	0.62	0.11–1.14	**0.018**	−0.06	−0.67–0.55	0.85	0.22	−0.38–0.82	0.465
Random Effects									
σ_2	0.93			1.33			0.75		
τ_{00}	0.25_{task}			0.41_{stress}			0.33_{group}		
ICC	0.21			0.24			0.30		
N	7_{task}			5_{stress}			20_{group}		
Observations	80			80			80		
Marginal/Conditional R^2	0.059/0.258			0.039/0.266			0.028/0.323		

small proportion of the variance in their task performance (τ_{task} = 0.10) and pre-stress (τ_{stress} = 0.09). Once again, much of the variances in students' collaboration performance was explained by the group differences (τ_{group} = 0.77). On the contrary, significant effects were found for each post-survey items in Scenario B. Specifically, students with higher PA_{group} were significantly more satisfied with their task performance (γ_{21} = 0.18, p = 0.028) after accounting for the variance explained by their task self-efficacy (τ_{task} = 0.25). Students with higher PS_{group} were significantly less stressed post-simulation (γ_{11} = −0.97, p = 0.017) after accounting for the variance explained by their pre-stress level (τ_{stress} = 0.39). Interestingly, students with higher PS_{group} were also significantly less satisfied with their group collaboration (γ_{11} = −1.02, p = 0.015) after accounting for the variance explained by their simulation group (τ_{group} = 0.18). This negative relationship can potentially be explained by the level of agency and control students had during Scenario B. For example, the students in high PS_{group} groups could all work together on the overtly deteriorating patient. However, they may have felt little agency in this situation as they were forced to collaborate on this task by the patient relative, potentially resulting in lower self-evaluation on their collaboration performance. Students in low PS_{group} groups may have actively distributed the tasks assigned to them and had the agency to decide task responsibilities. Thus, they could feel more satisfied in terms of collaboration (Tables 5 and 6).

Table 5. Group-level physiological synchrony and arousal (Scenario A).

	Task			Stress			Collaboration		
Predictors	*Estimates*	*CI*	*p*	*Estimates*	*CI*	*p*	*Estimates*	*CI*	*p*
(Intercept)	3.98	3.53–4.42	**<.001**	5.96	5.59–6.34	**<.001**	5.47	4.99–5.96	**<.001**
PS_{group}	−0.10	−1.06–0.85	0.828	0.43	−0.19–1.05	0.173	0.28	−1.02–1.58	0.670
PA_{group}	0.03	−0.05–0.11	0.445	0.02	−0.03–0.08	0.349	0.00	−0.11–0.11	0.968
Random Effects									
σ_2	2.11			0.89			1.09		
τ_{00}	0.10_{task}			0.09_{stress}			0.72_{group}		
ICC	0.05			0.09			0.40		
N	7_{task}			6_{stress}			23_{group}		
Observations	92			92			92		
Marginal/Conditional R^2	0.006/0.053			0.037/0.125			0.005/0.400		

Table 6. Group-level physiological synchrony and arousal (Scenario B).

	Task			Stress			Collaboration		
Predictors	*Estimates*	*CI*	*p*	*Estimates*	*CI*	*p*	*Estimates*	*CI*	*p*
(Intercept)	5.30	4.66–5.93	**<.001**	5.86	5.03–6.68	**<.001**	6.39	5.83–6.95	**<.001**
PS_{group}	-0.34	−1.00–0.33	0.315	−0.97	−1.77–−0.18	**0.017**	-1.03	−1.85–−0.20	**0.015**
PA_{group}	0.18	0.02–0.34	**0.028**	0.05	−0.14–0.24	0.577	0.16	−0.03–0.36	0.104
Random Effects									
σ_2	0.93			1.31			0.77		
τ_{00}	0.25_{task}			0.39_{stress}			0.18_{group}		
ICC	0.21			0.23			0.19		
N	7_{task}			5_{stress}			20_{group}		
Observations	80			80			80		
Marginal/Conditional R^2	0.052/0.250			0.057/0.275			0.124/0.288		

5 Discussion

This study investigated the significance of PS and PA measures as indicators of students' stress and learning performance in ECL after accounting for their individual and group differences. For RQ1, the findings suggested that individual-level PS and PA effects on students' stress, task, and collaboration performance vary across learning scenarios. While individual-level PS significantly reduced students' post-stress and PA significantly increased task performance in Scenario B, none of these effects were found in Scenario A. This finding contributes empirical evidence to support the importance of considering the differences in learning design when developing context-aware educational technologies and learning analytics [9]. These findings also contribute empirical evidence regarding the potential value contained within individual-level PS and PA as indicators of students' stress and learning performance, as they have rarely been investigated in prior studies on collaborative learning.

The findings of RQ2 were similar to those of RQ1, with significant effects of group-level PS and PA on students' stress, task, and collaboration performance in Scenario B. However, such effects were non-significant in Scenario A. Once again, adding empirical evidence to support the necessity of considering learning design differences [9]. The positive association between physiological measures and task performance is similar to the prior research but with a nuanced difference, as we found a significant relationship between PA and task performance, whereas prior research were between PS and task performance [7,28]. This nuanced difference could be due to the physically active nature of ECL compared to the more static collaborative learning tasks investigated in prior studies. Specifically, in ECL, PA may represent cognitive and physical engagement with different hands-on tasks, leading to higher perceived task performance.

The non-significant effects of group-level PS on task performance resonates with the findings of Dindar et al. [6], indicating the limitation of PS as an effective indicator of task performance in ECL. The significant effects of PS on reducing students' perceived stress resonate with prior psychophysiology studies [4], indicating the potential value of using group-level PS as an indicator of stress in ECL. On the other hand, the negative effect of group-level PS on students' collaboration performance is inconsistent with prior studies on the positive association between PS and collaboration quality [5, 21]. This difference may result from the different learning contexts (as explained in the Results RQ2) and the consideration of the variance explained by group differences in our study, which was not accounted for in prior studies.

Implications. The current findings have several implications for future AIED research and practice. As shown in this study, physiological indicators, such as PS and PA, may have different effects on students' stress and learning performance in ECL compared to collaborative learning contexts with a relatively low level of physical activity, suggesting the need to consider the influences of the learning design on students' cognitive and affective responses in future AIED research. Such considerations are particularly important for developing teacher and student-facing analytics user interfaces as generalised modelling methodologies could potentially lead to biased and inaccurate representation of students' learning experience [9]. The empirical evidence generated from modelling and analysing physiological data could provide the opportunity to advance knowledge and understanding of collaborative learning in the age of artificial intelligence, especially in novel and technological-integrated settings (e.g., ECL), where traditional approaches (e.g., interview and direct observation) could potentially become less effective due to the increased complexity [16]. This novel evidence could also contribute to the development of context-specific educational theories, such as using these physiological features as indicators of students' cognitive and affective states during collaborative learning and elaborating on the potential changes across different learning designs [17]. The potential of using heart rate to model stress and learning performance in ECL could signal the room for improvements in the scalability of sensor-based analytic tools, as more affordable wearables that only capture heart rate can be used to model higher-level learning constructs (e.g., performance and collaboration) in ECL without relying on EDA data that can only captured with the more expensive wearables [34].

Modelling stress and learning performance from physiological data has several ethical implications. Using physiological data could expose students to under-the-skin tracking, which may induce ethical issues where teachers, through the analytics, have more knowledge about students' cognitive and affective states than themselves, as identified in prior works on MMLA [3]. Ethical regulations must be established on the cognitive and affective insights that can be captured, modelled, and made available to educational stakeholders. Informed consents are essential for these kinds of analytics [10].

Limitations. The current findings should be interpreted with a limitation in mind. We only investigated PS and PA as indicators of students' perceived stress and learning performance of ECL in healthcare simulations. Our findings have already shown that these indicators are context-sensitive (e.g., the differences between Scenario A and B). Thus, future studies should validate the identified relationships in other learning contexts before using physiological features as indicators of students' stress and learning performance in collaborative learning.

6 Conclusion

With the opportunities enabled by artificial intelligence and MMLA, the current study contributed additional empirical evidence to support the role of PS and PA modelled from heart rates as significant indicators of students' stress and learning performance in ECL. The findings also emphasised the need for MMLA and analytics tools to have context sensitivity.

References

1. Bates, D., Mächler, M., Bolker, B., Walker, S.: Fitting linear mixed-effects models using lme4. J. Stat. Softw. **67**(1), 1–48 (2015)
2. Campbell, L., Kashy, D.A.: Estimating actor, partner, and interaction effects for dyadic data using PROC MIXED and HLM: a user-friendly guide. Pers. Relat. **9**(3), 327–342 (2002)
3. Cukurova, M., Giannakos, M., Martinez-Maldonado, R.: The promise and challenges of multimodal learning analytics. BJET **51**(5), 1441–1449 (2020)
4. Denk, B., et al.: Influence of stress on physiological synchrony in a stressful versus non-stressful group setting. J. Neural Transm. **128**(9), 1335–1345 (2021). https://doi.org/10.1007/s00702-021-02384-2
5. Dich, Y., Reilly, J., Schneider, B.: Using physiological synchrony as an indicator of collaboration quality, task performance and learning. In: Penstein Rosé, C., et al. (eds.) AIED 2018. LNCS (LNAI), vol. 10947, pp. 98–110. Springer, Cham (2018). https://doi.org/10.1007/978-3-319-93843-1_8
6. Dindar, M., Järvelä, S., Haataja, E.: What does physiological synchrony reveal about metacognitive experiences and group performance? Br. J. Educ. Technol. **51**(5), 1577–1596 (2020)
7. Elkins, A.N., Muth, E.R., Hoover, A.W., Walker, A.D., Carpenter, T.L., Switzer, F.S.: Physiological compliance and team performance. Appl. Ergon. **40**(6), 997–1003 (2009)
8. Ferreira, J.M.: What if we look at the body? an embodied perspective of collaborative learning. Educ. Psychol. Rev. **33**(4), 1455–1473 (2021)
9. Gašević, D., Dawson, S., Rogers, T., Gasevic, D.: Learning analytics should not promote one size fits all: the effects of instructional conditions in predicting academic success. Internet High. Educ. **28**, 68–84 (2016)
10. Giannakos, M., Cukurova, M., Papavlasopoulou, S.: Sensor-based analytics in education: Lessons learned from research in multimodal learning analytics. In: Giannakos, M., Spikol, D., Di Mitri, D., Sharma, K., Ochoa, X., Hammad, R. (eds.) The Multimodal Learning Analytics Handbook, pp. 329–358. Springer, Cham (2022). https://doi.org/10.1007/978-3-031-08076-0_13
11. Gordon, I., Wallot, S., Berson, Y.: Group-level physiological synchrony and individual-level anxiety predict positive affective behaviors during a group decision-making task. Psychophysiology **58**(9), e13857 (2021)
12. Haataja, E., Malmberg, J., Järvelä, S.: Monitoring in collaborative learning: co-occurrence of observed behavior and physiological synchrony explored. Comput. Hum. Behav. **87**, 337–347 (2018)
13. Hausser, J., Strimmer, K.: Entropy inference and the james-stein estimator, with application to nonlinear gene association networks. J. Mach. Learn. Res. **10**(7) (2009)

14. Ioannou, M., Georgiou, Y., Ioannou, A., Johnson, M.: On the understanding of students' learning and perceptions of technology integration in low-and high-embodied group learning. In: CSCL, vol. 1, pp. 304–311. ISLS (2019)
15. Janssen, J., Kirschner, P.A.: Applying collaborative cognitive load theory to computer-supported collaborative learning: towards a research agenda. Educ. Technol. Res. Dev. **68**(2), 783–805 (2020). https://doi.org/10.1007/s11423-019-09729-5
16. Järvelä, S., Dindar, M., Sobocinski, M., Nguyen, A.: Multimodal research for studying collaborative learning in higher education. In: Handbook of Digital Higher Education, pp. 199–210. Edward Elgar Publishing (2022)
17. Järvelä, S., Gašević, D., Seppänen, T., Pechenizkiy, M., Kirschner, P.A.: Bridging learning sciences, machine learning and affective computing for understanding cognition and affect in collaborative learning. BJET **51**(6), 2391–2406 (2020)
18. Järvelä, S., Järvenoja, H., Malmberg, J.: Capturing the dynamic and cyclical nature of regulation: methodological progress in understanding socially shared regulation in learning. Int. J. Comput. Support Collab. Learn. **14**(4), 425–441 (2019)
19. Joët, G., Usher, E.L., Bressoux, P.: Sources of self-efficacy: an investigation of elementary school students in France. J. Educ. Psychol. **103**(3), 649 (2011)
20. Leonard, M., Graham, S., Bonacum, D.: The human factor: the critical importance of effective teamwork and communication in providing safe care. BMJ Qual. Saf. **13**(suppl 1), i85–i90 (2004). https://doi.org/10.1136/qshc.2004.010033
21. Liu, Y., Wang, T., Wang, K., Zhang, Y.: Collaborative learning quality classification through physiological synchrony recorded by wearable biosensors. Front. Psychol. **12**, 674369 (2021)
22. Malmberg, J., Fincham, O., Pijeira-Díaz, H.J., Järvelä, S., Gašević, D.: Revealing the hidden structure of physiological states during metacognitive monitoring in collaborative learning. J. Comput. Assist. Learn. **37**(3), 861–874 (2021)
23. Malmberg, J., Haataja, E., Järvelä, S.: Exploring the connection between task difficulty, task perceptions, physiological arousal and learning outcomes in collaborative learning situations. Metacogn. Learn. **17**, 1–19 (2022)
24. Martinez-Maldonado, R., Echeverria, V., Fernandez Nieto, G., Buckingham Shum, S.: From data to insights: a layered storytelling approach for multimodal learning analytics. In: CHI, pp. 1–15 (2020)
25. Milstein, N., Gordon, I.: Validating measures of electrodermal activity and heart rate variability derived from the empatica e4 utilized in research settings that involve interactive dyadic states. Front. Behav. Neurosci. **14**, 148 (2020)
26. Nguyen, A., Järvelä, S., Rosé, C., Järvenoja, H., Malmberg, J.: Examining socially shared regulation and shared physiological arousal events with multimodal learning analytics. Br. J. Educ. Technol. **54**(1), 293–312 (2022)
27. Nokes-Malach, T.J., Richey, J.E., Gadgil, S.: When is it better to learn together? insights from research on collaborative learning. Educ. Psychol. Rev. **27**(4), 645–656 (2015)
28. Pijeira-Díaz, H.J., Drachsler, H., Järvelä, S., Kirschner, P.A.: Investigating collaborative learning success with physiological coupling indices based on electrodermal activity. In: LAK, pp. 64–73 (2016)
29. Schielzeth, H., et al.: Robustness of linear mixed-effects models to violations of distributional assumptions. Methods Ecol. Evol. **11**(9), 1141–1152 (2020)
30. Sobocinski, M., Malmberg, J., Järvelä, S.: Exploring adaptation in socially-shared regulation of learning using video and heart rate data. Technol. Knowl. Learn. **27**(2), 385–404 (2022)

31. Törmänen, T., Järvenoja, H., Saqr, M., Malmberg, J., Järvelä, S.: Affective states and regulation of learning during socio-emotional interactions in secondary school collaborative groups. Br. J. Educ. Psychol. **93**, 48–70 (2022)
32. Wallot, S., Roepstorff, A., Mønster, D.: Multidimensional recurrence quantification analysis (mdrqa) for the analysis of multidimensional time-series: a software implementation in matlab and its application to group-level data in joint action. Front. Psychol., 1835 (2016)
33. Yan, L., et al.: The role of indoor positioning analytics in assessment of simulation-based learning. Br. Educ. Res. J. **54**(1), 267–292 (2022)
34. Yan, L., Zhao, L., Gasevic, D., Martinez-Maldonado, R.: Scalability, sustainability, and ethicality of multimodal learning analytics. In: LAK, p. 13–23. LAK22, New York (2022)

Confusion, Conflict, Consensus: Modeling Dialogue Processes During Collaborative Learning with Hidden Markov Models

Toni V. Earle-Randell[1(✉)], Joseph B. Wiggins[1], Julianna Martinez Ruiz[1], Mehmet Celepkolu[1], Kristy Elizabeth Boyer[1], Collin F. Lynch[2], Maya Israel[1], and Eric Wiebe[2]

[1] University of Florida, Gainesville, FL, USA
{tearlerandell,jbwiggi3,juliannamartinez,mckolu,keboyer}@ufl.edu, misrael@coe.ufl.edu

[2] North Carolina State University, Raleigh, NC, USA
{cflynch,wiebe}@ncsu.edu

Abstract. There is growing recognition that AI technologies can, and should, support collaborative learning. To provide this support, we need models of collaborative talk that reflect the ways in which learners interact. Great progress has been made in modeling dialogue for high school and college-age learners, but the dialogue processes that characterize collaborative talk between elementary learner dyads are not currently well understood. This paper reports on a study with elementary school learners (4th and 5th grade, ages 9–11 years old) coded collaboratively in dyads. We recorded dialogue from 22 elementary school learner dyads, covering 7594 total utterances. We labeled this corpus manually with dialogue acts and then induced a hidden Markov model to identify the underlying dialogue states and the transitions between these states. The model identified six distinct hidden states which we interpret as Social Dialogue, Confusion, Frustrated Coordination, Exploratory Talk, Directive & Disagreement, and Disagreement & Self-Explanation. The HMM revealed that when students entered into a productive exploratory talk state, the primary way they transitioned out of this state is when they became confused or reached an impasse. When this occurred, the learners then moved into states of disputation and conflict before re-entering the Exploratory Talk state. These findings can inform the design of AI agents who support young learners' collaborative talk and help agents determine when students are conflicting rather than collaborating.

Keywords: Collaborative Dialogue · Elementary School Learners · Hidden Markov Models · Dialogue Acts · Pair Programming

1 Introduction and Related Work

In recent years, there has been growing attention to modeling collaborative talk between elementary school learners in the AIED community by detecting speakers [8], investigating interactive reading companions [15], and providing emo-

N. Wang et al. (Eds.): AIED 2023, LNAI 13916, pp. 615–626, 2023.
https://doi.org/10.1007/978-3-031-36272-9_50

tional support to learners [16]. There is a tremendous need to foster good collaboration among young learners [13,17]. In this paper, we investigate the following research question: *What are the dialogue states that characterize collaborative talk between elementary school learner dyads, and how do these dyads transition between these dialogue states?*

We analyzed collaborative talk between learner dyads in an elementary school classroom during a 45-min pair programming activity where students interacted with a block-based programming environment. As part of our analysis, we induced a hidden Markov model (HMM), which allows us to map the observed utterances to an underlying set of hidden dialogue states that drive the actions [19]. This approach, where we induce a model from the observable data and then interpret the hidden states to characterize different modes of talk, has shown success in prior work where it has been used to evaluate the importance of collaborative planning discussion [20]; to identify the ways in which learners resolve errors [6], and to detect confusion [22]. While this prior work has illustrated the utility of HMMs for dialogue analysis, they have focused primarily on high school or adult learners. To date, there has been limited research on modeling collaborative talk between elementary-aged learners.

In this work we leverage a leading discourse theory by Mercer et al. that deconstructs collaborative talk into three main components: Exploratory talk, Disputational talk, and Cumulative talk [11,12]. *Exploratory talk* is an embodiment of collaborative critical thinking [1]. In Exploratory talk, learners may express incomplete thoughts as they forge their own understanding [14]. In *Disputational talk*, learners disagree with each other, make assertions and counter-assertions, and make their own decisions instead of collaborating with their partner [11]. Finally, *Cumulative talk* is characterized by learners constructing shared knowledge by positively and uncritically building on each other's contributions [12]. In our view, a promising approach to moving toward AI-augmented support of collaboration for young learners is to build bottom-up models of their dialogue and explore the ways in which these learners move among states or modes of collaborative talk.

The HMM reported here identified six hidden states, which we interpreted as Social Dialogue, Confusion, Frustrated Coordination, Exploratory Talk, Directive & Disagreement, and Disagreement & Self-Explanation. These six states were derived from Mercer's three main components [11,12]. The model suggests that the dialogue states that occur between learner dyads are cyclical, and while Exploratory Talk can be interrupted when the learner dyad becomes confused, Disputational Talk can serve as an avenue back to Exploratory Talk once consensus has been reached. To the best of the authors' knowledge, this work is the first to build HMMs on elementary learners' dialogue, and the findings suggest implications for the ways in which AI-augmented technologies can foster productive collaboration among these young learners.

2 Methods

We analyzed a corpus of dialogue between upper-elementary school students collected in spring 2022 during a study in an elementary school in the southeastern

United States, consisting of 44 students that provided assent and parental consent. The school's student body was approximately 72% White/ Caucasian, 15% Hispanic/Latinx, 9% Black/African American, 4% multiracial, and 1% other. The school served a large percentage of economically disadvantaged learners, with 74% of the student body eligible for free or reduced meals. All students were in grade 4 and the mean age was 9.73, with ages ranging from 8 to 11 years old. Of the 44 students, 23 were female, 16 were male, and 5 preferred not to report their gender.

Fig. 1. The collaborative block-based learning environment.

Learning Environment. This study was conducted in our own collaborative block-based learning environment (Fig. 1), a system that utilizes the Netsblox block-based coding environment with embedded virtual learning companions. Students are paired into dyads and tasked with completing programming activities using the pair programming framework [24], in which one student is the "driver" controlling the mouse and keyboard and changing the code, and the other student is the "navigator", contributing ideas and suggestions and helping the driver. These students switched halfway through the session. The interface contains instructions for the coding task, a block-based scripting space, and a stage where they can view their code. In the environment, two virtual learning companions, Viviana and Jeremy, model exploratory talk in the form of vignettes, and brief dialogues between the agents that sometimes directly address the learners. These vignettes are designed to influence the uptake of positive collaborative behaviors between the learner dyads. The virtual learning companions stay on screen throughout the session.

Table 1. Dialogue act scheme used in this analysis.

Dialogue Act	Rel. Freq.	Description	Examples
Self-Explanation/ Justification (SE)	14.1%	Explaining the steps they are taking, or thoughts	*"We can put this here to make it happen twice"* *"I did that because of how slow it is"*
Suggestion/ Alternative Idea (SU)	11.5%	Any suggestions when directly talking to their partner.	*"Maybe we should put two of those"* *"How about doubling that?"*
Directive (DI)	8.10%	Telling partner to do something	*"Give me the keyboard"* *"Click that one"*
Agreement/ Acknowledgement (AG)	10.9%	Agreement on any opinion/edit	*"Good job"* *"Ok"*
Confusion/ Help-Seeking (C)	3.94%	Learner directly or indirectly seeking help from their partner	*"I'm confused"* *"I don't know"*
Question - Higher Order (QH)	1.45%	Asks a why question or a question that challenges a partner's idea.	*"Why is he moving like that?"* *"What happens if you keep it that way?"*
Question - Other (QO)	10.4%	Asks anything other than a why question	*"What does that block do?"*
Disagreement/ Negative Feedback (D)	7.12%	Disagreement on any opinion/edit	*"No no no"* *"That is wrong"*
Disagreement w/ Justification (DJ)	0.51%	Disagrees but provides reasoning	*"No, that won't work because it needs to go in a square."*
Antagonistic Action (AN)	1.51%	Actions that cause tension including harmful comments, putting down partner contributions.	*"You are being ridiculous"* *"Stupid"* *"You don't know anything"*
Social (S)	10.6%	Social dialogue	*"Did you hear about James?"* *[Quoting meme]* *"Thank you"*
Directed at Agent (DA)	1.24%	It was said to the agent, not the partner.	*"Thanks, we know we are great"*
Other (O)	18.8%	Something not covered by any of the other tags	*[Reading instructions]* *"Are my headphones working?"* *[Observations, comments on activity]*

Dialogue Tagging. Our corpus contains 22 sessions and 7594 utterances, with an average of 416 utterances per 45-min session (SD = 114, min = 274, max = 617). We developed a dialogue act taxonomy (Table 1) by drawing upon the exploratory talk framework [12] and a dialogue act taxonomy by Zakaria et al. [25] that was designed for a closely related application with elementary school learners in the classroom.

We modified Zakaria et al.'s [25] taxonomy to isolate exploratory talk dialogue moves, which required collapsing some tags into broader exploratory talk ideas (eg. combining "Self-explanation" with "Justification", and "Suggestion" with "Alternative idea"). Isolating exploratory talk moves was imperative because of their role in supporting student learning through collaboration. Mercer explains that exploratory talk enables partners to achieve a better mutual understanding of the problem [14]. We also added a tag to capture utterances that were directed at the virtual agents[1] to separate them from conversation within the dyad. This scheme, reported in Table 1, was applied by two independent annotators who achieved a Cohen's kappa of 0.816, a strong agreement [21].

Analysis. We implemented a hidden Markov model (HMM) to model learners' collaborative talk. An HMM is a probabilistic graphical model that can be used to describe hidden processes or states that influence the sequence of observable symbols. HMMs are defined by a set of hidden internal states and a set of evidence states representing observable symbols or actions. Taken together, the states define a set of Bayesian variables with transitions between the hidden states, and emission of the evidence states being governed by probability distributions. In our model, the observable states are the 13 labeled dialogue acts, shown in Table 1. We represented each of the 22 collaborative learning sessions as a sequence of these symbols (dialogue acts) and trained an HMM on these sequences. Our model does not consider the time between the actions or the speaker in each case. In order to identify the best set of hidden states, we trained a series of models ranging from 3 to 12 hidden states using leave-one-out cross-validation over the 22 sessions. We then compared the generated models using the average Akaike information criterion (AIC) score for each number of hidden states. We found that models with six states produced the best AIC scores, on average. We then trained a final model, which we report here.

3 Results

The resulting model (Fig. 2) revealed the following six hidden states. The most frequent of the dialogue states (32% of the time) was characterized by a high degree of *Self-Explanation (28%)*, *Suggestion (17%)*, and *Agreement (15%)* constituted 60% of observable symbols. *Higher-Order Questions (3%)* were also most likely to occur in this state out of all six states. These behaviors are consistent with Mercer's characterization of **Exploratory Talk** [14] and we will use that name to refer to this state below.

The next most common state (20% of dialogue) included *Question - Other, Agree, and Suggestion*, and it also had the highest likelihood of *Antagonistic Action (5%)* of all of the states. We label this state **Frustrated Coordination** because a qualitative examination of the dialogue in this state indicates that students are often negotiating computer control, effort, or focus within the dyad. Other studies of elementary pair programming treat this coordination as a distinct dialogue act in its own right [25], and research has long shown that the balance of control can be a source of conflict in pair programming [24].

Sessions were most likely to start in a state characterized by the *Confusion (18%)*, *Suggestion (17%)*, and *Question-Other (10%)* dialogue acts. This state accounted for 13% of the conversations, and we label this state **Confusion**, in line with D'Mello's theory on the dynamics of cognitive disequilibrium [4].

[1] While the agents were not designed to elicit verbal responses from the learners and could not listen or respond, some learners spoke to them nonetheless.

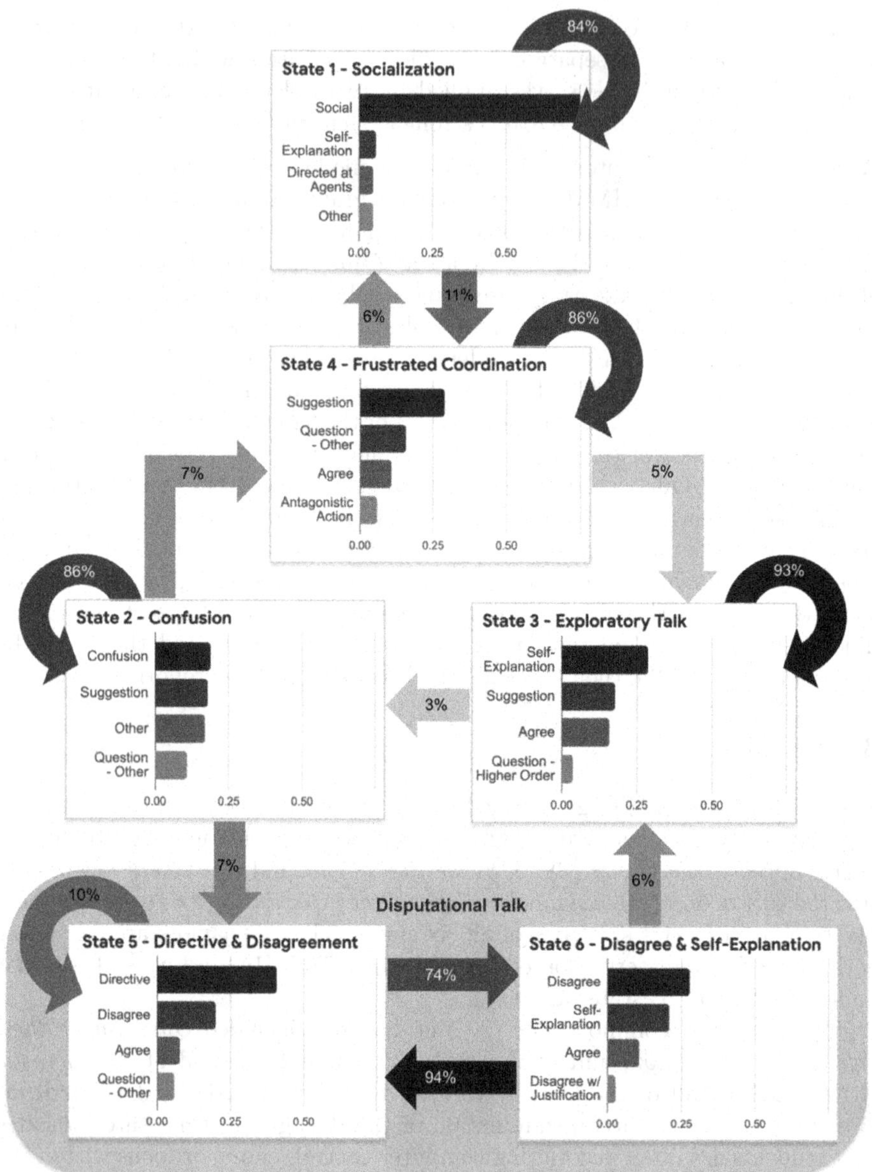

Fig. 2. High-level overview of states and transitions between them.

In contrast to the aforementioned states, two states showed a much higher probability of transition to each another than self-transition. The first such state (11% of conversations) is characterized by learners *Directing each other (40%)* and *Disagreeing with each other (19%)*, so we label it **Directive and Disagreement**, while the second state (9% of conversations) is mostly students

Disagreeing with each other (27%) and *Explaining themselves (20%)*, so we call it **Disagreement and Self-Explanation**. (Interestingly, *Disagreement with Justification is higher in this state (2%)* than in any other state.) Together, we label this tightly alternating pair of states **Disputational Talk**, because this kind of behavior is consistent with Mercer's definition of this type of collaborative talk. Finally, we found a hidden state where observations are almost exclusively *(75%) Social* dialogue acts. This hidden state accounted for 10% of the conversations.

4 Discussion

4.1 Dialogue States

There were five distinct states of conversation in these collaborative programming dialogues: Socializing, Exploratory Talk, Frustrated Coordination, Confusion, and Disputational Talk (representing both *Directive and Disagreement* and *Disagreement and Self-explanation*). The Disputational Talk state cycle was created as a collective because as Fig. 2 shows, the two states cycle back and forth instead of dwelling, and they contain many of the same observable symbols.

The **Socializing** state involves off-task discussions and rapport building. It occurs most frequently at the end of the session and at the halfway point when students switch pair programming roles. Other work has found that rapport building has a positive impact on collaborative learning: in work by Madaio et al. (2018), it was discovered that peer tutors and tutees with high rapport demonstrated greater performance than their non-rapport-building counterparts [9]. **Frustrated Coordination** is characterized by students refocusing the dyad on the current task or negotiating control of the computer. This state primarily occurs after the Socialization or Confusion states. A typical example of this dialogue occurs in one session after students were joking around (*"Where's my money?", "There's your $2000."*) Immediately afterward, one of the students dismissed the joke, saying *"Whatever. Okay, let's get back to coding."* This state often includes some antagonism, starting when a student appears to express frustration with their partner (*"No, well, you do it. Since you think you can do everything."*)

Utterances in the **Exploratory Talk** state are characterized by students asking higher-level questions, sharing partial understanding of the work, and listening and providing criticism of their partner's ideas [12]. They are typically focused on the task during this state. Higher-level questions in the Exploratory Talk state usually include the word *why:* for example, *"But why instead of going like this, back, down, back ... Isn't it supposed to be, like, going in a square?"* This is the most frequent state in our corpus and has the highest dwell time distribution (93%). **Confusion** occurs when students hit a roadblock or error in their code, and dialogue generally consists of statements like *"What the heck?"* and *"I don't know what to do now."* Sessions were most likely to begin in this state. While confusion may be perceived as negative, research shows that the moment of uncertainty has a positive impact on learning, because it compels students to stop and revise their understanding to resolve an impasse. [3,5,10]. Finally, **Disputational Talk** consists of students making decisions individually and disagreeing with each other, with few attempts to offer constructive criticism or make suggestions (*"Stop, stop. Let me do my work"*, *"No, no, no."*) Researchers consider this dialogue unproductive, as no effort is being made between the students to share their knowledge and work together [11,25]. This state cycle encompasses two states with a very high probability of back-and-forth transition, *Directive and Disagreement* and *Disagreement and Self-explanation.* While the states have distinct differences, they characterize very similar dialogue. The *Directive and Disagreement* state is perhaps less desirable than *Disagreement and Self-explanation* because the latter state involves more justification for decisions and is the predominant way for students to transition out of the Disputational talk and into Exploratory talk. When both partners accept the justification of an action taken, the dyad is able to move forward.

4.2 Transitions into and Out of Exploratory Talk

Exploratory talk is thought of as a gold standard of collaborative talk, with students openly sharing and enhancing each others' ideas. In our model, it was the most frequent state. However, we also found that learners spent substantial time in other kinds of talk during collaboration. This section discusses the most common ways students move into Exploratory Talk (Frustrated Coordination and Disputational Talk) and out of it (through Coordination).

Table 2. Annotated dialogue excerpt showing a dyad's transition from Exploratory Talk to Confusion.

S1	*But when we get him into a square I think they know when we put him into a square.*	Suggestion	Exploratory Talk
S2	*Okay. So move 10 steps and then we're gonna repeat that. We're gonna repeat that like...*	Self- Expl.	
S1	*(laughs)*	Social	
S1	*He left.*	Other	
S2	*Um, is he back?*	Question - Other	
S1	*He's not coming back.*	Other	Confusion
S2	*Okay. Repeat.*	Self- Expl.	
S2	*Repeat it four times maybe? Five?*	Self- Expl.	
S1	*Or where is he going?*	Question - Other	
S2	*I don't know where he went.*	Confusion	

Exploratory Talk is the most frequent state, and in our study, it was very unlikely for students to transition out of this kind of dialogue. When a dyad is demonstrating Exploratory Talk and working collaboratively, they build on each other's ideas and work together on completing the task, feeding a positive feedback loop that supports a continuation in this state. For the 3% of the time our learners transitioned out of Exploratory Talk, it was through Confusion; encountering a problem with their code, or getting stuck on the next step in their assignment. Confusion occurs when students reach an impasse and are confronted with an error or contradiction to their expectations and do not know how to move forward [23]. It has been found to positively correlate with learning because it gives students the opportunity to revise their misconceptions about the material [2]. However, for confusion to have a positive benefit in this context, the dyad needs to resolve the confusion to the satisfaction of (ideally) both students.

Confusion can lead to frustration and disengagement during this period of cognitive disequilibrium if the impasse is not resolved [4,18]. We give an example of this transition in Table 2, an excerpt in which two students reach an impasse while moving the sprite around the screen. When the sprite disappears from the screen due to an error they do not understand yet, the dialogue transitions from Suggestion (SU) to Confusion (C).

Once students entered the Confusion state, there was an 86% chance they would dwell there. In the event (14% probability) of leaving the Confusion state, our dyads transitioned to *Frustrated Coordination (7%)* or *Disputational talk (7%)*. This finding is consistent with literature showing that persistent confusion leads to frustration, among other things [4].

Table 3. Annotated dialogue excerpt showing a dyad's transition from Disputational Talk to Exploratory Talk.

S1	*Why are we taking it out? We already have it working.*	Question - Higher Order	Disputational Talk
S2	*No, take these two out and put this back in there.*	Directive	
S1	*No, we're not scrapping this.*	Disagree	
S2	*No, I'm saying, not, wait, no, look*	Disagree	
S1	*grab that, say "Hello" "Hi, Alonzo". Take that out. Then take the 'touching' out. And then take the 'if-else', put it in there, then go back to control and say 'if', then...*	Directive	
S1	*If. Grab the regular 'if'.*	Directive	
S2	*Okay.*	Agree	
S2	*Wait a minute, we still have to use some blocks right here, so...*	Suggestion	Exploratory Talk
S1	*That's what I'm saying, yeah, that's...*	Agree	
S1	*Else, motion...*	Self- Expl.	
S2	*Oh, wait, you're right.*	Agree	

The transition from Disputational Talk to Exploratory Talk is characterized by students justifying their disagreement and beginning to work cooperatively.

Table 3 provides an example in which the students are directing and disagreeing with each other, not making any progress toward a solution, until the dialogue changes from directives (DI) and disagreements (D) to suggestions (SU) and self-explanation (SE), which prompts the dyad to reach a consensus and return to exploratory talk.

Littleton and Light [7] suggest that disputational talk often does not offer constructive criticism, instead making counter-assertions, while exploratory talk offers justified challenges and alternative hypotheses. Students transitioning from Disputational Talk to Exploratory Talk in these sessions specifically utilize justification and self-explanation to challenge their partner, which indicates that the transition between the states occurs because students begin to make justified arguments rather than unfounded assertions.

Frustrated Coordination is the other entry point into Exploratory Talk. A hallmark of Frustrated Coordination is one or both members of the dyad attempting to refocus the conversation on the task, and in exiting this state, students either reach a consensus and transition to Exploratory Talk (5%), or they continue their conflict and transition to Disputational Talk (2%).

4.3 Design Implications

These models of collaborative talk in upper elementary school learners provide implications for determining when to intervene in support of collaborative learning. For example, the models suggest that when students became confused, they almost never returned to productive Exploratory Talk without first having Disputational Talk. In practice, students in the Confused state generally could not come to a consensus on the next step they should take, and a dispute would arise where the dyad would either express their frustration with who is in control (*Frustrated Coordination*) or they would oppose each other's suggestions without considering them (*Disputational Talk*). These frustrated and argumentative dialogues constituted conflict within the dyad, but it appears to be an important part of the cycle that brings students back to Exploratory Talk. If an AI-augmented system can detect when learners enter a Frustrated Coordination or Disputational Talk state from the Confusion state, tailored feedback could guide them to a productive resolution of the unresolved confusion, preventing these conflict states from becoming self-perpetuating. The feedback would likely need to be different depending on whether the goal was for students to move from Frustrated Talk back to Exploratory Talk or for them to move through Disputational Talk, entering the more productive Disagreement & Self-Explanation state within that cycle before returning to Exploratory Talk.

5 Conclusion

Supporting collaborative talk is an important direction for AI-augmented technologies to move into. This paper has presented an analysis of the states and flow of dialogue between elementary students who are collaboratively learning. The results highlight several promising directions for future work. The state of Frustrated Coordination and its relationship to Disputational Talk and Exploratory

Talk needs to be further investigated in future work, especially regarding antagonism between partners. Disputational dialogue appears to be an important component of larger cycles of Exploratory Talk, and this cycle of disputational and exploratory talk suggests that conflict between elementary-aged partners is a natural part of reaching a consensus. Tailored feedback could guide these young learners in navigating a dispute productively rather than devolving into unproductive antagonism. With a deeper understanding of collaborative talk between elementary-aged learners, researchers have the potential to improve adaptive feedback in the context of dialogue and ultimately support a better collaborative learning experience.

Acknowledgements. This work is supported by the National Science Foundation through grant DRL-1721160. Any opinions, findings, conclusions, or recommendations expressed in this report are those of the authors and do not necessarily represent the views of the National Science Foundation.

References

1. Barnes, D.: Exploring Talk in School: Inspired by the Work of Douglas Barnes. SAGE Publications Ltd, New York (2008). https://doi.org/10.4135/9781446279526
2. Craig, S., Graesser, A., Sullins, J., Gholson, B.: Affect and learning: an exploratory look into the role of affect in learning with AutoTutor. J. Educ. Media **29**, 241–250 (2004). https://doi.org/10.1080/1358165042000283101
3. D'Mello, S.K., Lehman, B., Person, N.: Monitoring affect states during effortful problem solving activities. Int. J. Artif. Intell. Educ. **20**(4), 361–389 (2010). https://doi.org/10.3233/JAI-2010-012
4. D'Mello, S., Graesser, A.: Dynamics of affective states during complex learning. Learn. Instr. **22**(2), 145–157 (2012). https://doi.org/10.1016/j.learninstruc.2011.10.001
5. Graesser, A.C., Olde, B.A.: How does one know whether a person understands a device? the quality of the questions the person asks when the device breaks down. J. Educ. Psychol. **95**, 524–536 (2003). https://doi.org/10.1037/0022-0663.95.3.524
6. Griffith, A.E., et al.: Discovering co-creative dialogue states during collaborative learning. In: Roll, I., McNamara, D., Sosnovsky, S., Luckin, R., Dimitrova, V. (eds.) AIED 2021. LNCS (LNAI), vol. 12748, pp. 165–177. Springer, Cham (2021). https://doi.org/10.1007/978-3-030-78292-4_14
7. Light, P., Littleton, K. (eds.): Learning with Computers: Analysing Productive Interactions. Routledge, Milton Park (1998)
8. Ma, Y., Wiggins, J.B., Celepkolu, M., Boyer, K.E., Lynch, C., Wiebe, E.: The challenge of noisy classrooms: Speaker detection during elementary students' collaborative dialogue. In: Roll, I., McNamara, D., Sosnovsky, S., Luckin, R., Dimitrova, V. (eds.) AIED 2021. Lecture Notes in Computer Science, vol. 12748, pp. 268–281. Lecture Notes in Computer Science, Springer, Cham (2021)
9. Madaio, M., Peng, K., Ogan, A., Cassell, J.: A climate of support: a process-oriented analysis of the impact of rapport on peer tutoring. In: Rethinking Learning in the Digital Age: Making the Learning Sciences Count, 13th International Conference of the Learning Sciences (ICLS) 2018. International Society of the Learning Sciences (2018)

10. McQuiggan, S.W., Robison, J.L., Lester, J.C.: Affective transitions in narrative-centered learning environments. J. Educ. Technol. Soc. **13**(1), 40–53 (2010)
11. Mercer, N.: The Guided Construction of Knowledge: Talk amongst Teachers and Learners, p. viii, 135. Multilingual Matters (1995)
12. Mercer, N.: Words and Minds: How We Use Language to Think Together. Routledge, Milton Park (2000). https://doi.org/10.4324/9780203464984
13. Mercer, N., Howe, C.: Explaining the dialogic processes of teaching and learning: the value and potential of sociocultural theory. Learn. Cult. Soc. Interact. **1**(1), 12–21 (2012). https://doi.org/10.1016/j.lcsi.2012.03.001
14. Mercer, N., Wegerif, R.: Is 'exploratory talk' productive talk? In: Learning with Computers: Analysing Productive Interaction, p. 23. Routledge, Milton Park (1998)
15. Misra, A., Loukina, A., Beigman Klebanov, B., Gyawali, B., Zechner, K.: A good start is half the battle won: unsupervised pre-training for low resource children's speech recognition for an interactive reading companion. In: Roll, I., McNamara, D., Sosnovsky, S., Luckin, R., Dimitrova, V. (eds.) Artificial Intelligence in Education. Lecture Notes in Computer Science, pp. 306–317. Springer, Cham (2021). https://doi.org/10.1007/978-3-030-78292-4_25
16. Morales-Urrutia, E.K., Ocaña Ch., J.M., Pérez-Marín, D., Pizarro-Romero, C.: Promoting learning and satisfaction of children when interacting with an emotional companion to program. In: Bittencourt, I.I., Cukurova, M., Muldner, K., Luckin, R., Millán, E. (eds.) Artificial Intelligence in Education, vol. 12164, pp. 220–223. Springer, Cham (2020). https://doi.org/10.1007/978-3-030-52240-7_40
17. Nystrand, M., Wu, L.L., Gamoran, A., Zeiser, S., Long, D.A.: Questions in time: Investigating the structure and dynamics of unfolding classroom discourse. Discourse Process. **35**(2), 135–198 (2003). https://doi.org/10.1207/S15326950DP3502_3
18. Piaget, J.: Part i: cognitive development in children: Piaget development and learning. J. Res. Sci. Teach. **2**(3), 176–186 (1964). https://doi.org/10.1002/tea.3660020306
19. Rabiner, L., Juang, B.: An introduction to hidden Markov models. IEEE ASSP Mag. **3**(1), 4–16 (1986)
20. Rodríguez, F.J., Boyer, K.E.: Discovering individual and collaborative problem-solving modes with hidden Markov models. In: Conati, C., Heffernan, N., Mitrovic, A., Verdejo, M.F. (eds.) AIED 2015. Lecture Notes in Computer Science, pp. 408–418. Springer, Cham (2015). https://doi.org/10.1007/978-3-319-19773-9_41
21. Sun, S.: Meta-analysis of Cohen's Kappa. Health Serv. Outcomes Res. Methodol. **11**, 145–163 (2011). https://doi.org/10.1007/s10742-011-0077-3
22. Tiam-Lee, T.J., Sumi, K.: Adaptive feedback based on student emotion in a system for programming practice. In: Nkambou, R., Azevedo, R., Vassileva, J. (eds.) Intelligent Tutoring Systems, vol. 10858, pp. 243–255. Springer, Cham (2018). https://doi.org/10.1007/978-3-319-91464-0_24
23. VanLehn, K., Siler, S., Murray, C., Yamauchi, T., Baggett, W.B.: Why do only some events cause learning during human tutoring? Cogn. Instr. **21**(3), 209–249 (2003). https://doi.org/10.1207/S1532690XCI2103_01
24. Williams, L., Wiebe, E., Yang, K., Ferzli, M., Miller, C.: In support of pair programming in the introductory computer science course. Comput. Sci. Educ. **12**(3), 197–212 (2002). https://doi.org/10.1076/csed.12.3.197.8618
25. Zakaria, Z., et al.: Two-computer pair programming: exploring a feedback intervention to improve collaborative talk in elementary students. Comput. Sci. Educ. **32**(1), 3–29 (2022). https://doi.org/10.1080/08993408.2021.1877987

Unsupervised Concept Tagging of Mathematical Questions from Student Explanations

K. M. Shabana[1](✉) and Chandrashekar Lakshminarayanan[2]

[1] Indian Institute of Technology Palakkad, Palakkad , India
shabana.meethian@gmail.com
[2] Indian Institute of Technology Madras, Chennai, India
chandrashekar@cse.iitm.ac.in

Abstract. Assigning concept tags to questions enables Intelligent tutoring systems (ITS) to efficiently organize resources, help identify students' strengths and weaknesses, and recommend suitable learning materials accordingly. Manual tagging is time-consuming, and inefficient for large question banks, and could lead to consistency issues due to differences in the perspectives of individual taggers. Automatic tagging techniques can efficiently generate consistent tags at lower costs. Generating automatic tags for mathematical questions is challenging as the question text is usually short and concise, and the question as well as the answer text contains mathematical symbols and formulas. However, prior works have not studied this problem extensively. In this context, we conducted a study in a graduate-level linear algebra course to understand if student explanations to solving mathematical problems can be employed to generate concept tags associated with those questions. In this paper, we propose a method called Unsupervised Skill Tagging (UST) to extract concept tags associated with a given question from explanation text. Using UST on the explanations generated, we show that the explanations indeed contain the expert-specified concept tags.

Keywords: question-tagging · text analysis · unsupervised concept-labeling

1 Introduction

Educational technology systems such as intelligent tutoring systems (ITS) consist of a domain model that stores the relations or dependencies between various assessment items such as questions, exercises and their associated concepts. In particular, the questions which are used to evaluate the student knowledge level are mapped with the concepts/skills that are required to solve them. This enables organization of content wherein questions associated with the same set of concepts can be grouped together. Assigning concept tags to questions also helps in assessing students knowledge level, identifying the concepts/skills that the

N. Wang et al. (Eds.): AIED 2023, LNAI 13916, pp. 627–638, 2023.
https://doi.org/10.1007/978-3-031-36272-9_51

student is unable to master, and thereby recommend suitable exercises/learning resources that facilitate learning. Therefore generating the correct tags for questions is important as incorrect tagging could adversely affect the performance of automatic tutoring systems.

Manual generation of question tags is tedious and costly in terms of the time involved in tagging and the labour costs incurred. Manual tagging could also lead to consistency issues due to the difference in perspectives of individual taggers with respect to a question. Automatic tagging procedures can generate consistent concept tags faster at a lower cost. Several techniques to generate automatic question tags for language related educational content have been proposed in literature [8]. However, automatic tagging of mathematical questions has not been studied extensively as compared to that of language questions. Generating tags for mathematical questions is demanding as the question text is mostly succinct and not rich enough to support tagging. The answers to such questions are mostly numerical values, which do not add any value to tag generation.

In this context, we study if student explanations on solving a mathematical problem could be utilized for unsupervised automatic tag generation. The key idea in this paper is that, when asked explicitly to explain their response to a given mathematical problem/question, students who have the right approach to solving a problem would mention the right concepts related to the question in their explanation. We tested our idea in a graduate-level linear algebra course where we asked the students to explain their problem-solving strategy for a selected set of questions in their weekly quizzes. It was observed that the most frequent unigrams and bigrams in the processed explanation text provided by students had good matching with the important concepts used in solving the problem. We then developed a technique called *Unsupervised Skill Tagging (UST)* to tag questions with labels from a master concept list based on the most frequent n-grams in the set of explanations for each problem. UST demonstrates good performance on our dataset comprising the student explanations, which reinforces the idea that student explanations indeed describe the right concepts required to solve a problem, which is exploited by *UST* to generate the correct question tags.

The rest of the paper is organized as follows: Sect. 2 discusses some of the related work on existing question tagging methods for educational content. Section 3 describes the experiment setup used to collect student explanations. In Sect. 4, we present *UST*, our unsupervised approach to match questions with concept labels by processing student explanations on solving the questions. Section 5 describes the results obtained when UST was applied on a test dataset collected during our experiment. In the final section, we summarize our work and discuss the scope of future work.

2 Related Work

B. Sun et al. [8] proposed a position-based attention model and keywords-based model to automatically associate questions with tags from a knowledge map.

Their work has shown that employing words of answers with the question text boosts the performance of automatic tagging methods for multiple choice questions (MCQs) belonging to English course. Wankerl et al. [9] proposed a method to perform tagging of mathematical questions solely based on the textual and visual representations of mathematical formula provided along with the question. Both the proposed techniques are supervised learning methods that require a large amount of labeled data for training.

A semi-automatic method to perform tagging was proposed by Ramesh et al. [6] to tag questions with metadata corresponding to cognitive level, question type, content and difficulty level. The tags were generated using techniques such as n-grams keyword matching, semantic dictionary and domain ontology.

Next we describe the studies that propose techniques to generate knowledge components (KC) for mathematical and English writing problems using crowdsourcing. Knowledge components represent the skills or knowledge needed to solve a problem. Moore et al. [4] studied whether KCs identified by domain experts could be generated through crowd-sourcing. Participants in this study were given a problem in mathematics or English writing and were asked to concisely list three skills required to solve the problem at hand. For each problem, three KCs were identified by a domain expert. The participants' responses were coded, using a codebook developed manually from the participants' responses, and compared to the KCs identified by the expert. It was found that roughly 33% of the crowdsourced KCs directly matched those generated by domain experts for the same problem. While the participants for this study were selected through the Amazon Mechanical Turk (AMT) platform, a similar study was conducted with learners from two online courses in chemistry and programming [5]. In this case, it was found that half of the crowdsourced KCs matched expert-generated KCs in each problem. Both these works involved a substantial amount of human intervention to code the participant responses and perform the analysis.

The efficacy of using topic modeling approaches for identifying relevant KCs from student explanations has also been studied [3]. It was observed that interpreting the topics as KCs was difficult and none of the topics were found to be directly related to any of the expert generated KCs, which made it unsuitable for automatic tag generation.

Supervised techniques for labelling educational content with skill tags have also been proposed in literature. Shen et al. [7] adopted a task adaptive pretrained BERT model that was fine tuned on three input types including KC descriptions, instructional video titles, and problem text, for automatic labeling of educational content. Recently Matsuda et al. [2] proposed an unsupervised method to mine latent skills from instructional text of assessment items and tag the discovered skills with human-friendly labels. The given text is clustered using k-means such that each cluster represents a unique skill and the most influential keyword extracted from each cluster is used to label the skill. This technique is used to generate skill model from an available body of text and it is not clear how the method can be extended to tag new content (instructional text/assessment item). Our method UST is an unsupervised technique for automatic tag generation and does not make use of any additional input other than the student

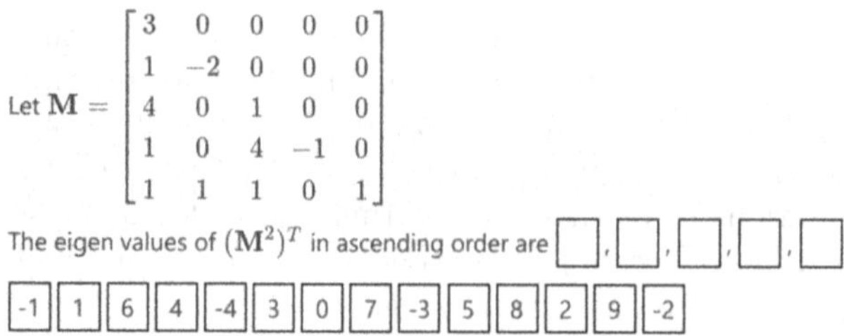

Fig. 1. A question asked in one of the weekly quizzes.

Table 1. Concept-labels included in the master tag-list.

orthonormal/orthogonal matrix, trace, transpose, frobenius norm, orthogonal projection, rank, range space, idempotent matrix, dimension, symmetry of inner product, linearity of inner product, positive definiteness of inner product, subspace addition, subspace intersection, linear independence, outer product, rank nullity theorem, null space, linear map composition, nullity, eigenvalues triangular matrix, eigenvalues matrix powers, eigenvalues matrix transpose, eigenvalue, eigenvector, invertibility, function space, derivative, inner product, linear transformation, nilpotent matrix, span

explanation text and the master tag list from which concepts are matched. It can be used to generate tags for any questions for which student explanations are available.

3 Experiment Setup

In this section, we describe the experimental setup used to collect data from the students. The study was conducted during the Linear Algebra for Engineers course in the Fall 2020 semester at the Institute of Technology Palakkad. The course was offered in online mode due to the pandemic and was credited by 30 Indian masters students. As a part of continuous evaluation, five weekly/regular quizzes $Q_1, Q_2, ..., Q_5$ were held through Moodle. Each quiz consisted of about 5–7 questions based on the topics covered in each week. The quizzes consisted of a mix of multiple-choice and numerical questions that were auto-evaluated. An example question has been provided in Fig. 1.

Each of the regular quizzes $Q_1, Q_2, ..., Q_5$ were accompanied by supplementary quizzes $SQ_1, SQ_2, ..., SQ_5$. In each supplementary quiz SQ_i, the students were asked to perform additional activities for two selected questions included in Q_i. Student responses for a total of 10 questions were collected as described below. First, the concepts required to solve each of the regular quiz questions included in the supplementary quiz were identified by the course instructor. These are referred to as the *expert-defined concept labels*, and these expert labels over all the questions is referred to as the *master tag-list* given in Table 1. Then,

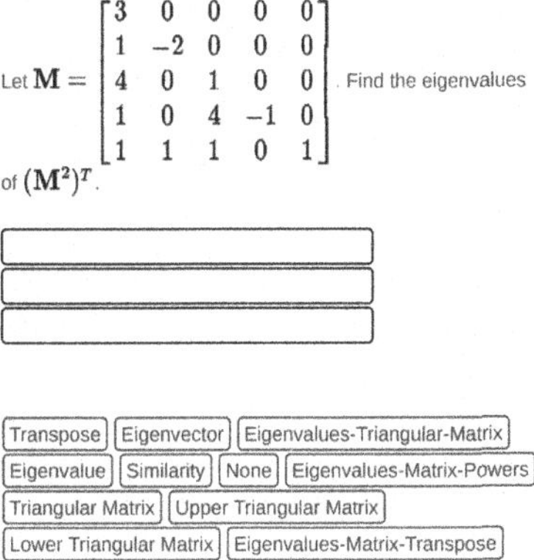

Fig. 2. A drag-and-drop activity used in one of the supplementary quizzes.

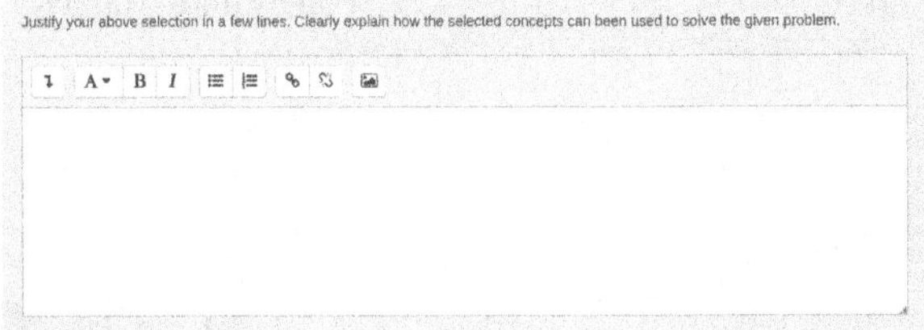

Fig. 3. A sample question from one of the supplementary quizzes for the student to explain how they solved the quiz question.

for each question selected from the regular quiz, two activities were included in the supplementary quiz.

1. Manual Tagging using a drag-and-drop activity where the student selects the labels of the concepts which they think are required to solve the given problem from a given list of concept labels (Fig. 2). The drag-and-drop activity was scored based on the level of agreement of the labels selected by the student with those decided by the instructor.
2. Justification of the choices they made in the earlier part by explaining how they solved the problem (Fig. 3).

Separate time and marks were allotted for the supplementary quizzes to motivate the students to attempt the quiz earnestly. The responses to these questions were evaluated by the teaching assistant for the course. The explanations were also intended for the students to learn by reflecting on their problem solving approach. Majority of the students who participated in the feedback survey at the end of the course stated that they found the supplementary quizzes useful and that it helped them in reflecting on their conceptual understanding.

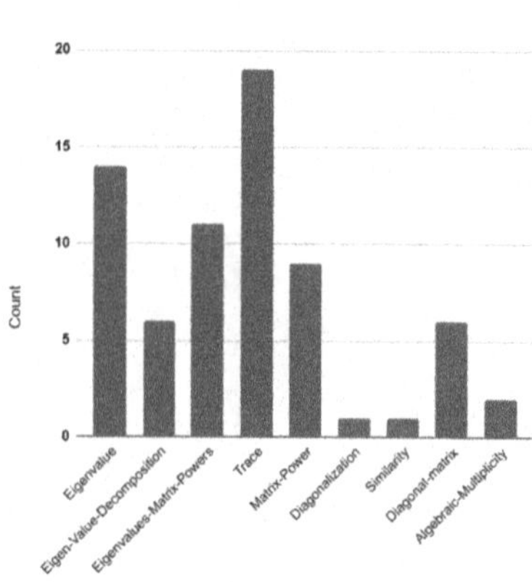

(a) Q1: Eigenvalue, Eigenvalues-Matrix-Powers, Trace

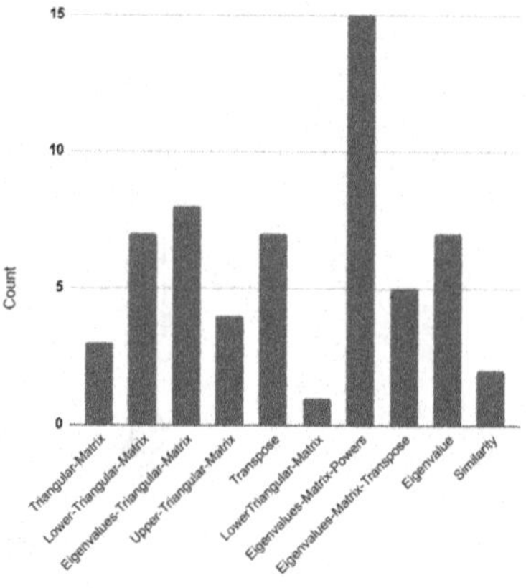

(b) Q2: Eigenvalues-Triangular-Matrix, Eigenvalues-Matrix-Powers, Eigenvalues-Matrix-Transpose

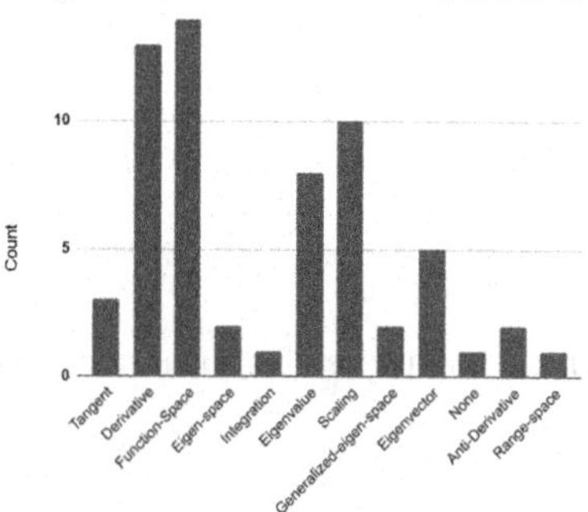

(c) Q3: Eigenvalue, Eigenvector, Function-space, Derivative

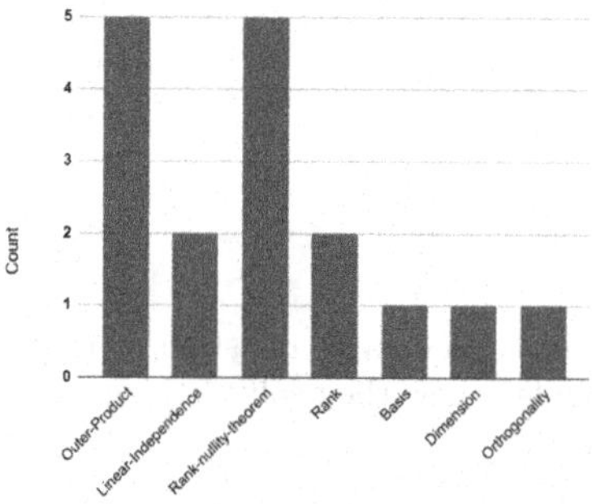

(d) Q4: Outer-Product, Rank, Dimension, Rank-nullity-theorem

Fig. 4. Bar plots of concept labels selected by students who correctly solved the problems in Table 2.

4 Unsupervised Question Tagging Based on Student Explanations

In this section, we first describe the performance of the manual tagging performed by students using the drag-and-drop activity and discuss why it does not always work well in practice. Next, we present *UST*, an unsupervised algorithm to generate concept tags for questions based on student explanations of their problem-solving strategy.

4.1 Manual Tagging Based on Drag-and-Drop Activity

Based on the student responses to the drag-and-drop activity, we attempted to perform manual tagging by assigning the most frequently chosen concept labels

Table 2. Questions associated with the bar plots in Fig. 4.

Q1	If the eigen values of $\mathbf{M}$ are $-2, 0, 1, 1, 3, 4$, then the trace of $(3\mathbf{I} - 2\mathbf{M})^2$ is
Q2	Let $\mathbf{M} = \begin{bmatrix} 3 & 0 & 0 & 0 & 0 \\ 1 & -2 & 0 & 0 & 0 \\ 4 & 0 & 1 & 0 & 0 \\ 1 & 0 & 4 & -1 & 0 \\ 1 & 1 & 1 & 0 & 1 \end{bmatrix}$ The eigen values of $(\mathbf{M}^2)^T$ in ascending order are ☐, ☐, ☐, ☐, ☐. -1 1 6 4 -4 3 0 7 -3 5 8 2 9 -2
Q3	The differentiation of $f(x)$ in function space w.r.t x is a linear transformation in x, which of the following is an eigen function of this linear transformation?
Q4	The nullity of the outer product between the normal of the hyperplanes x+y-z=0 and 2x-y-3z=0 in $\mathbb{C}^3$ is

to each question. Figure 4 shows four instances where this technique did not work well. Each bar chart displays the number of selections of each concept tag in the drag-and-drop activity for the questions in Table 2, by students who answered the question correctly in the regular quiz. The expert-defined concept labels for each question are specified in the caption of each bar chart.

It could be observed from the plots that except for one or two expert-defined concepts none of the other tags received a strong mandate that distinguishes them from the non-relevant labels. Note that during the drag-and-drop activity, the students were presented with around 10 concept tags to choose from, which also contained the expert-defined labels. Therefore the activity was easier as compared to an actual tagging procedure where the tags are to be selected from the list of all available concept labels. This perhaps points to a flaw of manual tagging where multiple taggers may not have a consensus regarding the concept-tags associated with a question. This surprisingly poor performance of manual tagging might be of interest of future study from a behavioral perspective.

4.2 Our Method: Unsupervised Skill Tagging (UST)

It has been observed that manual tagging does not always work well for mathematical questions. Automatic tagging methods that employ question text and words of answers to generate tags might not work well in the case of mathematical questions as: (i) the questions for mathematical problems are usually concise and contains various mathematical symbols and formula which makes tagging difficult. For instance, the concepts required to solve the question provided in Fig. 1 would depend on the properties of the matrix M, i.e., whether it is triangular, orthogonal, etc., which cannot be inferred from the question text, (ii) the answers to such questions are mostly numerical values, which do not add any value to tag generation.

Therefore, we study if student explanations on solving a problem could be used to generate tags for mathematical questions. Table 3 shows explanations

Table 3. Student explanations on solving a problem.

Let $\mathbf{A}$ be the matrix for the orthogonal projection of vectors in $\mathbb{R}^3$ onto the line spanned by the vector [-1 0 1]. Then $rank(\mathbf{A}^7)$ =
It is projecting R^3 to a line, so rank (A)=1. And the projection is orthogonal. which means it is just a rotation, not changing magnitude of vector. so the vector in the line will be projected there itself. $rank(A^7)$=1
A will be a idempotent matrix and have full rank since A will be identity
The projection matrix is idempotent $A^2 = A$, Hence $A^7 = A$. The rank of the projection matrix is the same as that of input space, in our case, the line spanned by [-1 0 1] Projection does not change the span of the input space, and the matrix is idempotent.
Projection matrices are idempotent. Here the matrix is projecting onto a vector space basis having dimension 2. Hence rank of A will be 2 which is the dimension of its range space. Since projection matrices are idempotent $A^7 = A$ itself and hence $rank(A^7) = rank(A) = 2$.

provided by four students on solving the given problem. Unlike essays wherein words are linked by underlying syntactic/semantic structures, explanations to math problems are linked via the underlying procedural relationships between high-level concepts. Language models such as Bidirectional Encoder Representations from Transformers (BERT) are of little help in this task because such models are trained for language tasks and they learn the contextual relations between words in a text, which could be used for grading essays, short answers, etc.

In this section, we describe our proposed method called *Unsupervised Skill Tagging* (UST), an unsupervised technique for tagging mathematical questions by linking them with an existing list of concept labels, referred to as the *master tag-list*. It is based on the idea that explanations by students with the right problem-solving approach contain the correct concepts needed to solve the problem. The steps of the algorithm are detailed below:

1. Pre-processing: The student explanation text are processed by first performing the basic text preprocessing steps such as removing punctuation, extra spaces and digits, conversion to lower case and removal of stop words, followed by tokenization and lemmatization.

2. Extracting the most frequent unigrams and bigrams: For each question, the most frequent unigrams and bigrams from all the student explanations were extracted. The unigrams and bigrams with a frequency of occurrence less than a *threshold* were filtered out to form the *frequent n-gram list.* The best value of the *threshold* parameter is determined during the training phase.

3. Matching the frequent n-grams with concept labels in the *master tag-list*: A string $s1$ is said to have an *exact match* in another string $s2$ if the words in $s1$ are present as exact continuous words in $s2$. For instance, 'norm' does not have an exact match in 'orthonormal' whereas 'nullity' has an exact match in 'rank nullity theorem'. The exact match of a concept-label *cLabel* in the master tag-list and a frequent n-gram *fGram* is given by:

$$contains(cLabel, fGram) = \begin{cases} True & \text{if } cLabel \text{ has an exact match of } fGram \\ False & \text{otherwise} \end{cases} \tag{1}$$

Levenshtein distance between two strings is defined as the minimum number of single-character edits (insertions, deletions or substitutions) required to change one string into the other. For instance, the Levenshtein distance between 'test' and 'tent' is 1, while that between 'test' and 'ten' is 2. A concept-label *cLabel* in the master tag-list is matched with a frequent n-gram *fGram* if *fGram* has an exact match in *cLabel* or if the Levenshtein distance between *fGram* and *cLabel* (given by $levD(cLabel, fGram)$) is less than the threshold γ_{dist}.

$$match(fGram, cLabel) = \begin{cases} True & \text{if } contains(cLabel, fGram) \text{ is True or} \\ & \quad levD(cLabel, fGram) < \gamma_{dist} \\ False & \text{otherwise} \end{cases} \tag{2}$$

4. Scoring Function: A scoring function is then defined to assign a numerical value to the level of matching as follows:

$$score(fGram, cLabel) = \begin{cases} \frac{\text{numWords(fGram)}}{\text{numWords(cLabel)}} & \text{if } match(fGram, cLabel) \text{ is True} \\ 0 & \text{otherwise} \end{cases} \tag{3}$$

$numWords(term)$ is the number of words in *term*. For each question, a score is computed for every tag in the *master tag-list* against each of the frequent n-grams using the scoring function. The tags in the master tag-list with a score greater than or equal to γ_{score} are determined to be the tags associated with that question. The values of γ_{dist} and γ_{score} are determined during training.

5 Results and Discussion

UST is an unsupervised tagging method. However, we make a train-test split so as to choose the values of the thresholds used in the scoring function (γ_{dist} and γ_{score}) and for selecting the frequent n-grams. Hence the dataset was randomly split into a training-test set with the training set consisting of 6 questions and the test set with 4 questions.

The text preprocessing for all the questions was performed using the NLTK library [1]. The master tag-list was constructed by combining the expert-defined concept labels associated with all the questions along with some additional linear-algebra concepts (Table 1). A total of 32 concept-tags were included in the master tag-list. UST was used to generate concept-tags for each question by including explanations given by:

Table 4. The expert-defined concept labels along with the labels generated by our method for two questions in the test set.

Let $\mathbf{M}$ be a rank-3 matrix such that $\mathcal{R}(\mathbf{M}) = \mathcal{N}(\mathbf{M})$, then $rank(\mathbf{M}^T)$= ____, and the dimension of the input and output vectorspaces of this linear map are ____ and ____.
Expert-defined: subspace addition, subspace intersection, dimension, linear independence **Generated by our method:** dimension, subspace addition, subspace intersection, span
Let $\mathbf{v}_1, \mathbf{v}_2, \mathbf{v}_3, \mathbf{v}_4$ be the basis of a vectorspace, $\mathcal{S}_1 = span(\mathbf{v}_1 + \mathbf{v}_3, \mathbf{v}_4)$ and $\mathcal{S}_2 = span(\mathbf{v}_3, \mathbf{v}_2 + \mathbf{v}_4)$. The dimension of $\mathcal{S}_1 + \mathcal{S}_2$ is
Expert-defined: range space, null space, linear map composition, rank nullity theorem, rank, nullity **Generated by our method:** rank, range space, dimension, rank nullity theorem, null space, linear map composition, nullity

Setting 1: all students
Setting 2: students who scored non-zero marks (partially correct answers) for the question in regular quiz
Setting 3: students who scored full marks for the question in regular quiz

For each of these settings, the values of γ_{dist}, γ_{score} and the frequency threshold for n-gram selection were chosen through parameter tuning on the training set. The best parameter values discovered under all the three settings were the same: frequency threshold for n-gram selection as one-tenth of the total number of student explanations for that question, $\gamma_{dist} = 1$ and $\gamma_{score} = 1$.

UST was applied on the test dataset with the chosen parameters. The concept labels generated for two questions in the test set have been provided in Table 4. As each question is tagged with multiple concept labels, this is an instance of multi-label classification and therefore the precision, recall and F_1-score of our method were computed for the test set as follows:

$$precision = \frac{1}{n}\sum_{i=1}^{n}\frac{|y_i \bigcap y_i'|}{|y_i'|} \quad (4)$$

$$recall = \frac{1}{n}\sum_{i=1}^{n}\frac{|y_i \bigcap y_i'|}{|y_i|} \quad (5)$$

$$F_1 score = 2 \times \frac{precision \times recall}{precision + recall} \quad (6)$$

n: number of questions
y_i: expert defined concept labels for question i
y_i': labels generated by UST for question i

Table 5. Evaluation metrics computed for the 3 settings under a) UST and b) Manual tagging.

Metric	UST			Manual Tagging		
	Setting 1	Setting 2	Setting 3	Setting 1	Setting 2	Setting 3
Precision	0.713	0.742	0.775	0.895	0.742	0.742
Recall	0.937	0.875	0.792	0.789	0.707	0.707
F1-score	0.810	0.802	0.783	0.838	0.724	0.724

Precision is the fraction of expert-defined concept labels among the labels retrieved by our method. *Recall* is the fraction of expert-defined concept labels retrieved by our method. F_1 *score* combines the precision and recall into a single metric by taking their harmonic mean.

The precision, recall and F_1 score for our method in each of the three cases is shown in Table 5. We have also compared the performance of UST against manual tagging where the question tags were chosen to be the ones selected by at least half of the students under each setting. It could be observed that for both the methods, the highest F_1 score is obtained when the explanations of all students were considered. This could be attributed to students making a mistake in solving the problem, despite having some idea about the right problem-solving strategy. In this setting, manual tagging performs slightly better than UST in terms of the F1 score. Note that for manual tagging, a small subset of tags in the *master tag-list* were presented to the students which also included all of the expert-defined labels, which contributed to the higher precision value and subsequently larger F1 score. UST on the other hand utilized the entire *master tag-list* and produced a better recall than manual tagging.

Note that while precision improves for UST from Setting 1 to 2 and 3, a drop in recall is simultaneously observed. This might be due to the students who correctly solved a problem listing mostly the right concept labels while missing out on a few other relevant ones. The students who have incorrectly solved a problem but with the right problem-solving strategy might still list out all the required concepts, while the ones with an incorrect strategy would specify irrelevant concepts in their explanations. Therefore including these explanations causes UST to lose on precision but gain on recall.

A limitation of UST is that it requires the master tag list, with all the concept-labels associated with the domain, to be defined and available.

6 Conclusion and Future Work

Question tags are utilized by educational technology systems to organize content as well as perform ability analysis of students that helps with intelligent question recommendation. It has been shown that including the words of answers along with the question text boosts the performance of automatic taggers. However for mathematical questions, the answers are mostly numerical values, that reveal no extra information to taggers. In this paper, we conducted a study in a graduate level linear algebra course to examine if students with the correct problem solving approach mentions the right concepts required to solve the problem, when asked to explain their problem solving strategy. Based on our observations, we have proposed an unsupervised method called unsupervised skill tagging (UST), to generate concept-tags associated with a question item by processing the student explanations. *UST* extracts frequent unigrams and bigrams from the text and uses a scoring function based on exact word matching and Levenshtein distance to perform matching with a master list of concept tags.

In spite of its simplicity, the scoring function demonstrates good performance on the test set. It would be interesting to study if the performance of *UST* could be enhanced with the help of a concept graph that models the relationships between the various concepts. It would also be interesting to study the performance of *UST* on questions from other domains.

Acknowledgement. We thank Dr. Lakshmi Narasimhan Theagarajan (Indian Institute of Technology Madras), the instructor of the linear algebra course, for his support and inputs during the study.

References

1. Loper, E., Bird, S.: Nltk: The natural language toolkit. CoRR cs.CL/0205028 (2002)
2. Matsuda, N., Wood, J., Shrivastava, R., Shimmei, M., Bier, N., et al.: Latent skill mining and labeling from courseware content. J. Educ. Data Min. **14**(2) (2022)
3. Moore, S., Nguyen, H.A., Stamper, J.: Evaluating crowdsourcing and topic modeling in generating knowledge components from explanations. In: Bittencourt, I.I., Cukurova, M., Muldner, K., Luckin, R., Millán, E. (eds.) AIED 2020. LNCS (LNAI), vol. 12163, pp. 398–410. Springer, Cham (2020). https://doi.org/10.1007/978-3-030-52237-7_32
4. Moore, S., Nguyen, H.A., Stamper, J.: Towards crowdsourcing the identification of knowledge components. In: Proceedings of the Seventh ACM Conference on Learning@ Scale, pp. 245–248 (2020)
5. Moore, S., Nguyen, H.A., Stamper, J.: Leveraging students to generate skill tags that inform learning analytics, pp. 791–798 (2022)
6. Ramesh, R., Mishra, S., Sasikumar, M., Iyer, S.: Semi-automatic generation of metadata for items in a question repository. In: 2014 IEEE Sixth International Conference on Technology for Education, pp. 222–228. IEEE (2014)
7. Shen, J.T., et al.: Classifying math knowledge components via task-adaptive pretrained BERT. In: Roll, I., McNamara, D., Sosnovsky, S., Luckin, R., Dimitrova, V. (eds.) AIED 2021. LNCS (LNAI), vol. 12748, pp. 408–419. Springer, Cham (2021). https://doi.org/10.1007/978-3-030-78292-4_33
8. Sun, B., Zhu, Y., Xiao, Y., Xiao, R., Wei, Y.: Automatic question tagging with deep neural networks. IEEE Trans. Learn. Technol. **12**(1), 29–43 (2018)
9. Wankerl, S., Götz, G., Hotho, A.: f2tag-can tags be predicted using formulas? In: 2020 19th IEEE International Conference on Machine Learning and Applications (ICMLA), pp. 565–571. IEEE (2020)

Robust Team Communication Analytics with Transformer-Based Dialogue Modeling

Jay Pande[1(✉)], Wookhee Min[1], Randall D. Spain[2], Jason D. Saville[1], and James Lester[1]

[1] North Carolina State University, Raleigh, NC 27695, USA
{jpande,wmin,jdsavill,lester}@ncsu.edu

[2] U.S. Army Combat Capabilities Development Command, Soldier Center, Orlando, FL 32826, USA
randall.d.spain.civ@army.mil

Abstract. Adaptive training environments that can provide reliable insight into team communication offer great potential for team training and assessment. However, traditional techniques that enable meaningful analysis of team communication such as human transcription and speech classification are especially resource-intensive without machine assistance. Additionally, developing computational models that can perform robust team communication analytics based on small datasets poses significant challenges. We present a transformer-based team communication analysis framework that classifies each team member utterance according to dialogue act and the type of information flow exhibited. The framework utilizes domain-specific transfer learning of transformer-based language models pre-trained with large-scale external data and a prompt engineering method that represents both speaker utterances and speaker roles. Results from our evaluation of team communication data collected from live team training exercises suggest the transformer-based framework fine-tuned with team communication data significantly outperforms state-of-the-art models on both dialogue act recognition and information flow classification and additionally demonstrates improved domain-transfer capabilities.

Keywords: Team Communication Analytics · Team Training · Dialogue Modeling · Text-to-Text Transformers · Deep Learning · Natural Language Processing

1 Introduction

Team-oriented and collaborative learning tasks require team members to actively communicate (e.g., share and provide information, request information) with one another in order to achieve shared goals [1, 2]. Research on the science of teamwork and team training has consistently shown that team communication is an important mediator of team performance [3]. Communication can impact team coordination [4], strategy formulation [5], team situation awareness [6], team cognition [7], and team cohesion [8], all of which are important processes and outcomes that can affect team performance.

N. Wang et al. (Eds.): AIED 2023, LNAI 13916, pp. 639–650, 2023.
https://doi.org/10.1007/978-3-031-36272-9_52

Given the critical role of communication in effective teamwork, it is important to develop methods for automatically analyzing verbal behaviors and using them to infer teamwork competencies. In particular, automatic recognition of dialogue acts and teamwork patterns in team communication would represent a significant step towards developing AI-driven team training systems [9, 10].

Recent dialogue modeling techniques rely on deep neural networks with a transformer architecture [11] such as BERT [12] and GPT-3 [13] that can perform a variety of natural language downstream tasks. Because it uses self-attention mechanisms, the transformer architecture possesses a substantial capacity for learning high-quality representations of complex data. To utilize this capacity, these models are trained with large-scale data corpora. The Text-to-Text Transfer Transformer (T5) has proven especially effective as a unified natural language processing (NLP) model because both inputs and outputs are in the form of text, which can more flexibly support various kinds of NLP tasks through prompt engineering [14]. Because of this flexibility and the rich representational capacity afforded by the transformer architecture, the creators of T5 were not only able to train the model on an enormous corpus of text, but also fine-tune the model on a broad range of NLP tasks such as machine translation, question answering, and sentiment analysis. T5 thus possesses unique versatility while also being more generalizable than previous approaches.

In this paper, we introduce a novel NLP framework for team communication analysis. This framework builds on T5 to recognize dialogue acts that occur between team members as they complete a complex training exercise. The framework also classifies how information flows directionally between team members. We evaluate the effectiveness of T5-based team communication analysis models utilizing natural language representations of speakers' utterances combined with the speakers' roles (a descriptor of a speaker's responsibilities within the team), use cross-validation to select the best-performing hyperparameters for our team communication analysis models, and report the generalization performance of the framework, as defined by accuracy of dialogue act recognition and information flow classification, on a held-out test set. In addition, we further explore the framework's domain-transfer capacity by evaluating the framework's performance on data from a different but related team training exercise.

2 Related Work

Recent advances in NLP along with the increasing availability of education-related text and speech data have contributed to the creation of effective learning technologies [15]. Research has examined NLP's ability to support a wide range of learning analytics tasks, such as short answer grading [16], automated essay scoring [17], learning outcome and goal prediction [18, 19], and discourse analysis [20]. Additionally, deep neural networks have been widely investigated as a computational approach for NLP-based learning analytics tasks due to their robust capacity for distributed representation learning in an end-to-end manner, thereby eliciting state-of-the-art performance [e.g., 17]. Notably, deep neural networks have shown significant promise for NLP tasks with small datasets through data augmentation [21] and few-shot and zero-shot learning [22] by effectively utilizing syntactics, semantics, and linguistic regularities learned from both labeled and unlabeled large-scale external language resources. Deep learning-based NLP has also been investigated to detect disruptive behaviors [20] and off-task behaviors [23] by analyzing the multi-party student dialogue in computer-supported collaborative learning environments, where individuals work together to solve problems and accomplish shared learning goals.

Much of the work to date on NLP for learning analytics has addressed individual learning tasks, but natural language data can also be collected for tasks completed by teams. Analyzing dialogue produced during team activities has potential to inform deeper understanding of the actions and behaviors that enable teams to perform more effectively. Moreover, if team training environments can provide appropriate feedback when a team demonstrates or does not demonstrate effective teamwork behavior, these environments can provide better adaptive support to improve team performance [3, 24]. Accordingly, team training researchers have developed several approaches using team communication data to produce insights into not only team performance, but also into the quality of team coordination and other team behaviors during training [25]. Computational approaches to such problems have included the use of latent semantic analysis to identify socio-cognitive roles in interaction between multiple parties [26] as well as the use of Markov models [27] to better understand how team communication changes over the course of an activity. Additionally, hierarchical regression models have been used to study the relationship of natural language utterances to a team's social development [28]. Deep learning approaches have been less common, but prior work has investigated models combining ELMo word embeddings with conditional random fields for dialogue act recognition and information flow classification [9].

Our current work investigates deep learning for team communication analytics by introducing a team communication analysis framework that leverages the versatility and transfer learning capacity of T5 to achieve high accuracy on both dialogue act recognition and information flow classification. This framework thus provides key functionality necessary to support adaptive team training, which may then be utilized to foster effective teamwork and improved team performance.

3 Team Communication

Team communication is instrumental for effective team performance [29] as it allows information to be exchanged as necessary and mediates essential team action processes. When teams maintain accurate shared mental models, members are able to actively recognize and anticipate the needs of other members without having to be prompted [30]. Further, teams whose members frequently and deliberately share information with one another rather than explicitly request information are better able to coordinate and respond to demanding situations [31].

To investigate this information exchange within teams, labeling schemes are sometimes applied to transcripts of team communication. Through the use of a labeling scheme, individual utterances may be classified according to their function in the information exchange process. Utilizing labeled transcripts, researchers may perform frequency or sequence analyses for insight into team performance [32]. For example, research involving flight crews conducting simulated flights categorized crew member speech into eight categories, such as uncertainty, planning, and acknowledgement statements [33]. Frequency analysis results indicated that high performing flight crews exhibited lower rates of non-task related communication and were also more inclined to use planning statements to respond to air traffic control than low performing flight crews.

Investigating how teams exchange information is critical for understanding the processes that teams use to coordinate [34]. However, human transcription and labeling of team communication is resource-intensive and often not well-suited to providing timely insight or feedback for team performance. While initial research into automating team communication analysis has shown potential [35], the framework presented in the current study (Fig. 1) substantially extends the existing work by achieving highly accurate recognition of team communication labels that are critical to support adaptive team training.

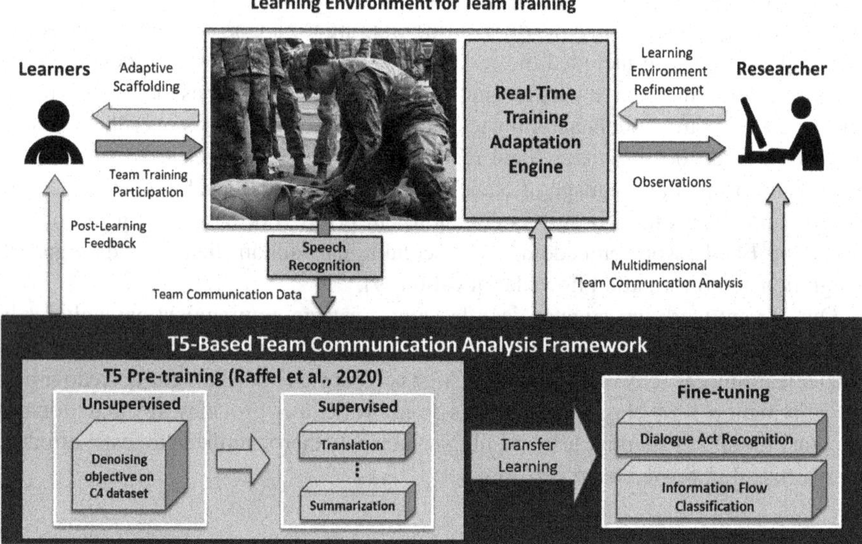

Fig. 1. Team communication analysis framework (at bottom, in dark blue) supporting adaptive team training environments. (Color figure online)

4 Dataset

The implementation and evaluation of our framework was conducted using a dataset of transcribed utterances from verbal communication captured during a live team training exercise. Six U.S. Army squads each participated in one simulation-based exercise followed by two live training exercises. The current work focuses on the final 45-min training exercise, during which each squad was tasked with completing a series of objectives, such as contacting local community leaders, neutralizing active threats, and administering tactical combat casualty care, while reacting to scripted events [36]. While accomplishing these objectives, team members were expected to maintain advanced situational awareness and appropriately inform the rest of their squad about any identified tactical threats and persons of interest. Squads were composed of approximately 10 squad members, each fulfilling a distinct role within a hierarchical structure, or chain of command, which guided their interaction with other squad members and role players in the environment. Leadership roles consisted of the squad leader, who was responsible for coordinating and directing the activities of the squad's subordinate teams, and the alpha and bravo team leaders, who each reported to the squad leader in the hierarchy and directed activities for their respective teams. The alpha and bravo teams each had four to five squad members.

During a mission or training event, a team member may gather and observe critical information that needs to be shared vertically or horizontally within the squad. Each squad member wore an individual microphone during the exercise to capture their utterances, which were later transcribed and annotated. The annotation coding scheme included 27 dialogue act labels and 18 information flow labels. The dialogue act labels describe the intent of each utterance, such as to give a command, make a suggestion, or request information. The information flow labels describe the type of information contained in an utterance and the direction in which it was directed (i.e., up the chain of command, down the chain of command). While each utterance was assigned a dialogue act label, an information flow label was only assigned to applicable utterances when squad members communicated information amongst each other. Interrater reliability for human labeling was substantial for both dialogue act labels (Krippendorff's alpha = 0.679, raw agreement = 0.715, expected agreement = 0.113) and information flow labels (Krippendorff's alpha = 0.710, raw agreement = 0.817, expected agreement = 0.335) [37].

Following annotation, a mapping was produced to create a reduced set of dialogue act labels with the dual goals of making the labels more useful for analysis and intervention and supporting better recognition accuracy. The nine dialogue act labels resulting from mapping are: ACKNOWLEDGEMENT, ACTION REQUEST, ACTION STATEMENT, COMMAND, ATTENTION, GREETING, PROVIDE INFORMATION, REQUEST INFORMATION, and OTHER, where OTHER captures non-meaningful utterances (e.g., incomplete statements, expletives, "hmm"). Eighteen information flow labels were used for classification. Examples of the information flow labels include COMMAND COMING FROM THE SQUAD LEADER, PROVIDE INFORMATION UP THE CHAIN OF COMMAND, and REQUEST INFORMATION FROM DOWN THE CHAIN OF COMMAND [9]. Labeled utterance examples captured from the actual dataset are shown in Table 1. One additional information flow label, "N/A", was created to enable classification of utterances to which the information flow labels did not apply. According to frequency analyses, out of 4,315 tagged utterances across all six squads, PROVIDE INFORMATION (n = 1,109) was the most frequent dialogue act label, and COMMAND (n = 805) was second-most frequent. The three most common information flow labels were N/A (n = 1,978), PROVIDE INFORMATION UP THE CHAIN OF COMMAND (n = 550), and COMMAND COMING FROM THE SQUAD LEADER (n = 362).

Prior research has demonstrated relationships between label frequencies and squad performance [38, 39]. Performance was rated for each of the U.S. Army squads during the live training exercises by trained observers who completed behaviorally-anchored rating scales, objectively capturing squad actions and inactions. Within a study utilizing the same live training exercise as the current study, researchers used performance ratings to categorize squads as either high or low performance. T-tests conducted between performance groups indicated that squads exhibiting high performance had significantly more utterances labeled with the PROVIDE INFORMATION dialogue act label ($t = -14.708, p < .01$) and utterances labeled with the COMMAND COMING FROM THE TEAM LEADER information flow label ($t = -21.0, p < .01$) than low performing squads [38]. More recent research shows that the communication patterns of high and low performing teams differ during stressful events. Mixed-model ANOVA results indicated that high performing teams were more likely to maintain consistent rates of information exchange whether or not a stressful event was occurring, whereas low performing teams showed significant increases in commands and information exchange during a stressful event [39]. Thus, accurate classification of utterances according to dialogue act and information flow labels can support more rigorous assessment of team performance and richer adaptive team training systems.

Table 1. Example utterances and their dialogue act (DA) and information flow (IF) labels.

Speaker	Example Utterances	DA	IF
Squad Leader	Hey, what side of the building is that?	REQUEST INFORMATION	REQUEST INFORMATION FROM DOWN THE CHAIN OF COMMAND
Alpha Team Leader	Two two alpha say again you're coming in broken and unreadable	ACTION REQUEST	REQUEST INFORMATION FROM UP THE CHAIN OF COMMAND
Squad Leader	Moving to conduct KLE with Father Romanov	ACTION STATEMENT	PROVIDE INFORMATION UP THE CHAIN OF COMMAND
Bravo Team Leader	Do you have eyes on the marketplace from here?	REQUEST INFORMATION	REQUEST INFORMATION FROM DOWN THE CHAIN OF COMMAND
Squad Leader	Make sure you get eyes on route black	COMMAND	COMMAND COMING FROM THE SQUAD LEADER

5 Team Communication Analysis Framework

T5 is a text-to-text transformer model which has demonstrated exceptional performance on several NLP benchmarks including CoLA (a sentence acceptability benchmark), MNLI (natural language inference), COPA (sentence completion), and GLUE (general language understanding) [14]. The transformer architecture uses self-attention to capture complex relationships between tokens regardless of their position in the sequence. Standard T5 implementations have undergone unsupervised pre-training using a "denoising" objective (predicting masked tokens) on the Colossal Clean Crawled Corpus (C4) dataset and supervised learning on NLP benchmarks.

Of the several available sizes of T5 models, we chose to use the "small" implementation (60 million parameters) as our base model because it can be trained relatively quickly and is thus well-suited for deployment in a low-resource setting. This model has a vocabulary size of 32,128 distinct tokens. There are 6 layers each in the encoder and decoder of the transformer. The output dimensionality of the feedforward sublayers is 2,048. The multi-head attention sublayers each have 8 heads, with queries, keys, and values projected down to 64 dimensions. The remaining sublayers have output dimensionality 512. For all layers with dropout, the dropout rate was set to 0.1, and the AdamW optimizer was used during training [40].

As noted above, T5 has been fine-tuned on many different NLP tasks. We took advantage of the flexible input format used for fine-tuning to encode both the speaker role and speaker utterance in each input from our dataset. In our encoding, the speaker role takes the place of the "task-specific prefix" [14], and the speaker utterance is placed after the colon. For example, an utterance from the alpha team leader would be encoded as *a_tmldr*: *Hey we're getting ready to move.*

We created two separate T5 models for dialogue act recognition and information flow classification. We conducted squad-level cross-validation to determine the best-performing model hyperparameter values for each task. The data from one of the six squads was reserved as a held-out test set and was not used in cross-validation. The other data was divided into five folds, where each fold consisted of the data from one squad. The training procedure thus used four folds of data. We used early stopping with a patience of 10 epochs, where 10% of the four folds was used for validation as part of the early stopping procedure, and the other 90% was used to train the model. The hyperparameters evaluated were learning rate (using values 1e-4 and 3e-4), warm-up steps (using values equal to 5% and 10% of the number of data instances used to train the model), and batch size (using values 4 and 8). Our performance metric for cross-validation was the macro average of classification accuracy across all five folds.

6 Evaluation

The cross-validation results for accuracy across all folds show that our framework significantly outperforms the majority-class baselines (25.7% for dialogue act recognition and 45.8% for information flow classification) for all hyperparameter settings. As shown in Table 2, the best-performing dialogue act recognition model used learning rate 3e-4, batch size 8, and warm-up steps equal to 10% of the number of training data instances as hyperparameter values. Dialogue act recognition accuracy for this model ranged from 70.97% to 77.41% across all folds ($M = 73.55\%$). The best-performing information flow classification model used learning rate 1e-4, batch size 4, and warm-up steps equal to 10% of the number of training data instances as hyperparameter values. Information flow classification accuracy for this model ranged from 58.62% to 71.30% across all folds ($M = 65.72\%$).

After identifying the best-performing hyperparameter values for each task, we created new models using the same hyperparameter values and tested the new models on the held-out test set. The data instances in the held-out test set were not seen by any of the models created during cross-validation to avoid data leakage. For this evaluation, the data from the other five squads was split with 90% of the instances used for training, and 10% used for validation as part of the early stopping procedure. Accuracy on the held-out test set was 75.92% for dialogue act recognition and 67.23% for information flow classification. The results for both tasks are higher than the average accuracy results reported during cross-validation, and more importantly, exceed the previous state-of-the-art performance based on this dataset [9] (Table 3). These results suggest that our team communication analysis framework, driven by T5's learned representations of a large corpus of natural language data and exposure to many types of NLP tasks, outperforms prior approaches using conditional random fields with ELMo word embeddings (CRF-ELMo).

Table 2. Cross-validation accuracy rates (%) averaged across folds for dialogue act (DA) recognition and information flow (IF) classification. The highest accuracy rate for each task is marked in bold.

{Learning rate, batch size, warm-up steps}	DA	IF
{1e-4, 4, 5%}	72.69	63.51
{1e-4, 4, 10%}	72.77	**65.72**
{1e-4, 8, 5%}	72.91	61.62
{1e-4, 8, 10%}	72.28	64.99
{3e-4, 4, 5%}	71.90	63.63
{3e-4, 4, 10%}	71.98	61.83
{3e-4, 8, 5%}	73.33	62.99
{3e-4, 8, 10%}	**73.55**	63.29

Table 3. Accuracy rates (%) for dialogue act (DA) recognition, information flow (IF) classification on a held-out test set, and the domain-transfer dialogue act recognition task (DA-Transfer). Results from prior work (CRF-ELMo [9]) are shown in addition to those for current work based on T5, and the higher number for each task is marked in bold.

	DA	IF	DA-Transfer
CRF-ELMo	69.42	64.92	67.35
T5	**75.92**	**67.23**	**70.60**

The marginal improvements made when comparing the T5 models' classification accuracy on the held-out test set to the majority-class baselines are 195.4% (dialogue act recognition) and 46.8% (information flow classification). The overwhelming prevalence of N/A labels makes information flow classification for our data more difficult than dialogue act recognition, for which class imbalance is less severe, since information flow classification models are more likely to overfit to the N/A label. This is also evidenced by comparatively lower performance on information flow classification across all hyperparameter settings in cross-validation (72.68% for dialogue act recognition vs. 63.45% for information flow classification, averaged across all hyperparameter value combinations) and on the held-out test set. These label distribution patterns may also explain the difference in best-performing hyperparameter values for each task. Information flow classification may require a smaller batch size because, if batches are large, more batches would consist predominantly of N/A-labelled inputs, and the model could overfit to these examples at the expense of others. A lower learning rate also makes such overfitting less likely. It is also notable that for both tasks, using a number of warm-up steps equal to 10% of the number of training data instances worked better than using a number of warm-up steps equal to 5% of the number of training data instances. This likely indicates that the small size of our dataset made it challenging for models to recover from mistakes made early in training without the benefit of a relatively long warm-up period in which a very low learning rate was used to slowly optimize the models.

Finally, we evaluated our framework's performance on a domain-transfer task. Specifically, we created another T5 model to perform dialogue act recognition on utterances from a separate dataset collected from a different but related team training exercise. Whereas our original dataset consists of utterances captured during the third of three team training exercises, this separate dataset consists of utterances captured during the second exercise. This model again used the best-performing hyperparameter values identified from cross-validation for dialogue act recognition, and it was trained with data from all six squads in our original dataset. The reported accuracy of this model was 70.60%, demonstrating that our framework effectively generalizes to data whose characteristics differ to some extent from those of the data used for training (e.g., different data distributions as a function of how many exercises a squad had completed together). Our T5-based communication analysis model showed improved classification accuracy for this domain-transfer task compared to the prior state-of-the-art result (67.35%) [9] (Table 3).

7 Conclusion

Team communication data captures information beneficial for evaluating teams' performance and skill level, providing a basis for adaptive feedback. However, rich team communication data presents a challenge for these evaluations. Our work addresses this challenge with a transformer-based team communication analysis framework that utilizes a natural language encoding of both team members' utterances and team members' hierarchical role within the team. Our framework exhibits high accuracy in dialogue act recognition and information flow classification on a dataset of utterances from live team training exercises, outperforming majority class baselines as well as prior work completed using the same dataset. The framework also exceeded benchmarks from prior work for generalizing to data from a different training event.

Results from this study encourage the continued exploration of transformer-based models such as T5 for developing the next generation of computational approaches for team communication analytics. In particular, future work should investigate an approach which allows past utterances to provide additional context to improve model performance for classifying subsequent utterances. It may also be useful to consider alternative prompt engineering approaches to represent team communication tasks as similarly as possible to other NLP tasks on which transformer-based models are known to perform well. A multitask model which can simultaneously perform dialogue act recognition and information flow classification could improve performance on one or both tasks and be deployed in the team communication analysis framework efficiently as a unified model. Additionally, error analyses could suggest ways to improve performance on data which these models most frequently misclassify. These investigations will deepen our understanding of how team communication relates to team performance as we work towards increasingly effective adaptive team training environments.

Acknowledgements. The research described herein has been sponsored by the U.S. Army DEVCOM, Soldier Center under cooperative agreement W912CG-19-2-0001. The statements and opinions expressed in this article do not necessarily reflect the position or the policy of the United States Government, and no official endorsement should be inferred.

References

1. Kaliisa, R., Rienties, B., Mørch, A.I., Kluge, A.: Social learning analytics in computer-supported collaborative learning environments: a systematic review of empirical studies. Comput. Educ. Open **3**, 100073 (2022)
2. Meslec, N., Duel, J., Soeters, J.: The role of teamwork on team performance in extreme military environments: an empirical study. Team Perform. Manage.: an Int. J. **26**(5/6), 325–339 (2020)
3. Marlow, S.L., Lacerenza, C.N., Paoletti, J., Burke, C.S., Salas, E.: Does team communication represent a one-size-fits-all approach?: A meta-analysis of team communication and performance. Organ. Behav. Hum. Decis. Process. **144**, 145–170 (2018)
4. Grimm, D.A.P., Gorman, J.C., Robinson, E., Winner, J.: Measuring adaptive team coordination in an enroute care training scenario. In: Proceedings of the Human Factors and Ergonomics Society Annual Meeting, vol. 66, no. 1, pp. 50–54. SAGE Publications, Los Angeles (2022)
5. Waring, S., Alison, L., Shortland, N., Humann, M.: The role of information sharing on decision delay during multiteam disaster response. Cogn. Technol. Work **22**(2), 263–279 (2019)
6. Endsley, M.R.: Situation awareness. In: Salvendy, G., Karwowski, W. (eds.) Handbook of Human Factors and Ergonomics, 5th edn., pp. 434–455. John Wiley & Sons Inc, Hoboken (2021)
7. Gorman, J.C., Grimm, D.A., Stevens, R.H., Galloway, T., Willemsen-Dunlap, A.M., Halpin, D.J.: Measuring real-time team cognition during team training. Hum. Factors **62**(5), 825–860 (2020)
8. Stucky, C.H., De Jong, M.J., Kabo, F.W.: Military surgical team communication: implications for safety. Mil. Med. **185**(3–4), e448–e456 (2020)
9. Min, W., Spain, R., Saville, J.D., Mott, B., Brawner, K., Johnston, J., et al.: Multidimensional team communication modeling for adaptive team training: a hybrid deep learning and graphical modeling framework. In: Roll, I., McNamara, D., Sosnovsky, S., Luckin, R., Dimitrova, V. (eds.) AIED 2021, LNCS, vol. 12748, pp. 293–305. Springer International Publishing, Cham (2021)
10. Foltz, P.W., Martin, M.J.: Automated communication analysis of teams. In: Salas, E., Goodwin, G.F., Burke, C.S. (eds.) Team Effectiveness in Complex Organizations, 1st edn., pp. 445–466. Routledge, New York (2008)

11. Vaswani, A., Shazeer, N., Parmar, N., Uszkoreit, J., Jones, L., Gomez, A.N., et al.: Attention is all you need. In: von Luxburg, U. et al., (eds.) Advances in Neural Information Processing Systems 30 (NIPS 2017), pp. 5999–6009. Curran Associates Inc, Red Hook (2017)
12. Devlin, J., Chang, M., Lee, K., Toutanova, K.: BERT: Pre-training of deep bidirectional transformers for language understanding. In: Burstein, J., Doran, C., Solorio, T. (eds.) Proceedings of the 2019 Conference of the North American Chapter of the Association for Computational Linguistics: Human Language Technologies, Volume 1 (Long and Short Papers), pp. 4171–4186. Association for Computational Linguistics, Stroudsburg (2019)
13. Brown, T.B., Mann, B., Ryder, N., Subbiah, M., Kaplan, J., Dhariwal, P., et al.: Language models are few-shot learners. In: Larochelle, H., Ranzato, M., Hadsell, R., Balcan, M.F., Lin, H. (eds.) Advances in Neural Information Processing Systems 33 (NeurIPS 2020), pp. 1877–1901. Curran Associates Inc., Red Hook (2020)
14. Raffel, C., Shazeer, N., Roberts, A., Lee, K., Narang, S., Matena, M., et al.: Exploring the limits of transfer learning with a unified text-to-text transformer. J. Mach. Learn. Res. **21**(1), 5485–5551 (2020)
15. Litman, D.: Natural language processing for enhancing teaching and learning. In: AAAI'16: Proceedings of the Thirtieth AAAI Conference on Artificial Intelligence, pp. 4170–4176. AAAI Press, Washington (2016)
16. Sawatzki, J., Schlippe, T., Benner-Wickner, M.: Deep learning techniques for automatic short answer grading: predicting scores for english and german answers. In: Cheng, E.C.K., Koul, R.B., Wang, T., Yu, X. (eds.) Artificial Intelligence in Education: Emerging Technologies, Models and Applications. LNDECT, vol. 104, pp. 65–75. Springer, Singapore (2022)
17. Ramesh, D., Sanampudi, S.K.: An automated essay scoring systems: a systematic literature review. Artif. Intell. Rev. **55**(3), 2495–2527 (2021)
18. Emerson, A., Min, W., Azevedo, R., Lester, J.: Early prediction of student knowledge in game-based learning with distributed representations of assessment questions. Br. J. Edu. Technol. **54**(1), 40–57 (2022)
19. Gupta, A., Carpenter, D., Min, W., Rowe, J., Azevedo, R., Lester, J.: Enhancing multimodal goal recognition in open-world games with natural language player reflections. In: Ware, S.G., Eger, M. (eds.) Proceedings of the Eighteenth Annual AAAI Conference on Artificial Intelligence and Interactive Digital Entertainment, pp. 37–44. AAAI Press, Palo Alto (2022)
20. Park, K., Sohn, H., Min, W., Mott, B., Glazewski, K., Hmelo-Silver, C., et al.: Disruptive talk detection in multi-party dialogue within collaborative learning environments with a regularized user-aware network. In: Lemon, O., et al. (eds.) Proceedings of the 23rd Annual Meeting of the Special Interest Group on Discourse and Dialogue, pp. 490–499. Association for Computational Linguistics, Stroudsburg (2022)
21. Lun, J., Zhu, J., Tang, Y., Yang, M.: Multiple data augmentation strategies for improving performance on automatic short answer scoring. In: Proceedings of the AAAI Conference on Artificial Intelligence, vol. 34, ed. 09, pp. 13389–13396. AAAI Press, Palo Alto (2020)
22. Bragg, J., Cohan, A., Lo, K., Beltagy, I.: FLEX: unifying evaluation for few-shot NLP. In: Ranzato, M., Beygelzimer, A., Dauphin, Y., Liang, P.S., Wortman Vaughan, J. (eds.) Advances in Neural Information Processing Systems 34 (NeurIPS 2021), pp. 15787–15800. Curran Associates, Inc., Red Hook (2021)
23. Carpenter, D., Emerson, A., Mott, B., Saleh, A., Glazewski, K., Hmelo-Silver, C., et al.: Detecting off-task behavior from student dialogue in game-based collaborative learning. In: Bittencourt, I.I., Cukurova, M., Muldner, K., Luckin, R., Millán, E. (eds.) AIED 2020, LNCS, vol. 12163, pp. 56–66. Springer, Heidelberg (2020)
24. Sottilare, R.A., Burke, C.S., Salas, E., Sinatra, A.M., Johnston, J.H., Gilbert, S.B.: Designing adaptive instruction for teams: a meta-analysis. Int. J. Artif. Intell. Educ. **28**(2), 225–264 (2018)

25. Foltz, P.W.: Automating the assessment of team collaboration through communication analysis. In: Sottilare, R.A., Graesser, A.C., Hu, X., Sinatra, A.M. (eds.) Design Recommendations for Intelligent Tutoring Systems, vol. 6, pp. 179–185. US Army Research Laboratory, Orlando (2018)
26. Dowell, N.M.M., Nixon, T.M., Graesser, A.C.: Group communication analysis: a computational linguistics approach for detecting sociocognitive roles in multiparty interactions. Behav. Res. Methods **51**(3), 1007–1041 (2019)
27. Ayala, D.F.M., Balasingam, B., McComb, S., Pattipati, K.R.: Markov modeling and analysis of team communication. IEEE Trans. Syst. Man, and Cybern. Syst. **50**(4), 1230–1241 (2020)
28. Yu, M., Litman, D., Paletz, S.: Investigating the relationship between multi-party linguistic entrainment, team characteristics, and the perception of team social outcomes. In: Barták, R., Brawner, K. (eds.) Proceedings of the 32nd International Florida Artificial Intelligence Research Society Conference (FLAIRS-32), pp. 227–232. AAAI Press, Palo Alto (2019)
29. Goodwin, G.F., Blacksmith, N., Coats, M.R.: The science of teams in the military: contributions from over 60 years of research. Am. Psychol. **73**(4), 322–333 (2018)
30. Cannon-Bowers, J.A., Salas, E., Converse, S.A.: Cognitive psychology and team training: training shared mental models of complex systems. Hum. Factors Soc. Bull. **33**(12), 1–4 (1990)
31. Gontar, P., Fischer, U., Bengler, K.: Methods to evaluate pilots' cockpit communication: cross-recurrence analyses vs. speech act–based analyses. J. Cogn. Eng. Decis. Making **11**(4), 337–352 (2017)
32. Nonose, K., Kanno, T., Furuta, K.: An evaluation method of team communication based on a task flow analysis. Cogn. Technol. Work **17**(4), 607–618 (2015)
33. Bowers, C.A., Jentsch, F., Salas, E., Braun, C.C.: Analyzing communication sequences for team training needs assessment. Hum. Factors **40**(4), 672–679 (1998)
34. Mesmer-Magnus, J.R., DeChurch, L.A.: Information sharing and team performance: a meta-analysis. J. Appl. Psychol. **94**(2), 535–546 (2009)
35. Spain, R.D., Min, W., Saville, J.D., Emerson, A., Pande, J., Brawner, K., et al.: Leveraging advances in natural language processing to support team communication analytics in GIFT. In: Sinatra, A. M. (ed.) Proceedings of the 10th Annual GIFT Users Symposium, pp. 147–156. US Army Combat Capabilities Development Command–Soldier Center, Orlando (2022)
36. Johnston, J.H., Phillips, H.L., Milham, L.M., Riddle, D.L., Townsend, L.N., DeCostanza, A.H., et al.: A team training field research study: extending a theory of team development. Front. Psychol. **10**, 1480 (2019)
37. Traum, D.: USC Institute for Creative Technologies, personal communication, July 25 (2022)
38. Saville, J.D., Spain, R.D., Johnston, J.H., Lester, J.: Exploration of team communication behaviors from a live training event. In: Nazir, S., Ahram, T.Z., Karwowski, W. (eds.) AHFE 2021. LNNS, vol. 269, pp. 101–108. Springer, Cham (2021)
39. Saville, J.D., Spain, R.D., Johnston, J.H., Lester, J.C.: An analysis of squad communication behaviors during a field-training exercise to support tactical decision making. In: Wright, J., Barber, D. (eds.) Human Factors and Simulation, vol. 30, pp. 109–116. AHFE Open Access (2022)
40. config.json t5-small at main, https://huggingface.co/t5-small/blob/main/config.json. Last accessed 16 Jan 2023

Teacher Talk Moves in K12 Mathematics Lessons: Automatic Identification, Prediction Explanation, and Characteristic Exploration

Deliang Wang[1](✉), Dapeng Shan[2], Yaqian Zheng[3], and Gaowei Chen[1]

[1] Faculty of Education, The University of Hong Kong, Hong Kong, China
wdeliang@connect.hku.hk

[2] Faculty of Engineering, The University of Hong Kong, Hong Kong, China

[3] School of Educational Technology, Faculty of Education, Beijing Normal University, Beijing, China

Abstract. Talk moves have been shown to facilitate enriched discussion and conversation in classrooms, leading to improved student learning outcomes. To support teachers in enhancing discursive techniques and providing timely feedback on their classroom discourse, this paper proposed two BERT-based deep learning models for automatically identifying teacher talk moves in K12 mathematics lessons. However, the proposed discourse models have complex structures and cannot offer clear explanations on the prediction of teacher utterance, potentially leading teachers to distrust the model. To address this issue, this paper employed three model-agnostic interpreting methods from explainable artificial intelligence and transparently explained to teachers how the model predictions were made by computing and displaying the word relevance. The analysis results confirmed the validity of these explanations. Further, the paper investigated the interpreting results to uncover key characteristics of each type of teacher talk moves. The findings indicated that simple words and phrases could serve as representative indicators of talk moves (e.g., phrases centering on "agree" and "disagree" in the case of talk move *getting students to relate to another's idea*), which shows the potential to assist teachers in mastering discursive techniques. We believe that the paper built a solid step towards building an automated classroom discourse analysis system and fully addressing the interpretability concerns of deep learning-based discourse models.

Keywords: Classroom discourse · Artificial intelligence · Explanations

1 Introduction

Rich discussion and conversation in the classroom can promote students' critical thinking, knowledge construction, and learning achievements [15]. In order to encourage teachers to use productive dialogic practices in their teaching,

N. Wang et al. (Eds.): AIED 2023, LNAI 13916, pp. 651–664, 2023.
https://doi.org/10.1007/978-3-031-36272-9_53

researchers have developed various pedagogical approaches related to classroom discourse [2,16]. For example, the academically productive talk (APT) framework has been designed to advocate for teachers employing talk moves (i.e., specific dialogic acts) to elicit learners' participation in classroom discussion [19]. However, not all teachers can possess the discursive techniques to effectively facilitate classroom conversation [8]. To address this issue, researchers have implemented programs using classroom videos to help teachers reflect on their lessons and improve the quality of dialogue [4,10]. However, the large amount of information contained in these videos can be overwhelming for teachers, making it difficult for them to focus on relevant segments [4]. Additionally, the reflection process requires teachers to manually extract and analyze key information from the dialogue, which is laborious and time-consuming [21].

The advancement of artificial intelligence (AI) techniques has enabled researchers to make significant progress in developing automated models for analyzing classroom dialogue [10]. However, the progress has also brought about some challenges. For instance, traditional machine learning techniques have been predominantly used in existing studies to model classroom discourse, leading to relatively limited performance [7,9,12]. While a small number of studies have employed deep learning techniques to achieve powerful performance [10,24], the resulting automated classroom discourse models often have complex structures, operate as black boxes, and do not offer clear explanations for their predictions of teachers' or students' utterances. On the one hand, this lack of transparency may lead teachers to distrust the models and even refuse to use them [17]. On the other hand, incorrect analysis of classroom discourse without explanations may cause teachers to make inappropriate adjustments to their teaching strategies, potentially negatively impacting their teaching practice. As [13] pointed out, it is important to focus on accountability and transparency issues when using educational AI tools. Researchers in the field of explainable AI (XAI) have proposed various interpreting methods to provide transparent explanations and increase users' trust and understanding of such deep and complex models.

Thus, this study aims to build a deep learning model for automatically analyzing teacher classroom discourse (i.e., talk moves) and provide explanations to teachers on how the model predicts their utterances. To achieve this goal, we propose two BERT-based deep learning models on the publicly available *TalkMoves* dataset and apply three generic interpreting methods to explain the prediction of teachers' talk moves. Our investigation of the interpreting results further reveals distinctive characteristics of each talk move and provides teachers with helpful advice on using specific words and phrases to function as talk moves.

2 Related Work

2.1 Automated Models on Classroom Discourse

Traditional machine learning models have frequently been used to automatically analyze classroom discourse, particularly teacher utterances. For instance, the random forest has been utilized to detect the type of support provided by teachers' utterances to students (e.g., emotional support and learning support) [9] and the presence of questions in classroom dialogue [7]. Decision trees and regression trees have been employed to identify authentic questions (i.e., questions without prescripted answers) [5]. Naive Bayes has been adopted to classify teachers' instructional activities in live classrooms (e.g., small group work and lecture) [12]. Additionally, linear regression and the k-nearest neighbor method have also been explored in modeling teacher discourse [7]. While these traditional machine learning models can achieve basic automated analysis of classroom discourse, their performance can be improved.

Currently, deep neural network-based models are preferred due to their powerful performance. Recurrent neural networks (RNNs) [24], convolutional neural networks (CNNs) [21], and attention-based networks [11] have been utilized to identify specific features in teachers' classroom discourse (e.g., instructional activities), with transformer-based models being the most competitive. As an important vehicle where learning occurs, it is necessary to model classroom discourse using more powerful deep learning models to provide teachers with better automatic analysis and feedback.

2.2 Explainable Artificial Intelligence

Interpretability, defined as the degree to which humans comprehend the cause of a decision or properly reproduce the decision-making process of the AI model [13,17], can be classified as ante-hoc and post-hoc interpretability. Ante-hoc interpretability refers to self-explainable models in which humans can understand the working mechanism and individual decisions without exterior messages. Self-explainable models are usually limited to traditional machine learning models with simple structures [3]. Post-hoc interpretability refers to models with complex structures that people can hardly understand, such as many deep neural network-based models [18]. The interpreting methods are usually needed to figure out what happens inside the deep models. Specifically, rule extraction, knowledge distillation, and activation maximization are typical global interpreting methods for understanding the fundamental logic. Backpropagation, sensitivity analysis, and local approximation are commonly adopted local interpreting methods to explain individual decisions, where the relevance of each input feature to the prediction is calculated. Additionally, the interpreting methods can be classified as model-agnostic or model-specific based on the types of neural networks they can interpret [1]. Model-agnostic methods can interpret any deep learning model, regardless of its internal structure, while model-specific methods can only be used for models with specific structures.

Researchers have started leveraging interpreting methods to explain AI-supported educational models to increase users' trust (e.g., deep knowledge tracing models [25]). However, few studies focus on the interpretability issue of deep learning-based models for classroom discourse which may impact teachers' teaching practice. Thus, this paper aims to address this gap by proposing the use of interpreting methods to explain the predictions of the models.

3 Method

3.1 Dataset

This paper employed *TalkMoves* as the dataset for automatically building and explaining deep learning-based teacher discourse models. *TalkMoves* [23] is a corpus that contains 567 transcripts of classroom discourse in K12 mathematics lessons, with a total of 174,186 teacher utterances and 59,874 learner utterances. Based on the APT framework [19], each teacher and student utterance is annotated with a talk move (i.e., specific dialogic acts) or non-talk-moves label.

Talk moves can facilitate rich discussions in which students share their understanding, consider others' perspectives, and engage in reasoning [10]. Therefore, this paper focuses on building models for teacher talk moves. The description of teacher talk moves in the *TalkMoves* dataset is shown in Table 1. When building automated teacher talk move models, 90% of the transcripts are randomly selected as the training data, while the remaining 10% serve as the test data.

Table 1. The description of teacher talk moves included in the TalkMoves dataset [10,23]

Talk move	Description
Keeping everyone together	Eliciting learners to listen and orient them to each other
Getting students to relate to another's idea	Encouraging students to respond to their peer's utterances
Restating	Asking students to repeat or rephrase the ideas of others
Revoicing	Repeating what a student said but extending his/her ideas
Pressing for accuracy	Prompting students to use mathematical language or make a mathematical contribution
Pressing for reasoning	Requesting students to provide evidence or explain the reasoning

3.2 Model Construction

Identifying teacher talk moves in the *TalkMoves* dataset is a 7-way sequence classification task that classifies teacher utterances as one of six talk moves or no talk moves. This paper selected two deep learning models (i.e., BertForSequenceClassification and BERT-BiLSTM) to achieve this goal due to their strong performance in many NLP tasks. Both of them used a pre-trained language model called Bidirectional Encoder Representations from Transformers (BERT) [6] to encode input token sequences. Specifically, the model input was set as student-teacher utterance pairs - a combination of the teacher sequence concatenated with its previous student sequence, considering the importance of context information for detecting some talk moves (e.g., restating) [23]. The encoded representation of the input utterances was then fed into the distinctive modules of the two deep learning-based models.

BertForSequenceClassification was composed of BERT and an additional linear layer. The encoded input representation was fed into the final linear layer and the softmax function was used to output the predicted talk move. The batch size, optimizer, epoch, and learning rate were set to 32, Adam, 10, and 2e–5.

BERT-BiLSTM was composed of BERT and a Bidirectional Long Short-Term Memory (BiLSTM) module. The encoded input representation was fed into the BiLSTM module followed by a linear layer, and the softmax function was also used to obtain the prediction. During training, the batch size, optimizer, and epoch were set to 32, Adam, and 10, respectively. The learning rate was set to 2e–5 for BERT and 5e–4 for BiLSTM.

After training, BertForSequenceClassification achieved an F1 score of 0.735 and an accuracy of 0.868. BERT-BiLSTM achieved an F1 score of 0.736 and an accuracy of 0.866.

3.3 Model Explanation

To address the interpretability issue in the proposed models, this paper leveraged three local interpreting methods to explain their predictions by computing the relevance of each token (or word) in the input utterance. Specifically, considering the different structures of *BertForSequenceClassification* and *BERT-BiLSTM*, this paper selected three model-agnostic (i.e., generic) methods, namely first-order Taylor expansion, saliency, and integrated gradient.

Interpreting Methods. Formally, suppose there is a teacher utterance x whose embedding is e, $e \in R^m$. And x is composed of n tokens and the embedding of each token is e_i, where i denotes the token position. The proposed model f predicts that the teacher utterance belongs to talk move c with a probability of $f_c(e)$. To compute the relevance of each token to the prediction, first-order Taylor expansion [14], saliency [20], and integrated gradient [22] work in the following ways.

First-order Taylor expansion [14] decomposes the non-linear prediction $f_c(e)$ into the linear sum of the contribution of each token, as shown in Eq. 1. The dot product of i-th token's embedding e_i and its derivative $\frac{\partial f_c(e)}{\partial e_i}$ is treated as the token's relevance, as shown in Eq. 2, where j represents the j-th dimension in e_i and m is the total number of dimensions. In contrast, the saliency method [20] takes the derivative $\frac{\partial f_c(e)}{\partial e_i}$ as a measure of the token's importance to the prediction, as shown in Eq. 3.

$$f_c(e) \approx \sum_{i=1}^{n} \frac{\partial f_c(e)}{\partial e_i} \cdot e_i + b \tag{1}$$

$$R_{taylor}(e_i) \approx \sum_{j=1}^{m} \frac{\partial f_c(e)}{\partial e_{ij}} e_{ij} \tag{2}$$

$$R_{saliency}(e_i) \approx \sum_{j=1}^{m} \frac{\partial f_c(e)}{\partial e_{ij}} \tag{3}$$

To explain the prediction $f_c(e)$, the integrated gradient method selects a baseline sample x' whose embedding and predicted probability of talk move c are e' and $f_c(e')$, respectively. The method then compares $f_c(e)$ and $f_c(e')$ and attributes their difference to the difference in their token embeddings, as shown in Eq. 4, where $R_{ig}(e_i)$ refers to the relevance of token e_i to its prediction. The integrated gradient method considers the straight-line path from the baseline embedding e' to the input embedding e and calculates the gradients at all points along the path [22]. $R_{ig}(e_i)$ is obtained by accumulating all of these gradients, as shown in Eq. 5. In this study, the baseline sample for the teacher discourse model was set as a sample whose tokens were all zeros.

$$\sum_{i=1}^{n} R_{ig}\left(e_i\right) = f_c(e) - f_c\left(e'\right) \tag{4}$$

$$R_{ig}(e_i) \approx \sum_{j=1}^{m} \left(e_{ij} - e'_{ij}\right) \times \int_{\alpha=0}^{1} \frac{\partial f\left(e' + \alpha \times \left(e - e'\right)\right)}{\partial e_{ij}} d\alpha \tag{5}$$

It is worth noting that if the summation term $\sum_{j=1}^{m}$ is removed from Eq. 2, 3, and 5, the equations represent the relevance of j-th dimension in token e_i to the prediction.

A Interpreting Case. The interpreting methods are used to explain the predictions from the proposed deep learning models for teacher utterances. To illustrate this, we present an example using the BertForSequenceClassification model and the integrated gradient method. Specifically, given the teacher utterance "can all of you repeat it?", the BertForSequenceClassification model classifies it as "keeping everyone together" with a probability of 0.84. The integrated gradient method is then used to calculate the relevance of each word to this prediction, as shown in Table 2. The larger the relevance value, the more important the token is considered to be in the prediction. In this case, "all of you" contributes the largest share to the prediction *keeping everyone together*, which is logical and reasonable. To further clarify the explanation, we have visualized it in Fig. 1. Instead of listing the relevance values, we use the background color to represent the importance of the tokens. Green and orange colors represent positive and negative contributions, respectively, and the darkness of the color indicates the level of contribution. From the figure, we can easily and clearly see that "all of you" is highlighted in a dark green color, indicating a strong positive relevance to the prediction.

Table 2. The explanation of a teacher utterance in a table form.

token	can	all	of	you	repeat	it	?
relevance	0.088	0.395	0.369	0.399	−0.126	0.158	0.707

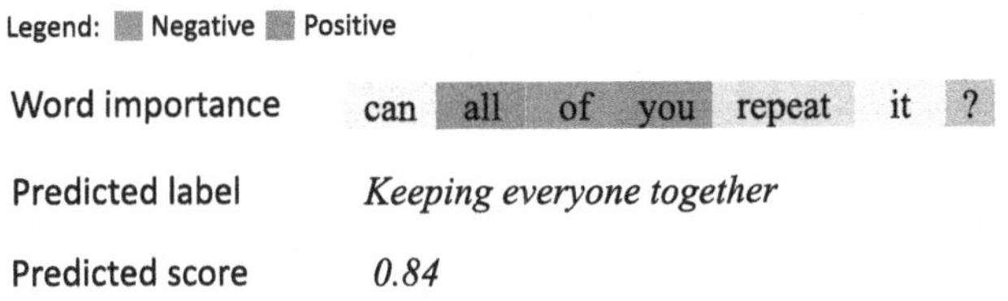

Legend: Negative Positive

Word importance can all of you repeat it ?

Predicted label *Keeping everyone together*

Predicted score *0.84*

Fig. 1. The explanation of a teacher utterance in a visualized form.

4 Experiments and Results

In this section, we applied the interpreting methods to explain the built deep learning-based teacher discourse models and further explored the interpreting results to examine the distinctive features of different types of teacher talk moves.

4.1 Interpreting Results Validation

Considering that the *BertForSequenceClassification* model shared comparable performance with the *Bert-BiLSTM* model, we chose *BertForSequenceClassification* for the subsequent experiments. Specifically, to verify the effectiveness of the token relevance value obtained from the interpreting methods, we conducted a validation experiment. The goal of this experiment was to determine whether the identified important tokens were indeed crucial for predicting the talk move of teacher utterances. To do this, we selected teacher utterances that had been correctly predicted with 100% accuracy from the *TalkMoves* dataset, and then removed important tokens (i.e., set them to zeros) in decreasing order of their relevance, one by one, until five tokens had been removed. By examining the changes in accuracy, we were able to confirm the effectiveness of the token relevance. However, we recognized that removing too many tokens from a short utterance could significantly affect prediction accuracy. Therefore, we set a minimum length of 10 and also included a random deletion for comparison.

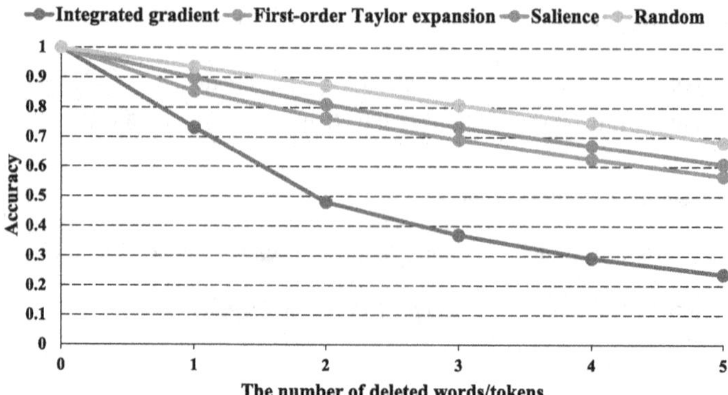

Fig. 2. The prediction accuracy changes of removing tokens from correctly-predicted teacher utterances based on their relevance.

Figure 2 shows the prediction accuracy change among the originally correctly-predicted teacher utterances. We can see that removing tokens based on their relevance computed from the integrated gradient, first-order Taylor expansion, and salience, causes the accuracy to drop from 100% to around 23%, 56%, and 60%, respectively, In contrast, the random deletion only causes the accuracy to drop to 68%. This demonstrates the effectiveness of the interpreting results. As the integrated gradient method was the most effective of the three interpreting methods, we used its results for further investigation.

4.2 Talk Move Characteristics Exploration

Many teachers struggle to create and sustain productive discussions in their classrooms due to the lack of discursive techniques such as talk move skills [10]. To help teachers improve their talk move skills, we analyzed the interpreting results to identify the words and phrases that characterize each type of talk move. Specifically, we used correctly-predicted sequences as the data source for our investigation, but excluded utterances labeled as no-talk-move. For sequences in each type of talk move, we selected the most important word (i.e., the word with the highest token relevance) and the most important phrases of different lengths (i.e., ranging from 2-token length to 4-token length). The most important phrase at a length of n was defined as the phrase with the highest sum of token relevance among all phrases composed of n consecutive tokens in a sequence. Among the identified indicative words and phrases, we selected and displayed those with the top 15 occurrences. We also invited experts proficient in the APT framework to highlight the words and phrases that could function best as talk moves. Note that we did not investigate the indicators for *restating* and *revoicing* due to the small number of teacher utterances labeled with these talk moves.

Token Level. Table 3 lists the most important tokens in talk moves. For *pressing for reasoning*, the identified and highlighted words include "why", "how", "because", "reason", "explain", and "justify". All of these words are related to the reasoning process, which suggests that they may be derived from utterances where teachers ask students to elaborate on their reasoning. "What", "how", "where", "who", and "which" are all question words that reflect teachers asking questions, and represent *pressing for accuracy.* "Agree", "disagree", "who", and "you" seem to imply that teachers are asking students whether they agree or disagree with someone, which also reflects the purpose of *getting students to relate to another's idea.* In contrast, "you", "your", "any", "what", and "who" are not specific enough to function as indicators of *keeping everyone together*, as these words may also be used in other talk moves.

Table 3. The most important token with frequent occurrences in talk moves

Talk move	The most important tokens with top 15 frequency
Keeping Everyone Together	'?', 'right', '**you**', 'okay', '.', 'say', '**your**', 'it', 'up', '**any**', '**what**', 'questions', 'read', '**who**', 'hand'
Getting Students to Relate to another's idea	'?', '**agree**', '**you**', 'do', 'with', 'up', 'something', '**who**', 'else', 'on', 'think', 'different', 'have', '**disagree**', 'if'
Pressing for accuracy	'?', '**what**', 'you', '**how**', 'do', '', 'many', 'or', 'me', '-', 'your', '**where**', '**who**', 'we', '**which**'
Pressing for reasoning	'**why**', '**how**', '?', 'come', 'you', 'know', 'not', '**because**', '**reason**', 'who', 'do', '', '**explain**', '**justify**', 'can'

Table 4. The most important 2-token phrases with frequent occurrences in talk moves

Talk move	The most important 2-token phrases with top 15 frequency
Keeping Everyone Together	'**right?**', '**okay?**', '**ready?**', 'sense?', 'it?', '**any questions**', 'do you', 'that?', '##s', 'one?', '**correct?**', 'are you', 'your hand', '**done?**', 'you?'
Getting Students to Relate to another's idea	'**agree?**', 'do you', '**agree with**', 'that?', '**disagree?**', 'thumbs up', 'do we', '**disagree with**', '**add on**', 'stand up', 'up if' 'differently?', 'you think', 'or have', '**agreed?**'
Pressing for accuracy	'**what**'', 'do you', '**what?**', '**what do**', '**what is**', '**how many**', 'is?', '**what are**', 'be?', 'did you', 's', '**what would**', '##s?' '**what does**'
Pressing for reasoning	'**why?**', '**why do**', '**why did**', '**how do**', '**why would**', '**why does**', '**why can**', 'you know', '**why not**', '**why are**', '**how come**', 'not?', '**how did**', '**why.**'

Phrase Level. Table 4, 5, and 6 display representative phrases of different lengths. In general, as the length of the phrases increases, more of the identified phrases can function as talk moves. For example, when the length of the tokens is three or above, almost all of the selected phrases in *pressing for accuracy* and *pressing for reasoning* are identified as talk moves by the experts. Specifically, the indicators in "pressing for accuracy" mainly include the first part of question sentences that begin with "what," such as "what are the", "what do you think", and "what did you get", through which teachers expect students to express their answers more academically. The indicators in *pressing for reasoning* also contain the beginnings of question sentences, but these start with "why" and "how," which elicit reasoning. Typical phrases include "why?", "why is it", "why does that work", "how come?", "how do I know", and "how did you know".

Table 5. The most important 3-token phrases with frequent occurrences in talk moves

Talk move	The most important 3-token phrases with top 15 frequency
Keeping Everyone Together	**', right¿, ', okay?', 'make sense?', 'you ready?', 'talk to your', 'raise your hand'**, 't it?', 'so far?', **'any questions on', 'many of you', 'any questions?', 'makes sense?'**, ', correct?', 'this one?', **'any questions about'**
Getting Students to Relate to another's idea	**'we agree?', 'you agree?', 'do you agree', 'do we agree'**, 'with that?', 'you think about', **'like to add'**, 'stand up if', **'or disagree?', 'do you guys'**, 'phone a friend', **'something to add'**, 'thumbs up if', **'to add on'**
Pressing for accuracy	**'what's', 'what do you', 'what did you', 'what do we', 'what is the', 'how did you', 'what are you', 'what do I', 'what are the', 'is what?', 'what are we'**, 'to be?', **'how would you', 'what does that', 'how do I'**
Pressing for reasoning	**'why do you', 'why did you', 'how do you', 'why not', 'why is it', 'and why?', 'why do we', 'why can', 'why would I', 'why did you', 'how come?', 'why are you', 'why does that', 'but why?'**

The indicators for *keeping everyone together* and *getting students to relate to another's idea* also have distinctive characteristics. For *keeping everyone together*, simple phrases such as "right?", "okay?", "done?", and "you ready?", can be used to connect all students in the class. Other phrases, such as "any questions", "make sense?", and "raise your hand", can also serve this purpose. For *getting students to relate to another's idea*, the indicators often focus on "disagree" and "agree," such as "you agree?" and "agree or disagree", which allow teachers to encourage students to elaborate on their ideas and evaluate the thinking of others. "Add on" is another frequently identified indicator, which also promotes connections between students by asking them to build upon others' ideas.

Table 6. The most important 4-token phrases with frequent occurrences in talk moves

Talk move	The most important 4-token phrases with top 15 frequency
Keeping Everyone Together	**'that make sense?'**, **'##s, right?'**, **'are you ready?'**, "t it?', **'are you done'**, **'questions so far?'**, **'raise your hand if'**, **'one, right?'**, **'how many of you'**, **'raise your hand.'**, **', all right?'**, 'two, right?', 'half, right?', ', is it?', 'stand up'
Getting Students to Relate to another's idea	**'do we agree?'**, **'do you agree with'**, **'do you agree?'**, **'agree with that?'**, **'do you think about'**, **'agree or disagree?'**, **'do we agree with'**, **'you like to add'**, 'a different way?', **'you guys agree?'**, 'stand up if you', 'a thumbs up.', **', do you agree?'**, **'want to add on'**, 'thumbs up if you'
Pressing for accuracy	**'what do you think'**, **'what's the'**, **'do you think?'**, **'what are you thinking'**, **'what did you do'**, 'going to be?', **'how did you get'**, **'what is it'**, **'what's a'**, **'what is your'**, 'den##omi##nator?', **', what do you'**, **'what did you get'**, **'what'd you'**
Pressing for reasoning	**'how do you know'**, **'why do you think'**, **'how did you know'**, **'why can't'**, **'or why not'**, **'why did you do'**, **'why do you say'**, **'how do we know'**, **'how do I know'**, **', why do you'**, **'why is it not'**, **'why does that work'**, **', but why?'**, **'why can we not'**, **'why did you change'**

5 Discussion and Conclusion

This paper built two deep learning-based models (i.e., BertForSequenceClassification and BERT-BiLSTM) to automatically identify teacher talk moves in mathematics lessons and provide teachers with timely feedback. Considering the complicated structures in the proposed models may cause teachers difficulty comprehending the decision, this paper proposed three interpreting methods to explain to teachers how the predicted talk moves were determined. Specifically, the paper utilized three model-agnostic interpreting methods to compute the relevance of each word to the prediction, and verified the validity of the obtained explanations. In addition, the paper further analyzed the interpreting results to uncover the characteristics of each talk move. The findings showed that certain simple words and phrases can function as talk moves. For *keeping everyone together*, phrases like "right?", "ready?", and "any questions" are effective at capturing the attention of all students. For *getting students to relate to another's idea*, representative indicators center around "agree", "disagree", and "add on". For *pressing for accuracy*, the typical indicators are the beginning part of question sentences that start with "what". For *pressing for reasoning*, question sentences starting with "why" and "how" can prompt students to elaborate on their reasoning process.

The findings of this study have implications for improving teachers' professional development. Firstly, the proposed automated deep learning-based models can be integrated into professional systems (e.g., the Classroom Discourse Analyzer [4]) to provide teachers with timely feedback on their classroom discourse. By uploading recordings of their lessons, teachers can quickly receive automated discourse analysis that allows them to reflect on and improve their dialogic teaching. Secondly, the explanations partially address the interpretability issue of "black box" models, making it clear to teachers how the prediction of their utterances was made. These explanations have the potential to increase teachers' understanding, acceptance, and trust in AI-powered discourse models and systems. Thirdly, the talk move characteristics extracted from the interpreting results offer useful guidance, particularly for novice teachers unfamiliar with discursive techniques that facilitate classroom discussion. By incorporating indicative words and phrases into their teaching, novice teachers can more easily learn and apply productive dialogic strategies.

There are a few limitations to this study. Firstly, the participants in the classroom included both teachers and students. Automatically analyzing student discourse is also important, as it can help teachers understand students' affective and cognitive states and intervene as needed. Secondly, this study used data from mathematics lessons to build the models. These models may not be generalized to other subjects, such as English or Chinese, as different subjects may have distinctive patterns of classroom discourse. Thirdly, while explaining model predictions can help make the deep learning-based discourse model more transparent, it is not clear what impact the explanations have on teachers. Future research could address these limitations by building automated classroom discourse models for both teachers and students in various subjects and evaluating the effects of the automated analyses and their explanations in practice.

Acknowledgements. This work was supported by Hong Kong Research Grants Council, University Grants Committee (Grant No.: 17605221), and by the Innovation and Technology Commission of the Government of the HKSAR (Grant No.: ITB/FBL/7026/20/P).

References

1. Adadi, A., Berrada, M.: Peeking inside the black-box: a survey on explainable artificial intelligence (XAI). IEEE Access **6**, 52138–52160 (2018)
2. Alexander, R.J.: Towards Dialogic Teaching: Rethinking Classroom Talk, 5th edn. Dialogos (2017)
3. Alvarez Melis, D., Jaakkola, T.: Towards robust interpretability with self-explaining neural networks. In: Advances in Neural Information Processing Systems, vol. 31. Curran Associates, Inc. (2018)
4. Chen, G., Chan, C.K.K., Chan, K.K.H., Clarke, S.N., Resnick, L.B.: Efficacy of video-based teacher professional development for increasing classroom discourse and student learning. J. Learn. Sci. **29**(4–5), 642–680 (2020)
5. Cook, C.: An open vocabulary approach for detecting authentic questions in classroom discourse. In: International Conference on Educational Data Mining (2018)

6. Devlin, J., Chang, M.W., Lee, K., Toutanova, K.: BERT: pre-training of deep bidirectional transformers for language understanding. In: International Conference of the North American Chapter of the Association for Computational Linguistics), pp. 4171–4186. Association for Computational Linguistics (2019)
7. Donnelly, P.J., Blanchard, N., Olney, A.M., Kelly, S., Nystrand, M., D'Mello, S.K.: Words matter: automatic detection of teacher questions in live classroom discourse using linguistics, acoustics, and context. In: International Learning Analytics and Knowledge Conference, pp. 218–227. ACM (2017)
8. Howe, C., Hennessy, S., Mercer, N., Vrikki, M., Wheatley, L.: Teacher-student dialogue during classroom teaching: does it really impact on student outcomes? J. Learn. Sci. **28**(4–5), 462–512 (2019)
9. Hunkins, N.C., Kelly, S., D'Mello, S.: "Beautiful work, you're rock stars!": teacher analytics to uncover discourse that supports or undermines student motivation, identity, and belonging in classrooms. In: 12th International Learning Analytics and Knowledge Conference (LAK 2022), pp. 230–238. ACM (2022)
10. Jacobs, J., Scornavacco, K., Harty, C., Suresh, A., Lai, V., Sumner, T.: Promoting rich discussions in mathematics classrooms: using personalized, automated feedback to support reflection and instructional change. Teach. Teach. Educ. **112**, 103631 (2022)
11. Jensen, E., Pugh, S.L., D'Mello, S.K.: A deep transfer learning approach to modeling teacher discourse in the classroom. In: 11th International Learning Analytics and Knowledge Conference (LAK 2021), pp. 302–312. ACM (2021)
12. Kelly, S., Olney, A.M., Donnelly, P., Nystrand, M., D'Mello, S.K.: Automatically measuring question authenticity in real-world classrooms. Educ. Res. **47**(7), 451–464 (2018)
13. Khosravi, H., et al.: Explainable artificial intelligence in education. Comput. Educ. Artif. Intell. **3**, 100074 (2022)
14. Li, J., Chen, X., Hovy, E., Jurafsky, D.: Visualizing and understanding neural models in NLP. In: International Conference of the North American Chapter of the Association for Computational Linguistics, pp. 681–691 (2016)
15. Mercer, N., Hennessy, S., Warwick, P.: Dialogue, thinking together and digital technology in the classroom: some educational implications of a continuing line of inquiry. Int. J. Educ. Res. **97**, 187–199 (2019)
16. Michaels, S., O'Connor, C., Resnick, L.: Deliberative discourse idealized and realized: accountable talk in the classroom and in civic life. Stud. Philos. Educ. **27**, 283–297 (2008)
17. Miller, T.: Explanation in artificial intelligence: insights from the social sciences. Artif. Intell. **267**, 1–38 (2019)
18. Montavon, G., Samek, W., Müller, K.R.: Methods for interpreting and understanding deep neural networks. Digit. Signal Process. **73**, 1–15 (2018)
19. O'Connor, C., Michaels, S.: Supporting teachers in taking up productive talk moves: the long road to professional learning at scale. Int. J. Educ. Res. **97**, 166–175 (2019)
20. Simonyan, K., Vedaldi, A., Zisserman, A.: Deep inside convolutional networks: visualising image classification models and saliency maps. arXiv preprint arXiv:1312.6034 (2013)
21. Song, Y., Lei, S., Hao, T., Lan, Z., Ding, Y.: Automatic classification of semantic content of classroom dialogue. J. Educ. Comput. Res. **59**(3), 496–521 (2021)
22. Sundararajan, M., Taly, A., Yan, Q.: Axiomatic attribution for deep networks. In: International Conference on Machine Learning. Proceedings of Machine Learning Research, vol. 70, pp. 3319–3328. PMLR (2017)

23. Suresh, A., Jacobs, J., Harty, C., Perkoff, M., Martin, J.H., Sumner, T.: The Talk-Moves dataset: K-12 mathematics lesson transcripts annotated for teacher and student discursive moves (2022)
24. Suresh, A., Sumner, T., Jacobs, J., Foland, B., Ward, W.H.: Automating analysis and feedback to improve mathematics teachers' classroom discourse. In: The Thirty-Third AAAI Conference on Artificial Intelligence (AAAI 2019), pp. 9721–9728. AAAI Press (2019)
25. Wang, D., Lu, Y., Zhang, Z., Chen, P.: A generic interpreting method for knowledge tracing models. In: Rodrigo, M.M., Matsuda, N., Cristea, A.I., Dimitrova, V. (eds.) Artificial Intelligence in Education (AIED 2022). LNCS, vol. 13355, pp. 573–580. Springer, Cham (2022). https://doi.org/10.1007/978-3-031-11644-5_51

Short Papers

Ethical and Pedagogical Impacts of AI in Education

Bingyi Han[1(✉)], Sadia Nawaz[2,3], George Buchanan[1], and Dana McKay[4]

[1] School of Computing and Information Systems, University of Melbourne, Parkville, Australia
bingyi.han@unimelb.edu.au

[2] Melbourne Assessment, MGSE, University of Melbourne, Parkville, Australia

[3] Faculty of IT, Monash University, Clayton, Australia

[4] School of Computing Technologies, RMIT University, Melbourne, Australia

Abstract. Artificial Intelligence is becoming pervasive in higher education. While these tools can provide customized intervention and feedback, they may cause ethical risks and sociotechnical implications. Current ethical discussions often focus on established technical issues and overlook further implications from students' perspectives, which may increase their vulnerability. Taking a student-centered view, we apply the story completion method to understand students' concerns about the future adoption of various analytics-based AI tools. 71 students elaborated on the provided story stems. A qualitative analysis of their stories reveals students' perceptions that AIEd may disrupt aspects of the pedagogical landscape such as learner autonomy, learning environments and approaches, interactions and relationships, and pedagogical roles. This study provides an initial insight into student concerns about AIEd and a foundation for future research.

Keywords: Learner Autonomy · Ethics · Pedagogical Roles · Student Perceptions

1 Introduction

The application of artificial intelligence in education (AIEd)[1] and learning analytics (LA) is becoming popular for improving teaching and learning. For instance, analytics-based tools and adaptive technologies have been used for providing customized recommendations and timely intervention to meet individual needs [1]. However, while AIEd strives to enhance positive educational outcomes, it also raises concerns regarding the unintended ethical impacts on students' welling-being and learning experiences.

Existing studies have limited coverage on ethical themes: they predominately align with technical principles of AI ethics such as privacy, bias, and fairness [2], but overlook the broader sociotechnical issues and pedagogical impacts that can arise due to students' direct interaction with AIEd. Typically, the ethical discussions are in department-specific silos. The technical community generally focuses on improving the functionality and

[1] Our study design sees the future implementation of AIEd techniques, where AIEd would entail AI, LA (Learning Analytics), and EDM (Educational Data mining techniques) tools. Therefore, we may use these terms interchangeably in the current context.

N. Wang et al. (Eds.): AIED 2023, LNAI 13916, pp. 667–673, 2023.
https://doi.org/10.1007/978-3-031-36272-9_54

addressing established technology-related ethical problems especially privacy and algorithmic bias [2]. This limited ethical focus has been repeated in AIEd and LA frameworks for governance and strategic adoption (e.g., [3]). However, consequential ethics such as pedagogical implications have rarely been considered. Tsai et al. [4] has critiqued the limited research on the impact of AIEd on student experience and learning. Similar arguments have been raised by the education community concerning learner autonomy, while it lacks student-based evidence.

It is further complicated by having fewer students who have actual AIEd experience, and they may struggle to predict the consequences of AIEd adoption. We are, therefore, left with a challenging dilemma: empirical evidence gained before exposure to AIEd may be unreliable, but identifying consequences after they occur can lead to unintended harm. Compliance alone may not guarantee student trust and confidence. To build trust in these deployments, it is essential to consider students' perspectives.

This paper aims to understand students' responses to and the pedagogical implications of AIEd. We focus on the question, "*What are the sociotechnical consequences that happen to students' learning experience due to AIEd adoption*?" Using qualitative methods and grounding our findings in learning theories, we contribute to existing work on ethics in AIEd by revealing likely impacts on student engagement and pedagogy: the Leaner Status, Learning Experience and Environment, Learning Process and Approach, Educational Interaction and Relationship, and Pedagogical Roles. This paper presents the pedagogical impacts, while impacts of other dimensions including individual [5], technical and system-design are presented in other publications. These insights can inform future work and supplement existing ethical frameworks.

2 Research Method

To reveal the possible impacts of adopting AI in higher education from students' perspective, we collected data through the Story Completion Method (SCM). It leverages speculative design and research fiction which can effectively probe the ethical and moral implications of future technologies [6].

Story Prompts: Participants were invited to create narratives based on given story openings as Story Completion Tasks[2] (SCT) that elicit their responses to specific contexts and the values that matter to them. We developed three futuristic scenarios of university classrooms based on proposed technologies in the AIEd research literature. We described the AIEd functions with a deliberate ambiguity to encourage students' imagination. We then hinted them by asking how the character would behave.

Data Collection: Stories were collected from 71 university students (54% males and 46% females). Considering the diverse background of students in real-world learning situations, we focused on broader concerns for all students, regardless of their AIEd experience or AI literacy.

Data Processing: The first author analysed data while disputed themes were discussed with the other authors. We adopted grounded theory (GT) techniques to tease out relationships, particularly consequences, among identified codes [7]. GT analysis included

[2] The full scenarios are presented at: https://sites.google.com/view/storystems

three stages: Open Coding, Axial Coding, and Selective Coding, which helped in identifying concepts, high-level phenomena, and interpretation. We then identified Causal Conditions (the specific function, e.g., customization, intervention), Phenomena (e.g., the affected learner status), and Consequences (the affected learning experiences).

3 Results, Discussion and Practical Implication

The analysis of the 3 story scenarios from the SCT tasks highlighted the potential impact of AI adoption under four dimensions: Individual, Pedagogical, Technological, and System-design. These impacts seemed to interact with each other. In a prior study [5], we demonstrated the impact on individuals due to their direct interaction with AI monitoring. This study presented only the pedagogical implications, categorized into five core categories, see Table 1. In the text, we would refer to Table 1 core categories in **bold** and the open codes in *italics*. Students' stories would be quoted.

Table 1. Codes developed based on participants' responses – Pedagogical Dimension

<table>
<tr><th>Core Category:Axial Code</th><th colspan="4">Sub-Category: Open Code (Mentioned Frequency)</th></tr>
<tr><td colspan="5">IMPACTS ON PEDAGOGICAL ELEMENTS</td></tr>
<tr><td rowspan="2">Learner Status</td><td>Decreased Learner Autonomy (8)</td><td>Passive Learning Pathway (10)</td><td>Forced Learning Initiatives (10)</td><td>Limited Exploration Opportunities - follow own choices (16)</td></tr>
<tr><td>Overly Relying on AIEd (9)</td><td>Decreased Learning Independence (8)</td><td>Restricted Freedom (13)</td><td>System Manipulation (11)</td></tr>
<tr><td rowspan="2">Learning Experience & Environment</td><td colspan="2">Scary and Intense Learning Environment (40)
- Students generally live with fear (22)
- Assume strict rules and supervisions (18)</td><td>More Engaged Class (13)</td><td>Assume more effort is needed (for the imagined standard) (6)</td></tr>
<tr><td colspan="2">Result-Oriented (Scores) (13)
Losing Learning/Exploration Enjoyment (6)</td><td>Educational Disruption (27)</td><td>Over-Digitized Learning (6)</td></tr>
<tr><td rowspan="2">Educational Processes & Approaches</td><td colspan="2">Equity (Equal Attention) for Individuals (6)</td><td colspan="2">Improve Effectiveness for Self-Learning (49)</td></tr>
<tr><td colspan="3">Provides Analytical Power and Insights to inform education (21)</td><td>Active Help for Introverts (5)</td></tr>
<tr><td>Educational Interaction & Relationship</td><td colspan="2">Affected Communication:
- Among students (12); Students & teachers (3)</td><td colspan="2">Affected Social & Educational Relationship:
- Among students (9); Students & the teaching body (12)</td></tr>
<tr><td>Pedagogical Roles</td><td>Student Role (3)</td><td>Teacher Role (5)</td><td colspan="2">Role of Other Teaching Agents (e.g., AIEd) (17)</td></tr>
</table>

3.1 Learner Status

Based on Table 1, the most frequently anticipated change by students is **learner status**. Students perceived that AI monitoring and its associated functions (e.g., feedback provision and adaptivity) could pose a threat to their *Autonomy* and reduce their *Learning Independence.* They expressed discomfort while being monitored by a fear of showing poor performance under scrutiny. This can lead to students' behavioral changes, e.g., engagement with a *Forced Learning Initiative.* Students also reported that automated scaffolding could leave them with *Passive Learning Pathways*, inhibiting them from their *Own Choice for Learning Process* and from the *Freedom of Self-Exploration.* e.g., "students are *forced* to be incorporated into this system" as the automated learning pathways are based on a predetermined standard. Students perceived that such learning design could cause forced compliance and disrupt their original choice, especially when one's curiosity or passion locates somewhere else. The intervention was perceived to restrict learning exploration and *Manipulative* due to the *Forced Learning Pathway.*

By comparison, some students viewed forced engagement as a positive push, e.g., to concentrate on the lecture. They expressed trust on AIEd's analytical power with a willingness to receive system-allocated learning resources, believing it could prevent stress or anxiety by filtering out content deemed unsuitable for them.

Discussion: Our findings mirror existing concerns of undermined learner autonomy [8] and students' choices being manipulated or "nudged". Nudge theory suggests using subtle methods to influence people's behaviors [9], e.g., using indirect methods of reinforcement where certain options can be made more accessible. In digital environments, students expect to be self-reliant and control their learning [10]. To address these concerns, individuals should be offered options and encouraged to take decision-making autonomy, even when their choice might disagree with the AIEd recommendations.

3.2 Learning Environment and Experience

The influenced **learning environment and experiences** are often expressed after describing the affected **learner status**. Many participants believed that being instructed on what and how to learn discouraged their enthusiasm for learning and *self-exploration of knowledge.* Consequently, when the system removes independence, the learning experience "loses its *enjoyment*". They also criticized that the instant help provided by intervention could cause *Result-Oriented Education* (score-focused), while neglecting the value of time and thinking in knowledge acquisition. Students also perceived that data-driven tools may lead to *Over-Digitized Learning*, making their experience "less personal and more transactional". Also, AI monitoring with compliance expectations can create an intimidating instead of encouraging environment that disrupts learning concentration. Nevertheless, for some, surveillance positively *Enhances Class Engagement* due to forced performativity.

Discussion: Those concerns echo existing arguments such as spoon-fed learning [2]. Most stories that mentioned the disruptions also expressed a preference for self-directed learning. These students perceived that interventions could hinder their learning as they felt being treated as passive learners who can only receive information rather than create knowledge actively. This is against one of the fundamental pedagogical theories - constructivism [11]. Hence, AIEd should encourage students to take charge of their learning through exploratory behaviors. Teachers' future role should assist students in developing self-regulatory behaviors where students set goals for themselves and adjust learning processes accordingly [12].

3.3 Educational Processes and Approaches

Students often expressed positive attitudes toward the affected **educational processes and approaches**, anticipating *Improved Learning Effectiveness* by targeting their weaknesses. Individuals can *receive undivided Equal Attention* from AIEd, whereas teachers' time and energy per student are limited. Some participants also believe customized intervention is more *Friendly for Introverts* as the system actively offers help and identifies struggles. However, quite a few respondents expressed an unwillingness for AIEd to detect their learning weaknesses, due to concerns about privacy and its exposure to others, including teachers and other students. On the contrary, some favored the weakness analysis for its shortcoming-targeted service in *Enhancing Self-Learning Reflection.* For instance, a story described a student's increased learning confidence while studying with recommended study plans that targeted his weakness and agreed that the captured weakness and its identification were accurate and effective.

Discussion: We now see students' opposing attitudes of acceptance due to their different expectations of customization. This divergence echoes a recent study [13]: low expectations of LA are linked to the perceived stifled independence, while high expectations are linked with hopes of LA to assist particular needs. This then raises the consideration of balancing the benefits and harms of LA in light of varying student needs.

3.4 Interaction, Pedagogical Relationship, and Roles

During the actual learning process, students frequently anticipated the disruption to **interaction, relationships, and roles**. While being monitored, students would find informal interactions such as small talk uncomfortable and become self-cautious of their expressions. They would *Decrease Learning-Related and Social Interactions* and may act inauthentically. It results from their fear of inappropriate performance being recorded. Those changes are often shown with forced performativity under surveillance, where students feel the need to pander to expectations of tone and expression, such as having solemn and concise communications to avoid low evaluation scores. *Students' Pedagogical Relationships* are thereby affected compared with when "students are used to learning from conversations with each other". Since the intervention can provide automatic learning assistance, struggling students might decrease their help-seeking from peers. When both struggling student and their peers could benefit from the problem-solving process, some now turn to technology alone. This further influences students' pedagogical roles: peers could once educate each other, but now "students are more like competitors against each other". Similarly, the *Interaction between Students and Teachers* might decrease when students trust and overly rely on the help and intervention from LA. In response, students were often concerned about *teacher-student relationships being impersonal*: the system's administrative tools, such as attendance checks and support, could lead to a lack of contact with teachers, raising questions about whether teachers would still make an effort to know them. Further, a story described that the student might game the LA system to obtain attention and receive teacher assistance: "He will play up his weaknesses to make the AI think that he needs further help and then hint the teacher…He won't ask the teacher directly as the system will inform the teacher either way". Nevertheless, in this way, LA might provide indirect communication between students and teachers, thereby *becoming a medium* that delivers students' wishes for more attention.

Discussion: The automatic learning assistance may weaken students' pedagogical roles such as peer learners. Peer learning is powerful, moving learners "beyond independent to interdependent or mutual learning" [14]. Once students perceive LA assistance as efficient for problem-solving, established interactions among students would be reduced. This might covertly subvert their pedagogical relationships of peer-learning. LA-assisted learning environment should create spaces for facilitating human relationships (in the classroom) by enhancing human interactions.

4 Conclusion

This study demonstrates the likely disruptions that can happen to students in the adoption of AIEd. Instructors and learning designers should not overlook students' interaction with the changed learning environment. The education landscape should be considered as a whole: including not only students and educators, but also all changes brought to students' behaviors, status, and the educational system. Automated systems should respect differences and provide flexibility and control (e.g., options) to mitigate concerns due to varying students' expectations and needs, so that to encourage the acceptance towards AIEd-afforded personalized learning.

References

1. Nawaz, S., et al.: Moments of confusion in simulation-based learning environments. Journal of Learning Analytics **7**(3), 118–137 (2020)
2. Tzimas, D., Demetriadis, S.: Ethical issues in learning analytics: A review of the field. Education Tech. Research Dev. **69**(2), 1101–1133 (2021)
3. Clark, J.-A., et al.: Critical success factors for implementing learning analytics in higher education: A mixed-method inquiry. Australasian Journal of Educational Technology (2020)
4. Tsai, Y.-S., Gasevic, D.: Learning analytics in higher education---challenges and policies: a review of eight learning analytics policies. In: Proceedings of 7th LAK conference (2017)
5. Han, B., Buchanan, G., Mckay, D.: Learning in the Panopticon: Examining the Potential Impacts of AI Monitoring on Students. In: Proceedings of the 34th OzCHI Confernce (2023)
6. Dunne, A., Raby, F.: Speculative everything: design, fiction, and social dreaming (2013)
7. Walker, D., Myrick, F.: Grounded theory: An exploration of process and procedure. Qual. Health Res. **16**(4), 547–559 (2006)

8. Schumacher, C., Ifenthaler, D.: Features students really expect from learning analytics. Comput. Hum. Behav. **78**, 397–407 (2018)
9. Southgate, E., et al.: Artificial intelligence and emerging technologies in schools (2019)
10. Nawaz, S., et al.: Analysis of task difficulty sequences in a simulation-based POE environment. In: Artificial Intelligence in Education. Ifrane, Morocco (2020)
11. Piaget, J.: The development of thought: Equilibration of cognitive structures. Viking (1977)
12. Van Der Graaf, J., et al.: Do instrumentation tools capture self-regulated learning?. In: International learning analytics and knowledge (LAK) conference (2021)
13. Whitelock-Wainwright, A., et al.: An exploratory latent class analysis of student expectations towards learning analytics services. The Internet and Higher Education, p. 100818 (2021)
14. Boud, D., Cohen, R.: Peer learning in higher education: Learning from and with each other. Routledge (2014)

Multi-dimensional Learner Profiling by Modeling Irregular Multivariate Time Series with Self-supervised Deep Learning

Qian Xiao(✉), Breanne Pitt, Keith Johnston, and Vincent Wade

Trinity College Dublin, College Green, Dublin 2 D02PN40, Ireland
{xiaoq,pittb,keith.johnston,vincent.wade}@tcd.ie

Abstract. Personalised or intelligent tutoring systems are being rapidly adopted because they enable tailored learner choices in, for example, exercise materials, study time, and intensity (i.e., the number of chosen exercises) over extended periods of time. This, however, poses significant challenges for profiling the characteristics of learner behaviors, mostly due to the great diversity in each individual's learning path, the timing of exercise accomplishments, and varying degrees of engagement over time. To address this problem, this paper proposes an innovative approach that uses self-supervised deep learning to consolidate learner behaviors and performance into compact representations via irregular multivariate time series modeling. These representations can be used to highlight learners' multi-dimensional behavioral characteristics on a massive scale for self-directed learners who can freely pick exercises and study at their own pace. With experiments on a large-scale real-world dataset, we empirically show that our approach can effectively reveal learner individuality as well as commonality in characteristics.

Keywords: Learner profiling · LSTM autoencoder · Self-supervised learning

1 Introduction

Analyzing learning behaviors together with learning performance is crucial for tutors and intelligent tutoring systems (ITSs) to understand distinct learner needs. Profiling learner behaviors to identify different learner personas based on learning traces, also referred to as learner profile representation [3], has been extensively studied. Prior works have demonstrated that learner profiling can help provide personalized feedback [3] and identify at-risk learners [2,3,5].

Recent works for learner profiling have primarily focused on identifying learner prototypes (also referred to as personas [3] or stereotypes) from multi-dimensional aspects (e.g. effort, consistency, regularity) using clustering approaches (e.g. k-means [3,7], spectral clustering [6]). Although useful, these

N. Wang et al. (Eds.): AIED 2023, LNAI 13916, pp. 674–680, 2023.
https://doi.org/10.1007/978-3-031-36272-9_55

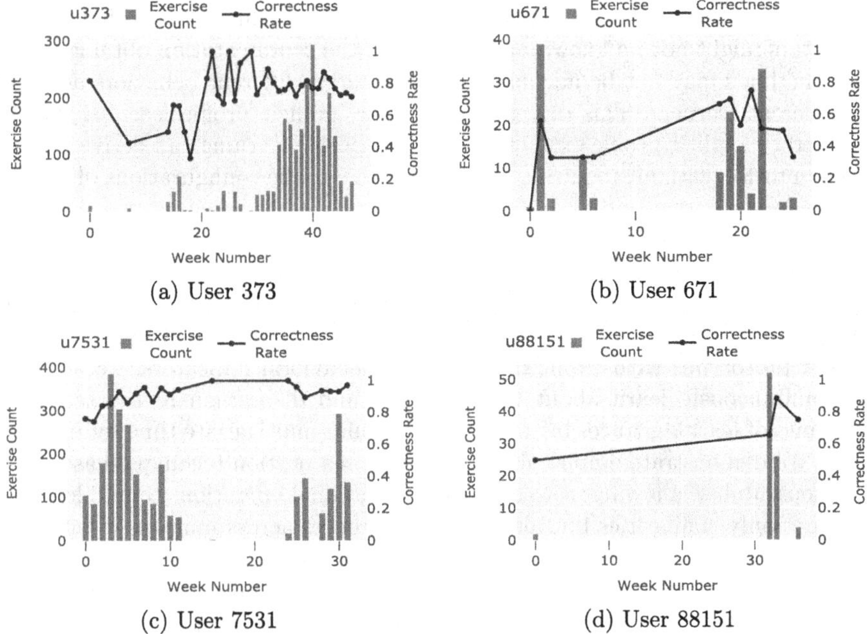

(a) User 373

(b) User 671

(c) User 7531

(d) User 88151

Fig. 1. Historgrams of four learners' exercises that cover one knowledge concept labeled with Tag 183, collected from Riiid's public dataset, Ednet-KT1 [4]. Each bar chart denotes the numbers of a learner's attempts to answer the questions in the ITS, measured by the primary y-axis on the left. Each line chart indicates correctness rates of a learner's exercises, measured by the secondary y-axis on the right side. Both are on a weekly basis. Each x-axis shows the week numbers in the calendar since the learner's first exercise associated with Tag 183.

predefined prototypes may overlook the impact of subtle behavioral differences between individual learners. Prior studies [1] in educational psychology have shown that profiling individual differences and distinct characteristics is also important. However, using pre-defined behavioral indicators (e.g., mean session duration [6], periodicity of week day [6], student shape [5]), each behavioral dimension of a learner is cast into one category (e.g. high/lower effort [6]). As a result, one learner is limited to belonging to a large cluster group (if it exists) with a single prototype (defined by an ensemble of behavioral indicator categories), and thereby the learner profile obtained ignores all behavioral differences at the individual level within one group. Moreover, these indicators are not always effective, or even known apriori for characterizing a learner's behavior. In fact, the effectiveness of the ensemble of indicators varies depending on the course setting and structure [5], as well as the learner groups [3]. For example, recent work in [5] found that the predictive capability of behavior indicators varies for different learner groups. As demonstrated in [5], the learner group at medium-risk is the most difficult on which to perform prediction, while it is much easier to predict successful or drop-out learner groups [3]. It is thereby highly possible that certain indicators might not be known apriori. Hence, the representation obtained with such methods may not be adequate to characterize different behaviors adopted by different learners. This drives the need for implicit profile learning [8]. As opposed to explicit profile extraction, implicit profile learning can derive learner representation without requiring prior knowledge of the configurations of behavioral indicators.

To overcome the challenges above, we propose a self-supervised deep learning approach that uses irregularly-observed time series modeling to derive implicit learner profile representations based on learning traces (e.g., choices of learning intensity and pace). Unlike previously cited clustering approaches that find learner prototypes with explicitly extracted behavioral indicators, we attempt to simultaneously learn about the similarity and distinction in the temporal dynamics of learning traces by leveraging irregular multivariate time series modeling. We demonstrate empirically that our representation technique has statistical separability, allowing us to derive computational operations to find learners with not only similarities but also subtle differences across multiple dimensions at the same time. In this way, a learner is not forced to conform to a fixed prototype. Instead, each learner is represented by a vector that encodes their multi-dimensional learning traces and identifies the unique properties of their behaviors and performance together.

We conducted empirical experiments on a large-scale dataset `Ednet` [4]. By putting learners' representations together in one visual view (visualization) and evaluating using the average score as the probing metric, our experiment shows that we can use vector representations visually in the latent space to outline comparative characteristics of learner performance and behaviors together. The representations can provide a comparative visualization of each learner's behaviors, highlighting the overall performance achieved by each learner and comparing a learner's performance against similarly behaving learners.

2 Approach

For each knowledge concept, we consider learner profile representation as a encoding of a multivariate time series that represents the temporal information on user behaviors and learning performance. Each representation is a fixed-length vector in the latent space derived by the LSTM autoencoder, which captures the unique dynamics of learner behaviors over time. In this study, we consider three types of learning traces: exercise correctness rates, the number of exercises attempted, and the actual exercising time, all on a weekly basis. Formally, we define *multivariate weekly performance time series* as below:

Definition 1 (Multivariate Weekly Performance Time Series). *A multivariate weekly performance time series for learner u, regarding a knowledge concept θ, is a series of weekly observations about u's performance on all exercises covering the concept θ that learner u attempted, since u's first attempt regarding that concept θ. It is represented as $\mathbf{x} = (\mathbf{x_1}, \mathbf{x_2}, \mathbf{x_3})^T$, where $\mathbf{x_i}$ =*

$(x_i^{w_0}, x_i^{w_1}, ..., x_i^{w_{\ell-1}}) \in \mathbb{R}^\ell$ *is a univariate series in a length of* ℓ*, and* ℓ *is the variable length of a learner's active exercising weeks. A weekly observation can be represented as* $(x_1^{w_k}, x_2^{w_k}, x_3^{w_k})^T$*, where* $x_1^{w_k}$ *denotes the weekly exercise correctness rate in the week* w_k*,* $x_2^{w_k}$ *counts the number of weekly exercises covering the concept* θ *and* $x_3^{w_k}$ *records the active exercising week number in the week* w_k*.*

3 Experiments

We use the `Ednet-KT1` dataset [4]. `Ednet-KT1` consists of learners' question-solving logs and records their interactions with `Santa` over two years. In `Ednet-KT1`, each question may cover multiple types of knowledge concepts. Each knowledge concept is labeled by one unique tag ID. Unless noted otherwise, *tags* and knowledge concepts are used interchangeably in the rest of the paper.

We implemented the autoencoder model with Tensorflow. For each tag, we filtered out all learners who interacted with the ITS for less than 12 weeks in the calendar, avoiding including new or inactive learners with too few behavioral characteristics. Then, for each learner, we kept all observations within 48 weeks in the calendar, since the learner's first attempt in the ITS regarding each tag. We used Mean-Squared Error (MSE) loss and ran for 20 epochs for each tag during training. We trained all models with the Adam optimizer, using a learning rate of 0.001, batch size of 32, and latent dimension size of 16.

Profiling High-Volume User Behaviors in One Visual View. Visualizing representations in the latent space is a useful way to understand the characteristics of the distribution of learning behaviors. We use Principle Component Analysis (PCA) to compress the latent representations in a 3D space. We next detail the visualization for the most frequent tag (i.e., Tag 183) in `Ednet-KT1`.

Figure 2a reports a scatter plot of representations after PCA for Tag 183. Here, we use a learner's average score of all exercises under observation as a probing metric to evaluate the learner's level. As can be seen, the majority of learners are located together on one side, leaving only a small number of learners that form a tail shape and stay far apart. This is because most learners did a considerable number of exercises associated with Tag 183, while just some learners did a very limited number of exercises with low intensity. To validate this point, in Fig. 2a, we also plot the positions where the four learners in the previous example (see Fig. 1) are located in the compressed latent space. For each learner, we use line charts in Figs. 2b–2d to compare the original time series and the reconstructed series. The blue lines in the charts indicate the actual records of their multivariate performing time series, while the red lines are reconstructed by the LSTM autoencoder. Among these four learners, only learner $u88151$ is located in the tail shape. This aligns with the fact that learner $u88151$ practised much less compared to the other three learners. Moreover, as shown in Fig. 2a, the positioning of representations in the compressed latent space also reflects the level of learner performance in terms of average scores. Learners located at the

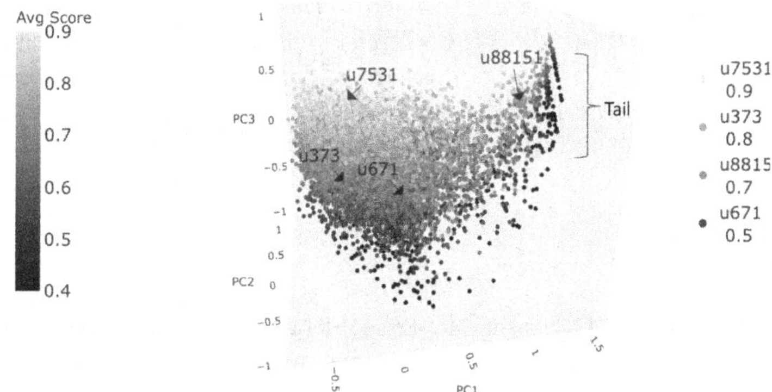

(a) PCA 3D Compressed View of the Latent Space for Tag 183. The annotations show that the four learners listed in Fig. 1 are located in different areas in this view, which indicates they interacted with the ITS in different ways. The legend on the right side explains each learner's average score.

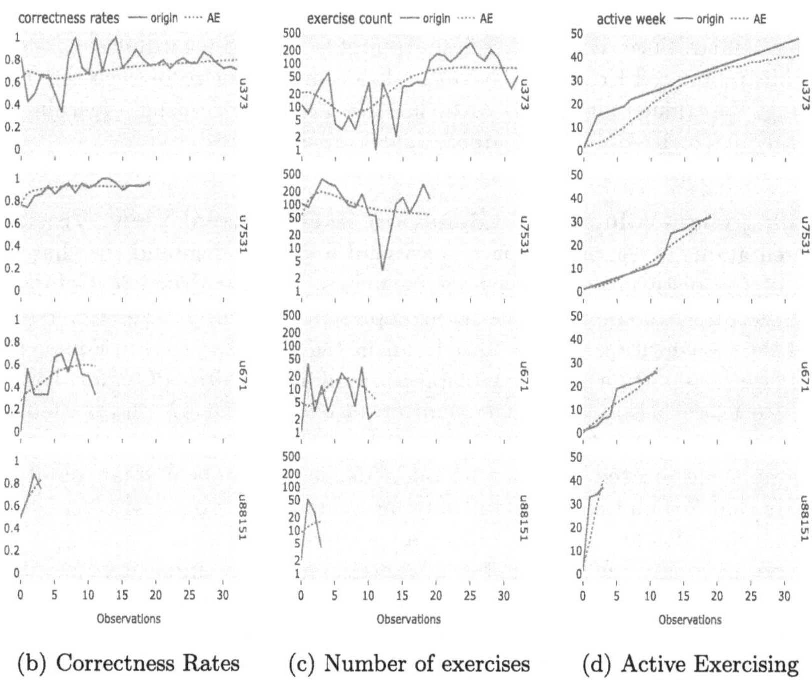

(b) Correctness Rates (c) Number of exercises (d) Active Exercising Week Numbers

Fig. 2. Profiling high-volume learner behavior data in one view. (a) presents the compressed 3D view of all learners regarding exercises that cover Tag 183, with the annotations highlighting the locations of four individual learners. (b)-(d) plot multivariate time series of these four learners. Blue lines are the original time series, and red dotted lines are the reconstructed time series by the LSTM autoencoder. Learner ID is annotated on the right side in each line chart. (Color figure online)

higher positions in this figure, in general, have higher average scores than learners located at the lower positions. Note that average scores are used as descriptive statistics to verify the effectiveness of the latent representations. Training LSTM autoencoders does not require average scores as inputs.

As it turns out, the scatter plot shows that color changes gradually in the distributions. These color gradients indicate smooth changes in the average scores of the learners. In this way, the plot of the latent space presents the overall views for understanding learner behaviors and performance at scale. Using average score as an evaluating metric, we can look at the latent spaces and quickly spot *marker* learners, such as learners that are at high risk of dropping out (e.g., learners near the tail at the right side in Fig. 2a, due to the attempted exercises are too few, compared to the majority), or underperforming learners but with high effort (e.g., learners at the left bottom in Fig. 2a). In addition, the gradients also show that the distribution of learners at medium-risk is in a continuum. As such, our model assesses not only performance but also their learning habits.

4 Conclusion

We proposed a novel self-supervised deep learning framework for learner profiling and examined the quality of learned vector representations. We empirically showed that it is possible to train high-quality representations using our framework. Coupled with the designed visualization methods, our approach opens new possibilities for learner profiling and offers the potential for learning intervention.

Acknowledgments. This project has received funding from the European Union's Horizon 2020 Research and Innovation Programme under the HUMAN+ COFUND Marie Skłodowska-Curie grant agreement No. 945447.

References

1. Alexander, P.A., Murphy, P.K.: Profiling the differences in students' knowledge, interest, and strategic processing. J. Educ. Psychol. **90**(3), 435 (1998)
2. Ben Soussia, A., Roussanaly, A., Boyer, A.: An in-depth methodology to predict at-risk learners. In: De Laet, T., Klemke, R., Alario-Hoyos, C., Hilliger, I., Ortega-Arranz, A. (eds.) EC-TEL 2021. LNCS, vol. 12884, pp. 193–206. Springer, Cham (2021). https://doi.org/10.1007/978-3-030-86436-1_15
3. Ben Soussia, A., Treuillier, C., Roussanaly, A., Boyer, A.: Learning profiles to assess educational prediction systems. In: Rodrigo, M.M., Matsuda, N., Cristea, A.I., Dimitrova, V. (eds) Artificial Intelligence in Education (AIED 2022). LNCS, vol. 13355, pp. 41–52. Springer, Cham (2022). https://doi.org/10.1007/978-3-031-11644-5_4
4. Choi, Y., et al.: EdNet: a large-scale hierarchical dataset in education. In: Bittencourt, I.I., Cukurova, M., Muldner, K., Luckin, R., Millán, E. (eds.) AIED 2020. LNCS (LNAI), vol. 12164, pp. 69–73. Springer, Cham (2020). https://doi.org/10.1007/978-3-030-52240-7_13
5. Marras, M., Vignoud, J.T.T., Kaser, T.: Can feature predictive power generalize? Benchmarking early predictors of student success across flipped and online courses. In: 14th International Conference on Educational Data Mining, pp. 150–160 (2021)

6. Mejia-Domenzain, P., Marras, M., Giang, C., Käser, T.: Identifying and comparing multi-dimensional student profiles across flipped classrooms. In: Rodrigo, M.M., Matsuda, N., Cristea, A.I., Dimitrova, V. (eds.) Artificial Intelligence in Education (AIED 2022). LNCS, vol. 13355, pp. 90–102. Springer, Cham (2022). https://doi.org/10.1007/978-3-031-11644-5_8
7. Mojarad, S., Essa, A., Mojarad, S., Baker, R.S.: Data-driven learner profiling based on clustering student behaviors: learning consistency, pace and effort. In: Nkambou, R., Azevedo, R., Vassileva, J. (eds.) ITS 2018. LNCS, vol. 10858, pp. 130–139. Springer, Cham (2018). https://doi.org/10.1007/978-3-319-91464-0_13
8. Wang, P., Fu, Y., Xiong, H., Li, X.: Adversarial substructured representation learning for mobile user profiling. In: Proceedings of the 25th ACM SIGKDD International Conference on Knowledge Discovery and Data Mining, pp. 130–138 (2019)

Examining the Learning Benefits of Different Types of Prompted Self-explanation in a Decimal Learning Game

Huy A. Nguyen[1(✉)], Xinying Hou[2], Hayden Stec[1], Sarah Di[1], John Stamper[1], and Bruce M. McLaren[1]

[1] Carnegie Mellon University, Pittsburgh, PA 15213, USA
hn1@cs.cmu.edu

[2] University of Michigan, Ann Arbor, MI 48104, USA

Abstract. While self-explanation prompts have been shown to promote robust learning in several knowledge domains, there is less research on how different self-explanation formats benefit each skill set in a given domain. To address this gap, our work investigates 214 students' problem-solving performance in a learning game for decimal numbers as they perform self-explanation in one of three formats: multiple-choice (N = 52), drag-and-drop (N = 72) and open-ended (N = 67). We found that self-explanation performance in all three formats was positively associated with problem-solving performance. At the same time, we observed better outcomes with the drag-and-drop format than the open-ended format for solving decimal addition problems that do not remind students about carrying, but worse outcomes than the multiple-choice and open-ended format for other problem types. These results point to the nuanced interactions between the problem type and self-explanation format that should be further investigated to identify when and how self-explanation is most helpful for learning.

Keywords: Decimal learning · Problem-solving · Self-explanation format

1 Introduction

Prompted self-explanation, where the student self-explains what they have learned or how they solved a problem, is a highly robust and effective learning strategy [2]. There are different ways to implement this activity in a learning system, such as multiple-choice, drag-and-drop and open-ended, which vary in their constructive or active nature [2]. Yet there is mixed evidence regarding their relative effectiveness. While [2] and [8] pointed to open-ended prompts being more beneficial than multiple-choice prompts, findings from [4] showed that the multiple-choice prompts led to better transfer. Drag-and-drop self-explanation is less explored, with one study showing it did not differ in learning benefits from the other formats [8], but drag-and-drop questions have become increasingly popular in computer-based assessments, with the advantage of promoting interactivity and reducing random guessing [1].

N. Wang et al. (Eds.): AIED 2023, LNAI 13916, pp. 681–687, 2023.
https://doi.org/10.1007/978-3-031-36272-9_56

Our conjecture regarding these mixed results is that the effectiveness of each self-explanation format could differ across knowledge domains, or even across problem types within a domain. To examine this possibility, we performed a post-hoc analysis of our prior study [8] which compares the learning outcomes from multiple-choice, drag-and-drop and open-ended self-explanation prompts in a digital learning game for decimal numbers. In the game, students solve a variety of decimal problem types (e.g., locate decimal numbers on a number line, sort sequences of decimal numbers) and perform self-explanation after every two rounds of problem-solving [7]. While we have found that open-ended self-explanation prompts led to higher delayed posttest scores than multiple-choice prompts [8], our prior work did not investigate whether the relative benefits of these formats also differ by the decimal problem types. Thus, our current work focuses on addressing this gap.

2 A Digital Learning Game for Decimal Numbers

Decimal Point is a web-based single-player game that teaches decimal numbers and operations to middle-school students [7]. The game features an amusement park metaphor with 8 theme areas and 24 mini-games. Each mini-game consists of two problem-solving activities, designed to help students practice decimal procedures, and a self-explanation activity, designed to reinforce their learning [2]. Each problem-solving activity belongs to one of five problem types listed in Table 1. As an example, in a *Sequence* mini-game called *Escape the Aliens*, the two problem-solving activities involve filling in the 4th and 5th numbers in the sequences 0.0, 0.35, 0.70, ___, ___ and 0.0, 4.4, 8.8, ___, ___. After solving these problems, students are asked to self-explain how they would find the 6th number in the second sequence (Table 2).

Table 1. Descriptions and examples of the five decimal problem types in *Decimal Point*.

Problem type	Description	Example
Addition	Enter the carry and result digits when adding two decimal numbers	Enter the carry and result digits when adding 7.50 to 3.90
Bucket	Label each of the given decimals as "less than" or "greater than" a threshold	**Given decimals**: 0.51, 0.132, 0.9, -0.833. **Threshold**: -0.4
Number Line	Locate the position of a given decimal on a number line	Place 0.111 on a number line from -1 to 1
Sequence	Fill in the next two numbers in a given sequence of decimals	Fill in the next two numbers: 0.33, 0.66, 0.99, ___, ___
Sorting	Sort a given list of decimal numbers in ascending or descending order	Sort the list 0.9063, 0.39, 0.291, 0.7 from small to large

Students need to play through all 24 mini-games in a fixed order to finish the game. They also get immediate feedback about the correctness of their answer in both the

problem-solving and self-explanation activities. If they make a mistake, they can retry any number of times until arriving at the correct answer, and must do so to advance to the next mini-game. Additionally, in the study of *Decimal Point* which we base our analysis on [8], students are randomly assigned to play one of three versions of the game with different self-explanation formats. In the control multiple-choice condition, students select a correct answer from three given options. In the drag-and-drop condition, students fill in the blanks of an answer statement by dragging items from a given word bank with 4 or 5 phrases. In the open-ended condition, students enter their responses in an open text box. To deter students from providing unthoughtful answers, we implemented a preliminary grading mechanism, using a keyword matching technique commonly applied to automated grading of short answers [10]. In particular, the game considers the student's answer as acceptable if it contains more than three words, with at least one of the words belonging to a set of keywords that would be found in a correct explanation. The keywords are pre-determined for every mini-game and are not revealed to the students. Table 2 shows the self-explanation question for the *Escape the Aliens* mini-game described earlier in each of the three game versions.

Table 2. A self-explanation question example in each of the three formats.

Multiple-choice	Drag-and-drop	Open-ended
The next number in the pattern can be found by adding 17.6 + 4.4. What is the answer and how do you know? ▢ 22.00 because you should carry a 1 over to the ones place ▢ 21.10 because .6 + .4 = .10 and 4 + 17 = 21 ▢ 22.10 because .6 + .4 = .10 and you also carry a 1	The next number in the pattern is 17.6 + 4.4, which equals _____ because when we add, we _____ **Word bank**: 22.0, 21.0, 22.10, carry once, add the right side separately	The next number in the pattern can be found by adding 17.6 + 4.4. What is the answer and how do you know? _________

3 Methods

Our analysis involves 357 5th and 6th graders from a prior classroom study [8] in the spring of 2021, which includes a pretest, game play and immediate posttest in the first week, followed by a delayed posttest one week later. 143 students who either did not finish all of the materials or had already played through the game in a previous year were excluded. Among the remaining 214 students, there were 75 in the multiple-choice condition, 72 in the drag-and-drop condition, and 67 in the open-ended condition.

While students' self-explanations can be automatically graded in the multiple-choice and drag-and-drop formats, the open-ended responses require manual grading. Thus, we coded all of the 5,142 self-explanation responses from students as either *correct*, *incorrect* or *off-topic* (i.e., the response is unrelated to the question). Following [3], the coding process started with one member of the research team analyzing 1.5% of the responses to develop a rubric for the correct, incorrect and off-topic labels. Two other members then applied this rubric to independently code 20% of the responses, achieving an inter-rater reliability of $\kappa = 62.7\%$. Next, the two members met to resolve disagreements in their codes and refine the rubric, which was then applied by the first member to code the remaining 80% of the data. This coding process yielded 1,000 correct, 4,076 incorrect and 66 off-topic responses.

4 Results

Our first analysis investigated how students' self-explanation performance relates to their problem-solving performance. As we had reported the comparison of test scores by condition in [8], with significantly higher delayed posttest scores in the open-ended than in the multiple-choice condition, in this work we focus only on in-game performance. To measure student performance in the problem-solving and self-explanation activities, we computed the rate at which their first attempt is correct (i.e., the number of times their first attempt in a question is correct, divided by the number of questions). We only considered the first attempt because subsequent attempts may be influenced by the game's corrective feedback and therefore not reflective of the student's performance; this approach is also common in learning assessment on digital platforms [11]. Next, we built a regression model that predicts students' problem-solving performance (measured as the rate of correctness at the first problem-solving attempt, i.e., *PS-success-rate*) based on their prior knowledge (measured by pretest scores), study condition and self-explanation performance (measured as the rate of correctness at the first self-explanation attempt, i.e., *SE-success-rate*). The overall model achieved an R^2 value of .498, with the coefficient terms reported in Table 3. We observed that, when controlled for prior knowledge, students' self-explanation performance in all three formats had a significant and positive association with their overall problem-solving performance.

Table 3. Results of the regression model predicting *PS-success-rate*. Rows with (***) indicate significant coefficients ($p < .001$)

	Coefficient	Std Err	t
Intercept	0.405	0.022	18.284 (***)
Condition (drag-and-drop)	0.004	0.014	0.275
Condition (open-ended)	0.026	0.018	1.439
SE-success-rate	0.129	0.035	3.694 (***)
Pretest Score	0.007	0.001	9.981 (***)

Next, we examined how students' problem-solving performance in each decimal problem type differs across conditions. To measure problem-solving performance, we again used the rate of correctness at the first problem-solving attempt (*PS-success-rate*),

Table 4. Descriptive statistics for *PS-success-rate* by problem type and study condition. Rows with (**) indicate significant condition differences ($p < .01$).

Problem type	Multiple-choice *M (SD)*	Drag-and-drop *M (SD)*	Open-ended *M (SD)*
Addition	0.660 (0.255)	0.660 (0.279)	0.698 (0.248)
Bucket	0.683 (0.248)	0.661 (0.225)	0.716 (0.238)
Number Line (**)	0.649 (0.227)	0.530 (0.237)	0.626 (0.233)
Sequence (**)	0.706 (0.098)	0.736 (0.099)	0.688 (0.104)
Sorting	0.605 (0.191)	0.603 (0.181)	0.629 (0.186)

but computed a separate value for each of the five decimal problem types (see Table 4 for descriptive statistics). We then conducted a series of ANCOVAs comparing students' rate of correctness in each decimal problem type across the three study conditions, with pretest scores as covariates to account for prior knowledge. Our results showed significant condition effects on *PS-success-rate* in the *Number Line* ($F(2, 210) = 5.136$, $p = .007$) and *Sequence* ($F(2, 210) = 5.391$, $p = .005$) problem types. Within the *Number Line* problems, post-hoc (Tukey) comparisons revealed significant differences between the drag-and-drop and multiple-choice conditions ($p = .006$), as well as between the drag-and-drop and open-ended conditions ($p = .043$), with the drag-and-drop condition leading to the lowest *PS-success-rate*. Within the *Sequence* problems, we similarly observed a significant difference between the drag-and-drop and open-ended conditions ($p = .014$), but with higher *PS-success-rate* in the drag-and-drop condition.

5 Discussion and Conclusion

This work examined the effects of prompted self-explanation formats on problem-solving performance in a digital learning game, *Decimal Point*. Our results showed that, when controlling for prior knowledge, self-explanation performance across all three formats is a significant and positive predictor of problem-solving performance. This benefit is consistent with prior cognitive theories and empirical evidence of self-explanation [2]. In the case of *Decimal Point*, the self-explanation questions were designed to target common decimal misconceptions [7]. Therefore, students who answered these questions correctly, especially on the first attempt, had acquired a good understanding of decimal concepts, which translated to better problem-solving. This connection suggests that self-explanation could also be helpful in other domains where students' learning difficulties lie primarily in their common misconceptions.

On the other hand, our analysis of individual decimal problem types revealed varying effects of the self-explanation format. First, drag-and-drop self-explanation prompts led to higher *PS-success-rate* than open-ended prompts in the *Sequence* problem type, which focuses on decimal additions. Because *Sequence* problems do not require specifying the carry digits during addition, the most common error in these problems is forgetting to carry across the decimal point [7]. In that case, the drag-and-drop prompts are better at addressing this misconception, thanks to the given word bank that always contained the term "carry," whereas the open-ended prompts provided no such reminders (Table 2). At the same time, we found that the *PS-success-rate* was lowest in the drag-and-drop condition across all other problem types, especially the *Number Line* type where the condition differences were significant. One explanation is that the drag-and-drop format could cause reading difficulties due to the blanks in the given self-explanation statements. Combined with evidence that young students' reading skills have declined during the COVID-19 pandemic [5], this suggests students may not have learned effectively from drag-and-drop self-explanation prompts because they had to exert more cognitive efforts on reading comprehension [6]. In turn, they would perform worse in most decimal problem types, especially *Number Line* which was identified as the most difficult type [9]. Finally, we note that there were no differences in the benefits of multiple-choice and open-ended self-explanation for solving any decimal problem type. On the other hand, our prior results did report higher delayed posttest scores from students who performed open-ended self-explanation [8], suggesting that the benefit of the open-ended format only manifests in the long term, but not during immediate game play.

In conclusion, our findings support the benefits of prompted self-explanation in decimal problem-solving, but also point to variances across problem types and self-explanation formats. At the same time, this work presents certain limitations that future research should address. First, the grading rubric of open-ended self-explanations can be refined to improve the graders' agreement and more accurately evaluate the students' work. Second, while several mini-games share the same decimal problem type, there may still be inherent differences in the game design or narrative that could influence students' performance. A more nuanced understanding of these latent factors will help identify when and how each self-explanation format is best for learning.

References

1. Arslan, B., Jiang, Y., Keehner, M., Gong, T., Katz, I.R., Yan, F.: The effect of drag-and-drop item features on test-taker performance and response strategies. Educ. Meas. Issues Pract. **39**(2), 96–106 (2020)
2. Chi, M.T., Wylie, R.: The ICAP framework: linking cognitive engagement to active learning outcomes. Educational psychologist. **49**(4), 219–243 (2014)
3. DeCuir-Gunby, J.T., Marshall, P.L., McCulloch, A.W.: Developing and using a codebook for the analysis of interview data: an example from a professional development research project. Field Methods **23**(2), 136–155 (2011)
4. Johnson, C.I., Mayer, R.E.: Applying the self-explanation principle to multimedia learning in a computer-based game-like environment. Comput. Hum. Behav. **26**(6), 1246–1252 (2010)
5. Kuhfeld, M., Lewis, K., Peltier, T.: Reading achievement declines during the COVID-19 pandemic: evidence from 5 million US students in grades 3–8. Reading and Writing. 1–17 (2022)
6. McCray, G., Brunfaut, T.: Investigating the construct measured by banked gap-fill items: evidence from eye-tracking. Lang. Test. **35**(1), 51–73 (2018). https://doi.org/10.1177/0265532216677105
7. McLaren, B.M., Adams, D.M., Mayer, R.E., Forlizzi, J.: A computer-based game that promotes mathematics learning more than a conventional approach. Int. J. Game-Based Learn. (IJGBL). **7**(1), 36–56 (2017)

8. McLaren, B.M., Richey, J.E., Nguyen, H.A., Mogessie, M.: Focused self-explanations lead to the best learning outcomes in a digital learning game. In: Proceedings of the 17th International Conference of the Learning Sciences, pp. 1229–1232. ISLS (2022)
9. Nguyen, H., Hou, X., Stamper, J., McLaren, B.M.: Moving beyond test scores: analyzing the effectiveness of a digital learning game through learning analytics. In: Proceedings of the 13th International Conference on Educational Data Mining (2020)
10. Rokade, A., Patil, B., Rajani, S., Revandkar, S., Shedge, R.: Automated grading system using natural language processing. In: 2018 Second international conference on inventive communication and computational technologies (ICICCT), pp. 1123–1127. IEEE (2018)
11. Yu, H.-F., et al.: Feature engineering and classifier ensemble for KDD cup 2010. In: Proceedings of the KDD Cup 2010 Workshop (2010)

Plug-and-Play EEG-Based Student Confusion Classification in Massive Online Open Courses

Han Wei Ng(✉)

School of Computer Science and Engineering, Nanyang Technological University, 50 Nanyang Ave., Singapore 639798, Singapore
hanwei001@e.ntu.edu.sg

Abstract. Use of Electroencephalography (EEG)-based monitoring devices in classrooms has seen greater uptake with increased interest in Internet of Things (IOT) and human-computer interaction (HCI). The ability to interact directly with digital interfaces using brain signals offer significant advantage towards seamless and natural communication by simply thinking of the desired outcome. We propose the implementation of a new leave-one-subject-and-video-out paradigm alongside a plug-and-play lightweight EEG-based classification framework to accurately analyse the efficacy of EEG signals in determining students' confusion levels. The proposed methodology achieves state-of-the-art performance, reaching 95.75% classification accuracy.

Keywords: Plug-and-play EEG · EEG Education · Signal Processing · Deep Feature Learning

1 Introduction

BCIs are used to communicate between humans and computers [12], with non-invasive Electroencephalography (EEG) signals being one of the most common signals collected for analysis [11]. The EEG signals stem from the users' neural activity and therefore can be said to directly reflect the individuals' mental state in real-time. In the classroom, confusion of a group of students in class can difficult to quantize or obtain in real-time, especially in online class formats such as the Massive Online Open Courses (MOOCs) [10]. This lack of information results in less than ideal outcomes for the students as the teaching is not catered towards their learning rate. Thus, EEG-based monitoring devices such as Neurosky Mindset [4] may be utilized towards providing instructors with real-time feedback from students alongside providing further insights into education [1]. One of the biggest barrier to entry of such EEG-based devices is the classification accuracy and efficacy of the products [8]. Thus, this study aims to examine the effectiveness of EEG-based monitoring of students' confusion levels in an online classroom setting. Presently, decoding non-invasive EEG signals is a huge

N. Wang et al. (Eds.): AIED 2023, LNAI 13916, pp. 688–694, 2023.
https://doi.org/10.1007/978-3-031-36272-9_57

challenge due to high noise-to-signal ratio [3]. This results in EEG-based decoding having significantly reduced discriminative ability that prevents translation towards real-world usage. Furthermore, there is a lack of easily obtainable EEG datasets which reduces deep learning models' ability to leverage upon other data to perform pretraining [9].

2 Dataset

For this work, we use the confused student EEG brainwave on MOOC dataset collected by Wang et al. from Carnegie Mellon University [10]. The data was collected by first preparing 20 videos belonging to two main categories, topics which are familiar to a normal college student and topics which they might find challenging to understand. These videos are extracted from the Massive Online Open Course (MOOC) videos in two-minute clips spliced from the middle of the topic to ensure a more distinctive difference between the confusing and non-confusing videos. EEG data was subsequently collected from 10 students, with each student undergoing 10 sessions.

Contrary to previous research which utilized a randomized train-validation-split to train the classifier networks, we propose the use of a structured split by separating each subject and video from one another. The rationale behind splitting by subject and video is to prevent potential overfitting of the model, which when exposed to sufficient data drawn from the same distribution as the target, is likely to perform better than in real-life scenarios. Most importantly, any EEG-based monitoring system deployed in real-world systems would likely be required to feed in data in chronological order to perform inference, and there are multiple instances whereby labelled data from the target is not readily available.

3 Methodology

In this study, we remove the predefined confusion labels since these are arbitrarily labelled and may not necessarily be always available in real-life scenarios. Furthermore, the attention and mediation features are removed from the dataset before training the classifier. The main rationale for this is to ensure that the resultant classifier is a model that is able to be utilized under most settings. Since the attention and mediation measurements come from a proprietary software developed for solely the companys' products, it would limit the EEG devices that may be used by students in the future which would not be ideal. Lastly, the user-defined labels are separated from the features to form the test labels for confusion. The remaining features are then used as the training data that is fed into the classifier model to train the neural network to discriminate between confused and non-confused classes.

We utilize the EEGNet classifier framework as first proposed by Lawhern et al. in 2016 [7] which has previously shown significant success in performing motor imagery classification. This network was chosen due to its flexibility in

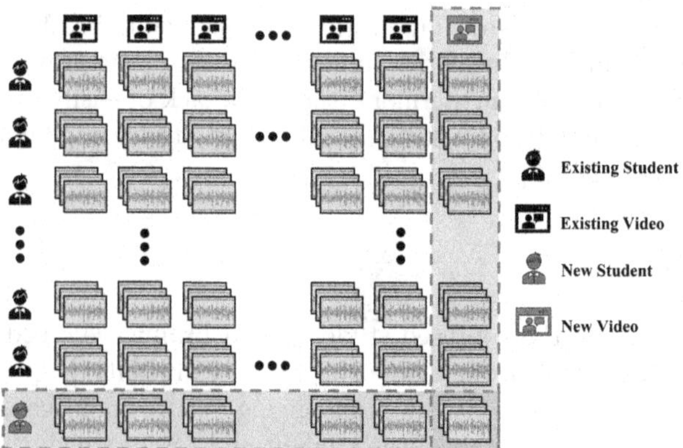

Fig. 1. In the proposed leave-one-subject-and-video-out (LOSVO) paradigm, the inference data is selected as the overlap between the chosen subject and video. The rest of the data is used as training and validation data.

adapting to different EEG paradigms while still retaining ease of training the classifier. In this work, we show that there is significant promise in performing model-level transfer learning to heavily reduce the amount of resources required to test a newly created model.

After the dataset has been split into their respective subjects and videos, normalization may be applied to each segment of subject and video specific EEG data. This helps to ensure that the neural network aims to capture the patterns of the EEG waves as opposed to simply memorizing values [2]. The resultant dataset is then further separated into the training-validation-test split by allocating each subject and video specific segment accordingly. As previous work done on the same dataset trains the classifier via randomizing all of the data without taking into consideration of scenarios involving unseen data from new subjects and videos, this work aims to explore the feasibility of a plug-and-play EEG-based confusion classifier that does not rely on target data. To achieve this, we introduce the leave-one-subject-and-video-out (LOSVO) paradigm (Fig. 1) whereby EEG data belonging to the same selected subject and video are fully removed from the training and validation data. This closely follows a well-establish leave-one-subject-out (LOSO) paradigm [5] but is modified to include the lack of unseen video information as well.

The training and validation split would thus contain the EEG data of students and videos that were not selected to be the simulated new student or video, while the test data would contain solely the EEG data of the simulated new student on the new video. For instance, if subject 10 and video 10 was selected to be the inference data, the EEG data of subject 10 for the other videos would be removed and data of video 10 for the other subjects would be removed from training. Considering that there are 10 students and 10 videos, this would allow us to compute a total of 100 individual simulated scenarios of new student and new video combinations. The initialized EEGNet model thus takes in a signal containing N number of EEG signal windows, with 15 channels of time length equivalent to the time window length.

4 Results

We show that the proposed confusion classifier achieves high classification accuracy in a variety of different paradigms, namely: (1) leave-one-subject-out (2) leave-one-video-out and (3) leave-one-subject-and-video-out. To ensure that the inference testing of the trained model closely mimics the real-world scenario, the inference data is fed as input to the network in chronological order. The accuracy is computed for each student for each video individually. The classification accuracy is determined by the total number of correct classified data points throughout the video clip for the student divided by the total number of individual data points corresponding to the number of EEG recordings throughout the video.

Table 1. Confusion Classification Accuracy (%) for Individual Video and Subjects using the proposed LOSVO-EEGNet methodology without normalization.

	Video Number										
Subject Number	1	2	3	4	5	6	7	8	9	10	Subject Average
1	100	100	100	100	100	100	100	100	100	100	100
2	100	72.00	100	100	100	100	100	0.00	100	97.00	86.90
3	100	91.00	100	100	100	100	100	100	100	83.00	97.40
4	100	100	100	100	100	100	100	100	100	86.00	98.60
5	100	100	100	100	100	100	100	100	100	100	100
6	100	100	100	100	100	100	100	100	100	100	100
7	100	100	100	100	100	100	100	100	0.00	100	90.00
8	100	100	100	64.00	100	100	100	100	0.00	100	86.40
9	100	100	100	100	100	100	100	100	100	100	100
10	100	100	100	100	100	100	100	100	100	82.00	98.20
Video Average	100	96.30	100	96.40	100	100	100	90.00	80.00	94.80	95.75

From the results in Table 1 and 2, we can observe that the proposed LOSVO paradigm with EEGNet is highly effective in achieving outstanding performance on new data which domain is not previously found in the training data. The proposed classification methodology with the best performing configuration was able to attain high average accuracy across both different videos (80.00–100%) and subjects (86.40–100%). Therefore, we show that the proposed EEGNet configuration is sufficient to achieve high accuracy confusion classification without much changes to the existing architecture, allowing instructors to easily utilize the network to perform confusion measurements. Between the use of normalization (Table 2) and without normalization (Table 1), it was discovered that the network has superior performance without normalization (95.75%) as compared to normalizing the signals (94.58%). This strongly suggests that beyond the relative shifts in the EEG signal, the deep neural network also considers the absolute numbers of the recorded signal. One plausible reason for the better performance of the non-normalized data is that when anomalous points exist within the data,

Table 2. Confusion Classification Accuracy (%) for Individual Video and Subjects using the proposed LOSVO-EEGNet methodology with normalization.

	Video Number										
Subject Number	1	2	3	4	5	6	7	8	9	10	Subject Average
1	62.00	100	100	48.00	65.00	100	100	95	100	100	87.00
2	95.00	100	100	100	97.00	100	100	88.00	76	87.00	94.30
3	97.00	95.00	100	100	100	100	100	100	100	100	99.20
4	100	100	100	97.00	100	100	100	100	100	100	99.70
5	35.00	100	100	100	100	82.00	100	100	88.00	100	90.50
6	95.00	100	100	100	100	100	100	100	100	100	99.50
7	70.00	100	100	100	100	100	100	100	100	94.00	96.40
8	100	100	100	96.00	100	42.00	100	100	100	100	93.80
9	100	100	100	46.00	100	100	100	100	100	100	94.60
10	13.00	100	100	100	100	100	100	100	100	95.00	90.80
Video Average	76.70	99.50	100	88.70	96.20	92.40	100	98.30	96.40	97.60	94.58

Table 3. Comparison of confusion classification accuracy (%) across different models against the propsosed method.

Methodology	Classification Accuracy (%)
Bi-LSTM (2017)	73.30
CF-BiLSTM (2019)	75.00
Bi-LSTM (2020)	88.00
ETC (2022)	66.60
LOSVO-EEGNet (Ours)	**95.75**

normalization might result in the over compression of the remaining data points which can lead to the network being unable to learn the associated features as the relative pattern of the signal has been diminished.

Despite using fewer data from the target subject and video and overall fewer data points from the original dataset, we show that the proposed methodology is able to achieve significantly superior classification performance over other proposed methods on the same dataset (Table 3). The proposed framework achieves a significantly superior mean accuracy of 95.75% as compared to previously known state-of-the-arts. This strongly indicates the ability for EEG-based neural networks such as EEGNet to be adapted towards other classification purposes such as mental state classification while retaining its key advantages in low computational complexity, minimal feature engineering and high classification capabilities. This allows for instructors to easily utilize largely available networks and tweak it towards their desired task, without having to manually optimize the architecture or perform feature engineering.

5 Conclusion

In conclusion, we propose the use of a leave-one-subject-and-video-out (LOSVO) paradigm that closely mimics real-life classroom scenarios where the subject and/or video data may not be available beforehand. By using the other available data, we present a method using EEGNet to establish a strong baseline EEG-based confusion classifier which readily generalizes towards new unseen data belonging to different domains. This enables a plug-and-play EEG-based confusion monitoring device to be deployed readily towards classrooms. As wearable devices become more commonplace in educational settings [6], the key benefit towards a plug-and-play system is that it enables easy integration of deep learning models into existing pipelines without much disruption to the workflow or increase in resource sinks. The use of a previously established deep learning neural network alongside the proposed training and inference paradigms displays the potential ease and effectiveness of integrating deep learning empowered wearables in classroom education even without the need of formal artificial intelligence training for the educators.

References

1. Davidesco, I., Matuk, C., Bevilacqua, D., Poeppel, D., Dikker, S.: Neuroscience research in the classroom: portable brain technologies in education research. Educ. Res. **50**(9), 649–656 (2021)
2. Giri, E.P., Fanany, M.I., Arymurthy, A.M., Wijaya, S.K.: Ischemic stroke identification based on EEG and EOG using ID convolutional neural network and batch normalization. In: 2016 International Conference on Advanced Computer Science and Information Systems (ICACSIS), pp. 484–491. IEEE (2016)
3. Goldenholz, D.M., et al.: Mapping the signal-to-noise-ratios of cortical sources in magnetoencephalography and electroencephalography. Hum. Brain Mapp. **30**(4), 1077–1086 (2009)
4. Katona, J., Farkas, I., Ujbanyi, T., Dukan, P., Kovari, A.: Evaluation of the NeuroSky MindFlex EEG headset brain waves data. In: 2014 IEEE 12th International Symposium on Applied Machine Intelligence and Informatics (SAMI), pp. 91–94. IEEE (2014)
5. Kunjan, S., et al.: The necessity of leave one subject out (LOSO) cross validation for EEG disease diagnosis. In: Mahmud, M., Kaiser, M.S., Vassanelli, S., Dai, Q., Zhong, N. (eds.) BI 2021. LNCS (LNAI), vol. 12960, pp. 558–567. Springer, Cham (2021). https://doi.org/10.1007/978-3-030-86993-9_50
6. Labus, A., Milutinovic, M., Stepanic, Đ., Stevanovic, M., Milinovic, S.: Wearable computing in e-education. RUO. Revija za Univerzalno Odlicnost **4**(1), A39 (2015)
7. Lawhern, V.J., Solon, A.J., Waytowich, N.R., Gordon, S.M., Hung, C.P., Lance, B.J.: EEGNet: a compact convolutional neural network for EEG-based brain–computer interfaces. J. Neural Eng. **15**(5), 056013 (2018)
8. Sawangjai, P., Hompoonsup, S., Leelaarporn, P., Kongwudhikunakorn, S., Wilaiprasitporn, T.: Consumer grade EEG measuring sensors as research tools: a review. IEEE Sens. J. **20**(8), 3996–4024 (2019)
9. Wan, Z., Yang, R., Huang, M., Zeng, N., Liu, X.: A review on transfer learning in EEG signal analysis. Neurocomputing **421**, 1–14 (2021)

10. Wang, H., Li, Y., Hu, X., Yang, Y., Meng, Z., Chang, K.M.: Using EEG to improve massive open online courses feedback interaction. In: AIED Workshops (2013)
11. Wang, Y., Nakanishi, M., Zhang, D.: EEG-based brain-computer interfaces. Neural Interface: Frontiers and Applications, pp. 41–65 (2019)
12. Wolpaw, J.R.: Brain-computer interfaces (BCIs) for communication and control. In: Proceedings of the 9th International ACM SIGACCESS Conference on Computers and Accessibility, pp. 1–2 (2007)

CPSCoach: The Design and Implementation of Intelligent Collaborative Problem Solving Feedback

Angela E. B. Stewart[1](✉), Arjun Rao[2], Amanda Michaels[3], Chen Sun[4], Nicholas D. Duran[3], Valerie J. Shute[5], and Sidney K. D'Mello[2]

[1] University of Pittsburgh, Pittsburgh, PA 15260, USA
angelas@pitt.edu
[2] University of Colorado Boulder, Boulder, CO 80309, USA
[3] Arizona State University, Glendale, AZ 85306, USA
[4] University of Manchester, Manchester, UK
[5] Florida State University, Tallahassee, FL 32306, USA

Abstract. We present the design of CPSCoach, a fully-automated system that assesses and provides feedback on collaborative problem solving (CPS) competencies during remote collaborations. We leveraged existing data to develop deep NLP models that automatically assess the CPS competencies from speech, achieving moderate to high accuracies (average area under the receiver operating characteristic curve of .78). We engaged 43 participants in an iterative process to design the feedback mechanism, resulting in the first prototype of CPSCoach. We conducted a user study with 20 dyads who engaged with CPSCoach over multiple rounds. Participants thought the system was usable, but they were mixed about the accuracy of the feedback. We discuss design considerations for feedback systems aimed at improving CPS competencies.

Keywords: Collaborative problem solving · Intelligent feedback

1 Introduction and Related Work

Collaborative problem solving (CPS) is a ubiquitous phenomenon, occurring when multiple people engage in a coordinated attempt to solve a problem [1]. Productive CPS involves combining socio-cognitive processes, such as maintaining a shared vision of the problem and coordinating teammate strengths to implement a solution [2]. Remote CPS is increasingly common [3]. However, remote collaborations are especially difficult, as the social signals from face-to-face communication are muted or non-existent [3]. Currently, students are expected to master CPS competencies through project-based group learning [2], yet they are evaluated on project outcomes, not CPS skill [2]. Further, they do not receive meaningful feedback on their CPS skills [2], making it difficult for them to acquire and improve these competencies.

Software systems do exist that support collaborative skills, although these largely are task-specific (e.g., interfaces to enhance team communication programming [4]). Most

N. Wang et al. (Eds.): AIED 2023, LNAI 13916, pp. 695–700, 2023.
https://doi.org/10.1007/978-3-031-36272-9_58

similar to our work are systems that support awareness of behavior in collaborative interactions. In this literature, low-level behaviors are synthesized into interpretable metrics, and visualized. For example, real-time visualizations of turn-taking behaviors (e.g., interruptions) have been aggregated from speech signals [5].

Taken together, there is a conspicuous absence of pathways to CPS proficiency. Accordingly, we developed CPSCoach, a system that uses NLP to automatically assess CPS skills [1] and provide feedback between short collaborations in an educational physics game. CPSCoach is based in theories on the importance of personalized and immediate feedback to support learning [6]. We extend current collaboration feedback by focusing on high-level CPS constructs, rather than low-level behaviors. Low-level behaviors are related to CPS, but CPS is far more than this. For example, ample verbal participation is related to problem solving ideation. However, this is insufficient as the team could be entirely off topic or discussing unproductive ideas. Accordingly, we move from low-level concrete and generic behaviors (e.g., amount of verbal participation) to abstract and specialized CPS behaviors (e.g., joint knowledge construction). In this work, our goal is to identify key lessons from a proof-of-concept prototype to inform future versions of CPSCoach.

2 Intervention Design and User Study

We conducted an iterative design study to explore effective designs for CPS feedback, focusing on instructional content, translation of model predictions to interpretable metrics, and the user interface. Participants in the design study were 43 college students from a large public university (female = 69%; male = 29%; 2% = non-binary; Asian = 23%, Black = 3%; Hispanic = 9%, white = 60%).

Our final prototype used a theoretically-grounded, empirically-validated CPS framework [1] which consists of three facets: shared knowledge construction, negotiation/coordination, and maintaining team function. This framework was applied in the context of Physics Playground [7], a physics problem-solving game. After collaborating in Physics Playground, teammates viewed their feedback in CPSCoach. The scores overview page included their scores for the current round and past scores. Participants could view a facet-specific page, which included a definition and indicative behaviors, and annotated videos of teams engaging in the behaviors.

Our CPS facet machine learning models were trained on a dataset of 94 triads playing Physics Playground [8]. We used the Bidirectional Encoder Representations from Transformers (BERT) model [9], with pre-trained word embeddings that were fine-tuned to our data. We chose hyper-parameters based on recommendations from [9] (e.g., fine-tuning over four epochs, batch size = 32, sequences padded or truncated at 300 words). We used team-level tenfold cross validation such that all utterances from a team were in the training set, or testing set, but never both. Using this approach, we achieved area under the receiver operating characteristic curve (AUROC) values of 0.88, 0.83, and 0.82 on our test data for shared knowledge construction, negotiation/coordination, and maintaining team function, respectively (chance = 0.5). To use the model for real-time assessment, we retrained the model with all the data, preserving the hyper-parameters. The final model yielded one probability score per facet that a given utterance was a positive example of that facet.

The probability scores were converted into interpretable metrics. Our scoring scheme used a normed-reference approach based on the training data. Given there are multiple utterances per CPS round, we averaged prediction probabilities across the round (per individual, per facet). We fit a probability density function on that average, after Winsorizing [10] the top and bottom 1.25% of outliers. We only used utterances in the training dataset from the first 10 min of the interaction as only 10 min of data is used in our CPSCoach. A final numeric score was determined by the cumulative distribution function where a score that is high, compared to the reference group, yields a higher value (range 0% to 100%). We also displayed ordinal feedback (low – bottom 25%, average – middle 50%, high – top 25%).

We recorded audio using Zoom, which records separate audio tracks for each teammate. After recording, we transcribed each participant's audio using IBM Watson [11] and the resultant utterances were submitted to the BERT model. We used only the first ten (of 15) minutes of a CPS interaction due to processing latency.

We conducted a user study with 40 students (20 dyads) from two large public universities in the United States (School One – N = 32, female = 55%, male = 42%, non-binary = 3%, Asian = 29%, Hispanic = 3%, white = 61%, other race = 3%; School Two – N = 8, female = 75%, male = 25%, Asian = 25%, Hispanic = 13%, Native American or Pacific Islander = 13%, white = 38%, other race = 13%). Participants completed a game tutorial and warmup, followed by a video orienting them to the upcoming task, CPS framework, and feedback. Teammates engaged in three to four rounds of collaborative gameplay and feedback, as time permitted (10 – 15 min each). They were given up to seven minutes to individually review feedback between rounds. They then completed a short survey on their perception of the accuracy of the scores, and usefulness of the tips for improving the scores. For example, they were asked: "Your score for sharing ideas and expertise was [score]%. How accurate was this score?" At the end of the study, participants completed the System Usability Scale [12], as a measure of general usability ($\alpha = 0.66$). All measures were completed on a five-point Likert scale. To conclude the study, we conducted semi-structured interviews to further understand participant's perception of CPSCoach. Other measures not germane to the present study were also collected.

3 Results and Discussion

To understand behavioral patterns as participants interacted with the feedback system, we examined average time spent on each element of CoachCPS (scores overview, facet text, and facet videos). We conducted two-tailed paired-samples t-tests between each round. Participants spent a similar amount of time on the score overview page across all three rounds (average of 52.84 s, $p > 0.05$). However, for the facet text, participants spent a third of the time on each subsequent round (round 1 = 61.86 s, round 2 = 41.44 s, round 3 = 21.99 s; $p < 0.05$ for round 1 to 2, $p < 0.01$ for round 2 to 3, $p < 0.001$ for round 1 to 3). There was a similar finding for the facet videos (round 1 = 159.29 s, round 2 = 110.55 s, round 3 = 42.88 s) where there were significant differences between rounds 1 and 3 ($p < 0.001$), and 2 and 3 ($p < 0.01$).

We also examined participants' subjective perceptions of the feedback system. First, we considered ratings of score accuracy and usefulness. For each participant, we averaged across all rounds, to obtain an overall accuracy and usefulness measure. We computed an overall usability rating for the CPS system as the average of items from the System Usability Scale. Descriptive statistics are shown in Table 1. We also conducted a thematic analysis [13] of the interviews, in order to gain a deeper understanding of participants' perceptions of CPSCoach.

Participants found the models to be somewhat accurate (a mean of 3.4 across all facets). A theme in the participant interviews was doubts about model accuracy, with 10% mentioning the accuracy of the shared knowledge construction and maintaining team function facets, and 10% expressing general doubt about score accuracy. For example, one user noted "I was kind of surprised that the sharing ideas and expertise score was pretty low for the first two rounds. I thought that we were sharing ideas pretty well for those first two rounds and it kind of surprised me that they were in the low area, especially when low is the bottom quarter." Participants also found the facet text and videos to be somewhat useful (average of 3.64 across facets). This sentiment was also reflected in the interviews, with 18% of participants mentioning they found the video examples helpful. For instance, one user noted: "Hearing the examples of what other people did I feel like was more helpful not so much reflecting on what I did last time, but what I can do next time instead." Ratings of usability of the feedback system were high (average of 4.24 out of 5), which was reflected in the interviews, as 20% of participants noted that the feedback was clear and easy-to-use.

Participants had suggestions for improving CoachCPS, such as examples of good and bad CPS from their own behavior (13%). 20% indicated they wanted the facet tips and videos to change each round, which gives them more examples to learn from, and 10% wanted indicator-level feedback. As an illustrative example of these two points, one participant made suggestions for how to provide indicator level feedback that changed each round: "I know it gives you the percentages, but in like positive team building, it said you did a really good job doing X but you could work on Y, something like that. So it feels like it's more based on the round instead of just the same ideas to improve every time…like within positive work environment asking for collaboration ideas you could say like oh like you did a good job asking for ideas but you could have done better with like specific solutions and stuff like that."

Our user study led us to two key lessons learned. First, *feedback should change each time a user views it, to encourage them to continuously improve their CPS.* We found that participants spent less time each round interacting with CPSCoach. This finding could be linked to increased familiarity with the content as rounds progressed. However, in interviews, participants requested different video examples each round, which could increase engagement and in turn support score improvement. Further feedback personalization, such as video examples extracted from the previous interaction, might aid engagement. This kind of personalization poses both technical and design challenges. Specific examples from the previous round of CPS require accurate models at fine-grained time intervals (e.g., 30 s rather than aggregated over a ten minutes). It is an open question as to how to automatically select examples from the interaction that are meaningful and can prompt positive behavior change.

Table 1. Descriptive statistics for subjective perception measures

Measure	M	SD	Min	Max
Construction of Shared Knowledge Accuracy	3.40	0.83	1	5
Construction of Shared Knowledge Usefulness	3.58	1.0	1	5
Negotiation/Coordination Accuracy	3.70	0.74	1.67	5
Negotiation/Coordination Usefulness	3.76	1.02	1	5
Maintaining Team Function Accuracy	3.05	0.84	1.33	4.33
Maintaining Team Function Usefulness	3.58	1.03	1	5
System Usability Scale	4.24	0.62	2.4	5

Our second lesson was that *increased transparency of the CPS feedback and underlying scoring mechanisms could increase users' trust in the feedback accuracy.* Participants rated usability of CPSCoach highly, but ratings of score accuracy were average. Facets with the two lowest accuracy ratings (shared knowledge construction and maintaining team function) were specifically mentioned in the interviews. Thus, there is a need for increased trust and transparency in the feedback. In order for participants to buy in to an AI system, the outcome of the system must be reasonably predictable [14]. Perhaps participants being surprised by the score resulted in the feedback being less effective. Additional training on the nuances of the CPS framework could help boost participant trust, by differentiating the facets they are being scored on from preconceived notions. Feedback designs proposed by participants (e.g., examples from their interactions) could also increase transparency of the underlying CPS model and in turn trust in the feedback.

Given that this is a first prototype CPSCoach, there are limitations that should be addressed. The purpose of this work was to design a prototype feedback system and conduct an initial user study. Accordingly, our study design did not include a control group, or formal efficacy evaluation. Findings from this study indicated several areas of improvement, thus it is prudent to make these changes before conducting an efficacy study. As collaborations increasingly move online, now, more than ever, we need to support teams CPS. Our work presents important steps towards this goal of developing systems that equip teammates with the skills they need to become more effective collaborative problem solvers.

This research was supported by the NSF National AI Institute for Student-AI Teaming (iSAT) under grant DRL 2,019,805. The opinions expressed are those of the authors and do not represent views of the NSF.

References

1. Sun, C., Shute, V.J., Stewart, A.E.B., Yonehiro, J., Duran, N., D'Mello, S.K.: Towards a generalized competency model of collaborative problem solving. Comput Educ. **143**, 103672 (2020). https://doi.org/10.1016/j.compedu.2019.103672
2. Graesser, A.C., Fiore, S.M., Greiff, S., Andrews-Todd, J., Foltz, P.W., Hesse, F.W.: Advancing the science of collaborative problem solving. Psychological Science in the Public Interest. **19**, 59–92 (2018). https://doi.org/10.1177/1529100618808244

3. Schulze, J., Krumm, S.: The “virtual team player”: a review and initial model of knowledge, skills, abilities, and other characteristics for virtual collaboration. Organizational Psychology Review. **7**, 66–95 (2017). https://doi.org/10.1177/2041386616675522
4. Čubraniundefined, D., Storey, M.A.D., Čubranić, D., Storey, M.A.D.: Collaboration support for novice team programming. In: Proceedings of the 2005 International ACM SIGGROUP Conference on Supporting Group Work, pp. 136–139. Association for Computing Machinery, New York, NY, USA (2005). https://doi.org/10.1145/1099203.1099229
5. Faucett, H.A., Lee, M.L., Carter, S.: I Should Listen More: Real-time Sensing and Feedback of Non-Verbal Communication in Video Telehealth. Proc. ACM Hum.-Comput. Interact. **1**, 44:1--44:19 (2017). https://doi.org/10.1145/3134679
6. Shute, V.J.: Focus on formative feedback. Rev Educ Res. **78**, 153–189 (2008). https://doi.org/10.3102/0034654307313795
7. Shute, V.J., Smith, G., Kuba, R., Dai, C.-P., Rahimi, S., Liu, Z., Almond, R.: The design, development, and testing of learning supports for the physics playground game. Int J Artif Intell Educ. **31**, 357–379 (2021). https://doi.org/10.1007/s40593-020-00196-1
8. Stewart, A.E.B., Keirn, Z., D'Mello, S.K.: Multimodal modeling of collaborative problem-solving facets in triads. User Model User-adapt Interact. **31**, 713–751 (2021). https://doi.org/10.1007/s11257-021-09290-y
9. Wolf, T., et al.: HuggingFace's Transformers: State-of-the-art Natural Language Processing (2019)
10. Dixon, W.J., Yuen, K.K.: Trimming and winsorization: A review. Statistische Hefte. **15**, 157–170 (1974). https://doi.org/10.1007/BF02922904
11. IBM: https://www.ibm.com/watson/services/speech-to-text/, last accessed 01 May 2018
12. Brooke, J.: others: SUS-A quick and dirty usability scale. Usability evaluation in industry. **189**, 4–7 (1996)
13. Blandford, A., Furniss, D., Makri, S.: Qualitative HCI Research: Goaing Behind the Scenes. Morgan & Claypool (2016). https://doi.org/10.2200/S00706ED1V01Y201602HCI034
14. Bostrom, N., Yudkowsky, E.: The ethics of artificial intelligence. The Cambridge handbook of artificial intelligence. **1**, 316–334 (2014)

Development of Virtual Reality SBIRT Skill Training with Conversational AI in Nursing Education

Jinsil Hwaryoung Seo[1(✉)], Rohan Chaudhury[1], Ja-Hun Oh[1], Caleb Kicklighter[1], Tomas Arguello[1], Elizabeth Wells-Beede[2], and Cynthia Weston[2]

[1] Soft Interaction Lab, School of Performance, Visualization and Fine Arts, Texas A&M University, College Station, TX 77843, USA
hwaryoung@tamu.edu

[2] School of Nursing, Texas A&M University, College Station, TX 77843, USA

Abstract. The paper presents the process of conversational AI patient design and development for SBIRT skills training. SBIRT (Screening, Brief Intervention, and Referral to Treatment) is a comprehensive public behavior health approach that is commonly used by nurses and social workers to detect potential substance abuse in their patients. In the VR exam room, a nursing student practices SBIRT skills with a virtual patient powered by a conversational AI system. The development of the VR patient system was started by collecting sample dialogs from a standardized patient and a nurse practitioner. In addition, extended conversations were collected through prototypes with different interaction modes. With the intelligent virtual patient, the "SBIRT VR Training Program" provides the user with a diverse selection of simulated environments and personalized training. Our research focuses on the efficacy of a VR-based conversational AI training on a student's acquisition and retention of the SBIRT training material.

Keywords: Virtual Reality · Conversational AI · SBIRT Training · Nursing Education

1 Introduction

The demand for nursing professionals trained in SBIRT (Screening, Brief Intervention, and Referral to Treatment) has been on the rise as substance abuse has become a prevalent issue in society. SBIRT is a comprehensive public behavioral health approach that nurses and social workers use to detect potential substance abuse in patients. However, the lack of qualified staff and standard patients present problems in the current medical education programs. Therefore, the development of new teaching solutions is essential to ease the burden and provide the necessary training to meet the increasing demand. In this paper, we concentrate on creating a virtual reality (VR) based conversational training application that supports nursing students to practice SBIRT skills with virtual patients (VPs). VPs are virtual interactive agents who simulate clinical situations with high realism [1]. Without arranging with actual patients, this enables trainees to exercise their

N. Wang et al. (Eds.): AIED 2023, LNAI 13916, pp. 701–707, 2023.
https://doi.org/10.1007/978-3-031-36272-9_59

abilities in a realistic setting. Smart VPs have the potential to be an invaluable resource because of their realistic scenarios, remote access capabilities, cost effectiveness, and individualized training, even though in-person training has shown to be the most productive way to learn healthcare-related topics. Studies have already shown the benefits of combining traditional classroom learning with virtual simulation tools [2]. This paper presents the interactive VR training application that allows novice behavioral health nurses to conduct an SBIRT training with a VP that emulates an adult male with alcohol abuse.

1.1 Conversational AI in Healthcare Education

Conversational Artificial Intelligence (AI) in healthcare education involves the incorporation of natural language processing (NLP) and machine learning (ML) techniques to facilitate communication between VPs and healthcare providers through chatbots or virtual assistants [3]. These AI-based tools usually assist with scheduling appointments, answering common medical questions, and providing symptom checker services. They also provide a safe and cost-effective way for students to gain clinical experience through simulations of a variety of symptoms and medical conditions. Additionally, AI-powered tutors can provide personalized feedback and guidance to students as they learn, helping them to improve their knowledge and skills. The use of conversational AI in healthcare education is a relatively new field of study, but it has been gaining significant attention in recent years due to advancements in NLP and ML technologies [4]. However, it's important to note that AI-powered conversational systems must be designed, implemented, and evaluated with a user-centered approach, to ensure they meet the needs and expectations of both students and healthcare professionals [5]. Moreover, the use of AI in healthcare education should be accompanied by ethical considerations, ensuring the protection of patient's privacy and safety [6].

2 Interaction Design and Implementation of the Application

The process of design and development of SBIRT VR training application consists of several steps: conversation data collection, interaction design, 3D environment/virtual character development, and AI system development. In this section, we are going to focus on the AI system development including data collection and interaction design processes.

2.1 SBIRT Conversation Data Collection

Collecting sufficient data for training a VP is significantly challenging, as NLP algorithms require large amounts of high-quality data to learn and improve. The development of a conversational AI system for VPs began by gathering sample dialogs from traditional video materials: recordings from a standardized patient and a nurse practitioner. Additionally, conversation data was collected from role-playing workshops with nursing faculty members, where they were assigned different roles, such as a virtual patient and a nurse. These sample dialogs were used to create prototypes to collect more conversation data. To further increase the amount of data for training, the conversation data was expanded using a paraphrasing text generation module, similar to the method proposed by Conneau et al. [7]. The text generation module used was an expansion of sequence-to-sequence text generation, implemented using a bidirectional encoder, which allows for more efficient and accurate text generation. Overall, the combination of different interaction methods and the paraphrasing text generation module helped overcome the challenge of collecting training data for the AI-powered VP.

2.2 Design of Interaction Modes

The research team experimented three interaction modes for SBIRT VR: Mode 1 - interacting with a human-controlled VP, Mode 2 - interacting a web-based VP, Mode 3 - interacting with a VP powered by conversational AI. This was initially planned to collect more data, but different interaction modes have unique characteristics that could benefit users. These different modes of interaction provide users to choose the one that best suits their learning situation and objectives. In Mode 1 & 3, multiple students can join the same virtual room together, so that a large group can experience the scenario at once (Fig. 1). In addition, several rooms can be run in parallel, pairing up any number of students with an instructor.

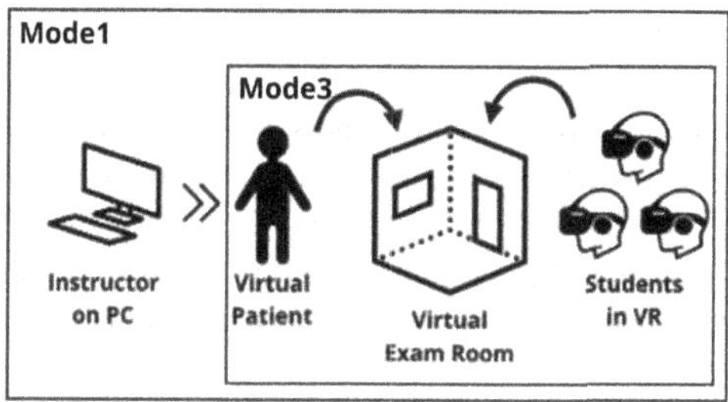

Fig. 1. Interaction structures for Mode 1 & 3

Mode 1: Human-controlled VP mode. In this mode, a nursing student interacts with a VP who is controlled by a remote instructor. The interaction is facilitated by two applications: an administrative application (PC-Admin) controlled by the instructor on a personal computer and a VR application (VR-User) for the Oculus Quest 2 used by the student. These two applications work together to create a simulated experience that allows an instructor and student to run through an SBIRT scenario together in a shared virtual space. The instructor controls the VP character using the PC-Admin application, while the student practices their SBIRT techniques by speaking directly to the "patient" in VR. The student experiences a simulated exam room, while the instructor observes from an omniscient camera view on the PC (Fig. 2). In this mode, the instructor adds value to interacting with a VP through the creative control that the instructor has in directing the lesson. The instructor can switch between roleplaying as the patient and intervening as the teaching expert. Through the PC-Admin application, the instructor is given a script for a pre-made scenario that they can follow, complete with the patient's background information and a menu of context-appropriate responses that they can choose from. When selected, each text response will cause the VP in the VR-User app to speak to the student in that character's voice using RT voice and Salsa units. However, if the script provides no suitable option, the instructor can type out a custom response for the character to say. In this interaction mode, conversations done by VR nurse users and instructors were collected and used to improve the AI system.

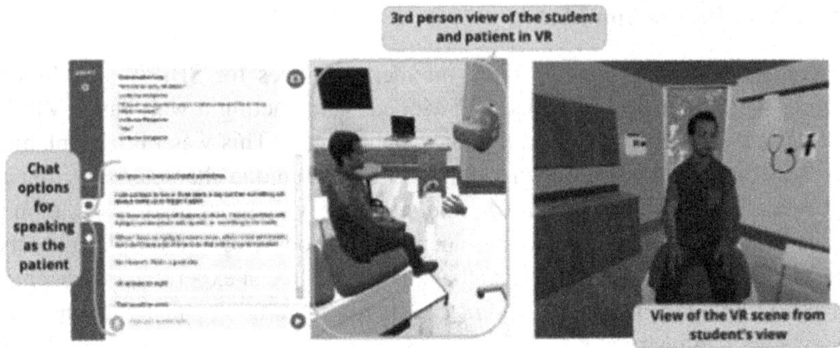

Fig. 2. Screen Shots of Nursing Virtual Reality Training Program's PC-Admin and VR-User

Mode 2: Web-based VP AI Mode. The Immersive Virtual System for education offers many benefits, such as a distraction-free and realistic environment. However, it requires VR equipment, which is not yet widely available. To address this issue, we developed a web-based system that utilizes conversational AI through a chat interface. This system also allows for the collection of a significant number of interactive dialogs (Fig. 3).

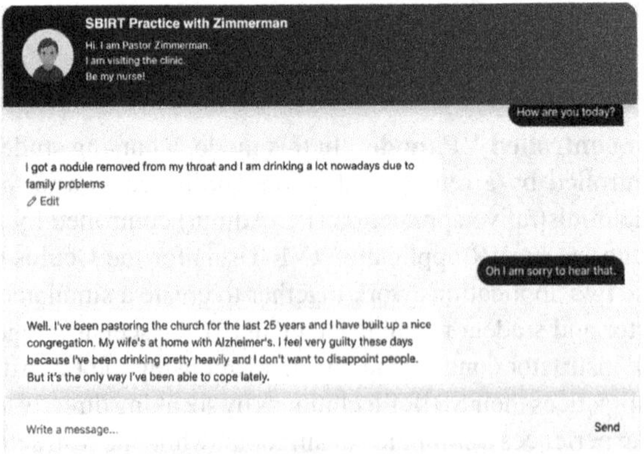

Fig. 3. Web-based VP

Chat Interface. This is a front-end interface that enables participants to interact with VP through a familiar chat format. It includes a text input field for students to type their message, as well as a dialog section where all chat conversations are displayed. When a message is submitted, it is sent to a response-generating computation to determine the most suitable response, which is then displayed on the website.

Prompt evaluation and modification. We have incorporated prompt evaluation to enable real-time modification of the chatbot's responses. This feature allows users to edit the AI agent's response if they find it to be misleading or out of context. These modifications are stored in a database that can be accessed by faculty and administrators. This human-in-the-loop architecture allows for timely modifications, preventing users from getting stuck in an interaction due to inappropriate responses from the conversational AI.

3 Mode 3: Virtual Patient Conversational AI Mode.

The team collected conversation data from over 100 training sessions to develop conversational AI models for Mode 3. The intent set includes over 300 questions and 100 responses. This mode enables nursing students to practice SBIRT skills directly with a VP powered by the conversational AI system in VR. Nursing students interact with the VP by speaking into the microphone of the VR headset, which records the audio. The audio is then sent to a Google speech recognition engine, which converts it to text. The text is then sent to the conversational AI models, which selects an appropriate response based on the intents and context of the scenario. The response is then sent to real-time voice and Salsa units, allowing the virtual character to speak with lip sync to the sound.

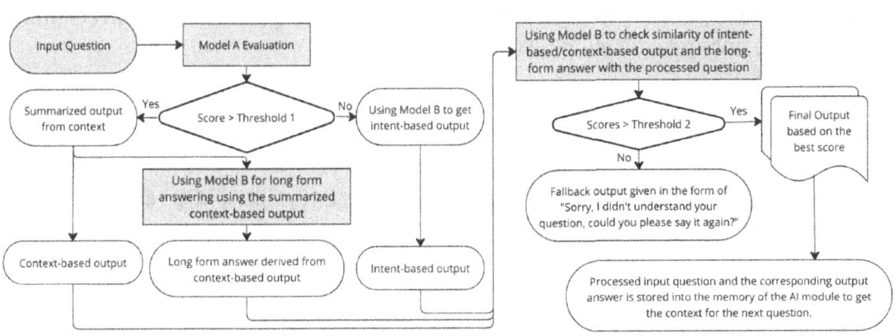

Fig. 4. Conversational AI model workflow

The response selection process is based on a conversational AI module that includes two AI models from Hugging Face [8]: (1) Model A (Deepset/Roberta-base-squad2) is a NLP model trained on question-answer pairs (SQuAD2.0 dataset [9]), including unanswerable questions, for the task of question answering [10]; (2) Model B (sentence-transformers/paraphrase-xlm-r-multilingual-v1) maps sentences and paragraphs to a 768-dimensional dense vector space and can be used for tasks like clustering or semantic search [11]. When a new question is received from the speech recognition module, it is fed to Model A, which gives a context-based output. If Model A's confidence score is greater than threshold 1, this context-based output is used to generate a long-form answer using Model B and a database of previously stored answers by comparing the cosine similarities of the vectorized outputs. If Model A's confidence score is less than threshold 1, the input question is fed to Model B, and the vectorized output is compared to a database of previously stored questions and answers. The answer corresponding to the question in the database with the highest cosine similarity with the input question is taken as a candidate output (intent-based output). Finally, the context-based output, the long-form answer, and the intent-based output are compared with the input question using Model B and cosine similarity, and the one with the highest cosine similarity is given as the final output, provided that it is greater than threshold 2. If these cosine similarity values are less than threshold 2, the model returns a variation of "unknown input" as the output. For the next turn of question-answering, the processed input question and output from the present turn is passed as memory into context for Model A. This allows the model to remember what questions were asked and what were the corresponding responses in previous turns and answer accordingly in present turns. The entire workflow is depicted in Fig. 4.

4 Conclusion

As we noticed during a preliminary user study, users of the SBIRT VR application show improvement in their decision-making and communication skills, as well as their ability to conduct an SBIRT screening. Our VP simulation system offers a practical and efficient resource for SBIRT education that allows interactive, collaborative, and immersive learning. The advancements achieved through this project will provide new ideas in how we approach nursing education, and the capabilities VR and Conversational AI have in the nursing education field.

References

1. Triola, M., Feldman, H., Kalet, A.L., Zabar, S., Kachur, E.K., Gillespie, C., et al.: A randomized trial of teaching clinical skills using virtual and live standardized patients. J. Gen. Intern. Med. **21**(5), 424–429 (2006)
2. Padilha, J.M., Machado, P.P., Ribeiro, A., Ramos, J., Costa, P.: Clinical virtual simulation in nursing education: randomized controlled trial. Journal of Medical Internet Research **21**(3) (2019)
3. de Achini Adikari, D., et al.: Empathic conversational agents for real-time monitoring and co-facilitation of patient-centered healthcare. Futur. Gener. Comput. Syst. **126**, 318–329 (2022)
4. Bohr, A., Memarzadeh, K.: The rise of artificial intelligence in healthcare applications. In: Artificial Intelligence in Healthcare, pp. 25–60 (2020). https://doi.org/10.1016/B978-0-12-818438-7.00002-2. Epub 2020 Jun 26
5. Nadarzynski, T., Miles, O., Cowie, A., Ridge, D.: Acceptability of artificial intelligence (AI)-led chatbot services in healthcare: A mixed-methods study. Digital health 5 (2019)
6. Gerke, S., Minssen, T., Cohen, G.: Ethical and legal challenges of artificial intelligence-driven healthcare. In: Artificial Intelligence in Healthcare, pp. 295–336. Academic Press (2020)

7. Conneau, J.S., Lample, G., Rault, L., Denoyer, L.: Paraphrasing for data augmentation in neural machine translation (2018)
8. Hugging Face: (n.d.). Retrieved from https://huggingface.co/
9. Hugging Face: SQUAD v2 (n.d.). Retrieved from https://huggingface.co/datasets/squad_v2
10. Hugging Face: deepset/roberta-base-squad2 (n.d.). Retrieved from https://huggingface.co/deepset/roberta-base-squad2
11. Hugging Face: sentence-transformers/paraphrase-xlm-r-multilingual-v1 (n.d.). Retrieved from https://huggingface.co/setence-transformers/paraphrase-xlm-r-multilingual-v1

Balancing Test Accuracy and Security in Computerized Adaptive Testing

Wanyong Feng(✉), Aritra Ghosh, Stephen Sireci, and Andrew S. Lan

University of Massachusetts Amherst, Amherst, USA
wanyongfeng@umass.edu

Abstract. Computerized adaptive testing (CAT) is a form of personalized testing that accurately measures students' knowledge levels while reducing test length. Bilevel optimization-based CAT (BOBCAT) is a recent framework that learns a data-driven question selection algorithm to effectively reduce test length and improve test accuracy. However, it suffers from high question exposure and test overlap rates, which potentially affects test security. This paper introduces a constrained version of BOBCAT to address these problems by changing its optimization setup and enabling us to trade off test accuracy for question exposure and test overlap rates. We show that C-BOBCAT is effective through extensive experiments on two real-world adult testing datasets.

1 Introduction

Compared to conventional testing/assessment mechanisms, computerized adaptive testing (CAT) is a form of personalized testing that adaptively selects the next question based on students' answers to previous questions to reduce the test length effectively [6,14,15]. A CAT system is generally comprised of the following components: a knowledge level estimator that estimates a student's current knowledge level given the answers to the previous questions, a response model that predicts how likely a student answers the question correctly given the knowledge level estimate and question features, and a question selection algorithm that selects the next most informative question given the output of response model [2]. Despite the fact that CAT has been widely applied to real-world tests, an important limitation is that most of the existing question selection algorithms are static, which means that they cannot be improved over time as more and more students go through tests. Recently, researchers proposed several conceptual ideas for learning a data-driven question selection algorithm [8].

Bilevel Optimization-Based Computerized Adaptive Testing (BOBCAT) [1] (and extensions [16]) learns a data-driven question selection algorithm while supporting different response models. BOBCAT learns the question selection algorithm by solving the following bilevel optimization problem: the inner level optimization updates the student's knowledge level estimate using questions selected by the question selection algorithm, while the outer level optimization updates the question selection algorithm and response model by evaluating the latest

N. Wang et al. (Eds.): AIED 2023, LNAI 13916, pp. 708–713, 2023.
https://doi.org/10.1007/978-3-031-36272-9_60

knowledge level estimate on held-out data, i.e., student responses to questions not seen by the inner optimization. BOBCAT has several advantages over conventional CAT: i) BOBCAT can achieve the same knowledge level estimate accuracy with fewer questions selected, and ii) the question selection algorithm is agnostic of the response model, which enables the test administrators to explore the best combination that meets their needs.

Although BOBCAT has several advantages over conventional CAT, the learned question selection algorithm suffers from high question exposure and test overlap rates. The high question exposure rate means some questions are constantly being selected and the high test overlap rate means different students may receive similar test contents. This limitation could potentially affect the validity and fairness of the test by nefarious test preparation activities that focus on harvesting items administered to previous test takers [5]. Existing methods that address this issue are designed for traditional CAT settings [13], by either introducing randomness in question selection [4] or using maximum clique algorithms [12] and cannot be applied to data-driven CAT settings.

Contributions. In this paper, we propose a constrained version of BOBCAT (C-BOBCAT) that can trade off test accuracy for question exposure and test overlap rates. C-BOBCAT i) uses a stochastic question selection algorithm instead of a deterministic one and ii) adds a penalty term to BOBCAT's optimization objective to promote the learned question selection algorithm to select diverse questions across different students. Therefore, we can trade off test accuracy for question exposure and test overlap rates via a temperature hyper-parameter. We validate the effectiveness of C-BOBCAT through experiments on two real-world adult test datasets and our implementation will be publicly available at: https://github.com/umass-ml4ed/C-BOBCAT.

2 Methodology

We first review BOBCAT and then detail the C-BOBCAT framework.

2.1 BOBCAT Background

The goal of BOBCAT is to solve the following bilevel optimization problem:

$$\underset{\gamma,\phi}{\text{minimize}} \ \frac{1}{N}\textstyle\sum_{i=1}^{N}\sum_{j\in\Gamma_i}\ell\Big(Y_{i,j}, g(j;\boldsymbol{\theta}_i^*)\Big) \tag{1}$$

$$\text{s.t. } \boldsymbol{\theta}_i^* = \underset{\boldsymbol{\theta}_i}{\text{argmin}} \textstyle\sum_{t=1}^{T}\ell\Big(Y_{i,j_i^{(t)}}, g(j_i^{(t)};\boldsymbol{\theta}_i)\Big) + \mathcal{R}(\gamma, \boldsymbol{\theta}_i) \tag{2}$$

$$\text{where } j_i^{(t)} \sim \Pi(Y_{i,j_i^{(1)}}, \ldots, Y_{i,j_i^{(t-1)}}; \boldsymbol{\phi}), \tag{3}$$

where $\Pi(\cdot)$ represents the question selection algorithm with parameters $\boldsymbol{\phi}$. The inputs are the responses of a particular student i to the previously selected questions $(Y_{i,j_i^{(1)}}, \ldots, Y_{i,j_i^{(t-1)}})$, and the output is the next selected question index $j_i^{(t)}$. $g(\cdot)$ represents the global response model with parameters γ. γ contains both the prior mean of student knowledge level estimate and question difficulties. The inputs are the student's knowledge level estimate $\boldsymbol{\theta}_i$ and question id j, and the output is the student's likelihood of responding to the question correctly. The global response model can either be IRT-based or neural-network-based. Within a batch of N students, BOBCAT splits each student i's response questions into training part Ω_i and held-out meta part Γ_i. For the inner-level optimization problem in (2), the question selection algorithm selects the next informative question from Ω_i. The current student knowledge level estimate $\boldsymbol{\theta}_i^*$, which is the local adaption to the prior mean of student knowledge level estimate, is calculated as a function of $\boldsymbol{\phi}$ and γ by minimizing the summation of binary cross-entropy loss $\ell(.)$ on the selected questions and penalty function $\mathcal{R}(\gamma, \boldsymbol{\theta}_i)$ that prevents large deviation. For the outer level optimization problem in (1), global response model and question selection algorithm are updated by minimizing the binary cross entropy loss $\ell(.)$ on the held-out meta data with the current student knowledge level estimate $\boldsymbol{\theta}_i^*$ as input. For further details, refer to [1].

2.2 C-BOBCAT

We now detail the modifications to BOBCAT. We turn BOBCAT's underlying question selection algorithm from deterministic to stochastic, which injects some randomness into the questions selected for each student. Specifically, we transform the original categorical question selection distribution to the Gumbel-Softmax distribution [3,7] with a fixed temperature hyperparameter. Moreover, since the entropy of a distribution increases as the distribution approaches the uniform distribution, maximizing the entropy of the categorical question selection distribution can further encourage the learned question selection algorithm to select a diverse set of questions for each student. Specifically, we add the negative summation of the entropy of categorical question selection distributions of all selected questions for each student in (5) to the outer level optimization function in (1) to create the new outer level function in (4).

$$\underset{\gamma,\boldsymbol{\phi}}{\text{minimize}} \ \frac{1}{N}\sum_{i=1}^{N}\sum_{j\in\Gamma_i}\ell\Big(Y_{i,j}, g(j;\boldsymbol{\theta}_i^*)\Big) - \lambda H_i(\boldsymbol{\phi}) \tag{4}$$

$$\text{where } H_i(\boldsymbol{\phi}) = \sum_{t=1}^{T} H\Big(\Pi(Y_{i,j_i^{(1)}}, \ldots, Y_{i,j_i^{(t-1)}}; \boldsymbol{\phi})\Big) \tag{5}$$

During the training process, the model needs to maximize the combination of both prediction accuracy on the held-out meta data and the uncertainty of the question selection algorithm, which is reflected in the entropy regularization term. This balance is controlled by the value of the hyperparameter λ: when $\lambda = 0$, the problem reduces to the original BOBCAT bi-level optimization problem; when $\lambda = \infty$, the solution to the problem is a question selection algorithm that selects each question with equal probability, i.e., the entropy is maximized when the question selection distribution is uniform distribution [10]. In practice, test administrators can explore different values of λ and select a suitable value, to achieve a desirable balance between maximizing the accuracy of the test and maintaining acceptable question exposure and test overlap rates, according to the requirement of each testing scenario.

3 Experiments

We now detail experiments we conducted on two real-world adult test datasets to validate C-BOBCAT's effectiveness.

3.1 Data, Experimental Setup, Baseline, and Evaluation Metrics

We adopt two new publicly available datasets collected from the Massachusetts Adult Proficiency Test (MAPT), the reading comprehension test (MAPT-Read), and the math test (MAPT-Math). In both datasets, there are a total of more than 90 K students, 1.7 K questions, and 4 M question responses [11]. We use these datasets since they are collected in real tests while datasets used in the original BOBCAT paper are collected during longer periods of student learning. As a result, the MAPT datasets are closer to the actual real-world situation addressed by CAT, i.e., a test taker's ability roughly remains unchanged during a short test. We choose the gradient-calculation-based method from BOBCAT to learn the question selection algorithm with C-BOBCAT framework (**C-BINN-Approx** and **C-BIIRT-Approx**). We split the datasets over students into train, validation, and test sets with the ratio of 6 : 2 : 2 respectively. For each student i, we partition the questions they answer (and their responses) into the training part (Ω_i, 80%) and the held-out meta part (Γ_i, 20%). We choose the IRT-based active learning method (**IRT-Active**), and the IRT-based random question selection method (**IRT-Random**) as the baseline models. We use the area under the receiver operating characteristics curve (**AUC**) as metrics to evaluate the predictive performance on the held-out meta data Γ_i. We use the Scaled Chi-square Statistics of the question exposure rate (**EXPOSE-CHI**) and the average value of test overlap rate between every two students (**OVERLAP-MU**) to measure question exposure and test overlap rates [9].

3.2 Results and Discussion

In Fig. 1, we use a series of plots to demonstrate how C-BOBCAT trades off test accuracy (AUC on held-out meta data) versus test security (EXPOSE-PHI and OVERLAP-MU). For both C-BINN-Approx and C-BIIRT-Approx, each point on the curve corresponds to a specific value of the hyperparameter λ. The baselines correspond to single points since they do not support this tradeoff. We observe that AUC and EXPOSE-PHI are positively correlated in the first column of the graph. Both metrics reach the minimum as λ approaches infinity, i.e., when the categorical question selection distribution's entropy is maximized. In this scenario, the question selection algorithm selects each question with equal

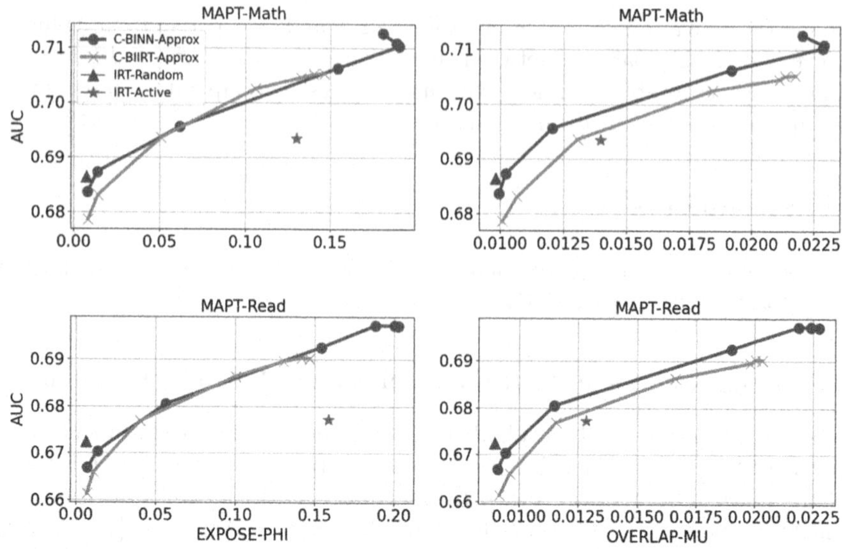

Fig. 1. C-BOBCAT can effectively trade off test accuracy (y-axis, AUC on held-out meta data) for test security (x-axes, EXPOSE-PHI and OVERLAP-MU) on both datasets.

probability, which decreases both the question exposure rate and test accuracy. Both metrics reach the maximum as λ approaches zero, i.e., when the categorical question selection distribution's entropy is minimized. In this scenario, the question selection algorithm greedily selects the question that leads to the highest test accuracy, which also results in a high question exposure rate. A similar relationship can also be found between OVERLAP-MU and AUC in the second column of the graph. More importantly, with the same EXPOSE-PHI and OVERLAP-MU values, both C-BINN-Approx and C-BIIRT-Approx achieve higher AUC values than the IRT-Active baseline. This observation implies C-BOBCAT is agnostic of the underlying response model and always achieves a better balance between test accuracy and question exposure and test overlap rates than the IRT-Active baseline.

4 Conclusions and Future Work

In this paper, we proposed C-BOBCAT, a framework that attempts to strike a balance between test accuracy and security in CAT settings, which we demonstrated to be effective via experiments on two real-world datasets. Avenues of future work include i) investigating the effect of combining questions' info with the responses from students as the new inputs to the question selection algorithm, ii) for multiple choice questions, investigating the effect of predicting the exact option that students select instead of correctness in both inner and outer level optimization, and iii) deploying C-BOBCAT in real-world tests.

Acknowledgements. The authors thank the NSF (under grants 1917713, 2118706, 2202506, 2215193) and Schmidt Futures for partially supporting this work.

References

1. Ghosh, A., Lan, A.: BOBCAT: bilevel optimization-based computerized adaptive testing. In: Proceedings of the Thirtieth International Joint Conference on Artificial Intelligence (IJCAI), pp. 2410–2417 (2021)
2. Han, K.C.T.: Components of the item selection algorithm in computerized adaptive testing. J. Educ. Eval. Health Prof. **15**, 1149084 (2018)
3. Jang, E., Gu, S., Poole, B.: Categorical reparameterization with Gumbel-Softmax. In: ICLR (2017)
4. Kingsbury, C.G., Zara, A.R.: A comparison of procedures for content-sensitive item selection in computerized adaptive tests. Appl. Measur. Educ. **4**(3), 241–261 (1991)
5. Lee, H., Dodd, B.G.: Comparison of exposure controls, item pool characteristics, and population distributions for cat using the partial credit model. Educ. Psychol. Measur. **72**(1), 159–175 (2012)
6. Linden, W.J., Glas, G.A.W. (eds.): Computerized Adaptive Testing: Theory and Practice. Springer, Dordrecht (2000). https://doi.org/10.1007/0-306-47531-6
7. Maddison, C.J., Mnih, A., Teh, Y.W.: The concrete distribution: a continuous relaxation of discrete random variables. In: ICLR (2017)
8. Nurakhmetov, D.: Reinforcement learning applied to adaptive classification testing. In: Veldkamp, B.P., Sluijter, C. (eds.) Theoretical and Practical Advances in Computer-based Educational Measurement. MEMA, pp. 325–336. Springer, Cham (2019). https://doi.org/10.1007/978-3-030-18480-3_17
9. Ozturk, N.B., Dogan, N.: Investigating item exposure control methods in computerized adaptive testing (2015)
10. Shannon, C.E.: A mathematical theory of communication. Bell Syst. Tech. J. **27**(3), 379–423 (1948)
11. Sireci, S.G., et al.: Massachusetts adult proficiency tests technical manual, version 2. Center for Educational Assessment Research Report No 677 (2008)
12. Ueno, M., Miyazawa, Y.: Uniform adaptive testing using maximum clique algorithm. In: Isotani, S., Millán, E., Ogan, A., Hastings, P., McLaren, B., Luckin, R. (eds.) AIED 2019. LNCS (LNAI), vol. 11625, pp. 482–493. Springer, Cham (2019). https://doi.org/10.1007/978-3-030-23204-7_40
13. Ueno, M., Miyazawa, Y.: Two-stage uniform adaptive testing to balance measurement accuracy and item exposure. In: Rodrigo, M.M., Matsuda, N., Cristea, A.I., Dimitrova, V. (eds.) Artificial Intelligence in Education (AIED 2022). LNCS, vol. 13355, pp. 626–632. Springer, Cham (2022). https://doi.org/10.1007/978-3-031-11644-5_59
14. Vats, D., Lan, A.S., Studer, C., Baraniuk, R.G.: Optimal ranking of test items using the Rasch model. In: 2016 54th Annual Allerton Conference on Communication, Control, and Computing (Allerton), pp. 467–473. IEEE (2016)
15. Vats, D., Studer, C., Lan, A.S., Carin, L., Baraniuk, R.: Test-size reduction for concept estimation. In: Educational Data Mining 2013 (2013)
16. Zhuang, Y., Liu, Q., Huang, Z., Li, Z., Shen, S., Ma, H.: Fully adaptive framework: neural computerized adaptive testing for online education (2022)

A Personalized Learning Path Recommendation Method for Learning Objects with Diverse Coverage Levels

Tengju Li[1], Xu Wang[1], Shugang Zhang[2], Fei Yang[3], and Weigang Lu[1,2](✉)

[1] Department of Educational Technology, Ocean University of China, Qingdao 266100, China
luweigang@ouc.edu.cn

[2] College of Computer Science and Technology, Ocean University of China, Qingdao 266100, China

[3] School of Mechanical, Electrical, and Information Engineering, Shandong University, Weihai 264200, China

Abstract. E-learning has resulted in the proliferation of educational resources, but challenges remain in providing personalized learning materials to learners amidst an abundance of resources. Previous personalized learning path recommendation (LPR) methods often oversimplified the competency features of learning objects (LOs), rendering them inadequate for LOs with diverse coverage levels. To address this limitation, an improved learning path recommendation framework is proposed that uses a novel graph-based genetic algorithm (GBGA) to optimize the alignment of features between learners and LOs. To evaluate the performance of the method, a series of computational experiments are conducted based on six simulation datasets with different levels of complexity. The results indicate that the proposed method is effective and stable for solving the LPR problem using LOs with diverse coverage levels.

Keywords: Personalized learning · Learning path recommendation · Learning object model · Genetic algorithms

1 Introduction

The rapid growth of e-learning has led to the proliferation of high-quality educational resources [9]. Without proper organization, the abundance of learning resources can lead to information overload and learning disorientation for e-learners, negatively impacting their learning experience. To optimize learning experience for e-learners, personalized learning path recommendation (LPR) is necessary [6]. Learning paths, composed of a carefully organized series of learning objects (LOs), enhance both learning effectiveness and efficiency [3].

While a number of methods have been proposed for LPR [8], it is worth noting that prior studies have oversimplified certain features of LOs, particularly their competency features, which may not accurately reflect the complexity of real-world e-learning environments. Competency features refer to both

N. Wang et al. (Eds.): AIED 2023, LNAI 13916, pp. 714–719, 2023.
https://doi.org/10.1007/978-3-031-36272-9_61

the prerequisites and learning outcomes for each individual learning unit [4]. In this paper, we define competency as the number of concepts that a learner can acquire by studying a particular LO. It is evident that educational resources may contain LOs covering varying numbers of concepts [5], but previous approaches have generally only considered LOs covering a single concept to simplify the LPR problem [1,2,7]. Consequently, LOs with diverse coverage levels in educational resources are incompatible with these methods. To address this issue, we propose an improved LPR framework that employs a graph-based genetic algorithm (GBGA) to generate learning paths consisting of diverse LOs.

2 The Learning Path Recommendation Framework

We have modified the LPR framework previously proposed by Bian et al. [2]. Figure 1 shows our proposed LPR process divided into three stages. The first stage focuses on the concept level, aiming to generate a target subgraph for the learner from the concept map. A target subgraph generation algorithm is applied, utilizing a breadth-first traversal [3] in the concept map, starting from the learning target set and ending at the prior knowledge set. The obtained concepts and relationships are added to the subgraph to generate a learner-specific target subgraph representing all the concepts that need to be learned.

In the second stage, LOs with diverse coverage levels are matched to the target subgraph to generate a LO map by a graph-based genetic algorithm. The generated LO map not only preserves the dependency relationships between concepts, but also maximizes the feature matching between learners and LOs. We describe this algorithm in Sect. 3.

During the third stage, the LO map is converted into a linear LO sequence via a classical topological sort algorithm. The algorithm primarily involves iteratively executing the following steps until all nodes are sorted: (1) selecting a node with an in-degree of 0 and outputting it, and (2) removing the selected node and all of its outgoing edges from the graph.

3 The Graph-Based Genetic Algorithm

The genetic algorithm (GA) is a metaheuristic algorithm. It is a powerful technique that has been applied with fair success to various optimization problems. In order to apply the GA to the problem of generating LO maps, we propose a graph-based genetic algorithm. The GBGA differs from the classic GA in two important aspects. First, a special encoding scheme is used in order to make the relevant structures of the LO map become genes in chromosomes. Second, given the encoding, special genetic operators are used.

3.1 Problem Modeling

GBGA is used to generate the LO map, maximizing matching with the learner's features. For learner u_i, the tuple $\left(f_{CL}^{(u_i)}, f_{MP}^{(u_i)}, f_{PT}^{(u_i)}\right)$ represents the features in

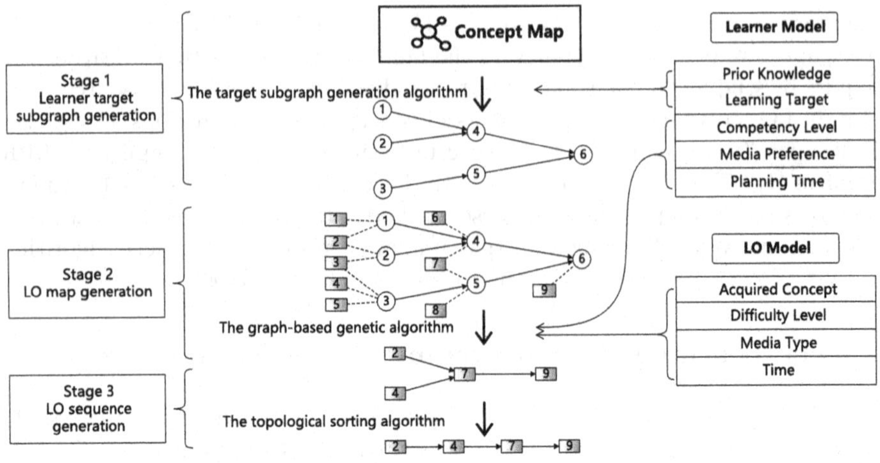

Fig. 1. The learning path recommendation framework

terms of Competency level, Media preference, and Planning time. The set of LOs is denoted by $X = \{x_1, x_2, \ldots, x_m\}$, and certain LO features need to be collected to match the features of learners. The tuple $\left(f_{DL}^{(x_i)}, f_{MT}^{(x_i)}, f_T^{(x_i)}\right)$ represents the features of x_i in terms of difficulty level, media type, and required time.

Assuming that there are N LOs in a LO map, the following objective functions are formulated to measure how well the LO map matches the learner:

$$F_1 = \frac{1}{N} \sum_{x_i \in X_{u_i}} \left| f_{CL}^{(u_i)} - f_{DL}^{(x_i)} \right| \tag{1}$$

$$F_2 = \frac{2}{3N} \sum_{x_i \in X_{u_i}} \operatorname{Pos}\left(f_{MP}^{(u_i)}, f_{MT}^{(x_i)}\right) \tag{2}$$

$$F_3 = \max\left(0, \sum_{x_i \in X_{u_i}} f_T^{(x_i)} - f_{PT}^{(u_i)}\right) \tag{3}$$

where $\operatorname{Pos}\left(f_{MP}^{(u_i)}, f_{MT}^{(x_i)}\right)$ refers to the position of the media type of x_i in the sequence of media preference for u_i. Then, generating the LO map becomes a constrained optimization problem. The mathematical formulation is as follows:

$$\min\ F = \sum_{i=1}^{3} w_i F_i \tag{4}$$

s.t.

$$\bigcup_{x_i \in X_{u_i}} C_{AC}^{(x_i)} = C_{u_i} \tag{5}$$

$$C_{AC}^{(x_i)} \cap C_{AC}^{(x_j)} = \emptyset \quad \forall x_i, x_j \in X_{u_i} \tag{6}$$

where F_i represents the degree of match between different characteristics, w_i is the corresponding weight, $C_{AC}^{(x_i)}$ represents the set of concepts associated with LO x_i, and C_{u_i} represents all the concepts the learner needs to learn.

Constraint (5) ensures that by learning all LOs in the LO map, the learner will be able to master all required concepts. Constraint (6) ensures that the concept sets corresponding to each LO do not overlap. In the process of learning the LOs, concepts will not be learned repeatedly. These two constraints prevent redundancy of concepts in the LO map, reducing cognitive load on the learner.

3.2 Chromosome Representation

In GBGA, the LO maps as candidate solutions should be encoded as chromosomes. We encode the information of matching between LOs in the LO map and concepts in the target subgraph as chromosomes. Suppose there are five concepts in the target subgraph, arranged in a fixed order. The LOs in the LO map will be encoded as genes of the chromosome, with their positions in chromosome indicating their corresponding concepts (note that LOs covering multiple concepts will be repeated). Figure 2 shows a chromosome representation of 3 LOs with 5 concepts. The matching relationships between LOs and concepts can be used to determine the dependency between LOs using the mapping rules. Hence, the chromosome can represent the LO map indirectly by representing the matching relationships between LOs and concepts.

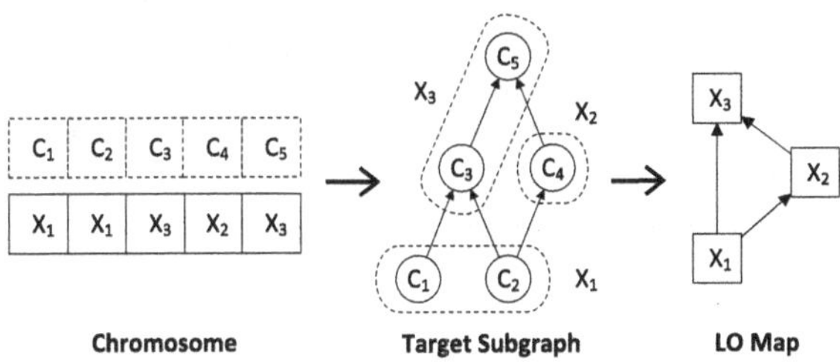

Fig. 2. The process of chromosome decoding

3.3 The Genetic Operators

In GBGA, the population is iteratively updated through the selection of high-fitness individuals and the use of genetic operators. In the following, two special genetic operators are introduced.

The crossover operator combines the genetic information of two parents to generate offspring, ensuring the search capability of the genetic algorithm in the solution space. In order to make offspring satisfy Constraints (5) and (6), a graph-based crossover operator is implemented, which has stronger local search capabilities. The graph-based crossover operator firstly generates a swappable concept set C'_{u_i}, which indicates the genes that can be swappable for crossover. In the two parent chromosomes, the genes covering the concepts in C'_{u_i} are independent of the other genes. Then, the corresponding genes will be swapped to generate two new individuals.

The mutation operator maintains the diversity across the populations by randomly changing certain genes in an individual. This operator randomly selects a LO within the individual and changes all genes at its positions. One or more LOs are randomly selected from the set of candidate LOs to replace them.

4 Experiments

To evaluate the validity of our proposed method, a series of simulation experiments were conducted and the performance of our proposed GBGA was compared with an improved genetic algorithm (IGA). The IGA, proposed by Benmesbah et al. [1], is utilized in a two-stage learning path generation framework and has demonstrated good performance in simplified datasets.

Six simulated datasets are generated. These datasets differ in concept coverage levels of LOs. Datasets 1, 2, and 3 are simplified datasets, in which each LO corresponds to a single concept, and datasets 4, 5, and 6 are non-simplified datasets, in which the coverage levels of LOs are uniformly distributed from 1 to 5. IGA and GBGA were run 20 times on the six simulation datasets. Table 1 presents the best fit (BF), average fit (AF), and standard deviation (SD) achieved by the two algorithms.

In datasets 1, 2, and 3, the performance of IGA and GBGA was similar. GBGA slightly outperformed IGA in finding global optimal solutions in datasets 2 and 3. In datasets 4, 5, and 6, the solutions found by GBGA were significantly better than those found by IGA. This indicates that GBGA is able to use LOs with diverse coverage levels to generate better solutions in the non-simplified

Table 1. Performance of IGA and GBGA on six simulation datasets

Datasets	Best Fitness		Average Fitness		Standard Deviation	
	IGA	GBGA	IGA	GBGA	IGA	GBGA
1	**0.37**	**0.37**	**0.37**	**0.37**	0.09	**0.01**
2	0.41	**0.33**	0.43	**0.34**	0.28	**0.02**
3	2.38	**2.21**	2.44	**2.31**	0.22	**0.07**
4	0.26	**0.12**	0.26	**0.13**	0.10	**0.03**
5	0.39	**0.28**	0.42	**0.29**	0.14	**0.03**
6	0.56	**0.37**	0.58	**0.42**	**0.09**	0.10

dataset. By using special genetic operators, GBGA avoids premature convergence and has stronger search capability. Comparison of standard deviation values indicates that GBGA is more robust.

5 Conclusion

The present work focuses on a crucial issue in e-learning contexts: the generation of personalized learning paths. Our main objective is to generate personalized learning paths using LOs with diverse coverage levels. To achieve this, we propose an improved LPR framework. Compared to previous studies, our method is compatible with LOs with diverse coverage levels and makes full use of various e-learning resources. In the proposed framework, GBGA was designed to solve the personalized LO map generation problem. The performance of GBGA was evaluated through computational experiments, which showed that it can obtain high-quality solutions efficiently. Overall, these findings suggest that the proposed approach is highly effective for the LPR problem.

Acknowledgements. This work has been supported by the Natural Science Foundation of China (No. 62277045) and the Natural Science Foundation of Shandong Province (No. ZR 2021MF011).

References

1. Benmesbah, O., Lamia, M., Hafidi, M.: An enhanced genetic algorithm for solving learning path adaptation problem. Educ. Inf. Technol. **26**(5), 5237–5268 (2021). https://doi.org/10.1007/s10639-021-10509-z
2. Bian, C.L., Wang, D.L., Liu, S.Y., Lu, W.G., Dong, J.Y.: Adaptive learning path recommendation based on graph theory and an improved immune algorithm. KSII Trans. Internet Inf. Syst. **13**(5), 2277–2298 (2019)
3. Durand, G., Belacel, N., LaPlante, F.: Graph theory based model for learning path recommendation. Inf. Sci. **251**, 10–21 (2013)
4. Machado, M.D.O.C., Bravo, N.F.S., Martins, A.F., Bernardino, H.S., Barrere, E., Souza, J.F.D.: Metaheuristic-based adaptive curriculum sequencing approaches: a systematic review and mapping of the literature. Artif. Intell. Rev. **54**(1), 711–754 (2021)
5. Martins, A.F., Machado, M., Bernardino, H.S., de Souza, J.F.: A comparative analysis of metaheuristics applied to adaptive curriculum sequencing. Soft. Comput. **25**(16), 11019–11034 (2021). https://doi.org/10.1007/s00500-021-05836-9
6. Meng, L., Zhang, W., Chu, Y., Zhang, M.: LD-LP generation of personalized learning path based on learning diagnosis. IEEE Trans. Learn. Technol. **14**(1), 122–128 (2021)
7. Nabizadeh, A.H., Goncalves, D., Gama, S., Jorge, J., Rafsanjani, H.N.: Adaptive learning path recommender approach using auxiliary learning objects. Comput. Educ. **147**, 103777 (2020)
8. Nabizadeh, A.H., Leal, J.P., Rafsanjani, H.N., Shah, R.R.: Learning path personalization and recommendation methods: a survey of the state-of-the-art. Expert Syst. Appl. **159**, 113596 (2020)
9. Siren, A., Tzerpos, V.: Automatic learning path creation using OER: a systematic literature mapping. IEEE Trans. Learn. Technol. **15**(4), 493–507 (2022)

Prompt-Independent Automated Scoring of L2 Oral Fluency by Capturing Prompt Effects

Ryuki Matsuura(✉) and Shungo Suzuki

Waseda University, Tokyo, Japan
{matsuura,ssuzuki}@pcl.cs.waseda.ac.jp

Abstract. We propose a prompt-independent automated scoring method of second language (L2) oral fluency, which is robust to different cognitive demands of speaking prompts. When human examiners assess L2 learners' oral fluency, they can consider the effects of different task prompts on speaking performance, systematically adjusting their evaluation criteria across prompts. However, conventional automated scoring methods tend to ignore such variability in speaking performance caused by prompt design and use prompt-specific features of speech. Their robustness is thus arguably limited to a specific prompt used in model training. To address this challenge, we operationalize prompt effects in terms of conceptual, linguistic and phonological features of speech and embed them, as well as a set of temporal features of speech, into a scoring model. We examined the agreement between true and predicted fluency scores in four different L2 English monologue prompts. The proposed method outperformed a conventional method which used only temporal features (κ = 0.863 vs. 0.797). The detailed analysis showed that the conceptual and phonological features improved the performance of automated scoring. Meanwhile, the effectiveness of the linguistic features was not confirmed possibly because it may largely reflect redundant information to capture the prompt demands. These results suggest that the robustness of the automated fluency scoring should be achieved by careful consideration of what characteristics of L2 speech reflect the prompt effects.

Keywords: Automated speech scoring · Fluency · Prompt-independent scoring · Foreign language learning · L2 speech

1 Introduction

It is crucial for second language (L2) learners to acquire an optimal level of fluency. In real-world communication, listeners tend to be distracted when speech is less fluent [4]. To encourage the acquisition of fluency, it is helpful to evaluate learners' current level and set realistic learning goals. However, the evaluation of L2 oral fluency by trained examiners requires time, cost and expert knowledge,

N. Wang et al. (Eds.): AIED 2023, LNAI 13916, pp. 720–726, 2023.
https://doi.org/10.1007/978-3-031-36272-9_62

which reduces learners' opportunities to attain fluency in an effective manner. Previous work has thus addressed automated speech scoring (ASS) for L2 fluency.

The existing ASS models have been trained to predict subjective fluency ratings based on temporal features of speech. The fluency rating is found to be significantly associated with speed of delivery, pausing behavior and disfluency phenomena (e.g., self-repair and repetition) [13]. Building on this finding, ASS models of fluency have commonly used the corresponding temporal features, such as speech rate, pause duration and repaired word frequency, as input (e.g., [6, 7,10]). While the models using the temporal features can approach human-level performance in a specific prompt, they have failed to maintain the same level of prediction accuracy in a multiple-prompt setting. This low applicability of the trained model to various speaking contexts may provide incorrect scores to learners and subsequently may hinder effective L2 speech learning. It is thus expected to develop a prompt-independent ASS system for fluency, which is robust to various speaking prompts.

The aforementioned problem is possibly due to an ignorance of variability in temporal features caused by different cognitive demands across prompts. L2 research consistently showed that variability in speaking performance across different prompts is explained by the fact that different speaking prompts impose different quality of demands on components of speech production processes, such as content planning and linguistic encoding [9]. Despite such variability of speech performance, human examiners can evaluate L2 fluency consistently across multiple prompts. Previous studies on L2 fluency found that the fluency evaluation is influenced not only by temporal features of speech but also by non-temporal ones (e.g., content, grammar and pronunciation), suggesting that human examiners adjust evaluation criteria by intuitively inferring the prompt effects from a whole range of speech characteristics [9].

To address this challenge of the prompt-specific ASS, the current study aims to operationalize the cognitive demands of prompts and incorporate conceptual, linguistic and phonological features of speech as well as temporal ones into the fluency scoring system. We examined the effectiveness of these features by comparing the performance of subjective fluency score predictions between the proposed scoring method and a conventional method based only on temporal features as the input. In an experiment, we used four monologue prompts to extract these features of speech and construct ASS models. We also conducted a follow-up analysis to examine the relative importance of them in relation to the prediction accuracy of the models.

2 Related Work

Previous research on ASS for fluency has focused on prompt-specific systems, which are developed and evaluated using the same prompt. For example, Shen et al. [10] developed a regression model to predict fluency scores in a picture description prompt, and it outperformed human performance (r = 0.956 vs. 0.910).

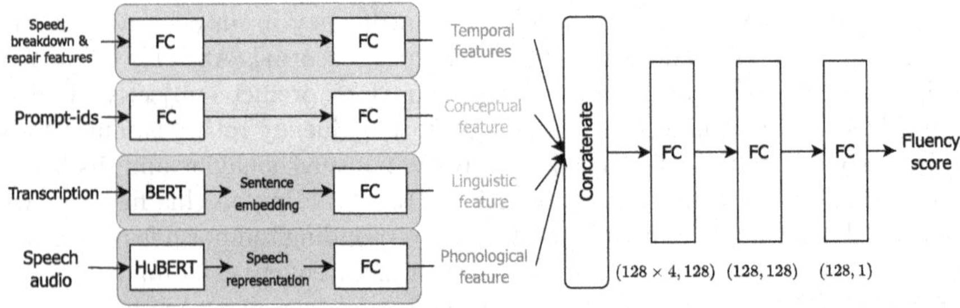

Fig. 1. Architecture of prompt-independent automated fluency scoring model

While the prompt-specific models have achieved high performance, the prediction accuracy often declines when the same method, which ignores prompt differences, is applied to a multiple-prompt setting. Matsuura et al. [6] adopted the prompt-specific model to predict human ratings of fluency in an oral proficiency interview task consisting of multiple question-format prompts. Their model showed low classification accuracy for novice and pre-advanced learners. An inability of the model to fully capture the prompt differences may have lowered the prediction accuracy of fluency scores, highlighting the necessity of prompt-independent ASS systems. To the best of our knowledge, such systems have not been developed in the domain of speaking assessment. Meanwhile, there are several successful scoring models for assessing L2 essays. Despite the difference in the modality between speaking and writing, one major issue is the methods for extracting and controlling for the effects of features that are highly subject to prompt design. In the automated essay scoring, semantic information of essays is considered to be prompt-specific, and thus alternatively some general features, such as grammatical structure and text length, are used as a proxy for the prompt-specific characteristics of essays [8]. When it comes to the oral fluency scoring, given temporal features of speech are, by nature, highly subject to the prompt design, an adaptation of the models to various prompts should be achieved by manipulating non-temporal features of speech. We therefore propose an incorporation of conceptual, linguistic and phonological features into the temporal ones to develop the prompt-independent fluency scoring system.

3 Prompt-Independent Automated Fluency Scoring

We propose the prompt-independent ASS model for fluency, which is compatible with the various prompt setting, using conceptual, linguistic and phonological features as well as temporal features (see Fig. 1). Three dimensions of temporal characteristics of speech are captured, following [6]. Speed feature is represented by articulation rate, speech rate and mean length of run; breakdown feature includes mid-clause pause ratio, end-clause pause ratio, filled pause ratio, mid-clause pause duration, end-clause pause duration and mean pause duration; and repair feature is measured by disfluency ratio, repetition ratio and self-repair ratio. To extract the conceptual feature, we assume that speech content is largely controlled by prompt design and thus use prompt-ids as an input of the ASS model. As for the linguistic feature, a sentence embedding is extracted by means of BERT [1], which can capture lexical and grammatical characteristics of sentences [3]. In our study, a hidden state of [CLS] token is used as the sentence embedding. For the phonological feature, we utilize the pre-trained HuBERT [2] and extract speech representation, which is reflective of phonetic characteristics of speech [14]. The extracted features are mapped to 128 dimension vectors by fully-connected (FC) layers and concatenated. This vector is further fed into three FC layers, and finally fluency score is predicted. Moreover, the dropout is adopted to avoid an overfitting to the training data. We employ Adam as an optimizer, and hyper-parameters are determined by the tuning.

4 Experiment

To examine the proposed prompt-independent fluency scoring model, we used English monologue speech elicited from four speaking tasks differing in the nature of cognitive demands [12]: an argumentative speech, a related picture narrative, a reading-to-speaking (RtoS) and a reading-while-listening-to-speaking (RwLtoS)[1]. The argumentative prompt demanded an ability of content planning, in which learners argued for or against a given statement with valid reasons. Meanwhile, with the picture narrative prompt, learners were demanded to use their ability to describe a cartoon, where the content of speech is largely predefined by the visual prompt. The RtoS was a prompt to give a oral summary of a given English written text and differs from the picture description in that the essential vocabulary was presented in the text. The requirement for the RwLtoS prompt was same for the RtoS, whereas the RwLtoS also provides phonological information of the vocabulary by the audio-recording of the text. All four tasks were completed by 128 Japanese learners of English, and speech duration was around two minutes on average. The first one-minute excerpts of 512 speech were scored for fluency by two Ph.D. students in Applied Linguistics independently. They evaluated L2 oral fluency on a nine-point scale, focusing on temporal aspects of speech, and the inter-rater reliability was confirmed (Cronbach's $\alpha = 0.819$). As the ground truth of fluency score, we removed raters' severity by conducting many-facet Rasch analysis [5], and the estimated fluency scores for each speech sample were re-scaled on a six-point scale. To train and evaluate the model, speech samples were split into 384 training, 64 validation and 64 test set, with a equal balance of four prompts.

Considering the ordinality of the fluency scores, we evaluated the fluency prediction performance using quadratic weighted κ (QWK) and Peasron's correlation coefficient (PCC) between the predicted and true fluency scores. To confirm the effectiveness of the non-temporal features for the prompt-independent ASS, we compared the proposed method with the conventional one which sorely

[1] https://osf.io/zrwmn/?view_only=0eeb1c966cb64afc9834acf80a42ad7e.

Table 1. QWK and PCC of fluency score predictions.

	model					QWK	PCC
(a)	temporal				(conventional)	0.797	0.904
(b)	temporal	+ conceptual	+ linguistic	+ phonological	(proposed)	**0.863**	**0.932**
(c)	temporal	+ conceptual				0.799	0.867
(d)	temporal		+ linguistic			0.830	0.917
(e)	temporal			+ phonological		0.859	0.931
(f)	temporal	+ conceptual	+ linguistic			0.853	0.929
(g)	temporal	+ conceptual		+ phonological		**0.896**	0.929
(h)	temporal		+ linguistic	+ phonological		0.870	0.919

depends on temporal features of speech. Moreover, we compared different combinations of the conceptual, linguistic and phonological features to investigate which features can capture the cognitive demands of the prompts. The basic architecture and training setting of each model was identical to the proposed method except for the type of features to be concatenated.

5 Results and Discussion

We summarized the results of the experiment in Table 1. First, the proposed method (b) outperformed the conventional one (a) in terms of QWK and PCC. This result indicates that the non-temporal features can increase the prediction accuracy of fluency scores when multiple prompts were included. We also compared the various combinations of the conceptual, linguistic and phonological features. When adding one of the non-temporal features ((c), (d) and (e)), all methods achieved higher performance than the conventional method (a) but lower than the proposed one (b). Meanwhile, the models which included two types of the features ((f), (g) and (h)) had the same or higher agreement than the one with one feature ((c), (d) and (e)). This result suggests that combining at least two non-temporal features is effective in improving the robustness of the fluency scoring model to multiple prompts. Moreover, the combination of the phonological feature and either conceptual (g) or linguistic (h) features approached the same level of performance of the proposed method. These findings suggest that the phonological feature strongly contribute to the prompt-independent prediction, whereas the effectiveness of the linguistic feature may need further investigation. However, these results should be interpreted carefully. According to Suzuki and Kormos [11], the listeners' perception of L2 oral fluency is related to the linguistic dimension of the speech. This discrepancy between the previous and current findings might be attributed to the information involved in the BERT's sentence embedding. BERT is known to capture the sentiment and semantic characteristics of the text as well as lexical and grammatical ones [3]. In contrast, raters in the current study may not have relied on such sentimental aspects of speech to adjust the effect of the cognitive demands by the prompts because they were instructed to focus solely on temporal aspects of speech. Therefore, it might be better to disentangle essential information from the information-rich features to capture more refined prompt effects which improve the performance of the ASS. In future work, it should be investigated how to extract the lexical and grammatical features, which can enhance the performance of the prompt-independent ASS system. Although the effectiveness of the prompt-independent ASS by capturing the prompt effects were confirmed, it should be noted that the prompt-ids were used to capture the conceptual dimension of speech. Since the prompt-ids are determined by the type of prompts used in the training set, and conceptual features of completely new prompts cannot be captured, the current method reduces the robustness of the ASS model to such prompts. To solve this problem, future works should design the extraction method of conceptual features from speech transcriptions using natural language processing techniques, such as speech act classification.

References

1. Devlin, J., Chang, M.W., Lee, K., Toutanova, K.: BERT: pre-training of deep bidirectional transformers for language understanding. In: Proceedings of the 2019 Conference of the North American Chapter of the Association for Computational Linguistics: Human Language Technologies, vol. 1, pp. 4171–4186 (2019). https://doi.org/10.18653/v1/N19-1423
2. Hsu, W.N., et al.: HuBERT: self-supervised speech representation learning by masked prediction of hidden units. IEEE/ACM Trans. Audio Speech Lang. Proc. **29**, 3451–3460 (2021). https://doi.org/10.1109/TASLP.2021.3122291
3. Jawahar, G., Sagot, B., Seddah, D.: What does BERT learn about the structure of language? In: Proceedings of the 57th Annual Meeting of the Association for Computational Linguistics, pp. 3651–3657. Association for Computational Linguistics, Florence, Italy (2019). https://doi.org/10.18653/v1/P19-1356
4. Lennon, P.: The lexical element in spoken second language fluency. In: Riggenbach, H. (ed.) Perspectives on Fluency, pp. 25–42. University of Michigan Press, Ann Arbor (2000)
5. Linacre, J.M.: Many-Facet Rasch Measurement. MESA Press, Chicago (1989)
6. Matsuura, R., Suzuki, S., Saeki, M., Ogawa, T., Matsuyama, Y.: Refinement of utterance fluency feature extraction and automated scoring of L2 oral fluency with dialogic features. In: 2022 Asia-Pacific Signal and Information Processing Association Annual Summit and Conference (APSIPA ASC), pp. 1312–1320 (2022). https://doi.org/10.23919/APSIPAASC55919.2022.9980148
7. Ramanarayanan, V., Lange, P.L., Evanini, K., Molloy, H.R., Suendermann-Oeft, D.: Human and automated scoring of fluency, pronunciation and intonation during human-machine spoken dialog interactions. In: Proc. Interspeech 2017, pp. 1711–1715 (2017). https://doi.org/10.21437/Interspeech.2017-1213
8. Ridley, R., He, L., Dai, X., Huang, S., Chen, J.: Prompt agnostic essay scorer: a domain generalization approach to cross-prompt automated essay scoring. arXiv preprint arXiv:2008.01441 (2020)
9. Segalowitz, N.: Cognitive Bases of Second Language Fluency. Routledge, London (2010). https://doi.org/10.4324/9780203851357

10. Shen, Y., Yasukagawa, A., Saito, D., Minematsu, N., Saito, K.: Optimized prediction of fluency of L2 English based on interpretable network using quantity of phonation and quality of pronunciation. In: 2021 IEEE Spoken Language Technology Workshop (SLT), pp. 698–704 (2021). https://doi.org/10.1109/SLT48900.2021.9383458
11. Suzuki, S., Kormos, J.: Linguistic dimensions of comprehensibility and perceived fluency: an investigation of complexity, accuracy, and fluency in second language argumentative speech. Stud. Second. Lang. Acquis. **42**(1), 143–167 (2020). https://doi.org/10.1017/S0272263119000421
12. Suzuki, S., Kormos, J.: The multidimensionality of second language oral fluency: interfacing cognitive fluency and utterance fluency. Stud. Second. Lang. Acquis. **45**(1), 38–64 (2022). https://doi.org/10.1017/S0272263121000899
13. Suzuki, S., Kormos, J., Uchihara, T.: The relationship between utterance and perceived fluency: a meta-analysis of correlational studies. Mod. Lang. J. **105**(2), 435–463 (2021). https://doi.org/10.1111/modl.12706
14. Wells, D., Tang, H., Richmond, K.: Phonetic analysis of self-supervised representations of English speech. In: Proc. Interspeech 2022, pp. 3583–3587 (2022). https://doi.org/10.21437/Interspeech.2022-10884

Navigating Wanderland: Highlighting Off-Task Discussions in Classrooms

Ananya Ganesh[1(✉)], Michael Alan Chang[2], Rachel Dickler[1], Michael Regan[1], Jon Cai[1], Kristin Wright-Bettner[1], James Pustejovsky[3], James Martin[1], Jeff Flanigan[4], Martha Palmer[1], and Katharina Kann[1]

[1] University of Colorado Boulder, Boulder, CO, USA
ananya.ganesh@colorado.edu
[2] University of California Berkeley, Berkeley, CA, USA
[3] Brandeis University, Waltham, MA, USA
[4] University of California Santa Cruz, Santa Cruz, CA, USA

Abstract. Off-task discussions during collaborative learning offer benefits such as alleviating boredom and strengthening social relationships, and are therefore of interest to learning scientists. However, identifying moments of off-task speech requires researchers to navigate massive amounts of conversational data, which can be laborious. We lay the groundwork for automatically identifying off-task segments in a conversation, which can then be qualitatively analyzed and coded. We focus on in-person, real-time dialog and introduce an annotation scheme that examines two facets of dialog typical to in-person classrooms: whether utterances are pertinent to the *lesson*, and whether utterances are pertinent to the *classroom*, more broadly. We then present two computational models for identifying off-task utterances.

Keywords: Collaborative learning · Classroom discourse · Off-task speech

1 Introduction

Towards supporting learning scientists in studying collaborative discourse, we hone in on the issue of *navigating* moments of collaboration that seemingly wander from the learning goals of an activity. Some previous work operates from the premise that such off-task remarks detract from collaborative effectiveness [4]. In contrast, others argue that off-topic behavior can be critical to collaborative learning because of the strengthening of the social relationships within a group [5]. Moreover, a better understanding of off-task behaviors is a key equity issue: Langer-Osuna demonstrates that collaborative wanderings often play a critical role towards elevating the voices of youth who are marginalized in the discussion because of their identity [5].

In this paper we propose a first step towards *automatically* spotlighting critical moments of interest where seemingly off-task interactions occur. Qualitative

N. Wang et al. (Eds.): AIED 2023, LNAI 13916, pp. 727–732, 2023.
https://doi.org/10.1007/978-3-031-36272-9_63

approaches towards analyzing interactions are usually constrained in speed and scale by a manual expert coding process [7]. Automated detection offers the opportunity to direct researchers to relevant segments of data for further examination [10]. However, prior work on computationally modeling off-task discussions [1,2,8] focuses on text-based environments where students communicate through chat messages, which can differ in quality from the spontaneous spoken dialog that qualitative researchers study. Additionally, prior work models on-task utterances from a "productivity" perspective, which may not reflect the nuances of real-time, in-person group work in classrooms.

Towards addressing these challenges, we make two main contributions in this paper. First, we introduce an **annotation scheme** suitable for in-person classroom environments, which examines two facets of collaborative dialog, which we call lesson-focused (LF) speech and classroom-focused (CF) speech. As shown in Table 1, lesson-focused speech focuses on discussions about the current learning activity, such as discussing ideas and solutions. Classroom-focused speech discusses any relevant classroom activity, including peripheral tasks such as team management. This distinction provides flexibility for modeling, such as using keywords, as well as for downstream analysis – for instance, lesson-focused speech could be examined for speech or skills related to problem-solving, while classroom-focused speech could shed light on team dynamics.

As our second contribution, we develop **two computational models** for detecting if utterances are lesson-focused, and classroom-focused respectively, which are trained on discussions from middle school science classrooms. Both models are transformer-based sequence classifiers, and perform classification at the utterance level to indicate if an utterance is focused, non-focused, or if it is undecidable given the input information. Experimenting with multiple input signals, we find that our best LF-classifier's performance is 0.59 F1, and our best CF-classifier's performance is 0.56 F1, both strongly outperforming random and majority-class baselines. Finally, we discuss the impact of design choices such as the amount of surrounding context used, in serving the needs of the user.

2 Methodology

2.1 Dataset

We use the dataset described in Southwell et al. [9], which consists of five-minute long transcripts of small-group discussions, collected from a middle-school science classroom in the United States. The subject of instruction is a curriculum unit called Sensor Immersion (SI) focused on "programmable sensor technology". Student interactions are recorded through desk-top mics, and manually transcribed and anonymized, yielding 32 transcripts with 2133 student utterances in total.

We then annotate utterances on a 3-point scale for the "lesson-focused" and "classroom-focused" facets, indicating *focused*, *non-focused*, and *undecidable* given the information available. As shown in Table 1, LF includes content such as discussing solutions, asking for help, discussing issues with the exercise, etc.,

Table 1. Examples of utterances in our classroom dataset, and the corresponding annotations for the lesson-focused facet and the classroom-focused facet.

Utterance	LF	CF
This is actually a different button	✓	✓
So maybe there's something that we put in a wrong spot or something we didn't add	✓	✓
So we're creating a greenhouse sensor?	✓	✓
What are you doing over there [partner]?	✗	✓
No, she's supposed to navigate you.	✗	✓
Teacher, their computer shut off.	✗	✓
Hold up. Aren't you in advanced math?	✗	✗
I got a tournament this weekend so I can't get quarantined	✗	✗
We should [inaudible]	Undecidable	Undecidable

whereas CF additionally includes activities like introductions, troubleshooting equipment, as well as meta-management of roles and responsibilities in groups. Each transcript is annotated by *two* annotators experienced with linguistic annotation tasks. They are provided access to the lesson plans to understand the context of discussions, and are instructed to evaluate every utterance in context of the entire transcript, by looking at past and future utterances. On the LF facet, inter-annotator agreement is 64.7% and on the CF facet, inter-annotator agreement is 71.3%. During disagreements, if only one of the annotators picks *undecidable*, we adjudicate to the other decisive label. When annotators disagree diametrically (focused and non-focused), we manually adjudicate using a third trained annotator. Diametric disagreements comprise 41% of all disagreements on the LF facet, and 42% of all disagreements on the CF facet. Due to the in-person, spontaneous nature of the interactions, our dataset contains phenomena like deictic references, fillers and interleaved conversations, which we believe make the annotation process challenging, thereby hindering high agreement.

We divide our data into three sets at the transcript level, using 18 for training, 6 for development, and 8 for testing. Of the 1254 utterances in the training set, 74% are *focused*, 22.5% are *non-focused*, and 2.6% are undecidable for the LF facet, and 86% are *focused*, 11.8% are *non-focused*, and 2% are undecidable for the CF facet.

2.2 Models

We now investigate the research question, *how well can state of the art natural language processing models identify off-task utterances?* We develop two classification models, to classify utterances on the LF, and the CF facet respectively. Since pretrained language representation models such as RoBERTa [6] perform competitively on classification tasks, we finetune them for sequence classification

Table 2. Performance on test set for both facets, on the labels focused (✓), non-focused (✗) and undecidable (*). RoBERTa models are given utterance history of window size w, RoBERTa-UL models are given utterance and gold label history.

Model	Lesson-focused				Classroom-focused			
	Macro F1	✓F1	✗F1	* F1	Macro F1	✓F1	✗F1	* F1
Random	0.29	0.41	0.36	0.10	0.29	0.43	0.33	0.11
Majority	0.25	0.74	0.00	0.00	0.26	0.78	0.00	0.00
RoBERTa ($w = 5$)	**0.59**	0.83	0.76	0.19	**0.56**	0.83	0.53	0.32
RoBERTa-UL ($w = 4$)	0.51	0.83	0.71	0.00	0.50	0.85	0.49	0.15

here. Specifically, each classifier is built by adding a separate classification head on top of the RoBERTa model, following which we train for LF or CF classification as described below. While our models are trained for predicting both facets separately, a multi-task training setup could be explored in future work.

Input: Each input instance consists of a single utterance, preceded by a speaker ID that is an anonymized unique identifier. We experiment with passing in a context history – a sequence of utterances in a window preceding the current utterance. The utterances in the history are all concatenated, with a special token to separate each utterance. We optionally provide the gold LF/CF label corresponding to each utterance in the history as input. At inference time, the label corresponding to each previous utterance in the history is *predicted.*

The number of preceding utterances passed to the model as context is a tunable hyperparameter, w. We experiment with ten window sizes, values ranging from 0 to 9, and also experiment with setting w to be the maximum length of a transcript, that is, using all previous utterances in a transcript as context.

3 Results

Table 2 shows the performance of our best performing models, along with comparisons to two baselines, Random and Majority. The random baseline randomly selects one out of the three labels, and the majority baseline always chooses the majority class, *focused.* We report class-wise and macro-averaged F1 scores.

For the lesson-focused models, Random achieves an F1 of 0.29, while Majority achieves an F1 of 0.25. Our best performing model, with an utterance history of size 5, outperforms both these baselines by a large margin, with an F1 of 0.59. However, the model that sees context as well as label history sees a drop in performance, to 0.51 F1, potentially due to noise in predicting labels (at test time) corresponding to the history. The impact of class imbalance can be seen in the class-wise F1s – high performance of 0.83 on the majority label *focused,* and very low performance of 0.19 on *undecidable.*

In the classroom-focused models, similarly, the RoBERTa models outperform the baselines by a strong margin, and the model that sees only the utterance context achieves an F1 of 0.56, outperforming the model with context+label history. We again observe a very high performance on the *focused* label, at 0.85 F1. In comparison to the LF models, performance on the *non-focused* or non-CF labels is lower, with the best model having an F1 of 0.53. One potential direction for improving performance in future is making use of multimodal signals, such as gestures and tone. As discussed in Sect. 2.1, this could help resolve instances when the utterance alone does not provide enough information (deictic references, short ambiguous utterances), which are challenging even for humans.

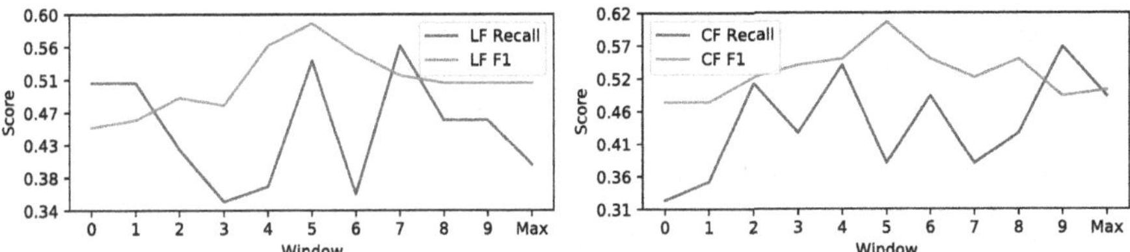

Fig. 1. Effect of context window size on recall

Recall Performance: An important indicator of the usefulness of our system to scaling qualitative coding of off-task discussions, is *recall* performance, which tells us how many off-task utterances were retrieved correctly by the model. As observed in Crowston et al. [3], a high recall is desirable in tools supporting qualitative research, to discover *all* relevant segments. Figure 1 shows recall performance on our development set for all window sizes.

We first note that maximum recall of 0.56 is observed at window size 7 for LF and 0.57 at window size 9 for CF. Further, recall performance does not necessarily follow the same trends as F1 score, which climbs until window size 5 before dropping off. Upon manually examining false negatives from models with differing window sizes, we find that the models with longer contexts tend to miss the *initiation* of an off-task segment, but correctly identify off-task utterances that are part of a longer sequence. The size of the context window must therefore serve as an important and modifiable design consideration for the researcher using our system, depending on the objective that needs to be optimized.

4 Conclusion

Towards assisting qualitative researchers in understanding the dynamics of collaboration between students, we propose using NLP to automatically highlight moments of wandering discussion. We present an annotation scheme that categorizes whether utterances are *lesson-focused* or *classroom-focused*, and annotate real-time spoken dialog. We then develop two transformer-based models for automatically classifying utterances accordingly. While our models outperform multiple baselines, our results show there is room for improvement, potentially using multi-modal information. In practice, we envision that qualitative researchers can filter large amounts of transcribed dialog based on our models' predictions, choosing to examine either of the two facets in greater detail. Further analysis can be facilitated in future through frameworks for researchers to query predictions, looking for specific types of interactions – for instance, off-task sequences that engage all speakers, or that result in increased subsequent on-task contributions.

Acknowledgments. This research was supported by the NSF National AI Institute for Student-AI Teaming (iSAT) under grant DRL 2019805. The opinions expressed are those of the authors and do not represent views of the NSF.

References

1. Baker, R.S., Corbett, A.T., Koedinger, K.R., Wagner, A.Z.: Off-task behavior in the cognitive tutor classroom: when students "game the system". In: Proceedings of the Conference on Human Factors in Computing Systems, pp. 383–390 (2004)
2. Carpenter, D., et al.: Detecting off-task behavior from student dialogue in game-based collaborative learning. In: Bittencourt, I.I., Cukurova, M., Muldner, K., Luckin, R., Millán, E. (eds.) AIED 2020. LNCS (LNAI), vol. 12163, pp. 55–66. Springer, Cham (2020). https://doi.org/10.1007/978-3-030-52237-7_5
3. Crowston, K., Allen, E.E., Heckman, R.: Using natural language processing technology for qualitative data analysis. Int. J. Soc. Res. Methodol. **15**(6), 523–543 (2012)
4. Godwin, K.E., et al.: Off-task behavior in elementary school children. Learn. Instr. **44**, 128–143 (2016)
5. Langer-Osuna, J., Gargroetzi, E., Chavez, R., Munson, J.: Rethinking loafers: understanding the productive functions of off-task talk during collaborative mathematics problem-solving. In: International Society of the Learning Sciences (2018)
6. Liu, Y., et al.: RoBERTa: a robustly optimized BERT pretraining approach. arXiv preprint arXiv:1907.11692 (2019)
7. Marathe, M., Toyama, K.: Semi-automated coding for qualitative research: a user-centered inquiry and initial prototypes. In: Proceedings of the 2018 CHI Conference on Human Factors in Computing Systems, pp. 1–12 (2018)
8. Segal, A., et al.: Keeping the teacher in the loop: technologies for monitoring group learning in real-time. In: André, E., Baker, R., Hu, X., Rodrigo, M.M.T., du Boulay, B. (eds.) AIED 2017. LNCS (LNAI), vol. 10331, pp. 64–76. Springer, Cham (2017). https://doi.org/10.1007/978-3-319-61425-0_6
9. Southwell, R., Pugh, S., Perkoff, E., Clevenger, C., Bush, J., D'Mello, S.: Challenges and feasibility of automatic speech recognition for modeling student collaborative discourse in classrooms. In: Proceedings of the 15th International Conference on Educational Data Mining. International Educational Data Mining Society (2022)
10. Worsley, M.: Multimodal learning analytics for the qualitative researcher. In: International Society of the Learning Sciences, Inc. [ISLS] (2018)

C^{2}Tutor: Helping People Learn to Avoid Present Bias During Decision Making

Calarina Muslimani[1,2(✉)], Saba Gul[1], Matthew E. Taylor[1,2], Carrie Demmans Epp[1], and Christabel Wayllace[1,2]

[1] University of Alberta, Edmonton, AB T6G 2R3, Canada
{musliman,matthew.e.taylor,demmanse,wayllace}@ualberta.ca
[2] Alberta Machine Intelligence Institute (Amii), Edmonton, AB T5J 3B1, Canada

Abstract. Procrastination can harm many aspects of life, including physical, mental, or financial well-being. It is often a consequence of people's tendency to prefer immediate benefits over long-term rewards (i.e., present bias). Due to its prevalence, we created C^{2}Tutor, an intelligent tutoring system (ITS) that can potentially reduce procrastination habits by teaching planning strategies. C^{2}Tutor teaches people how to make decisions aligned with long-term benefits. It will discourage present bias behavior while allowing for differences in user cognitive abilities. Our study found that C^{2}Tutor encourages far-sighted behavior while reducing maladaptive planning-strategy use.

Keywords: Procrastination · Behavior Change · Lifelong Learning

1 Introduction

Procrastination, the action of voluntarily and unnecessarily delaying something despite adverse consequences, is a failure in self-regulation, where a person is unable to regulate their thoughts, emotions, impulses, or behavior [1]. Several theories from economics and psychology suggest that procrastination can be a consequence of people's tendency to prefer immediate benefits over more critical long-term rewards, a phenomenon known as present bias [12].

Despite its prevalence, there is a substantial shortage of mental healthcare professionals [11]. This shortage creates an opportunity and the need for more virtual interventions. Recent work has developed intelligent tutoring systems (ITSs) to teach people strategies for reducing present bias behavior [2,8].

Therefore, the present work aims to develop an ITS, C^{2}Tutor, that teaches planning strategies to reduce present bias. Unlike prior work, we developed a flexible backwards planning strategy that discourages present bias behavior without assuming each individual has the same cognitive ability. We also developed a teaching methodology for C^{2}Tutor based on theories of formative feedback, meta-cognitive reinforcement learning, and behavior change. To that end, we performed a study to investigate the use of C^{2}Tutor in a simulated environment in which present bias is likely to occur. We address the following questions:

N. Wang et al. (Eds.): AIED 2023, LNAI 13916, pp. 733–738, 2023.
https://doi.org/10.1007/978-3-031-36272-9_64

1. Do people learn our flexible backwards planning strategy using C^{2}Tutor?
2. Can C^{2}Tutor improve people's decision-making in a simulated environment?
3. Do people find C^{2}Tutor useful for improving decision-making?

2 Literature Review and Research Gaps

Present bias or near-sightedness is a fundamental factor influencing how much people procrastinate [10,12]. Prior work [2,8] attempted to reduce procrastination using an ITS to discourage present bias in the Mouselab-MDP paradigm [3] (Fig. 1 Left). These ITSs teach an optimal planning strategy that obtains the best possible tradeoff between the expected cost of making a decision and the expected utility of that decision [7]. To teach this strategy, these systems use meta-cognitive reinforcement learning, where positive or negative reinforcement is given based on how a decision was made rather than what decision was made.

Although this prior work showed the potential of using ITSs to discourage present bias, the optimal planning strategy and teaching methodology were limited. A primary weakness is their reliance on a single optimal planning strategy for all individuals. This choice has two underlying assumptions: (1) all learners have the same cognitive abilities, and (2) the environment is consistent. For example, following this optimal planning strategy, users will start reviewing the rewards of different options in the future and finalize the plan when they find the first maximum possible reward. Any further exploration is discouraged. Therefore, this strategy is not flexible to how much exploration an individual might want to perform when planning. Moreover, relying on meta-cognitive reinforcement learning does not fully account for what is known about feedback delivery [5,9]. For example, these prior ITSs [2,8] used praise as the primary source of positive reinforcement, even though praise can discourage learning [5].

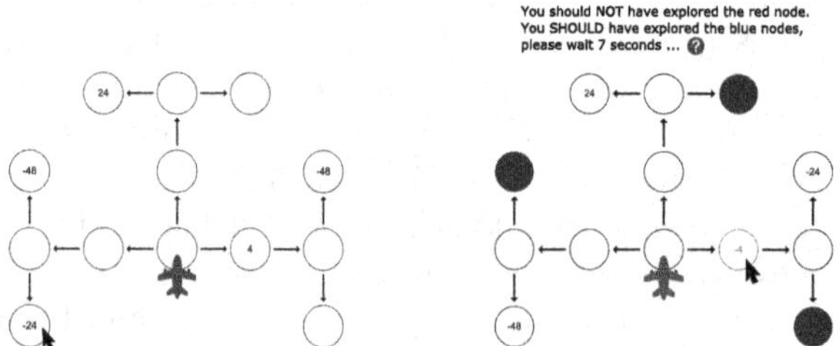

Fig. 1. Mouselab-MDP - Participants click to reveal the value at future states (Left). C^{2}Tutor incorporated in Mouselab-MDP (Right).

3 C^{2}Tutor - Reducing Present Bias

To teach people how to plan, we must be able to observe how people make decisions. To that end, we employed an implementation of Mouselab-MDP since it externalizes people's planning behavior [3] - See Fig. 1 Left. Mouselab-MDP represents planning problems using simple directed acyclic graphs (DAGs). Along each path, there are nodes with hidden rewards. Users click on nodes to reveal their values. After revealing as many nodes as desired, the individual chooses the path they think is best (i.e., the one with the highest total reward). A node-reveal corresponds to a user evaluating the quality of a future state, which is a fundamental cognitive operation in planning. As this process requires mental effort, there is a cost of 1 per click operation. The participant's goal is to maximize their cumulative reward for a decision-making task.

To represent real-life settings where present bias is commonly a factor, we used a three-level DAG in which the rewards are randomly drawn from the following distributions: In level 1, closest to the root, $\{-4, -2, +2, +4\}$; in level 2 $\{-8, -4, +4, +8\}$; and in level 3 $\{-48, -24, +24, +48\}$. The order in which nodes are explored reveals the planning strategy used. Common near-sighted strategies include only inspecting the immediate nodes (i.e., those in level 1) or inspecting an immediate node first. Other poor planning strategies include zero-planning and random exhaustive search. In zero-planning, no nodes are inspected and a random path is taken. For random exhaustive search, all nodes are inspected before taking the path with the maximum cumulative reward [6]. For a full description of all planning strategies, we refer readers to Jain et al. [6].

C^2Tutor aims to teach a general planning strategy that does not assume individuals' cognitive limits are fixed. To that end, we develop our flexible backwards planning strategy that explicitly considers future rewards first and immediate rewards last. As in typical backwards planning, this strategy explores node rewards by level, going from the leaf nodes to the root. Thus, it first explores the reward of all leaf nodes. If there is not enough information to choose a path, it explores nodes of the upper levels. As the range of rewards is higher for the leaf nodes and decreases towards the root, one can prune paths with low rewards.

A limitation of exhaustive backwards planning (i.e., reviewing all nodes from the leaves to the root) is that it can be time-consuming or infeasible for large graphs. For example, if a person has limited time or has more cognitive constraints, they can make a decision (i.e., choose a path) based solely on reviewing the rewards of the leaf nodes. Otherwise, they can plan backwards for more than one step. Critically, C^2Tutor does not assume there is one optimal planning strategy, making our planning strategy flexible to individual differences.

C^2Tutor also provides theoretically-grounded formative feedback to support behavior change. It uses twice as many practice rounds as prior systems [2,8] since an individual's performance on a task is directly related to the amount of deliberate practice performed [4]. To guide design decisions about the timing and types of feedback provided, we used meta-cognitive reinforcement learning and a computer-based feedback framework [9]. To determine the type of feedback to provide, we first assumed that learners have low prior knowledge of effective planning strategies. Therefore, we provided correct response and response-contingent feedback after each planning operation [9]. To alleviate the potential for unnecessarily elaborate feedback, we gave learners control over response-contingent feedback. Learners could choose whether to view this type of feedback (see the green question mark in Fig. 1 Right). This allows learners to obtain clarification if needed or wanted. We also included negative reinforcement as a time penalty because it has been effective when teaching planning strategies [2,8].

4 Experimental Design

Following an institutional ethics review, we conducted a between-subjects study with two conditions: a Tutor experimental condition and a No Tutor control.

We had 15 participants (10 male and 5 female): 7 in the Tutor condition and 8 in the No Tutor condition. Age ranges were collected: 7 were aged 18–24, 4 were 25–30, 2 were 31–40, and 2 were aged 41–50. Participants self-identified their membership in racial groups: 6 were Asian, 5 were White, 3 were Middle Eastern or North African, and 1 was multi-racial.

Following consent procedures, participants were given instructions on how to complete tasks in Mouselab-MDP. In the Tutor condition, participants also read an instructional page describing C^{2}Tutor. Participants were then given the instructions quiz, for which they had to answer all questions correctly to continue. Qualifying participants interacted with the Mouselab-MDP for 1 pre-test round, 20 practice rounds, and 10 test rounds. In the Tutor condition, participants had access to C^{2}Tutor during the practice rounds only. After the test rounds, participants completed an ITS preference and demographics questionnaire.

For all rounds, we collected data on which nodes users revealed, the order of node reveals, and which path they chose. We used this information and the planning strategies defined by Jain et al. [6] to identify the strategies used. To compare performance across groups, we used a one-tailed Mann-Whitney U Test ($\alpha = .05$). We report r as the effect size.

5 Findings

Do People Learn the Backwards Planning Strategy Using C^{2}Tutor? Figure 2 Right shows the number of times participants used various planning strategies during the test rounds. Those in the Tutor condition used our backwards planning strategy close to twice as much ($U = 43.5$, $p = .038$, $r = 0.47$). Meanwhile, participants in the No Tutor condition had significantly higher use of maladaptive planning strategies during both the training ($U = 8.5$, $p = .012$, $r = 0.59$) and the test rounds ($U = 14.0$, $p = .021$, $r = 0.545$).

Can C^{2}Tutor Improve People's Decision-Making in a Simulated Environment? We compared the participants' cumulative training and test scores across conditions. We found a significant difference ($U = 52.0$, $p = .003$, $r = 0.72$) in scores during training between the Tutor ($M = 735.57$, $SD = 39.24$) and No Tutor conditions ($M = 560.25$, $SD = 278.61$). None was not found during testing ($U = 33.0$, $p = .301$, $r = 0.15$) across the Tutor ($M = 466.43$, $SD = 15.56$), and No Tutor ($M = 403.88$, No Tutor $SD = 124.08$) groups.

Across all rounds, participants in the Tutor condition achieved significantly higher scores compared to those in the No Tutor condition ($U = 658.0$, $p = .006$, $r = 0.32$). This highlights that using C^{2}Tutor resulted in consistently higher performance during both the training and test phases (Fig. 2 Left).

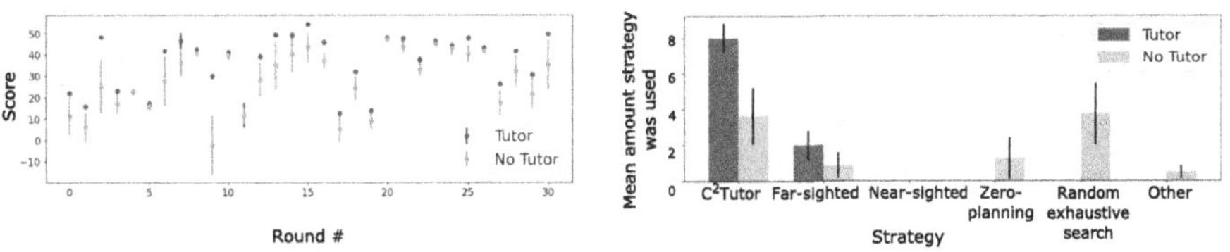

Fig. 2. Left: Average score per round. Right: Average number of times participants used a planning strategy (test rounds). Error bars show standard error for both plots.

Do People Find C^{2}Tutor Useful for Improving Decision-Making? Overall, participants in the Tutor condition found C^{2}Tutor helpful. They also found the feedback easy to understand and use (Fig. 3). Moreover, approximately a third of participants believed that the concepts they learned during the study would be helpful for future decision-making tasks, suggesting the potential for the learned planning strategies to transfer to future decision-making tasks.

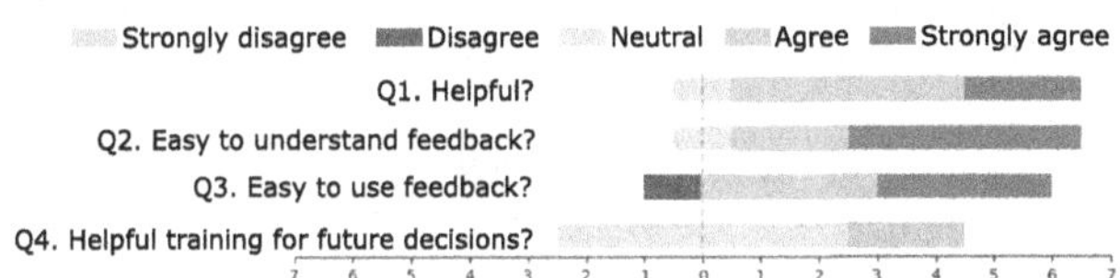

Fig. 3. Participant perceptions of C^{2}Tutor.

6 Conclusion and Future Work

In this work, we sought to create an intervention that has the potential to reduce procrastination habits. To that end, we developed an ITS that teaches people a planning strategy that can reduce present bias tendencies. Unlike prior work, C^{2}Tutor teaches a planning strategy that discourages present bias behavior in a manner that enables individualization. In addition, C^{2}Tutor uses a principled teaching methodology that is based on theories of formative feedback, metacognitive reinforcement learning, and behavior change.

We found that participants who practiced decision-making with C^{2}Tutor were significantly more likely to use our flexible backwards planning strategy. Meanwhile, participants who did not use our ITS were significantly more likely to use maladaptive strategies. While these results are promising, we note that C^{2}Tutor only focused on reducing present bias in a simple simulated environment. Future work should consider how educational technologies can be used to tempter other contributors to procrastination in more complex environments.

Acknowledgements. This work was supported in part by grants from Amii; a Canada CIFAR AI Chair, Amii; Compute Canada; Huawei; Mitacs; and NSERC.

References

1. Baumeister, R.F., Heatherton, T.F., Tice, D.M.: Losing control: how and why people fail at self-regulation (1994)
2. Callaway, F., et al.: Leveraging artificial intelligence to improve peoples planning strategies. In: Proceedings of the National Academy of Sciences, vol. 119, no. 12 (2022)
3. Mouselab-MDP: A new paradigm for tracing how people plan (2017)
4. Ericsson, K.A., Krampe, R.T., Tesch-Römer, C.: The role of deliberate practice in the acquisition of expert performance. Psychol. Rev. **100**(3), 363–406 (1993)
5. Hattie, J., Timperley, H.: The power of feedback. Rev. Educ. Res. **77**(1), 81–112 (2007)
6. Jain, Y.R., et al.: A computational process-tracing method for measuring people's planning strategies and how they change over time. Behav. Res. **55**, 2037–2079 (2023)
7. Lieder, F., Griffiths, T.L.: Resource-rational analysis: understanding human cognition as the optimal use of limited computational resources. Behav. Brain Sci. **43**, e1 (2020)
8. A cognitive tutor for helping people overcome present bias (2019)
9. Mason, B.J., Bruning, R.: Providing feedback in computer based instruction: what the research tells us. Tech. rep. 9 (2001)
10. O'Donoghue, T., Rabin, M.: Doing it now or later. Am. Econ. Rev. **89**(1), 103–124 (1999)
11. Psychologists struggle to meet demand amid mental health crisis. Tech. rep. DOE-SLC-6903-1. American Psychological Association (2022)
12. Steel, P., König, C.J.: Integrating theories of motivation. Acad. Manag. Rev. **31**(4), 889–913 (2006)

A Machine-Learning Approach to Recognizing Teaching Beliefs in Narrative Stories of Outstanding Professors

Fandel Lin[1(✉)], Ding-Ying Guo[2], and Jer-Yann Lin[3]

[1] USC Information Sciences Institute, Marina del Rey, CA, USA
fandel.lin@usc.edu

[2] Department of Education, National University of Tainan, Tainan, Taiwan
ying@mail.nutn.edu.tw

[3] Department of Material Science, National University of Tainan, Tainan, Taiwan
yann@mail.nutn.edu.tw

Abstract. The coding of text information to recognize the teaching beliefs of outstanding professors is crucial research to enhance teaching performance in university. Most previous studies adopted manual coding, and thus text information was limited to briefly descriptive statements or questionnaires, rather than full narrative stories of outstanding professors, owing to the time-consuming of manual coding. However, outstanding professors' narrative stories, which contained more detailed information about the outstanding professors' thinking and behaviors, were valuable text information to recognize the types of teaching beliefs of outstanding professors. Therefore, to overcome the time-consuming obstacle of manual coding, this study proposes a machine-learning-based approach, which exploits BERT with convolutional LSTM, to code narrative stories of outstanding teachers for the identification of the types of teaching beliefs of outstanding professors. In this study, the text information used for coding was a series of fourteen books published across fourteen years, namely The Stories of Outstanding Professors in National Taiwan University (NTU), which contained one million words describing the stories of three hundred NTU outstanding professors. Towards identifying six categories and thirty subcategories of teaching beliefs revealed in the narrative stories, our approach outperforms comparative methods by 1% to 86% in the F1 score. Comprehensive evaluations validate the effectiveness of our approach in assisting in not only recognizing the teaching beliefs from stories of outstanding professors, but also the process of coding text information from narrative stories.

D.-Y. Guo and J.-Y. Lin are supported by National Science and Technology Council of Taiwan under Grants MOST 110-2410-H-024-022; while Fandel Lin is partially supported by Ministry of Education of Taiwan under Grant Pilot Project on Scholarships for Taiwanese Studying in the Focused Fields at Top Foreign University, and partially supported by USC Information Sciences Institute and Viterbi School of Engineering from University of Southern California.

N. Wang et al. (Eds.): AIED 2023, LNAI 13916, pp. 739–745, 2023.
https://doi.org/10.1007/978-3-031-36272-9_65

Keywords: Educational data mining · Teaching beliefs · Text mining · Multi-label text classification

1 Introduction

The exploration of the outstanding professors' teaching beliefs, which is defined as teachers' ideas or concepts concerning teaching, is important to achieve excellent teaching in universities [6]. Professors' excellent teaching promotes students' effective learning [3], and hence elevates national competitiveness [5]. However, Clark and Peterson [1] argued that teachers' teaching beliefs dominate their teaching behaviors. Thus, investigating outstanding professors' teaching beliefs is an effective approach to understanding what teachers' concepts construct their excellent teaching.

To investigate the teaching beliefs of outstanding professors, the manual coding analysis of text information was frequently adopted as a research method. Because of time consumption, briefly descriptive statements, rather than full narrative stories, were chosen as text information in most manual coding studies to obtain the teaching beliefs of outstanding professors. However, it can take coders up to three years to complete the coding of one million words from a series of 341 narrative stories of outstanding professors [4].

Therefore, a machine-learning-based method is an excellent approach to overcoming the time-consuming obstacle of manual coding. For instance, Tseng, Chou, and Tsai [14] used the SnowNLP to quantify the textual statements in questionnaires for the selection of outstanding professors. However, they analyzed the textual statements, rather than a full narrative story of the professors.

Consequently, this study formulates the recognizing teaching beliefs in narrative stories as a multi-label text classification problem. Given the narrative stories of outstanding professors and the hierarchical taxonomy of teaching beliefs, our goal is to identify the labels of teaching beliefs for each sentence.

We propose a dual-attention convolutional LSTM with BERT in narrative classification and depict the workflow of our approach in Fig. 1. It applies BERT [2] with focal loss [8] to retrieve preliminary sentence representation, and exploits SA-Net [15] with convolutional LSTM to synergize the retrieved labels and contextual patterns among narrative sentences. Evaluation results on a publicly accessible book series show that our approach surpasses comparative and baseline methods by 1% to 86%.

Problem Statement. Given the narrative stories of outstanding professors and the hierarchical taxonomy (category) of teaching beliefs, our goal is to find the corresponding labels of teaching beliefs for each sentence. A sentence from the narrative stories can correspond to either zero, one, or even multiple categories of teaching beliefs.

Besides, we formulate three goals that solutions to the targeted research problem shall achieve as follows.

- **G1** Support multi-label text classification from narrative stories effectively.

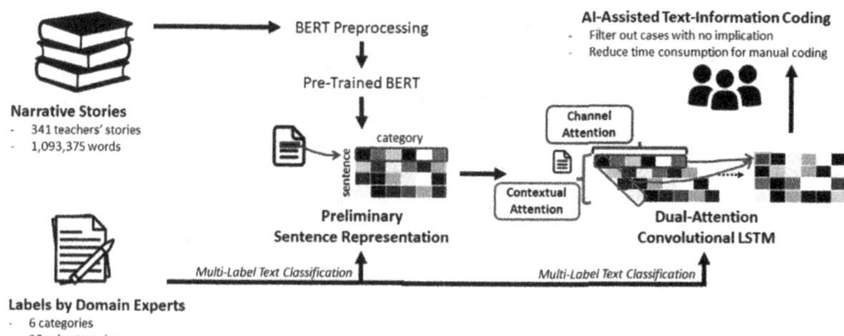

Fig. 1. Workflow of our approach to the targeted research problem.

– **G2** Provide assistance in manual text-information coding by filtering out cases with no implication (falling under no hierarchical taxonomy).
– **G3** Support efficient classification over manual text-information coding.

2 Related Work

In educational data mining in teaching beliefs and outstanding teachers, some targets the perspective of students and apply content analysis of statements from questionnaires [9,14]. In which attentional Long Short-Term Memory (LSTM) [14], k-nearest neighbor [9] are adopted to identify factors related to outstanding teachers. In contrast, some takes the perspective of teachers. In which content analysis for statements of award-winning professors [12], thematic analysis with a semi-structured interview [10], or self-report qualitative questionnaires [11] are adopted. Nevertheless, all the above-mentioned studies analyzed the textual statements from semi-structured interviews or pre-defined questionnaires, rather than a full narrative story. To the best of our knowledge, this study is the first to apply machine-learning-based techniques in identifying the teaching beliefs from narrative stories of outstanding teachers.

3 Dataset

We obtain the stories of outstanding professors from the book series named 'The Stories of Outstanding Professors in National Taiwan University (NTU).' This book series is written based on interviews with outstanding professors each year at NTU, the top university in Taiwan, from 2007 to 2021. With a time span of 14 years, it has 341 stories about outstanding professors from various departments in the university. The book series is written in Chinese and ends up having 25,490 sentences and 1,093,375 words.

We apply the hierarchical taxonomy of teaching beliefs proposed by Guo and Lin [4]. The adopted hierarchical taxonomy has 6 categories with 30 subcategories and 246 sub-subcategories, and can be mapped to other taxonomies proposed in the domain of categorizing teaching beliefs. A sentence may be labeled with zero or multiple subcategories of teaching beliefs. With a 0.97 inter-rater reliability towards the labeling results, these labels are used as the ground truth for our targeted multi-label text classification problem. For the cases in that domain-expert labels do not agree, we use the labels given by the first expert.

4 Methodology

We illustrate the workflow for our approach in Fig. 1. The idea is to first emphasize the sentence representation towards targeted taxonomies, and then refine it according to the taxonomic and contextual relationships.

First, we apply BERT [2] with focal loss [8] for sentence representation to derive the preliminary probabilities towards taxonomies for each sentence. Next, we apply convolutional LSTM [13] with channel and contextual attention [15] to perform multi-label classification based on preliminary probabilities.

5 Evaluation

To demonstrate the efficacy of our approach, we implement the ideas in a system named DECAL (***D***ual-att***E***ntion ***C***onvolutional LSTM with BERT in n***A***rrative c***L***assification). We conduct the evaluation on the book series and aim to answer the three goals in accord with **G1** to **G3** from Sect. 1.

5.1 Evaluation Setting

We adopt the (i) *F1 Score* and (ii) *Area Under Curve (AUC)* as the quantitative evaluation matrices for the targeted multi-label text classification problem. F1 score is the harmonic mean of precision and recall; while AUC is bounded by 0 and 1, and a 0.5 AUC can be viewed as randomization. All the evaluation matrices including F1 score, precision, recall, and AUC are to be maximized.

The narrative stories are split into training, validation, and testing datasets with a ratio of 5:2:3. All methods are implemented in Python based on an Acer desktop computer equipped with an Intel i7-1165G7 CPU at 2.80GHz and 32 GB of RAM. We run all methods 10 times and report the average along with the standard deviation. Source code is available on Figshare[1].

5.2 Comparative Method

We adopt six comparative and baseline methods. For all comparative methods, we employ BERT preprocessing for the text representation of sentences. The comparative methods include: BERT [2], LSTM [7], LSTM with SA-Net (AT-LSTM) [14,15], Convolutional LSTM (CV-LSTM) [13], Multi-Label Random Selection (ML-RS), and Single-Label Random Selection (SL-RS).

[1] https://dx.doi.org/10.6084/m9.figshare.21908154.

Table 1. Overall performance among comparative methods for multi-label text classification towards 30 subcategories.

Method	F1	Precision	Recall	AUC
DECAL	**0.277** ± 0.005	0.207 ± 0.005	0.421 ± 0.019	**0.827** ± 0.003
BERT	0.158 ± 0.000	**0.297** ± 0.000	0.107 ± 0.000	0.787 ± 0.000
LSTM	0.266 ± 0.011	0.205 ± 0.006	0.380 ± 0.038	0.820 ± 0.011
AT-LSTM	0.259 ± 0.004	0.199 ± 0.007	0.370 ± 0.021	0.815 ± 0.005
CV-LSTM	0.274 ± 0.005	0.204 ± 0.004	0.420 ± 0.024	0.825 ± 0.006
ML-RS	0.087 ± 0.001	0.047 ± 0.001	**0.500** ± 0.006	0.500 ± 0.004
SL-RS	0.039 ± 0.001	0.048 ± 0.002	0.033 ± 0.001	0.500 ± 0.004

Table 2. Reliability assessment of comparative methods in decreasing candidate size while preserving instances that may be labeled with incorrect taxonomies.

Task	6 Categories		30 Subcategories	
Method	Recall*	Decrease Rate	Recall*	Decrease Rate
DECAL	0.994 ± 0.002	0.115 ± 0.005	0.989 ± 0.006	0.125 ± 0.008
BERT	0.969 ± 0.000	0.163 ± 0.000	0.455 ± 0.000	0.623 ± 0.000
LSTM	0.998 ± 0.002	0.106 ± 0.013	0.972 ± 0.017	0.127 ± 0.024
AT-LSTM	0.993 ± 0.004	0.111 ± 0.009	0.989 ± 0.007	0.120 ± 0.015
CV-LSTM	0.995 ± 0.003	0.113 ± 0.014	0.986 ± 0.005	0.128 ± 0.010
ML-RS	0.984 ± 0.002	0.016 ± 0.001	1.000 ± 0.000	0.000 ± 0.000
SL-RS	1.000 ± 0.000	0.000 ± 0.000	1.000 ± 0.000	0.000 ± 0.000

5.3 Evaluation Result

Overall Quantitative Performance. For **G1**, we compare the overall quantitative performance for classifying 30 subcategories in Table 1. Our proposed DECAL retains the highest F1 score and AUC among all comparative methods.

Since we employ BERT preprocessing for the text representation of sentences, BERT acts as a baseline that retrieves the highest precision among all methods. All the other methods tend to sacrifice precision in exchange for a higher recall.

Reliability Assessment. For **G2**, we present the reliability assessment for 6 and 30 (sub)categories in Table 2. In Table 2, recall* is the fraction of sentences, with their labels corresponding to more than one hierarchical taxonomies, that are successfully recognized despite assigning incorrect taxonomies. On the other hand, the decrease rate refers to the fraction of instances that are filtered out.

According to the procedure of manual text-information coding and interview with domain experts, we find that a higher recall* is preferred over precision for experts. Since recall* and decrease rate is to be maximized, there is no comparative method that can retrieve better results in both criteria over DECAL.

Running Time Performance. For **G3**, the time required in the training and testing process of DECAL is less than 6 h based on the provided computational resources. Considering the recall* and decrease rate in Table 2, the running time for DECAL is reasonable compared to manual text-information coding by domain experts, which takes around three years for the targeted book series.

6 Conclusion

This paper studies the problem of recognizing teaching beliefs from the narrative stories of outstanding professors. We take the book series named The Stories of Outstanding Professors in National Taiwan University (NTU) as the dataset for multi-label text classification. The book series contains more than three hundred narrative stories of outstanding professors from NTU across fourteen years. To extract sentences that reveal teaching beliefs, we propose a dual-attention convolutional LSTM with BERT in narrative classification. Evaluation results show that our approach surpasses comparative methods by 1% to 86% in the F1 score. Besides, our approach can provide reliable assistance in manual text-information coding by filtering out a decent amount of cases while maintaining high recall.

References

1. Clark, C.M., Peterson, P.L.: Teachers' thought processes.In: Wittrock, M.C. (Ed.), Handbook of Research on Teaching, 3rd edn, pp. 255–296. Macmillan (1986)
2. Devlin, J., Chang, M.W., Lee, K., Toutanova, K.: Bert: Pre-training of deep bidirectional transformers for language understanding. arXiv preprint arXiv:1810.04805 (2018)
3. Feldman, K.A.: Identifying exemplary teachers and teaching: evidence from student ratings. In: Perry, R.P., Smart, J.C. (eds.) The scholarship of teaching and learning in higher education: An evidence-based perspective, pp. 93–143. Springer, Cham (2007). https://doi.org/10.1007/1-4020-5742-3_5
4. Guo, D.Y., Lin, J.Y.: The analysis of teaching ideas in the stories of outstanding teachers in national Taiwan university. Under Review (2023)
5. Hardy, I.: Education as a 'risky business': theorising student and teacher learning in complex times. Br. J. Soc. Educ. **36**(3), 375–394 (2015)
6. Hativa, N., Barak, R., Simhi, E.: Exemplary university teachers: knowledge and beliefs regarding effective teaching dimensions and strategies. J. High. Educ. **72**(6), 699–729 (2001)
7. Hochreiter, S., Schmidhuber, J.: Long short-term memory. Neural Comput. **9**(8), 1735–1780 (1997)
8. Lin, T.Y., Goyal, P., Girshick, R., He, K., Dollár, P.: Focal loss for dense object detection. In: Proceedings of the IEEE International Conference on Computer Vision, pp. 2980–2988 (2017)
9. Okoye, K., Arrona-Palacios, A., Camacho-Zuñiga, C., Achem, J.A.G., Escamilla, J., Hosseini, S.: Towards teaching analytics: a contextual model for analysis of students' evaluation of teaching through text mining and machine learning classification. Educ. Inf. Technol. **27**(3), 3891–3933 (2022)
10. Pychyl, T.A., Flett, G.L., Long, M., Carreiro, E., Azil, R.: Faculty perceptions of mattering in teaching and learning: a qualitative examination of the views, values, and teaching practices of award-winning professors. J. Psychoeducational Assess. **40**(1), 142–158 (2022)

11. Reichert, F., Torney-Purta, J.: A cross-national comparison of teachers' beliefs about the aims of civic education in 12 countries: a person-centered analysis. Teach. Teach. Educ. **77**, 112–125 (2019)
12. Rossetti, J., Fox, P.G.: Factors related to successful teaching by outstanding professors: an interpretive study. J. Nurs. Educ. **48**(1), 11–16 (2009)
13. Shi, X., Chen, Z., Wang, H., Yeung, D.Y., Wong, W.K., Woo, W.C.: Convolutional lstm network: A machine learning approach for precipitation nowcasting. In: Advances in Neural Information Processing Systems, vol. 28 (2015)
14. Tseng, C.W., Chou, J.J., Tsai, Y.C.: Text mining analysis of teaching evaluation questionnaires for the selection of outstanding teaching faculty members. IEEE Access **6**, 72870–72879 (2018)
15. Zhang, Q.L., Yang, Y.B.: Sa-net: Shuffle attention for deep convolutional neural networks. In: ICASSP 2021–2021 IEEE International Conference on Acoustics, Speech and Signal Processing (ICASSP), pp. 2235–2239 (2021)

BETTER: An Automatic feedBack systEm for supporTing emoTional spEech tRaining

Adam Wynn(✉) and Jingyun Wang

Durham University, Durham DH1 3LE, UK
adam.t.wynn@durham.ac.uk

Abstract. Feedback is a crucial process in education because it helps learners identify their weaknesses whilst motivating them to continue to learn. Existing systems often only provide a score or rating with basic explanations. Although some systems provide detailed feedback, they require manual input from teachers. This paper proposes a real-time feedback visualisation system (called BETTER) for supporting emotional speech training which uses a visual dashboard to provide the learner with immediate feedback in the form of written, audio, and visual feedback. The AI-based feedback system utilises pitch tracking, transcriptions, and audio modifications in addition to one-dimensional convolutional neural networks (CNNs) to categorise speech into emotional states. A preliminary experiment was conducted involving a speech expert and 8 non-native speakers to assess their cognitive load, technology acceptance, and satisfaction while using the system.

Keywords: Automatic visualisation feedback · Emotion recognition · Speech prosody · Emotional speech training · System Evaluation

1 Introduction

In education, feedback is an important process as it enables students to identify misconceptions and reflect on their learning progress whilst reducing instructor effort [3]. The primary methods and techniques used to generate automatic feedback include comparisons with the desired solution, interactive dashboards and visualisations, and text generated by natural language processing [3]. Due to advances in machine learning and audio feature extraction techniques, new feedback mechanisms involving emotion detection, virtual worlds feedback, game-mediated feedback, and oral feedback have been proposed [2].

It can be challenging to successfully convey emotion in a public speaking environment. Therefore, this paper proposes providing automatic feedback for emotional speech training and seeks to address the following research questions: *1) Does providing automatic feedback help learners improve their speech?*; *2) What are the learner perception (including cognitive load, technology acceptance,*

N. Wang et al. (Eds.): AIED 2023, LNAI 13916, pp. 746–752, 2023.
https://doi.org/10.1007/978-3-031-36272-9_66

and satisfaction for the system) differences between training one and multiple sentences? ; 3) What are the learner perception differences between real-time feedback addressing a specific discrete emotion and addressing positive and negative emotions?

To explore the above questions, BETTER, an AI-based feedBack systEm for supporTing emoTional spEech tRaining, which gives real-time automatic feedback via a visualisation dashboard is proposed. Our main contributions are supporting speech training by providing detailed feedback on a visual dashboard including not only the transcription and pitch information but also emotional information, and expanding the system based on the feedback of participants so that for the first time, visual strategies related to pitch changes and modified audio are provided automatically to assist the learners in conveying their speech with more accurate emotions.

2 Related Work

Although there are some systems that include techniques for acquiring features from audio signals, they usually only focus on single features at a time such as pronunciation, pitch, and emotion and most of these systems do not have feedback for the learner as their focus and focus more on how to abstract the features accurately. For example, Sztaho, et al. [14] aimed to improve the intonation and rhythm of children with hearing loss using a visual feedback system displaying the learner's pitch and the correct pitch in an easy-to-understand way using a line graph. Aucouturier, et al. [1] and Rachman, et al. [13] provide audio feedback by modifying speech signals in real-time by applying audio effects including pitch shifting, inflection and vibrato, and by using high-pass and low-pass filters. However, these effects can made speech sound unnatural and in [1], not all participants detected any manipulation. Zhou, et al. [16] use sequence-to-sequence encoders with attention to reconstruct speech which conveys a combination of emotions, however, the challenges of unnatural emotional expressions remain.

Detecting emotion involving classifying speech into emotional states is one research direction. Using Convolutional Neural Networks (CNNs), an accuracy of 71% was achieved by Franti, et al. and it was identified that someone who was speaking faster which a higher and wider pitch range were more likely to be experiencing emotions of fear, anger of joy, whereas emotion such as sadness generate slow and lower pitched speech. The framework by Issa, et al. [8] using CNNs obtains 71.61% for the RAVDESS dataset [11] with 8 classes. Using a reinforcement learning network to keep track of emotional changes in every utterance throughout the speech is another research direction. Huang et al. [7] propose a method where a sliding window would be used to collect emotional information using Mel-frequency cepstral coefficients (MFCCs), which approximate the human auditory system's response. Using this information, an agent would decide which emotion to classify the sentence as.

3 Methodology

3.1 BETTER Version 1.0

Both visual and basic written feedback is provided by BETTER 1.0 [15] built using the Panel [5] library. Two modes are provided: Mode 1) a positive and negative emotions mode and Mode 2) a 5 discrete emotions mode. In mode 1, the learner is presented with a graph which shows their audio emotion and content emotion on a 5-point Likert scale (very positive, positive, neutral, negative, very negative), and their pitch. In mode 2, as shown in Fig. 1, the user is able to see how the intensity of the focused emotion changes throughout the speech, how long each sentence is, and how long the pauses between each sentence are. BETTER shows which emotion was most prevalent out of fear, happy, anger, sad, and neutral along with the level of the emotion detected indicated by the colour of the bar. The "Content Emotion" is also provided in green which is the emotion of the text produced by the transcript. The learner can look at the emotion identified for the content of their speech, and try to match that emotion when speaking. Finally, BETTER shows how the pitch changes throughout the speech which can be used to infer how this might affect the emotion detected. The pitch is not displayed when there is a pause between sentences to make it clearer to the user how the pitch in each individual sentence fluctuates.

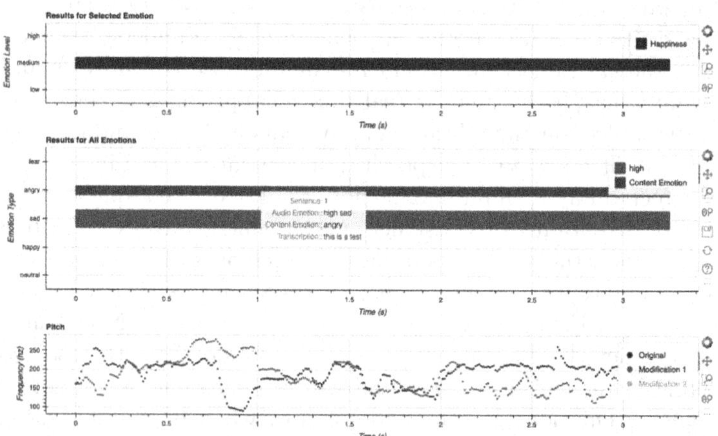

Fig. 1. Part of the system output when in Discrete Emotions Mode showing how the user's emotion and pitch changes throughout their speech.

1D CNNs are used to classify each sentence and were programmed using the Keras library [4] and TensorFlow. 40 MFCC features are extracted from the data using the Librosa package [12] and used as input into the models. To classify emotions on a 5-point Likert scale, a model was trained using batches of size 64 over 100 epochs. The model architecture consists of 4 1-dimensional convolutional layers, a batch normalisation layer, dropout layers and 3 dense layers. The model was trained using the RAVDESS [11], TESS [6], and SAVEE [9] datasets. These datasets were manually relabelled by a native speaker using the same 5-point likert scale as above to avoid inconsistencies between datasets. This model achieved a validation accuracy of 89.82% using 5-fold validation. To classify discrete emotions (anger, fear, sadness, happiness, and neutral), another model was trained with the same architecture as above, using data from the RAVDESS [11], TESS [6], and SAVEE [9] datasets which include a total of 3876 audio recordings of English speakers. This model achieved a validation accuracy of 90.05% using 5-fold validation.

3.2 Preliminary Experiment

A preliminary experiment involving a female expert in speech training and eight male Chinese students who study in a university in the UK, was conducted using BETTER 1.0. During the experiment, participants were asked to use Mode 1 followed by Mode 2. Once the audio was analysed, the participant was presented with visual feedback showing how their emotion, pronunciation, and pitch changes (using Parselmouth [10]) throughout their speech.

After the speech training activity, the students were interviewed and completed a questionnaire consisting of 31 questions related to mental load, technology acceptance, and satisfaction. 62.5% and 12.5% of participants agreed and strongly agreed respectively that BETTER's feedback on their emotions was easy to understand, and 87.5% of participants found the feedback given on their emotions effective and useful. Participants had mixed opinions on whether viewing positive and negative emotions or discrete emotions was easier to understand and more useful, with 37.5% and 25% preferring each, respectively. Overall, 75% of participants found the combination of content and audio emotion helpful for improving their speech. Overall, 87.5% of participants found BETTER easy to use and navigate, and 62.5% felt confident using it to improve their speech. However, only 50% agreed that the overall feedback provided by the system was effective and useful, and 75% expressed a desire to use the system frequently. For the questions about cognitive load, participants answered on a scale from 1–7 where 1 is "very little" and 7 is "a lot". Participants rated the effort required to understand the learning activity with a mean of 3.5 (SD = 1.41). In general, participants felt more distracted when using mode 2 (Mean = 3.625, SD = 1.30) compared to mode 1 (Mean = 3.5, SD = 1.60), and more stressed or irritated when using mode 2 (Mean = 4.125, SD = 1.25) compared to mode 1 (Mean = 3, SD = 1.60).

Six of the students found the transcription and audio playback helpful for revising and fixing pronunciation mistakes. However, participants suggested that the feedback could be more detailed, such as providing an explanation of how to match the detected emotion or giving feedback on energy, speed, and other speech aspects. The expert suggested providing written feedback and a comparison with a native speaker's recorded sample sentence for easier interpretation.

Participants found it difficult how to interpret the curve of the sentence's pitch and suggested an explanation of how pitch impacts the conveyed emotion.

3.3 BETTER Version 2.0

Based on the results of the preliminary study, BETTER was extended to version 2.0 and enhanced with more detailed feedback by adding written explanations of how to match the selected emotion along with a simple diagram. The explanation consisted of how the user can adapt their pitch throughout the sentence and their rate of speech, as shown in the left part of Fig. 2.

Moreover, an audio playback function was added, as shown in the right part of Fig. 2, to provide the user with 2 modified versions of their speech, together with their original recording. The audio clips were modified by applying effects including speed changes, pitch shifting, inflection, vibrato, and filtering [1,13], and by adjusting parameters to create subtle and exaggerated versions.

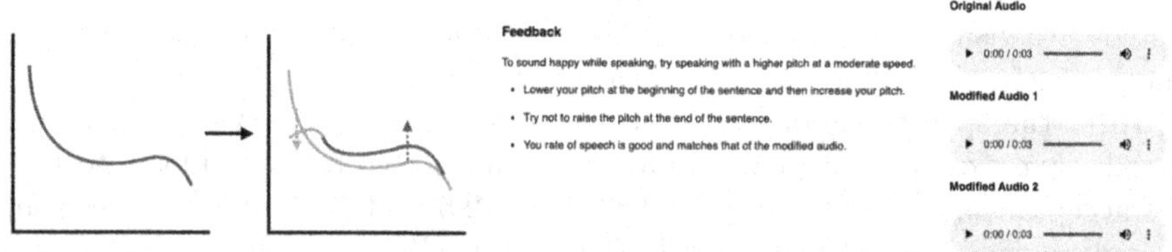

Fig. 2. Part of the system output when in Discrete Emotions Mode showing written feedback and audio modifications

A further experiment is currently being conducted to evaluate the effectiveness of BETTER 2.0. Until now, 31 students aged 18–35 consisting of 18 male and 13 female speakers have participated in a 1 h long experiment and were required to interact with BETTER 2.0 during the experiment. It is found that most of the participants agreed that automatic feedback is effective in improving the speech of learners and thought that training one sentence at a time is slightly easier and more useful. The participants acknowledge that both modes of BETTER 2.0 are easy to use and useful for speech training and the results suggest that mode 2 is more useful despite mode 1 being easier to use. The experiment will involve more female participants to ensure a more balanced representation. The final results will be analysed in the future.

4 Conclusion

The main contribution of this paper was to present a system that has been implemented from experimental results to support emotional speech training. For the first time, speech training is supported by detailed feedback provided on a visual dashboard including not only the transcription and pitch information but also emotional information with visual strategy related to pitch changes and modified audio, to support the learners in adjusting their speech.

With reference to future improvement, we plan to provide more specific personalised feedback tailored towards each user considering differences between different languages and cultures. Furthermore, we plan to carry out further experiments with more diverse participants and analyse and explore the speaker's audio collected during the experiments.

References

1. Aucouturier, J.J., Johansson, P., Hall, L., Segnini, R., Mercadié, L., Watanabe, K.: Covert digital manipulation of vocal emotion alter speakers' emotional states in a congruent direction. Proc. Natl. Acad. Sci. **113**(4), 948–953 (2016). https://doi.org/10.1073/pnas.1506552113
2. Bahari, A.: Computer-mediated feedback for l2 learners: challenges versus affordances. Jouney Comput. Assist. Learn. **37**, 24–38 (2020). https://doi.org/10.1111/jcal.12481
3. Cavalcanti, A.P., Barbosa, A., Carvalho, R., Freitas, F., Tsai, Y.S., Gašević, D., Mello, R.F.: Automatic feedback in online learning environments: a systematic literature review. Comput. Educ. Artif. Intell. **2**, 100027 (2021). https://doi.org/10.1016/j.caeai.2021.100027
4. Chollet, F., et al.: Keras (2015). https://github.com/fchollet/keras
5. contributors, H.: Panel (2022). https://panel.holoviz.org
6. Dupuis, K., Pichora-Fuller, M.: Toronto emotional speech set (TESS) Collection (2010). https://doi.org/10.5683/SP2/E8H2MF
7. Huang, X., et al.: Emotion detection for conversations based on reinforcement learning framework. IEEE Multimedia **28**(2), 76–85 (2021). https://doi.org/10.1109/MMUL.2021.3065678
8. Issa, D., Fatih Demirci, M., Yazici, A.: Speech emotion recognition with deep convolutional neural networks. Biomed. Signal Process. Control **59**, 101894 (2020). https://doi.org/10.1016/j.bspc.2020.101894
9. Jackson, F.P., Haq, S.: Surrey Audio-Visual Expressed Emotion (SAVEE) Data (2015). http://kahlan.eps.surrey.ac.uk/savee/Introduction.html
10. Jadoul, Y., Thompson, B., de Boer, B.: Introducing parselmouth: a python interface to praat. J. Phon. **71**, 1–15 (2018). https://doi.org/10.1016/j.wocn.2018.07.001
11. Livingstone, S.R., Russo, F.A.: The Ryerson audio-visual database of emotional speech and song (RAVDESS) (2018). https://doi.org/10.5281/zenodo.1188976
12. McFee, B., et al.: Thassilo: librosa/librosa: 0.8.1rc2 (2021). https://doi.org/10.5281/zenodo.4792298
13. Rachman, L., et al.: DAVID: an open-source platform for real-time transformation of infra-segmental emotional cues in running speech. Behav. Res. Methods **50**(1), 323–343 (2018)
14. Sztahó, D., Kiss, G., Vicsi, K.: Computer based speech prosody teaching system. Comput. Speech Lang. **50**, 126–140 (2018). https://doi.org/10.1016/j.csl.2017.12.010

15. Wynn, A.T., Wang, J., Umezawa, K., Cristea, A.I.: An AI-based feedback visualisation system for speech training. In: Rodrigo, M.M., Matsuda, N., Cristea, A.I., Dimitrova, V. (eds.) Artificial Intelligence in Education. Posters and Late Breaking Results, Workshops and Tutorials, Industry and Innovation Tracks, Practitioners' and Doctoral Consortium, vol. 13356, pp. 510–514. Springer, Cham (2022). https://doi.org/10.1007/978-3-031-11647-6_104
16. Zhou, K., Sisman, B., Rana, R., Schuller, B.W., Li, H.: Speech synthesis with mixed emotions (2022). https://doi.org/10.48550/ARXIV.2208.05890

Eliciting Proactive and Reactive Control During Use of an Interactive Learning Environment

Deniz Sonmez Unal[1(✉)], Catherine M. Arrington[2], Erin Solovey[3], and Erin Walker[1]

[1] University of Pittsburgh, Pittsburgh, PA 15260, USA
{d.sonmez,eawalker}@pitt.edu

[2] Lehigh University, Bethlehem, PA 18015, USA
kate.arrington@lehigh.edu

[3] Worcester Polytechnic Institute, Worcester, MA 01609, USA
esolovey@wpi.edu

Abstract. The dual mechanisms of control framework describes two modes of goal-directed behavior: proactive control (goal maintenance) and reactive control (goal activation on task demands). Although these mechanisms are relevant to learner behaviors during interaction with intelligent tutoring systems (ITS), their relation to ITSs is under-researched. We propose a manipulation to induce proactive or reactive control during interaction with an online tutoring system. We present two experiments where students solved problems using either proactive or reactive control. Study 1 validates the manipulation by investigating behavioral measures that reflect usage of the intended strategy and assesses whether either mode impacted learning. Study 2 investigates if alternating between control modes during problem solving affects student performance.

Keywords: Cognitive control · problem solving · learning environments

1 Introduction

Cognitive control is the set of processes that enable behavior adjustments based on current goals [2]. It is explored within cognitive science with its roles in functions such as attention, task switching, and goal maintenance [13], self regulation [8]. Previous work has explored these functions at an abstract level in ITSs, through detection of mind wandering [9], divided attention [17], and multi-tasking [11] as well as self-regulatory behaviors such as resisting the urge to game the system [5] and self regulated learning [1]. Detecting these functions to adapt instruction is the main feature of ITSs and it is in interest of AIED community. Although these functions all share the underlying mechanism of cognitive control, cognitive control itself has been explored less in AIED.

N. Wang et al. (Eds.): AIED 2023, LNAI 13916, pp. 753–759, 2023.
https://doi.org/10.1007/978-3-031-36272-9_67

The Dual Mechanisms of Cognitive Control Framework [3] suggests cognitive control has two modes: proactive and reactive control. Proactive control involves anticipating and preparing for future events by biasing attention and actions towards task goals (e.g. a student preparing a strategy for an expected problem in advance). Reactive control involves retrieving goal-relevant information just-in-time when triggered by external stimuli (e.g. a student relies on a keyword in the problem to use a strategy). Even though prior work shows a positive relationship between proactive control and student performance [12,16], proactive control has not been explored in realistic learning settings. In this work, we aim to gain insight into the underlying mechanisms of higher-order functions studied in ITSs and assess if these functions' benefits can be attributed to cognitive control. Our research questions are: (1) How can we induce proactive or reactive modes of cognitive control within an ITS? (2) What is the relationship between cognitive control and learning?

2 Task Design

The relative usage of proactive and reactive control can be determined using the AX Continuous Performance Task (AX-CPT) [3]. Participants are presented with cue-probe pairs of letters and their response time and error rates indicate their bias towards proactive or reactive control. Previous research has successfully shifted this bias through training strategies encouraging the use of one mode over the other [6]. Our task design creates a goal maintenance scenario similar to this manipulation in problem solving. Our task had 9 probability problems in the ASSISTments tutoring system [7]. Problems were divided into 3 to 4 substeps. First, participants were shown the full problem and asked to report their confidence level. Then, they were shown the problem's substeps, which they solve sequentially (see Fig. 1). This design was used in a thinkaloud study where subjects verbalized their thought processes [15] showing success in inducing proactive and reactive control. Here, we do not use the thinkaloud protocol to eliminate confounding factors, and to get more accurate behavioral data.

The initial presentation of the full problem serves to establish a task goal. The substeps of the problem serve to make goal maintenance more challenging. To induce proactive or reactive control, we have participants use an instructed control strategy. The task had two different versions of instructions. In the "proactive" version, in each problem step, participants were instructed: "Think about how this step relates to the goal of the problem,". In the "reactive" version, the instruction was: "Think about how you are solving this step,". The proactive instruction encourages the student to maintain the main problem's goal while solving its substeps and the reactive instructions made the student only focus on the current step and not the overall goal until they need it.

3 Study 1: Inducing Proactive and Reactive Control

We recruited 29 undergraduate students (21 female, 5 male, 3 non-binary; 14 Asian, 13 White/Caucasian, 1 two or more races) between 18 and 22 years old

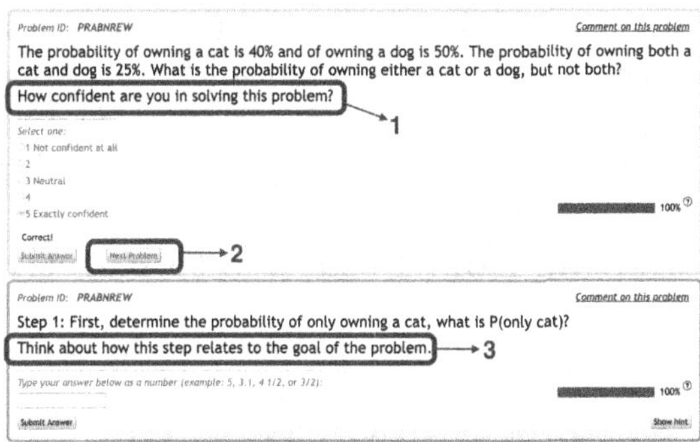

Fig. 1. Screenshot of a sample problem. 1: Participants rate their confidence level. 2: When this button is clicked, the second box with the first problem step unfolds. 3: Prompt to trigger proactive control.

(M = 19.8, SD = 1.13) at a university in the Northeastern US. We recruited participants via email and online bulletin boards. Eligibility criteria included reporting completion of no more than two university-level math courses. Participants first completed a sample problem and a pre-test on ASSISTments. Then, they practiced using proactive or reactive control with another sample problem. They then solved the probability problems using the instructed control strategy, followed by a post-test. Finally they completed a demographic questionnaire.

Data and Measures. One participant's data were excluded as they scored 100% on the pre-test. We also excluded a participant and problem step pair due to a logging error. We had 1035 rows in the final dataset from 28 participants and 37 problem steps. To evaluate participants' compliance with the instructions, we analyzed response times, as previous research showed participants instructed to use proactive control spend more time on problem steps [15]. Studies with AX-CPT report similar results, where response times were significantly different between when proactive and reactive control were used [6]. We measured the time spent on problem steps as the duration between viewing a problem step and the next step. Learning gains were assessed by the difference in the percentages of correctly answered questions between the pre-test and the post-test.

Results. Results of a two-sample t-test show participants in the proactive condition spent significantly more time on the problem steps ($Proactive : M = 21.36, SD = 6.07, Reactive : M = 16.59, SD = 4.53, t(24.05) = 2.36, p < .05$), and marginally more time on the initial step with the full problem ($Proactive : M = 16.50, SD = 6.30, Reactive : M = 12.67, SD = 2.99, t(18.58) = 2.05, p = .05$). In line with the results from [15], the extra time may reflect time spent on proactive strategies. Next, we tested if control strategy affected learning. We first confirmed students' pre-test scores did not differ significantly between the conditions ($Proactive : M = 0.42, SD = 0.20, Reactive : M = 0.52, SD = 0.18$), ($t(25.81) = -0.51, p > 0.5$). Then we used a repeated-measures ANOVA with condition as the between-group variable and the tests (pre and post) as the within-group variable. The significant main effect of test ($F(1, 26) = 51.33, p < .001$) indicated that participants improved from pre to post test. We found no significant interaction between the condition and test ($F(1, 26) = 1.45, p > .05$) indicating no significant difference in learning gain between conditions.

4 Study 2: Shifting from Proactive to Reactive Control

The proactive control manipulation in our task promotes self-explanation by requiring participants to consider the problem steps' relation to the problem goals. Prior work suggests these activities become extraneous to learning once the skill associated with the problem step was acquired [14]. Therefore, in a second study, we investigate if proactive control helps in problem solving when it is used only in the first time that the participant was introduced to a skill.

Adjustments to Task Design. We introduced another condition which we called 'Mixed'. Participants in this condition were given the proactive instructions at first time they see a problem type. In the next occurrences of the same problem type, reactive instructions were given. This condition aims to facilitate skill acquisition while the students' cognitive resources are still available by prompting them to use proactive control at the first occurrence of such skill. In later occurrences, the students will use reactive control to avoid cognitive load proactive control may cause and to speed up solving the problems. Another addition was participants were prompted to indicate their knowledge on each problem step after giving their answer (Fig. 2). This was for ongoing work to study the stages of rule learning.

We recruited 56 undergraduate students (40 female, 14 male, 2 non-binary; 22 Asian, 21 White/Caucasian, 4 African-American, 4 Two or More Races, 3 Hispanic/Latino, 2 did not disclose), ranging from 18 to 27 years old (M = 20.36, SD = 1.41). Participants followed the same procedure as in Study 1.

Data and Measures. Our dataset had 3380 rows consisting of 52 participants (4 participants were excluded as they scored 100% on the pre-test) and 37 problem steps and 28 self-report knowledge prompts. We labeled each row in the dataset with a "step condition" to account for participants' proactive and reactive modes at different steps of a problem. This new column enabled us to compare the time spent between proactive and reactive steps across all participants, including those in the mixed condition who could switch between modes.

Planned Analyses. We build a mixed effects model with time spent on problem steps (excluding the confidence rating and knowledge indication steps, Fig. 2 A and C, as there were no strategic instructions on these steps) as the dependent variable, participant, problem, and problem steps as random factors and step condition as the fixed factor. We found step condition was not effective on time spent as the coefficient of the fixed effect explains time spent on the problem steps decrease by 1.65 s from proactive to reactive steps (Table 1).

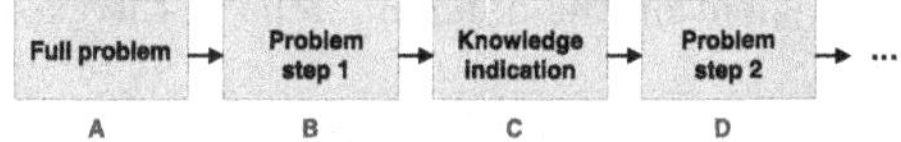

Fig. 2. Sequence of steps in problem solving in Study 2.

To investigate learning, we first confirmed the pre-test scores did not differ significantly ($F(2, 49) = 1.26, p = 0.29$). We conducted a repeated-measures ANOVA with condition as the between-group variable and the tests (pre and post) as the within-group variable. The main effect of test was significant ($F(1, 49) = 116.07, p < .001$) indicating participants improved from pre-test to post-test. The condition and test interaction was not significant ($F(2, 49) = 0.5, p = 0.61$) meaning no significant difference in learning among the conditions.

Table 1. Summary of the planned and post-hoc mixed effects models.

	Planned Analyses			Post-hoc Analyses		
Fixed effects						
(Intercept)	Coefficient	Std. Error	*t* value	Coefficient	Std. Error	*t* value
Step condition (reactive)	28.322	3.687	7.681	19.751	1.926	10.257
	−1.645	1.883	-0.874	−2.716	1.362	−1.993
Random effects						
Groups		Variance	Std. Dev		Variance	Std. Dev
Participant ID		63.52	7.970		15.23	3.902
Problem step: Problem		147.18	12.132		32.35	5.688
Problem		51.49	7.176		10.63	3.261
Residual		372.94	19.312		328.29	18.119

Post-hoc Analyses. We investigated if task design adjustments caused the non-significant result on response time as it contradicts previous results. We first assessed if participants in the mixed condition complied with the instructions. We conducted a paired samples t-test to compare the time spent on problem steps with proactive and reactive instructions. Results showed a significant difference ($t(14) = 9.88, p < 0.001, Proactive : M = 42.91, SD = 10.78, Reactive : M = 26.73, SD = 11.51$). A lack of significant difference in time spent on problem steps between the general proactive and reactive conditions (Table 2) suggests that participants in the proactive condition may not be engaging in the instructed reflection or they may be spending more time on the initial step with the full problem goal they were instructed to maintain. Additionally, they may still engage in reflection during the knowledge indication prompts since those appear before a new problem step. We repeat the mixed effects model including the time spent on the confidence rating and the knowledge indication steps (Fig. 2 A, B, C, D). Results (Table 1) show a stronger effect of step condition as the time spent was decreasing by 2.716 s between proactive and reactive steps. A post-hoc Tukey test shows the difference was significant ($p = 0.04$).

Table 2. Average time spent on the different steps in the experiment in seconds.

Condition	Problem substeps	Confidence rating	Knowledge indication
Proactive	25.7 (22.1)	20.0 (17.7)	7.47 (5.49)
Reactive	26.6 (24.7)	17.3 (12.2)	6.64 (4.65)
Mixed	30.5 (28.3)	16.7 (8.43)	7.70 (5.02)

5 Discussion

Our experiments confirm proactive and reactive control can be induced in a realistic learning setting in line with the prior findings [15]. Even though we obtained mixed results regarding the time spent in study 2, post-hoc analyses showed participants who received the proactive instructions still spent more time when the additional steps in study 2 were considered. These results can indicate the proactive control prompt may encourage other strategies such as reflecting on the solution [10]. We found no significant difference in learning. We argue proactive prompts encourage self-explanations and therefore provide benefits. However, it is possible that the quality of self-explanations was not sufficient for improving learning [4]. Another factor may be the limited problem set we used to ensure consistency with previous work. Participants may have adjusted to the few problem types quickly and therefore there is no visible differences in learning. Another explanation may be the benefits of learner behaviors that involve proactive control could not be isolated to this underlying mechanism, and there may be no direct causal connection between proactive control and learning. Our studies provide preliminary insights into low-level cognitive mechanisms that support higher-level cognitive states studied to understand learner behaviors.

Acknowledgments. This work was supported by the National Science Foundation Award Nos 185307 and 1912474.

References

1. Azevedo, R., Hadwin, A.F.: Scaffolding self-regulated learning and metacognition-implications for the design of computer-based scaffolds. Instr. Sci. **33**(5/6), 367–379 (2005)
2. Botvinick, M.M., Braver, T.S., Barch, D.M., Carter, C.S., Cohen, J.D.: Conflict monitoring and cognitive control. Psychol. Rev. **108**(3), 624 (2001)
3. Braver, T.S.: The variable nature of cognitive control: a dual mechanisms framework. Trends Cogn. Sci. **16**(2), 106–113 (2012)
4. Conati, C., Vanlehn, K.: Toward computer-based support of meta-cognitive skills: a computational framework to coach self-explanation. Int. J. Artif. Intell. Educ. (IJAIED) **11**, 389–415 (2000)
5. Dang, S., Koedinger, K.: Exploring the link between motivations and gaming. Int. Educ. Data Min. Soc. (2019)
6. Gonthier, C., Macnamara, B.N., Chow, M., Conway, A.R., Braver, T.S.: Inducing proactive control shifts in the AX-CPT. Front. Psychol. **7**, 1822 (2016)

7. Heffernan, N.T., Heffernan, C.L.: The ASSISTments ecosystem: building a platform that brings scientists and teachers together for minimally invasive research on human learning and teaching. Int. J. Artif. Intell. Educ. **24**(4), 470–497 (2014). https://doi.org/10.1007/s40593-014-0024-x
8. Hofmann, W., Schmeichel, B.J., Baddeley, A.D.: Executive functions and self-regulation. Trends Cogn. Sci. **16**(3), 174–180 (2012)
9. Hutt, S., Krasich, K., R. Brockmole, J., K. D'Mello, S.: Breaking out of the lab: Mitigating mind wandering with gaze-based attention-aware technology in classrooms. In: Proceedings of CHI 2021 (2021)
10. Kramarski, B., Mevarech, Z.R.: Enhancing mathematical reasoning in the classroom: the effects of cooperative learning and metacognitive training. Am. Educ. Res. J. **40**(1), 281–310 (2003)
11. Lottridge, D.M., et al.: The effects of chronic multitasking on analytical writing. In: Proceedings of CHI 2015 (2015)
12. Meijer, A.M., Aben, B., Reynvoet, B., Van den Bussche, E.: Reactive and proactive cognitive control as underlying processes of number processing in children. J. Exp. Child Psychol. **215**, 105319 (2022)
13. Monsell, S., Driver, J.: Control of Cognitive Processes: Attention and Performance XVIII, vol. 18. MIT Press, Cambridge (2000)
14. Renkl, A., Atkinson, R.K.: Structuring the transition from example study to problem solving in cognitive skill acquisition: a cognitive load perspective. Educ. Psychol. **38**(1), 15–22 (2003)
15. Unal, D.S., Arrington, C.M., Solovey, E., Walker, E.: Using Thinkalouds to understand rule learning and cognitive control mechanisms within an intelligent tutoring system. In: AIED (2020)
16. Wang, C., Li, B., Yao, Y.: Proactive control mediates the relationship between working memory and math ability in early childhood. Front. Psychol. **12**, 1612 (2021)
17. Xiao, X., Wang, J.: Understanding and detecting divided attention in mobile MOOC learning. In: Proceedings of the CHI 2017 (2017)

How to Repeat Hints: Improving AI-Driven Help in Open-Ended Learning Environments

Sébastien Lallé[1(✉)], Özge Nilay Yalçın[2], and Cristina Conati[3]

[1] LIP6, CNRS, Sorbonne University, Paris, France
sebastien.lalle@sorbonne-universite.fr
[2] School of Interactive Arts and Technology, Simon Fraser University, Burnaby, Canada
[3] Department of Computer Science, University of British Columbia, Vancouver, Canada

Abstract. AI-driven personalized support can help students learn from *Open-Ended Learning Environments* (OELEs). In this paper, we focus on how to effectively provide repeated hints in OELEs, when students repeat a sub-optimal behavior after receiving a hint on how to recover from the first occurrence of the behavior. We formally compare two repeated hint designs in UnityCT, an OELE that fosters *Computational Thinking* (CT) via free-form game design in K-6 education, with the long-term goal of providing AI-driven personalized hints.

Keywords: Open-Ended Environment · Real-Time Hint · Hint Design

1 Introduction

There is evidence (e.g., [4,6,8]) that AI-driven personalized support can help students learn from OELEs, namely, educational software that is specifically designed to let the students learn via exploration of the learning material with minimal constraints. However, delivering such help in an effective way is challenging for two reasons: 1) *student modeling*, namely, how to capture and recognize students' behaviors that indicate the need for help and 2) *how to deliver the help effectively* without interfering with the interaction with the OELE. Our long-term goal is to study how to design effective personalized help for OELEs that support game-based activities for K-6 students. This context is especially challenging from the point of view of delivering help that is effective and not intrusive. For this purpose, we leverage UnityCT, an OELE designed by UME Academy (*www.ume.academy*) to foster CT among young learners. CT, or the ability to express problems and their solutions computationally, has become an essential part of today's K12 curricula [1]. Previous work [7] addressed the challenge of building student models in UnityCT, by inferring from data non-obvious sub-optimal student behaviors that call for personalized help.

In this paper, we address the challenge of *how to deliver this personalized help effectively* once the need for it has been established by the student model.

N. Wang et al. (Eds.): AIED 2023, LNAI 13916, pp. 760–766, 2023.
https://doi.org/10.1007/978-3-031-36272-9_68

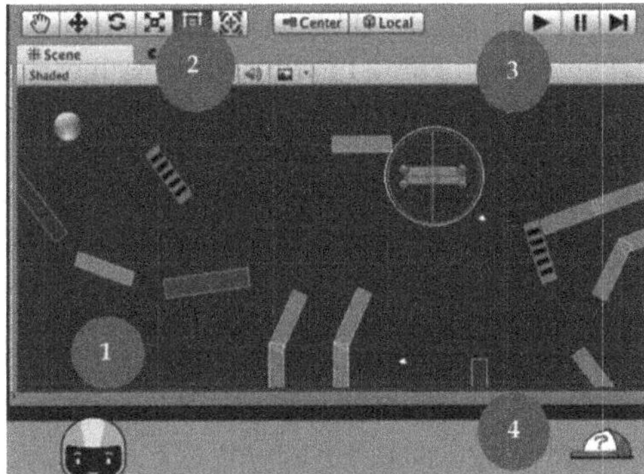

Fig. 1. UnityCT environment.

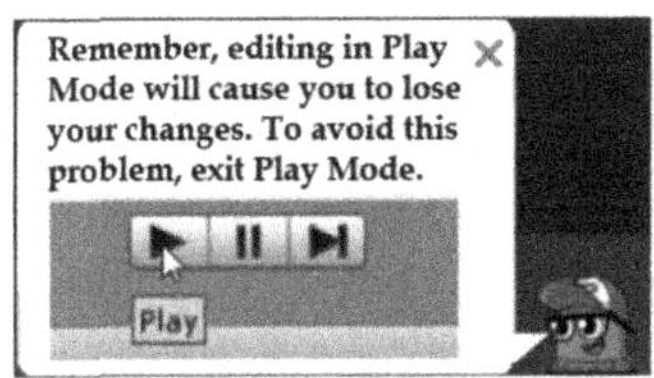

Fig. 2. New design of the repeated hints with combined texts.

We focus on a specific form of help provision, namely how to effectively provide *repeated hints* when students repeat sub-optimal behaviors and they have already received a hint on how to recover from the first occurrence of the behavior without interfering with the exploratory nature of OELEs. Some research on this issue has targeted OELEs for university students, by providing a more explicit justification as for why students should follow a hint when an error is repeated, with positive results on the students' experience [4,6]. However, these findings with college students may not translate to younger K-6 students who are less experienced with computers and learning environments. Two works targeted OELEs for K-6 students, but they just repeated hints without changing their content or format, in an OELE for model building activities (e.g., modeling engine force) [2] and in UnityCT [8]. Here, we build on the results of [8] by exploring how to alleviate the issues of increased confusion and lack of impact on error correction they found with repeated hints. Specifically, we address these research questions (RQs): "Compared to [8], does changing the design of repeated hints in UnityCT: RQ1/ improve students' error correction behavior? RQ2/ reduce students' reported confusion with these hints?" The evaluation of the new hints was conducted in real-life settings aiming for high ecological validity, and our results show that the new design successfully answered both RQs.

2 UnityCT

UnityCT is built on the Unity game engine and used in online lessons to foster CT skills as suitable for young audiences. During each 1-h lesson, an instructor teaches the CT skills covered in that lesson, and then assigns a 30-min challenge to design incremental game components that meet specific constraints. Students can freely explore how to build these components in the Unity scene (Fig. 1) in which they can manipulate game objects with different manipulators (e.g., move, rotate, Fig. 1). Students can enter into Play Mode by clicking the

Run button (Fig. 1) to execute and test their game. The feedback from UME Academy is in line with research showing that game-based activities increase student engagement in CT classes [3,5]. However, UnityCT introduces specific challenges due to students having to both operate in the complex Unity interface and learn the CT material. Although students can ask the instructor for help, not all students do it. Instructors generally try to watch for students needing help, but this is challenging without continuously monitoring what each student is doing. Thus, UME Academy has been focusing on augmenting UnityCT with personalized hints delivered by an *Intelligent Pedagogical Agent* (*IPA* from now on), as reported in [8]. Figure 1 shows the IPA when it is not providing a hint.

In [8], authors worked with the introductory lesson with UnityCT as a proof of concept focusing on errors commonly addressed by the instructors. Here, we focus on one of them, the *PlayMode* error, for which the hint repetition in [8] yielded some positive results. This error occurs when students enter "Play Mode" to execute their current game and make changes to their scene while in this mode. This is problematic because UnityCT does not record changes while in Play Mode, and can be corrected by pressing the Play button (Fig. 1) to exit the mode or by avoiding making changes to the scene. In [8], the PlayMode hint was designed to provide help at incremental levels of detail. This incremental support is consistent with how the UnityCT instructors provide help: they first flag the problem and why it exists (e.g., your changes will be lost because of editing in play mode), followed by an explanation of how to fix the problem if the student asks (e.g., exit play mode). This 2-step design was used to format the hints via speech bubbles. UnityCT and the hints are fully described in [8].

An evaluation of this work [8] showed that repeating the same PlayMode hint helped students make fewer errors and perform more correct behaviors, as compared to delivering the hints only once (1-Shot) or having no hints. However, hint repetition did not help students correcting their errors compared to 1-Shot and no hints, even though the hints explicitly showcase how to do so. The repeated hints were also judged by the students to be more confusing than the 1-shot hints, showing that there is room for improvement. Hence, we designed a new version of the repeated hint, described next.

3 User Study

To address the aforementioned limitations of previous repeated hints (*repeated-two-levels* version from now on), we designed a new version of these hints, in collaboration with UME Academy's UX/UI expert and three experienced instructors who used the previous IPAs. The key of this new design (*repeated-one-level* version from now on, see Fig. 2) is that the two levels of the repeated hint are combined into one bubble, as opposed to two bubbles in the previous study [8], so when students repeat the error they see a reminder of the nature of the error and how to correct it all at once. The rationale for this new design is two-fold. First, showing the solution right away when a hint is repeated is meant to increase error correction, especially for those students who, with a two-level hints, would

Table 1. Summary statistics (mean and st. dev.) of the corrrection measures.

Performance measure	Repeated-one-level	Repeated-two-levels	No-hint
Prop. of 2nd error correction	0.92	0.38	0.13
Rate of spont. error correction	0.92 (±0.14)	0.29 (±0.41)	0.12 (±0.16)

choose to only see the first speech bubble and thus never see how to correct the error in the second bubble. Second, the instructors suggested that receiving the same speech bubbles twice may cause students to wonder why they are receiving the hint again and generating the confusion with the repeated-hints IPA reported in [8]. Hence, changing the repeated hint so that it provides all the information at once may help reduce this source of confusion by diversifying it from the first occurrence of the hint. Importantly, these changes apply only to the repeated hints, i.e., we did not modify the first hint because the previous study [8] found that the students who got only one hint did not report confusion. The phrasing of the hints was finalized by the instructors and UX/UI designer.

The repeated-one-level version of the hints was deployed in 10 on-line classes for a total of 53 students. The study procedure and participant population was the same as in [8]: the classes covered exactly the same content, were open to students in grade 4 to 6 living in North America and used the same enrolling process. The inclusion of the hints did not alter the structure of the lessons, which took place in fully ecological settings; the students received no training nor specific instructions about the hints At the end of class, we collected the students' levels of perceived confusion with the hints using the same usability survey as in [8]. Confusion is rated on a 3-level Likert scale, coded as 0 for "Never", 0.5 for "Sometimes" and 1 for "Always". Maintaining the exact same study process allows us to compare the results of this evaluation of the repeated-one-level hints against the results obtained in the study in [8].

4 Analysis and Results

To answer our RQs, we compare the new *repeated-one-level* condition, against the *repeated-two-level* and *no-hint* conditions from [8] (each having 56 students).

Results for RQ1. We use the two measures of error correction from [8]:

M1. *Second error correction*: the proportion of students who corrected their second PlayMode error. For students in the conditions with an IPA, in which the second error triggers a repeated hint, this measure evaluates the *immediate effect* of the repeated hint on correcting the error they just made. For the no-hint condition, it measures if the students spontaneously correct their second error.

M2. *Rate of spontaneous error corrections* after the second error: the ratio of errors corrected over errors made after the second one. For students who received hints, it shows if the hints are successful in teaching the students to spontaneously detect and fix errors, when they don't receive hints anymore.

We retain students who made a second PlayMode error, namely 13 students from the repeated-one-level condition, 21 students from the repeated-two-levels condition and 38 students for the no-hint condition. Table 1 shows descriptive statistics for (M1-M2). We report statistical significance at the 0.05 level, and use Cohen's d for effect sizes (reported as large for $d > 0.8$, medium for $d > 0.5$, small otherwise.) We include *class_id* as a random effect in all models, to account for potential differences in the classes involved in the study.

Second Error Correction. We ran a generalized linear mixed model (GLMM) with a logistic link with this measure as dependent variable, condition (repeated-one-level, repeated-two-level, no-hint) as a factor, class_id as a random effect and first error correction (corrected, non-corrected) as covariate. This model estimates the probability that the students corrected their second error depending on their hint condition. The covariate captures if students could correct their error after receiving the first hint. This GLMM revealed a significant main effect of conditions ($p = .02$, $d = .68$). Pairwise comparisons using Holm-adjusted t-tests reveal a significant difference between the repeated-one-level condition and the no-hint conditions ($p = .03$, $d = .6$). There was no other significant pairwise comparisons ($p = .2$, $d = .4$), i.e., only the one-level hints outperformed no-hint.

Spontaneous Error Correction. For M2, we look at those students who continued to make errors after the second one (7 in repeated-one-level, 14 in repeated-two-levels and 37 in no-hint). We fit a GLMM with spontaneous correction rate as the dependent variable, condition (three levels) as the factor, and class_id as a random effect. This GLMM revealed a significant main effect of condition ($p < .01$, $d = 2.02$). Pairwise comparisons with Holm-adjusted t-tests show that the repeated-one-level condition outperforms both the no-hint condition ($p < .001$, $d = 2.46$) and the repeated-two-levels condition ($p < 0.001$, $d = 1.32$) with large effect sizes, highlighting an additional benefit of our new hint design[1].

Results for RQ2. We examine the self-reported levels of confusion with the hints in the usability survey, which was completed by 19 students from the repeated-one-level and 23 from the repeated-two-level condition. We use Ordinal Logistic Regression (OLR) to compare the Likert-scale ratings among the conditions. The OLR shows that the repeated-one-level hints are rated as significantly less confusing ($p = .01$, $d = .74$) than the repeated-two-levels hints. This indicates that the new hints were successful in reducing the confusion seen in the repeated-two-levels condition, with an average rating of 0.13 and a mode of 0, which corresponds to"never confusing" in the survey, as compared to a mode of 0.5 for repeated-two-levels ("sometimes confusing"). In addition, the students' perception of the hints was very positive for all other usability items, with almost all students liking the hints and wanting to reuse them.

[1] Note that we also confirmed that the repeated-one-level hints do maintain the positive results found for the repeated-two-levels hints in [8] as for fostering less errors and more correct behaviors (we did so with the exact same analysis as in [8].).

5 Discussion and Future Work

Our results provide evidence that the repeated-one-level condition can improve student error correction behavior (RQ1), while eliminating the confusion generated by the repeated-two-levels hints in [8] (RQ2). The repeated-one-level hints also maintain all of the positive results already found in [8] related to the performance and experience of the students. These results are encouraging as there has been no research on OELEs for K-6 students that formally investigated how to format a repeated hint, and here we propose a new promising design that we tested in regular classes, thus demonstrating high ecological validity. While we focus on one OELE, we see our study as a proof of concept for the value of personalized help to support young learners in game-based learning activities. This opens the possibility of testing the hint design in other game-based OELEs, which are getting popular to engage children in CT.

As a next step, we plan to leverage our IPA to address other problematic behaviors using the existing data-driven student models in UnityCT mentioned in the introduction [7]. We will focus on providing personalized support to the sub-optimal behaviors mined in [7] that are similar in nature to the PlayMode error targeted in this paper. In particular, in [7], they found that students failed to remediate other errors (e.g., deleting a key object, panning outside the scene), or did not perform important actions (e.g., not interacting with objects). Applying our repeated hint design to these behaviors would be a important step toward evaluating fully AI-driven hints in game-based OELEs for K-6 learners.

Acknowledgement. This work was supported by UME Academy Ltd. (*ume. academy*) and Mitacs (Grant #IT13336).

References

1. Barr, V.: Stephenson, C: bringing computational thinking to K-12: what is Involved and what is the role of the computer science education community? ACM Inroads **2**(1), 48–54 (2011)
2. Basu, S., Biswas, G., Kinnebrew, J.S.: Learner modeling for adaptive scaffolding in a Computational Thinking-based science learning environment. User Model. User-Adap. Inter. **27**(1), 5–53 (2017)
3. Biswas, G., Segedy, J.R., Kinnebrew, J.S.: Smart open-ended learning environments that support learners cognitive and metacognitive processes. In: Holzinger, A., Pasi, G. (eds.) HCI-KDD 2013. LNCS, vol. 7947, pp. 303–310. Springer, Heidelberg (2013). https://doi.org/10.1007/978-3-642-39146-0_27
4. Borek, A., McLaren, B.M., Karabinos, M., Yaron, D.: How much assistance is helpful to students in discovery learning? In: Cress, U., Dimitrova, V., Specht, M. (eds.) EC-TEL 2009. LNCS, vol. 5794, pp. 391–404. Springer, Heidelberg (2009). https://doi.org/10.1007/978-3-642-04636-0_38
5. Çakır, N. A., Gass, A., Foster, A., Lee, F.J.: Development of a game-design workshop to promote young girls' interest towards computing through identity exploration. Comput. Educ. **108**, 115–130 (2017)

6. Kardan, S., Conati, C.: Providing adaptive support in an interactive simulation for learning: an experimental evaluation. In: Proceedings of the ACM Conference on Human Factors in Computing Systems, pp. 3671–3680 (2015)
7. Lallé, S., Yalçın, Ö. N., Conati, C.: Combining data-driven models and expert knowledge for personalized support to foster computational thinking skills. In: 11th Inter. Learning Analytics and Knowledge Conferencepp. 375–385. ACM (2021)
8. Yalçin, Ö. N., Lallé, S., Conati, C.: An intelligent pedagogical agent to foster computational thinking in open-ended game design activities. In: International Conference on Intelligent User Interfaces, pp. 633–645. ACM (2022)

Automatic Detection of Collaborative States in Small Groups Using Multimodal Features

Mariah Bradford, Ibrahim Khebour, Nathaniel Blanchard, and Nikhil Krishnaswamy(✉)

Colorado State University, Fort Collins, CO 80523, USA
{mbrad,nkrishna}@colostate.edu

Abstract. Cultivating collaborative problem solving (CPS) skills in educational settings is critical in preparing students for the workforce. Monitoring and providing feedback to all groups is intractable for teachers in traditional classrooms but is potentially scalable with an AI agent who can observe and interact with groups. For this to be feasible, CPS moves need to first be detected, a difficult task even in constrained environments. In this paper, we detect CPS facets in relatively unconstrained contexts: an in-person group task where students freely move, interact, and manipulate physical objects. This is the first work to classify CPS in an unconstrained *shared* physical environment using multimodal features. Further, this lays the groundwork for employing such a solution in a classroom context, and establishes a foundation for integrating classroom agents into classrooms to assist with group work.

Keywords: Collaborative Problem Solving · Multimodal · Natural Language Processing · Small Groups

1 Introduction

Working in teams is an essential skill in the workforce, which the education system needs to prepare students for. Such practices have been formalized into pedagogical techniques of *collaborative problem solving* (CPS) wherein students learn by working together to achieve a shared goal. With CPS, peers can develop a "positive interdependence" [8], but doing so depends on having an effective group dynamic that does not fall into dysfunction. With proper facilitation, this can be avoided, but this role usually falls to the teacher, and with a single teacher and many small groups, such facilitation becomes intractable.

For some group-facilitation tasks, such as aligning group goals, an artificially intelligent agent can be a useful tool to help teachers manage groups. A prerequisite to agent interaction with a group is that the agent must be able to observe the group and detect its state. Sun *et al.* [13] recently proposed a novel coding

M. Bradford, I. Khebour—These authors contributed equally to this work.

N. Wang et al. (Eds.): AIED 2023, LNAI 13916, pp. 767–773, 2023.
https://doi.org/10.1007/978-3-031-36272-9_69

scheme where each component of CPS could occur simultaneously, rather than being treated as distinct phenomena (e.g., social or cognitive indicators). They define an alternative framework composed of three main facets: *construction of shared knowledge*, *negotiation/coordination*, and *maintaining team function*, which are defined by *indicators* (e.g., "proposes specific solutions"). In this work, we adopt the Sun *et al.* framework from [13] and use it to annotate a novel collaborative problem-solving task: the Weights Task (Sect. 2.1).

A number of works have trained machine learning models to detect CPS in controlled, virtual environments [1,12]. We extend these findings, showing that CPS facets can still be automatically detected in in-person group work, despite the increased complexity. Several works have explored what features are important for detecting in-person CPS states [4,5,9], and we took inspiration from these works in the design of our study. Our work presents a novel physical, in-person shared task, and we utilized many of the recommended features to establish our baseline on this data.[1]

2 Methods

2.1 Data Collection

Weights Task. We collected audiovisual data of small groups collaborating on an in-person, shared, physically-grounded problem-solving task, known as the Weights Task. An example still can be seen in Fig. 1. In this task, triads are given five colored cubes of different weights, a balance scale, and a worksheet to track answers. We identify the weight of one block, and ask them to use the balance scale to identify the weights of the remaining blocks. When participants have identified the weights of all five blocks, we remove the balance scale and provide a new block of unknown weight to participants. They must then try to identify the weight of the mystery block. To successfully do this, they must infer the pattern in the block weights. They have two attempts. We then ask participants for the weight of a hypothetical next block in the sequence, according to the pattern. They again have two attempts. Recording ends at the end of this second attempt. A prior version of this task was described in [3]—our version extends those methods in several ways to elicit more collaborative moves.

Thirty participants were recruited for this study. All participants were over the age of 18, spoke fluent English, and were drawn from the student population of Colorado State University. Participants were 20% female and 80% male. When asked to identify their ethnicity, 60% of participants identified as Caucasian, 10% identified as Hispanic or Latino, and 30% identified as Asian. Participants indicated a range of native languages including English, Hindi, Assamese, Gujarati, Bengali, Telugu, Persian, Malayalam, Urdu, and Spanish. The full dataset consists of ten triads completing the Weights Task. Recordings average 16 min.

[1] Supplemental material can be found here: https://github.com/Blanchard-lab/aied_2023_suppmat.

Fig. 1. The Weights Task

Audio recording used an MXL AC-404 Procon microphone as advised by findings from [3]. Each audio recording was processed using Google's Voice Activity Detection (VAD) [11] to automatically segment audio files into utterances, with only one speaker per segment. Next, we transcribe the segmented audio files using Google's automatic speech recognition (ASR). After preprocessing, there were a total of 1,822 utterances, with an average length of 4.26 s.

2.2 Annotations

Utterances from each group were annotated for collaborative problem solving (CPS) in accordance with the CPS framework developed by Sun *et al.* [13]. Each utterance was annotated by two trained coders. Table 1 shows the average number of occurrences of each facet per group.

Table 1. Descriptive statistics of all 1,822 utterances across all groups

	Average	SD	Min	Max
All utterances	182.20	80.51	90	380
# None	62.10	30.50	40	141
# Construction of shared knowledge	69.20	25.49	33	131
# Negotiation/Coordination	68.10	27.12	28	126
# Maintaining team function	37.10	23.59	9	73
Time (s)	4.26	2.84	0.84	23.64

2.3 Verbal Features

Verbal features comprised features corresponding to the words in each utterance as transcribed by Google's ASR. Each group's utterance-level transcripts were preprocessed for formatting (including removing newlines and periods and surrounding the utterance with BERT's required `[CLS]` and `[SEP]` tokens), and then fed into the BERT Transformer model [6] to retrieve the sentence embedding for each utterance. To expedite computation, we use the BERT-SMALL model first published in [14] and made available on the HuggingFace platform. Therefore the embedding size is 512 dimensions.

2.4 Prosodic Features

Prosodic features here refers to the non-linguistic features of speech. Each group's audio files were processed using openSMILE to extract prosodic features of speech—e.g., features relating to frequency, amplitude, and balance. We used the extended feature set predefined by Eyben *et al.* [7]. This feature set aims to be minimalist while still effective. After processing, each utterance has an associated total of 88 prosodic features, such as loudness and spectral flux.

2.5 Model Training

All models were trained and evaluated using leave-one-group-out cross-validation. We trained and evaluated three types of models for evaluation: a random forest (RF) and AdaBoost (AB) classifier where optimal hyperparameters were identified with Hyperopt's [2] guided TPE search, and a neural network (NN) classifier where hyperparameters were identified with a grid search.

3 Results

Table 2 shows the results of binary CPS facet classification with respective standard deviations given in Table 3. As discussed in our related works, binary classification (presence or absence of a CPS facet) is important since some utterances may contain multiple collaborative components [13]. Area Under the Receiver Operating Characteristic Curve (AUROC) was computed using test results from every utterance.

Table 2. Weighted average AUROC for binary classification

	Construction of shared knowledge			Negotiation/ Coordination			Maintaining team function		
Modalities	RF	AB	NN	RF	AB	NN	RF	AB	NN
Verbal	.814	.804	.829	.788	.783	.791	.712	.689	.678
Prosodic	.832	.796	.714	.730	.710	.595	.661	.649	.598
Verbal + Prosodic	.840	.818	.794	.785	.794	.760	.720	.699	.645

Table 3. Standard deviations of weighted average AUROC across all 10 groups for binary classification

	Construction of shared knowledge			Negotiation/ Coordination			Maintaining team function		
Modalities	RF	AB	NN	RF	AB	NN	RF	AB	NN
Verbal	.044	.037	.040	.054	.052	.057	.082	.079	.079
Prosodic	.038	.051	.118	.055	.056	.094	.077	.074	.091
Verbal + Prosodic	.035	.044	.143	.054	.052	.099	.076	.088	.095

4 Discussion

In many cases, we achieve results comparable to or even exceeding those reported in [12], even though our shared environment and task are noisier and our data size is smaller (30 participants compared to 111).

We often observe that performance with feature combinations does not significantly exceed that with verbal (linguistic) features alone. In these cases, the utterances or numerical representations thereof usually carry sufficient information to classify or detect CPS facets most of the time.

4.1 Qualitative Error Analysis

In order to identify instances where multimodal features helped with CPS facet detection, we examined 50 random samples where predictions by the random forest model were wrong using verbal features only and correct when using verbal+prosodic. 32% of utterances required prosody to clarify intent (e.g., one participant said “that’s 100” to indicate the team had done well, not to posit that a block weighed 100 g)—half of this set were instances of participants asking questions. 16% of the 50 sample utterances contained interruptions (e.g., Participant 2: “So even when they’re equal it leans—”; Participant 1: “It leans slowly that way”). For the remaining samples, it was not immediately clear how the inclusion of prosodic information led to correct CPS identification, but within this subset, we noted that 78.9% of utterances were correctly classified by the prosodic classifier, indicating the prosodic signal alone was sufficient and they did not benefit from the combination of verbal and prosodic features. Finally, we found 374 out of 1,822 (20%) utterances were misclassified by all models—these utterances clearly require additional features or modalities.

5 Limitations, Future Work, and Conclusion

In this study, we have used several tools for automatic feature extraction and trained multiple machine learning classifier models to detect collaborative problem solving (CPS) in small groups. We achieve promising results on multimodal detection of CPS in a challenging in-person setting: a task that requires real-time interaction with physical objects.

There are several limitations to this effort that future work could address. While this is in line with other efforts, we are currently identifying CPS at the facet-level (the most coarse-grained) and future work will need to identify CPS at the sub-facet and indicator levels. While our participants exhibit a range of ethnic, national, and linguistic backgrounds, all participants still tend to satisfy most conditions of the WEIRD paradigm [10]. We are not utilizing any visual features but the weights task itself is ripe to take advantage of body pose and block interactions—some work on incorporating gesture was recently conducted by [15]. We experiment with only one CPS task and it would behoove future research to explore task-agnostic CPS classification. Our dataset is collected outside of classrooms—classroom environments will inevitably be subject to additional noise. Future work will need to consider how robust our solutions are to classroom noise. Finally, the ultimate goal of this work is to create an AI agent to assist small groups—this work is an important milestone, but much remains to be done before an agent can actively interact with a group.

References

1. Avci, U., Aran, O.: Predicting the performance in decision-making tasks: from individual cues to group interaction. IEEE Trans. Multimedia **18**(4), 643–658 (2016). https://doi.org/10.1109/TMM.2016.2521348
2. Bergstra, J., Yamins, D., Cox, D.D.: Making a Science of model search: hyperparameter optimization in hundreds of dimensions for vision architectures. In: Proceedings of the 30th International Conference on Machine Learning (2013)
3. Bradford, M., Hansen, P., Beveridge, J.R., Krishnaswamy, N., Blanchard, N.: A deep dive into microphone hardware for recording collaborative group work. In: Proceedings of the International Conference on Educational Data Mining (2022)
4. Castillon, I., VanderHoeven, H., Bradford, M., Venkatesha, V., Krishnaswamy, N., Blanchard, N.: Multimodal features for group dynamic-aware agents. In: Interdisciplinary Approaches to Getting AI Experts and Education Stakeholders Talking Workshop at AIEd. International AIEd Society (2022)
5. Cukurova, M., Zhou, Q., Spikol, D., Landolfi, L.: Modelling collaborative problem-solving competence with transparent learning analytics: is video data enough? In: Proceedings of the Tenth International Conference on Learning Analytics & Knowledge, pp. 270–275 (2020)
6. Devlin, J., Chang, M.W., Lee, K., Toutanova, K.: BERT: pre-training of Deep Bidirectional Transformers for Language Understanding (May 2019). https://doi.org/10.48550/arXiv.1810.04805
7. Eyben, F., et al.: The geneva minimalistic acoustic parameter set (GeMAPS) for voice research and affective computing. IEEE Trans. Affect. Comput. **7**(2), 190–202 (Apr 2016). https://doi.org/10.1109/TAFFC.2015.2457417
8. Graesser, A.C., Fiore, S.M., Greiff, S., Andrews-Todd, J., Foltz, P.W., Hesse, F.W.: Advancing the science of collaborative problem solving. Psychol. Sci. Public Interest **19**(2), 59–92 (2018). https://doi.org/10.1177/1529100618808244

9. Grover, S., Bienkowski, M., Tamrakar, A., Siddiquie, B., Salter, D., Divakaran, A.: Multimodal analytics to study collaborative problem solving in pair programming. In: Proceedings of the Sixth International Conference on Learning Analytics & Knowledge, pp. 516–517 (2016)
10. Henrich, J., Heine, S.J., Norenzayan, A.: The weirdest people in the world? Behav. Brain Sci. **33**(2–3), 61–83 (2010)
11. Karrer, R.: Google WebRTC Voice Activity Detection (VAD) module (2022). https://www.mathworks.com/matlabcentral/fileexchange/78895-google-webrtc-voice-activity-detection-vad-module
12. Stewart, A.E.B., Keirn, Z., D'Mello, S.K.: Multimodal modeling of collaborative problem-solving facets in triads. User Model. User-Adap. Inter. **31**(4), 713–751 (2021). https://doi.org/10.1007/s11257-021-09290-y
13. Sun, C., Shute, V.J., Stewart, A., Yonehiro, J., Duran, N., D'Mello, S.: Towards a generalized competency model of collaborative problem solving. Comput. Educ. **143**, 103672 (2020). https://www.sciencedirect.com/science/article/pii/S0360131519302258
14. Turc, I., Chang, M.W., Lee, K., Toutanova, K.: Well-read students learn better: on the importance of pre-training compact models. arXiv preprint arXiv:1908.08962 (2019)
15. Vanderhoeven, H., Blanchard, N., Krishnaswamy, N.: Robust motion recognition using gesture phase annotation. In: International Conference on Human-Computer Interaction (HCII). Springer (2023)

Affective Dynamic Based Technique for Facial Emotion Recognition (FER) to Support Intelligent Tutors in Education

Xingran Ruan[1(✉)], Charaka Palansuriya[1,2], and Aurora Constantin[1]

[1] The University of Edinburgh, Edinburgh, UK
{xruan,aurora.constantin}@ed.ac.uk
[2] Edinburgh Parallel Computing Centre, Edinburgh, UK
charaka@epcc.ed.ac.uk

Abstract. Facial expressions of learners are relevant to their learning outcomes. The recognition of their emotional status influences the benefits of instruction or feedback provided by the intelligent tutor in education. However, learners' emotions expressed during interactions with the intelligent tutor are mostly detected by self-reports of learners or judges who observe them in manually. The automated Facial Emotion Recognition (FER) task has been a challenging problem for intelligent tutors. The state-of-art automated FER methods target six basic emotions instead of learning-related emotions (e.g., neutral, confused, frustrated, and bored). Thus our research contributes to training a machine learning (ML) model to recognise learning-related emotions for intelligent tutors automatically, based on an Affective Dynamics (AD) model. We implement the AD model into our loss function (AD-loss) to fine-tune the ML model. In the test scenario, the AD-loss method improves the performance of state-of-art FER algorithms.

Keywords: Facial emotion recognition · Intelligent tutors · Epistemic emotion · Affective dynamics model

1 Introduction

Self-regulated learning (SRL) is a complex educational construct that describes how students as regulated learners manage cognitive functions to achieve educational goals [15]. Under the framework of SRL, a considerable number of variables that influence learning outcomes are studied. In terms of emotions, research has shown that emotions can have positive and negative influences on learning [7,9]. Significantly, the analysis of D'Mello et al. [7] points out that engagement, boredom, confusion, frustration, and curiosity are the most frequent learning-related emotions that predict learning. Researchers have proposed a number of intelligent tutors that are adaptive to learners' emotions to provide support during

N. Wang et al. (Eds.): AIED 2023, LNAI 13916, pp. 774–779, 2023.
https://doi.org/10.1007/978-3-031-36272-9_70

their learning [3,5]. Thus, emotion is a critical component of learning activity and FER of learners has become an essential task in education [9].

However, the state-of-art FER techniques in intelligent tutors are recognising emotions manually which limits to support learners adaptively [3,8]. In [8], subjects are invited to learn about computer literacy with AutoTutor and researchers record subjects' facial expressions during their learning. The expressed affective statuses of subjects in videos are recognised by themselves in a post-learning session. In the study of Craig et al. [3], learners' emotions (e.g. boredom, flow, and confusion) during learning activities are recognised by five observers who were given a training session. On the other hand, the state-of-art automatic FER techniques mostly focus on six basic emotions (sadness, happiness, fear, anger, surprise and disgust) [12], but affective studies have shown that basic emotions occurred considerably rare when compared to epistemic emotions (boredom, confusion, frustration, flow/engagement) during learning. For example, D'Mello et al. [5] analysed studies of affective status of learning activity and claimed that these epistemic emotions were the most frequent. Thus our research is motivated to propose the FER technique to support intelligent tutors in recognising the epistemic emotions (neutral, confusion, frustration, and boredom) of learners during their learning process automatically.

In this paper, Sect. 2 introduces the affective dynamic model and explains using AD-Loss to fine-tune the ML model. Experimental results are presented in Sect. 3. Finally, conclusions and further works are provided in Sect. 4.

2 Proposed Method

In this section, we describe an AD-Loss framework which helps to improve and solve FER to support intelligent tutors to interpret learners' epistemic emotions. Our design is inspired by a model of affective dynamics during learning [6] which proposed that the experience of learners' emotions during SRL depends on their cognitive disequilibrium status. To imitate this affective dynamics model in our FER approach, we simultaneously take a pair of joint images as input to a pre-trained network then sequentially apply Multilayer perception (MLP), and an affective dynamic transition operation. The whole framework is shown in Fig. 1.

2.1 Affective Dynamics Model

D'Mello and Graesser's Model of Affective Dynamics [6] aims to highlight the critical role of cognitive disequilibrium in the learning activity and explain the trajectory of cognitive-affective states that are spawned by cognitive disequilibrium. The model emphasizes that cognitive disequilibrium occurs when a learner is confronted with obstacles whilst trying to achieve learning goals (e.g., interruptions, contradictions, unexpected feedback etc.) and claimed that cognitive disequilibrium is the key to understanding the affective states that underlie complex learning. In the form of the transition network, a learner's affective state of

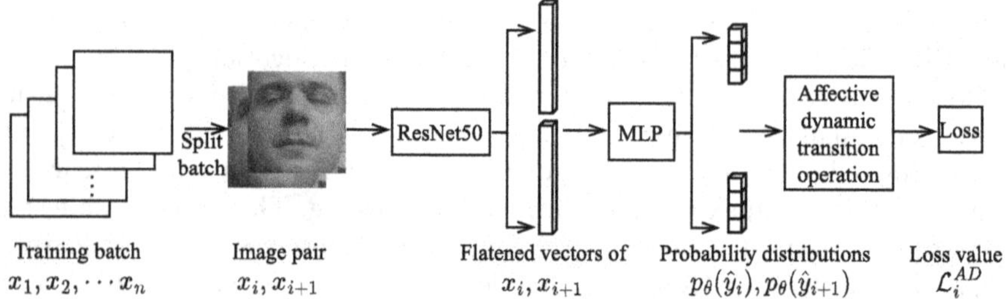

Fig. 1. AD-Loss framework. We split each incoming training batch of size n into $n-1$ pairs and feed them into pre-trained ResNet50 (pre-trained on ImageNet) and MLP. Each pair of frames will be converted to two probability distributions and be fed into affective dynamic transition operation to compute loss value.

confusion is one key signature of cognitive disequilibrium. Learners must engage in effortful problem-solving activities in order to resolve the impasse and resolve equilibrium. A hopeless confusion occurs when the impasse can not be resolved, the learner will get lost and experience frustration. Furthermore, persistent frustration may transition into boredom.

2.2 Affective Dynamics Based FER Technique

Using a pre-trained deep convolutional neural network (DCNN) model through appropriate transfer learning is the main baseline for solving the FER task. Since training a DCNN model from scratch (with all weights randomly initialized) is a computing intensity workload, a DCNN which already been pre-trained on another large dataset (e.g. ImageNet [11]) can be fine-tuned on the target dataset for emotion classification. The following subsection describes implementing the proposed AD-Loss to fine-tune a pre-trained network for emotion classification.

We employ a pre-trained ResNet50 network [10] as the foundation of our work, incorporating flatten layers, dropout layers, and dense layers based on empirical evidence [1]. The first convolutional layer of ResNet50 takes the input of size $248 \times 248 \times 3$ for input RGB colour images with size 248×248; after successive convolutional blocks and pooling average layers, the output size is $9 \times 9 \times 2048$ (2048 volumes with the size of 9×9). Following the ResNet50 network, we add a flatten layer which converts the output into a linear vector of size $165{,}888 (= 9 \times 9 \times 2048)$. After that, we add three dense layers with lengths of 1000, 128, and 4 respectively behind the flatten layer and one dropout layer with the rate of 0.3 between each of the two dense layers to increase the robustness. The last dense layer with size 4 represents 4 different emotion statuses (neutral, confused, frustrated, and bored).

During fine-tuning on learners' facial expressions during learning $\mathcal{X} = \{x_1, x_2, \ldots x_n\}$[1], we freeze all parameters in the pre-trained ResNet50 and

[1] $\mathcal{X}$ is a sequence of frames of facial emotion expressions of learners.

update the parameters in three dense layers. The ResNet50-based network converts each frame pair (x_i, x_{i+1}) to two conditional probability distributions $p_\theta(\hat{y}_i \mid x_i)$ and $p_\theta(\hat{y}_{i+1} \mid x_{i+1})$[2], where $\hat{y}_i$ is the prediction of classifier given x_i. Each of these distributions is a four-dimensional vector, with each element (e.g. the element j) denoting the probability of the classifier in class j given input x_i. The θ are the parameters in the network and the final AD-Loss of x_i and x_{i+1} is $\mathcal{L}_i^{AD}$.

Now we will explain how to adapt the affective dynamics model in AD-Loss. Because the transition of a learner's emotional status mostly follows a partition that can transfer from equilibrium (neutral) to confusion when encountering impasses; transfer from confusion to frustration if the confronted impasses can not be resolved; furthermore, a persistent frustration can transition into boredom, we create a transition matrix $\mathcal{T}$ that describes the emotion status trajectory based on the affective dynamics model, see below. $\mathcal{T}(i, j)$ describes current emotional status $i \in [1, 4]$ transfers to the next (predicted) emotional status $j \in [1, 4]$ ($\mathcal{T}(i, j) = 0$ or 1). When $\mathcal{T}(i, j) = 0$, a penalty λ_{ad} will be aggregated on $\mathcal{L}_i^{AD}$, otherwise the penalty is zero. In order to implement the transition matrix into AD-Loss, we create the loss function $\mathcal{L}^{AD}$ as described in Eq. 1 and set a parameter $\lambda_{ad} \in \mathcal{R}$ to control the penalty from affective dynamics transition operation.

$$\mathcal{T} = \begin{array}{c} \\ \\ \\ \\ \\ \end{array}\begin{array}{cccc} \text{Neutral} & \text{Confused} & \text{Frustrated} & \text{Boredom} \end{array}$$

$$\mathcal{T} = \begin{bmatrix} 1 & 1 & 0 & 0 \\ 1 & 1 & 1 & 0 \\ 1 & 0 & 1 & 1 \\ 1 & 0 & 0 & 1 \end{bmatrix} \begin{array}{l} \text{Neutral} \\ \text{Confused} \\ \text{Frustrated} \\ \text{Boredom} \end{array}$$

$$\mathcal{L}_i^{AD} = \sum_{k=i}^{i+1} \mathcal{L}_k^{CE} + \lambda_{ad}(1 - \mathcal{T}(\lfloor argmax(p_\theta(\hat{y}_i)) \rfloor, \lfloor argmax(p_\theta(\hat{y}_{i+1})) \rfloor)) \quad (1)$$

$$\mathcal{L}_i^{CE} = p_\theta(\hat{y}_i)^T log(y_i) \quad (2)$$

We proceed to learn parameters θ^* with the following learning objective function for a sequence of emotional frames $\mathcal{X} = \{x_1, x_2, \dots x_n\}$:

$$\theta^* = argmin \sum_{i=1}^{n-1} \mathcal{L}_i^{AD} \quad (3)$$

This objective function can be explained as: for ith frame in the training sequence, we select its successive frame (the $i + 1$th), then search for an optimal θ^* to minimize the cross-entropy loss $\mathcal{L}^{CE}$ and the penalty of AD-Loss $\mathcal{L}^{AD}$.

[2] We note $p_\theta(\hat{y}_i)$ as the abbreviation of $p_\theta(\hat{y}_i \mid x_i)$ in the rest of paper.

3 Experimental Evaluations

To evaluate our method on recognising the epistemic emotions of learners during their learning activity, we tested our AD-Loss on PUZZLED [13][3] and compared with other state-of-art FER methods that support intelligent tutors in education.

3.1 Results

In the pre-processing phase, we first recognised and cropped face areas from video frames. The transformed face was considered as the face region and we resized the face region to 248 × 248 pixels. In the test scenario, PUZZLED was split into training, and test sets by subject ID. We randomly chose 10% of subjects as the test set and the rest of the data as the training set in each test round. To increase the robustness of the network, we augmented the origin training dataset by performing colour shift (maximum value of 20), rotation, smoothing (maximum window size 5), mirror, and multi-crop [2] before the training.

In our test experiment, we 'plugged-in' AD-Loss (with $\lambda_{ad} = 1$) on three different state-of-art FER methods and compared the accuracy and F1 score with their origin. The results were summarized in Table 1. It can be seen from Table 1 that on average our method improved Cross-Entropy (CE) loss [1] 5% on accuracy and 9% on F1 score, respectively; improved Additive Margin Softmax (AM) loss [14] 6% on accuracy and 5% on F1 score, respectively; improved ArcFace (AF) loss [4] 2% on accuracy and 1% on F1 score, respectively.[4]

Table 1. Recognition Rate and F1 score of testing on PUZZLED. **CE**: Cross Entropy; **AD**: AD-Loss (we proposed); **AM**: Additive Margin Softmax; **AF**: ArcFace.

Base Model	Loss Function	Weighted F1 score (Ave)	Accuracy (Ave)
ResNet50	CE	0.4866	0.5299
ResNet50	**CE + AD**	**0.5296**	**0.5541**
ResNet50	AM	0.5194	0.6376
ResNet50	**AM + AD**	**0.5216**	**0.6419**
ResNet50	AF	0.5055	0.6027
ResNet50	**AF + AD**	**0.5320**	**0.6414**

4 Conclusion

In this paper, we proposed an ML-based FER technique and showed the affective dynamics model was able to support intelligent tutors to recognise learners' epistemic emotions during their learning. We implemented the affective dynamics

[3] The PUZZLED consists of 10 videos of students when they are watching educational videos (contains neutral, confused, frustrated, and boredom).

[4] To assess the performance, we utilized CE to measure the loss of the output from both AM and AF.

model into the AD-Loss to fine-tune the pre-trained ResNet50 on the learners' emotion dataset. We tested our proposed technique on the PUZZLED dataset and compared it with other state-of-art FER techniques. It shows that AD-Loss improves the recognition rate of baseline FER methods. Our research introduced psychological aspects into the ML framework to recognise learners' emotional expressions during learning activities more accurately. This will aid in providing effective real-time support for learners during learning. In future work, our research will focus on implementing the proposed FER technique for intelligent tutors to interpret learners' (especially neural typical children's) emotions and provide real-time dynamic feedback to promote them reaching educational goals.

References

1. Akhand, M., Roy, S., Siddique, N., Kamal, M.A.S., Shimamura, T.: Facial emotion recognition using transfer learning in the deep CNN. Electronics **10**(9), 1036 (2021)
2. Caron, M., Misra, I., Mairal, J., Goyal, P., Bojanowski, P., Joulin, A.: Unsupervised learning of visual features by contrasting cluster assignments. arXiv preprint arXiv:2006.09882 (2020)
3. Craig, S., Graesser, A., Sullins, J., Gholson, B.: Affect and learning: an exploratory look into the role of affect in learning with AutoTutor. J. Educ. Media **29**(3), 241–250 (2004)
4. Deng, J., Guo, J., Xue, N., Zafeiriou, S.: ArcFace: additive angular margin loss for deep face recognition. In: Proceedings of the IEEE/CVF Conference on Computer Vision and Pattern Recognition, pp. 4690–4699 (2019)
5. D'mello, S., Graesser, A.: AutoTutor and affective AutoTutor: learning by talking with cognitively and emotionally intelligent computers that talk back. ACM Trans. Interact. Intell. Syst. (TiiS) **2**(4), 1–39 (2013)
6. D'Mello, S., Graesser, A.: Dynamics of affective states during complex learning. Learn. Instr. **22**(2), 145–157 (2012)
7. D'Mello, S., Kappas, A., Gratch, J.: The affective computing approach to affect measurement. Emot. Rev. **10**(2), 174–183 (2018)
8. Graesser, A., Chipman, P., King, B., McDaniel, B., D'Mello, S.: Emotions and learning with auto tutor. Front. Artif. Intell. Appl. **158**, 569 (2007)
9. Graesser, A.C.: Emotions are the experiential glue of learning environments in the 21st century. Learn. Instr. **70**, 101212 (2020)
10. He, K., Zhang, X., Ren, S., Sun, J.: Deep residual learning for image recognition. In: Proceedings of the IEEE Conference on Computer Vision and Pattern Recognition, pp. 770–778 (2016)
11. ImageNet: ImageNet (2021). https://www.image-net.org/. Accessed 28 Dec 2022
12. Li, S., Deng, W.: Deep facial expression recognition: a survey. IEEE Trans. Affect. Comput. **13**, 1195–1215 (2020)
13. Linson, A., Xu, Y., English, A.R., Fisher, R.B.: Identifying student struggle by analyzing facial movement during asynchronous video lecture viewing: Towards an automated tool to support instructors. In: Rodrigo, M.M., Matsuda, N., Cristea, A.I., Dimitrova, V. (eds.) AIED 2022. LNCS, vol. 13355, pp. 53–65. Springer, Cham (2022). https://doi.org/10.1007/978-3-031-11644-5_5
14. Wang, F., Liu, W., Liu, H., Cheng, J.: Additive margin softmax for face verification. arXiv preprint arXiv:1801.05599 (2018)
15. Winne, P.H.: A cognitive and metacognitive analysis of self-regulated learning. Handbook of self-regulation of learning and performance, pp. 15–32 (2011)

Real-Time Hybrid Language Model for Virtual Patient Conversations

Han Wei Ng[1,2], Aiden Koh[1(✉)], Anthea Foong[1], and Jeremy Ong[1]

[1] MediVR, 10 Central Exchange Green, Singapore 138649, Singapore
{hanwei,aiden}@medivr.io

[2] Nanyang Technological University, School of Computer Science and Engineering, 50 Nanyang Ave, Singapore 639798, Singapore
http://www.medivr.io

Abstract. Advancements in deep learning have enabled the development of online learning tools for medical training, which is important for remote learning. However, face-to-face interaction is essential for practicing human-centric skills such as clinical skills. Presently, in medical training, such interactions can be mimicked using deep learning methodologies. However, the understanding of such models is often limited whereby lightweight models are unable to generalize beyond scope while large language models tend to produce unexpected responses. To overcome this, we propose a hybrid lightweight and large language model for creating virtual patients, which can be used for real-time autonomous training of trainee doctors in clinical settings using online platforms. This ensures high-quality and standardized learning for all individuals regardless of location and background.

Keywords: Natural language understanding · Natural language generation · Simulated virtual human · Knowledge graph

1 Introduction

Deep learning has made large advances towards its ability to process a large variety of inputs. However, the applications of deep learning within the education sector still lags behind. In the medical education sector, there has been cases whereby the usage of such technology has enabled better learning outcomes for trainee doctors [6,10]. In clinical training, Ng et al. showed how deep learning can augment learning, whereby doctors learn to communicate with digital patients in order to diagnose the underlying condition [8]. Traditional training is often done through in-person lessons, thus proving to be a challenge towards digitalization of such an experience such that other individuals from different backgrounds may gain access to top quality medical education as well [2].

Thus, digital recreations of patient and doctor conversations have been attempted [8]. However, in the instance of training the clinical skills of trainee doctor, a highly immersive and robust system has yet to be achieved. Although

N. Wang et al. (Eds.): AIED 2023, LNAI 13916, pp. 780–785, 2023.
https://doi.org/10.1007/978-3-031-36272-9_71

lightweight models have shown great promise in handling a large number of intents as shown by Ng et al. [8], the data collection and labelling process of many intents is highly resource intensive and may still be insufficient to capture all possible intents during a clinical session between doctor and patient.

In this work, we propose the novel creation of a hybrid language model capable of both natural language understanding and generation. The hybrid model comprises of a lightweight model to handle domain-specific intents, whereas the large language model is used towards handling out-of-scope intents which have not been previously defined. We show the application of the resultant model in both virtual reality, augmented reality and webpage application platforms.

2 Related Works

Present clinical teaching methods often rely on highly manual and expensive methods such as training an actor patient to mimic the underlying disease [7]. The use of actor patients in medical training has several disadvantages, including high costs and limited scalability, inability to visually mimic illness symptoms, and limited training time for students due to one-on-one interactions. Additionally, students often face anxiety during initial training sessions. To address these issues, a simulated virtual patient can be created to replace the actor patient.

There are various limitations in the current solutions that strongly limits their retention and uptake rate among students. Most solutions utilize either text-based [9] or multiple choice-based inputs [3] in their application. The lack of immersion is further exacerbated by models which operate on rule-based logic, failing to take into account conversational history when determining the appropriate response to the student [11]. Therefore, we propose a hybrid lightweight and large language model to handle diverse conversational topics which strongly imitates real-world conversations, allowing for students to obtain higher immersion in online learning experiences.

3 Dataset

The dataset comprises of three main parts: the possible intention categories, example sentences for each intention and the expected response towards each intention. The data is collected from medical experts in the form of recorded and written conversations. From there, the sentences are extracted into text form. Subsequently the sentences may be grouped into overarching intentions with their respective expected responses.

Conventional natural language generation typically assumes that each intention would have a single response [1]. Although this would work well in cases whereby intentions are highly separable, a common issue arises in real world applications whereby intentions with similar semantic structures are confused with one another. Furthermore, recent work in knowledge graphs have shown that there are implicit links between different topics, which may make it difficult for NLP models to discriminate into explicitly defined classes [12]. As such, rather than relying on single intentions to discriminate between sentences of similar meanings, we utilize secondary intentions as well to capture underlying relationships between classes.

4 Methodology

For the lightweight model, we utilize the open-source RASA NLU model [5]. The RASA NLU model operates using the dual intent and entity transformer (DIET) classifier to consider both intents and entities. For the large language model, GPT-3 is utilized [4]. The GPT-3 model was trained upon large amount of data from various sources, giving it the ability to understand and respond to any input given. However, one of the limitations of having such a large number of data is that the specificity of the response given would be more limited in nature. Thus, the usage of the GPT-3 model within the proposed system architecture is entirely limited towards the instances whereby the original domain-specific trained DIET classifier fails to recognise an intention.

To ensure that the responses given by the GPT-3 model still fall in line with the general conversational history, we first initialize the GPT-3 model through fine-tuning the model using the domain-specific data. However, given the size of the GPT-3 model, the fine-tuning acts to encourage the model to give responses within the context of the simulated patient.

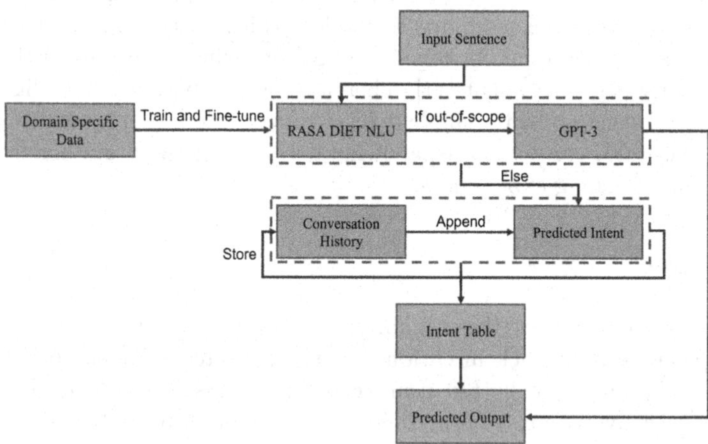

Fig. 1. Overall proposed framework for handling open-ended conversational input from students.

The hybrid model architecture is achieved via combining both the DIET classifier and GPT-3 model into a single conversational pipeline (Fig. 1). The trained RASA NLU classifier model outputs specific pre-defined intents which are indicated in the training data. These intents are then brought forward to future conversations to be utilized as conversational history to provide contextual information to the system. The overall determined intent(s) are subsequently compared against a pre-defined table containing the explicitly defined knowledge graph containing the relationship between different intents. However, if no suitable intent or permutation exists, the input sentence is then sent to the finetuned GPT-3 model to obtain the most suitable response given that the sentence has been determined to be outside of the domain context. Overall, the output model outputs an appropriate text response which may then be relayed back to the student via text or audio (Fig. 2).

Fig. 2. From left to right: Classroom implementations of the fully immersive virtual patient conversation in (1) Augmented Reality, (2) Virtual Reality and (3) Web-based Application.

This study proposes a hybrid language model architecture for a medical training scenario where a simulated patient avatar is created and hosted on digital online platforms such as virtual reality, augmented reality, and web applications. The student users interact with the simulated patient through natural spoken language, which is transcribed into text using Text-to-Speech (TTS) models. The response is then played out as sound using a Speech-to-Text (STT) model alongside the associated avatar movements, enabling a free-flow conversation that simulates a clinical environment. This proposed system can help students practice their clinical skills and communication in a fully autonomous way without any breaks in conversation.

5 Results and Discussion

The proposed hybrid language model combines the strengths of lightweight and heavy models, achieving state-of-the-art natural language processing and generation in a medical clinical simulation. It is able to handle content not previously possible, perform natural language understanding and generation without losing its ability to handle out-of-context inputs, and consistently perform well on domain-specific user intents. The system was tested in real-world classroom scenarios.

The proposed hybrid language model architecture is observed to be more effective in imitating a real-life clinical patient compared to a single NLU model like RASA DIET. The RASA DIET model fails to recognize certain questions and responds with a null fallback response, causing frustration for users. This failure could be prevented using the proposed architecture, which allows for maximum immersion in the training scenario for students.

Table 1. Comparison of performance across different methods to handle conversational systems.

Methodology	Metric		
	Mean Accuracy (%)	Model Failures (%)	Model Confusion (%)
RASA NLU	89.3	10.5	12.4
+ Knowledge Graph	93.2	7.5	6.2
+ **GPT-3 Hybrid (Ours)**	**98.7**	**2.0**	**0.5**

A total of 100 students utilized the solution across a period of 10 separate trial sessions for a total of 200 trials. We see greater overall accuracy (Table 1) in discerning the users' true intention. To summarize, the proposed virtual patient conversational system in this work greatly enables state-of-the-art medical practical lessons to be recreated into a digital format. The proposed method achieved high mean accuracy (98.7%) since the model is now able to recognise intents that was previously not included in the predefined intention labels, thus leading to an increase in the overall accuracy.

Furthermore, a significant decrease in model failure was observed (Table 1), whereby failure is deemed when the model is unable to proceed further into the conversation after three attempts by the student. The model failure rate is then computed as the number of model failures over the total number of trials (n = 200). In the previous two models, the failure rate was significantly higher due to their inability to parse sentences which may contain information that the model was not trained upon. This is further compounded by the fact that some individuals fail to rephrase their question in a manner that is recognizable by the trained model, thus leading to conversational failure.

Finally, it was found that the students' ability to converse with the model patient was greatly increased as indicated by the reduction of repetitions required to be uttered by the users throughout the clinical scenario. The model confusion rate measures the smoothness of the conversation and is defined by the number of times a clarification is required by the user over the total number of conversational turns. In the case of our proposed method, the confusion rate was 0.5% due to the usage of GPT-3 to handle instances whereby the domain-specific NLU model failed. This allows the proposed system to bypass the clarification step that is normally seen in conventional chatbot agents.

6 Conclusion

In conclusion, the combination of two language models can create state-of-the-art conversational capabilities as demonstrated in a medical clinic scenario. The proposed system of using simulated patients achieves significant cost savings compared to traditional actor-patient training, making quality medical education more accessible to schools in poorer socioeconomic backgrounds. The digitalization of training scenarios allows for clinical training to be practiced at any time and place, increasing accessibility. The proposed system is scalable to a large number of different scenarios, unlike traditional teaching pedagogy, and offers greater realism through virtual and augmented reality platforms. While simulated patients cannot completely replace the authenticity of human-to-human interactions, this work presents a first step towards creating standardized training for such scenarios.

References

1. Abro, W.A., Qi, G., Gao, H., Khan, M.A., Ali, Z.: Multi-turn intent determination for goal-oriented dialogue systems. In: 2019 International Joint Conference on Neural Networks (IJCNN), pp. 1–8. IEEE (2019)
2. Almousa, O., et al.: Virtual reality technology and remote digital application for tele-simulation and global medical education: an innovative hybrid system for clinical training. Simul. Gaming **52**(5), 614–634 (2021)
3. Aubart, F.C., et al.: Progressive clinical case-based multiple-choice questions: an innovative way to evaluate and rank undergraduate medical students. Rev. Med. Interne **42**(5), 302–309 (2021)
4. Brown, T., et al.: Language models are few-shot learners. Adv. Neural. Inf. Process. Syst. **33**, 1877–1901 (2020)
5. Bunk, T., Varshneya, D., Vlasov, V., Nichol, A.: Diet: lightweight language understanding for dialogue systems. arXiv preprint arXiv:2004.09936 (2020)
6. Carin, L.: On artificial intelligence and deep learning within medical education. Acad. Med. **95**(11S), S10–S11 (2020)
7. Crofts, J., et al.: Patient-actor perception of care: a comparison of obstetric emergency training using manikins and patient-actors. BMJ Quality & Safety **17**(1), 20–24 (2008)
8. Ng, H.W., Koh, A., Foong, A., Ong, J., Tan, J.H., Khoo, E.T., Liu, G.: Real-time spoken language understanding for orthopedic clinical training in virtual reality. In: Rodrigo, M.M., Matsuda, N., Cristea, A.I., Dimitrova, V. (eds.) AIED 2022. LNCS, vol. 13355, pp. 640–646. Springer, Cham (2022). https://doi.org/10.1007/978-3-031-11644-5_61
9. Oliven, A., Nave, R., Gilad, D., Barch, A.: Implementation of a web-based interactive virtual patient case simulation as a training and assessment tool for medical students. In: User Centred Networked Health Care, pp. 233–237. IOS Press (2011)
10. Sanal, M.G., Paul, K., Kumar, S., Ganguly, N.K.: Artificial intelligence and deep learning: the future of medicine and medical practice. J. Assoc. Physicians India **67**(4), 71–73 (2019)
11. Xu, P., Sarikaya, R.: Contextual domain classification in spoken language understanding systems using recurrent neural network. In: 2014 IEEE International Conference on Acoustics, Speech and Signal Processing (ICASSP), pp. 136–140. IEEE (2014)
12. Yu, D., He, L., Zhang, Y., Du, X., Pasupat, P., Li, Q.: Few-shot intent classification and slot filling with retrieved examples. arXiv preprint arXiv:2104.05763 (2021)

Towards Enriched Controllability for Educational Question Generation

Bernardo Leite[1,2(✉)] and Henrique Lopes Cardoso[1,2]

[1] Faculty of Engineering of the University of Porto (FEUP), Porto, Portugal
{bernardo.leite,hlc}@fe.up.pt
[2] Artificial Intelligence and Computer Science Laboratory (LIACC), Porto, Portugal

Abstract. *Question Generation* (QG) is a task within Natural Language Processing (NLP) that involves automatically generating questions given an input, typically composed of a text and a target answer. Recent work on QG aims to control the type of generated questions so that they meet educational needs. A remarkable example of *controllability* in educational QG is the generation of questions underlying certain *narrative elements*, e.g., causal relationship, outcome resolution, or prediction. This study aims to enrich controllability in QG by introducing a new guidance attribute: *question explicitness*. We propose to control the generation of explicit and implicit (*wh*)-questions from children-friendly stories. We show preliminary evidence of controlling QG via question explicitness alone and simultaneously with another target attribute: the question's narrative element. The code is publicly available at https://github.com/bernardoleite/question-generation-control.

Keywords: Natural Language Processing · Question Generation · Controllability · Question Explicitness

1 Introduction

In the educational context, Question Generation (QG) can potentially automate and assist the teacher in what can be a time-consuming and effortful task. QG may also be helpful for the learner's formative assessment via self-study and engagement with computer-generated practice questions. However, automatic QG tools are not widely used in classrooms [2,8], namely because generated questions are generally limited in types and difficulty levels [2]. As pointed by Wang *et al.* [8], there is a strong desire for user control, where humans provide input to QG systems and can decide when to use their output. Inspired by this need, this study proposes a QG framework for controlling the generation of explicit and implicit questions, using question explicitness as a guidance attribute during the generation process. Generally, explicit questions center on a particular story fact, whereas implicit questions rely on summarizing[1] and draw-

[1] Summarization skills have been used to assess and improve students' reading comprehension ability [8].

N. Wang et al. (Eds.): AIED 2023, LNAI 13916, pp. 786–791, 2023.
https://doi.org/10.1007/978-3-031-36272-9_72

ing inferences from implicit information in the text. As stated by Xu *et al.* [9], explicit and implicit questions are formally defined as follows:

- **Explicit** *questions ask for answers that can be directly found in the stories. In other words, the source of answer are spans of text.*
- **Implicit** *questions ask for answers that cannot be directly found in the text. Answering the questions requires either reformulating language or making inferences. In other words, the answer source is "free-form", meaning that the answers can be any free-text, and there is no limit to where the answer comes from.*

Noteworthy, prior research [6,9,11] suggests that a combination of explicit and implicit questions contributes to a more balanced difficulty in the assessments. To achieve our goal, we use a recent dataset called FAIRYTALEQA [9], which contains question-answering (QA) pairs derived from children-friendly stories. Each question is categorized as "explicit" or "implicit" by expert annotators.

Some previous work has addressed controllability in educational QG. For instance, Ghanem *et al.* [1] control the reading comprehension skills required by the question, e.g., figurative language and summarization. Similarly, Zhao *et al.* [10] control the narrative elements underlying the generated questions, such as causal relationship, outcome resolution, or prediction. They use the same dataset as this study, FAIRYTALEQA, where each question, beyond explicitness, is also categorized according to the referred narrative elements.

2 Generating Explicit and Implicit Questions

In this study, we fine-tune the T5 pre-trained model [5] with the controllable mechanism for generating explicit and implicit questions. T5 is a text-to-text generation model which has achieved state-of-the-art results on multiple natural language generation benchmarks, including QA and summarization. We train the model to generate both questions and answers for a particular story text. To control the explicitness of the generated questions, we prepend a special token <EX> followed by "explicit" or "implicit" attribute at the beginning of the input, before the story text. This attribute guides the system to generate a question of the desired type. Other special tokens (<SECTION>, <QUESTION> and <ANSWER>) are used to delimit the input and output information of the model. This technique is based on a recent study [10] with the purpose of controlling QG conditioned on another target attribute: the question's narrative elements. We also investigate controlling simultaneously the question's explicitness along with that target attribute. To that end, beyond <EX>, we prepend <NAR> followed by the narrative attribute name.

3 Experimental Setup

Data: We use FAIRYTALEQA [9], in which educational experts have manually created 10,580 QA pairs from 278 children-friendly stories. Each question is

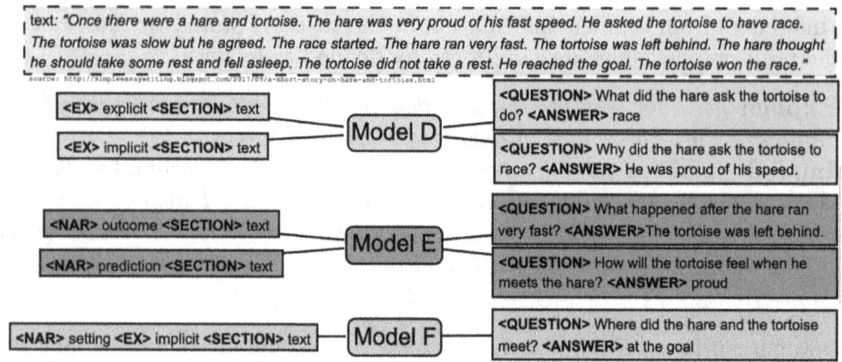

Fig. 1. An illustrative example of question-answer pairs generated by different models.

annotated with an explicitness label, which can be "explicit" or "implicit". Also, each question is labeled with one of the following narrative elements[2]: "character", "setting", "action", "feeling", "causal relationship", "outcome resolution", or "prediction". Statistically, each story has ≈15 sections and each section (composed of multiple sentences) has ≈3 questions. Explicit questions represent ≈75% of all questions. We use the original train/val/test splits composed of 8,548/1,025/1,007 QA pairs.

Models: From the original dataset, we have trained different models[3]: (A) question-section:answer; (B) answer-section:question; (C) section:question-answer; (D) ex-section:question-answer; (E) nar-section:question-answer; and (F) nar-ex-section:question-answer. Models A and B will serve as a baseline comparison with the QA and QG models from the FairytaleQA paper. Model C only contains the section text as input, so its purpose is to serve as a baseline to compare with models D-F, which include control attributes. Model D includes the question's explicitness attribute in the input. Model E includes the narrative attribute in the input. Model F has both control attributes included. Figure 1 shows an illustrative example of the models with controllability prompts.

***Controlled Test* Set:** For assessing the effectiveness of controllability along models D-F, we have prepared a reorganized version from the original *test* set which we call *controlled test*: each example includes a section and all ground-truth QA pairs regarding that section, being that these QA pairs belong to one explicitness type (explicit or implicit) and narrative element. Also, for comparability between models C and D-F, each section only appears once.

Implementation Details: We use the *t5-base*[4] model version. We have set 512 and 128 for the maximum token input and output, respectively. We train the models with a maximum of 10 epochs, early stopping with a patience of 2, and a batch size of 32. For inference, we use beam search with a beam width of 5.

[2] Detailed information of each aspect is described in the FairytaleQA paper [9].

[3] A colon separates the input and output information used by the models.

[4] https://huggingface.co/t5-base.

4 Results

Baselines: FairytaleQA authors have reported n-gram similarity ROUGE$_L$-F1 [3] values of 0.536 (QA) and 0.527 (QG) on the *test* set. Using our baseline models (A and B) we correspondingly obtained 0.559 (QA) and 0.529 (QG). This shows that our baseline models are quantitatively aligned with previously obtained results.

QA Results by Question Explicitness: More on baseline model A for QA, our ROUGE$_L$-F1 QA results for explicit and implicit questions are 0.681 and 0.194, respectively. This notable difference is also observe by Xu *et al.* [9]. According to the authors, this situation is expected since the answers to explicit questions can be directly found in the text. In contrast, implicit questions call for in-depth inference and summarization. We use this rationale to evaluate the controllability of the question's explicitness. We hypothesize that the QA model obtained in setup A will perform significantly better on explicit than implicit questions generated from models D and F.

Controllability: We look for evidence of the question's controllability by employing both QA and QG tasks. For QA, we use the ROUGE$_L$-F1 metric and EXACT MATCH, which is a strict all-or-nothing score between two strings. For QG, we use n-gram similarity ROUGE$_L$-F1 and BLEU-4 [4]. Also, we use BLEURT [7], which is a more recent text generation performance metric.

Table 1 refers to the QA results, which have been obtained as follows. We use the QA model (obtained in setup A) for answering the generated questions from models D and F. Then, the answers obtained from the QA model are compared against the answers generated from models D and F, yielding the reported results. For both evaluation metrics, the QA model performs significantly better on explicit than implicit generated questions (confirming our hypothesis). Thus, we conclude that these scores indicate compelling evidence that it is possible to control the question's explicitness using the proposed controllable mechanism.

Table 2 presents the obtained QG results. Here the traditional evaluation procedure in QG is employed, which is to directly compare the generated questions with the ground-truth[5]. We find no significant differences in the QG scores obtained by model D compared to C, which can be explained as follows: controlling the question's explicitness has more influence on the type of answer required to respond to the generated question than on the syntax of that generated question. Therefore, we consider the non-significant differences between models C and D in the QG results to be expected. In contrast, a significant improvement is observed in models E and F (which receive narrative controllability prompts) compared to model C. This can be explained as follows: controlling the question's narrative elements strongly influences the syntax of the generated questions. For

[5] Note that the drop in QG ROUGE$_L$-F1 values relative to baseline model B is expected, since in these models the answer is not included in the input. The generated questions may thus focus on target answers that are not part of the gold standard.

instance, we empirically observe that when requesting the model to generate questions about the "causal relationship" element, it generates (in many cases) questions starting with "Why did...?". As for "outcome resolution", the model generates "What happened...?" questions. As for "prediction", the model generates "How will...?" questions.

Finally, it should be noted that model F (which receives both explicitness and narrative controllability prompts) is shown to be effective for controlling *simultaneously* question's explicitness and question's narrative elements.

Table 1. QA results (0–1) for assessing the question's controllability (*controlled test*).

	ROUGE$_L$-F1			**EXACT MATCH**		
Models	Overall	Explicit	Implicit	Overall	Explicit	Implicit
ex-section:question-answer (D)	0.656	0.741	0.431	0.434	0.483	0.306
nar-ex-section:question-answer (F)	0.671	0.730	0.514	0.449	0.489	0.343

Table 2. QG results (0–1) for assessing the question's controllability (*controlled test*).

Models	**ROUGE$_L$-F1**	**BLEU-4**	**BLEURT**
section:question-answer (C)	0.305	0.099	0.370
ex-section:question-answer (D)	0.303	0.104	0.369
nar-section:question-answer (E)	0.432	0.189	0.432
nar-ex-section:question-answer (F)	0.432	0.195	0.424

5 Conclusion

In this study, we work towards enriched controllability for educational QG. Through automatic evaluation, the results show preliminary evidence that it is possible to (1) control the question's explicitness and (2) *simultaneously* control both the question's explicitness and question's narrative elements. We argue that the next developments in educational QG should involve enriching (even more) the controllability process with multiple guidance and educationally relevant attributes. Looking for additional effective control mechanisms is also an interesting route. For future work, we intend to perform a large-scale human evaluation focusing on QG controllability in an actual educational environment.

Acknowledgments. This work was financially supported by Base Funding - UIDB/00027/2020 of the Artificial Intelligence and Computer Science Laboratory - LIACC - funded by national funds through the FCT/MCTES (PIDDAC). Bernardo Leite is supported by a PhD studentship (with reference 2021.05432.BD), funded by Fundação para a Ciência e a Tecnologia (FCT).

References

1. Ghanem, B., Lutz Coleman, L., Rivard Dexter, J., von der Ohe, S., Fyshe, A.: Question generation for reading comprehension assessment by modeling how and what to ask. In: Findings of the Association for Computational Linguistics: ACL 2022, pp. 2131–2146. Association for Computational Linguistics, Dublin (2022). https://doi.org/10.18653/v1/2022.findings-acl.168
2. Kurdi, G., Leo, J., Parsia, B., Sattler, U., Al-Emari, S.: A systematic review of automatic question generation for educational purposes. Int. J. Artif. Intell. Educ. **30**(1), 121–204 (2019). https://doi.org/10.1007/s40593-019-00186-y
3. Lin, C.Y.: ROUGE: a package for automatic evaluation of summaries. In: Text Summarization Branches Out, pp. 74–81. ACL, Barcelona (2004). https://www.aclweb.org/anthology/W04-1013
4. Papineni, K., Roukos, S., Ward, T., Zhu, W.J.: Bleu: a method for automatic evaluation of machine translation. In: Proceedings of the 40th Annual Meeting of the Association for Computational Linguistics, pp. 311–318. ACL, Philadelphia (2002). https://doi.org/10.3115/1073083.1073135
5. Raffel, C., et al.: Exploring the limits of transfer learning with a unified text-to-text transformer. J. Mach. Learn. Res. **21**(140), 1–67 (2020). http://jmlr.org/papers/v21/20-074.html
6. Raphael, T.E.: Teaching question answer relationships, revisited. Read. Teach. **39**(6), 516–522 (1986)
7. Sellam, T., Das, D., Parikh, A.: BLEURT: learning robust metrics for text generation. In: Proceedings of the 58th Annual Meeting of the Association for Computational Linguistics, pp. 7881–7892. Association for Computational Linguistics (2020). https://doi.org/10.18653/v1/2020.acl-main.704
8. Wang, X., Fan, S., Houghton, J., Wang, L.: Towards process-oriented, modular, and versatile question generation that meets educational needs. In: Proceedings of the 2022 Conference of the North American Chapter of the Association for Computational Linguistics: Human Language Technologies, pp. 291–302. Association for Computational Linguistics, Seattle (2022). https://doi.org/10.18653/v1/2022.naacl-main.22
9. Xu, Y., et al.: Fantastic questions and where to find them: FairytaleQA - an authentic dataset for narrative comprehension. In: Proceedings of the 60th Annual Meeting of the Association for Computational Linguistics (vol. 1: Long Papers), pp. 447–460. Association for Computational Linguistics, Dublin, (2022). https://doi.org/10.18653/v1/2022.acl-long.34
10. Zhao, Z., Hou, Y., Wang, D., Yu, M., Liu, C., Ma, X.: Educational question generation of children storybooks via question type distribution learning and event-centric summarization. In: Proceedings of the 60th Annual Meeting of the Association for Computational Linguistics (vol. 1: Long Papers), pp. 5073–5085. Association for Computational Linguistics, Dublin (2022). https://doi.org/10.18653/v1/2022.acl-long.348
11. Zucker, T.A., Justice, L.M., Piasta, S.B., Kaderavek, J.N.: Preschool teachers' literal and inferential questions and children's responses during whole-class shared reading. Early Child. Res. Q. **25**(1), 65–83 (2010). https://doi.org/10.1016/j.ecresq.2009.07.001

A Computational Model for the ICAP Framework: Exploring Agent-Based Modeling as an AIED Methodology

Sina Rismanchian(✉) and Shayan Doroudi

University of California, Irvine, Irvine, CA 92697, USA
{srismanc,doroudis}@uci.edu

Abstract. Recently, researchers have advocated for using complex systems methodologies including agent-based modeling in education. This study proposes using agent-based models to simulate teaching and learning environments. Specifically, we present ABICAP, an agent-based model that simulates learning in accordance with the ICAP framework, which defines four levels of cognitive engagement: Interactive, Constructive, Active, and Passive. The ICAP hypothesis suggests a higher level of engagement results in improved learning outcomes. To show how ABICAP can support running hypothetical studies in a risk-free and inexpensive environment, we present two simulations examining different pedagogical scenarios. We show how our model can surface counterintuitive results which may lead to a more nuanced understanding of ICAP. More generally, this paper provides a concrete example of how agent-based modeling can be used as a methodology for advancing education research.

Keywords: agent-based models · ICAP · computational models · cognitive modeling

1 Introduction

In recent years, researchers have suggested that education and learning are best viewed as complex systems, and therefore education research needs to embrace complex systems methodologies [8,9]. For example, Jacobson et al. "regard learning not as something that is—such as receiving instruction or acquiring knowledge—but rather, as something that emerges" (p. 212) from a complex system of interactions [7]. Such a perspective is useful in understanding learning as an inherently social and situated process that is mediated by teachers, peers, and tools [1,4], but it can also be useful in explaining how learning takes place in an individual's mind—for example, as a complex system of interacting neurons or a network of "p-prims" [6,7]. A prominent method for exploring complex systems is agent-based modeling. Despite the call to use such methods in education, there are only a few examples of agent-based models (ABMs) of learning in the literature, mostly modeling specific learning scenarios [1,2]. In this paper,

N. Wang et al. (Eds.): AIED 2023, LNAI 13916, pp. 792–797, 2023.
https://doi.org/10.1007/978-3-031-36272-9_73

we show the potential promise of agent-based modeling as a methodology for studying how people learn by developing an ABM for a prominent theoretical framework in the learning sciences: the ICAP framework.

Our modeling effort builds on the cognitive science tradition of developing computational models of learning [3,11,12]. However, what differentiates the present work is that rather than focusing on accurately modeling low-level information-processing mechanisms, we focus on modeling different kinds of learning, including interactions between learners, to see how modeling these kinds of learning in tandem with one another can give rise to broader inferences about how people learn and how to best structure pedagogical activities.

The ICAP framework [5] claims that learners' behaviors fall under one of four modes of cognitive engagement: Interactive, Constructive, Active, and Passive. For instance, in a hypothetical lecture, corresponding learning activities to the modes of engagement are "arguing a position in dyads or small groups", "drawing concept maps", "taking notes", and "listening without doing anything else" [5], respectively. The ICAP hypothesis suggests that as the level of cognitive engagement increases (from passive to interactive), learning outcomes are also expected to increase. This hypothesis is supported by a large body of prior studies, when they are interpreted in light of the ICAP framework [5], and has been further confirmed with studies conducted to explicitly test the hypothesis [10]. The ICAP framework has gained increasing popularity in recent years; it is the most cited paper in *Educational Psychologist* since it was published in 2014. The ICAP hypothesis is a simple but powerful takeaway from the framework; however, the framework can also be used to inquire into interesting questions about the nature of learning and teaching. In this paper, we report on preliminary simulations to show how ABMs could be used to compare different pedagogical scenarios (e.g., ways of sequencing pedagogical activities).

One of the affordances of using an ABM is being able to simulate long trajectories of student learning in a short amount of time without subjecting some students to potentially counterproductive learning conditions. As such, agent-based modeling can complement existing quantitative and qualitative research methods used in education. It can be used to gain preliminary insights that help refine theory and make suggestions for practice.

2 ABICAP

The ABICAP model assumes each learner is an agent that consists of two main parts: 1) knowledge and 2) learning. We describe ABICAP at a high-level below, but for the exact implementation we refer readers to our Jupyter Notebook[1], which also contains additional simulations and robustness checks.

To represent knowledge, we use a knowledge graph which is a weighted undirected graph comprised of nodes and edges. Each node represents a knowledge component. Each edge shows the connection between two nodes. Each edge has a weight between 0 and 1 to indicate the extent of the connection between corresponding nodes. We assume there is a source knowledge graph G which shows

[1] https://github.com/sinaris76/ABICAP.

the nodes that are being taught to the learners and the relationships between them. G can and likely should include many nodes that are not explicitly taught in a curriculum; in this sense, while we assume there is a universal knowledge graph, this does not imply that instructors necessarily know the entire graph or that all students will learn all the nodes in the graph. Specifically, we assume learners will be taught a set of focal nodes, but they can also learn additional nodes in G based on their mode of engagement. Each learner has an evolving knowledge graph G_l which determines their current learned nodes and edges.

2.1 Learning

According to ICAP, learning outcomes are determined by the corresponding level of cognitive engagement. The framework assumes the existence of different underlying processes for learning, namely knowledge-change processes [5]. Thus, learners in a specific mode of engagement evoke the respective processes. The learning functions in ABICAP were implemented to match the descriptions of the corresponding knowledge-change processes. The exact specification of ABICAP is the result of many back-and-forth discussions between the authors while interpreting preliminary simulation results. Furthermore, we appreciate feedback from two experts on ICAP to make our model more compatible with the framework. However, we note that how to instantiate the knowledge-change processes in using specific procedures and functions is open to interpretation. One advantage of using concrete computational models to describe a theoretical framework is that it can open up the details of the framework to discussion [2]. As Abrahamson and Wilensky state, "the *lingua franca* of ABM enables researchers who otherwise use different frameworks, terminology, and methodologies to understand and critique each others' theory and even challenge or improve the theory by modifying and/or extending the computational procedures that underlie the model." Therefore, we invite researchers to challenge ABICAP's specifications.

We simulate the learning session as a sequence of steps where at each step, all learners attempt to learn the same focal node. Each learner has a chance to learn some nodes and edges according to rules corresponding to their mode of cognitive engagement. In **passive** learning, "new information is stored in an isolated way" [5]. Therefore, in ABICAP, a learner's probability of learning the focal node is a fixed amount that does not depend on prior knowledge. In **active** learning, "new information activates relevant prior knowledge and while storing, new information is integrated with activated prior knowledge" [5]. In ABICAP, the probability of active learning increases as a sigmoidal function of the sum of the weights of edges in G_l that are connected to the focal node. If a learner learns the focal node, they also learn all edges connecting the focal node to G_l, unlike the passive learner who only learns nodes and not edges. In the **constructive** mode, "new information is integrated with activated prior knowledge, and new knowledge is inferred from activated and integrated knowledge" [5]. Therefore, the constructive learner learns the focal node and its edges just as the active learner; however, they also have the ability to infer up to two extra nodes connected to the focal node. If the constructive learner successfully learns the focal node, the additional nodes are learned in the same fashion but with the probability reduced by some amount for each successive node. Finally, **interactive** learning occurs between two learners who each learn constructively: "Each learner infers new knowledge from activated and integrated knowledge and iteratively infers knowledge with new inputs from conversational partner(s)" [5]. Two interactive learners are paired with each other; this is done by picking a peer randomly in the simulations. Unlike in other modes, the knowledge of the peer plays a role in increasing the probability of learning for a learner. In particular, the probability of learning each node is a sigmoidal function of the sum of the weights of edges connected to the focal node in either learners' knowledge graphs (with edges only in the partners' knowledge graph given half as much weight). If either partner learns the node, each partner can learn an additional node in a similar fashion, and if they do, their partner can learn the same node as well.

3 Simulations and Results

We contend that ABICAP can be used to ask many different kinds of questions (e.g., what is the relative advantage of different strategies for pairing students? how efficient does interactive learning need to be in order to maintain its advantage?). As an initial case study, we present two simulations focused on comparing different pedagogical situations where learners change their mode of engagement over time. Both simulations below use 50 time steps, a knowledge graph with 100 nodes (including 50 focal nodes). We repeat each simulation ten times in order to run a t-test to see if there are significant differences in certain measures.

In the first simulation, we compare two groups of agents: I) a group of 50 agents that learn passievly in the first half and in the interactive mode in the second half of the session, and II) a group of 50 agents learning in the reverse order. Group I could resemble a flipped classroom setting, where the learner might go from watching videos passively at home to engaging in interactive activities in the classroom. Group II resembles an alternative strategy where learners might do interactive group work before attending a lecture. To measure the performance of these groups, we report the average number of nodes in the learners' knowledge graphs (knowledge amount; KA) and the average number of nodes in the largest connected subgraph in the learners' knowledge graphs (largest subgraph; LS). KA could be seen as a measure of breadth of knowledge while LS could be seen as a measure of depth of knowledge. As seen in Fig. 1, which shows one run of the simulation, we see that group I learns more (KA = 50.49 in I vs. KA = 44.47 in II, $t(9) = 18.65, p < 0.001$) and deeper (LS = 49.54 in I vs. LS = 35.09 in II, $t(9) = 41.78, p < 0.001$) than group II. We believe the main reason for the difference is that when group I starts learning interactively, they have already gained some knowledge passively; although this knowledge is isolated, once they learn neighboring nodes they establish links to their prior knowledge that helps increase their partner's probability of learning new nodes.

To more accurately reflect what happens in a flipped classroom setting, we ran a second simulation with two new types of learners. A **diligent** learner is

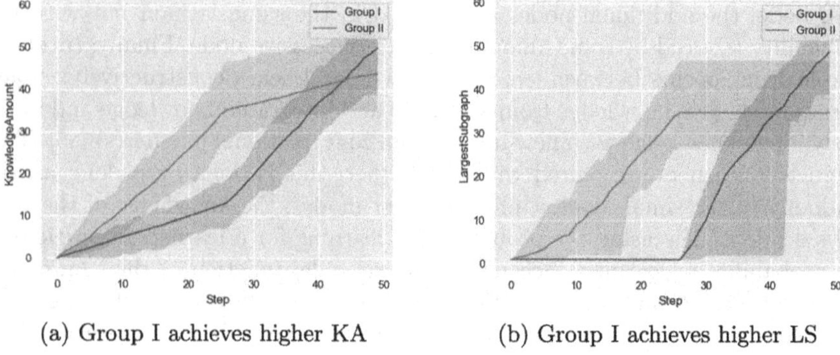

(a) Group I achieves higher KA

(b) Group I achieves higher LS

Fig. 1. Comparison of two group types

a regular student in a flipped classroom who learns passively in the first half of a learning session, then learns interactively for the rest of the session. A **lazy** learner learns nothing in the first half (i.e., does not watch videos or read papers at home) and then learns interactively in the second half. We aimed to study the effect of a greater proportion of lazy learners in a classroom. We compared the results of two classroom settings: I) a classroom where 15% of students are lazy, and II) a classroom where 85% of students are lazy. In each classroom of 200 students, students were paired randomly with any other student in the class, regardless of type.

The performance of diligent learners was almost identical between the two settings (KA = 66.28 in I vs. KA = 66.71 in II, $t(9) = 1.67, p = 0.11$) and similarly for lazy students (KA = 57.34 vs. 58.25; $t(9) = 1.66, p = 0.11$). We initially found this result to be counterintuitive since we expected that both types of students would benefit more from having diligent partners (which is more likely in classroom I). When investigating this result, we realized that if a learner has learned all the neighboring nodes, then they do not have a chance to construct new nodes; therefore the partner also loses one of their two chances to learn a new node. Based on the model assumptions, which are rooted in how interactive learning was defined in the ICAP framework, an interactive agent only shares new knowledge that it constructs with their partner. If this is true, learners are better off being partnered with someone who knows somewhat less to take advantage of their learning process. However, it seems reasonable to us that at least some peers will share their prior knowledge with their partner acting as a peer tutor, a phenomenon that ICAP does not seem to account for since it assumes both partners are constructive.

4 Discussion

ABMs in education can help translate theories into concrete computational models, which the process can uncover theoretical limitations and hidden assumptions, such as variations in the amount of time each step takes for different modes of engagement, allowing researchers to test and interrogate the ICAP hypothesis under different time scenarios. Once the model is specified, ABM simulations can reveal counterintuitive results that shed light on theoretical limitations and hidden assumptions, such as discovering that partnering with knowledgeable learners is not always advantageous, which could lead to new questions about learning scenarios and potential extensions to ICAP. Agent-based models, provide a risk-free environment to test hypotheses about teaching and learning, allowing researchers, instructional designers, and instructors to run "what-if" scenarios that may be difficult, costly, or counterproductive to test in the real world, and fine-tune simulations to specific contexts. We acknowledge that ABICAP is not a perfect simulation of how people learn, but rather a tool to advance educational theory and practice by bringing precision to ICAP, uncovering hidden assumptions, and simulating various learning scenarios and pedagogical practices. We hope that this work can serve as a case study for how ABMs can bring new insights into the complex nature of learning.

References

1. Abrahamson, D., Blikstein, P., Wilensky, U.: Classroom model, model classroom: Computer-supported methodology for investigating collaborative-learning pedagogy. In: Proceedings of the 8th International Conference on Computer Supported Collaborative Learning, pp. 49–58 (2007)
2. Abrahamson, D., Wilensky, U.: Piaget? Vygotsky? I'm game!-Agent-based modeling for psychology research. In: Annual Meeting of the Jean Piaget Society (2005)
3. Anderson, J.R., Lebiere, C.J.: The Atomic Components of Thought. Psychology Press, London (2014)
4. Carley, K.: Knowledge acquisition as a social phenomenon. Instr. Sci. **14**(3), 381–438 (1986)
5. Chi, M.T., Wylie, R.: The ICAP framework: linking cognitive engagement to active learning outcomes. Educ. Psychol. **49**(4), 219–243 (2014)
6. DiSessa, A.A.: Toward an epistemology of physics. Cogn. Instr. **10**(2–3), 105–225 (1993)
7. Jacobson, M.J., Kapur, M., Reimann, P.: Conceptualizing debates in learning and educational research: toward a complex systems conceptual framework of learning. Educ. Psychol. **51**(2), 210–218 (2016)
8. Jacobson, M.J., Levin, J.A., Kapur, M.: Education as a complex system: conceptual and methodological implications. Educ. Res. **48**(2), 112–119 (2019)
9. Jacobson, M.J., Wilensky, U.: Complex systems in education: scientific and educational importance and implications for the learning sciences. J. Learn. Sci. **15**(1), 11–34 (2006)
10. Menekse, M., Stump, G.S., Krause, S., Chi, M.T.: Differentiated overt learning activities for effective instruction in engineering classrooms. J. Eng. Educ. **102**(3), 346–374 (2013)
11. Rafferty, A.N., Jansen, R.A., Griffiths, T.L.: Assessing mathematics misunderstandings via Bayesian inverse planning. Cogn. Sci. **44**(10), e12900 (2020)
12. VanLehn, K., Jones, R.M., Chi, M.T.: A model of the self-explanation effect. J. Learn. Sci. **2**(1), 1–59 (1992)

Automated Program Repair Using Generative Models for Code Infilling

Charles Koutcheme[1(✉)], Sami Sarsa[1], Juho Leinonen[2], Arto Hellas[1], and Paul Denny[2]

[1] Aalto University, Espoo, Finland
{charles.koutcheme,sami.sarsa,arto.hellas}@aalto.fi
[2] The University of Auckland, Auckland, New Zealand
{juho.leinonen,p.denny}@auckland.ac.nz

Abstract. In educational settings, automated program repair techniques serve as a feedback mechanism to guide students working on their programming assignments. Recent work has investigated using large language models (LLMs) for program repair. In this area, most of the attention has been focused on using proprietary systems accessible through APIs. However, the limited access and control over these systems remain a block to their adoption and usage in education. The present work studies the repairing capabilities of open large language models. In particular, we focus on a recent family of generative models, which, on top of standard left-to-right program synthesis, can also predict missing spans of code at any position in a program. We experiment with one of these models on four programming datasets and show that we can obtain good repair performance even without additional training.

Keywords: Program repair · Large Language Models · Computer Science Education

1 Introduction

Novice programmers will eventually have bugs in their programs which require debugging [9,14]. One possible stream of research for providing support in debugging is the use of large language models (LLMs) [3,4]. Contrary to the use of proprietary LLMs through an API at a cost, we explore *open large language models* that are freely accessible and that can be run locally.

For the present work, we focus on Generative Models for Code Infilling (GMCI) [8] which can be used to fill in missing sections within a given input. Our overarching research goal is to explore the applicability of GMCI for repairing novice programs. We address the following two research questions with respect to GMCI-based program repair techniques: *(RQ1) How do GMCIs perform in fixing programs with a single bug?; (RQ2) with multiple problems and more diverse issues?* In answering the research questions, we also discuss the effect of including metadata information in the prompt and fine-tuning the model on educational data.

N. Wang et al. (Eds.): AIED 2023, LNAI 13916, pp. 798–803, 2023.
https://doi.org/10.1007/978-3-031-36272-9_74

2 Background

Automated program repair (APR) research seeks to find ways to automatically fix bugs in programs. Traditional approaches to program repair have utilised test suites to identify defects to fix, followed by the generation of candidate patches that are then validated against the test suite (see e.g. [11,13]). An example in this category is the work of Hu et al. [10], whose tool – Refactory – generates code by refactoring existing solutions to a problem. Then, given an incorrect program, its control flow structure is analysed to find a closely matching solution which is then used to isolate the buggy components of the program.

A variety of machine learning-based methods for program repair exist [5,19], and the recent emergence of LLMs has led to suggestions on using them to localize and fix bugs [15]. In an educational context, Zhang et al. [20] leveraged Codex to fix bugs in Python programming assignments, using correct solutions, test cases, and assignment descriptions for prompting. They evaluate their method on 286 Python programs produced by novices and show that their approach can repair up to 96.5% of the programs, and with a smaller edit distance compared to other automated program repair approaches [10].

One challenge of the recent proprietary LLMs [18,20] is that the underlying models are opaque and their access is only available through APIs and at a cost. A recent advance that is intuitively applicable for code repair is Generative Models for Code Infilling (GMCI) [2,8] that extend LLMs with infilling capabilities that allow completion within text. This provides an intuitive approach for fixing bugs in place. Such models have been explored and evaluated recently in the context of program repair, where Xia et al [18] showed that LLM-based approaches outperform traditional APR tools.

3 Methodology

In our work, we study GMCIs for program repair in an educational context using the InCoder model [8] available on HuggingFace. The preprocessed data and code for our experiments are released online[1]. In this section, we introduce our data and present the experiments performed to answer our research questions.

Data. We use the QuixBugs dataset [12] and three datasets of student solutions to programming assignments written in Python: (1) Dublin City University data (DB) [1]; (2) University of New Caledonia data (NC) [6]; and (3) National University of Singapore data [10]. We scope our evaluation to assignments that require writing a single function that takes fixed inputs and produces one output and use a subset of assignments to balance diversity and complexity. Additionally, we remove duplicate solutions that could bias evaluation by comparing ASTs.

[1] https://github.com/KoutchemeCharles/aied2023.

Experiments. To answer our first research question, we use our model for repairing programs on the QuixBugs dataset, where each code contains a single function with a unique bug on a single line. Although this dataset is not strictly a dataset of student programs, many of the bugs present, such as "incorrect assignment operator" and "missing condition", are typical mistakes that students make. We report the number of programs that we can repair, and compare our performance against the work of Prenner and Robbes [15] who used OpenAI Codex with the same data.

To answer our second research question, we evaluate our program repair strategy using InCoder on our student datasets. We report our performance in terms of success rate (i.e. the ratio of the number of programs that we can repair for each assignment in each dataset), and we contrast our results against the Refactory automated program repair tool [10].

Technical Details. To repair a given buggy program, we adapt the multi-line infilling strategy of Fried et al. [8]. Given a function with N lines, our method systematically selects a span of n ($n \leq N$) consecutive lines and asks the model to complete the code with the missing lines. When completing a program, we generate ten candidate completions using top-p nucleus sampling with $p = 0.95$ and a fixed temperature of 0.6 [4]. A program is considered to be repaired if one of the model-generated solutions passes all automated tests associated with the buggy program. We evaluate up to 50 different spans to find a repair [16]. We refer the reader to the implementation for details of the algorithm.

To allow the model to better understand the buggy program's intended functionality, we add a docstring and we expose the test cases that need to be passed. Figure 1 illustrate our approach. The QuixBugs dataset contains properly formatted docstrings, whereas the Dublin and New Caledonia datasets do not contain docstrings. In these cases, we add single-line docstrings which summarize briefly what the code is supposed to do. For the Singapore dataset, we format the original assignment description as a docstring.

4 Results

Repairing Programs with a Single Bug. Table 1 shows a comparison of InCoder [8] and Codex [15] on the QuixBugs dataset [12]. Our model managed to repair 13 out of the 28 programs. In contrast, Codex performs better with 23 bugs repaired in total.

Repairing Student Programs. Table 2 breaks down our results per assignment, for a subset of selected assignments, when repairing programs in our student datasets. Overall, although performance varies greatly between different assignments, our results are competitive with the Refactory repair tool (RF), both in terms of the number of repairs found and in terms of distance to the original buggy program.

Original incorrect student code with one error

```
def mean(arr):
    """Compute the mean of an array."""
    if len(arr) == 0:
        return None
    sum = 0
    for i in range(0, len(arr)-1):
        sum = sum + arr[i]
    avg = sum / len(arr)
    return avg

if __name__ == "__main__":
    assert mean([2, 4]) == 3
    assert mean([]) == None
```

Completed student code (after inference)

```
def mean(arr):
    """Compute the mean of an array."""
    if len(arr) == 0:
        return None
    sum = 0
    for i in range(0, len(arr)):
        sum = sum + arr[i]
    avg = sum / len(arr)
    return avg

if __name__ == "__main__":
    assert mean([2, 4]) == 3
    assert mean([]) == None
```

Fig. 1. Prompting our GMCI for repairing student programs. We add a short description of the program functionality as docstring as well as example test cases. We remove part of the buggy code (in orange) and prompt the model to complete the code (in green). (Color figure online)

Table 1. Incoder vs Codex on QuixBugs. ✓(resp. ✗) marks bugs which could be successfully (resp. unsuccessfully) repaired. We highlight in gray the results for the assignments which we found were typical in CS1 courses.

	Codex	InCoder
bitcount	✓	✓
bucketsort	✓	✓
find-first-in-sorted	✓	✓
flatten	✓	✗
gcd	✓	✓
get_factors	✓	✓
hanoi	✓	✗
is_valid_parenthes.	✓	✓
kheapsort	✓	✗
knapsack	✓	✗
kth	✓	✓
lcs_length	✓	✗
levenshtein	✓	✗
lis	✗	✗
long_com_subseq	✓	✗
max_sublist_sum	✓	✓
next_palindrome	✗	✗
next_permutation	✓	✓
pascal	✗	✓
possible_change	✓	✓
powerset	✓	✗
quicksort	✓	✓
shunting_yard	✗	✓
sieve	✓	✗
sqrt	✓	✗
subsequences	✗	✗
to_base	✓	✗
wrap	✓	✗

Table 2. Comparing program repair performance between our language model (GMCI) and Refactory (RF) in terms of success rate (SR) and average sequence distance (SD).

dataset	assignment_id	RF_SR	GMCI_SR	RF_SD	GMCI_SD
DB	append2list	1.00	0.93	11.43	15.19
DB	fibonacci_iter	1.00	0.94	56.03	33.81
DB	fibonacci_recur	1.00	0.95	16.79	25.41
DB	index_iter	1.00	0.93	21.36	23.83
DB	index_recur	0.98	0.52	26.11	23.66
DB	maximum	0.96	1.00	26.60	33.19
DB	merge_lists	1.00	0.28	52.45	27.51
DB	minimum	1.00	0.97	24.78	25.56
DB	reverse_iter	1.00	0.88	17.29	14.86
DB	reverse_recur	1.00	0.90	15.79	16.92
DB	search_iter	0.91	0.82	19.65	27.67
DB	search_recur	1.00	0.91	17.41	23.26
NC	decreasing_list	1.00	0.68	26.63	31.68
NC	is_palindrome	1.00	1.00	43.32	25.97
NC	maximum	0.40	0.96	22.24	26.35
NC	mean	0.77	0.99	33.29	20.61
NC	minimum	0.43	0.96	17.13	22.56
NC	sum	1.00	1.00	11.56	8.53
NC	sum_even_numbers	1.00	0.48	23.14	15.86
NC	sum_n_first_even	1.00	0.96	26.96	17.96
NC	symetrical_list	1.00	0.96	32.70	18.09
SP	remove_extras	1.00	0.42	44.77	32.69
SP	search	0.99	0.81	25.79	24.13
SP	sort_age	0.99	0.67	82.48	40.49
SP	top_k	1.00	0.91	51.30	30.39

5 Discussion and Conclusion

It seems that generative models for code infilling offer good potential for supporting novice programmers and using open large language models straight out of the box performs relatively well for fixing buggy programs. We notice in general that the model's performance depends heavily on the program's complexity, its functionality, and the type of issue(s) encountered in it. Our results are in line with similar work evaluating LLMs for program repair [18,20]. We do not obtain state-of-the-art results, so there is still much room for improvement.

Practical Tips. In our preliminary experiments, we experimented with the effect of adding a docstring. Overall, we found that adding a docstring does improve performance. In particular, the more precise the description of the code functionality, the better the results. We also experimented with finetuning the InCoder model on programming data from a previous semester for one of our datasets. Although we notice minor performance improvements on some assignments, we see a degradation in overall performance across all datasets. Instead of fine-tuning, selecting the right generation parameters for the specific dataset can be a better strategy to improve the model performance [8].

Limitations, and Future Work. One limitation of our work is that we did not compare closed-source state-of-the-art LLMs against InCoder on the student solutions. However, because of the closed nature of these, such a large-scale evaluation is costly. We are also working on developing repair algorithms more adapted for GMCIs, using, for instance, better enumeration strategies [16]. In the spirit of developing more sustainable models, we will also investigate how to create smaller, but perhaps even more efficient, LLMs for program repair. We believe that with a better pre-training strategy, combined with more sophisticated repair algorithms, we can obtain improved performance sufficient for practical use. In the long term, we see great potential in deploying these models in the classroom to address long-standing debugging challenges faced by novices. Indeed, between working on our article in late 2022 and preparing the final submission in May 2023, there has been a growing emphasis on the use and availability of open large language models (e.g. [7,17]).

References

1. Azcona, D., Smeaton, A.: +5 Million Python & Bash Programming Submissions for 5 Courses & Grades for Computer-Based Exams Over 3 Academic Years (2020). https://doi.org/10.6084/m9.figshare.12610958.v1
2. Bavarian, M., et al.: Efficient training of language models to fill in the middle (2022). https://doi.org/10.48550/ARXIV.2207.14255
3. Bommasani, R., et al.: On the opportunities and risks of foundation models (2021). https://doi.org/10.48550/ARXIV.2108.07258

4. Chen, M., et al.: Evaluating large language models trained on code (2021). https://doi.org/10.48550/ARXIV.2107.03374
5. Chen, Z., Kommrusch, S., Tufano, M., Pouchet, L., Poshyvanyk, D., Monperrus, M.: SequenceR: sequence-to-sequence learning for end-to-end program repair. IEEE Trans. Softw. Eng. **47**(09), 1943–1959 (2021). https://doi.org/10.1109/TSE.2019.2940179
6. Cleuziou, G., Flouvat, F.: Learning student program embeddings using abstract execution traces. In: 14th International Conference on Educational Data Mining, pp. 252–262 (2021)
7. Dey, N., et al.: Cerebras-GPT: open compute-optimal language models trained on the Cerebras wafer-scale cluster. arXiv preprint arXiv:2304.03208 (2023)
8. Fried, D., et al.: InCoder: a generative model for code infilling and synthesis (2022). https://doi.org/10.48550/ARXIV.2204.05999
9. Hirsch, T., Hofer, B.: A systematic literature review on benchmarks for evaluating debugging approaches. J. Syst. Softw. **192**, 111423 (2022). https://doi.org/10.1016/j.jss.2022.111423
10. Hu, Y., Ahmed, U.Z., Mechtaev, S., Leong, B., Roychoudhury, A.: Re-factoring based program repair applied to programming assignments. In: 2019 34th IEEE/ACM International Conference on Automated Software Engineering (ASE) (2019)
11. Le Goues, C., Nguyen, T., Forrest, S., Weimer, W.: GenProg: a generic method for automatic software repair. IEEE Trans. Softw. Eng. **38**(1), 54–72 (2012). https://doi.org/10.1109/TSE.2011.104
12. Lin, D., Koppel, J., Chen, A., Solar-Lezama, A.: QuixBugs: a multi-lingual program repair benchmark set based on the Quixey challenge. In: Proceedings Companion of the 2017 ACM SIGPLAN International Conference on Systems, Programming, Languages, and Applications: Software for Humanity, pp. 55–56. SPLASH Companion 2017, ACM (2017). https://doi.org/10.1145/3135932.3135941
13. Long, F., Rinard, M.: Automatic patch generation by learning correct code. In: Proceedings of the 43rd Annual ACM SIGPLAN-SIGACT Symposium on Principles of Programming Languages. POPL 2016, pp. 298–312. ACM (2016)
14. McCauley, R., et al.: Debugging: a review of the literature from an educational perspective. Comput. Sci. Educ. **18**(2), 67–92 (2008)
15. Prenner, J.A., Babii, H., Robbes, R.: Can OpenAI's codex fix bugs? An evaluation on QuixBugs. In: Proceedings of the Third International Workshop on Automated Program Repair, pp. 69–75 (2022)
16. Pu, Y., Narasimhan, K., Solar-Lezama, A., Barzilay, R.: Sk_p: a neural program corrector for MOOCs. In: Companion Proceedings of the 2016 ACM SIGPLAN International Conference on Systems, Programming, Languages and Applications: Software for Humanity, pp. 39–40. ACM (2016). https://doi.org/10.1145/2984043.2989222
17. Touvron, H., et al.: Llama: open and efficient foundation language models. arXiv preprint arXiv:2302.13971 (2023)
18. Xia, C.S., Wei, Y., Zhang, L.: Practical program repair in the era of large pre-trained language models (2022). https://doi.org/10.48550/ARXIV.2210.14179
19. Yasunaga, M., Liang, P.: Graph-based, self-supervised program repair from diagnostic feedback (2020). https://doi.org/10.48550/ARXIV.2005.10636
20. Zhang, J., et al.: Repairing bugs in python assignments using large language models (2022). https://doi.org/10.48550/ARXIV.2209.14876

Measuring the Quality of Domain Models Extracted from Textbooks with Learning Curves Analysis

Isaac Alpizar-Chacon[1,2](✉), Sergey Sosnovsky[1], and Peter Brusilovsky[3]

[1] Utrecht University, Utrecht, The Netherlands
{i.alpizarchacon,s.a.sosnovsky}@uu.nl
[2] Instituto Tecnológico de Costa Rica, ATI, Compus Cartago, Costa Rica
ialpizar@itcr.ac.cr
[3] University of Pittsburgh, Pittsburgh, USA
peterb@pitt.edu

Abstract. This paper evaluates an automatically extracted domain model from textbooks and applies learning curve analysis to assess its ability to represent students' knowledge and learning. Results show that extracted concepts are meaningful knowledge components with varying granularity, depending on textbook authors' perspectives. The evaluation demonstrates the acceptable quality of the extracted domain model in knowledge modeling.

Keywords: Knowledge Extraction · Learning Curves · Textbooks

1 Introduction

Automation of the creation of domain models (DMs) has been a long-standing practical and research problem in the field of Artificial Intelligence in Education. Most types of adaptive educational systems require high-quality composite fine-grained representations of domain knowledge to be able to model students' abilities and provide meaningful support of their learning. Development of such representations has been traditionally a manual task demanding a great deal of time and expertise. Over the last decade, a range of approaches have been introduced capable of extracting different elements of domain semantics from domain-oriented documents and user data [6,7,11]. However, very few studies examined the applicability of automatically extracted domain semantics to the task of modeling student knowledge [13].

This paper presents an evaluation of a DM that has been automatically extracted from a collection of textbooks in the same domain—Python programming. Section 2 briefly outlines the approach developed to produce such a model. The quality of the DMs extracted from textbooks with this approach has been evaluated before in terms of accuracy [2], semantic completeness [1], coverage [1], and domain specificity [4]. This paper explores the quality of the concepts

N. Wang et al. (Eds.): AIED 2023, LNAI 13916, pp. 804–809, 2023.
https://doi.org/10.1007/978-3-031-36272-9_75

extracted from textbooks as Knowledge Components (KCs) assessing their cognitive validity and applicability for knowledge modeling and assessment. Additionally, the concepts are evaluated to see if they cover too much or too little knowledge (granularity).

The best approach to validate the extracted concepts from this perspective is the *learning curve analysis* [9][1]. Learning curves are graphs that plot performance on a task versus the number of attempts to practice. Performance is usually measured using the proportion of incorrect responses (the error rate) for a KC that is being practiced. Learning curve analysis is used to qualify learning performance. If learning occurs for the KC being measured, the learning curve should follow the power law [10]. That is, the error rate of a KC should decrease as a power function of the number of attempts involving this component. A positive slope (α) indicates a decreasing curve and, therefore, a learning effect. A high fit (R^2) indicates that the KC successfully identifies the student's learning.

Motivated by the mentioned learning curve analysis, this paper describes an experiment to assert the cognitive validity and granularity of concepts extracted from textbooks. The analysis of the learning curves showed the DMs automatically extracted from textbooks consist of cognitively valid knowledge components for domain knowledge modeling. Additionally, textbooks provide both fine- and coarse-grained concepts; smaller concepts are shown to support more accurate student modeling.

2 Background

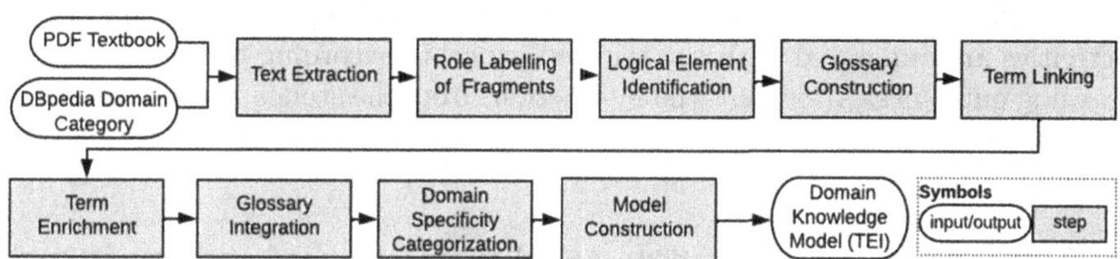

Fig. 1. Stages for the extraction of DMs from textbooks.

We have developed a workflow for the automated extraction of DMs from textbooks [1–4]. Figure 1 shows the main stages of this approach. The first three stages use an extensive set of rules that capture common conventions and guidelines for textbook formatting, structuring, and organization. The textbook's structure (chapters and subchapters), content (words, lines, text fragments, pages, and sections), and domain terms (terminology used in the textbook and the domain) are extracted. In the following three stages, the domain terms are used as a bridge to link the textbooks to entities in DBpedia[2]. The linking with

[1] The reader is directed to this source for a comprehensive introduction to learning curves.

[2] https://www.dbpedia.org/.

DBpedia allows for the enrichment of the domain terms with semantic information (e.g., abstracts and categories). In the seventh stage, terms from multiple textbooks are integrated into a single model to get better coverage of the target domain. Then, in the next stage, terms are categorized according to their relevance to the target domain (main, related, or unrelated domain). Finally, all the extracted knowledge is serialized as a descriptive XML file[3].

3 Experiment

This experiment examines the conceptual representation of knowledge in DMs from textbooks, exploring the validity of concepts as cognitive KCs and analyzing their granularity. Learning curve analysis is used to quantify learning performance for these concepts.

Data. Concepts extracted from three introductory Python programming textbooks[4] are analyzed using learning activities and learner data from PythonGrids [5], a personalized practice system for Python programming. PythonGrids learning activities are grouped into 15 ordered topics, from simple (e.g., "Variables and Operations") to advanced (e.g., "Classes/Objects"). Eleven datasets of students' interactions with PythonGrids are used for this experiment[5]. A final combined dataset containing 57929 interactions of 465 students with 85 activities was used in this experiment.

Procedure. The experiment has four steps. The first step is extracting the domain knowledge from textbooks and selecting the relevant concepts. Then, the learning activities are annotated with the selected concepts according to their expected learning outcomes. After that, the interactions from the students are aggregated, filtered out, and augmented. Finally, learning curves are generated for each concept.

1. Domain Knowledge Generation. A DM is extracted and enriched for each textbook (see Sect. 2). The models are combined into a single model to merge repeated terms. The model contains 600 terms, with 266 considered relevant for Python programming, which are used to annotate learning activities.

2. Content Annotation. Learning activities are annotated with the selected concepts by experts to indicate expected learning outcomes. Annotators assess topics, prerequisite/outcome relations, and activity outputs to choose the relevant concepts. In total, 54 concepts are used in annotations (the KCs), including "variable", "function call", "for loop", "function", and "exception".

[3] https://tei-c.org/.

[4] Python for everybody, Think Python, and Introduction to computation and programming using Python.

[5] Provided through the PSLC DataShop at http://pslcdatashop.web.cmu.edu.

3. Data Preparation. The used dataset comes from students using the Python-Grids system in real and diverse settings with no control over the environment. Therefore, the data have to be treated with caution. Reliable learning sequences that help to evaluate the concepts need to be extracted from the data. Sequences with no evidence of learning have to be regarded as noise. First, the interactions in the dataset are grouped into sequences containing all student attempts per concept [12]. In total, 10946 student-concept-attempts are generated. Correct outcomes are marked with 1's, incorrect ones with 0's. After that, we identify rapid-guessing [8] to reduce the number of incorrect attempts (smoothing) in the sequences of activities where the trial-and-error strategy works exceptionally well. Then, noise in the student-concept-attempts sequences is filtered out. Sequences are labeled with four tags: *known*, *understood_strong*, *understood_weak*, and *not_understood*. These tags identify students who already know the concept, have practiced until mastery, practiced until getting a correct answer, or stopped before showing any learning, respectively. Sequences without learning (*known* and *not_understood*) are filtered out, leaving 8079 student-concept-attempts (73.8% of all sequences). After filtering, new attempts are generated by augmenting student-concept-attempts sequences to maintain the learning evidence for students who stopped practicing earlier. This ensures the same number of concept-attempts for each student. Correct attempts (1's) are inserted in *understood_strong* sequences, while the average of the original attempts are inserted in *understood_weak* sequences until the maximum number of attempts is reached.

4. Learning Curves Generation. The learning curves are generated using the processed student-concept-attempts sequences. For each concept, the error rates at each attempt are calculated using $1 - \frac{sum\ of\ all\ outcomes}{total\ number\ of\ outcomes}$. After generating and analyzing the learning curves, a cut-off point was selected when the number of attempts was less than 25% of the first attempts. This threshold maintains a good balance between the number of attempts and the fit of the learning curves.

4 Results and Analysis

In total, 46 unique learning curves have been generated (eight concepts have had learning curves identical to other concepts'). Figure 2 displays representative examples of learning curves to guide the discussion of the results. All 46 learning curves are available online[6].

Cognitive Validity. Five[7] of 46 concepts show no learning, while the remaining 41 display a positive learning trend. The mean fit (R^2) of 0.65 ($SD = 0.27$) for these 41 positive curves indicates that **the assessed textbook concepts are cognitively valid units of knowledge**. The results align with similar literature evaluating fine-grained knowledge components [9,12]. Learning curves are classified using the power law parameters, with higher fit indicating more

[6] https://github.com/isaacalpizar/learning_curves.

[7] "Exception", "conditional statement", "value", "variable", and "iteration".

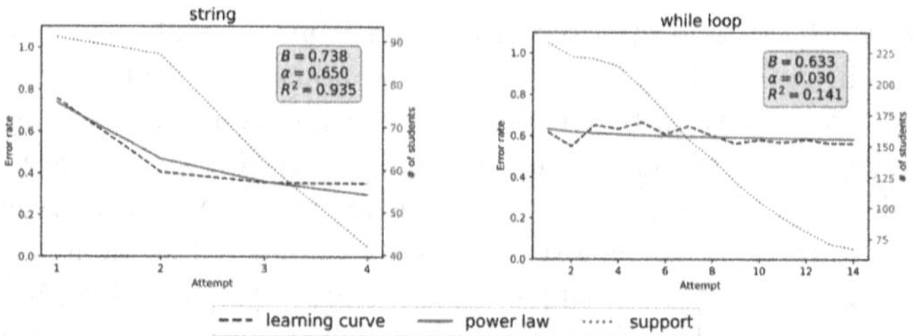

Fig. 2. Examples of learning curves.

reliable learning and steeper slope suggesting faster learning. Of the 41 downward learning curves, four are high-quality, 28 are medium-quality, and nine are low-quality. High-quality curves, such as "string" ($R^2 = 0.94$, $\alpha = 0.65$, Fig. 2-left), demonstrate more and faster learning compared to the other categories, with other concepts in this category including "hello, world", "optional parameter", and "counter". A low-quality learning curve corresponds to the "while loop" concept (Fig. 2-right), with a fit ($R^2 = 0.14$) indicating some learning but a low slope ($\alpha = 0.03$) suggesting slow progress. This may imply students struggle with while loops and that activities could be improved. Concepts "instance" and "reference" have similar curves.

Granularity. Regarding granularity, concepts in only one topic in PythonGrids are considered fine-grained (e.g., "float" in "Variables and Operations"), while those in multiple topics are coarse-grained (e.g., "iteration" in "While Loops" and "For Loops"). Fine-grained concepts have acceptable learning curves, while coarse-grained concepts are a special case. Out of these concepts, all but "iteration" produce downward learning curves, though not smooth. We can analyze the "indentation" concept. In Python, indentation is semantically meaningful and is used to indicate a block of code in many statements or expressions. This concept is linked to the "conditional statement", "while loop", "for loop", and "function" concepts. The learning curve for "indentation" is uneven with multiple upticks (bumps). The other coarse-grained concepts show similar learning curves. "Bumpy" learning curves may result from concepts being learned in the context of associated concepts. When a new associated concept is introduced, the probability of making a mistake is high, but the central concept eventually shows a downward trend after all associated concepts have been trained. This finding aligns with previous studies observing worse learning curves for more general groupings [9,12].

5 Conclusion and Future Work

This paper explored the quality of DMs extracted from textbooks regarding their ability to model students' knowledge and learning. Learning curve analysis showed textbook concepts in *Python programming* measure students' learning (cognitive validity) and displayed different granularity levels. In conclusion, this paper provided strong evidence of the richness of the extracted models for domain modeling and assessment. Future work includes a more complex experiment with an educational system designed with textbook concepts at the core.

References

1. Alpizar-Chacon, I., Sosnovsky, S.: Expanding the web of knowledge: one textbook at a time. In: Proceedings of the 30th on Hypertext and Social Media, HT 2019, ACM, New York, NY, USA (2019)
2. Alpizar-Chacon, I., Sosnovsky, S.: Order out of chaos: construction of knowledge models from pdf textbooks. In: Proceedings of the ACM Symposium on Document Engineering 2020, pp. 1–10 (2020)
3. Alpizar-Chacon, I., Sosnovsky, S.: Knowledge models from pdf textbooks. New Rev. Hypermed. Multimed. **27**(1–2), 128–176 (2021)
4. Alpizar-Chacon, I., Sosnovsky, S.: What's in an index: extracting domain-specific knowledge graphs from textbooks. In: 2022 Proceedings of the ACM Web Conference (WWW 2022) (2022)
5. Brusilovsky, P., et al.: An integrated practice system for learning programming in python: design and evaluation. Res. Pract. Technol. Enhanced Learn. **13**(18), 1–40 (2018)
6. Chaplot, D.S., Yang, Y., Carbonell, J., Koedinger, K.R.: Data-driven automated induction of prerequisite structure graphs. Int. Educ. Data Min. Soc. (2016)
7. Chau, H., Labutov, I., Thaker, K., He, D., Brusilovsky, P.: Automatic concept extraction for domain and student modeling in adaptive textbooks. Int. J. Artif. Intell. Educ. **31**(4), 820–846 (2021)
8. Kong, X.J., Wise, S.L., Bhola, D.S.: Setting the response time threshold parameter to differentiate solution behavior from rapid-guessing behavior. Educ. Psychol. Measur. **67**(4), 606–619 (2007)
9. Martin, B., Mitrovic, A., Koedinger, K.R., Mathan, S.: Evaluating and improving adaptive educational systems with learning curves. User Model. User-Adapt. Interact. **21**(3), 249–283 (2011). https://doi.org/10.1007/s11257-010-9084-2
10. Newell, A., Rosenbloom, P.: Mechanisms of skill acquisition. Cognitive skills and their acquisition (1981)
11. Pan, L., Li, C., Li, J., Tang, J.: Prerequisite relation learning for concepts in MOOCs. In: The 55th Annual Meeting of the Association for Computational Linguistics, Association for Computational Linguistics (2017)
12. Sosnovsky, S., Brusilovsky, P.: Evaluation of topic-based adaptation and student modeling in QuizGuide. User Model. User-Adapt. Interact. **25**(4), 371–424 (2015)
13. Wang, M., Chau, H., Thaker, K., Brusilovsky, P., He, D.: Knowledge annotation for intelligent textbooks. Technol. Knowl. Learn. **28**, 1–22 (2021). https://doi.org/10.1007/s10758-021-09544-z

Predicting Progress in a Large-Scale Online Programming Course

Vincent Zhang[1(✉)], Bryn Jeffries[1,2], and Irena Koprinska[1]

[1] School of Computer Science, The University of Sydney, Sydney, NSW 2006, Australia
vzha9726@uni.sydney.edu.au, irena.koprinska@sydney.edu.au

[2] Grok Academy, PO Box 144, Broadway, Sydney, NSW 2007, Australia
bryn.jeffries@grokacademy.org

Abstract. With vast amounts of educational data being generated in schools, educators are increasingly embracing data mining techniques to track student progress, especially in programming courses, a growing area of computer science education research. However, there are few accurate and interpretable methods to track student progress in programming courses. To bridge this gap, we propose a decision tree approach to predict student progress in a large-scale online programming course. We demonstrate that this approach is highly interpretable and accurate, with an overall average accuracy of 88% and average dropout accuracy of 82%. Additionally, we identify important slides such as problem slide which significantly impact student outcomes.

Keywords: Student progress · Decision trees · Computer programming

1 Introduction

Computational skills are essential in advanced economies, but mastering them is difficult, particularly for K-12 students with no prior programming experience. Many schools use online coding platforms to teach computational skills and gather student programming data. Educational data mining techniques can help analyse data, monitor progress and prevent dropout. However, student data is often complex and difficult to understand. This paper proposes a decision tree approach that uses slide interaction features to predict student progress and identify important slides for students to focus on. Overall, we aim to answer the question: *How can we make predicting K-12 student progress in programming courses more accurate and interpretable using educational data mining techniques?* This is guided by three research questions:

1. Are slide interaction features a useful predictor of end of module outcomes?
2. How can we represent student progress in a useful format for educators?
3. What are the important slides in each module?

N. Wang et al. (Eds.): AIED 2023, LNAI 13916, pp. 810–816, 2023.
https://doi.org/10.1007/978-3-031-36272-9_76

2 Background and Related Work

Improving learning outcomes for student programmers requires understanding their progress and behaviour. Past research has largely focused on knowledge gaps to explain poor programming performance, including difficulty with recursion [2] and confusing equality with assignment [1]. However, student interactions with course activities has received little attention, despite its positive impact on performance in other subjects [4].

Additionally, there is a growing body of research on data mining to predict student progress. Progress networks [7] provide a visualisation of student progression through learning tasks. Approaches based on extracting features and using clustering techniques to identify typical student behaviours have also been investigated in [6,8]. Predicting final marks based on past assessment results and student participation has been studied in [5,9], while [3] examined the effect of prior knowledge, problem solving skills and engagement on student performance.

Overall, there has been little research on using course content features (represented as log data) to predict programming performance, and doing so in an interpretable way for educators. In this paper we aim to address these gaps by:

1. Using course content features such as each student's interactions with course slides (i.e. slide interactions) to predict student performance;
2. Using decision tree classifiers with feature selection to represent programming progress in a more interpretable and useful format for educators;
3. Identifying key points in course modules where teachers can intervene to improve student performance and prevent dropout.

3 Data

We use a dataset from the National Computer Science School Challenge from 2018. Four courses were offered:

1. **Newbies**: Designed for young students (10–14 years old) with no prior programming experience, and conducted in the Blockly visual environment.
2. **Beginners Blockly**: Designed for students (12–16 years old) with no prior programming experience, and conducted in the Blockly visual environment.
3. **Beginners**: Same problems and target group as Beginners Blockly, but conducted in Python.
4. **Intermediate**: Designed for high school students (14–18 years old) with prior programming experience. Conducted in Python.

There are ~35,000 students in total. Each course has 10 modules containing a number of content and problem slides. Content slides provide prerequisite knowledge, interactive exercises and examples that students can do to help them solve the tasks on the problem slides which test programming skills and understanding. Problem slides occur less frequently than content slides are typically located at the middle and end of each module. We used student log data to extract slide interaction events such as slide visits, running code and submitting code. Test outcomes ("Pass" if all test cases passed, otherwise "Fail") were also recorded.

4 Method

Our approach predicts performance on the last problem in each module using interactions on previous content and problem slides in the same module. There are three stages outlined below.

4.1 Slide Interaction Data Extraction and Train-Test Split

We represent student slide interactions for each module as a vector, where each entry is a content or problem slide event and the last entry is the outcome of the last problem. The length of the vector is the number of slides in the module. Content slide events are "Completed" / "Not completed". Problem slide events are "No submission" / "Failed submission" / "Passed submission". Separate datasets are constructed for each module, and split into training and test sets using 10-fold cross-validation with stratification.

4.2 Feature Selection and Ranking

We use two feature selection algorithms: Correlation-Based Feature Selection (CFS) and Information Gain Ratio (GR). CFS with best-first search starts with an empty feature set and picks features highly correlated with the target class but uncorrelated with each other. GR feature selection evaluates features by their gain ratio with respect to the target class. We selected the top 2–3 features based on GR's rankings.

4.3 Classification

We investigated the performance of a Decision Tree (DT) classifier under three conditions - without feature selection, with CFS and with GR. We compared results with a baseline which predicts the majority class.

5 Results

Table 1 presents results for each course, including overall and per-class accuracy, number of leaves and tree size. DT with and without feature selection has consistently high overall accuracy (85.2 to 90.7% on all four courses), surpassing the baseline (Table 1). Passed submission and no submission accuracy are also high, ranging from 82% to 97%, and at least 70% respectively. However, predicting failed submissions remains challenging due to limited representation in the dataset. This is because the coding problems are designed so that most students are able to solve them. Figure 1 shows overall accuracy and no submission accuracy for each module. Overall accuracy remains consistently high (80% and above) for most modules up to Module 6, after which it decreases (Fig. 1a), possibly because later modules tend to have fewer students due to dropout.

Table 1. Comparison of average accuracies and tree characteristics across all courses

	Overall Accuracy	Accuracy "Passed"	Accuracy "Failed"	Accuracy "No Submission"	Number of leaves	Size of tree
Newbies						
Baseline	58.5					
DT	85.3	74.3	4.5	91.5	6.1	9.9
DT + CFS	85.2	74.4	2.0	90.8	4.3	6.3
DT + GR	85.3	82.5	2.0	91.0	4.1	5.9
Beginners Blockly						
Baseline	60.4					
DT	88.8	96.5	2.6	87.0	7.6	11.9
DT + CFS	88.9	96.8	1.8	87.2	4.2	5.9
DT + GR	86.9	89.1	0.2	86.7	3.9	5.4
Beginners						
Baseline	71.7					
DT	90.8	98.0	5.9	77.6	13.9	23.6
DT + CFS	90.8	97.8	3.4	77.9	4.7	6.8
DT + GR	90.7	97.7	5.4	78.0	5.0	7.0
Intermediate						
Baseline	71.8					
DT	88.9	97.2	0.6	72.5	7.8	13.3
DT + CFS	88.9	97.0	0.1	73.3	3.4	4.7
DT + GR	88.9	96.9	0.0	73.5	3.2	4.3
Average	88.3	91.5	2.4	82.2	5.7	8.8

5.1 Evaluation by Educators

Results from 14 modules across 3 courses were evaluated by an educator, who found "the trees help identify key slides where students may be struggling, and slides with little pedagogical value that could be removed or improved". One drawback was that "nested trees could be misleading as educators should not think that slides that were not ranked should be removed".

6 Discussion

6.1 Accuracy in Predicting End of Module Outcomes

Q1: Are slide interaction features a useful predictor of end of module outcomes? The DT algorithm has high overall accuracy on all courses, accurately predicting passed and no submissions (Table 1), demonstrating that slide interaction features can be a useful and accurate predictor of student outcomes. The high "No submission" accuracy means this method can identify students who are likely to drop-out, giving educators a chance to intervene (Fig. 1b).

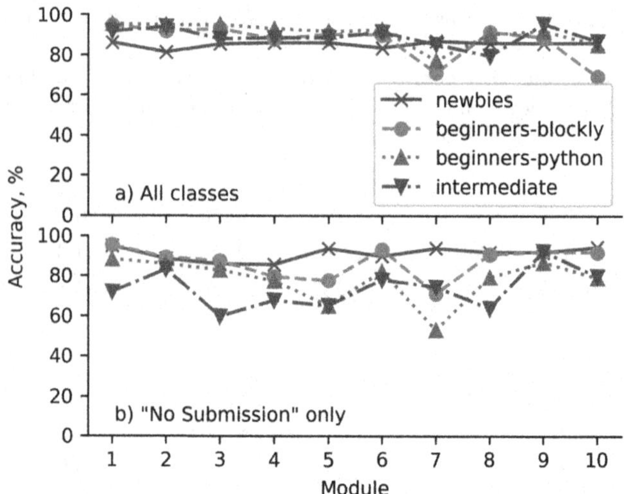

Fig. 1. Classifier accuracies (with GR feature selection) across all courses

6.2 Effects of Feature Selection

Q2: How can we represent student progress in a useful format for educators? The two feature selection methods (CFS and GR) created simpler decision trees with fewer leaf nodes while maintaining accuracy. This makes it easier for educators to identify key areas where students may need intervention. For instance, the DT in Fig. 2a shows that passing Slide 5 or 10 is necessary to achieve a "Pass" (Fig. 2).

6.3 Key Observations on Chosen Slides

Q3: What are the important slides in each module? Completing problem slides is critical for passing final module problems as they provide a chance to practice key skills. On average, they make up 75% of selected features across modules. Interactive content slides are also selected sometimes. These results are consistent with the doer effect [4] which posits that interactive practice is more beneficial than passive reading and video watching.

(a) Intermediate w1p1 (with GR feature selection)

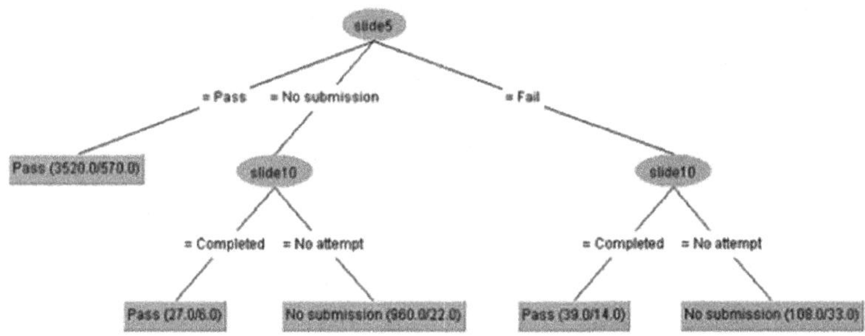

(b) Newbies w1p1 (with GR feature selection)

Fig. 2. Example decision trees

7 Conclusion

The proposed approach accurately predicts student progress and identifies intervention points, demonstrating the benefits of using course content features and interpretable methods like decision trees. However, our data is limited by the lack of sufficient failed submissions for the later courses. Future work will examine the impact of content slides, slide attempt order, and timing on progression, so we can gain a better understanding of progression at each step of the module and different student approaches.

References

1. Bonar, J., Soloway, E.: Preprogramming knowledge: a major source of misconceptions in novice programmers. Hum.-Comput. Interact. **1**(2), 133–161 (1985)
2. Dale, N.B.: Most difficult topics in CS1: results of an online survey of educators. SIGCSE Bull. **38**(2), 49–53 (2006)
3. Kennedy, G., Coffrin, C., de Barba, P., Corrin, L.: Predicting success: how learners' prior knowledge, skills and activities predict MOOC performance. In: Learning Analytics and Knowledge, pp. 136–140 (2015)
4. Koedinger, K.R., McLaughlin, E.A., Jia, J.Z., Bier, N.L.: Is the doer effect a causal relationship? In: Learning Analytics and Knowledge, pp. 388–397 (2016)

5. Koprinska, I., Stretton, J., Yacef, K.: Predicting student performance from multiple data sources. In: Artificial Intelligence in Education, pp. 678–681 (2015)
6. McBroom, J., Jeffries, B., Koprinska, I., Yacef, K.: Mining behaviors of students in autograding submission system logs. In: Educational Data Mining (2016)
7. McBroom, J., Paassen, B., Jeffries, B., Koprinska, I., Yacef, K.: Progress networks as a tool for analysing student programming difficulties. In: Australasian Computing Education, pp. 158–167 (2021)
8. Perera, D., Kay, J., Koprinska, I., Yacef, K., Zaïane, O.: Clustering and sequential pattern mining of online collaborative learning data. IEEE Trans. Knowl. Data Eng. **21**(6), 759–772 (2009)
9. Romero, C., López, M.I., Luna, J.M., Ventura, S.: Predicting students' final performance from participation in on-line discussion forums. Comput. Educ. **68**, 458–472 (2013)

Examining the Impact of Flipped Learning for Developing Young Job Seekers' AI Literacy

Hyo-Jin Kim[1,2], Hyo-Jeong So[2(✉)], and Young-Joo Suh[3]

[1] Institute of Artificial Intelligence, POSTECH, Pohang, South Korea
hyojinkim@postech.ac.kr

[2] Department of Educational Technology, Ewha Womans University, Seoul, South Korea
hyojeongso@ewha.ac.kr

[3] Graduate School of Artificial Intelligence, POSTECH, Pohang, South Korea
yjsuh@postech.ac.kr

Abstract. While AI literacy is regarded as an essential competency to become a citizen in a rapidly changing society, it is challenging for people without computer science (CS) backgrounds to develop a sufficient level of AI competency. The main goal of this research is to examine the impact of the flipped learning approach to equip non-CS major students who intend to pursue careers in AI-related fields with basic AI literacy. Among various learner-centered methods, flipped learning was chosen as the main pedagogical frame to design an AI literacy curriculum. The participants were 80 adult learners who enrolled in the AI education program in Korea. The control group (N = 40) was taught in traditional instructor-centered methods whereas the experimental group (N = 40) was taught with a flipped learning method. Our research results indicate that AI literacy education with flipped learning improves the learning achievements of both CS majors and non-majors, especially effective for higher-order problem-solving skills.

Keywords: AI literacy · Flipped learning · Non-majors · Job seekers

1 Introduction

As AI rapidly emerges as a key technology in the era of the 4th industrial revolution, the demand for education on understanding and using AI is explosively increasing. AI literacy is regarded as an essential competency to become a citizen in a rapidly changing society [1]. To develop AI literacy competency, it is necessary to understand the basic technique and concepts underlying AI and the impact and ethics of AI in our society and human life [2]. As AI literacy is valued in the changing job structure, young job seekers are increasingly interested in attaining AI literacy. Demand for AI literacy education is high not only for computer science (CS) majors but also for non-CS majors due to the nature of AI that can be applied in various fields. However, non-CS majors often experience relative difficulties compared to CS majors [3] due to various reasons. For instance, AI education for job seekers is often offered as an intensive program for a limited period targeting a large number of students. Thus, it is challenging to provide individual scaffolding for students with various levels of prior knowledge in the classroom.

N. Wang et al. (Eds.): AIED 2023, LNAI 13916, pp. 817–823, 2023.
https://doi.org/10.1007/978-3-031-36272-9_77

The current situation of AI literacy education indicates the need for a learner-centered pedagogy that focuses on individual learners to enhance the motivation and academic achievement of individual students from various majors [4]. To this end, this study was motivated by the assumption that when the flipped learning method is adopted to provide relevant scaffolding in and outside the classroom, non-CS major students can gain benefits by being more cognitively prepared to participate in problem-solving activities. Flipped learning supports effective learner-centered AI education by securing class time that can be devoted to high-order learning activities such as in-depth discussion, hands-on programming, and complex problem-solving tasks [5]. The main goal of this study, thus, was to implement flipped learning as a learner-centered pedagogical approach in AI education for both non-CS majors and CS majors who have difficulties learning about AI. In particular, we investigated whether flipped learning is effective for achieving high-level learning goals and improving learner satisfaction. This study also suggests ways to improve the instructional design of AI literacy education programs for young job seekers.

2 Methodology

2.1 Participants and Research Design

In this study, the participants were 80 students enrolled in an AI education program designed for young job seekers at a university in South Korea. We randomly assigned 40 students to the control group and 40 students to the experimental group. Both the control and experimental groups included 20 CS-related major students and 20 non-major students, respectively. We used the posttest-only nonequivalent control group design. As presented in Table 1, the control group was taught in traditional instructor-centered methods whereas the experimental group was taught in a flipped learning method that incorporated online learning tasks out of class. The instructor created self-paced learning materials with video content about fundamental concepts and theories in AI and uploaded them to the Learning Management System (LMS). Individual text messages were sent to students to inform them of their pre-learning progress. Before each session, students posted their learning points and question on Padlet.

Table 1. Learning activity design in control and experimental groups

	Control	Experimental
Out of class	No required activity	• Lecture video for conceptual learning (30 min) • Q&A on Padlet
In class	• Theory lesson and discussion (2 h) • Hands-on programming training (2 h) • Project work (2 h)	• Mini lecture and discussion (1 h 30 min) • Hands-on programming training (2 h) • Project work (2 h)

2.2 Data Collection and Analysis

The main topic of the class was AIoT (Artificial Intelligence of Things). The AI literacy program was for 30 h and each class was designed to combine theoretical and practical sessions. Both the control and experimental groups received the same lecture time, learning content, tasks, and assignments. After completing the AIoT course, students took a test that measured academic achievement. The test included items that measured both low-order (e.g., knowledge, comprehension) and higher-order (e.g., analysis, synthesis, and evaluation learning skills based on Bloom's taxonomy. The students also completed the survey that measured their satisfaction with the AI education program. The survey for the experimental group included additional items to measure student satisfaction with flipped learning. In addition, the instructor and selected students in the experimental group participated in the focus group interview (FGI) to share their perceptions and experiences with flipped learning. SPSS 25.0 was used to analyze the test and survey data at the significance level of .5. The FGI data were content-analyzed to identify common themes.

3 Results

3.1 Academic Achievement

Skewness and kurtosis were calculated to determine whether the assumption of normality was satisfied for the academic achievement data. An absolute value of less than 3 for skewness and an absolute value of less than 10 for kurtosis were judged to be close to a normal distribution [6], meeting the normality assumption. As seen in Table 2, an independent sample t-test reveals that the experimental group shows significantly higher academic achievement than the control group in both lower-order and higher-order learning goals.

Table 2. Comparison of academic achievement

Learning skills	Group	*N*	Mean	SD	*t*	*p*
Lower-order	Control	40	27.00	13.39	11.136*	< .001
	Experimental	40	51.65	4.08		
Higher-order	Control	40	16.20	11.66	8.396*	< .001
	Experimental	40	35.75	9.00		
Total	Control	40	43.20	23.57	10.719*	< .001
	Experimental	40	87.40	11.16		

* $p < .05$

3.2 Educational Satisfaction

Program Satisfaction. The program satisfaction survey consisted of eight multiple-choice questions on a 5-point Likert scale to measure students' overall satisfaction with the AI education program. As seen in Table 3, the independent sample t-test reveals that there was no significant difference between the control and experimental groups. Also, there was no significant difference in program satisfaction for both CS-related major and non-major students.

Table 3. Comparison of program satisfaction

	Group	*N*	Mean	SD	*t*	*p*
All	Control	39	4.44	0.72	0.635	.528
	Experimental	40	4.53	0.57		
Non-major	Control	20	4.41	0.76	0.809	.424
	Experimental	20	4.58	0.60		
Major	Control	19	4.47	0.69	0.040	.968
	Experimental	20	4.48	0.54		

Flipped Learning Satisfaction. The satisfaction survey on the flipped learning experience was additionally conducted in the experimental group. The survey consisted of 16 multiple-choice items using a 5-point Likert scale to measure flipped learning experiences such as attention, interaction, and satisfaction. Table 4 shows that students in the experimental group were highly satisfied with their flipped learning (FL) experiences, as seen in the high scores across all areas. Regarding prior experiences, students with previous FL experiences (n = 13) showed significantly higher satisfaction than those without prior FL experience (n = 27) in all areas except interaction.

Table 4. Flipped learning satisfaction

Area	*N*	Mean	SD	Prior FL	Mean	SD	*t*	*p*
Attention	40	4.38	0.77	O	4.77	0.44	2.362*	.023
				X	4.19	0.83		
Contents	40	4.48	0.61	O	4.82	0.32	3.318*	.002
				X	4.31	0.65		
Pre-class	40	4.63	0.47	O	4.87	0.29	2.942*	.006
				X	4.51	0.49		

(*continued*)

Table 4. (*continued*)

Area	*N*	Mean	SD	Prior FL	Mean	SD	*t*	*p*
Interaction	40	4.59	0.42	O	4.69	0.29	1.246	.221
				X	4.54	0.46		
Satisfaction	40	4.37	0.70	O	4.79	0.42	3.418*	.002
				X	4.17	0.73		
Overall	40	4.48	0.52	O	4.79	0.26	3.562*	.001
				X	4.33	0.55		

* $p < .05$

3.3 Focus Group Interview

Focus group interviews (FGI) were conducted to qualitatively examine flipped learning experiences with 10 students (5 majors and 5 non-majors) in the experimental group. The interview data were analyzed using Consensual Qualitative Research (CQR), which is an inductive qualitative research method to explain the phenomenon with core ideas. The flipped learning experiences derived from the FGI were organized according to the four pillars of flipped learning proposed by the Flipped Learning Network [7], as shown in Table 5.

Table 5. Flipped learning experiences in AI literacy education

Pillars	Group	Flipped learning experience
Flexible environment	All	Group discussions were beneficial in class (General) Useful to organize what I have learned on Padlet (Typical) Pre-learning took a lot of time (Variant)
	Major	Short training period (Typical) Pre-learning videos required fast pace function (Typical) Pre-learning reminder texts were burdensome (Variant)
	Non-major	Studied with pre-learning videos repeatedly (Typical) Pre-learning reminder texts were helpful (Variant)
	Instructor	Students' understanding of learning increased through pre-learning and discussions in the classroom
Learning culture	All	Flipped learning is effective (General) The class progress rate was fast (Typical)

(*continued*)

Table 5. (*continued*)

Pillars	Group	Flipped learning experience
	Major	Need to avoid too many Q&A in the classroom (Typical) Study group activities were helpful (Variant)
	Non-major	Helped from peers (Typical)
	Instructor	Students had an active learning attitude
Intentional content	All	AI education is useful for self-development and employment (General)
	Major	Easy to understand learning contents (General) Need more explanation of basic theory such as AI terms (Variant)
	Non-major	Unfamiliar AI terms (General) Had learning difficulties (Typical)
	Instructor	Use video examples for pre-learning Explaining basic AI terms in detail in pre-learning
Professional educator	Non-major	The teaching assistants were nice (Typical)
	Instructor	Systematic instructional design Instructor's understanding of flipped learning

4 Discussion and Conclusion

This study uses mixed methods to examine the effect of flipped learning in the AI education program for young job seekers with various levels of prior knowledge and backgrounds. The key findings of this study are as follows. First, AI education with flipped learning improved academic achievement in both lower- and higher-order learning levels for both CS-related majors and non-majors. Second, both the control and experimental groups were highly satisfied with the program. This study confirms that as AI is a broad and rapidly evolving field and it is challenging to address all contents in limited face-to-face class time, FL can be an effective learning method for students with diverse backgrounds [8]. Third, the FGIs revealed seven ways to improve flipped learning in the AI curriculum design: (a) the importance of explaining basic terms, (b) the use of multimedia, (c) classroom environments for group activities, (d) students' active participation in and out of the classroom, (e) the professional preparation of lectures by instructors, (f) the LMS support for students, and (g) the use of online collaboration tools.

The study has certain limitations. First, the sample size is rather small and findings are limited to the Korean learners and context. Therefore, further research with a larger sample size is necessary to validate the educational impact of flipped learning. Additionally, the classification of CS-related majors and non-majors was based solely on undergraduate majors. It is possible that some non-majors had taken CS courses during their undergraduate studies. Future research could consider collecting more comprehensive academic background data to better classify majors and non-majors. Despite these limitations, this study provides empirical evidence of the positive impact of flipped learning in teaching AI literacy, particularly for those without CS backgrounds.

Acknowledgment. This research was supported by Basic Science Research Program through NRF of Korea funded by the Ministry of Education (2022R1A6A1A0305295412). This research is part of the master's thesis of the first author at Ewha Womans University.

References

1. Long, D., Magerko, B.: What is AI literacy? competencies and design considerations. In: Proceedings of the 2020 CHI Conference on Human Factors in Computing Systems, pp. 1–16. Association for Computing Machinery, New York, NY, USA (2020)
2. Park, S.H., Suh, E.K.: A study on artificial intelligence education design for business major students. The Journal of Industrial Distribution & Business **12**(8), 21–32 (2021)
3. Thakkar, D., Ismail, A., Kumar, P., Hanna, A., Sambasivan, N., Kumar, N.: When is machine learning data good?: Valuing in public health datafication.In: Proceedings of the 2022 CHI Conference on Human Factors in Computing Systems, pp.1–13. Association for Computing Machinery, New York, NY, USA (2022)
4. McCombs, B.L., Whisler, J.S.: The learner-centered classroom and school. Jossey-Bass, San Francisco (1997)
5. Anderson, L.W., Krathwohl, D.R.: A taxonomy for learning, teaching, and assessing: A revision of Bloom's taxonomy of educational objectives. Longman, New York (2001)
6. Kline, T.J.B.: Psychological testing: A practical approach to design and evaluation. Sage, Thousand Oaks, CA (2005)
7. Flipped Learning Network: The four pillars of F-L-I-P, https://flippedlearning.org/wp-content/uploads/2016/07/FLIP_handout_FNL_Web.pdf, last accessed 16 January 2023
8. Kong, S.C., Man-Yin Cheung, W., Zhang, G.: Evaluation of an artificialintelligence literacy course for university students with diverse study backgrounds. Computers & Education: Artificial Intelligence 2 (2021)

Automatic Analysis of Student Drawings in Chemistry Classes

Markos Stamatakis[1(✉)], Wolfgang Gritz[1,2], Jos Oldag[3], Anett Hoppe[1,2], Sascha Schanze[3], and Ralph Ewerth[1,2]

[1] Leibniz University Hannover, L3S Research Center, Appelstraße 9A, 30167 Hannover, Germany
markos.stamatakis@tib.eu

[2] TIB – Leibniz Information Centre for Science and Technology, Welfengarten 1B, 30167 Hannover, Germany

[3] Leibniz University Hannover, Institute of Science Education, Am Kleinen Felde 30, 30167 Hannover, Germany

Abstract. Automatic analyses of student drawings in chemistry education have the potential to support classroom teaching. To date, related work on handwritten chemical structures or formulas is limited to well-defined presentation formats, e.g., Lewis structures. However, the large variety of possible illustrations in student drawings in chemical education has not been addressed yet. In this paper, we present a novel approach to identify visual primitives in student drawings from chemistry classes. Since the field lacks suitable datasets for the given task, we introduce a method to synthetically create a dataset for visual primitives. We demonstrate how detected visual primitives can be used to automatically classify drawings according to a taxonomy of drawing characteristics in chemistry and physics. Our experiments show that (1) the detection of visual primitives in student drawings, and (2) the subsequent classification of chemistry- and physics-specific drawing characteristics is possible.

Keywords: Computer Vision · Chemistry Education · Visual Analysis · Visual Primitives · Synthetic Data

1 Introduction

Drawing tasks allow students to express their conceptions in individual ways. The analysis of student drawings arouses interest in education due to their diversity [1,15], but the manual analysis can be challenging for teachers. Specifically, no approaches for an overall automatic analysis are available, except for some that work for chemical structures [6,14]. The problem is that such approaches

This work has been mainly supported by the Ministry of Science and Culture of Lower Saxony, Germany, through the PhD Program "LernMINT: Data-assisted classroom teaching in the MINT subjects".

N. Wang et al. (Eds.): AIED 2023, LNAI 13916, pp. 824–829, 2023.
https://doi.org/10.1007/978-3-031-36272-9_78

are designed for specific tasks and not applicable in general. Automatic analysis is not trivial for several reasons: (1) drawing tasks can be very diverse, and (2) the illustrations can vary between students, e.g., using circles, text, or triangles to represent a water molecule. Thus, no fixed set of objects can be defined and visual primitives must be considered. Tang et al. [13] introduced a suitable framework to analyze drawings specifically from chemistry and physics. The therein-identified drawing characteristics can enable a systematized analysis of the expressed conceptions.

In this paper, we propose an approach for the domain-independent detection of visual primitives and text in student drawings created on paper or in digital environments. To address the lack of a publicly available dataset that covers all required visual primitives, we introduce a method to synthetically create a large-scale dataset by leveraging the *"Quick, Draw!"* and *IAM* datasets [5,7]. Using these datasets, we train Convolutional Neural Network [11] to predict bounding boxes for visual primitives and text. The predictions are used to categorize the drawings according to Tang et al.'s taxonomy [13]. We evaluate our approach on a dataset of N = 306 hand-drawn student drawings from chemistry classes.

The remainder of the paper is structured as follows: In Sect. 2 the novel approach for visual primitive detection is presented. The experimental results are reported in Sect. 3, while we discuss conclusions and future work in Sect. 4.

2 Automatic Categorization of Student Drawings

2.1 Automatic Dataset Creation

To the best of our knowledge, there is currently no dataset of hand-drawn images encompassing all visual primitives. To address this limitation, we synthetically generate training images that consist of objects from the classes `oval`, `square`, `line`, `triangle`, and `text`, with the aim of covering a maximum number of commonly used drawing elements. Contrary to the other object classes, we have not found a collection of hand-drawn images with distinct arrows and they are therefore neglected in our synthetically created dataset. For this reason, we additionally include datasets with hand-drawn diagrams [2–4,8,12] that contain arrows. We build on the *Quick, Draw!* and *IAM* datasets [5,7]. Preprocessing steps include (1) conversion of images into vector representation using a pre-trained *ResNet-101*, (2) reduction of dimensions by performing principal component analysis, and (3) removal of outliers in the *Quick, Draw!* [5] dataset using *DBSCAN*. Additionally, the text from the *IAM* [7] dataset is extracted using *Otsu's method* [9]. Up to five elements of an object class are randomly used to create a synthetic image, with each object that belongs to one of the classes `oval`, `square` or `triangle` having a chance of 50% that text is placed centrally within it. If the text overlaps the margins of the object, it is resized to 75%, 50%, or 25%. The colors and sizes of objects are randomly chosen. Finally, a random text phrase from the *IAM* dataset is placed without overlapping with any other object if possible, otherwise it is discarded.

Table 1. Selected characteristics for the study of student drawings with their corresponding category and description of the framework of Tang et al. [13].

Characteristic	Category	Description
Connection	Association	Objects are connected to each other, e.g., by lines
Adjoining	Association	Objects are positioned exactly next to each other
Intersection	Association	Objects overlap
Independent	Association	Assigned, if no features from the same category apply
Movement	Movement	Movements are displayed with arrows
Inclusion	Textual Contextualisation - Label	Objects contain text
Indexial	Textual Contextualisation - Label	Text and objects are connected with lines or arrows

2.2 Detection of Objects in Student Drawings

The specialized object detection needs to detect the visual primitives (*oval*, *rectangle*, *triangle*, *line*, *arrow*) and text. We adopt *Arrow R-CNN* architecture [11] which was developed for the similar problem of object detection in diagrams. It consists of a *Faster R-CNN* [10] object detector with an additional head that locates and classifies key points of arrows, i.e., their head and tail points. We decided to re-implement this approach as it achieved promising results in similar tasks. The original *Arrow R-CNN* [11] was trained on diagrams consisting of a subset of visual primitives (i.e., *ovals*, *rectangles* and *arrows*) and text. Instead of the original classes for diagrams, we map the output to visual primitives and text and obtain comparable results with our classes. For each drawing, this approach yields bounding boxes and, in the case of arrows, head and tail points.

2.3 Classification of Drawing Characteristics

We use the predicted bounding boxes and key points for the classification of the drawing characteristics according to Tang et al. [13]. Since not all characteristics exist in our set of student drawings, we focus on eight of them with a minimum number of appearances (Table 1).

3 Experimental Evaluation

3.1 Datasets

For the evaluation, 306 student drawings from chemistry classes of different grade levels and schools are available. The drawings consist of photos and scans covering two topics: (1) the combustion of coal, and (2) weak and strong acids at the submicroscopic level. We annotated bounding boxes for visual primitives, text, and associated key points for arrows. From this data set, 33 drawings are used as a validation set; the remaining 273 drawings are used as unseen examples for evaluation. Besides, we use 6000 synthetically generated samples (see Sect. 2.1) for the training of the object detection approach. For better generalization and completion, we combine the synthetic data with 1372 training samples from diagram datasets [2–4,8,12].

Table 2. Mean average precision (mAP) and average precision [%] without and with fine-tuning (using 5-fold cross-validation) for object classes and keypoints.

Cross-Validation	mAP	Oval	Rect	Text	Arrow	Triangle	Line	Points
Yes	42.6	81.0	48.7	42.2	34.4	26.8	22.6	35.4
No	25.6	48.3	38.4	17.7	0.1	45.1	4.1	9.4

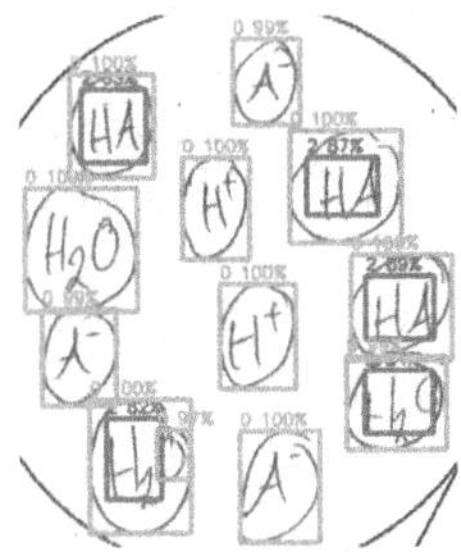

(a) Ovals with partially detected text

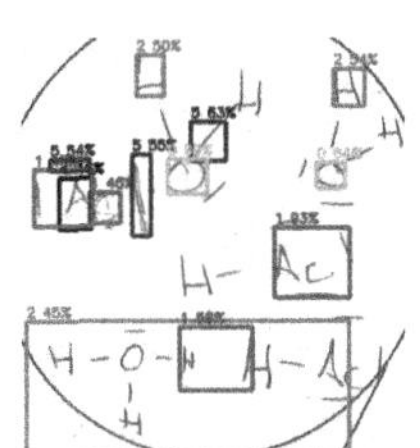

(b) Barely detected text and lines

Fig. 1. Qualitative examples (non-fine-tuned), where color represents ovals (*yellow*), text (*blue*), squares (*red*), triangles (*black*), and lines (*purple*). The numbers above a box indicate the class (left) and the model's confidence (right) [%]. (Color figure online)

3.2 Detection of Objects

To evaluate the performance of the object detection, we train our model on (1) the synthetic and diagram data, and (2) fine-tune the model on a subset of student drawings using a five-fold cross-validation. We report the results on the remaining 273 student drawings in Table 2.

It is apparent that there is a significant performance boost after fine-tuning. Especially after fine-tuning on student drawings the highest precision is achieved for ovals, which is reasonable, as it is the dominant class in the data. For the other classes, the performance varies. This is possibly caused by their lower frequency in the training data. Figure 1 shows the detected objects for two example drawings without fine-tuning. Our manual analysis of the results showed that often the diverse object representations impair their successful detection. Thus, the fine-tuned model would benefit from training with a larger and more diverse set of student drawings.

3.3 Classification of Drawing Characteristics Used by Students

As described in Sect. 2.3, the analysis of the characteristics is performed based on the framework of Tang et al. [13]. The experiment is divided into two sub-experiments: (1) using the model for object detection without fine-tuning (see Sect. 3.2) ; (2) using manually annotated bounding boxes, to focus on the potential for classifying the drawing characteristics. A chemistry didactic expert annotated 60 of the drawings as ground-truth data. Ten of them are used to find

Table 3. *Accuracy (Acc)*, *precision (P)*, *recall (R)*, and F_1*-score (F1)* for estimating the appearance of drawings characteristics [%] using the baseline, predicted, and ground truth objects.

		Baselines				Predicted				Ground Truth			
Characteristic	N	Acc	P	R	F1	Acc	P	R	F1	Acc	P	R	F1
Connecting	14	72.0	28.0	**100**	43.8	50.0	29.6	57.1	39.0	**90.0**	**73.7**	**100**	**84.8**
Adjoining	17	66.0	34.0	**100**	50.7	64.0	48.6	100	65.4	**94.0**	**85.0**	**100**	**91.9**
Intersecting	9	82.0	18.0	**100**	30.5	56.0	15.8	33.3	21.4	**86.0**	**62.5**	55.6	**58.8**
Independent	19	62.0	38.0	**100**	55.1	60.0	40.0	10.5	16.7	**94.0**	**100**	84.2	**91.4**
Movement	2	**96.0**	4.0	**100**	7.7	90.0	20.0	50.0	28.6	92.0	**95.8**	95.8	**95.8**
Inclusion	10	**80.0**	20.0	**100**	33.3	58.0	21.1	40.0	27.6	**80.0**	**50.0**	30.0	**37.5**
Indexial	5	**90.0**	10.0	**100**	18.2	56.0	9.5	40.0	15.4	88.0	**44.4**	80.0	**57.1**

thresholds for predicting the different drawing characteristics; the remaining 50 images are the test set. To compare the results, we define two baselines: 1) the majority vote for *accuracy*, and, contrary 2) the prediction of the corresponding class for *precision*, *recall*, and F_1-score. The results are shown in Table 3.

For the non-fine-tuned model, we achieve only a higher F_1-score in two cases. For a practical application in teaching, the detection results have to be improved significantly. However, the predominantly high F_1-scores when using our ground-truth annotated objects signal that basically our approach is suitable for determining the drawing characteristics. Overall, the results on ground-truth data show that with better detection, the characteristics could be well classified.

4 Conclusions

In this paper, we have presented an approach for the detection of visual primitives and text in student drawings to simplify educational assessment in chemistry classes. We re-implemented *Arrow R-CNN* [11] and trained it on a novel, synthetically generated dataset. We consequently used the detected objects for the classification of drawing characteristics using Tang et al.'s framework [13]. Our results show that the recognition of visual primitives can be a useful first step for automated didactic analyses. Further, commonly used training materials for objection detection methods are mainly created digitally, while our student drawings were created on paper and later scanned, making the task harder.

Future plans include extending the used datasets: The synthetic set is to be extended by arrow drawings. Moreover, we plan to collect additional, more diverse student drawings; including drawings to cover all of the categories described by Tang et al. [13], to enable practical use in educational settings. On the side of computational model design, transfer learning seems promising.

References

1. Ainsworth, S., Prain, V., Tytler, R.: Drawing to learn in science. Science **333**(6046), 1096–1097 (2011). https://doi.org/10.1126/science.1204153
2. Bresler, M., Phan, T.V., Průša, D., Nakagawa, M., Hlavác, V.: Recognition system for on-line sketched diagrams. In: International Conference on Frontiers in Handwriting Recognition, ICFHR 2014, Crete, Greece, pp. 563–568. IEEE Computer Society (2014). https://doi.org/10.1109/ICFHR.2014.100
3. Bresler, M., Průša, D., Hlaváč, V.: Online recognition of sketched arrow-connected diagrams. Int. J. Doc. Anal. Recogn. (IJDAR) **19**(3), 253–267 (2016). https://doi.org/10.1007/s10032-016-0269-z
4. Bresler, M., Průša, D., Hlavác, V.: Recognizing off-line flowcharts by reconstructing strokes and using on-line recognition techniques. In: International Conference on Frontiers in Handwriting Recognition, ICFHR 2016, Shenzhen, China, pp. 48–53. IEEE Computer Society (2016). https://doi.org/10.1109/ICFHR.2016.0022
5. Ha, D., Eck, D.: A neural representation of sketch drawings. In: International Conference on Learning Representations, ICLR 2018, Vancouver, Canada. OpenReview.net (2018). https://openreview.net/forum?id=Hy6GHpkCW
6. Hagag, A., Omara, I., Alfarra, A.N.K., Mekawy, F.: Handwritten chemical formulas classification model using deep transfer convolutional neural networks. In: International Conference on Electronic Engineering, ICEEM 2021, Menouf, Egypt, pp. 1–6 (2021). https://doi.org/10.1109/ICEEM52022.2021.9480627
7. Marti, U., Bunke, H.: The IAM-database: an English sentence database for offline handwriting recognition. Int. J. Doc. Anal. Recogn. **5**(1), 39–46 (2002). https://doi.org/10.1007/s100320200071
8. Mouchère, H.: Online handwritten flowchart dataset (OHFCD). http://tc11.cvc.uab.es/datasets/OHFCD_1
9. Otsu, N.: A threshold selection method from gray-level histograms. IEEE Trans. Syst. Man Cybern. **9**(1), 62–66 (1979). https://doi.org/10.1109/TSMC.1979.4310076
10. Ren, S., He, K., Girshick, R.B., Sun, J.: Faster R-CNN: towards real-time object detection with region proposal networks. IEEE Trans. Pattern Anal. Mach. Intell. **39**(6), 1137–1149 (2017). https://doi.org/10.1109/TPAMI.2016.2577031
11. Schäfer, B., Keuper, M., Stuckenschmidt, H.: Arrow R-CNN for handwritten diagram recognition. Int. J. Doc. Anal. Recogn. **24**(1), 3–17 (2021). https://doi.org/10.1007/s10032-020-00361-1
12. Schäfer, B., Stuckenschmidt, H.: DiagramNet: hand-drawn diagram recognition using visual arrow-relation detection. In: Lladós, J., Lopresti, D., Uchida, S. (eds.) ICDAR 2021. LNCS, vol. 12821, pp. 614–630. Springer, Cham (2021). https://doi.org/10.1007/978-3-030-86549-8_39
13. Tang, K.S., Won, M., Treagust, D.: Analytical framework for student-generated drawings. Int. J. Sci. Educ. **41**(16), 2296–2322 (2019). https://doi.org/10.1080/09500693.2019.1672906
14. Weir, H., Thompson, K., Woodward, A., Choi, B., Braun, A., Martínez, T.J.: ChemPix: automated recognition of hand-drawn hydrocarbon structures using deep learning. Chem. Sci. **12**, 10622–10633 (2021). https://doi.org/10.1039/D1SC02957F
15. Wu, S.P.W., Rau, M.A.: How students learn content in science, technology, engineering, and mathematics (STEM) through drawing activities. Educ. Psychol. Rev. **31**(1), 87–120 (2019). https://doi.org/10.1007/s10648-019-09467-3

Training Language Models for Programming Feedback Using Automated Repair Tools

Charles Koutcheme(✉)

Aalto University, Espoo, Finland
charles.koutcheme@aalto.fi

Abstract. In introductory programming courses, automated repair tools (ARTs) are used to provide feedback to students struggling with debugging. Most successful ARTs take advantage of context-specific educational data to construct repairs to students' buggy codes. Recent work in student program repair using large language models (LLMs) has also started to utilize such data. An underexplored area in this field is the use of ARTs in combination with LLMs. In this paper, we propose to transfer the repairing capabilities of existing ARTs to open large language models by finetuning LLMs on ART corrections to buggy codes. We experiment with this approach using three large datasets of Python programs written by novices. Our results suggest that a finetuned LLM provides more reliable and higher-quality repairs than the repair tool used for finetuning the model. This opens venues for further deploying and using educational LLM-based repair techniques.

Keywords: Program Repair · Large Language Models · Computing Education

1 Introduction

Debugging is particularly challenging and time-consuming for novice programmers [6], and thus educators are eagerly looking for tools and techniques for assisting with the process. One possibility is the use of Automated Repair Tools (ARTs), which are given erroneous codes, which the ARTs then seek to repair. One of the challenges with ARTs, however, is that the proposed fixes can be relatively far from the original buggy code.

Pre-trained large language models (LLMs) - deep neural networks trained on enormous quantities of data - have presented new avenues for program repair. Existing LLMs [10] have shown promising results when finetuned on datasets comprising buggy codes and their corrections. However, employing these models in an educational context requires having access to such annotated datasets. Although many institutions collect and keep track of students' submissions over time, privacy concerns have often limited the possibility of sharing such data

N. Wang et al. (Eds.): AIED 2023, LNAI 13916, pp. 830–835, 2023.
https://doi.org/10.1007/978-3-031-36272-9_79

with the public (and thus engaging in any annotation process). Consequently, improving the performance of LLM-based program repair techniques has often been limited to seeking effective prompts [11].

In this paper, we introduce a framework to deploy non-propriety/open large language models for program repair in educational contexts. As a workaround for the lack of available annotated datasets, we propose to finetune LLMs on pairs of buggy student code and their repaired versions created by existing ARTs. Figure 1 shows an overview of the framework. In essence, this method facilitates and allows the creation of neural automated repair tools (NARTs) for programming feedback. In this work, we evaluate the proposed framework to understand *to what extent NARTs can address the challenges around repairing student programs?* In particular, we are interested in two real-life deployment scenarios: (1) a scenario in which data collected over one semester in an introductory programming course can be used for providing feedback in another semester and (2) a scenario where only data from another course is available. We thus address the following two research questions concerning NARTs: (RQ1) How do NARTs perform for repairing familiar buggy programs; (RQ2) How do NARTs perform for repairing unfamiliar buggy programs?

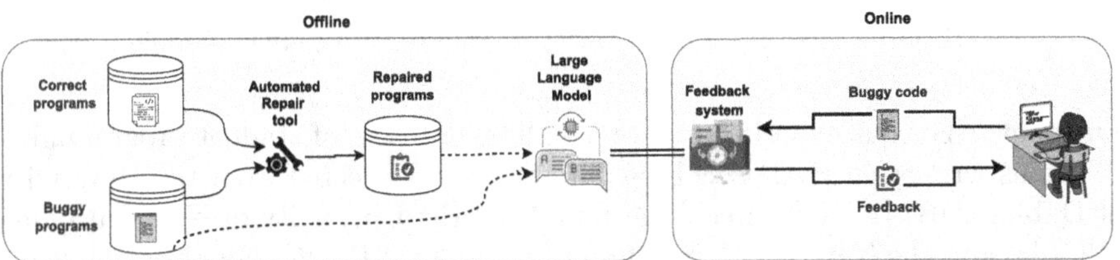

Fig. 1. Overview of the proposed framework. Offline: We use an automated repair tool (ART) to create a dataset of buggy codes and their associated repairs. We use this correction dataset to train an LLM for program repair. Online, a feedback system queries the LLM, providing quality feedback to students.

2 Related Work

Automated Repair Tools have been classically built using rule-based systems to construct working solutions from existing data. Most approaches developed for educational contexts use heuristics to match a student's buggy code to a set of correct programs written by other students, or to the instructor's reference solution [4,5,8,9]. An example in this category is the work of Hu et al. [5], whose tool (Refactory) generates a suite of semantically equivalent code by refactoring all available correct solutions to a problem. Then, given an incorrect program, its control flow structure is analysed to find a closely matching working program that can be compared to isolate the buggy components of the program. Previous work has also employed available data with sequence-to-sequence (SEQ2SEQ) models for repairing programs, training a Recurrent Neural Network -based SEQ2SEQ on students' correct solutions [7].

Recently, the large pre-trained language models have gained more attention. For instance, Zhang et al. [11] introduced MMAPR, an automated repair technique for introductory programming assignments. Their approach uses correct solutions, test cases, and assignment descriptions to prompt OpenAI Codex. They evaluate their method on 286 Python programs produced by novices and show that their approach can repair up to 96.5% of the programs, with a smaller edit distance compared to Refactory [5]. Their work is similar in spirit to ours. The differences are that they use Codex (a proprietary system), and their repairing technique uses only a few correct solutions to prompt the system. In contrast, we propose to *finetune* open (i.e. non proprietary) models (such as those available on HuggingFace[1]) using all available data.

3 Methodology

To answer our two research questions, we perform two experiments with three python introductory programming datasets. We use the semantic automated repair tool Refactory [5] to train the language model CodeT5 [10]. CodeT5 is a sequence-to-sequence model pre-trained on a large quantity of bimodal natural and programming language data. As an additional contribution, we release the preprocessed dataset as well as the code to conduct our experiments[2].

Datasets. For our experiments, we use three datasets of student programming solutions written in Python. These datasets were collected from the University of Dublin (DB) [1] over three semesters, from the University of New Caledonia (NC) [3] and the National University of Singapore (SP) [5]. For these datasets, we select the assignments which require writing a single function that takes fixed inputs and produces one output. We further select a subset of assignments from the student datasets to maintain a good ratio of diversity and complexity level.

Experiments. To answer our first research question, we finetune our LLM on the corrections to the solutions of two semesters of data before using it subsequently to find repairs to buggy programs submitted in the following semester. The Dublin dataset contains assignment data from three academic years (2015–2016 to 2017–1018). We use Refactory to find corrections to the buggy programs submitted in the first two academic years (using all the correct solutions of the same two years), forming a training dataset to finetune our LLM. We create a validation set by keeping the submissions of a subset (10%) of students [7]. Finally, we use the finetuned LLM model to repair the buggy solutions submitted in the academic year 2017–2018. We compare our results against those of Refactory when the latter has only access to the original correct programs in the first two academic years. We note that the test set contains exclusively assignments

[1] https://huggingface.co/.
[2] https://github.com/KoutchemeCharles/aied2023.

seen before during training, while the training set contains data from additional assignments not present in the academic year 2017–2018.

To answer our second research question, we evaluate the model trained on the Dublin dataset to repair the buggy solutions in the Singapore and New Caledonia datasets. We note that the Dublin dataset contains no similar assignments with the Singapore dataset and only two common assignments with New Caledonia dataset. In order to compare our model against Refactory, we provide the ART with the reference solutions for each exercise in the two test datasets.

Technical Details. As input to the model, we concatenate the assignment description and the buggy code into a single sequence [10]. For each experiment, we search for the best training hyperparameters using the validation loss. Additionally, for validation and testing, we generate 10 candidate repairs, and we select the best generation parameters on the validation set using the pass@10 metric [2]. The pass@10 metric tells us the probability that *at least one* of the 10 generated repairs passes *all* the unit tests. When testing, out of the ten model outputs, we choose the closest generation to the input buggy program in terms of sequence edit distance. We report our performance on the test sets in terms of success rate, that is, the percentage ratio of programs we manage to repair successfully, and in terms of average sequence edit distance [11].

4 Results and Discussion

Statistics. In the two first academic years at the University of Dublin, students submitted 4269 correct and 3217 buggy programs. Refactory managed to repair 2840 of these buggy programs. We finetuned CodeT5 on the resulting correction dataset aiming to correct 1593, 563, and 933 buggy programs of the test split of the Dublin, New Caledonia and Singapore datasets, respectively.

Results. Table 1 details our results for a subset of the selected assignments. Answering our first research question, when data from the same distribution is available, we observe that the LLM consistently finds higher quality repairs than Refactory (judged in terms of distance) while remaining competitive in terms of the number of repairs found. Answering our second research question, even having never seen most of the assignments before, our model can repair up to 97 % of the buggy programs. We believe the pretraining strategy of LLMs allows them to generalize well to the task of program repair, even when the finetuning data might not be optimal.

Surprisingly, we can observe that even when only given a reference solution (RQ2), Refactory seems to be able to correct almost all buggy codes. Still, these repairs seem to be worse than our model when considering the distance from the original buggy code. When investigating that effect further, we found that, for *both experiments*, Refactory often outputs the reference solution as a correction to buggy codes, even when this reference-based correction does not match the problem-solving intent of the student buggy code. Figure 2 illustrates this issue.

In the example, students had to write a function that removed the duplicates from a list. The original buggy code contains a typo in the second variable "occurrences". The (optimal) repair found by our model correctly identifies this typo but Refactory outputted the reference solution. Although outputting the reference solutions remains a valid functional correction to the buggy code, we argue that these kinds of "repairs" do not align with the spirit of providing valuable feedback for debugging. On the other hand, our model only produces repairs when these can adequately match the student's problem-solving intent. In a nutshell, our model prioritizes quality over quantity.

Table 1. Per assignment experiments results. The left (resp. right) table shows the results for the first (resp. second) experiment. Legend: LLM (Large Language Model), RF (Refactory), SR (success rate), SD (average sequence edit distance), DB (Dublin), NC (New Caledonia), SP (Singapore).

dataset	assignment	RF_SR	LLM_SR	RF_SD	LLM_SD
DB	append2list	1.00	0.96	11.43	12.19
DB	fibonacci_iter	1.00	0.76	56.03	48.65
DB	fibonacci_recur	1.00	1.00	16.79	18.07
DB	index_iter	1.00	0.86	21.36	19.76
DB	index_recur	0.98	1.00	26.11	24.86
DB	maximum	0.96	0.85	26.60	11.36
DB	merge_lists	1.00	0.18	52.45	8.00
DB	minimum	1.00	0.63	24.78	15.00
DB	reverse_iter	1.00	1.00	17.29	17.75
DB	reverse_recur	1.00	0.97	15.79	14.16
DB	search_iter	0.91	0.95	19.65	12.10
DB	search_recur	1.00	1.00	17.41	13.12

dataset	assignment	LLM_SR	RF_SR	LLM_SD	RF_SD
NC	decreasing_list	1.00	0.02	26.63	16.00
NC	is_palindrome	1.00	0.23	43.32	22.86
NC	maximum	0.40	0.19	22.24	15.33
NC	mean	0.77	0.29	33.29	5.29
NC	minimum	0.43	0.20	17.13	10.36
NC	sum	1.00	0.97	11.56	6.91
NC	sum_even_numbers	1.00	0.48	23.14	5.21
NC	sum_n_first_even	1.00	0.21	26.96	2.17
NC	symetrical_list	1.00	0.43	32.70	25.40
SP	remove_extras	1.00	0.01	44.77	22.50
SP	search	0.99	0.37	25.79	5.75
SP	sort_age	0.99	0.17	82.48	4.87
SP	top_k	1.00	0.27	51.30	4.58

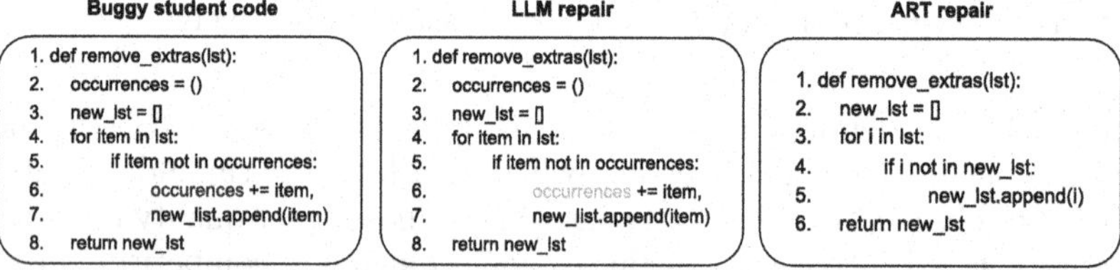

Fig. 2. Comparing a correction found by a language model against a correction found by an automated repair tool.

5 Conclusion

In this work, we showed that, compared to a classical automated repair tool, using a neural automated repair tool provides: more quality repairs when used in a familiar context (RQ1), and more reliable corrections when used in unfamiliar settings (RQ2).

Limitations and Future Work. One limitation of our work is that we only used and compared one APR tool with one large language model. Moreover, we did not contrast the performance improvements of this approach against other repairing techniques. On top of addressing these issues, we are currently working on improvements to this framework, namely, using the corrections of multiple repairing tools or other sources to train our models, and experimenting with other types of large language models. We believe that using automated program repair tools as means for training large language models can open new opportunities for applying program repair techniques in education.

References

1. Azcona, D., Smeaton, A.: +5 Million Python & Bash Programming Submissions for 5 Courses & Grades for Computer-Based Exams Over 3 Academic Years (2020). https://doi.org/10.6084/m9.figshare.12610958.v1
2. Chen, M., et al.: Evaluating large language models trained on code (2021). https://doi.org/10.48550/ARXIV.2107.03374
3. Cleuziou, G., Flouvat, F.: Learning student program embeddings using abstract execution traces. In: 14th International Conference on Educational Data Mining, pp. 252–262 (2021)
4. Gulwani, S., Radiček, I., Zuleger, F.: Automated clustering and program repair for introductory programming assignments (2018). http://arxiv.org/abs/1603.03165, arXiv:1603.03165 [cs]
5. Hu, Y., Ahmed, U.Z., Mechtaev, S., Leong, B., Roychoudhury, A.: Re-factoring based program repair applied to programming assignments. In: 2019 34th IEEE/ACM International Conference on Automated Software Engineering (ASE), pp. 388–398. IEEE/ACM (2019)
6. McCauley, R., et al.: Debugging: a review of the literature from an educational perspective. Comput. Sci. Educ. **18**(2), 67–92 (2008). https://doi.org/10.1080/08993400802114581
7. Pu, Y., Narasimhan, K., Solar-Lezama, A., Barzilay, R.: sk_p: a neural program corrector for MOOCs. arXiv:1607.02902 [cs] (2016). http://arxiv.org/abs/1607.02902
8. Singh, R., Gulwani, S., Solar-Lezama, A.: Automated feedback generation for introductory programming assignments. In: Proceedings of the 34th ACM SIGPLAN Conference on Programming Language Design and Implementation, pp. 15–26. PLDI 2013, Association for Computing Machinery, New York, NY, USA (2013). https://doi.org/10.1145/2491956.2462195
9. Wang, K., Singh, R., Su, Z.: Data-driven feedback generation for introductory programming exercises. arXiv:1711.07148 [cs] (2017). http://arxiv.org/abs/1711.07148
10. Wang, Y., Wang, W., Joty, S.R., Hoi, S.C.H.: Codet5: Identifier-aware unified pre-trained encoder-decoder models for code understanding and generation. In: EMNLP, pp. 8696–8708. Association for Computational Linguistics (2021)
11. Zhang, J., et al.: Repairing bugs in python assignments using large language models (2022). https://doi.org/10.48550/ARXIV.2209.14876

Author Index

N. Wang et al. (Eds.): AIED 2023, LNAI 13916, pp. 837–840, 2023.
https://doi.org/10.1007/978-3-031-36272-9

Printed in the United States
by Baker & Taylor Publisher Services